Unified Design of Steel St…

Louis F. Geschwindner
Former Vice President of Engineering and Research, American Institute of Steel Construction; Professor Emeritus of Architectural Engineering, The Pennsylvania State University; and Senior Consultant, Providence Engineering

Judy Liu
Professor, School of Civil and Construction Engineering, Oregon State University

Charles J. Carter
President, American Institute of Steel Construction

State College, Pennsylvania

Dedication

This book is dedicated to our mentors, teachers and former students, each of whom taught us something and, in the process, contributed to who we are today.

Cover photo: University of Rhode Island, Fascitelli Center for Advanced Engineering
Photo courtesy of Ballinger, Philadelphia, PA, © Albert Vecerka/Esto

•

Available from Amazon.com

For additional information see www.SteelStuff.com

ISBN: 9798351317908

Preface

INTENDED AUDIENCE

This book presents the basics of design of steel building structures and is based on the 2022 unified specification, ANSI/AISC 360-22 *Specification for Structural Steel Buildings*. It is intended primarily as a text for a first course in steel design for civil and architectural engineers. Such a course usually occurs in the third or fourth year of an engineering program. The book can also be used in a second, building-oriented course in steel design, depending on the coverage in the first course. In addition to its use as a textbook, it provides a good review for practicing engineers looking to learn the provisions of the latest specification or to convert their practice from any of the old specifications to the new specification. Users are expected to have a firm knowledge of statics and strength of materials and have easy access to the AISC *Steel Construction Manual*, 16th Edition.

UNIFIED ASD AND LRFD

A preferred approach to the design of steel structures had been elusive over a 20-year period from 1986 to 2005. In 1986, the American Institute of Steel Construction (AISC) issued its first *Load and Resistance Factor Design (LRFD) Specification for Structural Steel Buildings*. This specification came after almost 50 years of publication of an Allowable Stress Design (ASD) specification. Unfortunately, LRFD was accepted by the academic community but not by the professional engineering community. Although AISC revised the format of the ASD specification in 1989, it had not updated its provisions for over 25 years. This use of two specifications was seen as an undesirable situation by the professions related to the building industry, and in 2001 AISC began the development of a combined ASD and LRFD specification. In 2005, AISC published its first unified specification, combining the provisions of both the LRFD and ASD specifications into a single standard for the design of steel building structures. That specification, ANSI/AISC 360-05 *Specification for Structural Steel Buildings*, reflected a major change in philosophy by AISC, one that made the use of ASD and LRFD equally acceptable as approaches for the design of steel buildings. That has been the case for the past 17 years. Now, the newest specification, ANSI/AISC 360-22 *Specification for Structural Steel Buildings* continues that philosophy of equal status for design by ASD and LRFD.

The reader familiar with past editions of the separate ASD and LRFD specifications but not with the 2005, 2010 or 2016 editions will undoubtedly question how these two diverse design philosophies could be effectively combined into one specification. This is a reasonable question to ask. The primary answer is that the 2005 specification was not a combination of the old ASD and LRFD provisions. It was a new approach with a new ASD and a new LRFD that used the best of both previous approaches. The first unified specification took a different approach. It was based on the understanding that the strength of an element or structure, called the nominal strength in the specification, can be determined independent of the design philosophy. Once that nominal strength is determined, the available strength for ASD or LRFD is determined as a function of that nominal strength. Thus, the available strength of the element is always based on the same behavior and no inconsistency in behavior results from the use of ASD or LRFD. This important aspect of the unified specification is further explained in Chapter 1.

The 2022 *Specification* shows AISC's commitment to maintaining both design approaches. With only one strength equation for both methods, any updates are applied equally to the ASD and LRFD provisions. The designer continues to be given the opportunity to apply engineering judgment in deciding what approach to use for design, either by ASD or LRFD. AISC has also issued ANSI/AISC 341-22 *Seismic Provisions for Structural Steel Buildings*, the standard that guides the design of steel building structures to resist seismic loads.

CHANGES IN BUILDING LOADS

In addition to the provisions for steel design issued by AISC, structural engineering has seen many changes in the area of loads for which buildings must be designed. The American Society of Civil Engineers (ASCE) is continually revising ASCE 7 *Minimum Design Loads and Associated Criteria for Buildings and Other Structures*, its standard for building loads. The International Code Council (ICC) continues to issue its International Building Code (IBC) on a three-year cycle, and it in turn is adopted by a local jurisdiction on the schedule chosen by that jurisdiction. The IBC adopts ASCE 7, ANSI/AISC 360, ANSI/AISC 341, and other structural standards. For the calculation of loads within this text, ASCE 7-22 provisions are used. For any actual design, the designer must use the loadings established by the governing building code. ANSI/AISC 341 is discussed in Chapter 13.

UNITS

ANSI/AISC 360-22 is, as much as possible, a unitless specification. In those rare instances where equations could not be written in a unitless form, two equations are given, one in U.S. customary units and one in SI units. The *Manual* presents all of its material in U.S. customary units. The construction industry in this country has not adopted SI units in any visible way, and it is not clear that they will in the foreseeable future. Thus, this book uses only U.S. customary units.

TOPICAL ORGANIZATION

Chapters 1 through 3 present the general material applicable to all steel structures. This is followed in Chapters 4 through 9 with a presentation of member design. Chapters 10 through 12 discuss connections and Chapter 13 provides an introduction to seismic design.

In Chapter 1, the text addresses the principles of limit states design upon which all steel design is based. It shows how these principles are incorporated into both LRFD and ASD approaches and shows how reliability varies with live-to-dead load ratio for both approaches. Chapter 1 also provides a description of ANSI/AISC 360-22, the *Steel Construction Manual* and the AISC web site. These are the major source documents for all that is presented in this book. Chapter 2 introduces the development of load factors, resistance factors, and safety factors. It discusses load combinations and compares the calculation of required strength for both LRFD and ASD. The calibration of ASD and LRFD and the resulting system reliability is also addressed. Chapter 3 discusses steel as a structural material. It describes the availability of steel in a variety of shapes and the grades of steel available for construction.

Once the foundation for steel design is established, the various member types are considered. Tension members are addressed in Chapter 4, compression members in Chapter 5, and bending members in Chapter 6. Chapter 7 covers plate girders, which are simply bending members made from individual plates. Chapter 8 treats members subjected to combined axial load and bending as well as design of bracing. Chapter 9 deals with composite members, that is, members composed of both steel and concrete working together to provide the available strength. Each of these chapters begins with a discussion of that particular member type and how it is used in buildings. This is followed by a discussion of the specification provisions and the behavior from which those provisions have been derived. The LRFD and ASD design philosophies of the 2022 *Specification* are used throughout. Design examples that use the specification provisions directly are provided along with examples using the variety of design aids available in the AISC *Steel Construction Manual*. All examples that have an LRFD and ASD component are provided for both approaches. Throughout this book, ASD examples, or portions of examples that address the ASD approach, are presented with shaded background for ease of identification.

The member-oriented chapters are followed by chapters addressing connection design. Chapter 10 introduces the variety of potential connection types and discusses the strength of bolts, welds, and connecting elements. Chapter 11 addresses simple connections. This includes simple beam shear connections, light bracing connections and direct bearing connections. Chapter 12 deals with moment-resisting connections. As with the member-oriented chapters, the basic principles of limit states design are developed first. This is followed by the application of the provisions to simple shear connections and beam-to-column moment connections through extensive examples in both LRFD and ASD.

The text concludes in Chapter 13 with an introduction to steel systems for seismic force resistance. It discusses the variety of structural framing systems available and approved for inclusion in the seismic force resisting system.

NEW TO THIS EDITION

This fourth edition is based on ANSI/AISC 360-22 *Specification for Structural Steel Buildings*, the 2022 edition of the AISC specification. The committee responsible for developing the specification was charged by its chair to develop a new specification that would show minimal changes. Thus, changes in the specification upon which this book is based are limited to those areas where the committee felt important knowledge had been gained since the 2016 edition. As a result, the changes in this book related to the specification provisions are also somewhat limited. Other changes have been implemented here to better explain some topics and to expand others while revising the entire book wherever needed to reflect the latest specification. Throughout, the use of *Specification* and *Manual* equation numbers has been implemented to assist the reader in navigating the *Specification* and *Manual*.

Example Problems for ANSI/AISC 360-22: Examples have been updated to the new specification and the new table of member properties, as well as being revised to improve clarity of intent. In addition, selected examples that had used A36 material have been changed to use material with F_y = 50 ksi to be consistent with similar changes in the *Manual*.

New Homework Problems: Over 110 homework problems have been added and 180 have been revised. This edition provides over 590 homework problems intended to provide meaningful activities for student learning. Problems continue to be included that carry over from one chapter to another so that an opportunity exists to link concepts of design to one or two specific structures.

The following addresses the changes to be found in each chapter:

Chapter 1 includes updated references to new and revised Design Guides, an expanded discussion of the benefits of student membership in AISC, and reference to Volume 2 of the Companion to the AISC *Steel Construction Manual*. It also includes discussion of AISC 303-22, the *Code of Standard Practice*. The presentation of reliability analysis, the foundation of the *Specification*, has also been revised.

Chapter 2 contains a new section on reliability and probability of failure. Graphs are provided to illustrate how these two indicators of safety are different for ASD and LRFD and how different load combinations lead to different levels of reliability. The relationship between reliability and probability of failure is also explained. To support these calculations, statistical data are presented for several load types. Additionally, several examples have been revised to include calculation of loads that include roof loads.

Chapter 3 discusses the changes in preferred materials and provides updates on those material specifications that have been revised.

Chapter 4 addresses tension members. As was the case for the 3rd edition of this book, the provisions have not changed. However, it has been clarified that when holes are discussed, they are understood to be bolt holes. Examples have been revised to be for 50 ksi material, consistent with the preferred materials discussed in Chapter 3. The limit states that control the strength of pin connected members are expanded upon.

Chapter 5 addresses compression members. The terminology change in the *Specification* where nominal stress is now used in place of critical stress is reflected throughout the chapter. The limitation on "as fabricated" slenderness is discussed. The change from 36 to 50 ksi material as the preferred material for angles results in a need to confirm which provisions apply that might not have applied for earlier designs. A new example is added to address constrained axis torsional buckling.

Chapter 6 on bending members includes a change in how the most economical member is selected when using *Manual* Table 3-10. This is illustrated in examples for beams with unbraced length. In addition, more attention is given to dealing with the self-weight of members throughout the examples.

Chapter 7 addresses plate girders and contains a new section on proportioning of those plate girders. The discussion of tension field action in end panels is expanded and reference is made to the shear design aids in the *Manual*.

Chapter 8 on combined compression and bending incudes a somewhat expanded discussion of the various methods of stability analysis and design. The section on initial trial member selection is expanded to address shapes deeper than W14 and reference is added to a more refined approach found in the literature.

Chapter 9 examples addressing composite design have been adjusted to match the tables in the 16th edition *Manual*. In addition, guidance is provided for selection of camber for composite beams.

Chapter 10 has been revised to be consistent with new bolt group designations based on bolt strength and all examples similarly revised. The directional strength increase for welds has been expanded and the weld sizes in several examples have been changed to more realistic sizes.

Chapter 11 now includes direction for locating connection design examples beyond those presented here. The chapter has been revised throughout to provide additional clarity in discussion and examples. Many examples have been revised to use 50 ksi preferred materials. Discussion of eccentricity on welds, single angles and shear tabs is expanded. Derivation of the prying action equations have been expanded.

Chapter 12 continues to address moment connections and has had examples revised to incorporate 50 ksi connecting elements, reflecting the revised preferred materials presented in the *Manual*. As result, examples have been modified to describe the new limit states that will now control.

Chapter 13 continues to outline the application of the *Seismic Provisions*.

Throughout the book, new figures are included, and others revised, to better illustrate the corresponding material.

EXAMPLES AND HOMEWORK PROBLEMS IN LRFD AND ASD

The LRFD and ASD design philosophies of the 2022 *Specification* are used throughout the book. It is anticipated, however, that instructors and students will concentrate on only one design approach. This will be the most effective use of time and resources except for situations where it will be valuable to illustrate the differences and similarities of the LRFD and ASD philosophies. ASD is presented in this book primarily for those who are looking to compare methods in order to better understand the concepts of the unified specification and designers who have been practicing with ASD and wish to move their practice to LRFD. LRFD should be the primary approach used to teach structural steel design to new students and in current day professional practice.

LRFD and ASD in Examples

Design examples that use the specification provisions directly are provided along with examples using the variety of design aids available in the AISC *Steel Construction Manual*. All examples that have an LRFD and ASD component are provided for both approaches. Throughout this book, ASD examples, or portions of examples that address the ASD approach, are presented with shaded background for ease of identification.

EXAMPLE 4.10b
Tension Member Design by ASD

Goal: Select a double-angle tension member for use as a web member in a truss and determine the maximum area reduction that would be permitted for bolt holes and shear lag.

Given: The member must carry a dead load of $P_D = 67.5$ kips and a live load of $P_L = 202.5$ kips. For the load combination $P_D + P_L$, the ASD required strength is $P_a = 270$ kips. Use equal leg angles of A992 steel.

SOLUTION

Step 1: Determine the minimum required gross area based on the limit state of yielding where $\Omega = 1.67$:

$$A_{g\ \min} = P_a/(F_y/\Omega) = 270/(36/1.67) = 12.5\ \text{in.}^2$$

LRFD and ASD in Homework Problems

Each chapter includes homework problems at the end of the chapter. These problems are organized to follow the order of presentation of the material in the chapters. Several problems are provided for each general subject. Problems are provided for both LRFD and ASD solutions. There are also problems designed to show comparisons between ASD and LRFD solutions. These problems show that in some instances one method might give a more economical design, whereas in other instances the reverse is true.

WEBSITE

Additional resources are available from the book website at www.SteelStuff.com. The following resources are available on the student section of the website.

- **Answers:** Selected homework problem answers are available on the student section of the website.
- **Videos:** Links to several example problem videos are presented through instructor lead YouTube videos.
- **Qnect:** A guide to the use of the Qnect connection software in conjunction with the integrated design project is available
- **Errata:** We have reviewed the text to make sure that it is as error-free as possible. However, if any errors are discovered, they will be listed on the book website as a reference.

If you encounter any errors as you are using the book, please send them directly to the authors (LFG@psu.edu) so we may include them on the website and correct these errors in future editions.

RESOURCES FOR INSTRUCTORS

All resources for instructors are available through an Instructor link on the website at www.SteelStuff.com.

The following resources are available to instructors who adopt the text:

- **Solutions Manual:** Solutions for all homework problems in the text.
- **Integrated Building Project RAM Structural System:** Computer model and example output for one solution of integrated Building project.
- **Textbook Figures:** Select figures available in PowerPoint format.

Visit the Instructor link on the website at www.SteelStuff.com to register and request access to these resources.

ACKNOWLEDGEMENTS

The first two editions of this book were under the sole authorship of the first author. Many individuals contributed to those editions, and we want to thank them for their contributions to those works as those contributions clearly are carried over to this new edition. This fourth edition has been produced through the cooperative efforts of three individuals who have a long standing common academic and professional relationship. For the many individuals who have motivated and guided each of us, we acknowledge your contribution to all that we do. We thank those who have provided input and clarity to our understanding of structural steel design through our involvement in the development of the AISC *Specification* and *Manual*. Finally, we acknowledge the contributions of our spouses and families through whose support we are able to accomplish this project.

Louis F. Geschwindner
Judy Liu
Charles J. Carter

(This page intentionally left blank)

Contents

Chapter 1

Introduction

University of Rhode Island, Fascitelli Center for Advanced Engineering
Photo courtesy of Ballinger, Philadelphia, PA
© James Ewing/JBSA

1.1 Scope

A wide variety of designs can be characterized as *structural steel design.* This book deals with the design of steel structures for buildings as governed by the ANSI/AISC 360-22 *Specification for Structural Steel Buildings*, published by the American Institute of Steel Construction (AISC) in 2022, and referred to as the *Specification* in this book. The areas of application given throughout this book specifically focus on the design of steel building structures. The treatment of subjects associated with bridges and industrial structures, if addressed at all, is kept relatively brief.

The book addresses the concepts and design criteria for the two design approaches detailed by the *Specification*: Load and Resistance Factor Design (LRFD) and Allowable Strength Design (ASD). Both methods are discussed later in this chapter.

In addition to the *Specification*, the primary reference for this book is the 16th edition of the AISC *Steel Construction Manual.* This reference handbook contains tables of the basic values needed for structural steel design, design tables to simplify actual design, and the complete *Specification*. Throughout this book, this is referred to as the *Manual*.

1.2 The Specification

The ANSI/AISC 360-22 *Specification for Structural Steel Buildings* is the latest in a long line of standard specifications published by the American Institute of Steel Construction for the design and construction of structural steel buildings. The first edition was published in 1923. For the reader interested in the historical aspects of these specifications, AISC has two resources that provide detailed guidance on the historical structural steel standards. The first is AISC Design Guide 15, *Rehabilitation and Retrofit,* 2nd edition. This Design

Guide provides outline comparisons of the provisions in the different editions of the *Specification*. The second resource is found on the AISC web site, www.aisc.org, *Historic Specifications for Steel Buildings*, which contains downloadable versions of all of the AISC *Specifications for Structural Steel Buildings* produced from 1923 through 2016. Like most AISC documents, these are available as free downloads to AISC members. Note also that AISC offers free membership to students (see further information in Section 1.3).

Current design is carried out under the provisions published in the 2022 edition of the AISC *Specification*. In addition to the detailed provisions, the *Specification* contains User Notes and a detailed Commentary that provides insights into the source and application of the provisions. The reader interested in additional background on the provisions discussed in this book is encouraged to investigate the materials cited in the appropriate sections of the Commentary. The *Specification* contains 14 chapters and 8 appendices. To provide a concise guide to the use of the *Specification*, a brief description is given here.

Chapter A: General Provision. This chapter provides the scope of the *Specification* and summarizes all referenced specifications, codes, and standards. It also provides the requirements for materials to be used in structural steel design and the design documents necessary to communicate that design.

Chapter B: Design Requirements. This chapter gives the general requirements for analysis and design that are applicable throughout the entire *Specification*. It provides the charging language needed for application of the subsequent chapters and it addresses how tolerances are treated throughout the *Specification*.

Chapter C: Design for Stability. This chapter, along with Appendix 7, addresses the requirements for the design of structures to ensure stability. It details those factors that must be taken into consideration in any analysis and design.

Chapter D: Design of Members for Tension. This chapter applies to the design of members subjected to axial tension.

Chapter E: Design of Members for Compression. This chapter addresses members subjected to axial compression.

Chapter F: Design of Members for Flexure. This chapter applies to members loaded in a plane parallel to a principal axis that passes through the shear center or is restrained against twisting at load points and reactions. This is referred to as simple bending about one axis.

Chapter G: Design of Members for Shear. This chapter addresses webs of singly or doubly symmetric members subject to shear in the plane of the web. It also addresses other shapes such as single angles and hollow structural sections (HSS).

Chapter H: Design of Members for Combined Forces and Torsion. This chapter addresses design of members subject to an axial force in combination with flexure about one or both axes, with or without torsion. It also applies to members subjected to torsion only.

Chapter I: Design of Composite Members. This chapter addresses the design of members composed of steel shapes and concrete working together as a member. It addresses compression, flexure, and combined forces.

Chapter J: Design of Connections. This chapter addresses the design of connections, including the connecting elements, the connectors, and the connected portions of members.

Chapter K: Additional Requirements for HSS and Box-Section Connections. This chapter addresses requirements in addition to those given in Chapter J for the design of connections to hollow structural sections and built-up box sections of uniform thickness and connections between HSS and box members.

Chapter L: Design for Serviceability. This chapter summarizes the performance requirements for the design of a serviceable structure.

Chapter M: Fabrication and Erection. This chapter addresses the requirements for fabrication and erection documents along with the detailed requirements for fabrication, shop painting, and erection.

Chapter N: Quality Control and Quality Assurance. This chapter addresses the minimum requirements for ensuring quality of the constructed project.

Appendix 1: Design by Advanced Analysis. The body of the *Specification* addresses design based on an elastic analysis. This appendix addresses design by alternative methods generally referred to as advanced methods. It includes the classical plastic design method and design by direct modeling of imperfections.

Appendix 2: Design of Filled Composite Members (High Strength). This appendix provides methods to design filled composite members with material strength greater than that permitted by Chapter I.

Appendix 3: Fatigue. This appendix provides requirements for addressing the influence of high-cycle loading within the elastic range on members and connections that could lead to cracking and progressive failure. For most building structures, fatigue is not an issue of concern.

Appendix 4: Structural Design for Fire Conditions. This appendix provides the criteria for evaluation of structural steel subjected to fire conditions, including (1) the prescriptive approach provided for in the model building code and most commonly used in current practice and (2) the engineered approach.

Appendix 5: Evaluation of Existing Structures. This appendix provides guidance on the determination of the strength and stiffness of existing structures by load tests or a combination of tests and analysis.

Appendix 6: Member Stability Bracing. This appendix details the criteria for ensuring that column, beam and beam-column bracing has sufficient strength and stiffness to meet the requirements for member bracing assumed in the provisions of the *Specification* for design of those members.

Appendix 7: Alternative Methods of Design for Stability. This appendix reiterates the general stability requirements of Chapter C and provides methods of designing structures to ensure stability. Two alternative methods are provided here, including the method most commonly used in past practice.

Appendix 8: Approximate Analysis. This appendix provides an approximate method for obtaining second-order effects by an amplified first-order analysis. The provisions are limited to structures supporting load primarily through vertical columns. It also provides an approximate method for inelastic moment redistribution.

Each chapter of this book will identify those chapters of the *Specification* that are pertinent to that chapter. The reader is encouraged to become familiar with the organization of the *Specification*.

1.3 THE MANUAL

The AISC *Steel Construction Manual*, 16th edition, is the latest in a series of manuals published to assist the building industry in designing safe and economical steel building structures. The first edition was published in 1928 and the ninth edition in 1989. These manuals addressed design by the allowable stress method. In 1986 the first edition of the load and resistance factor design method manual was published, with the third edition published in 1999. The next in this unbroken string of manuals published in support of steel design and construction was the first manual to unify these two design methods and was published in 2005 as the 13th edition. The current edition of the *Manual* is the 16th. Students who purchase the *Manual* through the AISC Student Manual Discount Program also have an opportunity to apply for a free AISC Student Membership at the same time. Students are encouraged to become AISC Student Members in order to take full advantage of all free member benefits including PDF versions of the AISC Design Guides, scholarships, student competitions, and more.

As is the case for the *Specification*, AISC has two resources to assist in addressing the historic aspects of steel design and construction. The first is, again, AISC Design Guide 15, *Rehabilitation and Retrofit,* 2nd edition. This Design Guide provides properties of beam and column sections as old as the wrought iron shapes produced as early as 1873. The second resource is the electronic *AISC Shapes Database*. This database is available through the AISC web site www.aisc.org. It is a searchable database with properties for all shapes produced since 1873, consistent with the printed data in Design Guide 15. Access to the electronic shapes database is free to AISC members.

The *Manual* is presented in 17 parts as follows:

Part 1: Dimensions and Properties

Part 2: General Design Considerations

Part 3: Design of Flexural Members

Part 4: Design of Compression Members

Part 5: Design of Tension Members

Part 6: Design of Members Subject to Combined Forces

Part 7: Design Considerations for Bolts

Part 8: Design Considerations for Welds

Part 9: Design of Connecting Elements

Part 10: Design of Simple Shear Connections

Part 11: Design of Moment Connections

Part 12: Design of Simple Connections for Combined Forces

Part 13: Design of Bracing Connections and Truss Connections

Part 14: Design of Beam Bearing Plates, Column Base Plates, Anchor Rods, and Column Splices

Part 15: Design of Hanger Connections, Bracket Plates, and Crane-Rail Connections

Part 16: Specifications and Codes

Part 17: Miscellaneous Data and Mathematical Information

Each part of the *Manual* includes tables useful for designing steel structures. Because tables can only provide data in incremental form, it is important to understand the basis of each table before interpolating between the incremental values provided. When interpolation is used in this book, it is understood that the accuracy of that interpolation is sufficient for the design being carried out.

Each chapter of this book identifies those parts of the *Manual* that will be used with the material to be addressed. In many instances, the user will need to look in several parts of the *Manual* to fully understand the topics or solve the problems presented.

1.4 AISC Web Site Resources

Another primary resource is the AISC web site, where there is information that is free to all visitors and additional electronic resources that are free to members only. Students will find a great deal of useful information on the AISC publications web site, www.aisc.org/publications. The primary resources under the Steel Construction Manual link include electronic versions of the *Specification*, the Shapes Database, the Steel Construction Manual References, and the Steel Construction Manual Companion which includes Design Examples and Design Tables not included in the printed *Manual*. The *Specification*, as described in Section 1.2 and the historic Shapes Database, as mentioned in Section 1.3, are available free to all through the web site. The 16th edition Steel Construction Manual Shapes Database is also available free to all. The AISC web site also includes an extensive array of journal and proceedings papers. All of the references cited in the Commentary and the *Manual*, for which AISC owns the copyright, are accessible under Steel Construction Manual Resources Interactive Reference List.

Probably the most valuable aspect of the AISC web site for readers of this book is the complete set of the 16th edition Steel Construction Manual Design Examples found in Volume 1 of the *Companion to the AISC Steel Construction Manual*. These examples are presented in three sections.

Section I: Examples Based on the AISC *Specification*. This section contains examples demonstrating the use of the specific provisions of the *Specification*, organized by *Specification* chapter.

Section II: Examples Based on the AISC *Steel Construction Manual*. This section contains examples of connection design using the *Specification* and the tables found in the *Manual*.

Section III: System Design Examples. This section contains examples associated with the design of a specific building and the application of the system-wide requirements.

Volume 2 of the *Companion to the AISC Steel Construction Manual* contains design tables that compliment those available in the printed *Manual*. It provides design tables for filled composite HSS compression members, W-shape and HSS sections with higher-strength steels than published in the printed *Manual*, and a table for the plastic section modulus for coped W-shapes.

Although the topics covered in this book are supported by calculated example problems, the reader might find the electronic Steel Construction Manual Design Examples helpful for further understanding of some of the specific provisions or design aids described in the book. In addition, some of the Design Examples go beyond the coverage in this book and provide additional useful information regarding typical design or detailing. The reader is encouraged to investigate what the AISC web site has to offer through both free and member-only publications.

1.5 PRINCIPLES OF STRUCTURAL DESIGN

From the time an owner determines a need to build a building, through the development of conceptual and detailed plans, to completion and occupancy, a building project is a multi-faceted task that involves many professionals. The owner and the financial analysis team evaluate the basic economic criteria for the building. The architects and engineers form the design team and prepare the initial proposals for the building, demonstrating how the users' needs will be met. This teamwork continues through the final planning and design stages, where the design documents, specifications, and contract documents are readied for the construction phase. During this process, input may also be provided by the individuals who will transform the plans into a real-life structure. The steel detailer, fabricator, and erector all have a role in that process and add their respective expertise to make the design constructible. Thus, those responsible for the construction phase of the project often help improve the design by taking into account the actual on-site requirements for efficient construction.

Once a project is completed and turned over to the owner, the work of the design teams is normally over. The operation and maintenance of the building, although major factors in the life of the structure, are not usually within the scope of the designer's responsibilities, except when significant changes in building use are anticipated. In such cases, a design team should verify that the proposed changes can be accommodated.

The basic goals of the design team can be summarized by the words *safety*, *function*, and *economy*. The building must be safe for its occupants and all others who may come in contact with it. It must neither fail locally nor overall, nor exhibit behavioral characteristics that test the confidence of rational human beings. To help achieve that level of safety, building codes and design specifications are published that outline the minimum criteria that any structure must meet.

The building must also serve its owner in the best possible way to ensure that the functional criteria are met. Although structural safety and integrity are of paramount importance, a building that does not serve its intended purpose will not have met the goals of the owner.

Last, but not least, the design, construction, and long-term use of the building should be economical. The degree of financial success of any structure will depend on a wide range of factors. Some are established prior to the work of the design team, whereas others are determined after the building is in operation. Nevertheless, the final design should, within all reasonable constraints, produce the lowest combined short- and long-term expenditures.

The AISC *Specification* follows the same principles. The mission of the AISC Committee on Specifications is to "develop the practice-oriented specification for design, fabrication, and erection of structural steel buildings that provide for life safety, economical building systems, predictable behavior and response, efficient use, durability, and facilitates speed of project delivery." Thus, this book emphasizes the practical orientation of this *Specification*.

Another important AISC document is ANSI/AISC 303-22 *Code of Standard Practice for Steel Buildings and Bridges* which relates to the process of buying and selling structural steel. It is oriented toward the trade practices in the industry that have developed over time among those that are involved in the design, purchase, fabrication, and erection of structural steel. It provides the standard basis for these activities in addition to some information, like tolerances for fabrication and erection, that our design requirements are based on. The Code can be found in Part 16 of the Manual.

1.6 PARTS OF THE STEEL STRUCTURE

All structures incorporate some or all of the following basic types of structural components:

1. Tension members
2. Compression members
3. Bending members
4. Combined force members
5. Connections

The first four items represent structural members. The fifth, connections, represents the contact regions between the structural members, which ensure that all components work together as a structure.

Detailed evaluations of the strength, behavior, and design criteria for these members are presented in the following chapters:

Tension members:	Chapter 4
Compression members:	Chapter 5
Bending members:	Chapters 6 and 7
Combined force members:	Chapter 8
Connections:	Chapters 10, 11, and 12

The strength and behavior of structural frames composed of a combination of these elements are covered in Chapters 8 and 13, and the special considerations that apply to composite (steel and concrete working together) construction are presented in Chapter 9. An introduction to the design of steel structures for earthquake loading is presented in Chapter 13. The properties of structural steel and the various shapes commonly used are discussed in Chapter 3, and a brief discussion of the types of loads and load combinations is presented in Chapter 2.

Tension members are typically found as web and chord members in trusses and open-web steel joists; as diagonal members in structural bracing systems; and as hangers for balconies, mezzanine floors, and pedestrian walkways. They are also used as sag rods for purlins and girts in many building types, as well as to support platforms for mechanical equipment and pipelines. Figures 1.1 and 1.2 illustrate typical applications of tension members in actual structures.

In the idealized case, tension members transmit concentric tensile forces only. In certain structures, reversals of the overall load may change the tension member force from tension to compression. Some members will be designed for this action; others will have been designed with the assumption that they will carry tension only.

The idealized tension member is analyzed with the assumption that its end connections are pins, which prevent any moment or shear force from being transmitted to the member. However, in an actual structure, the type of connection normally dictates that some bending may be introduced to the tension member. This is also the case when the tension member is directly exposed to some form of transverse load. Moments will also be introduced if the element is not perfectly straight, or if the axial load is not applied along the centroidal axis of the member.

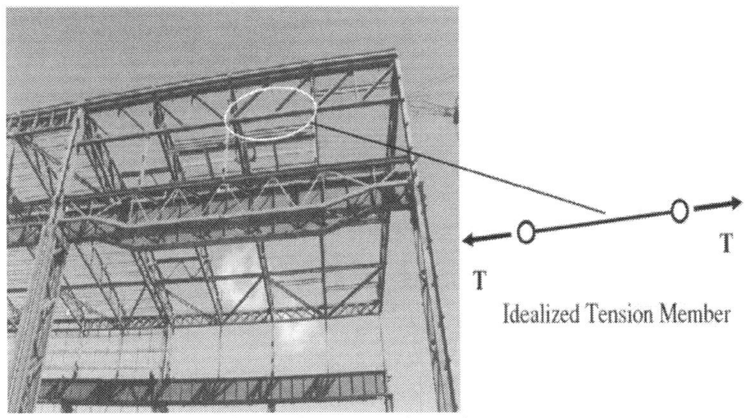

Figure 1.1 Use of Tension Members in a Truss
Photo courtesy Ruby + Associates

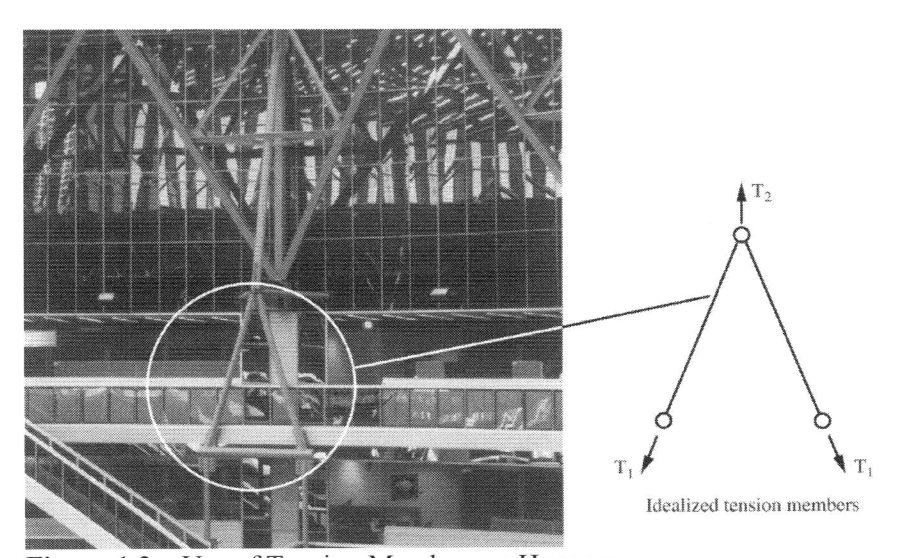

Figure 1.2 Use of Tension Members as Hangers

The primary load effect in the tension member is a concentric axial force, with bending and shear considered as secondary effects.

Compression members are also known as columns, struts, or posts. They are used as web and chord members in trusses and joists and as vertical members in all types of building structures. Figure 1.3 shows a typical use of structural compression members.

The idealized compression member carries only a concentric, compressive force. Its strength is heavily influenced by the distance between the supports, as well as by the support conditions. The basic column is therefore defined as an axially loaded member with pinned ends. Historically, design rules for compression members have been based on the behavior and strength of this idealized compression member.

Figure 1.3 Use of Columns in a Building Frame

The basic column is practically nonexistent in real structures. Realistic end supports rarely resemble perfect pins; the axial load is normally not concentric, due to the way the surrounding structure transmits its load to the member; and beams and similar components are likely to be connected to the column in such a way that moments are introduced. All of these conditions produce bending effects in the member, making it a combined force member or beam-column, as distinct from the idealized column.

The primary load effect in the pinned-end column is therefore a concentric axial compressive force accompanied by the secondary effects of bending and shear.

Bending members are known as beams, girders, joists, spandrels, purlins, lintels, and girts. Although all of these are bending members, each name implies a certain structural application within a building:

1. Beams, girders, and joists form part of common floor systems. The beams are most often considered as the members that are directly supported by girders, which in turn are usually supported by columns. Joists are beams with fairly close spacing. A girder may generally be considered a higher-order bending member compared with a beam or joist. However, variations to this basic scheme are common.

2. The bending members that form the perimeter of a floor or roof plan in a building are known as spandrels or spandrel beams. Their design may be different from other beams and girders because the load comes primarily from one side of the member.

3. Bending members in roof systems that span between other bending members are usually referred to as rafters or purlins.

4. Lintels are bending members that span across the top of openings in walls, usually carrying the weight of the wall above the opening as well as any other load brought into that area. They typically are seen spanning across the openings for doors and windows.

5. Girts are used in exterior wall systems. They transfer the lateral load from the wall surface to the exterior columns. They may also assist in supporting the weight of the wall.

Figure 1.4 shows beams and girders in an actual structure under construction. The idealized beam is shown in the figure as a member with a uniform load and simple supports. This member, with its placement within the bay and not on a column line, is sometimes called a "filler beam".

The basic bending member carries transverse loads that act in a plane containing the longitudinal centroidal axis of the member. The primary load effects are bending moment and shear force. Axial forces and torsion may occur as secondary effects.

The most common *combined force member* is known as a beam-column, implying that this structural element is simultaneously subjected to bending and axial compression. Although the presence of both bending and axial tension represents a potential loading case for the combined force member, this case is not as critical or common as the beam-column loading case.

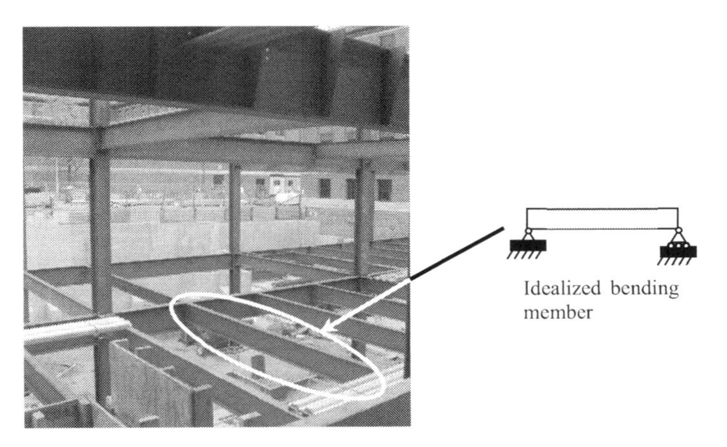

Figure 1.4 Building Structure Showing Bending Members

Figure 1.5a is a schematic illustration of a multi-story steel frame where the beams and columns are joined with rigid connections. Because of the geometric configuration, the types of connections, and the loading pattern, the vertical members are subjected to axial loads and bending moments. This is a typical application of practical beam-columns; other examples are the members of the gable frame shown in Figure 1.5b and the vertical components of a single-story portal frame shown in Figure 1.5c.

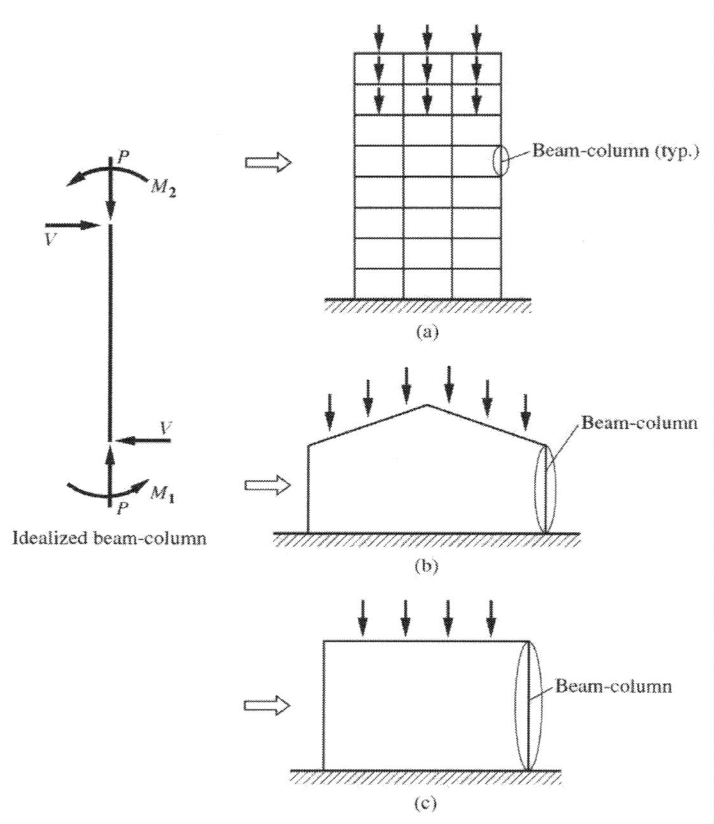

Figure 1.5 Schematic Representation of Steel Frames in Which the Vertical Members Are Subjected to Axial Loads and Bending Moments

The beam-column may be regarded as the general structural element, where axial forces, shear forces, and bending moments act simultaneously. Thus, the basic column may be thought of as a special case, representing a beam-column with no moments or transverse loads. Similarly, the basic bending member may be thought of as a beam-column with no axial load. Therefore, the considerations that must be accounted for in the design of both columns and beams must also be applied to beam-columns.

Because of the generalized nature of the combined force element, all load effects are considered primary. However, when the ratio of axial load to axial load strength in a beam-column becomes high, column behavior will overshadow other influences. Similarly, when the ratio of applied moment to moment strength is high, beam behavior will outweigh other effects. The beam-column is an element in which a variety of different force types interact. Thus, practical design approaches are normally based on interaction equations.

Connections are the collection of elements that join the members of a steel structure together. Whether they connect the axially loaded members in a truss or the beams and columns of a multi-story frame, connections must ensure that the structural members function together as a unit, consistent with the assumptions made in the design.

(a) Tee connection (b) Shear end plate

(c) Shear tab (d) Moment end plate

Figure 1.6 Building Connections

The fasteners used in structural steel connections today are almost entirely limited to bolts and welds. The load effects that the various elements of the connection must resist are a function of the specific connection type being considered. They include all of the possible forces and moments. Figure 1.6 illustrates a variety of connections. The idealized representations for connections are presented in Chapters 10, 11, and 12.

1.7 Types of Steel Structures

It is difficult to classify steel structures into neat categories, due to the wide variety of systems available to the designer. The elements of the structure, as defined in Section 1.6, are combined to form the total structure of a building, which must safely and economically carry all imposed loads. This combination of members is usually referred to as the framing system.

Steel-framed buildings come in a wide variety of shapes and sizes and in combinations with other structural materials. A few examples are given in the following paragraphs, to set the stage for the application of structural design presented in subsequent chapters.

1.7.1 Bearing Wall Construction

Bearing wall construction is primarily used for one- or two-story buildings, such as storage warehouses, shopping centers, office buildings, and schools. This system normally uses brick or concrete block masonry walls, on which are placed the ends of the flexural members supporting the floor or roof. The flexural members are usually hot-rolled structural steel shapes, alone or in combination with open web steel joists or cold-formed steel shapes. An example of a bearing wall application is shown in Figure 1.7.

Figure 1.7 Bearing Wall
Photo courtesy Providence Engineering

1.7.2 Beam-and-Column Construction

Beam-and-column construction is the most commonly used system for steel structures today. It is suitable for large-area buildings such as schools and shopping centers, which often have no more than two stories but may have a large number of spans. It is also suitable for buildings with many stories. Columns are placed according to a regular, repetitious grid that supports the beams, girders, and joists, which are used for the floor and roof systems. The regularity of the floor plan lends itself to economy in fabrication and erection, because most of the members will be of the same size. An example of this type of structure is shown in Figure 1.8.

For multi-story buildings, the use of composite steel and concrete flexural members affords additional savings. Further advances can be expected as designers become more familiar with the use of composite columns and other elements of mixed construction systems.

Figure 1.8 Beam-and-Column Building
Photo courtesy Douglas Steel Fabricating Corporation

Beam-and-column structures rely on either their connections or a separate bracing system to resist lateral loads. A frame in which all connections are moment resistant provides resistance against the action of lateral loads, such as wind and earthquakes, and overall structural stability, through the bending stiffness of the overall frame.

A frame without member-end restraint needs a separate lateral load resisting system, which is often afforded by having the elements along one or more of the column lines act as braced frames, as seen in Figure 1.9. One of the most common types of bracing is the vertical truss, which is designed to take the loads imposed by wind and seismic action. Both types of frames may also be referred to as *bents*. Other bracing schemes involve shear walls and reinforced concrete cores. The latter type may also be referred to as a braced core system and can be highly efficient because of the rigidity of the box-shaped cross section of the core. The core serves a dual purpose in this case: In addition to providing the bracing system for the building, it serves as the vertical conduit in the completed structure for all of the necessary services, including elevators, staircases, electricity, and other utilities.

Combinations of these types of construction are also common. For example, frames may have been designed as moment resistant in one direction of the building and as truss braced in the other. Of course, such a choice recognizes the three-dimensional nature of the structure.

Figure 1.10 shows an idealized representation of several types of beam-and-column framed structures.

Figure 1.9 Braced Beam-and-Column Building
Photo courtesy Douglas Steel Fabricating Corporation

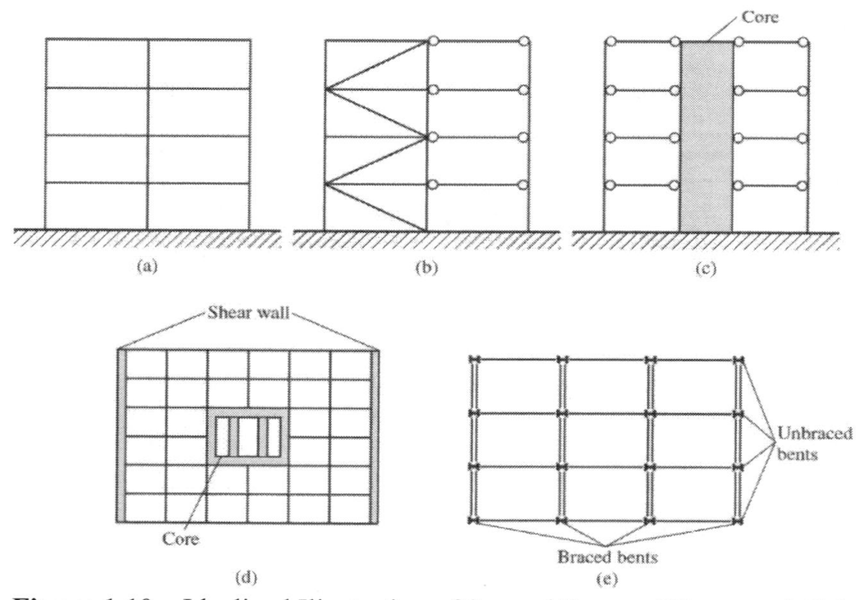

Figure 1.10 Idealized Illustration of Several Types of Beam-and-Column Framed Structures (a) moment-resistant frame; (b) truss-braced frame; (c) core-braced frame; (d) floor plan of shear wall and core-braced building; (e) floor plan of building with a combination of braced and unbraced frames or bents.

1.7.3 Long-Span Construction

This type of construction encompasses steel-framed structures with long spans between the vertical load-carrying elements, such as covered arenas. The long distances may be spanned by one-way trusses, two-way space trusses, or plate and box girders. A long-span structure is shown in Figure 1.11. Arches or cables could also be used, although they are not considered here.

Figure 1.11 Long-Span Structure
Photo courtesy Douglas Steel Fabricating Corporation

Long-span construction is also used in buildings that require large, column-free interiors. In such cases the building may be a core- or otherwise braced structure, where the long span is the distance from the exterior wall to the core.

Many designers would also characterize single-story rigid frames as examples of long-span construction systems. Depending on the geometry of the frame, such structures can span substantial distances, often with excellent economy.

1.7.4 High-Rise Construction

High-rise construction refers to multi-story buildings of significant height; an example is shown in Figure 1.12. The large heights and unique problems encountered in the design of such structures warrant treating them independently from typical beam-and-column construction. In addition, over the past 40 years several designers have developed a number of new concepts in multi-story frame design, such as the super composite column, the steel plate shear wall, and the SpeedCore system.

Figure 1.12 High-Rise Building Structure
Photo courtesy Douglas Steel Fabricating Corporation

Particular care must be exercised in the choice and design of the lateral load resisting system in high-rise construction. It is not just a matter of extrapolating from the principles used in the analysis of lower-rise structures, because many effects play a major role in the design of high-rise buildings but have significantly less impact on frames of smaller height. These effects are crucial to the proper design of the high-rise structure.

Some of these effects may be referred to as second-order effects, because they cannot be quantified through a normal, linearly elastic analysis of the frame. Although second-order effects are present in all structures, they may be more significant in high-rise structures. For example, when a structure is displaced laterally, additional moment is induced in a column due to the eccentricity of the column loads. When added to the moments and shears produced by gravity and lateral loads, the resulting effects may be significantly larger than those computed without considering the second-order effects. A designer who does not incorporate both types of effects will be making a serious and perhaps unconservative error.

Framing systems for high-rise buildings reflect the increased importance of lateral load resistance. Thus, attempts at making the perimeter of a building act as a unit or tube have proven quite successful. This tube may be in the form of a truss, as with 875 North Michigan Avenue in Chicago, Illinois (formerly known as the John Hancock Building), shown in Figure 1.13a, or a frame, as in the former World Trade Center in New York City, shown in Figure 1.13b; a solid wall tube with cutouts for windows, as used in the Aon Center in Chicago (formerly known as the Standard Oil Building), shown in Figure 1.13c; or several interconnected or bundled tubes, such as in the Willis Tower in Chicago (formerly known as the Sears Tower), shown in Figure 1.13d.

Figure 1.13 High-Rise Buildings (a) 875 North Michigan Avenue; (b) the World Trade Center; (c) the Aon Center; (d) the Willis Tower. Photo (b) courtesy Leslie E. Robertson Associates, RLLP

1.7.5 Gable-Frame Construction

Many designers include the single-story frame as part of the long-span construction category. The metal building industry has capitalized on the use of this system through fine-tuned designs of frames for storage warehouses, industrial buildings, temporary and permanent office buildings, and similar types of structures. The gable frames are typically built-up, web tapered members that can be optimized for the forces present. Members are connected in the field using bolted end plate connections. Metal buildings, commonly referred to as pre-engineered metal buildings, are now available as custom engineered for specific applications. An example of a gable-frame metal building is seen in Figure 1.14.

Figure 1.14 Gable-Frame Metal Building
Photo courtesy Metal Building Manufacturers Association

1.8 DESIGN PHILOSOPHIES

A successful structural design results in a structure that is safe for its occupants, can carry the design loads without overstressing any components, does not deform or vibrate excessively, and is economical to build and operate for its intended life span. Although economy may appear to be the primary concern of an owner, safety must be the primary concern of the engineer. Costs of labor and materials will vary from one geographic location to another, making it almost impossible to design a structure that is equally economical in all locations. Because the foremost task of the designer is to produce a safe and serviceable structure, design criteria such as those published by the American Institute of Steel Construction are based on technical models and considerations that predict structural behavior and material response. The use of these provisions by the designer will dictate the economy of a particular solution in a particular location and business climate.

To perform a structural design, it is necessary to quantify the causes and effects of the loads that will be exerted on each element throughout the life of the structure. This is generally termed the *load effect* or the *required strength*. It is also necessary to account for the behavior of the material and the shapes that compose these elements. This is referred to as the *nominal strength* or *capacity* of the element.

In its simplest form, structural design is the determination of member sizes and their corresponding connections, so that the strength of the structure is greater than the load effect. The degree to which this is accomplished is often termed the *margin of safety*. Numerous approaches for accomplishing this goal have been used over the years.

Although past experience might seem to indicate that the structural designer knows the exact magnitude of the loads that will be applied to the structure, and the exact strength of all of the structural elements, this is usually not the case. Design loads are provided by many codes and standards and, although the values given are specific, significant uncertainty is associated with those magnitudes. Loads, load factors, and load combinations are discussed in Chapter 2.

As is the case for loading, significant uncertainty is associated with the determination of the behavior and strength of structural members. The true indication of load-carrying capacity is given by the magnitude of the load that causes the failure of a component or the structure as a whole. Failure may either occur as the physical collapse of part of the building, or be considered to have occurred if deflections, for instance, exceed certain predetermined values. Whether the failure is the result of a lack of strength (collapse) or stiffness (deflection), these phenomena reflect the limits of acceptable behavior of the structure. Based on these criteria, the structure is said to have reached a specific *limit state*. A strength failure is termed an *ultimate limit state,* whereas a failure to meet operational requirements, such as deflection, is termed a *serviceability limit state*.

Regardless of the approach to the design problem, the goal of the designer is to ensure that the load on the structure and its resulting load effect, such as bending moment, shear force, and axial force, in all cases are sufficiently below each of the applicable limit states. This ensures that the structure meets the required level of safety or reliability.

Three approaches to the design of steel structures are permitted by the AISC *Specification*:

1. Allowable strength design (ASD)

2. Load and resistance factor design (LRFD)

3. Design by advanced analysis

The design approaches represent alternative ways of formulating the same problem, and all have the same goal. All three are based on the nominal strength of the element or structure. The nominal strength, most generally expressed as R_n, is determined in exactly the same way, from the exact same equations, whether used in ASD or LRFD. Some formulations of design by advanced analysis, such as plastic design, also use these same nominal strength equations whereas other approaches to elastic and inelastic design addressed in Appendix 1 model in detail every aspect of the structural behavior and do not rely on the equations provided through the *Specification*. The use of a single nominal strength expression for both ASD and LRFD permits the unification of these two design approaches. It will become clear throughout this book how this approach has simplified steel design for those who have struggled in the past with comparing the two available philosophies. The following sections describe these design approaches, any one of which is an acceptable approach to structural steel design according to the AISC *Specification*.

1.9 FUNDAMENTALS OF ALLOWABLE STRENGTH DESIGN (ASD)

Prior to 2005, allowable strength design was referred to as allowable stress design. It is the oldest approach to structural design in use today and has been the foundation of AISC Specifications since the original provisions of 1923. Allowable stress design was based on the assumption that under actual load, stresses in all members and elements would remain elastic. To meet this requirement, a safety factor was established for each potential stress-producing state. Although historically ASD was thought of as a stress-based design approach, the allowable strength was always obtained by using the proper combination of the allowable stress and the corresponding section property, such as area or elastic section modulus.

The current allowable strength design approach is based on the concept that the required strength of a component is not to exceed a certain permitted or allowable strength under normal in-service conditions. The required strength is determined on the basis of specific ASD load combinations and an elastic analysis of the structure. The allowable strength incorporates a factor of safety, Ω, and uses the nominal strength of the element under consideration. This strength could be presented in the form of a stress if the appropriate section property is used. As a result of doing this, the resulting stresses will most likely again be within the elastic range, although this is not a preset requirement of the *Specification*.

The magnitude of the factor of safety and the resulting allowable strength depends on the particular governing limit state against which the design must produce a certain margin of safety. Safety factors are obtained from the *Specification*. This requirement for ASD is provided in Section B3.2 of the *Specification* as

$$R_a \leq \frac{R_n}{\Omega} \qquad \text{(AISC B3-2)}$$

which can be stated as

$$\text{Required Strength (ASD)} \leq \frac{\text{Nominal Strength}}{\text{Safety Factor}} = \text{Allowable Strength}$$

The governing strength depends on the type of structural element and the limit states being considered. Any single element can have multiple limit states that must be assessed. The safety factor specified for each limit state is a function of material behavior and the limit state being considered. Thus, it is possible for each limit state to have its own unique safety factor. For example, the limit state of yielding of a tension member is given by

$$P_n = F_y A_g$$

where F_y is the steel yield strength and A_g is the gross area of the member. The safety factor is $\Omega = 1.67$. Thus, for steel with a yield strength of 50 ksi, the allowable strength is

$$\frac{P_n}{\Omega} = \frac{50.0A_g}{1.67} = 30A_g$$

Design by ASD requires that the allowable stress load combinations of the building code be used. Loads and load combinations are discussed in detail in Chapter 2.

1.10 Fundamentals of Load and Resistance Factor Design (LRFD)

Load and resistance factor design explicitly incorporates the effects of the random variability of both strength and load. Because the method includes the effects of these random variations and formulates the safety criteria on that basis, it is expected that a more uniform level of reliability, and thus safety, for the structure and all of its components will be attained.

LRFD is based on the concept that the required strength of a component under LRFD load combinations is not to exceed the design strength. The required strength is obtained by increasing the load magnitude by load factors that account for load variability and load combinations. The design strength is obtained by reducing the nominal strength by a resistance factor, ϕ, that accounts for the many variables that impact the determination of member strength. Load factors for LRFD are obtained from the building codes for strength design and will be discussed in Chapter 2. As for ASD safety factors, the resistance factors are obtained from the *Specification*.

The basic LRFD provision is provided in Section B3.1 of the *Specification* as

$$R_u \leq \phi R_n \qquad \text{(AISC B3-1)}$$

which can be stated as

Required Strength (LRFD) ≤ Resistance Factor × Nominal Strength=Design Strength

Again, considering the limit state of yielding of a tension member,

$$P_n = F_y A_g$$

and the resistance factor is $\phi = 0.90$. For steel with a yield strength of 50 ksi, the design strength is

$$\phi P_n = 0.90(50)A_g = 45A_g$$

LRFD has been a part of the AISC Specifications since it was first issued in 1986.

1.11 Design by Advanced Analysis

The *Specification* permits a wide variety of formulations for the elastic and inelastic analysis of steel structures through the use of Appendix 1. Any analysis method will require that the structure and its elements be modeled in sufficient detail to account for all types of behavior. An analysis of this type must be able to track the structure's behavior from the unloaded condition through every load increment to complete structural failure. The only inelastic design approach that will be discussed in this book is plastic design (PD).

Plastic design is an approach that has been available as an optional method for steel design since 1961, when it was introduced as Part 2 of the then current *Specification*. The limiting condition for the structure and its members is attainment of the load that would cause the structure to collapse, usually called the ultimate strength or the plastic collapse load. For an individual structural member this means that its plastic moment capacity has been reached. In most cases, due to the ductility of the material and the member, the ultimate strength of the entire structure will not have been reached at this stage. The less stressed members can take additional load until a sufficient number of members have exhausted their individual capacities so that no further redistribution or load sharing is possible. At the point where the structure can take no additional load, the structure is said to have collapsed. This load magnitude is called the *collapse load* and is associated with a particular *collapse mechanism*.

The collapse load for plastic design is the service load times a certain load factor. The limit state for a structure that is designed according to the principles of plastic design is therefore the attainment of a mechanism. For this to occur, all of the structural members must be able to develop the yield stress in all fibers at the most highly loaded locations.

There is a fine line of distinction between the load factor of PD and the safety factor of ASD. The former is the ratio between the plastic collapse load and the service or specified load for the structure as a whole, whereas the latter is an empirically developed, experience-based term that represents the relationship between the elastic strength of the elements of the structure and the various limiting conditions for those components. Although numerically close, the load factor of plastic design and the factor of safety of allowable stress design are not the same parameter.

1.12 Structural Safety and Integrity

The preceding discussions of design philosophies indicate that although the basic goal of any design process is to ensure that the end product is a safe and reliable structure, the ways in which this is achieved may vary substantially.

In the past, the primary goal for safety was to provide an adequate margin against the consequences of overload. Load factor design and its offshoots were developed to take these considerations into account. In real life, however, many other factors also play a role. These include, but are not limited to the following:

1. Variations of material strength
2. Variations of cross-sectional size and shape
3. Accuracy of method of analysis
4. Influence of workmanship in shop and field
5. Presence and variation of residual stresses

6. Lack of member straightness

7. Variations of locations of load application points

These factors consider only some of the sources of variation of the strength of a structure and its components. An even greater source of variation is the loading, which is further complicated by the fact that different types of loads have different variational characteristics.

Thus, a method of design that does not attempt to incorporate the effects of strength and load variability will be burdened with unaccounted-for sources of uncertainty. The realistic solution, therefore, is to deal with safety as a probabilistic concept. This is the foundation of load and resistance factor design, where the probabilistic characteristics of load and strength are evaluated, and the resulting safety margins determined statistically. Each load type is given its own specific factor in each combination, and each material limit state is also given its own factor. This method recognizes that there is always a finite, though very small, chance that structural failure will actually occur. However, this method does not attempt to attach a specific value to this probability. No specific level of probability of failure is given or implied by the *Specification*.

In the historic ASD approach, the variabilities of load and strength were not treated explicitly as separate parameters. They were lumped together through the use of a single factor of safety. The factor of safety varied with each strength limit state but did not vary with load source. At that time, ASD load combinations were simply the summation of the applied loads. That approach to ASD could be thought of as LRFD with a single load factor of 1.0. Thus, when LRFD was first introduced in 1986, many engineers referred to ASD as design using "unfactored loads" and LRFD as design using "factored loads." In ASD there was no attempt to attain uniform reliability; rather, the goal was to simply have a safe structure, although some elements would be safer than others.

When originally introduced, LRFD designs were generally expected to have a more uniform level of reliability than ASD designs. That is, the probability of failure of each element in an LRFD design would be the same or very close, regardless of the type of load or load combination. However, a detailed analysis of reliability under the LRFD provisions showed that reliability still varied under different load combinations.
For the development of LRFD, load effect (member force), Q, and resistance (strength), R, are assumed to each have a variability that can be described by the normal distributions shown by the bell-shaped curves in Figure 1.15. Structures can be considered safe as long as the resistance is greater than the load effect, $R > Q$. If it were appropriate to concentrate solely on the mean values, Q_m and R_m, it would be relatively easy to ensure a structure's safety. However, the full representation of the data shows an area where the two curves overlap.

This area represents cases where the load effect exceeds the resistance and therefore identifies occurrences of failure. Safety of the structure is a function of the size of this region of overlap. The smaller the region of overlap is, the lower the probability of failure.

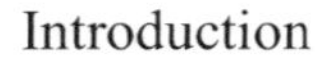

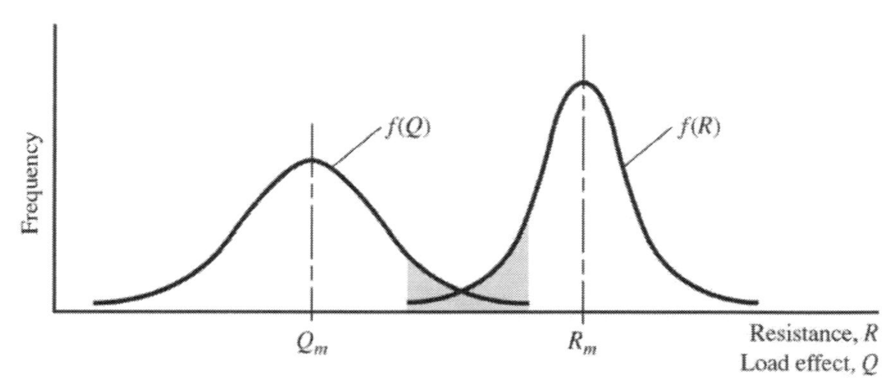

Figure 1.15 Probability Distribution, R and Q

Another approach to presenting the data is to look at the difference between resistance and load effect. Figure 1.16 shows the same data as Figure 1.15 but presents it as $(R - Q)$. For all cases where $(R - Q) < 0$, the structure is said to have failed, and for all cases where this difference is positive, the structure is considered safe. In this presentation of the data, the shaded area to the left of the origin represents the probability of failure. To limit that probability of failure, the mean value, $(R\text{–}Q)_m$, must be maintained at an appropriate distance from the origin. This distance is shown in Figure 1.16 as $\beta\sigma_{(R-Q)}$, where β is the reliability index and $\sigma_{(R-Q)}$ is the standard deviation of $(R\text{–}Q)$.

A third representation of the data is shown in Figure 1.17. In this case, the data is presented as $\ln(R/Q)$. The logarithmic form of the data is a well-conditioned representation and is more useful in the derivation of the factors required in LRFD. If we know the exact distribution of the resistance and load effect data, the probability of failure can be directly related to the reliability index β. Unfortunately, we know the actual distributions for relatively few resistance and load effect components. Thus, we must rely on other characteristics of the data, such as means and standard deviations.

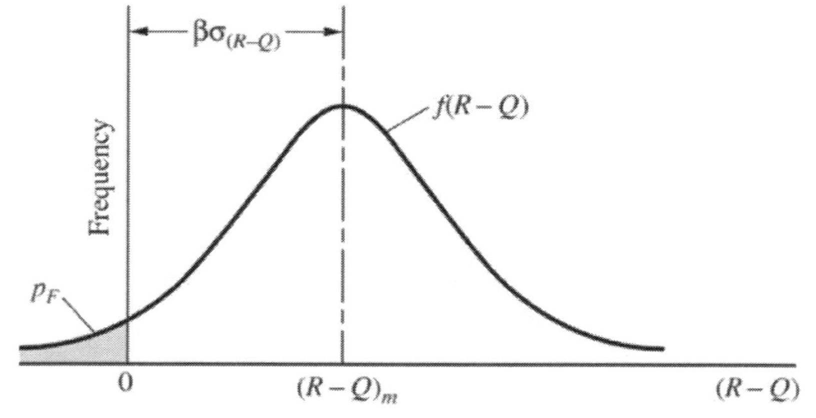

Figure 1.16 Probability Distribution, $(R - Q)$

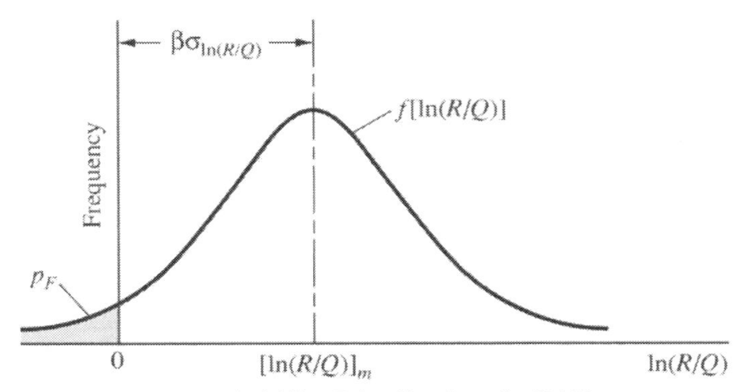

Figure 1.17 Probability Distribution, $\ln(R/Q)$

The statistical analyses required to establish an appropriate level of reliability have been carried out by the appropriate specification committees, and the resulting load factors, resistance factors, and safety factors have been established. Load factors are presented in the building codes whereas resistance factors and safety factors for each limit state are given in the *Specification*.

Since the load combinations and resistance and safety factors have been established, the reliability can be determined for specific design situations. The reliability index, β, is given in the *Specification* Commentary as

$$\beta = \frac{\ln(R_m/Q_m)}{\sqrt{V_R^2 + V_Q^2}} \qquad \text{(AISC C-B3-2)}$$

where R_m is the mean resistance, Q_m is the mean load effect, as discussed earlier, and V_R and V_Q are the coefficients of variation of resistance and load effect respectively. Design according to LRFD is given by

$$R_u \leq \phi R_n \qquad \text{(AISC B3-1)}$$

where the required strength, R_u is another term for the load effect, Q, R_n is the nominal strength and ϕ is the resistance factor. The reliability of design is determined when the required strength is exactly equal to the available strength. Thus, Equation B3-1 can be rewritten as $Q = \phi R_n$. The load effect will depend on the load combination being considered. Thus, for the LRFD live load plus dead load combination, written in terms of the live-to-dead load ratio, L/D,

$$Q = 1.2D + 1.6L = \left(1.2 + 1.6\left(L/D\right)\right)D = \phi R_n \qquad (1.1)$$

From Ravindra and Galambos[1] the mean resistance is given by

[1] Ravindra, M.K. and Galambos, T.V. (1978), "Load and Resistance Factor Design for Steel," *Journal of the Structural Division*, American Society of Civil Engineers, Vol. 104, No. ST9, September, pp. 1,337–1,353.

$$R_m = R_n M_m F_m P_m \tag{1.2}$$

and the coefficient of variation of the resistance is given by

$$V_R = \sqrt{V_M^2 + V_F^2 + V_P^2} \tag{1.3}$$

M_m is the mean of the ratio of the actual yield stress to the specified yield stress and V_M is the coefficient of variation; F_m is the mean of the ratio of the actual section property to the *Manual* value and V_F is the coefficient of variation; and P_m is the mean of the ratio of the test specimen strength to the predicted strength using the *Specification* equations and the actual material and geometric properties and V_P is the coefficient of variation.

Solving Equation 1.1 for R_n and substituting into Equation 1.2 yields

$$R_m = \frac{\left(1.2 + 1.6\left(L/D\right)\right)D}{\phi} M_m F_m P_m \tag{1.4}$$

Rearranging Equation 1.2 yields

$$M_m F_m P_m = \frac{R_m}{R_n} \tag{1.5}$$

Thus, combining Equations 1.4 and 1.5 gives

$$R_m = D\left(1.2 + 1.6\left(L/D\right)\right)\left(\frac{R_m}{\phi R_n}\right) \tag{1.6}$$

From Ravindra and Galambos the mean load effect for dead load plus live load is

$$Q_m = D_m + L_m \tag{1.7}$$

which can, with some manipulation, be rewritten as

$$Q_m = D_m + L_m = \left(D_m/D + \left(L_m/L\right)\left(L/D\right)\right)D = \gamma_{Q_m} D \tag{1.8}$$

Ravindra and Galambos also give the coefficient of variation of the load effect, which can be written as a function of the live-to-dead load ratio, as

$$V_Q = \frac{\sqrt{\left(\left(D_m/D\right)V_D\right)^2 + \left(\left(L_m/L\right)\left(L/D\right)V_L\right)^2}}{Q_m/D} \tag{1.9}$$

If Equations 1.6 and 1.8 are substituted into Equation C-B3-2 the reliability index, β, will be given in terms of the live-to-dead load ratio, L/D, and the resistance factor, ϕ. Thus,

$$\beta = \frac{1}{\sqrt{V_R^2 + V_Q^2}} \ln\left[\frac{R_m}{\phi R_n}\left(\frac{(1.2 + 1.6(L/D))D}{((D_m/D) + (L_m/L)(L/D))D}\right)\right] \tag{1.10}$$

For LRFD, the complete live plus dead load combinations are 1.4D and (1.2D + 1.6L). The discussion above used only one of those combinations. To include both, the effective dead load factor as a function of the live-to-dead-load ratio can be taken as

$$\gamma_{\text{LRFD}_i} = \max\begin{bmatrix} 1.4 \\ 1.2 + 1.6(L/D)_i \end{bmatrix} \tag{1.11}$$

and the mean load effect dead load multiplier from Equation 1.8 is

$$\gamma_{m_i} = D_m/D + (L_m/L)(L/D)_i \tag{1.12}$$

Thus, Equation 1.10 can be generalized to address other LRFD load combinations as follows:

$$\beta = \frac{1}{\sqrt{V_R^2 + V_Q^2}} \ln\left[\frac{R_m}{\phi R_n}\left(\frac{1.2 + 1.6(L/D)}{(D_m/D) + (L_m/L)(L/D)}\right)\right] = \frac{1}{\sqrt{V_R^2 + V_{Q_i}^2}} \ln\left[\frac{R_m}{\phi R_n}\left(\frac{\gamma_{\text{LRFD}_i}}{\gamma_{m_i}}\right)\right] \tag{1.13}$$

where γ_{LRFD} is the effective LRFD load factor for the load combination under consideration, γ_{m_i} is the mean load effect multiplier for that load combination, and V_{Q_i} is the coefficient of variation of the load effect, all as a function of the varying load ratio as indicated by the subscript *i*.

For ASD, the complete picture for dead plus live load is given by the load combination (D + L). To convert Equation 1.13 for use with ASD load combinations, γ_{LRFD} is replaced by γ_{ASD} and ϕR_n is replaced by R_n/Ω. Thus, Equation 1.13 becomes

$$\beta = \frac{1}{\sqrt{V_R^2 + V_{Q_i}^2}} \ln\left[\frac{R_m}{(R_n/\Omega)}\left(\frac{\gamma_{\text{ASD}_i}}{\gamma_{m_i}}\right)\right] \tag{1.14}$$

Based on extensive studies for A992 steel (Bartlett et al.[2]) and the original work for the development of the 1986 AISC LRFD *Specification* by Ravindra and Galambos, the following values for means and coefficients of variation can be used:

$$M_m = 1.028; \quad V_M = 0.058$$
$$F_m = 1.00; \quad V_F = 0.05$$
$$P_m = 1.02; \quad V_P = 0.06$$

Thus,

$$R_m = R_n(1.028)(1.00)(1.02) = 1.049R_n$$

$$V_R = \sqrt{(0.058)^2 + (0.05)^2 + (0.06)^2} = 0.097$$

Based on Galambos et al.[3], the ratio of mean to code specified dead and live loads and coefficients of variation can be taken as, for dead load

$$D_m/D = 1.05; \quad V_D = 0.10$$

and for live load

$$L_m/L = 1.00; \quad V_L = 0.25$$

The values for γ_{m_i} and V_{m_i} will be functions of the live-to-dead load ratio.

The reliability index, β, based on Equations 1.13 and 1.14, for a live-to-dead load ratio from 0 to 6.0 is presented in Figure 1.18 for a compact laterally supported wide-flange beam under uniform moment for both LRFD and ASD. The figure is based on the load combination of live load plus dead load and the statistical variations consistent with those used in the development of the *Specification* and presented above. It is seen that the reliability of design by LRFD is somewhat more uniform for this condition than design by ASD and that at a live to dead load ratio of approximately 3, the two approaches yield the same reliability. The higher the reliability index, the safer the structure. Regardless of the numerical value of *R*, any structure that meets the requirements of the *Specification* will be sufficiently safe. The reliabilities presented in Figure 1.18 with today's materials and load combinations are very close to what they were in 1986 when first developed. Reliability for structural steel under load combinations that include wind will be presented later in Chapter 2. A more detailed discussion of the statistical basis of steel design is available in *Load and Resistance Factor Design of Steel Structures*.[4] Since the introduction of the 2005 AISC *Specification*, design by ASD and LRFD have essentially been equivalent and differ

---

[2] Bartlett, R.M., Dexter, R.J., Graeser, M.D., Jelinek, J.J., Schmidt, B.J. and Galambos, T.V. (2003), "Updating Standard Shape Material Properties Database for Design and Reliability," *Engineering Journal,* American Institute of Steel Construction, Vol. 40, No. 1, pp. 2–14.

[3] Galambos, T.V., Ellingwood, B., MacGregor, J.G. and Cornell, C.A. (1982), "Probability-Based Load Criteria: Assessment of Current Design Practice," *Journal of the Structural Division*, American Society of Civil Engineers, Vol. 108, No. ST5, May, pp. 959–977.

[4] Geschwindner, L. F., Disque, R. O., and Bjorhovde, R. (1994), *Load and Resistance Factor Design of Steel Structures.* Englewood Cliffs, NJ: Prentice Hall.

only by the effect of load combinations. This will be demonstrated in a discussion of calibration in Section 2.6.

General structural integrity requires a continuous load path to the ground for resisting all gravity and lateral loads that might be applied to the structure. The AISC *Specification* addresses structural integrity beyond these general requirements when required by the applicable building code in Section B3.9. These requirements are beyond normal strength requirements and are intended to improve the connectivity of the structure and thus the performance of the structure under undefined extraordinary events. These requirements apply only to a small set of structures where additional structural integrity is mandated.

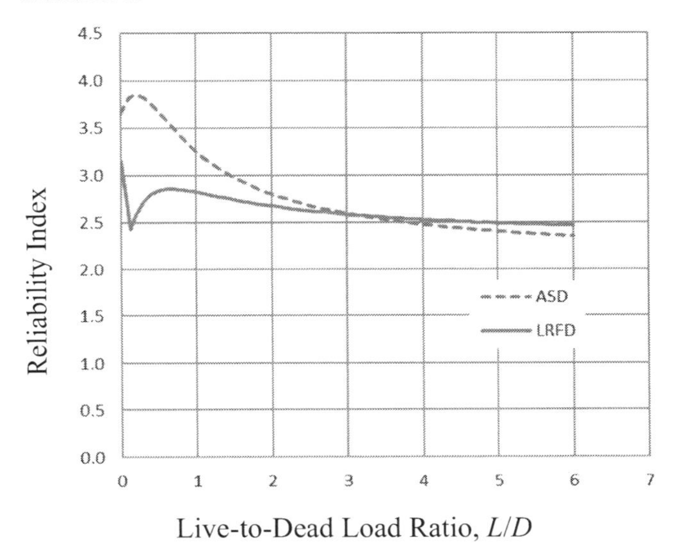

Figure 1.18 Reliability Index vs Live-to-Dead Load Ratio for Compact Laterally Supported Wide-Flange Beam with Uniform Moment

1.13 Limit States

Regardless of the design approach, ASD or LRFD, or the period in history of the design's execution, 1923 or 2024, all design is based on the ability of a structure or its elements to resist load. This ability is directly related to how an element carries that load and how it might be expected to fail, which is referred to as the element's limit state. Each structural element can have multiple limit states, and the designer is required to determine which of these limit states will actually limit the structure's capacity.

There are two types of limit states to be considered: strength limit states and serviceability limit states. Strength limit states are those limiting conditions that, if exceeded, will lead to collapse of the structure or a portion of the structure, or to such serious deformations that the structure can no longer be expected to resist the applied load. Strength limit states are identified by the *Specification*, and guidance is provided for determination of the nominal strength, R_n, the safety factor, Ω, and the resistance factor, ϕ.

Examples of the more common strength limit states found in the *Specification* are yielding, rupture, and buckling.

Serviceability limit states are not as well defined as strength limit states. If a serviceability limit state is exceeded, it usually means that the structure has reached some performance level that someone would find objectionable. The *Specification* addresses design for serviceability in Chapter L and defines serviceability in Section L1 as "a state in which the function of a building, its appearance, maintainability, durability, and the comfort of its occupants are preserved under typical usage." Chapter L lists deflections, drift, vibration, wind-induced motion, thermal expansion and contraction, and connection slip as items to be considered, although no specific limitations are given for any of these limit states.

Strength and serviceability limit states will be addressed throughout this book as appropriate for the elements or systems being considered.

1.14 Building Codes and Design Specifications

The design of building structures is regulated by a number of official, legal documents that are known commonly as *building codes*. These cover all aspects of the design, construction, and operation of buildings and are not limited to just the structural design aspects.

The model code currently in use in the United States is the ICC International Building Code. Model codes are published by private organizations and have been adopted, in whole or in part, by state and local governments as the legal requirements for buildings within their area of jurisdiction. In addition to the model codes, cities and other governmental entities have written their own local building codes. Unfortunately, since the adoption of a building code is in great part a political activity, the regulations in use across the country are not uniform. A new International Building Code is published every 3 years but not uniformly adopted as quickly as issued across all jurisdictions. Thus, building codes with effective dates from 2003 to 2021 are still in use. In addition, governmental bodies will often adopt a model code with local amendments. Because of the technical nature of the AISC *Specification*, local amendments normally do not affect those aspects of steel design, but they often do modify the loading definitions and thus do ultimately affect steel design.

To the structural engineer, the most important sections of a building code deal with the loads that must be used in the design, and the requirements pertaining to the use of specific structural materials. The load magnitudes are normally taken from *Minimum Design Loads and Associated Criteria for Buildings and Other Structures*, a national standard published by the American Society of Civil Engineers, Structural Engineering Institute as ASCE/SEI 7. The loads presented in ASCE/SEI 7 may be altered by the model code authority or the local building authority upon adoption, although this practice adds complexity for designers who may be called upon to design structures in numerous locations under different political entities. Throughout this book, ASCE/SEI 7 will be referred to simply as ASCE 7 as it is most commonly referred to in the profession.

The AISC *Specification* is incorporated into the model building code by reference. The *Specification*, therefore, becomes part of the code, and thus part of the legal requirements of any locality where the model code is adopted. Locally written building codes also exist and the AISC *Specification* is normally adopted within those codes by reference also. Through these adoptions the AISC *Specification* becomes the legally binding standard by which all structural steel buildings must be designed. However, regardless of the *Specification* rules, it is always the responsibility of the engineer to ensure that their structure can carry the intended loads safely, without endangering the occupants.

1.15 INTEGRATED DESIGN PROJECT

This section introduces a building to be used in subsequent chapters of this book as an integrated design project. It is a relatively open-ended design project in that only a limited set of design parameters are set at this point. Several options will be presented in subsequent chapters so that the project can be tailored at the desire of the instructor.

The building is a four-story office building with one story below grade. It is located in Downers Grove, Illinois, at approximately 42°N latitude and 88°W longitude. This is a 102,000 ft^2 building with approximately 25,500 ft^2 per above-grade floor. For the first three floors, the floor-to-floor height is 13 ft 6 in. For the top floor, the floor-to-roof height is 14 ft 6 in. The below-grade floor-to-floor height is 15 ft 6 in. The façade is a lightweight metal curtain wall that extends 2.0 ft above the roof surface, and there is a 6.0 ft screen wall around the middle bay at the roof to conceal mechanical equipment and roof access. All steel will receive spray-applied fireproofing as necessary.

Based on preliminary discussions with the architectural design team, the design will start with bay sizes of 30.0 ft in the east-west direction and 45.0 ft, 30.0 ft, and 45.0 ft in the north-south direction, as shown in Figure 1.19. Representative floor and roof framing plans are shown in Figures 1.20 through 1.22. To accommodate a two-story atrium on the first floor, the second-floor framing plan shows an opening bounded by column lines A, C, 4, and 5. The lateral load resisting system consists of a pair of three-bay moment frames in the east-west direction and a pair of one-bay chevron braced frames in the north-south direction.

Much of today's structural analysis and design is accomplished through the use of integrated analysis and design software. Figure 1.23 shows an example of a complete three-dimensional model of the given preliminary framing system developed using RAM Structural System. Figure 1.24 shows the results of the same computer model with the gravity-only structural elements removed.

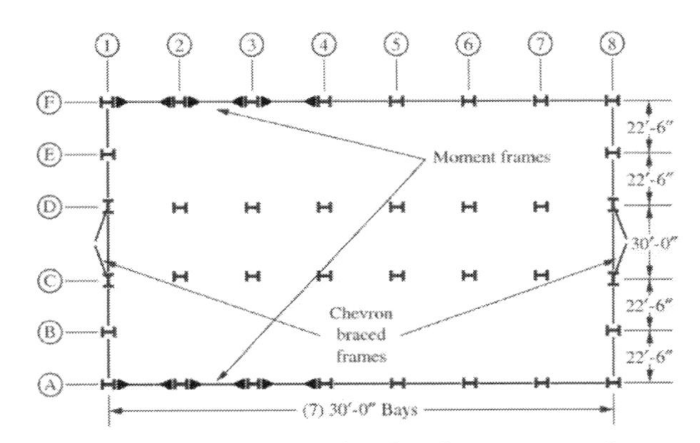

Figure 1.19 Schematic Plan for Integrated Design Problem

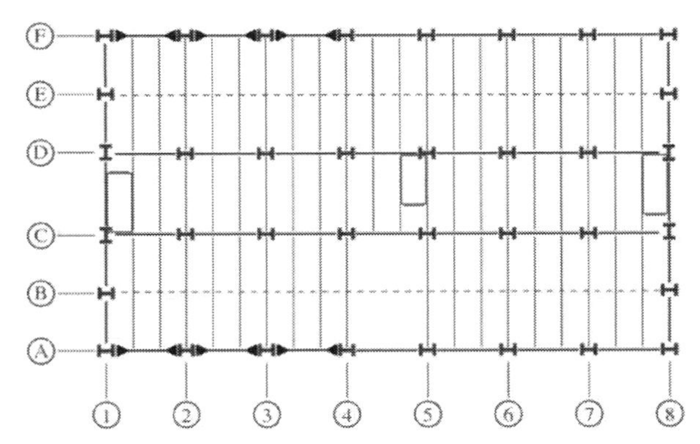

Figure 1.20 Representative Second-Floor Framing Plan

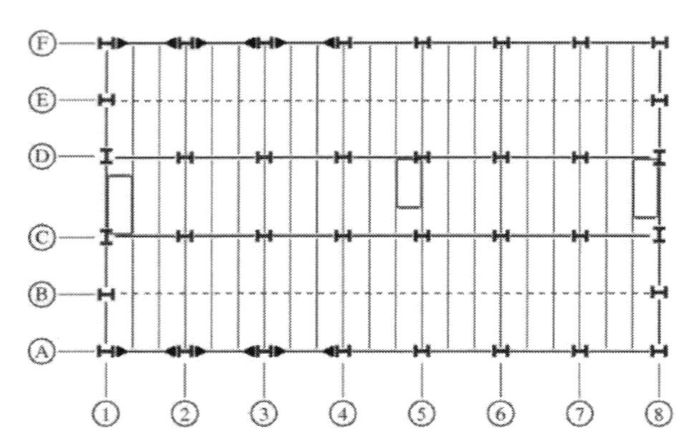

Figure 1.21 Representative First-, Third-, and Fourth-Floor Framing Plan

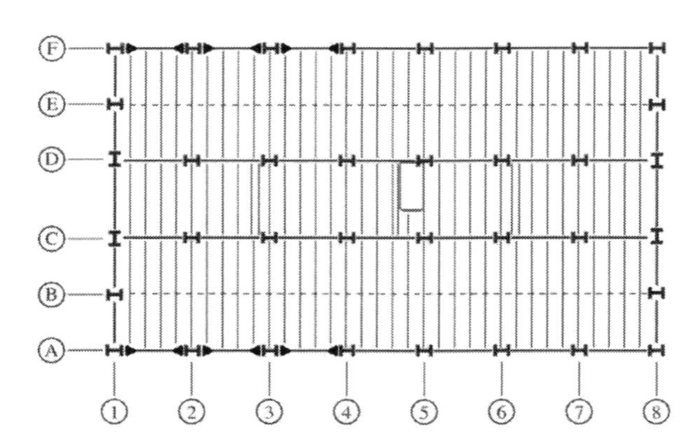

Figure 1.22 Representative Roof Framing Plan

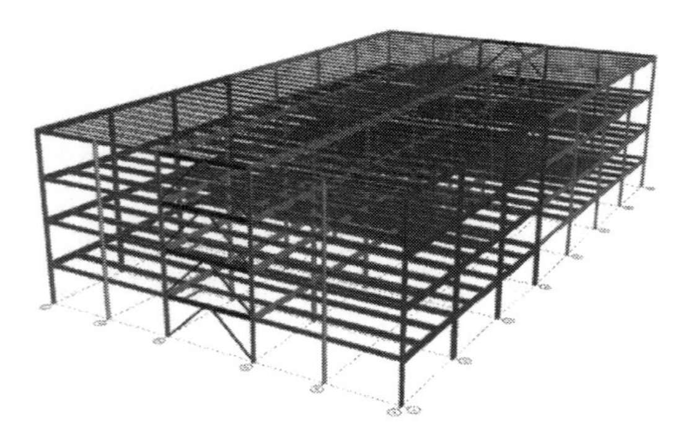

Figure 1.23 Three-Dimensional Computer Model of Complete Structure from RAM Structural System

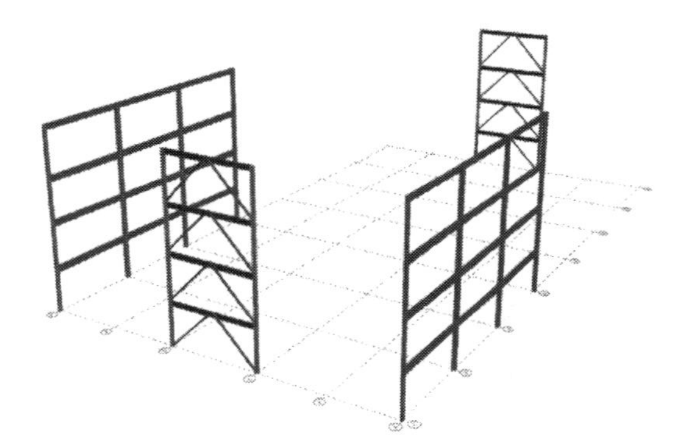

Figure 1.24 Three-Dimensional Computer Model Showing Only Lateral Load Resisting Systems from RAM Structural System

1.16 PROBLEMS

1. Where could one find information about the provisions of the 1961 AISC *Specification*?

2. What resource would be most likely to assist in the determination of properties of a steel member found in a building built in 1954?

3. Which chapter of the AISC *Specification* provides information about:

- **a.** general requirements for analysis and design
- **b.** design of members for flexure
- **c.** design of connections

4. Discuss the types of members addressed in AISC *Specification* Chapter H.

5. What types of members are addressed in AISC *Specification* Chapter I?

6. In the *AISC Steel Construction Manual*, where can one find:

- **a.** the AISC *Specification*
- **b.** design aids for bolts
- **c.** dimensions and properties for structural steel shapes
- **d.** design aids for compression members

7. What types of members does Part 5 of the *Manual* address?

8. What is the primary purpose of *Manual* Part 16?

9. List and define the three basic goals of a design team for the design of any building.

10. All structures are composed of some or all of five basic structural types. List these five basic structural components and provide an example of each.

11. What is the difference between beams, girders, and joists?

12. Provide an example of each of the following types of construction. To the extent possible, identify specific buildings in your own locale.

- **a.** Bearing wall
- **b.** Beam-and-column
- **c.** Long-span
- **d.** High-rise
- **e.** Gable-frame

13. What type of structural system uses the combined properties of two or more different types of materials to resist the applied loads?

14. List and describe two types of lateral load resisting systems commonly used in beam-and-column construction.

15. In designing a steel structure, what must be the primary concern of the design engineer?

16. Provide a simple definition of structural design.

17. Describe the difference between a strength limit state of a structure and a serviceability limit state.

18. Give a description of both the LRFD and ASD design approaches. What is the fundamental difference between the methods?

19. Provide a brief description of plastic design (PD).

20. Identify three sources of variation in the strength of a structure and its components.

21. Define the terms in Equation 1.2.

22. Define the terms in Equation 1.3.

23. It was shown that designs by ASD and LRFD result in the same reliability index for a specific Live-to-Dead load ratio. What is that ratio and describe the ramifications for other Live-to-Dead load ratios.

24. Provide three examples of strength, or ultimate, limit states.

25. Provide three examples of serviceability limit states.

Chapter 2

Loads, Load Factors, and Load Combinations

Boston University Center for Computing & Data Sciences
Photo courtesy of LeMessurier Consultants
© John Cannon

2.1 Introduction

Material design specifications, like the AISC *Specification*, do not normally prescribe the magnitudes of loads that are to be used as the basis for design. These loads vary based on the usage or type of occupancy of the building, and their magnitudes are dictated by the applicable local, regional, or state laws, as prescribed through the relevant building code.

Building code loads are given as nominal values. These values are to be used in design, even though it is well known that the actual load magnitude will differ from these specified values. This is a common usage of the term *nominal*, the same as will be used for the nominal depth of a steel member, to be discussed later. These nominal values are determined on the basis of material dimensions and densities for dead load, load surveys for live loads, weather data for rain, snow and wind loads, and geological data for earthquake or seismic loads. These loads are further described in Section 2.2. To be reasonably certain that these loads are not exceeded in a given structure, code load values have generally been higher than the actual loads on a random structure at an arbitrary point in time. This somewhat higher load level also accounts for the fact that all structural loads will exhibit some random variations as a function of time and load type.

To properly address this random variation of load, an analysis reflecting time and space interdependence should be used. This is called a *stochastic analysis*. Many studies have dealt with this highly complex phenomenon, especially as it pertains to live load in buildings. However, the use of time-dependent loads is cumbersome and does not add significantly to the safety or economy of the final design. For most design situations the building code will specify the magnitude of the loads as if they were constant or unchanging. Their time and space variations are accounted for through the use of the maximum load occurring over a certain reference or return period. American live load criteria are based on a reference period of 50 years.

The geographical location of a structure plays an important role for several load types, such as those from snow, wind, or earthquake. These loads also are specified based upon a reference period appropriate for the type of load.

2.2 BUILDING LOAD SOURCES

Many types of loads may act on a building structure at one time or another, and detailed data for each are given later. Loads of primary concern to the building designer include:

1. Dead load
2. Live load
3. Snow load
4. Wind load
5. Seismic load
6. Special loads

These primary load types are characterized as to their magnitude and variability by the building code and are described in the ensuing paragraphs.

2.2.1 Dead Load

Theoretically, the dead load of a structure remains constant throughout its lifespan, unless modifications are made. The dead load includes the self-weight of the structure, as well as the weight of any permanent construction materials, such as stay-in-place formwork, partitions, floor and ceiling materials, machinery, and other equipment. The dead load may vary from the magnitude used in the design, even in cases where actual element weights are accurately calculated.

The weight of all dead load elements can be exactly determined only by actually weighing and/or measuring the various pieces that compose the structure. This is almost always an impractical solution, and the designer therefore usually relies on published data of building material properties to obtain the nominal dead loads to be used in design. These data can be found in such publications as ASCE 7, the model building codes, and product literature. Some variation will thus likely occur in the real structure. Similarly, differences are bound to occur between the weights of otherwise identical structures, representing another source of dead load variability. However, compared to other structural loads, dead load variations are relatively small, the actual mean values are quite close to the published data, and the coefficients of variation are quite small. This was used in Section 1.12 where the mean dead load-to-code dead load ratio, D_m/D, was given as 1.05 and the coefficient of variation, V_D, was given as 0.10.

2.2.2 Live Load

Live load is the load on the structure that results from all of the non-permanent installations. It includes the weight of the occupants, the furniture and moveable equipment, plus anything else that the designer could possibly anticipate might occur in the structure. The fluctuations in live load are potentially quite substantial. They vary from being essentially zero immediately before the occupants take possession to a maximum value at some arbitrary point in time during the life of the structure. The magnitude of the live load to be used in a design is obtained from the appropriate building code and is the maximum live load expected over a 50-year period. The actual live load on the structure at any given time may differ significantly from that specified by the building code. This is one reason why numerous attempts have been made to model live load and its variation and why measurements in actual buildings continue to be made. Although the nominal live loads found in modern building codes have not changed much over the years, the actual use of buildings has, and load surveys continue to show that the specified maximum live load levels are still an adequate representation of the loads the structure should be designed to resist.

The actual live load on a structure at any given point in time is called the *arbitrary point-in-time live load* (L_{apt}). Figure 2.1 shows the variation of the live load on a structure as might be obtained from a live load survey. The load specified by the building code is always higher than the actual load found in the building survey. In addition, a portion of the live load remains constant. This load comes from the relatively permanent fixtures and furnishings associated with a particular occupancy and can be referred to as the *sustained live load* (SLL). The occupants who enter and leave the space form another part of the live load, raising and lowering the overall live load magnitude with time. This varying live load is called a *transient live load* (TLL).

Since the code specified live load is the maximum anticipated live load over a 50-year span, the mean maximum live load-to-code live load, L_m/L, is taken as 1.0 as used in Section 1.12. The coefficient of variation, V_L, is greater than that for dead load and is taken as 0.25. When live load occurs in combination with other transient load, such as wind, the statistics for arbitrary point in time live load must be used. Thus, $L_{m\ apt}/L$ is taken as 0.40 with a coefficient of variation, V_L, taken as 0.55.

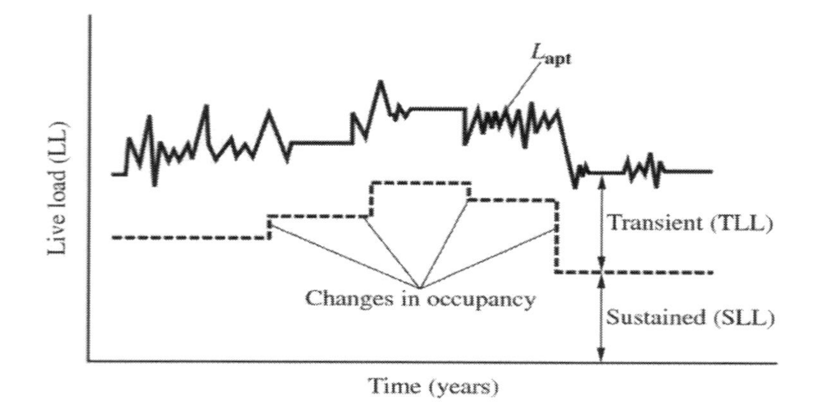

Figure 2.1 Variation of Live Load with Time

2.2.3 Snow Load

Although snow load might be considered a form of live load, unique conditions govern its magnitude and distribution, and it is treated separately in the load combinations used in design. It is the primary roof load in many geographical areas and heavily depends on local climate, building exposure, and building geometry.

Snow load data are normally based on surveys that result in isoline maps showing areas of equal depth of ground snow. Using this method, annual extreme snowfalls have been determined over a period of many years. These data have been analyzed through statistical models and the expected lifetime maximum snow loads estimated. The reference period is again the 50-year anticipated life of the structure.

A major difficulty is encountered in translating the ground snow load into a roof snow load. This is accomplished through a semi-empirical relationship whereby the ground snow load is multiplied by factors to account for properties such as roof geometry and the thermal characteristics of the roof. In addition to the uniform snow load determined using the ground snow load, drifts often lead to areas of roofs with significantly higher loads which must be accounted for. Work continues to be done to improve the method of snow load computation and to collect snowfall data.

The statistical data for snow load used to assess reliability is based on the maximum 50-year snow load. The mean maximum snow load-to-code snow load, S_m/S, is taken as 0.82 with a coefficient of variation, V_S, of 0.26. As with live load, when snow occurs in combination with wind, the arbitrary point in time data must be used. For snow, $S_{m\,apt}/S$ is taken as 0.20 with a coefficient of variation V_S taken as 0.73.

2.2.4 Wind Load

By its very nature, wind is a highly dynamic natural phenomenon. For this reason, it is also a complex problem from a structural perspective. Wind forces fluctuate significantly and are also influenced by the geometry of the structure, including the height, width, depth, plan and elevation shape, and the surrounding landscape. The basic building code approach to wind load analysis is to treat wind as a static load problem, using the Bernoulli equation to translate wind speed into wind pressure. In an approach similar to that used for snow load, a semi-empirical equation is used to give the wind load at certain levels as a function of a number of factors representing effects of features such as wind gusts, topography, and structural geometry.

The data used for determining wind loads are based on measured wind speeds. Meteorological data for 3-second wind speed gusts have been accumulated over the contiguous United States and corrected to a standard height of 33 ft. These data are then used to model the long-term characteristics for a mean recurrence interval of 300, 700, 1700 or 3000 years, as required for buildings in different risk categories. ASCE 7 and the model codes provide maps to be used as the foundation of wind force calculations. Because local site characteristics often dictate wind behavior, there are locations for which special attention must be given to wind load calculation. In addition, some buildings require (or can benefit from) special attention in determining wind load magnitude. In these cases, it might be valuable to conduct wind tunnel tests before the structural design for wind is carried out.

Since the development of the original LRFD load combinations, wind load magnitude has gone from being established at a service load level to an ultimate load level. Thus, wind load magnitude, load combinations including wind, and wind load statistics have changed. Current mean to specified wind load, W_m/W, is taken as 0.55 and the coefficient of variation, V_W, is taken as 0.35.

2.2.5 Seismic Load

The treatment of seismic load effects is extremely complicated because of the high variability of this natural phenomenon and the many factors that influence the impact of an earthquake on any particular structure. In addition, because the force the building is subject to is the result of the reaction of the mass of the building responding to the ground moving, inertia effects must be considered.

For most buildings it is sufficient to treat seismic effects through the use of equivalent static loads over the height of the building, provided that the magnitudes of these equivalent static loads properly reflect the dynamic characteristics of the seismic event. Many characteristics of the problem must be quantified in order to establish the correct magnitudes of these static loads. These characteristics include such factors as the ground motion and response spectra for the seismic event and the structural and site characteristics for the specific project. Meanwhile, the extension of seismic design requirements to all areas of the country through the current model building codes is making seismic design a requirement for many more structures. Earthquakes are no longer a design consideration only on the West coast.

Because of the complexity of seismic load determination and the acceptance of different levels of reliability for different regions of the country, a single mean to code load ratio and single coefficient of variation are not used. Thus, reliability under seismic loading is treated differently from that for wind and gravity loads only.

2.2.6 Special Loads

Several other loads are sometimes important. These include impact, blast, and thermal effects.

Impact: Most building loads are static or essentially so, meaning that their rate of application is so slow that the kinetic energy associated with their motion is insignificant. For example, a person entering a room is actually exerting a dynamic load on the structure by virtue of his or her motion. However, because of the relatively small mass and slow movement of the individual, the kinetic energy is essentially zero.

When loads are large and/or their rate of application is very high, the influence of the energy brought to bear on the structure as the movement of the load is suddenly restrained must be taken into account. This phenomenon, known as *impact*, occurs as the kinetic energy of the moving mass is translated into a load on the structure. Depending on the rate of application, the effect of the impact is that the structure experiences a load that may be as large as twice the static value of the same mass.

Impact is of particular importance for structures where machinery processes and similar actions occur. Cranes, elevators, and equipment such as repetitive-action industrial machinery could all produce impact loads that would need to be considered in a design. In addition, vibrations may be induced into a structure either by these high-magnitude impact loads or by the normally occurring occupancy loads. Although normal live load occupancy, such as walking, is not likely to produce increased design load magnitudes, the potential for vibration from these activities may require consideration in the design. Additional guidance may be found in AISC Design Guide 11, *Vibrations of Steel Framed Structural Systems Due to Human Activity*, 2nd edition.

Blast: Blast effects on buildings have become a more important design consideration during the first years of the twenty-first century. Prior to that time, blast effects were primarily considered to be accidental. These types of blast do not occur as often as impact for normal structures but should be considered under certain circumstances. Many structures designed for industrial installations, where products of a volatile nature are manufactured, are designed with resistance to blast as a design consideration. When the structure is called upon to resist the effects of blast, a great deal of effort must be devoted to determining the magnitude of the blast to be resisted.

The threat of terrorism has been increasingly recognized since the attacks on the World Trade Center and Pentagon on September 11, 2001. In order to take that threat into account, owners must determine the level of threat to be designed for and design engineers must establish the extent to which a particular threat will influence the design of a particular structure. Generally speaking, analysis and design data for blast effects are somewhat limited. Researchers continue to work toward establishing design guidelines that help determine blast effects and member strength in response to blast. Additional guidance may be found in AISC Design Guide 26, *Design of Blast Resistant Structures* and AISC Facts for Steel Buildings 2, *Blast and Progressive Collapse*.

Thermal Effects: Steel expands or contracts under changing temperatures and in so doing may exert considerable forces on the structure if the members are restrained from moving. For most building structures, the thermal effects are less significant than other loads for structural strength. Because the movement of the structure results from the total temperature change and is directly proportional to the length of the member experiencing the change, the use of expansion joints becomes important in long dimensions of structural framing. When expansion or contraction is not permitted, the resulting forces must be accommodated in the members. Additional guidance may be found in AISC Design Guide 3, *Serviceability Design Considerations for Steel Buildings*, 2nd edition and AISC Design Guide 7, *Industrial Building Design*, 3rd edition.

The AISC *Specification* includes guidance on the design of steel structures exposed to fire. Appendix 4 provides criteria for the design and evaluation of structural steel components, systems, and frames for fire conditions. In the current building design environment, design for fire is usually accomplished by means of a prescriptive approach defined in the *Specification* as design by qualification testing. If the actual thermal effects of a fire are to be addressed, the *Specification* permits design by engineering analysis. Additional guidance may be found in AISC Design Guide 19, *Fire Resistance of Structural Steel Framing*.

Other Effects: Building codes include special requirements for loads due to rain, ice, floods, and tsunamis. Although these loads may occur on a building structure, they are not the primary loads on most steel buildings and will not be discussed here.

2.3 BUILDING LOAD DETERMINATION

Once the appropriate building load sources are identified, their magnitudes must be determined. Methods to determine these magnitudes are set by the applicable building code for each load source. The following sections provide general guidance to determine the building load magnitudes, but for specific details, the applicable building code must be consulted.

2.3.1 Dead Load

Building dead load determination can be either quite straightforward or very complex. If the sizes of all elements of the structural system are known before an analysis is conducted, actual material weights may be determined and applied in the structural analysis. Selected unit weights of typical building materials are given in Table 2.1. *Manual* Table 17-13 and ASCE 7 Table C3.1-1 provide the weights of building materials, and product catalogs provide weights of things such as building mechanical equipment.

If the final sizes are not known, as is normally the case in the early stages of design, assumptions need to be made to estimate the self-weight of the structure. This necessitates an iterative process of refinement as the design and its corresponding weight are brought together.

Table 2.1 Unit Weights of Typical Building Materials

Material	Weight (lb/ft^3)
Aluminum	170
Brick	120
Concrete	
Reinforced, with stone aggregate	150
Block, 60 percent void	87
Steel, rolled	490
Wood	
Fir	32–44
Plywood	36

2.3.2 Live Load

As discussed earlier, live load magnitudes are established by the applicable building code. Table 2.2 provides values for the minimum uniformly distributed live loads for buildings for selected occupancies.

In design, how much of the load is supported by a particular element can be determined by multiplying the uniformly distributed load (lb/ft^2) by the tributary area, A_T (ft^2). The tributary area method is a simplified approach for visualizing the load on a structural element without performing the actual equilibrium calculations. It does, however, provide the same result because it is fundamentally based on an equilibrium analysis. Tributary areas for some structural members are given in Figure 2.2. For example, the tributary area for an interior column extends halfway to the next column in each direction (Figure 2.2).

Table 2.2 Minimum Uniformly Distributed Live Loads for Building Design[a]

Occupancy or use	Load (lb/ft^2)
Roof, flat	20
Residential dwellings, apartments, hotel rooms, school classrooms	40
Offices	50
Corridors above first floor	80
Auditoriums (fixed seats)	60
Retail stores	75–100
Bleachers	100
Library stacks	150
Heavy manufacturing and warehouses	250

[a]Data are taken from ASCE 7.

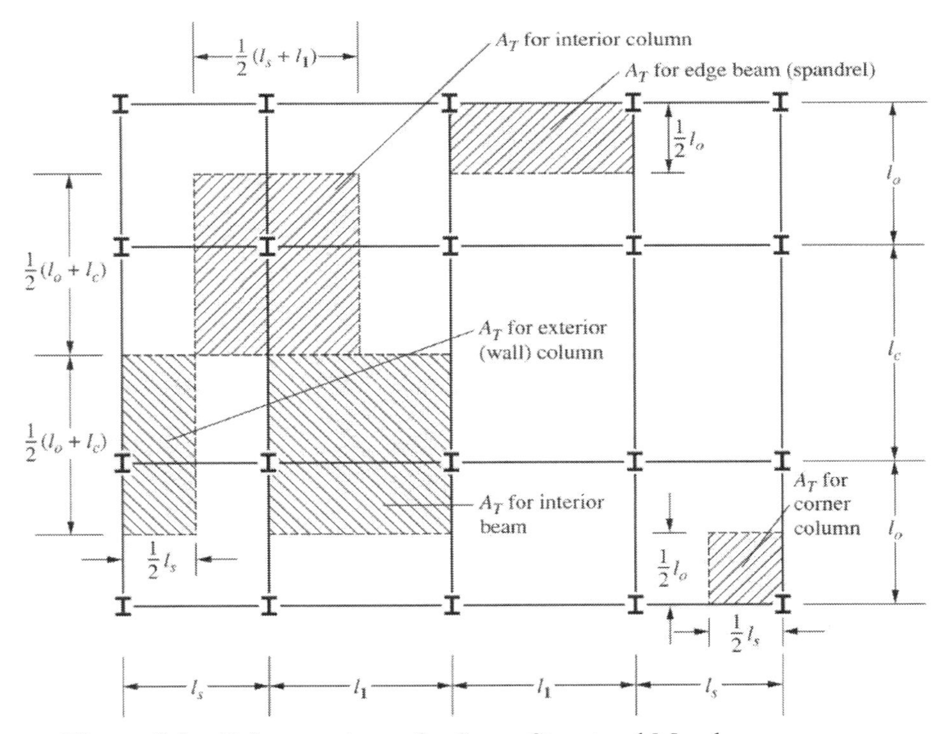

Figure 2.2 Tributary Areas for Some Structural Members

Although the concept of the tributary area can be used to determine the load on a member, an equally important concept is the *influence area*, A_I. The influence area is significant because it reflects the area over which any applied load has an influence on the member of interest. The member under consideration will experience no portion of the load applied outside of the influence area. Table 2.3 provides the relationship between tributary area and influence area, as defined by ASCE 7, for several specific structural elements, where $A_I = K_{LL}A_T$. For example, the influence area for an interior column extends all the way to the next column in each direction, resulting in an influence area that is four times the tributary area, or $K_{LL} = 4$. Several of the values in this table are simplified relative to the actual relationship, and it is always permissible to calculate the actual influence area.

Table 2.3 Live Load Element Factor, K_{LL} [a]

Element	K_{LL}
Interior columns	4
Exterior columns without cantilever slabs	4
Edge columns with cantilever slabs	3
Corner columns with cantilever slabs	2
Edge beams without cantilever slabs	2
Interior beams	2
All other members not identified above, including	1
Edge beams with cantilever slabs	
Cantilever beams	
One-way slabs	
Two-way slabs	
Members without provisions for continuous shear transfer normal to their span	

Note: In lieu of the values above, calculation of K_{LL} is permitted.
[a] Data are taken from ASCE 7.

As the influence area increases for a particular member, the likelihood of the full code-specified nominal live load actually occurring on it decreases. Because the code cannot predict the likelihood of that full area being loaded, the magnitude of the specified load is set without consideration of loaded area. Thus, the tabulated values are referred to as the unreduced nominal live load. To account for the size of the influence area and thereby provide a more realistic predictor of the actual live load on the structure, a live load reduction factor is introduced. For influence areas greater than 400 ft^2, the live load may be reduced according to the live load reduction equation:

$$L = L_o\left(0.25 + \frac{15}{\sqrt{A_I}}\right) \tag{2.1}$$

where

L = reduced live load
L_o = code-specified nominal live load
A_I = influence area = $K_{LL}A_T$

Limitations on the use of this live load reduction are spelled out in ASCE 7. The most important one is that L may not be taken less than $0.50L_o$ for members supporting one floor nor less than $0.40L_o$ for members supporting more than one floor. Other limitations and exceptions are given for loads over 100 lb/ft^2, passenger vehicle garages and assembly buildings.

2.3.3 Snow Load

Roof snow load calculations start with determination of the ground snow load for the building site. Table 2.4 provides typical ground snow load values for selected locations. The complete picture of ground snow load is provided in the appropriate building code. In many locations, however, the snowfall depth is a very localized phenomenon, and the variability is such that it is not appropriate to map those values. In these situations, local building officials should be consulted to determine what the local requirements are.

The determination of roof snow load is more complex and there are many acceptable approaches. Roof snow load on an unobstructed flat roof, as given in ASCE 7, is

$$p_f = 0.7 C_e C_{ts} p_g \tag{2.2}$$

where p_g is the ground snow load determined from the appropriate map for the appropriate risk category, as shown in Figure 2.3 for Risk Category II which is considered normal risk. Ground snow load by location may also be found using the website http://snowload.atcouncil.org/.

Table 2.4 Typical Ground Snow Loads, $p_g{}^a$

Location	Load (lb/ft^2)
Whittier, Alaska	300
Talkeetna, Alaska	120
Duluth, Minnesota	60
Minneapolis, Minnesota	50
Hartford, Connecticut	30
Chicago, Illinois	25
St. Louis, Missouri	20
Raleigh, North Carolina	15
Memphis, Tennessee	10
Atlanta, Georgia	5

[a] http://snowload.atcouncil.org/

The exposure factor, C_e, varies from a low of 0.7 for a windswept site above the tree line to a high of 1.2 for a sheltered site. The thermal factor, C_t, accounts for the melting effect of heat escaping through the roof. It varies from 0.85 for continuously heated greenhouses to 1.3 for freezer buildings. In the past there had been an importance factor for snow that varied from 0.8 for buildings that represent a low risk to human life, risk category I, to a high of 1.2 for buildings designated as essential, risk category IV. Buildings of normal risk fell into risk category II where the snow importance factor was 1.0. In the current ASCE 7, the influence of risk is included through use of the appropriate risk-based ground snow load map. Numerous other factors enter into the determination of roof snow load, including roof slope, roof configuration, snowdrift, and additional load due to rain on the snow. The applicable building code or ASCE 7 should be referred to for the complete provisions regarding snow load determination.

Figure 2.3 Ground Snow Load Map for Risk Category II from ASCE 7 (Used with permission from ASCE)

2.3.4 Wind Load

As with snow load and other geographically linked environmental loads, the starting point for wind load calculation is map based. In the case of wind loads, maps are provided based upon 3-second gust wind velocities in the building code (see Figure 2.4). Different wind speed maps are provided for the different risk categories of buildings. Thus, there are four basic wind speed maps for the four risk categories. Table 2.5 provides the wind speed data for several selected locations with varying wind velocities for risk category II. Wind speed by location may also be found using the website http://windspeed.atcouncil.org/.

These data must be transformed into wind pressures on a given building to determine the appropriate design wind loads. This transformation must take into account, in addition to the importance of the building included through selection of the appropriate map, such factors as height above the ground, relative sheltering of the site, topography, and the direction of the dominant winds.

ASCE 7 offers several approaches to the determination of design wind load for the main wind force resisting system (MWFRS). The directional procedure found in ASCE 7, Chapter 27 for buildings of all heights, where the load must be applied to the windward, leeward, and side walls converts the mapped data to velocity pressure at height z:

$$q_z = 0.00256 K_z K_{zt} K_e V^2 \tag{2.3}$$

where V is the wind speed obtained from the appropriate risk category map. The exposure coefficient, K_z, ranges from 0.57 at grade for the least critical exposure to 1.89 at a 500 ft elevation for the most critical exposure. The topographic factor, K_{zt}, is a calculated factor which is intended to account for the wind speed-up in locations of a hill, ridge or escarpment. For all other locations, $K_{zt} = 1.0$. The ground elevation factor, K_e, is intended account for air density at the site. It is permitted to take K_e =1.0 for all locations. For a site with ground elevation at 6,000 ft above sea level, K_e =0.80, its lowest value.

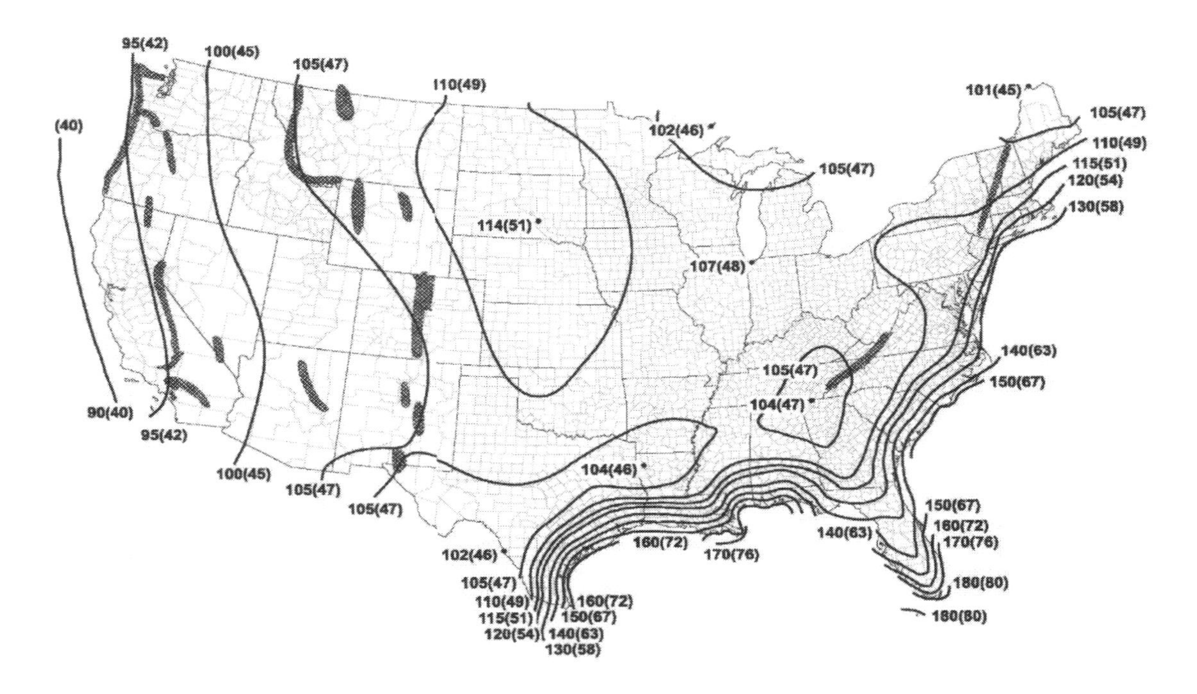

Figure 2.4 Basic Wind Speed, 3-Second Gust at 33 ft Above Ground for Risk Category II Exposure Category C from ASCE 7 (Used with Permission from ASCE)

Once the velocity pressure is determined through Equation 2.3, it must be converted to the external design wind pressure. For the MWFRS, this is given by

$$p = qGC_p - q_i(GC_{pi}) \tag{2.4}$$

where

p = design wind pressure
q = velocity pressure from Equation 2.3 as a function of wall location and height
q_i = velocity pressure from Equation 2.3 at mean roof height
G = gust-effect factor taken as 0.85 for rigid buildings
C_p = external pressure coefficient determined as a function of building geometry
GC_{pi} = internal pressure coefficient taken at ± 0.18 for enclosed buildings

The actual forces applied to the structure are then determined by multiplying the design wind pressures by the tributary areas at each level. Wind pressure distribution on the main wind force resisting system will vary with height according to the variation of the exposure coefficient, K_z. The pattern of distribution is illustrated in Figure 2.5. This pattern is commonly simplified to a "stepped" distribution of wind pressures. To account for the fact that wind may come from any direction, ASCE 7 requires that the wind load be determined for eight wind directions at 45° intervals in addition to three quarters of the load along the principal axis in conjunction with a torsional moment. This results in the wind load being applied to the structure in 12 different directions and magnitudes without consideration yet of load combinations. Because building codes have different requirements for wind load determination, the designer must review the provisions specified in the governing code. If there is no building code, ASCE 7 should be used.

Table 2.5 Approximate Representative Wind Velocities and Resulting Dynamic Pressures[a]

Location	Wind velocity (mph) V	Dynamic pressure (lb/ft^2) $= 0.00256V^2$
Miami, Florida	170	74.0
Houston, Texas	130	43.3
New York, New York	120	36.9
Chicago, Illinois	115	33.9
San Francisco, California	110	31.0

[a]Data are taken from ASCE 7 for Risk Category II.

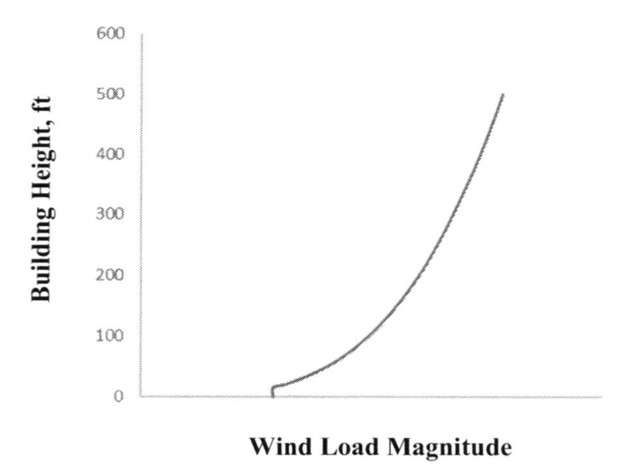

Figure 2.5 Wind Distribution over Building Height

2.3.5 Seismic Load

Perhaps the most rapidly fluctuating area of building load determination is that for seismic design. Although there have been many advances in the use of dynamic analysis for earthquake response, common practice is still to model the phenomenon using a static load. Seismic design requirements are based on the seismic design category, A through F, with A being the least severe and requiring no consideration of seismic loads. For a large majority of buildings in all seismic risk categories, ASCE 7 permits the equivalent lateral force procedure for determination of the building base shear through the expression

$$V = C_s W \tag{2.5}$$

where

V = seismic base shear
C_s = seismic response coefficient
W = effective seismic weight of the building

The seismic response coefficient determined by Method 2 is given by

$$C_s = \frac{S_{DS}}{\left(R/I_e\right)} \tag{2.6a}$$

but for $T \leq T_L$ need not be greater than

$$C_s = \frac{S_{D1}}{T(R/I_e)} \tag{2.6b}$$

and for $T > T_L$

$$C_s = \frac{S_{D1}T_L}{T^2(R/I_e)} \tag{2.6c}$$

where

S_{DS} = design spectral response acceleration for short period
S_{D1} = design spectral response acceleration for 1 sec period
T = fundamental building period
T_L = long period transition period
R = response modification factor
I_e = seismic importance factor which is 1.0 for risk categories I and II

The U.S. Geological Survey provides a web-based tool to determine site specific data for seismic design that precludes the need to use the maps found in ASCE 7. It is available at https://earthquake.usgs.gov/hazards/designmaps/.

Once the seismic base shear is determined it must be converted to horizontal loads to be applied at each story to the seismic force resisting system. For a building with uniform story heights and uniform seismic weight distribution over the height of the building, this results in a linear variation of load from a maximum at the top to zero at the bottom as illustrated in Figure 2.6.

Each seismic force resisting system is assigned a response modification factor, R, by the building code, according to its ability to resist seismic forces through a

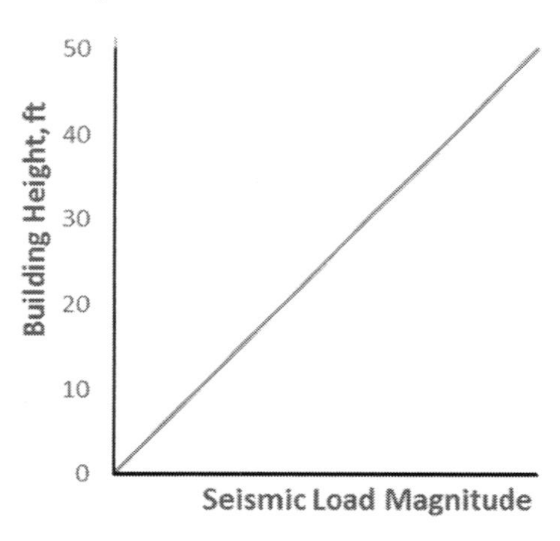

Figure 2.6 Seismic Force distribution over Height for Building with Uniform Story Height, Story Seismic Weight, and Period less than 0.5 sec.

ductile response. The most ductile steel systems, e.g., special moment frames, are assigned $R = 8$ while less ductile systems are assigned lower values. A higher R value leads, through Equation 2.6, to a lower required base shear and thus to lower lateral load on the structure. The resulting lower lateral load, however, leads to higher demand for ductility and corresponding special detailing requirements. In cases where appropriate, the selection of $R = 3$ usually permits a building structure to be designed according to the AISC *Specification* without use of any special seismic provisions. If a value of R greater than 3 is used in design of a building structure, or in cases where specifically required, the design must proceed according to the additional provisions of ANSI/AISC 341-22 *Seismic Provisions for Structural Steel Buildings*. This is discussed further in Chapter 13.

As with the other environmental loads discussed here, the details of load determination for seismic response must be found in the appropriate building code. ASCE 7 requires that the seismic load be applied to the structure in such a way as to produce the most critical load effects.

2.4 LOAD COMBINATIONS

In addition to specifying the load magnitudes for which building structures must be designed, building codes specify how the individually defined loads should be combined to obtain the maximum load effect for design. Care must be exercised in combining loads to determine the most critical combination because all loads are not likely to be at their maximum magnitude at the same time. For instance, it is unlikely that the maximum snow load and maximum wind load would occur simultaneously. Another unlikely occurrence would be a design earthquake occurring at the same time as the maximum design wind. Thus, building codes specify which loads are to be combined and at what magnitude they should be considered. The designer must exercise judgment when combining loads in situations where the normal expectations of the building code might not be satisfied or where some particular combination would result in a greater demand than previously identified.

The two design philosophies addressed in the AISC *Specification* are the direct result of the two approaches to load combinations presented in current building codes and ASCE 7.

2.4.1 ASD Load Combinations

ASD uses the load combinations defined in ASCE 7 as being for allowable stress design. The provisions in ASCE 7 for allowable stress design combine loads normally at or below their nominal or serviceability levels, the load magnitudes discussed in Section 2.3. These load combinations were historically used to determine the load effect under elastic stress distributions, and those stresses were compared to the allowable stresses established at some arbitrary level below failure, indicated by either the yield stress or the ultimate stress. Including only dead, live, wind, and roof live, the load combinations presented in ASCE 7 Section 2.4 for ASD can be written as

1. Dead
2. Dead + Live
3. Dead + Roof Live
4. Dead + 0.75 Live + 0.75 Roof Live
5. Dead + (0.6 Wind)
6. Dead + 0.75 Live + 0.75 (0.6 Wind) + 0.75 Roof Live
7. 0.6 Dead + (0.6 Wind)

In load combinations 3, 4, and 6, Roof Live refers to a code-specified roof live load, 0.7 times the snow load, or a rain load, whichever is greatest.

Used with the current AISC *Specification*, these load combinations are not restricted to an elastic stress distribution as was the practice in the distant past. Even recent past ASD editions of the AISC *Specification* had transitioned to use strength past the elastic level in many cases. The current *Specification* is a strength-based specification, not a stress-based one, and the requirement for elastic stress distribution is no longer applicable. This has no impact on the use of these load combinations but may have some historical significance to those who were educated primarily with this previously used interpretation.

2.4.2 LRFD Load Combinations

The second approach available in ASCE 7 combines loads at an amplified level. LRFD uses load combinations defined as being for strength design. These combinations permit one to investigate the ability of the structure to resist loads at its ultimate strength level. In this approach, loads are multiplied by a load factor that incorporates both the likelihood of the loads occurring simultaneously at their maximum levels and the margin against which failure of the structure is measured. Again, using only dead, live, wind, and roof live loads, the load combinations presented in ASCE 7 Section 2.3, if the live load is not greater than 100 lb/ft^2 (psf), for LRFD are

1. 1.4 Dead
2. 1.2 Dead + 1.6 Live + 0.5 Roof Live
3. 1.2 Dead + 1.6 Roof Live + (0.5 Live or 0.5 Wind)
4. 1.2 Dead + 1.0 Wind + 0.5 Live + 0.5 Roof Live
5. 0.9 Dead + 1.0 Wind

As was the case for ASD load combinations, Roof Live is intended to be taken as a code-specified roof live load, or the appropriate percentage of the snow or rain load, whichever produces the greatest load effect.

2.4.3 Controlling Load Combinations

For both ASD and LRFD, load combinations that include wind must be applied with the wind load acting independently in each of the directions specified in ASCE 7. For those combinations that include seismic load, that load must also be applied in the directions specified by ASCE 7. Load cases with a reduced dead load in combination with wind or seismic load are used to account for situations where the application of dead load might reduce the effect of the lateral load. This could become important if the actual dead load is less than anticipated. The design method to be used, and thus the choice of load combinations, is at the discretion of the designer. All current building codes permit either ASD or LRFD, and the AISC *Specification* provisions address all limit states for each approach. As discussed in Chapter 1, the resulting design may differ slightly for each design philosophy because the approach taken to ensure safety is different, but safety is assured when the appropriate building code and the AISC *Specification* are followed, regardless of the design approach. When there are no other important factors under consideration, it is recommended that design be carried out according to the requirements for LRFD.

It may be helpful to consider how these load combinations interact. This will be done by looking first at the ASD load combinations of dead, live, and wind as given earlier as combinations 2, 5, and 6. To determine which load combination controls, we will first ask, when is combination 2 greater than combination 6? Setting these combinations as

$$D+L \geq D+0.75L+0.45W$$

dividing by D and solving for L/D gives

$$L/D \geq 1.8W/D$$

Next, ask when is

$$D+0.75L+0.45W \geq D+0.6W$$

Again, dividing by D and solving for L/D gives

$$L/D \geq 0.2W/D$$

Figure 2.7 illustrates the relationships for determining the controlling load combinations for ASD.

For LRFD there are only two load combinations to consider, 2 and 4, thus we ask when is combination 2 greater than combination 4? Setting these as

$$1.2D+1.6L \geq 1.2D+0.5L+1.0W$$

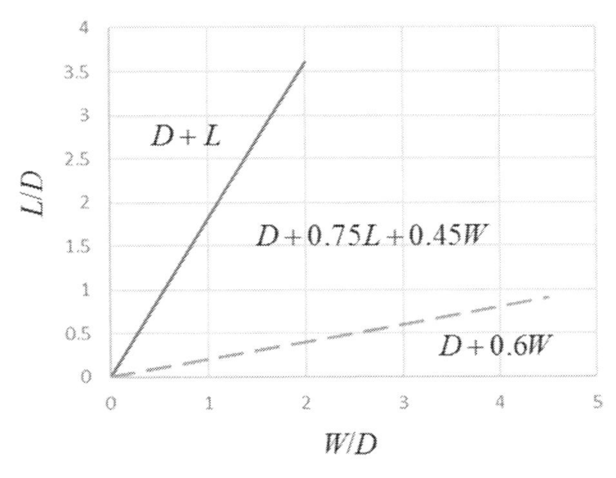

Figure 2.7 ASD Controlling Load Combinations

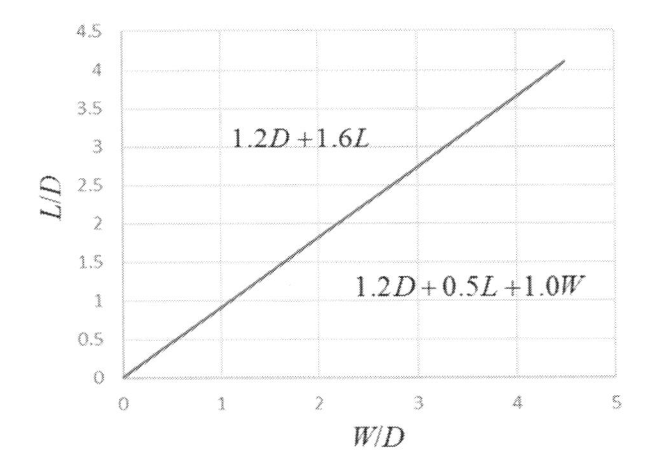

Figure 2.8 LRFD Controlling Load Combinations

Dividing by D and solving for L/D gives

$$L/D \geq 0.91W/D$$

Figure 2.8 illustrates the relationship for determining the controlling load combinations for LRFD.

Using Figures 2.7 and 2.8, for an arbitrary combination of $L/D = 3$ and $W/D = 3$, it is seen that for ASD the controlling combination is $D + 0.75L + 0.45W$ while for LRFD it is $1.2D + 1.6L$. In one case wind is involved, in the other it is not. Through comparisons like this it becomes clear that design by ASD and by LRFD have many differences.

2.5 LOAD CALCULATIONS

To understand the impact of these two approaches on analysis, it is helpful to compute the load effect for a variety of structural members according to both ASD and LRFD load combinations. The floor plan of a two-story office building is given in Figure 2.7. Load combination 2 for dead plus live load is considered for several beams and columns. The building is an office building with a nominal live load of 50 psf and a calculated dead load of 70 psf. Assume all beams and girders are simply supported and a deep floor system that can span 20 ft is assumed for simplicity.

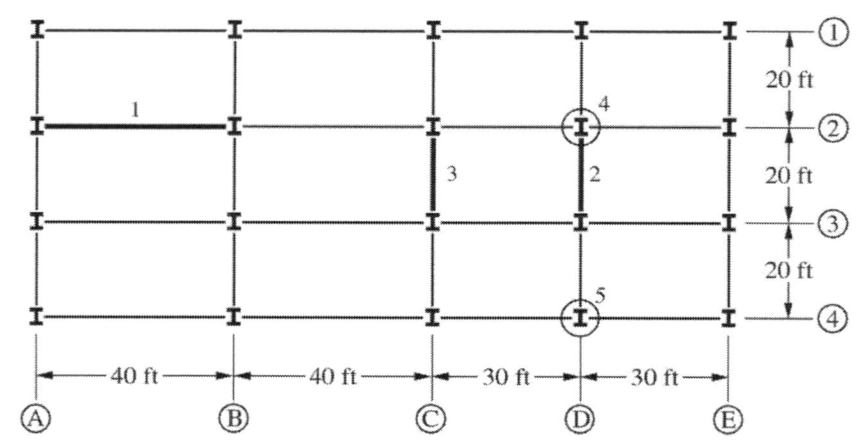

Figure 2.9 Floor Plan of a Two-Story Office Building

Example 2.1
Girder Load Calculation

Goal: Determine the required moment strength for a girder from the framing plan shown in Figure 2.9 due to live load of 50 psf and dead load of 70 psf.

Given: Girder A-B on line 2-2 if the floor deck spans from line 1-1 to 2-2 to 3-3, resulting in a uniformly distributed load on girder A-B:

SOLUTION

Step 1: Determine the live load reduction from Equation 2.1 using K_{LL} from Table 2.3

Tributary area: $A_T = (40)(20) = 800\ \text{ft}^2$
Influence area: $A_I = 2A_T = 2(800) = 1600\ \text{ft}^2 > 400\ \text{ft}^2$

Live load reduction:

$$0.25 + \frac{15}{\sqrt{A_I}} = 0.25 + \frac{15}{\sqrt{1600}} = 0.625 > 0.50$$

Since this member supports one floor the maximum reduction is 0.50.

For LRFD Step 2: Determine the required moment strength for the LRFD load combination

LRFD loads per foot:

1.2 Dead = 1.2(70 psf)(20 ft) = 1680 lbs/ft
1.6 Live = 1.6(0.625)(50 psf)(20 ft) = 1000 lbs/ft
w_u = 1.2 Dead + 1.6 Live
= 1680 + 1000 = 2680 lbs/ft = 2.68 kips/ft

Required moment strength (LRFD),

$$M_u = \frac{w_u l^2}{8} = \frac{2.68(40)^2}{8} = 536 \text{ ft-kips}$$

For ASD Step 2: Determine the required moment strength for the ASD load combination

ASD loads per foot:

Dead = (70 psf)(20 ft) = 1400 lbs/ft
Live = 0.625(50 psf)(20 ft) = 625 lbs/ft
w_a = Dead + Live = 1400 + 625 = 2030 lbs/ft = 2.03 kips/ft

Required moment strength (ASD),

$$M_a = \frac{w_a l^2}{8} = \frac{2.03(40)^2}{8} = 406 \text{ ft-kips}$$

Example 2.2 ***Bean Load Calculation***

Goal: Determine the required moment strength for a beam from the framing plan shown in Figure 2.9 due to live load of 50 psf and dead load of 70 psf.

Given: Floor beam 2-3 on line D-D if the floor deck spans from line C-C to D-D to E-E, resulting in a uniformly distributed load on floor beam 2-3:

SOLUTION

Step 1: Determine the live load reduction from Equation 2.1 using K_{LL} from Table 2.3

Tributary area: $A_T = (20)(30) = 600 \text{ ft}^2$
Influence area: $A_I = 2A_T = 2(600) = 1200 \text{ ft}^2 > 400 \text{ ft}^2$

Live load reduction:

$$0.25 + \frac{15}{\sqrt{1200}} = 0.683 > 0.50$$

Since this member supports one floor the maximum reduction is 0.50.

For LRFD Step 2: Determine the required moment strength for the LRFD load combination

LRFD loads per foot:

1.2 Dead = 1.2(70 psf)(30 ft) = 2520 lbs/ft
1.6 Live = 1.6(0.683)(50 psf)(30 ft) = 1640 lbs/ft
w_u = 1.2 Dead + 1.6 Live
= 2520 + 1640 = 4160 lbs/ft = 4.16 kips/ft

Required moment strength (LRFD),

$$M_u = \frac{w_u l^2}{8} = \frac{4.16(20)^2}{8} = 208 \text{ ft-kips}$$

For ASD Step 2: Determine the required moment strength for the ASD load combination

ASD loads per foot:

Dead = (70 psf)(30 ft) = 2100 lbs/ft
Live = 0.683(50 psf)(30 ft) = 1020 lbs/ft
w_a = Dead + Live = 2100 + 1020 = 3120 lbs/ft = 3.12 kips/ft

Required moment strength (ASD),

$$M_a = \frac{w_a l^2}{8} = \frac{3.12(20)^2}{8} = 156 \text{ ft-kips}$$

Example 2.3
Beam Load Calculation

Goal: Determine the required moment strength for a beam from the framing plan shown in Figure 2.9 due to live load of 50 psf and dead load of 70 psf.

Given: Floor beam 2-3 on line C-C if the floor deck spans from line B-B to C-C to D-D, resulting in a uniformly distributed load on floor beam 2-3:

SOLUTION

Step 1: Determine the live load reduction from Equation 2.1 using K_{LL} from Table 2.3

Tributary area: $A_T = (20)(30+40)/2 = 700 \text{ ft}^2$
Influence area: $A_I = 2A_T = 2(700) = 1400 \text{ ft}^2 > 400 \text{ ft}^2$

Live load reduction:

$$0.25 + \frac{15}{\sqrt{1400}} = 0.651 > 0.50$$

Since this member supports one floor the maximum reduction is 0.50.

For LRFD Step 2: Determine the required moment strength for the LRFD load combination

LRFD loads per foot:

1.2 Dead = 1.2(70 psf)(30 ft + 40 ft)/2 = 2940 lbs/ft
1.6 Live = 1.6(0.651)(50 psf)(30 ft + 40 ft)/2 = 1820 lbs/ft
w_u =1.2 Dead + 1.6 Live = 2940 + 1820 = 4760 lbs/ft = 4.76 kips/ft

Required moment strength (LRFD),

$$M_u = \frac{w_u l^2}{8} = \frac{4.76(20)^2}{8} = 238 \text{ ft-kips}$$

For ASD Step 2: Determine the required moment strength for the ASD load combination

ASD loads per foot:
Dead = (70 psf)(30 ft + 40 ft)/2 = 2450 lbs/ft
Live = 0.651(50 psf)(30 ft + 40 ft)/2 = 1140 lbs/ft
w_a = Dead + Live = 2450 + 1140 = 3590 lbs/ft = 3.59 kips/ft

Required moment strength (ASD),

$$M_a = \frac{w_a l^2}{8} = \frac{3.59(20)^2}{8} = 180 \text{ ft-kips}$$

Example 2.4
Column Load Calculation

Goal: Determine the required axial strength for a column from the framing plan shown in Figure 2.9 due to live load of 50 psf and dead load of 70 psf.

Given: Interior column D-2 to support one floor regardless of deck span direction:

SOLUTION

Step 1: Determine the live load reduction from Equation 2.1 using K_{LL} from Table 2.3.

Tributary area: $A_T = (30)(20) = 600 \text{ ft}^2$
Influence area: $A_I = 4A_T = 4(600) = 2400 \text{ ft}^2 > 400 \text{ ft}^2$

Live load reduction:

$$0.25 + \frac{15}{\sqrt{2400}} = 0.556 > 0.50$$

Since this member supports one floor the maximum reduction is 0.50.

For LRFD Step 2: Determine the required axial strength for the LRFD load combination

LRFD load entering column at this level:

1.2 Dead = 1.2(70 psf)(600 ft^2) = 50,400 lbs
1.6 Live = 1.6(0.556)(50 psf)(600 ft^2) = 26,700 lbs
1.2 Dead + 1.6 Live = 50,400 + 26,700 = 77,100 lbs = 77.1 kips

Required axial strength (LRFD), P_u = 77.1 kips

For ASD Step 2: Determine the required axial strength for the ASD load combination

ASD load entering column at this level:

Dead = (70 psf)(600 ft^2) = 42,000 lbs
Live = 0.556(50 psf)(600 ft^2) = 16,700 lbs
Dead + Live = 42,000 + 16,700 = 58,700 lbs = 58.7 kips

Required axial strength (ASD), P_a = 58.7 kips

Example 2.5
Column Load Calculation

Goal: Determine the required axial strength for a column from the framing plan shown in Figure 2.9 due to live load of 50 psf and dead load of 70 psf.

Given: Exterior column D-4 to support one floor regardless of deck span direction:

SOLUTION

Step 1: Determine the live load reduction from Equation 2.1 using K_{LL} from Table 2.3.

Tributary area: $A_T = (30)(10) = 300$ ft^2
Influence area: $A_I = 4A_T = 4(300) = 1200$ ft^2 >400 ft^2

Live load reduction:

$$0.25 + \frac{15}{\sqrt{1200}} = 0.683 > 0.50$$

Since this member supports one floor the maximum reduction is 0.50.

For LRFD Step 2: Determine the required axial strength for the LRFD load combination

LRFD load entering column at this level:

1.2 Dead = 1.2(70 psf)(300 ft^2) = 25,200 lbs
1.6 Live = 1.6(0.683)(50 psf)(300 ft^2) = 16,400 lbs

1.2 Dead + 1.6 Live = 25,200 + 16,400 = 41,600 lbs = 41.6 kips

Required axial strength (LRFD), P_u = 41.6 kips

For ASD Step 2: Determine the required axial strength for the ASD load combination

ASD load entering column at this level:

Dead = (70 psf)(300 ft^2) = 21,000 lbs

Live = 0.683(50 psf)(300 ft^2) = 10,200 lbs

Dead + Live = 21,000 + 10,200 = 31,200 lbs = 31.2 kips

Required axial strength (ASD), P_a = 31.2 kips

The floor plan of a 10-story office building is given in Figure 2.10. As with the previous examples, load combination 2 for dead plus live load is considered for several columns to illustrate the application of live load reduction to members that support loads from two or more floors.

The building is to be designed for flexibility in the floor layout, so the nominal live load of 80 psf for corridors above the first floor (Table 2.2) will be used. The calculated dead load is 100 psf.

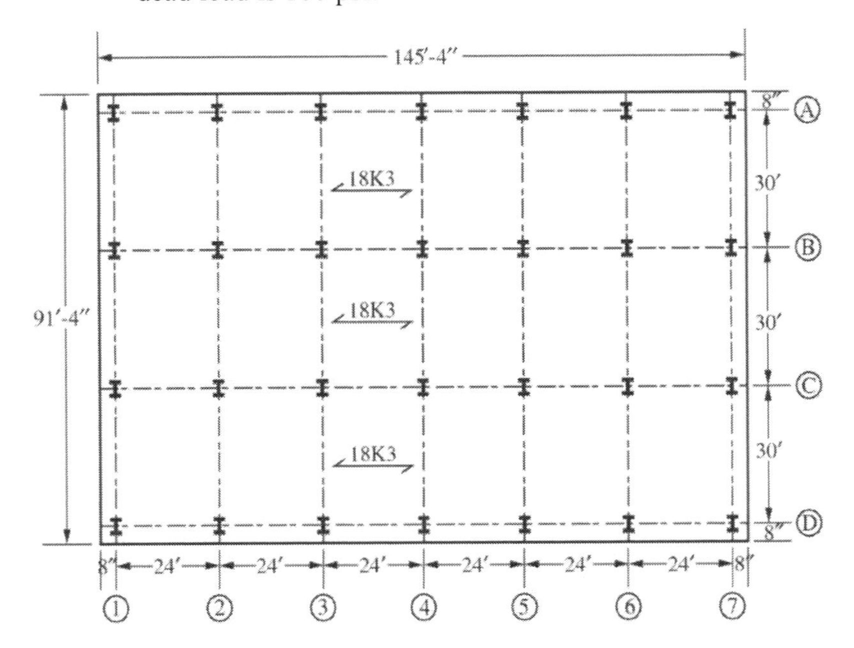

Figure 2.10 Floor Plan of a 10-story Office Building

Example 2.6
Column Load Calculation

Goal: Determine the required axial strength for a column from the framing plan shown in Figure 2.10 due to live load of 80 psf and dead load of 100 psf.

Given: Interior column C-3 supporting one level with 18K3 steel joists spanning as shown in all panels:

SOLUTION

Step 1: Determine the live load reduction from Equation 2.1 using K_{LL} from Table 2.3.

Tributary area per floor: $A_T = (30)(24) = 720$ ft^2
Influence area per floor: $A_I = 4A_T = 4(720) = 2880$ ft^2 >400 ft^2

Live load reduction:

$$0.25 + \frac{15}{\sqrt{2880}} = 0.530 > 0.50$$

Since this member supports one floor the maximum reduction is 0.50.

For LRFD Step 2: Determine the required axial strength for the LRFD load combination

LRFD load entering column at this level:
1.2 Dead = 1.2(100 psf)(720 ft^2) = 86,400 lbs
1.6 Live = 1.6(0.530)(80 psf)(720 ft^2) = 48,800 lbs
1.2 Dead + 1.6 Live = 86,400 + 48,800 = 135,000 lbs = 135 kips

Required axial strength (LRFD), $P_u = 135$ kips

For ASD Step 2: Determine the required axial strength for the ASD load combination

ASD load entering column at this level:
Dead = (100 psf)(720 ft^2) = 72,000 lbs
Live = 0.530(80 psf)(720 ft^2) = 30,500 lbs
Dead + Live = 72,000 + 30,500 = 103,000 lbs = 103 kips

Required axial strength (ASD), $P_a = 103$ kips

Example 2.7
Column Load Calculation

Goal: Determine the required axial strength for a column from the framing plan shown in Figure 2.10 due to live load of 80 psf and dead load of 100 psf.

Given: Interior column C-3 supporting five levels with 18K3 steel joists spanning as shown in all panels:

SOLUTION

Step 1: Determine the live load reduction from Equation 2.1

Tributary area per floor: $A_T = (30)(24) = 720\ \text{ft}^2$
Tributary area for five floors: $A_T = 5(720) = 3600\ \text{ft}^2$
Influence area per floor: $A_I = 4A_T = 4(720) = 2880\ \text{ft}^2$
Influence area for five floors: $A_I = 5(2880) = 14{,}400\ \text{ft}^2$

Live load reduction:

$$0.25 + \frac{15}{\sqrt{14,400}} = 0.375 < 0.40$$

Therefore, the live load reduction is limited to 0.40 since the member supports more than one level.

LRFD Step 2: Determine the required axial strength for the LRFD load combination

LRFD load in the column from five levels:
1.2 Dead = 1.2(100 psf)(3600 ft^2) = 432,000 lbs
1.6 Live = 1.6(0.40)(80 psf)(3600 ft^2) = 184,000 lbs
1.2 Dead + 1.6 Live = 432,000 + 184,000 = 616,000 lbs = 616 kips

Required axial strength (LRFD), P_u = 616 kips

ASD Step 2: Determine the required axial strength for the ASD load combination

ASD load in column from five levels:
Dead = (100 psf)(3600 ft^2) = 360,000 lbs
Live = 0.40(80 psf)(3600 ft^2) =115,000 lbs
Dead + Live = 360,000 + 115,000 = 475,000 lbs = 475 kips

Required axial strength (ASD), P_a = 475 kips

Example 2.8
Column Load Calculation

Goal: Determine the required axial strength for a first story column from the framing plan shown in Figure 2.10. The column supports nine floors above and the roof. Use a floor live load of 80 psf and dead load of 100 psf. For the roof use an unreducible live load, L_r, of 50 psf and dead load of 80 psf.

Given: Exterior column D-4 supporting nine floor levels and roof with 18K3 steel joists spanning as shown in all panels:

SOLUTION

Step 1: Determine the live load reduction from Equation 2.1 calculating the influence area without using the simplified approach from Table 2.3.

Note that the exterior column supports an 8 in. (0.67 ft) portion of slab outside the column center line. This must be included in the tributary and influence areas.

Tributary area: $A_T = 9(24)(15) + 9(24)(0.67) = 3380\text{ ft}^2$
Influence area: $A_I = 4(9)(24)(15) + 2(9)(24)(0.67) = 13{,}200\text{ ft}^2$

Live load reduction:

$$0.25 + \frac{15}{\sqrt{13{,}200}} = 0.380 < 0.40$$

Therefore, the live load reduction is limited to 0.40 since the member supports more than one level.

For LRFD Step 2: Determine the required axial strength for LRFD load combination 2.

1.2 Dead + 1.6 Live +0.5 Roof Live

LRFD load in column from nine levels of floor load:
1.2 Dead = 1.2(100 psf)(3380 ft^2) = 406,000 lbs
1.6 Live = 1.6(0.40)(80 psf)(3380 ft^2) = 173,000 lbs
1.2 Dead + 1.6 Live = 406,000 + 173,000 = 579,000 lbs = 579 kips

LRFD load in column from the roof level:
1.2 Dead = 1.2(80 psf)(376 ft^2) = 36,100 lbs
0.5 Roof Live = 0.5(50 psf)(376 ft^2) = 9,400 lbs
1.2 Dead + 0.5 Roof Live = 36,100 + 9,400 = 45,500 lbs = 45.5 kips

Required axial strength (LRFD), $P_u = 579 + 45.5 = 625$ kips

For ASD Step 2: Determine the required axial strength for ASD load combination 4.

Dead + 0.75 Live + 0.75 Roof Live

ASD load in column from ten levels of floor load:

Dead = (100 psf)(3380 ft^2) = 338,000 lbs

0.75Live = 0.75(0.40)(80 psf)(3380 ft^2) = 81,100 lbs

Dead + 0.75Live = 338,000 + 81,100 = 419,000 lbs = 419 kips

ASD load in column from the roof level:

Dead = (80 psf)(376 ft^2) = 30,100 lbs

0.75 Roof Live = 0.75(50 psf)(376 ft^2) = 14,100 lbs

Dead + 0.75 Roof Live = 30,100 + 14,100 = 44,200 lbs = 44.2 kips

Required axial strength (ASD), P_a = 419 + 44.2 = 463 kips

The 10-story building given in Figure 2.10 and just discussed for gravity load has uniform story heights of 13.5 ft. The wind load is determined to be 40 psf over the upper 40 ft, and the lateral load is to be resisted in each direction by perimeter moment frames. The 3-bay moment frames are located on grid lines A, D, 1, and 7. Use the load case that includes dead, live and wind loads to calculate resultant wind forces.

Example 2.9
Wind Load Calculation for Roof Level, Frames 1 and 7

Goal: Determine the wind force applied to roof level of a wind resisting frame for the framing plan shown in Figure 2.10 due to the load combination with live load, dead load and wind load. The wind load is 40 psf over the upper 40 ft.

Given: For the moment frames on grid lines 1 and 7, the force applied at the roof level of each of two frames:

SOLUTION

Step 1: Determine the area of loading:

Tributary area for upper half of the top story and one moment frame:

A_T = (145.3/2)(13.5/2)= 490 ft^2

For LRFD Step 2: Determine the lateral load for the LRFD load combination

LRFD wind load for load combination 4:

1.0(Wind) = 1.0(40 psf) = 40 psf

W_u = (40 psf)(490 ft^2) = 19,600 lbs = 19.6 kips

For ASD Step 2: Determine the lateral load for the ASD load combination

ASD wind load for load combination 6:

$0.75(0.6\text{Wind}) = 0.75(0.6(40 \text{ psf})) = 18 \text{ psf}$
$W_a = (18 \text{ psf})(490 \text{ ft}^2) = 8{,}820 \text{ lbs} = 8.82 \text{ kips}$

Example 2.10
Column Load Calculation for Top Floor Level, Frames A and D

Goal: Determine the wind force applied to the top floor level of a wind resisting frame for the framing plan shown in Figure 2.10 due to the load combination with live load, dead load and wind load. The wind load is 40 psf over the upper 40 ft.

Given: For the moment frames on grid lines A and D, the force applied at the first level below the roof level of each of two frames:

SOLUTION

Step 1: Determine the area of loading:

Tributary area for the first level below the roof and one moment frame:

$$A_T = (91.3/2)(13.5) = 616 \text{ ft}^2$$

For LRFD Step 2: Determine the lateral load for the LRFD load combination

LRFD wind load, load combination 4:
$1.0(\text{Wind}) = 1.0(40 \text{ psf}) = 40 \text{ psf}$
$W_u = (40 \text{ psf})(616 \text{ ft}^2) = 24{,}600 \text{ lbs} = 24.6 \text{ kips}$

For ASD Step 2: Determine the lateral load for the ASD load combination

ASD wind load, load combination 6:
$0.75(0.6\text{Wind}) = 0.75(0.6(40 \text{ psf})) = 18 \text{ psf}$
$W_a = (18 \text{ psf})(616 \text{ ft}^2) = 11{,}100 \text{ lbs} = 11.1 \text{ kips}$

2.6 Calibration

The basic requirements of the ASD and LRFD design philosophies were presented in Sections 1.9 and 1.10 and Equations B3-2 and B3-1. The required load combinations for ASD and LRFD, as found in ASCE 7, were presented earlier in the chapter. The reliability and probability of failure for ASD and LRFD will be discussed in the following section. This section establishes the relationship between the resistance factor, ϕ, and the safety factor, Ω, as found in the AISC *Specification*.

Early development of the LRFD approach to design concentrated on the determination of resistance factors and load factors that would result in a level of structural reliability consistent with previous practice but more uniform for different load combinations. Because the design of steel structures before that time had no particular safety-related concerns, the LRFD approach was calibrated to the then-current ASD approach. This calibration was carried out for the live load plus dead load combination at a live-to-dead-load ratio, *L/D*, of 3.0. This can be seen in Figure 1.18, where for a live-to-dead-load ratio of 3.0 the reliabilities for ASD and LRFD are essentially the same. It was well known that for any other load combination or live-to-dead-load ratio, the two methods could give different answers for the same design situation. This will be discussed in Section 2.7 and illustrated with Figures 2.10 and 2.11.

The current *Specification* has been developed with this same calibration, which results in a direct relationship between the resistance factor of LRFD and the safety factor of ASD. For the live load plus dead load combination in ASD, using Equation B3-2, and representing the load effect simply in terms of *D* and *L*,

$$(D+L) \leq \frac{R_n}{\Omega}$$

This same combination in LRFD, using Equation B3-1 is

$$\left(1.2D+1.6L\right) \leq \phi R_n$$

If it is assumed that the load effect is equal to the available strength and each equation is solved for the nominal strength, the results for ASD are

$$\Omega\left(D+L\right)=R_n$$

and for LRFD,

$$\frac{(1.2D+1.6L)}{\phi}=R_n$$

With *L/D* taken as 3, the above equations are set equal. They are then solved for the safety factor, which gives

$$\Omega=\frac{1.5}{\phi} \tag{2.7}$$

The resistance factors in the *Specification* were developed through a stochastic analysis to be consistent with the specified load factors and result in the desired reliability for each limit state. More detail on the development of these resistance factors can be found in Section B3.1 of the Commentary to the *Specification*. Once the resistance factors were established, the corresponding safety factors were determined using Equation 2.7. This relationship has been used throughout the *Specification* to set the safety factor for each limit state based on the previously determined resistance factor.

Although the relationship is simple, there is actually no reason for the engineer to use it to determine safety factors. This has already been done in the *Specification*, which explicitly defines resistance factors and safety factors for every limit state.

2.7 RELIABILITY AND PROBABILITY OF FAILURE

Load combinations, resistance and safety factors, and loading statistics have been discussed. However, establishing the reliability of steel design requires that one additional factor be determined. The variability that was discussed for loading must also be assessed for the resistance of the structure. In Section 1.12, the reliability of a compact, laterally supported wide-flange beam with uniform moment, for a load combination that included only dead load and live load, was presented. The resistance was presented there as

$$R_m = R_n M_m F_m P_m \tag{1.2}$$

and the coefficient of variation of the resistance as

$$V_R = \sqrt{V_M^2 + V_F^2 + V_P^2} \tag{1.3}$$

where M_m is the mean of the ratio of the actual yield stress to the specified yield stress and V_M is the coefficient of variation; F_m is the mean of the ratio of the actual section property to the *Manual* value and V_F is the coefficient of variation; and P_m is the mean of the ratio of the test specimen strength to the predicted strength using the *Specification* equations and the actual material and geometric properties and V_P is the coefficient of variation. Specific values were used there without discussion. The following will present a discussion of these factors and look at how they impact reliability.

2.7.1 Resistance Statistical Data

The required statistical data for *Material Factor*, M_m and V_M, will depend on what material property will impact the strength of the particular element being assessed. For the compact, laterally supported wide-flange beam with uniform moment previously discussed, the material property of interest is static yield stress. Table 2.6 gives some of the material property statistics for hot-rolled steel elements. Note that the mean and coefficient of variation for static yield stress are different from that used in Section 1.12 where, for A992 steel $M_m = 1.028$ and $V_M = 0.058$ were used.

Table 2.6 Material Property Statistics for Hot-Rolled Steel Elements [a]

Property	Mean	M_m	Coefficient of Variation, V_M
Static Yield Stress, Flanges	$1.05F_y$	1.05	0.10
Static Yield Stress, Webs	$1.10\,F_y$	1.10	0.11
Modulus of Elasticity	$1.00E$	1.00	0.06
Tensile Strength	$1.10F_u$	1.10	0.11
Static Yield Stress in Shear	$1.10\,F_y/\sqrt{3}$	1.10	0.10
Tensile Strength of Weld	$1.05F_{EXX}$	1.05	0.04
Tensile Strength of A325 Bolts	$1.20F_u$	1.20	0.07
Tensile Strength of A490 Bolts	$1.07F_u$	1.07	0.02

[a] Data taken from Ellingwood et at., *Development of a Probability Based Load Criterion for American National Standard A58*, SP577, National Bureau of Standards, 1980.

It is generally understood that the cross-sectional properties of hot-rolled steel elements are as given in the *Manual*. Thus, the *Fabrication Factor* is taken as $F_m = 1.0$, and the coefficient of variation is $V_F = 0.05$ for all steel elements and all cross-sectional properties.

Table 2.7 Professional Factor Statistics for Hot-Rolled Steel Elements [a]

Element	P_m	Coefficient of Variation, V_P
Tension member	1.00	0
Compact beams, uniform moment	1.02	0.06
Elastic lateral-torsional buckling beams	1.03	0.09
Inelastic lateral-torsional buckling beams	1.06	0.09
Beam-columns	1.02	0.10
Plate girders - flexure	1.03	0.05
Plate girders - shear	1.03	0.11

[a] Data taken from Ellingwood et at., *Development of a Probability Based Load Criterion for American National Standard A58*, SP577, National Bureau of Standards, 1980.

The ratio of the test specimen strength to the nominal strength predicted by the appropriate Specification equation, using the actual material and cross-sectional properties, is referred to as the *Professional Factor*. Since each limit state will have its own nominal strength equation, the statistics of the professional factor also must be established. Table 2.7 gives the professional factor statistics for hot-rolled steel elements. These factors are based on the strength equations used during the development of the 1986 LRFD Specification. Since the nominal strength equations in the current Specification, in some cases, have changed, it is likely that some new professional factor statistics were developed in conjunction with those new equations.

In Section 1.12, a compact, laterally supported wide-flange beam under uniform moment was evaluated. The statistics used there for the material factor were based on an assessment of the steels being used in modern construction, thus $M_m = 1.028$ and $V_M = 0.058$ were used. The fabrication factor and professional factor were the same as presented here. If the three statistical factors just discussed are properly combined using Equations 1.2 and 1.3, the ratio of the mean resistance to the nominal resistance and its coefficient of variation can be determined for each element and corresponding limit state. Table 2.8 lists several elements and the corresponding combined resistance statistics, based on the original development of LRFD. Again, note that for a compact beam with uniform moment these data differ from that used in Section 1.12 where $R_m/R_n = 1.049$ and $V_R = 0.097$ was used. Figure 2.9 illustrates the difference between the reliability using the original material property statistics and those based on current material statistics as used in Section 1.12.

2.7.2 Reliability under Wind Load

Reliability under gravity load was illustrated in Figures 1.18 and 2.11. It is also interesting to look at reliability under gravity plus wind. In order to address wind load combinations, it is first necessary to address the fact that we can no longer use the statistics for the static yield stress. It is usual practice in a reliability analysis to increase the mean for yield stress by a factor of 1.1 to account for the dynamic aspect of wind loading. Thus, the mean resistance becomes

$$R_m = R_n M_m F_m P_m = R_n\left(1.1(1.028)(1.0)(1.02)\right) = 1.15$$

while the coefficient of variation remains the same at

$$V_R = \sqrt{V_M^2 + V_F^2 + V_P^2} = \sqrt{(0.058)^2 + (0.05)^2 + (0.06)^2} = 0.097$$

Table 2.8 Resistance Statistics for Hot-Rolled Steel Elements [a]

Element	R_m/R_n	V_R
Tension member, yield	1.05	0.11
Tension member, ultimate	1.10	0.11
Compact beams, uniform moment	1.07	0.13
Elastic lateral-torsional buckling beams	1.03	0.12
Inelastic lateral-torsional buckling beams	1.11	0.14
Beam-columns	1.07	0.15
Plate girders – flexure	1.08	0.12
Plate girders – shear	1.14	0.16

[a] Data taken from Ellingwood et at., *Development of a Probability Based Load Criterion for American National Standard A58*, SP577, National Bureau of Standards, 1980.

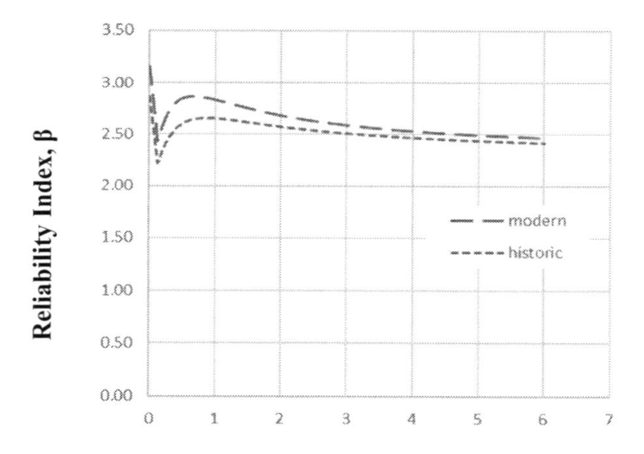

Figure 2.11 LRFD Reliability for Live plus Dead Load Based on Historic 1986 Material Property vs. Modern 2022 Material Property Data

In Section 2.2.2, the load statistics for arbitrary point in time live load were given as $L_{m\ apt}/L = 0.40$ and $V_L = 0.55$ and in Section 2.2.4 the statistics for wind load were given as $W_m/W = 0.55$ and $V_W = 0.35$. Equations 1.13 and 1.14 can now be used to determine the reliability index, β, for live plus dead plus wind load combinations, such as those given in ASCE 7. Figures 2.12 and 2.13 illustrate the reliability of a compact, laterally supported beam with uniform moment for two wind loading conditions, a wind-to-dead load ratio,

W/*D* = 4, and *W*/*D* = 10. For LRFD these results are for the load combination where the live load is less than 100 psf, load combination 4. The ASD curves show two segments reflecting the fact that there are two ASD wind load combinations that must be evaluated. The reliability of LRFD and ASD vary significantly over the range of data presented. The lowest reliability for LRFD given is $\beta = 2.15$ for $W/D = 4$ and the lowest value for ASD is $\beta = 2.12$ for $W/D = 10$. The reliability for other *W*/*D* and *L*/*D* ratios will clearly show different results. The engineer does not need to be concerned with these differences since the appropriate standards development committees, such as the ASCE 7-22 committees, have already taken this difference into account in the development of the appropriate load combinations and determined that they are acceptable.

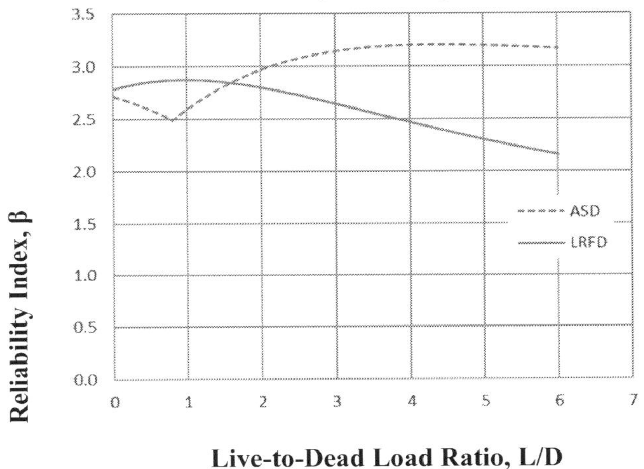

Figure 2.12 Reliability for Live Load plus Dead plus Wind for W/D = 4

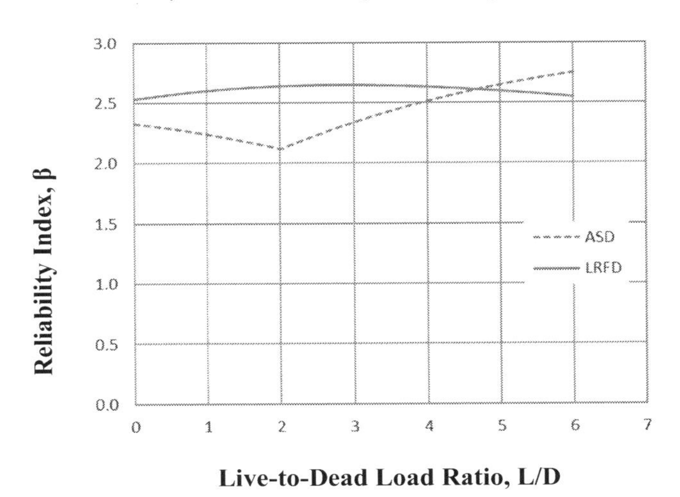

Figure 2.13 Reliability for Live plus Dead plus Wind for W/D = 10

2.7.3 Probability of Failure

Once the reliability of a system is determined it is possible to determine the probability of failure. Probability of failure was addressed in Section 1.12, but no calculations were carried out to investigate what that probability was, and no target probability of failure was established. ASCE 7-22 provides a table of target reliabilities and probabilities of failure for load conditions that do not include earthquake, tsunami, or other extraordinary events for its four risk categories. Normal buildings fall into risk category II. According to ASCE 7, Table 1.3-1, for failure that is not sudden and does not lead to widespread progression of damage, the target reliability is $\beta = 3.0$ and the target annual probability of failure is $P_F = 3.0 \cdot 10^{-5}$ per year. Examination of Figures 1.18, 2.10, and 2.11 clearly show that the target reliability is not a realistic representation of what happens in actual design. Since there is a direct relationship between the reliability index, β, and the probability of failure, P_F, it makes sense that the target probability of failure is also not likely to be met.

Calculation of the probability of failure is not a simple task but the relationship between β and P_F is fixed so there are several tables of values available in the literature. Also, an EXCEL spreadsheet can be used with the function LOGNORM.DIST()[1] and the arguments $\left(1, \beta\sqrt{V_R^2 + V_Q^2}, \sqrt{V_R^2 + V_Q^2}, \text{true}\right)$. Since the load statistics used to determine the reliability index were based on a 50-year recurrence interval, the probability of failure that results are for a 50-year interval, $P_F(50)$. To convert this to a yearly recurrence interval, the following relationship is used.

$$P_F(1) = 1 - \left(1 - P_F(50)\right)^{\frac{1}{50}} \tag{2.7}$$

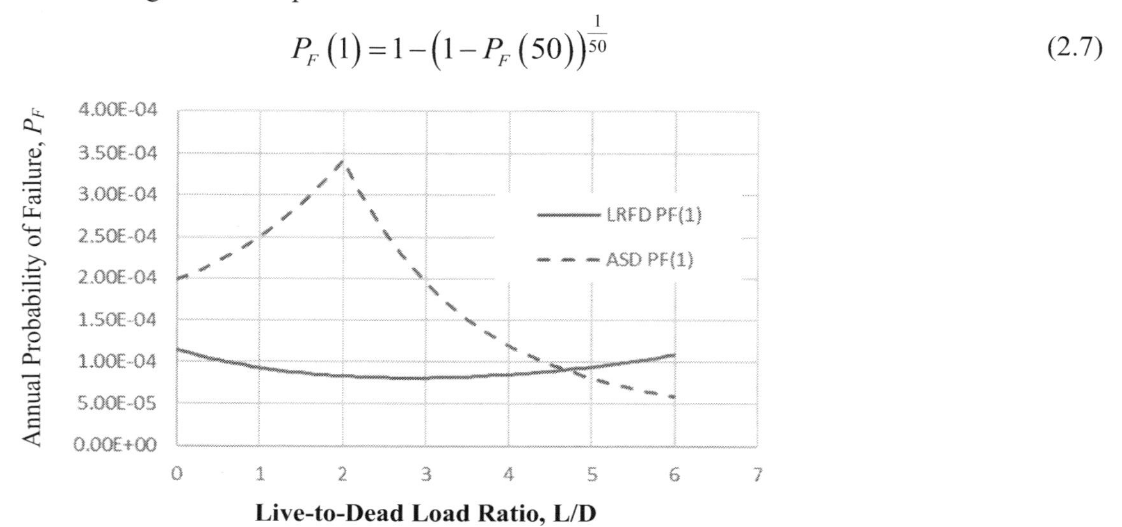

Figure 2.14 Probability of Failure for Live plus Dead plus Wind for W/D = 4

[1] Hamburger, R. O., "Is it Likely My Design Will Fail?," NASCC: The Steel Conference, AISC, 2018. https://www.aisc.org/education/continuingeducation/education-archives/is-it-likely-my-design-will-fail-current-views-from-past-higgins-award-winners-n33/

Comparison of the results presented in Figures 2.14 and 2.15 with the target probability of failure given in ASCE 7, Table 1.3-1 shows that, for the load ratios presented, the maximum probability of failure is approximately 10 times greater than the target. As when reliability index, β, does not meet the target, the engineer need not be concerned with this probability of failure since it has already been considered by the appropriate standards committees as mentioned previously. Thus, design carried out using the ASCE 7 load standard and the AISC Specification can be expected to result in a safe and reliable design.

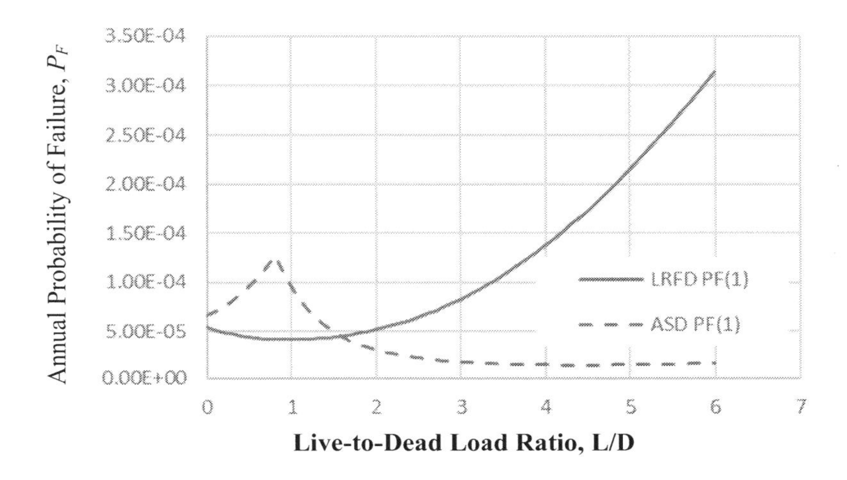

Figure 2.15 Probability of Failure for Live plus Dead plus Wind for W/D = 10

2.8 PROBLEMS

1. Name and describe five basic types/sources of building loads.

2. Categorize the following loads as dead load, live load, snow load, wind load, seismic load, or special load.

- **a.** Load on an office floor due to filing cabinets, desks, and computers.
- **b.** Load on a roof from a permanent air handling unit.
- **c.** Load on stadium bleachers from students jumping up and down during a college football game.
- **d.** Load on a building caused by an explosion.
- **e.** Weight on a steel beam from a concrete slab that it is supporting.
- **f.** Load experienced by an office building in California as it shakes during an earthquake.
- **g.** Load on a skyscraper in Chicago on a day with blustery conditions causing the building to sway back and forth.

3. What is one source you can consult to find the snow load data for a particular region as well as maps showing wind gust data to allow you to calculate wind loads?

4. Where in the AISC *Manual* can you find a table of selected unit weights of common building materials?

5. Give a source you can use to find the minimum uniformly distributed live load for buildings.

6. What analysis method allows the designer to visualize the load on a particular structural element without performing an actual equilibrium calculation?

7. What is the rationale for the live load reduction factor, when can this reduction factor be used, and what are the limitations on its use?

8. In determining the snow load on a structure, what value that can be obtained from the applicable building code is multiplied by a series of factors to obtain the roof snow load?

9. Identify and briefly describe three factors that affect the magnitude of the snow load on an unobstructed flat roof.

10. Name four factors that must be taken into account in converting wind speed data referenced by the building code into wind pressure on a given building.

11. If the local building code specifies a design load that differs from what is stated in ASCE 7, which document should you follow?

12. Special moment frames are assigned a response modification factor of 8. Why do they warrant a high value and what is its impact on the seismic base shear?

13. If a response modification factor of 3 is chosen in the design of a steel building to resist seismic loads, what design specification should be consulted?

14. Which design approach combines loads that are normally at their nominal or serviceability levels?

15. Strength load combinations that are incorporated by the LRFD method take into account what two factors?

16. Using ASCE 7, determine the minimum uniformly distributed live load for a hospital operating room.

17. Using ASCE 7, determine the minimum uniformly distributed live load for library stacks.

18. Using ASCE 7, determine the minimum uniformly distributed live load for an apartment building.

19. Using ASCE 7, determine the nominal uniformly distributed self-weight of a 6 in. thick reinforced concrete slab. If the slab has a 7/8 in. hardwood floor topping, what would its self-weight be?

20. ASCE 7-22 Table 1.3-1 provides target reliability values for various situations. It was shown here that these targets are not actually met in practice. Is that a problem for the designer? Why or why not?

21. If a design is carried out according to the applicable building codes and standards, is it guaranteed that a failure will not occur? Explain your answer.

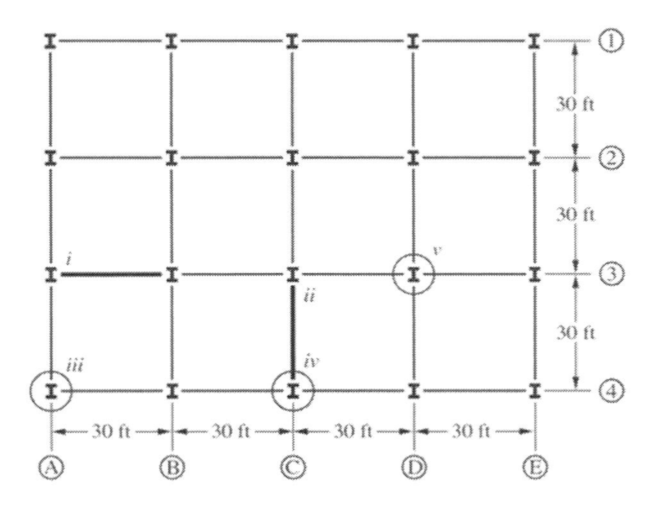

P2.22

22. A building has a column layout as shown in Figure P2.22 with 30 ft bays in each direction. It must support a uniform dead load of 90 psf and a uniform live load of 80 psf. Determine the required moment and shear strength for beams and girders, and axial strength for columns, as noted below for design by (a) LRFD and (b) ASD. Calculate with and without live load reduction, as appropriate.

i. The beam on column line 3 between column lines A and B if the deck spans from line 2-2 to 3-3 to 4-4.

ii. The girder on column line C between column lines 3 and 4 if the deck spans from line B-B to C-C to D-D.

iii. The column at the corner on lines 4 and A and supporting one floor.

iv. The column on the edge at the intersection of lines C and 4 and supporting one floor.

v. The interior column at the intersection of column lines D and 3 and supporting one floor.

23. If the framing plan shown in Figure P2.23 were for the roof of a structure that carried a dead load of 55 psf and a roof live load of 30 psf, determine the required moment and shear strength for beams and girders, and axial strength for columns, as required below for (a) design by LRFD and (b) design by ASD. Do not reduce the roof live loads.

i. The girder on column line A between column lines 1 and 2 if the deck spans from line A-A to B-B.

ii. The beam on column line 3 between column lines B and C if the deck spans from line 2-2 to 3-3 to 4-4.

iii. The column at the corner on lines 1 and E.

iv. The column on the edge at the intersection of lines 1 and B.

v. The interior column at the intersection of column lines C and 2.

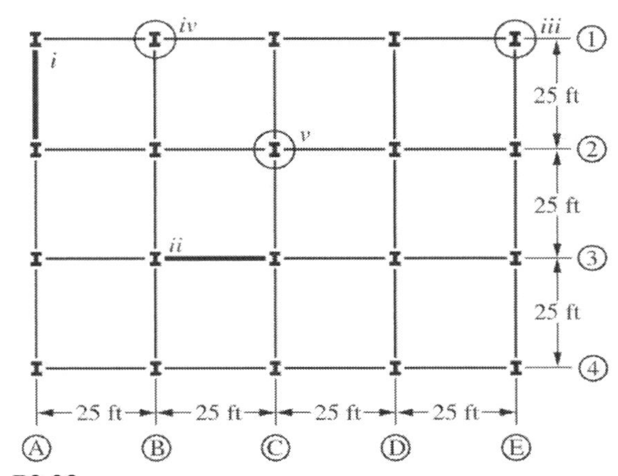

P2.23

24. The framing plan shown in Figure P2.24 is for an 18-story office building. It must support a floor and roof dead load of 80 psf, a floor live load of 50 psf, and a roof live load of 30 psf. In all cases, the decking spans in a direction from line A toward line E. Determine the required moment and shear strength for the beams and girders, and axial strength for the columns, as required below for (a) design by LRFD and (b) design by ASD. Calculate with and without live load reduction, as appropriate.

i. The beam between column lines 2 and 3 along line D.

ii. The girder on column line 3 between column line E and midway between lines D and C.

iii. The beam on the line between lines C and D and column lines 3 and 4.

iv. The column at the corner on lines 1 and E that supports eight floor levels plus the roof.

v. The column on the edge at the intersection of lines 4 and A that supports eight floor levels plus the roof.

vi. The interior column at the intersection of column line 4 and the point midway between lines C and D that supports three floor levels plus the roof.

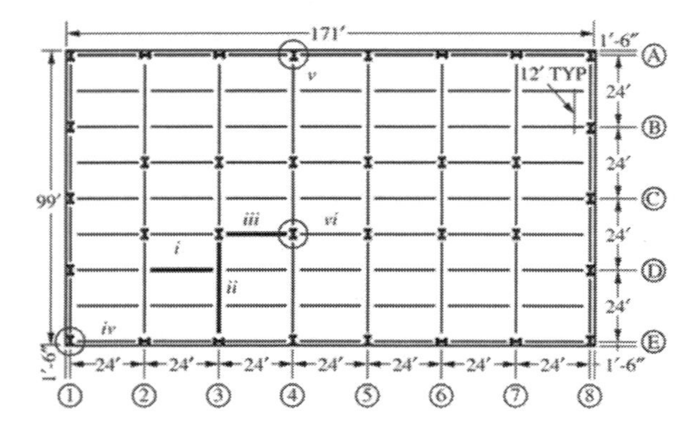

P2.24

25. The 18-story building given in Figure P2.24 must support a floor and roof dead load of 80 psf, a floor live load of 50 psf, and a roof live load of 30 psf as given in problem 24. The building has a uniform story heights of 14 ft. The wind load is determined to be 40 psf over the upper 50 ft, and the lateral load is to be resisted in each direction by perimeter moment frames. The 4-bay moment frames are located on grid lines 1 and 8 and the 7-bay moment frames are located along lines A and E. Use the load case that includes dead, live and wind loads to calculate resultant wind forces in each frame at the roof level and at the next level down for (a) design by LRFD and (b) design by ASD.

26. The floor plan given in Figure P2.24 is to be used on a 14-story building and will again support a floor and roof dead load of 80 psf, a floor live load of 50 psf, and a roof live load of 30 psf as given in problem 24. The building height comes from a 16 ft first story and uniform upper stories of 13.5 ft. The wind load is 50 psf over the upper 54 ft, and the lateral load is to be resisted in each direction by perimeter moment frames as for Problem 25. Use the load case that includes dead, live and wind loads to calculate resultant wind forces in each frame at the roof level and at the next level down for (a) design by LRFD and (b) design by ASD.

26. Integrated Design Project. For a Downers Grove, Illinois, office building, determine the dead, live, snow and wind loads based on ASCE 7 and the construction suggested below. A reasonable assumption for the construction of the roof is the following:

Roofing	5 psf
Insulation	2 psf
Deck	2 psf
Beams	3 psf
Joists	3 psf
Misc.	5 psf
Total	20 psf

Downers Grove is at an elevation of approximately 690 ft so a case-specific snow load calculation is not required. Because of the perimeter parapet and screen walls, snow drift loading must be considered. A reasonable assumption for the floor construction is

Slab and deck	60 psf
Beams and girders	8 psf
Misc.	12 psf
Total	80 psf

Since no information has been given regarding office layout, it is prudent to use the corridor loading over the entire floor plan. This will permit architectural flexibility. The two-story area bounded by column lines A, C, 4, and 5 should be treated as a lobby.

For wind load, this is a rigid enclosed building located on a site where there are no special terrain features to consider nor is air density a concern since the elevation is approximately 700 ft above sea level. Thus, to determine the velocity pressure, $K_d = 0.85$, $K_{zt} = K_e = 1.0$ and K_z at the top of the building is 1.09. To determine the external design wind pressure, use $G = 0.85$ for a rigid building, $C_p = 0.8$ for the windward wall and $GC_{pi} = +0.18$ for an enclosed structure.

a. Develop a table showing the gravity loads in all columns based on tributary area and live load reductions. Remember that the columns in the lateral load resisting systems will also carry wind or seismic forces. Wind loads will be determined in part c; seismic loads are in the Chapter 13 Integrated Design Project problem.

b. Develop a table showing beam loading based on tributary area and live load reductions. Remember that beams will see a uniform load and girders will see concentrated loads and some uniform load due to their self-weight. Beam and girder self-weight is included in the construction loads given above. Perimeter beams and girders will also support the lightweight metal curtain wall at each level. A reasonable assumption for the weight of the curtain wall is 18 psf.

c. Using a uniform wind pressure as determined at the top of the building and spread over the entire windward wall, determine the concentrated wind force at each level for each main wind force resisting system in each direction.

Chapter 3

Steel Building Materials

The Seattle Spheres
Photo courtesy of Magnusson Klemencic Associates

3.1 INTRODUCTION

Steel has been produced in the United States since the 1800s. Its first use in a bridge was for a railroad bridge across the Mississippi River in St. Louis, built in 1874 by James B. Eads. The bridge, known as the Eads Bridge and shown in Figure 3.1, is still an inspiring steel structure crossing the river in the shadow of the St. Louis Gateway Arch. The first skyscraper is generally considered to be the Home Insurance Building, designed by William LeBaron Jenney and erected at 135 South La Salle Street, Chicago. The building, shown in Figure 3.2, was started on May 1, 1884, and completed in the fall of 1885. It was originally a 10-story building but later had 2 stories added. The original structural design called for wrought iron beams bolted through angle-iron brackets to cast iron columns. As the framework reached the sixth floor, the Carnegie-Phipps Steel Company of Pittsburgh, Pennsylvania, indicated that they were now rolling "Bessemer Steel" and requested permission to substitute steel members for the wrought iron beams on all remaining floors. This was the first use of steel beams in a building. The Home Insurance Building was demolished in 1929.

The first all-steel skyscraper was the Rand-McNally Building at 165 West Adams Street in Chicago, designed by Daniel Burnham and John Root. This 10-story building, shown in Figure 3.3, was built from 1888–1890 and was constructed of built-up members made from standard rolled steel bridge shapes that were riveted together. It marked the beginning of a continuous evolution in steel building structures that is ongoing today as new ideas are brought into play by architects and engineers who build with steel.

This evolution in steel buildings encompasses the materials used, the applications of innovative designers, and the specifications that direct their designs.

3.2 APPLICABILITY OF THE AISC SPECIFICATION

The specification that guides the design of our modern steel buildings was first published by AISC in 1923. Its purpose was to promote uniform practice in the design of steel buildings. At that time, numerous approaches were being used across the industry. Steel producers each had their own standard for design, and the larger cities also required that their own standards be used. This multiplicity of standards meant no standard at all. It led to a confusion of approaches in which designers were continually called upon to change the way they designed, depending on where their current building project was to be located.

The 1923 *Specification* defined "the practice adopted by the American Institute of Steel Construction for the design, fabrication, and erection of structural steel buildings." It went on to provide direction on how to obtain a satisfactory structure. The following requirements were to be fulfilled:

1. The material used must be suitable, of uniform quality, and without defects affecting the strength or service of the structure.
2. Proper loads and conditions must be assumed in the design.
3. The unit stresses must be suitable for the material used.
4. The workmanship must be good, so that defects or injuries are not produced in the manufacture.
5. The computations and design must be properly performed so that the unit stresses specified shall not be exceeded, and the structure and its details shall possess the requisite strength and rigidity.

The *Specification* also provided guidance on the material to be used, stating, "Structural steel shall conform to the Standard Specifications of the American Society for Testing Materials for Structural Steel for Buildings, Serial Designation A9-21, as amended to date." These principles from 1923 are still important to steel construction a century later.

The 2022 AISC *Specification for Structural Steel Buildings* supersedes all previous AISC *Specification*s. Over the years, the *Specification* has lost the terms *fabrication* and *erection* from its title, nonetheless, the *Specification* continues to address issues pertinent to design, fabrication and erection of steel structures. In addition, the AISC *Specification* has regularly been used to guide the design of structures other than buildings. In recognition of this practice, and to ensure that the *Specification* is properly applied, the scope of the *Specification* states, "The specification sets forth criteria for the design, fabrication, and erection of structural steel buildings and other structures, where other structures are defined as structures designed, fabricated, and erected in a manner similar to buildings, with building-like vertical and lateral load resisting elements."

Figure 3.1 Eads Bridge, St. Louis, Missouri
Photo courtesy Luke Gatta

Figure 3.2 Home Insurance Building, Chicago, Illinois.

Figure 3.3 Rand McNally Building, Chicago, Illinois

Additionally, the *Specification* indicates that it "shall apply to the design, fabrication, erection, and quality of the structural steel system or systems with structural steel acting compositely with reinforced concrete, where the steel elements are defined in Section 2.1 of the AISC *Code of Standard Practice for Steel Buildings and Bridges* (ANSI/AISC 303)." In that document, structural steel is defined as "the elements of the structural frame that are shown and sized in the structural design documents, essential to support the design loads," and they are given here in Table 3.1. A quick review of the table contents indicates that elements are structural steel if they are part of or attached to the structural frame. Examples of the use of many of these elements are shown in Figure 3.4. All elements discussed in this text will meet the above definition.

3.3 STEEL FOR CONSTRUCTION

Since the introduction of the first AISC *Specification*, a variety of steels have been approved for use in steel construction. The specific steels approved at any given time has changed along with the techniques of manufacture and steel chemistry. Steels available for use in construction have increased in strength as manufacturing has become more refined. One important aspect of all steel is that it generally behaves in a uniform and consistent manner. Thus, although the strength might be different for different grades of steel, the steel can be expected to behave the same, regardless of grade, up to its various strength limits.

The characteristics of steel that are important to the structural engineer can be determined through a simple uniaxial tension test. This standard test is conducted according to the requirements of ASTM A370, *Standard Test Methods and Definitions for Mechanical Testing of Steel Products*. A specimen of specific dimensions is subjected to a tensile force, and the resulting stress and strain are plotted for the duration of the test. The stress, f, and strain, ε, are plotted in Figure 3.5 and defined as follows:

$$f = \frac{P}{A} \quad \text{and} \quad \varepsilon = \frac{\Delta L}{L}$$

where

A = cross-sectional area at start of test
L = length of specimen at start of test
P = tensile force
f = axial tensile stress
ΔL = change in length of specimen under stress
ε = axial strain

Table 3.1 Definitions of Structural Steel[a]

Anchor rods that will receive structural steel.
Base plates, if part of the structural steel frame.
Beams, including built-up beams, if made from standard structural shapes and/or plates.
Bearing plates, if part of the structural steel frame.
Bearings of steel for girders, trusses, or bridges.
Bracing, if permanent.
Canopy framing, if made from standard structural shapes and/or plates.
Columns, including built-up columns, if made from standard structural shapes and/or plates.
Connection materials for framing structural steel to structural steel.
Crane stops, if made from standard structural shapes and/or plates.
Door frames, if made from standard structural shapes and/or plates and if part of the structural steel frame.
Edge angles and plates, if attached to the structural steel frame or steel (open-web) joists.
Embedded structural steel parts, other than bearing plates, that will receive structural steel.
Expansion joints, if attached to the structural steel frame.
Fasteners for connecting structural steel items: permanent shop bolts, nuts, and washers; shop bolts, nuts, and washers for shipment; field bolts, nuts, and washers for permanent connections; and permanent pins.
Floor-opening frames, if made from standard structural shapes and/or plates and attached to the structural steel frame or steel (open-web) joists.
Floor plates (checkered or plain), if attached to the structural steel frame.
Girders, including built-up girders, if made from standard structural shapes and/or plates.
Girts, if made from standard structural shapes.
Grillage beams and girders.
Hangers, if made from standard structural shapes, plates, and/or rods and framing structural steel to structural steel.
Leveling nuts and washers.
Leveling plates.
Leveling screws.
Lintels, if attached to the structural steel frame.
Machinery supports, if made from standard structural shapes and/or plates and attached to the structural steel.
Marquee framing, if made from standard structural shapes and/or plates.
Monorail elements, if made from standard structural shapes and/or plates and attached to the structural steel frame.
Posts, if part of the structural steel frame.
Purlins, if made from standard structural shapes.
Relieving angles, if attached to the structural steel frame.
Roof-opening frames, if made from standard structural shapes and/or plates and attached to the structural steel frame or steel (open-web) joists.
Roof-screen support frames, if made from standard structural shapes.
Sag rods, if part of the structural steel frame and connecting structural steel to structural steel.
Shear stud connectors, if specified to be shop attached.
Shims, if permanent.
Struts, if permanent and part of the structural steel frame.
Steel plate shear walls and/or composite steel plate shear wall systems, and steel plate structures, if made from standard shapes and/or plates, and if part of the structural steel frame.
Tie rods, if part of the structural steel frame.
Trusses, if made from standard structural shapes and/or built-up members.
Wall-opening frames, if made from standard structural shapes and/or plates and attached to the structural steel frame.
Wedges, if permanent.

[a]From *Code of Standard Practice for Steel Buildings and Bridges*, ANSI/AISC 303-22.

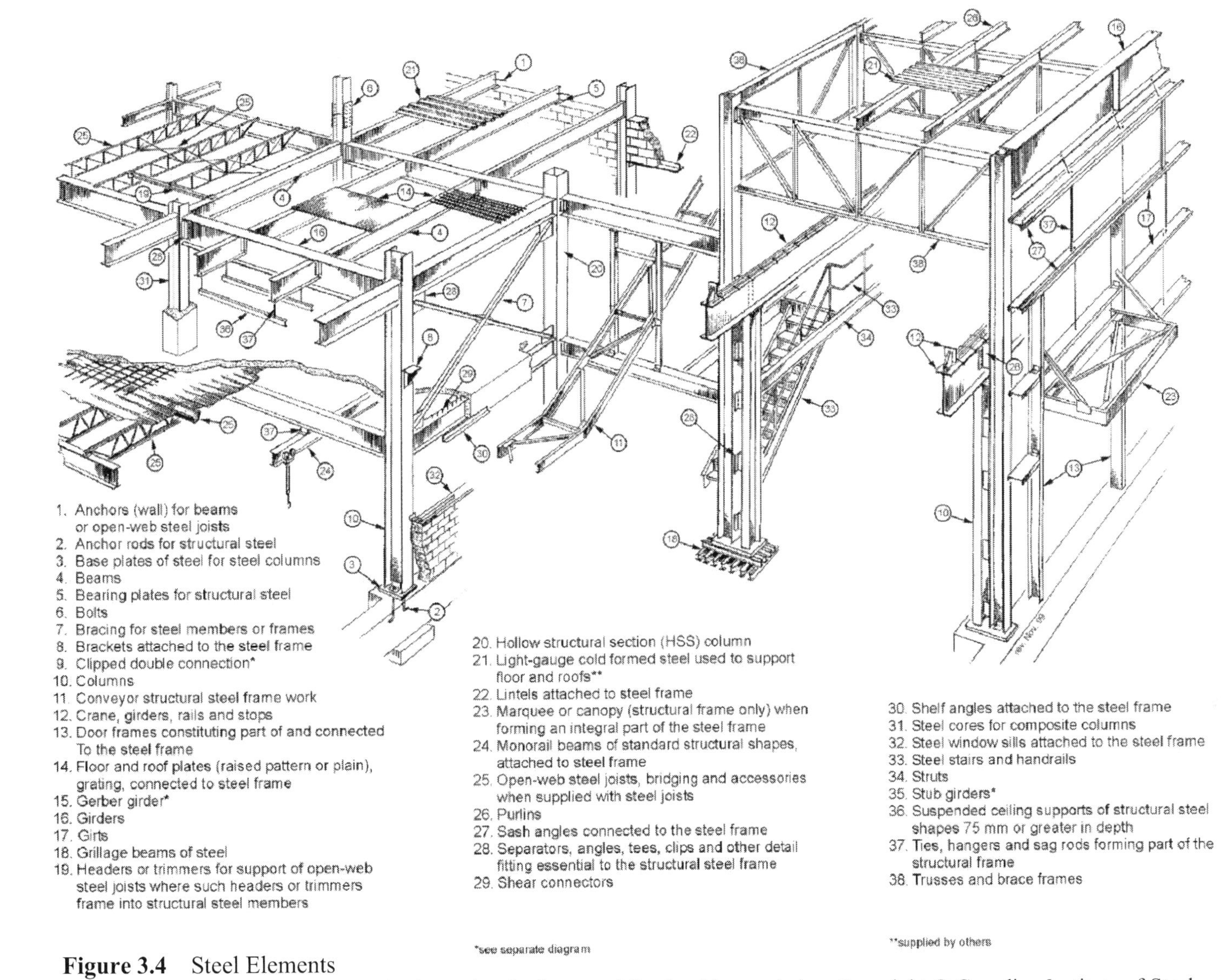

Figure 3.4 Steel Elements
Reprinted from CISC Code of Standard Practice for Structural Steel, with permission,

The curve shown in Figure 3.5 is typical of mild carbon steel. Several characteristics of this stress-strain curve are worth noting. First, the initial portion of the curve, which indicates the response that would be expected under most nominal or service loading conditions, follows a straight line up to a point called the *proportional limit*. For structural steel with yield stresses of 65 ksi or less, this proportional limit is the point where the curve first deviates from linear and is called the *yield point*. Some tests of structural steel exhibit a portion of the curve as indicated by the upper yield point followed by the lower yield point. The separation of these points appears to be affected by test techniques including speed of testing, shape of specimen, and accuracy of alignment. They do not indicate sustainable conditions and the variation of the upper yield point is not considered in design. The ratio of stress to strain in this region is constant and is called Young's modulus or the modulus of elasticity, E. All structural steels exhibit the same initial stress-strain behavior and thus have the same E value. The value of E obtained through a large number of tests is consistently between 29,000 ksi and 30,000 ksi. For all calculations based on the AISC *Specification*, $E = 29{,}000$ ksi has historically been used. Within the straight-line portion of the curve, the material is said to behave elastically. A load can be applied and then removed with the structure returning to its original configuration, showing no permanent deformation.

After reaching the yield stress, the stress-strain curve for mild carbon steel exhibits a long plateau where the stress remains essentially constant while the strain increases. This region is called the *plastic region* and indicates a stable characteristic of mild carbon steels. Any structure that is loaded into this region exhibits a permanent plastic deformation as shown by the unloading line in Figure 3.5. The length of this plastic region depends on the particular type of steel but typically is 15 to 20 times the strain at yield.

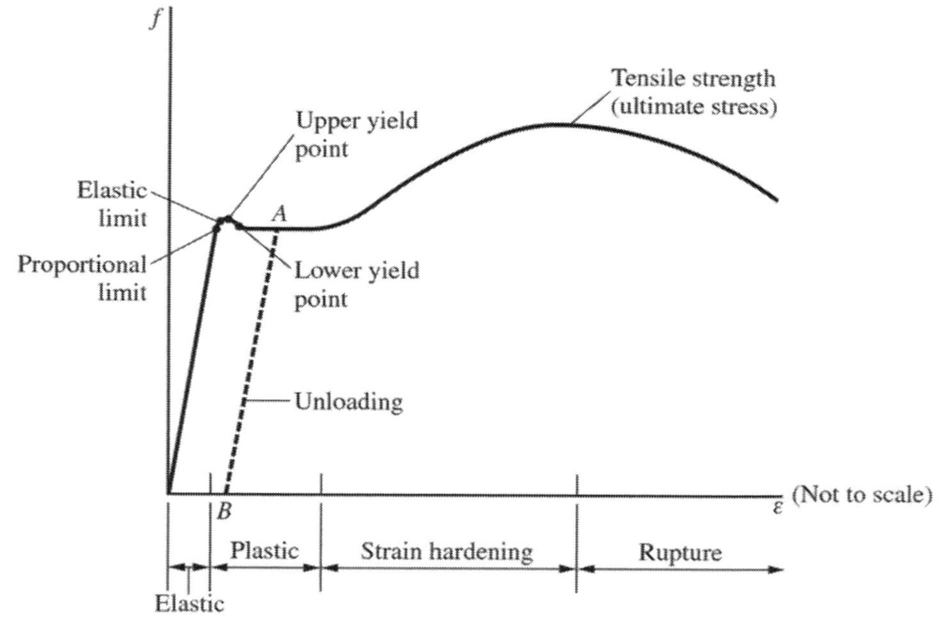

Figure 3.5 Typical Stress-Strain Plot for Mild Carbon Steel

At the end of the plastic region, the curve again rises with increasing stress and strain. This increase is called *strain hardening* and continues until the specimen reaches its tensile strength, or ultimate stress, F_u, at the peak of the stress-strain curve. Once the tensile strength is reached, the specimen rapidly sheds load and increases strain until complete rupture occurs.

Yield stress, tensile strength, and modulus of elasticity are the engineering data values used throughout design to fully describe the material and to determine the strength of the structural elements. The ratio of the tensile strength to the yield stress is also an important characteristic of steel. It is used to control the basic material behavior so that at various limit states, the expected behavior can be ensured.

Figure 3.6 shows the lower strain region of the stress-strain curves for three steels with different yield stresses: 36 ksi, 50 ksi, and 100 ksi. Elastic behavior for the higher-strength steels is the same as for the lower-strength steels. As already noted, $E = 29{,}000$ ksi for all steel. The differences occur after the proportional limit is reached. For steels with a yield stress less than or equal to 65 ksi, the plateau defining the plastic region can be expected to occur. However, for steels with a yield stress greater than 65 ksi, it is expected that no well-defined yield point will exist, and no well-defined plastic plateau will occur, as illustrated for the steel having $F_y = 100$ ksi. For these steels it is necessary to define yield strength by some other means. ASTM A370 provides for yield strength determination by the 0.2 percent offset method or the 0.5 percent elongation method. In either case, the stress-strain curve must be obtained, and the specified offset or elongation used to determine the appropriate stress value. The results of these two approaches are shown in Figure 3.6, and the two methods would yield slightly different yield strength values.

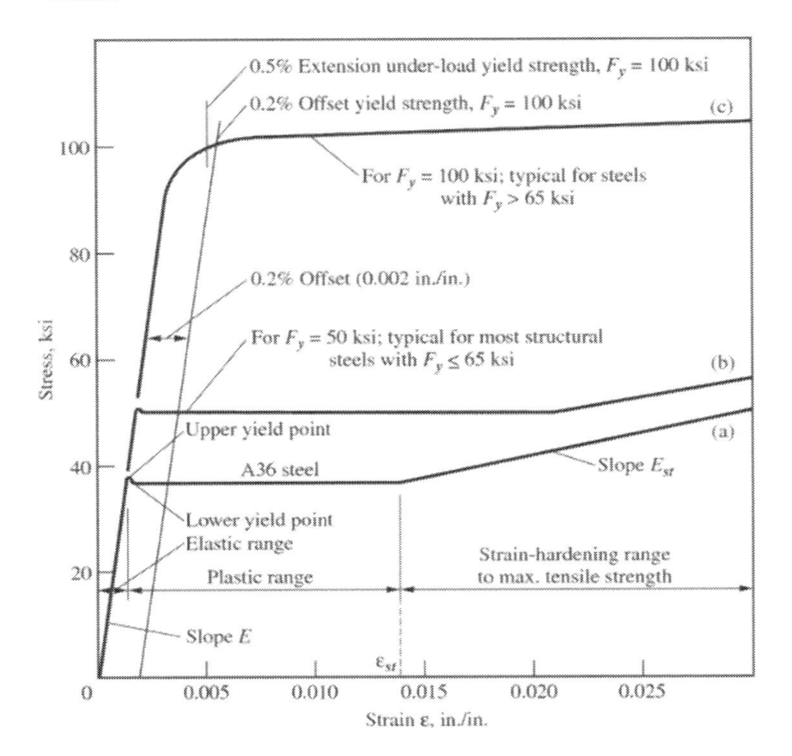

Figure 3.6 Enlarged Typical Stress-Strain Curves for Steels with Different Yield Stresses

3.4 STRUCTURAL STEEL SHAPES

Structural steel design serves to determine the appropriate shape and amount of steel needed to carry a given applied load. This is normally accomplished by selecting, from a predetermined list of available shapes, the lightest-weight member. However, options could also include a combination of steel elements in some particular desired form. The early days of steel construction witnessed very little standardization of available shapes. Although each mill would produce its own shapes, the variety of available shapes was limited and most structural members were composed of these available shapes riveted together. One of AISC's original goals was to standardize the shapes being produced. Over the years, shapes became standardized and more shapes, designed specifically for the needs of building construction, became available. Modern production practices now make a wide variety of shapes available to the designer so that design can almost always be accomplished by selecting one of these standard shapes. In situations where these standard shapes do not meet the needs of a project, members composed of plate material can be produced to carry the imposed loading.

3.4.1 ASTM A6 Standard Shapes

The first standard shapes to be discussed are those defined by ASTM A6: W-shapes, S-shapes, HP-shapes, M-shapes, C-shapes, MC-shapes, and L-shapes. Cross-sections of these shapes are shown in Figure 3.7, where it can be seen that W-, M-, S-, and HP-shapes all take the form of an I. C- and MC-shapes are called channels and take the form of a C, while L-shapes are called angles. Part 1 of the *Manual* contains tables of properties for all the standard shapes.

W-Shapes

W-shapes are usually referred to as wide-flange shapes and are the most commonly used shapes in buildings. They have two flanges with essentially parallel inner and outer faces and a single web located midway along the flanges. The overall shape of the wide flange may vary from a fairly deep and narrow section, as shown in Figure 3.7a, to an almost square section, as shown in Figure 3.7b. These shapes have two axes of symmetry; the x-axis is the strong axis and the y-axis is the weak axis. Wide-flange shapes can be as deep as 44 in. and as shallow as 4 in. A typical wide-flange shape is designated as a W16×26, where the W indicates it is a W-shape, the 16 indicates it has a nominal depth of 16 in., and the 26 indicates its weight is 26 pounds per foot. The nominal depth indicates an approximate member depth but not its actual depth. The production of wide-flange shapes results in their grouping in families according to the size of the rolls used to produce the shape. All shapes in a family have the same dimension between the inner faces of the flanges. The different weights are achieved by increasing the actual depth of the member and thus the flange thickness (the web thickness also generally increases as the flanges get thicker). *Manual* Table 1-1 provides the dimensions and section properties needed for design for all W-shapes.

HP-Shapes

HP-shapes are similar to wide-flange shapes and normally are used as bearing piles. These shapes also have parallel face flanges, but unlike the W-shapes, their webs and flanges are of the same nominal thickness, and they are all close to being square, as shown in Figure 3.7c. An HP14×117 is an HP-shape with a nominal depth of 14 in. and a weight of 117 pounds per foot. *Manual* Table 1-4 provides the dimensions and section properties needed for design for all HP-shapes.

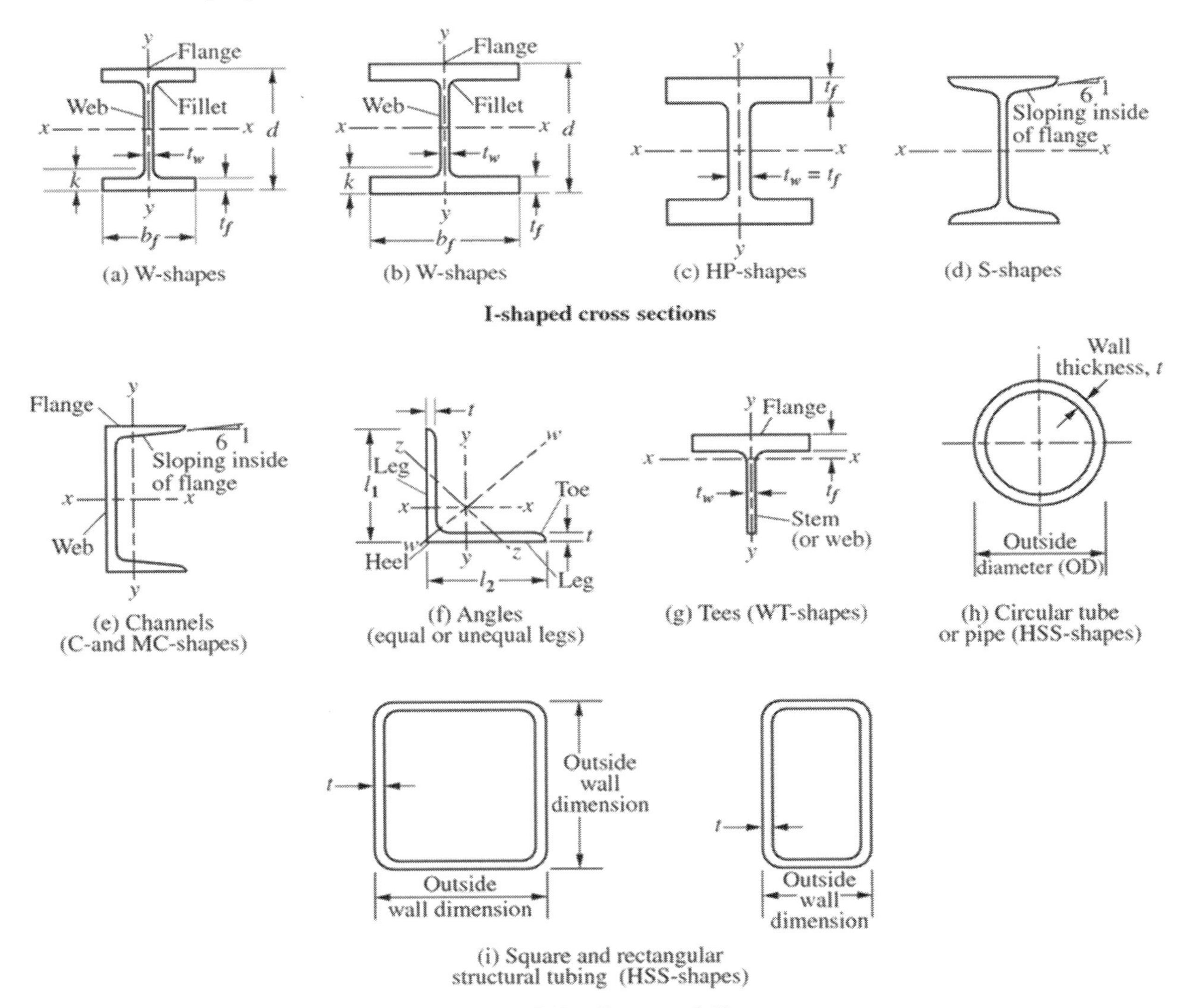

Figure 3.7 Structural Shapes

S-Shapes

S-shapes are American Standard beams and were historically referred to as I-beams. They were the standard shapes used in construction prior to the development of the rolling process that permitted the introduction of the wide-flange shapes. Although these shapes are still available, their use today is infrequent, and their availability should be confirmed prior to their specification. These shapes have narrow flanges in relation to their depth, and the flanges have a sloping interior face, as shown in Figure 3.7d. The *Manual* lists 28 S-shapes and gives their properties in Table 1-3. As with the shapes previously discussed, the numbers in the name refer to the nominal depth and the weight per foot. In all cases except for two S24s and two S20s, the nominal depth and the actual depth are the same.

M-Shapes

M-shapes are miscellaneous shapes that do not fit the definitions of W-, HP-, and S-shapes. The *Manual* lists 16 miscellaneous shapes. They are not particularly common in building design but are common in other applications. They should be used in design only after confirmation that they are economically available. A typical designation is M12×11.8; as with the other shapes, the 12 indicates the nominal depth and the 11.8 indicates the weight per foot. Dimensions and properties for M-shapes are found in *Manual* Table 1-2.

C-Shapes

C-shapes are American Standard channels and are produced using essentially the same process as for S-shapes. They have two flanges and a single web located at the end of the flanges, as shown in Figure 3.7e. These shapes have only one axis of symmetry and, like the W-shapes, the x-axis is the strong axis and the y-axis is the weak axis. As with the S-shapes, the flanges have sloping inner faces. One of the 32 C-shapes found in *Manual* Table 1-5 is a C8×18.75. All C-shapes have an actual depth equal to the nominal depth.

MC-Shapes

MC-shapes are miscellaneous channels that cannot be classified as C-shapes. Their designations follow the same rules as the previous shapes, with a typical shape being an MC6×18. *Manual* Table 1-6 lists 40 MC-shapes, and their sizes fall into the same overall range as the C-shapes.

L-Shapes

L-shapes are angles that can have equal or unequal legs. The largest angle legs are 12 in., and the smallest are 2 in., with the dimension taken from heel to toe of the angle. A typical angle designation is L6×4×7/8, where the first two numbers are the dimensions of the legs and the third is the leg thickness. Leg dimensions are actual dimensions, and the leg thickness is the same for both legs. For unequal leg angles, the longest leg is given first. Equal leg angles have one axis of symmetry, whereas unequal leg angles have no axis of symmetry.

All angles have two sets of two axes of interest to the designer: the geometric and principal axes, which are illustrated in Figure 3-7(f). The geometric axes are parallel to the faces of the legs with the *x*-axis, parallel to the short leg, and the *y*-axis, parallel to the long leg. For equal leg angles, the *x*- and *y*-axes are similarly oriented even though one cannot distinguish between longer and shorter legs. Alternatively, the principal axes can be used. The minor principal axis, which for equal leg angles is perpendicular to the axis of symmetry, is the *z*-axis; and the major principal axis, the axis of symmetry for equal leg angles, is the *w*-axis. The principal axes are similar for unequal leg angles, except there is no axis of symmetry.

Manual Table 1-7 provides the dimensions and section properties for the design for all angles. For some special cases, properties that may be needed for the *w*-axis are not available in Table 1-7. These can be found in the 16th edition electronic shapes database.

WT-Shapes

WT-shapes are tees that have been cut from W-shapes. Shown in Figure 3-7(g), these shapes are also called split tees because that is how they are produced – by splitting a W-shape. These shapes are designated as WT5×56, where both numbers are one-half of the corresponding numbers of the parent W-shape they were cut from. Dimensions and properties for WT-shapes are given in *Manual* Table 1-8.

MT-Shapes and ST-Shapes

MT-shapes and *ST-shapes* are tees that have been cut from parent M- and S-shapes. The properties and dimensions for these shapes are found in *Manual* Tables 1-9 and 1-10.

3.4.2 Hollow Shapes

Another group of shapes commonly used in building construction are the hollow shapes commonly referred to as tubes or pipes. These shapes are produced by bending and welding flat plates or by hot rolling to form a seamless section. For all hollow structural sections (HSS), ASTM specifications set the requirements for both the materials and the sizes.

Round HSS

Round hollow structural sections (Figure 3-7(h)) are manufactured through a process called "formed from round," which takes a flat strip of steel and gradually bends it around its longitudinal axis and joins the edges by welding. Once the weld has cooled, the round shape is passed through additional shaping and sizing rolls to fix the final diameter. An example of a round HSS is HSS5.563×0.258, where the first number is the outside diameter and the second is the nominal wall thickness. For round HSS, the diameter and nominal wall thickness are always shown as a decimal number to three places. These shapes are found in *Manual* Table 1-13.

Square and Rectangular HSS

Square and rectangular HSS (Figure 3-7(i)) may initially be formed from round, with the final sizing used to also change the formed shape into a rectangle, or formed from a flat plate through a "formed-square weld-square" process, in which the plate is gradually bent into its near final size. Another process starts with two flat pieces that are each bent, with the two half sections then joined to form the final shape. A typical rectangular HSS is the HSS12×8×1/2. The first number indicates the actual height of the section, the second the actual width, and the third the nominal thickness of the section wall. For square and rectangular HSS, the nominal wall thickness is always shown as a fraction. *Manual* Tables 1-11 and 1-12 provide the dimensions and section properties needed for the design of rectangular and square HSS-shapes, respectively.

Steel Pipes

Steel pipes are another hollow round section used in building construction. They are produced to different material standards than the round HSS. Pipes are available as standard weight (Std.), extra strong (x-Strong), and double extra strong (xx-Strong), which refer to the wall thickness for a given outside diameter. The standard designation for a pipe section is in the form Pipe 5 x-Strong, indicating that it must meet the pipe material standards, have a nominal 5 in. outside diameter, and a thickness corresponding to the "extra strong" designation. This particular pipe has an actual outside diameter of 5.56 in. and a nominal wall thickness of 0.375 in. *Manual* Table 1-14 provides the properties for steel pipes. Note that many round HSS cross sections are made to match the steel pipe cross sections. Although they are dimensionally interchangeable, it is important to remember that round HSS and steel pipes are produced to different material standards.

3.4.3 Plates and Bars

In addition to the shapes already discussed, steel is available as plates and bars, as shown in Figure 3.8. These elements are rarely used alone as shapes but are combined to form built-up shapes or used alone as connecting elements to join other shapes.

Table 3.2 Preferred Dimensions for Plates and Bars

Product	Range of Thicknesses/Diameters		
	$t \leq 3/8$ in.	3/8 in. $< t \leq 1$ in.	1 in. $< t$
Plates	1/16	1/8	1/4
Square and rectangular bars	1/8	1/8	1/8
Circular bars	1/8	1/8	1/8

Note: Table gives increments in thickness or diameter.

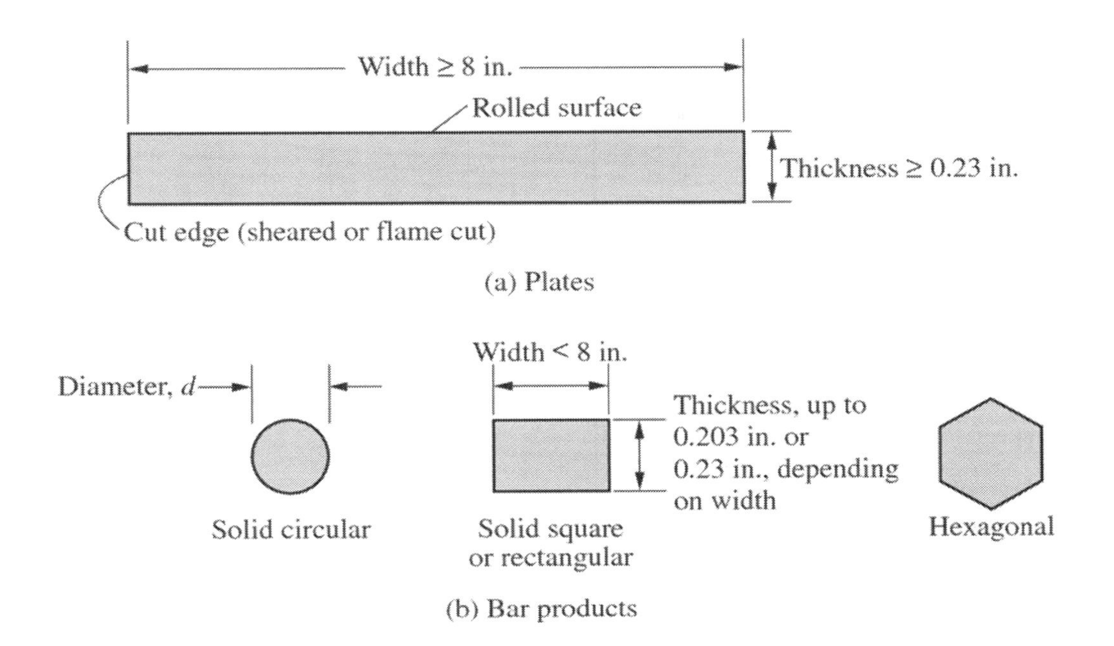

Figure 3.8 Plate and Bar Products

Plates

Plates are flat rectangular elements that are hot-rolled to a given thickness and sheared to the appropriate width. At one time, plates were also available that were rolled to a given width as well as a given thickness. These plates were called universal mill plates. Because of the manufacturing process, these plates had patterns of residual stresses that differed from the patterns in sheared plates and resulted in lower strength. Current manufacturing practice is to produce all plates as sheared plates. By industry definition, plates are a minimum of 8 in. in width and may vary in thickness from 3/16 in. up. The designation for a typical plate is PL 1/2×10×2 ft–4 in., where the first number is the thickness, the second the plate width, and the third the length. Table 3.2 reflects the preferred standard practice for plate thickness increments. For example, plates less than 3/8 in. thick can be obtained in 1/16 in. thickness increments, while plates greater than 1 in. thick are typically produced in 1/4 in. increments.

Bars

Bars are available in rectangular, circular, and hexagonal shapes, with the rectangular bar the most commonly used shape in building construction. The only difference between rectangular bars and plates is the width. Any rectangular solid element less than 8 in. in width is technically referred to as a bar. Because the distinction between bars and plates is not significant to the designer, the designation for these narrow elements is the same as for a plate. Thus, PL 1/2×6×2 ft–4 in. is a 6 in. wide bar.

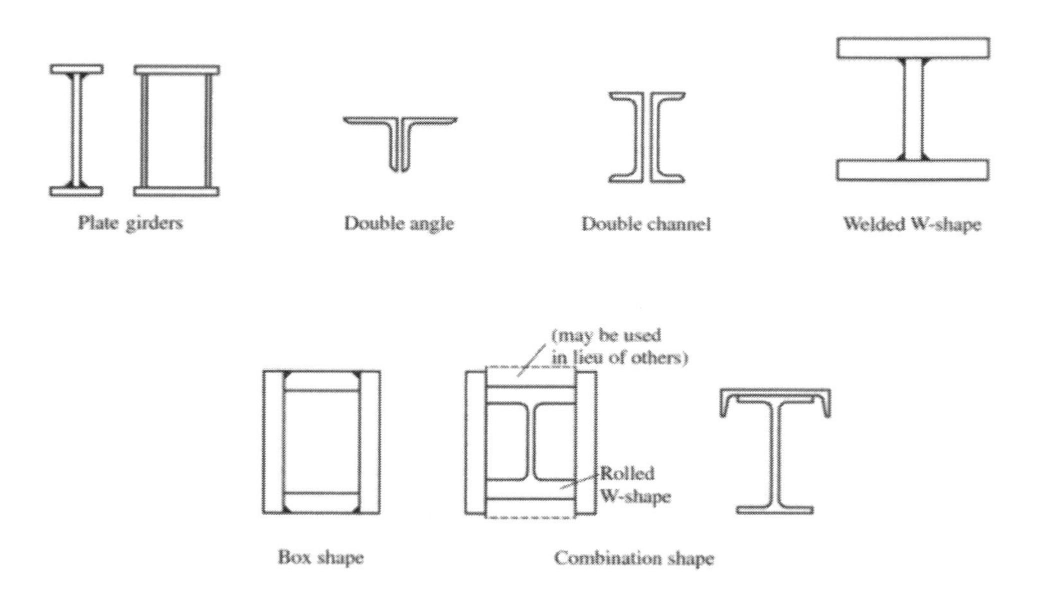

Figure 3.9 Examples of Built-up Shapes

3.4.4 Built-up Shapes

Other shapes are available and may be found in the *Manual*, but they are of limited application in building construction. The *Manual* also contains tables for combinations of standard shapes that have, over the years, been found to be useful to the designer. Figure 3.9 shows a variety of built-up shapes formed from combining plates and shapes. These include a double angle member often used in braced frames and trusses, plate girders for long building spans and bridges, and welded W-shape used when large W-shapes are called for but are not readily available.

3.5 CHEMICAL COMPONENTS OF STRUCTURAL STEEL

The basic mechanical properties of structural steel were presented in Section 3.3 with a limited discussion of the types of steel available for use by the building industry. Essentially three types of steel are used for shapes in the construction industry: carbon steel, sometimes referred to as carbon-manganese steel, high-strength low-alloy steel (HSLA), and corrosion-resistant high-strength low-alloy steel. For plates and bars, quenched and tempered steels are also available.

The chemical composition of steel significantly influences the properties that are of ultimate importance to the engineer. Steel is primarily made of iron but also contains such other elements as carbon, silicon, nickel, manganese, and copper. The primary element that influences the characteristics of steel, not counting iron, is carbon. The addition of carbon increases steel strength but decreases ductility and weldability. Even though carbon is the most significant component of steel (after iron), it still represents a very small percentage of the final product. Steels generally have a carbon content of not more than 0.3 percent by weight. Steels that contain strengthening elements in addition to carbon and manganese are referred to as HSLA steels although there are no strict rules that apply to that designation.

Although the formula for a specific steel might be different from that of any other steel, certain elements are required in order to meet a specific set of criteria. These specifications come from the ASTM standards for each steel type and are discussed in Section 3.6. However, the chemical elements that may be found in the most dominant steel for wide flange shapes are reviewed here. The specific percentage requirements for ASTM A992 steel are given in Table 3.3.

Carbon

Carbon (C) is the most common element, excluding iron, found in all steel. It is the most economical element used to increase strength. However, it also decreases ductility. Carbon content usually ranges from about 0.15 to 0.30 percent. Anything lower than 0.15 percent would produce steel with too low a strength, and anything higher than 0.30 percent would yield steel with poor characteristics for use in construction.

Manganese

Manganese (Mn) has an effect on strength similar to that of carbon. It is a necessary component because of the way it combines with oxygen and sulfur and its impact on the rolling process by decreasing the critical cooling rate. In addition, manganese improves the notch toughness of steel. It is added to steel to offset reductions in notch toughness due to the presence of other elements. It has a negative effect on material weldability.

Silicon

Silicon (Si) is an important element for removing oxygen from hot steel.

Vanadium

Vanadium (V) is another strengthening element. It refines the grain size and thus increases strength. Its biggest advantage is that while increasing strength, it does not negatively impact weldability or notch toughness.

Columbium

Columbium (Cb) is a strengthening element that, in small quantities, can increase the yield point and, to a lesser extent, the tensile strength. However, it has a significant negative impact on notch toughness.

Table 3.3 Chemical Requirements for A992 Steel

Element	Composition, %
Carbon, max	0.23
Manganese	0.50 to 1.60
Silicon, max	0.40
Vanadium, max	0.15
Columbium, max	0.05
Phosphorus, max	0.035
Sulfur, max	0.045
Copper, max	0.60
Nickel, max	0.45
Chromium, max	0.35
Molybdenum, max	0.15

Phosphorus

Phosphorus (P) increases strength and decreases ductility. It improves resistance to atmospheric corrosion, particularly when used in combination with copper. It has a negative impact on weldability that is more severe than that of manganese. It is generally an undesirable element but is permitted in very limited quantities in all steel.

Sulfur

Sulfur (S) is also permitted in very limited quantities in all steel. It has a negative impact on weldability comparable to that of phosphorus. Generally, steelmaking practice works to remove as much sulfur as possible.

Copper

Copper (Cu) in limited quantities is beneficial to steel. It increases strength with only a limited negative impact on ductility. If its content is held relatively low, it will have little effect on weldability. It is the most significant contributing element in the production of corrosion-resistant steel.

Nickel

Nickel (Ni) can provide a moderate improvement in strength and enhances corrosion resistance. It can also improve resistance to corrosion for steel subjected to seawater when used in combination with copper or phosphorus. It generally leads to a slight improvement in notch toughness.

Chromium

Chromium (Cr) is typically used in combination with copper to improve corrosion resistance. It also provides some strengthening of steels containing copper and vanadium. Chromium is an integral component of stainless steel.

Molybdenum

Molybdenum (Mo) increases strength but significantly decreases notch toughness. However, this negative impact can be controlled by appropriate processing or balancing with other elements.

3.6 GRADES OF STRUCTURAL STEEL

3.6.1 Steel for Shapes

Many more grades of steel are produced than are approved by AISC for use in structures. A unique ASTM number designates each type of approved steel. The steels approved for structural shapes are grouped as carbon steel (A36, A53, A500, A501, A529, A709, A1043 and A1085), high-strength low-alloy steel (A572, A618, A709, A913, A992, and A1065), corrosion-resistant quenched and tempered low-alloy steel (A709), and corrosion-resistant high-strength low-alloy steel (A588, A847, and A1065). Figure 3.10 lists these approved steels, their minimum yield and tensile stresses, and the shapes to which they are applicable. The table also identifies which combination of shape and material specification are preferred, such as A992 for W-shapes, and which material specifications do not apply for particular shapes.

A36 Steel

A36 steel was the most commonly available structural steel for many years. It was introduced in the 1961 AISC *Specification* and until the late 1990s was the steel of choice for most steel shapes (except for HSS and pipe) and plates. It is a mild carbon steel, so it is well suited for bolted or welded construction. Even when higher-strength steels were used for members, this steel had been the usual choice for connecting elements. It is no longer a preferred material for any shapes but continues to be available for W-, HP-, and L-shapes. It has a minimum yield stress, $F_y = 36$ ksi and a tensile stress, $F_u = 58$ to 80 ksi, although $F_u = 58$ ksi should be used for calculations throughout the *Specification*.

A53 Steel

A53 steel is the single standard for steel pipes approved for construction. It is available in three types and two grades; however, they are not all approved for structural applications. These pipes are generally intended for mechanical and pressure applications, and the only grade approved for construction is Grade B. This grade is available as Type E, which denotes electric-resistance welding of the seam, or Type S, which is a seamless pipe. A53

Grade B has a minimum yield stress, $F_y = 35$ ksi and a minimum tensile stress, $F_u = 60$ ksi. A53 Grade B steel pipe are treated as HSS in the *Specification*.

A500 Steel

A500 steel is a carbon steel used for structural tubing in round, square, and rectangular shapes, otherwise known as HSS. It comes in three grades approved by AISC for construction: Grade C, which is the preferred grade, and Grades B and D. The standard permits either welded or seamless manufacture. Grade C has a minimum yield stress, $F_y = 50$ ksi and a minimum tensile stress, $F_u = 62$ ksi.

A501 Steel

A501 steel is a carbon steel used for round and rectangular HSS. It is approved in Grade B, which has a minimum yield stress, $F_y = 46$ ksi, and a minimum tensile stress, $F_u = 65$ ksi.

A529 Steel

A529 steel is a carbon-manganese steel available in grades 50 and 55. It is approved for the smaller shapes with flange thickness no greater than 1.5 in. A529 Grade 50 has a minimum yield stress, $F_y = 50$ ksi and a tensile stress, $F_u = 65$ to 100 ksi whereas Grade 55 has a minimum yield stress, $F_y = 55$ ksi, and a tensile stress, $F_u = 70$ to 100 ksi.

A709 Steel

A709 steel represents three types of steel, carbon steel, high-strength low-alloy steel, and corrosion-resistant quenched and tempered low-alloy steel. Eight grades in four yield strengths are included. This standard is developed specifically for bridges where requirements necessary for bridge applications are added to the basic requirements for A36, A572, A992, A588, A1010, and A913 steels. In buildings, the corresponding grade of A709 steel may be used in place of these six steels.

A1043 Steel

A1043 steel is a carbon steel available in two grades. Grade 36 has a yield stress, $F_y = 36$ to 52 ksi and a minimum tensile strength, $F_u = 58$ ksi. Grade 50 has a yield stress, $F_y = 50$ to 65 ksi and a minimum tensile strength, $F_u = 65$ ksi. This steel is available when a maximum yield-to-tensile ratio of 0.80 is required. W-shapes with a flange width of 6 in. or greater and plates up to 5 in. thickness are included.

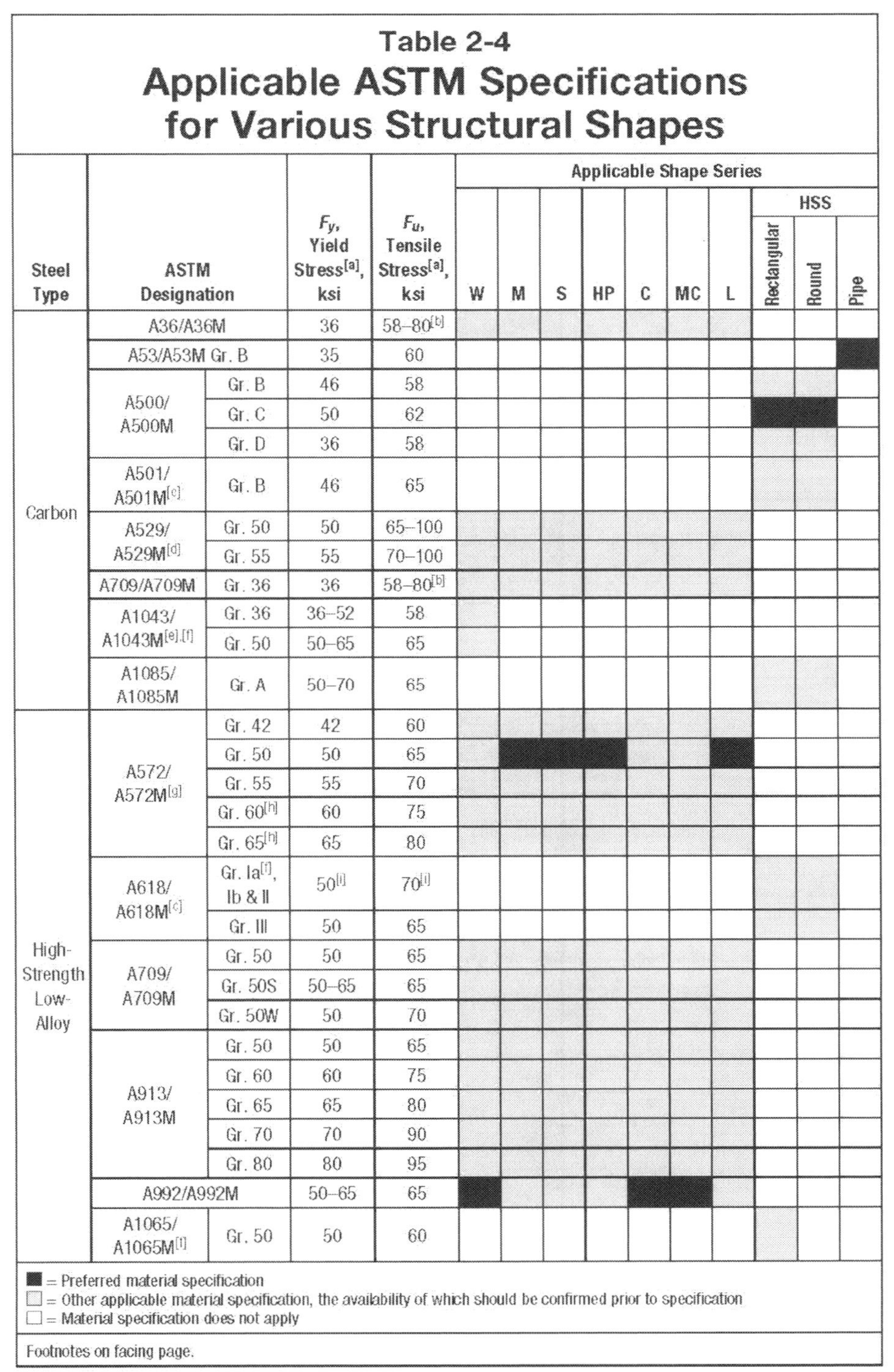

Table 2-4
Applicable ASTM Specifications for Various Structural Shapes

Steel Type	ASTM Designation		F_y, Yield Stress[a], ksi	F_u, Tensile Stress[a], ksi	Applicable Shape Series									
					W	M	S	HP	C	MC	L	HSS Rectangular	HSS Round	Pipe
Carbon	A36/A36M		36	58–80[b]	▒	▒	▒	▒	▒	▒	▒			
	A53/A53M Gr. B		35	60										■
	A500/ A500M	Gr. B	46	58								▒	▒	
		Gr. C	50	62								■	■	
		Gr. D	36	58								▒	▒	
	A501/ A501M[c]	Gr. B	46	65								▒	▒	
	A529/ A529M[d]	Gr. 50	50	65–100	▒	▒	▒	▒	▒	▒	▒			
		Gr. 55	55	70–100	▒	▒	▒	▒	▒	▒	▒			
	A709/A709M	Gr. 36	36	58–80[b]	▒	▒	▒	▒	▒	▒	▒			
	A1043/ A1043M[e],[f]	Gr. 36	36–52	58	▒									
		Gr. 50	50–65	65	▒									
	A1085/ A1085M	Gr. A	50–70	65								▒	▒	
High-Strength Low-Alloy	A572/ A572M[g]	Gr. 42	42	60	▒	▒	▒	▒	▒	▒	▒			
		Gr. 50	50	65	▒	■	■	■	▒	▒	■			
		Gr. 55	55	70	▒	▒	▒	▒	▒	▒	▒			
		Gr. 60[h]	60	75	▒	▒	▒	▒	▒	▒	▒			
		Gr. 65[h]	65	80	▒	▒	▒	▒	▒	▒	▒			
	A618/ A618M[c]	Gr. Ia[i], Ib & II	50[j]	70[j]								▒	▒	
		Gr. III	50	65								▒	▒	
	A709/ A709M	Gr. 50	50	65	▒	▒	▒	▒	▒	▒	▒			
		Gr. 50S	50–65	65	▒	▒	▒	▒	▒	▒	▒			
		Gr. 50W	50	70	▒	▒	▒	▒	▒	▒	▒			
	A913/ A913M	Gr. 50	50	65	▒	▒	▒	▒	▒	▒	▒			
		Gr. 60	60	75	▒	▒	▒	▒	▒	▒	▒			
		Gr. 65	65	80	▒	▒	▒	▒	▒	▒	▒			
		Gr. 70	70	90	▒	▒	▒	▒	▒	▒	▒			
		Gr. 80	80	95	▒	▒	▒	▒	▒	▒	▒			
	A992/A992M		50–65	65	■	▒	▒	▒	■	■	▒			
	A1065/ A1065M[l]	Gr. 50	50	60								▒		

■ = Preferred material specification
▒ = Other applicable material specification, the availability of which should be confirmed prior to specification
□ = Material specification does not apply

Footnotes on facing page.

Figure 3.10 Applicable ASTM Specifications for Various Structural Shapes

Table 2-4 (continued)
Applicable ASTM Specifications for Various Structural Shapes

Steel Type	ASTM Designation		F_y, Yield Stress[a], ksi	F_u, Tensile Stress[a], ksi	Applicable Shape Series									
												HSS		
					W	M	S	HP	C	MC	L	Rectangular	Round	Pipe
Corrosion Resistant Quenched and Tempered Low-Alloy	A709/ A709M	Gr. QST 50	50	65										
		Gr. QST 50S	50–65	65										
		Gr. QST 65	65	80										
		Gr. QST 70	70	90										
Corrosion Resistant High-Strength Low-Alloy	A588/A588M		50	70										
	A847/A847M		50	70										
	A1065/ A1065M[f]	Gr. 50W	50	70										

■ = Preferred material specification
□ = Other applicable material specification, the availability of which should be confirmed prior to specification
□ = Material specification does not apply

Note: Referenced standards are intended to be the versions specified in AISC *Specification* Section A2.
[a]Minimum, unless a range is shown.
[b]For wide-flange shapes with flange thicknesses over 3 in., only the minimum of 58 ksi applies.
[c]Electric resistance welded (ERW) or seamless HSS only.
[d]For shapes with a flange or leg thickness less than or equal to 1½ in. only.
[e]For shape profiles with a flange width of 6 in. or greater.
[f]This specification is not a prequalified base metal per AWS D1.1/D1.1M.
[g]Type 1, 2, or 3 only as given in AISC *Specification* Table A3.1.
[h]For shapes with a flange or leg thickness less than or equal to 2 in. only.
[i]Minimum applies for walls nominally ¾ in. thick and under. For wall thickness over ¾ in. and up to 1½ in., F_y = 46 ksi and F_u = 67 ksi.

Figure 3.10 (*continued*) Applicable ASTM Specifications for Various Structural Shapes

A1085

A1085 steel is a carbon steel for hollow structural sections. For HSS produced according to this standard, the wall thickness is permitted to be no more than 5% under the nominal wall thickness and the mass cannot be more than 3.5% under the nominal mass. This standard was developed so that the design properties for HSS could be developed based on the nominal dimensions of the section. For other HSS material specifications, the tolerance on thickness is so large that design must be carried out using properties determined by using 0.93 times the nominal thickness, rather than the nominal thickness. Grade A is the only grade provided for in the standard. It is available with a yield stress, F_y = 50 to 70 ksi, and a minimum tensile stress, F_u = 65 ksi.

A572 Steel

A572 is a high-strength low-alloy steel, also referred to as columbium-vanadium structural steel, available in five grades. It is a versatile high-strength steel with good weldability. Availability of shapes and plates is a function of grade, generally depending on element thickness. It is available in all shapes other than HSS and pipe. The full range of minimum yield stress is 42 to 65 ksi, depending on grade, and the minimum tensile stress ranges from 60 to 80 ksi, again depending on grade. A572 Grade 50 is the preferred steel for M-, S-, HP-, and L-shapes.

A618 Steel

A618 is a high-strength low-alloy steel used for HSS. Grades I, II, and III are approved for use in structures by AISC. It is one of two high-strength low-alloy steels available in HSS. Grade II has limited atmospheric corrosion resistance, and Grade III can be produced with increased corrosion resistance if required. The minimum yield stress depends on the particular product and may vary from 46 to 50 ksi. The minimum tensile stress varies from 65 to 70 ksi, again depending on grade and product wall thickness.

A913 Steel

A913 is a high-strength low-alloy steel produced by quenching and self-tempering. It is available in grades 50, 60, 65, 70, and 80. The minimum yield stress ranges from 50 to 80 ksi and the minimum tensile stress ranges from 65 to 95 ksi.

A992 Steel

A992 steel is a high-strength low-alloy steel that has become the steel of choice for wide-flange shapes as well as C- and MC-shapes. It was first approved for use in 1998 as a replacement for steel that had come to be dual-certified as meeting both A36 and A572 Grade 50. This standard was developed partly as the result of an improved understanding of the impact of material property variations on structural behavior, and partly as the result of the changes in properties caused by the use of scrap as the main resource for steel production. The chemical components for A992 steel were given in Table 3.3 and discussed in Section 3.5. It has a minimum yield stress, F_y = 50 ksi, and a minimum tensile stress, F_u = 65 ksi. An additional, and critical, requirement is that the yield-to-tensile ratio cannot exceed 0.85.

A1065 Steel

A1065 steel is a high-strength low-alloy steel specifically for HSS. This steel is produced to the standards for steels previously approved for use as plates. Thus, the tolerances are tighter than those for other common HSS steels such as A500 and A501. HSS produced according to this standard are first formed into two channels on a press brake and then welded together to form the tube. This material is produced in Grade 50 and in Grade 50W, a corrosion resistant grade, with a minimum yield stress, F_y = 50 ksi, and a minimum tensile stress, F_u = 60 ksi for Grade 50 and F_u = 70 ksi for Grade 50W. A1065 is also available in a corrosion-resistant grade for rectangular HSS. Before specifying A1065 steel, the designer should confirm its availability.

A588 Steel

A588 is a high-strength low-alloy corrosion-resistant steel with substantially better corrosion resistance than carbon steel with or without copper. It is available for all shapes, except HSS and pipe, as well as plate. For all shapes, and for plates up to 4 in., it has a minimum yield stress of 50 ksi and a minimum tensile stress of 70 ksi. Plates up to 8 in. are available at reduced stress values.

A847 Steel

A847 is a high-strength low-alloy corrosion-resistant steel used for HSS. It has a minimum yield stress of 50 ksi and a minimum tensile stress of 70 ksi.

3.6.2 Steel for Plates and Bars

Many of the steels already discussed for shapes are also available for plates and bars. Figure 3.11 shows the ASTM designations, the corresponding yield and tensile stresses, and the plate thickness for which they apply. The only steels available for plates and bars that are not also available for shapes are A283, A514, and A1066.

A283 Steel

A283 steel is a low to intermediate strength carbon steel that is available in Grade C and Grade D. Grade C has a minimum yield stress, F_y = 30 ksi, and a tensile stress, F_u = 55 to 75 ksi while Grade D has a minimum yield stress, F_y = 33 ksi, and a tensile stress, F_u = 60 to 80 ksi. Although there are no thickness limits specified, plates thicker than 2 in. may be more difficult to obtain, and the designer should check availability prior to specifying.

A514 Steel

A514 is a high-yield strength-quenched and tempered alloy steel suitable for welding. It is available as plate material up to 6 in. There are 14 different grades, which vary according to the chemical content and maximum thickness. The minimum yield stress is either 90 or 100 ksi, and the ultimate tensile stress ranges from 100 to 130 ksi. This is the highest-yield-stress steel approved for use, except for fasteners, according to the AISC *Specification*.

A1066 Steel

A1066 steel is a high-strength low-alloy steel produced in 5 grades, Grade 50, Grade 60, Grade 65, Grade 70, and Grade 80 with a minimum yield stress corresponding to the grade designation. Minimum tensile strength for Grade 50 is 65 ksi, Grade 60 is 75 ksi, Grade 65 is 80 ksi, Grade 70 is 85 ksi and Grade 80 is 90 ksi. Grades 50 and 60 are available up to 4 in. thickness, Grade 65 to 3 in. Grade 70 to 2 in. and Grade 80 to 1 in.

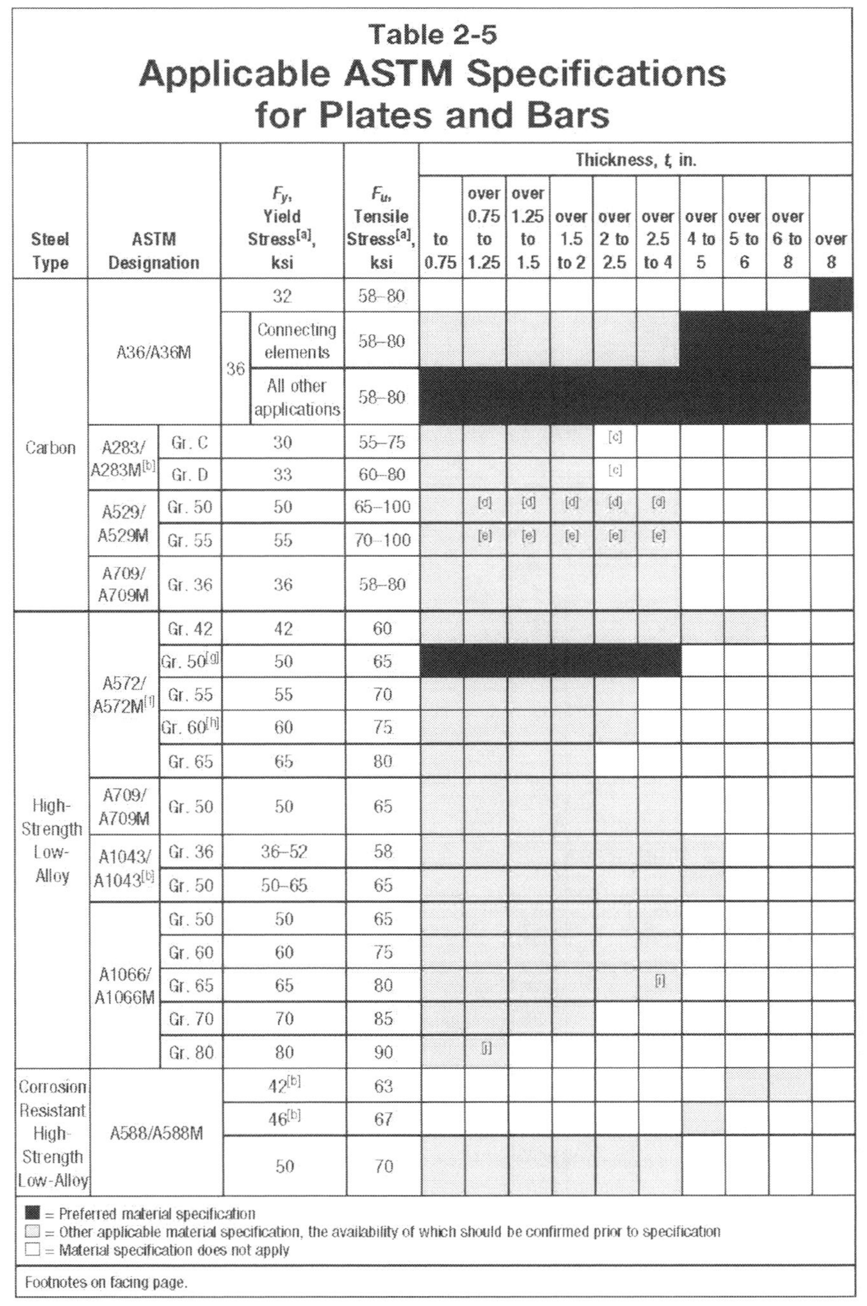

Table 2-5
Applicable ASTM Specifications for Plates and Bars

Steel Type	ASTM Designation		F_y, Yield Stress[a], ksi	F_u, Tensile Stress[a], ksi	Thickness, t, in.									
					to 0.75	over 0.75 to 1.25	over 1.25 to 1.5	over 1.5 to 2	over 2 to 2.5	over 2.5 to 4	over 4 to 5	over 5 to 6	over 6 to 8	over 8
Carbon	A36/A36M		32	58–80										■
			36 Connecting elements	58–80	▒	▒	▒	▒	▒	▒	■	■	■	
			36 All other applications	58–80	■	■	■	■	■	■	■	■	■	
	A283/ A283M[b]	Gr. C	30	55–75	▒	▒	▒	▒	[c]					
		Gr. D	33	60–80	▒	▒	▒	▒	[c]					
	A529/ A529M	Gr. 50	50	65–100	▒	▒ [d]	▒ [d]	▒ [d]	▒ [d]	▒ [d]				
		Gr. 55	55	70–100	▒	▒ [e]	▒ [e]	▒ [e]	▒ [e]	▒ [e]				
	A709/ A709M	Gr. 36	36	58–80	▒	▒	▒	▒	▒	▒				
High-Strength Low-Alloy	A572/ A572M[f]	Gr. 42	42	60	▒	▒	▒	▒	▒	▒	▒	▒		
		Gr. 50[g]	50	65	■	■	■	■	■	■				
		Gr. 55	55	70	▒	▒	▒	▒	▒					
		Gr. 60[h]	60	75	▒	▒	▒	▒	▒					
		Gr. 65	65	80	▒	▒	▒	▒						
	A709/ A709M	Gr. 50	50	65	▒	▒	▒	▒	▒	▒				
	A1043/ A1043[b]	Gr. 36	36–52	58	▒	▒	▒	▒	▒	▒	▒			
		Gr. 50	50–65	65	▒	▒	▒	▒	▒	▒	▒			
	A1066/ A1066M	Gr. 50	50	65	▒	▒	▒	▒	▒	▒				
		Gr. 60	60	75	▒	▒	▒	▒	▒	▒				
		Gr. 65	65	80	▒	▒	▒	▒	▒	▒ [i]				
		Gr. 70	70	85	▒	▒	▒	▒						
		Gr. 80	80	90	▒	▒ [i]								
Corrosion Resistant High-Strength Low-Alloy	A588/A588M		42[b]	63								▒	▒	
			46[b]	67							▒			
			50	70	▒	▒	▒	▒	▒	▒				

■ = Preferred material specification
▒ = Other applicable material specification, the availability of which should be confirmed prior to specification
□ = Material specification does not apply

Footnotes on facing page.

Figure 3.11 Applicable ASTM Specifications for Plates and Bars

Table 2-5 (continued)
Applicable ASTM Specifications for Plates and Bars

Steel Type	ASTM Designation		F_y, Yield Stress[a], ksi	F_u, Tensile Stress[a], ksi	Thickness, t, in.									
					to 0.75	over 0.75 to 1.25	over 1.25 to 1.5	over 1.5 to 2	over 2 to 2.5	over 2.5 to 4	over 4 to 5	over 5 to 6	over 6 to 8	over 8
Quenched and Tempered Alloy	A514/A514M[b]		90	100–130						▒	▒	▒		
			100	110–130	▒	▒	▒	▒	▒					
Corrosion Resistant Quenched and Tempered Low-Alloy	A709/A709M	Gr. 50W	50	70	▒	▒	▒	▒	▒	▒				
		Gr. HPS 50W	50	70	▒	▒	▒	▒	▒	▒				
		Gr. HPS 70W	70	85–110	▒	▒	▒	▒	▒	▒				
		Gr. HPS 100W[b]	90	100–130						▒				
			100	110–130	▒	▒	▒	▒	▒					

■ = Preferred material specification
▒ = Other applicable material specification, the availability of which should be confirmed prior to specification
□ = Material specification does not apply

Note: Referenced standards are intended to be the versions specified in AISC *Specification* Section A2.
[a]Minimum, unless a range is shown.
[b]This specification is not a prequalified base metal per AWS D1.1/D1.1M.
[c]Thickness is not limited to 2 in. in ASTM A283/A283M and thicker plates may be obtained but availability should be confirmed.
[d]Applicable for plates to 1 in. thickness and bars to 3½ in. thickness.
[e]Applicable for plates to 1 in. thickness and bars to 3 in. thickness.
[f]Type 1, 2, or 3 only as given in AISC *Specification* Table A3.1.
[g]Round bars up to and including 11 in. in diameter are permitted.
[h]Round bars up to and including 3½ in. in diameter are permitted.
[i]Applicable for plates to 3 in. thickness.
[j]Applicable for plates to 1 in. thickness.

Figure 3.11 (*continued*) Applicable ASTM Specifications for Plates and Bars

3.6.3 Steel for Fasteners

Fasteners for steel construction today include high-strength bolts, common bolts, threaded rods, and anchor rods. In addition, nuts, washers, and direct-tension indicators must be specified. The ASTM steels approved for these elements are listed in Figure 3.12. Many grades of steel are appropriate for the variety of mechanical fasteners used in steel construction, but only the three steels commonly specified for bolts are discussed here.

A307 Bolts

A307 bolts are also called *common bolts* or *black bolts*. Although the ASTM standard specifies three grades, only Grade A is approved for use as bolts in general applications. These bolts have an ultimate tensile strength of 60 ksi and are thus at a strength level similar to A36 steel. Although these bolts continue to be listed by AISC, they are rarely used in steel-to-steel structural connections.

F3125 Bolts

ASTM F3125 is a standard that consolidates what had previously been given in multiple separate standards for many years. It provides for quenched and tempered bolts manufactured in two strength grades, two types, and two styles. Thus, there are 6 grades specified, two of those six being metric equivalents to two other grades.

Grade A325 Bolts

Grade A325 bolts are heavy hex head bolts that are the predominant high-strength bolts used in construction. Two types are available: type 1, the normal medium carbon bolt; and type 3, which is the same bolt provided in a weathering steel. Bolts are available from 1/2 in. to 1-1/2 in. diameter and the minimum tensile strength is 120 ksi. These bolts are commonly referred to as A325 bolts.

Grade A490 Bolts

Grade A490 also designates a heavy hex head structural bolt. These fasteners are used when a higher tensile strength is required. As with the Grade A325 bolts, they are available as type 1 or type 3 with the same distinction. Again, the bolts are available from 1/2 in. to 1-1/2 in. diameter and have a minimum tensile strength of 150 ksi. These bolts are commonly referred to as A490 bolts.

Grade F1852 Bolts

Grade F1852 provides the standard specification for "twist-off" tension control (TC) bolt-nut-washer assemblies with a tensile strength of 120 ksi. These structural fasteners are unique in that they do not have a hex head but rather have a splined shank that permits installation to the required pretension force through the use of a special torque wrench which grabs the splined shank while the nut is turned. Grade F1852 connectors are essentially Grade A325 bolts but are manufactured as a different grade because their geometric characteristics differ from those of heavy hex head bolts. As with Grade A325, the tensile strength of these fasteners is 120 ksi for the diameters available, from 1/2 to 1-1/2 in. These bolts are commonly referred to as A325 TC bolts.

Grade F2280 Bolts

Grade F2280 provides the standard specification for "twist-off" tension control bolt-nut-washer assemblies, with a tensile strength of 150 ksi. These connectors are essentially Grade A490 bolts but, like Grade F1852 connectors, must be manufactured as a separate grade because their geometric characteristics differ from those of heavy hex head bolts. These fasteners are available in diameters of 1/2 to 1-1/2 in. These bolts are commonly referred to as A490 TC bolts.

Table 2-6
Applicable ASTM Specifications for Various Types of Structural Fasteners

ASTM Designation		F_y, Min. Yield Stress, ksi	F_u, Tensile Stress[a], ksi	Diameter Range, in.	Bolts: High-Strength: Conventional	Bolts: High-Strength: Twist-Off-Type Tension-Control	Bolts: High-Strength: Fixed-Spline Tension-Control	Bolts: Common Bolts	Nuts	Washers: Hardened	Washers: Plain	Washers: Direct-Tension Indicator	Threaded Rods	Anchor Rods: Hooked	Anchor Rods: Headed	Anchor Rods: Threaded and Nutted
F3125/F3125M	Gr. A325[b]	–	120	0.5 to 1.5	■											
F3125/F3125M	Gr. F1852[b]	–	120	0.5 to 1.5		■										
F3125/F3125M	Gr. A490[b]	–	150	0.5 to 1.5	■											
F3125/F3125M	Gr. F2280[b]	–	150	0.5 to 1.5		■										
F3111		–	200–215	1 to 1.25	▒											
F3043		–	200–215	1 to 1.25		▒										
F3148[b]		–	144	0.5 to 1.25			■									
A194/A194M Gr. 2H		–	–	0.25 to 4					▒							
A563/A563M		–	–	0.25 to 4					■							
F436/F436M		–	–	0.25 to 4[c]						■						
F844		–	–	any							■					
F959/F959M		–	–	0.5 to 1.5[c]								■				
A36/A36M		36	58–80	to 15									■	▒	▒	▒
A193/A193M Gr. B7		105	125	to 2.5									▒	▒	▒	▒
A193/A193M Gr. B7		95	115	over 2.5 to 4									▒	▒	▒	▒
A193/A193M Gr. B7		75	100	over 4 to 7									▒	▒	▒	▒
A307 Gr. A		–	60	0.25 to 4				■					▒	▒	▒	▒
A354	Gr. BC	109	125	0.25 to 2.5	▒ [d]								▒ [d]	▒	▒	▒
A354	Gr. BC	99	115	over 2.5 to 4	▒ [d]								▒ [d]	▒	▒	▒
A354	Gr. BD	130	150–173	0.25 to 4	▒ [d]								▒ [d]	▒	▒	▒
A449[b]		92	120	0.25 to 1	▒ [d]								▒ [d]	▒	▒	▒
A449[b]		81	105	over 1 to 1.5	▒ [d]								▒ [d]	▒	▒	▒
A449[b]		58	90	1.75 to 3	▒ [d]								▒ [d]	▒	▒	▒

■ = Preferred material specification
▒ = Other applicable material specification, the availability of which should be confirmed prior to specification
□ = Material specification does not apply

Footnotes on facing page.

Figure 3.12 Applicable ASTM Specifications for Various Types of Structural Fasteners

Table 2-6 (continued)
Applicable ASTM Specifications for Various Types of Structural Fasteners

ASTM Designation		F_y, Min. Yield Stress, ksi	F_u, Tensile Stress[a], ksi	Diameter Range, in.	Bolts: High-Strength: Conventional	Bolts: High-Strength: Twist-Off-Type Tension-Control	Bolts: High-Strength: Fixed-Spline Tension-Control	Bolts: Common Bolts	Nuts	Washers: Hardened	Washers: Plain	Washers: Direct-Tension Indicator	Threaded Rods	Anchor Rods: Hooked	Anchor Rods: Headed	Anchor Rods: Threaded and Nutted
A572/A572M	Gr. 42	42	60	to 6									▒	▒		▒
	Gr. 50	50	65	to 4[e]									▒	▒		▒
	Gr. 55	55	70	to 2.5									▒	▒		▒
	Gr. 60	60	75	to 3.5									▒	▒		▒
	Gr. 65	65	80	to 2									▒	▒		▒
A588/A588M		50	70	to 4									▒	▒		▒
		46	67	over 4 to 5									▒	▒		▒
		42	63	over 5 to 8									▒	▒		▒
F1554	Gr. 36	36	58–80	0.25 to 4									▒	■	■	■
	Gr. 55	55	75–95	0.25 to 4									▒	■	■	■
	Gr. 105	105	125–150	0.25 to 3									▒	▒	▒	▒

■ = Preferred material specification
▒ = Other applicable material specification, the availability of which should be confirmed prior to specification
□ = Material specification does not apply

Note: Referenced standards are intended to be the versions specified in AISC *Specification* Section A2.
– Indicates that a value is not specified in the material specification.
[a] Minimum unless a range is shown or maximum (max.) is indicated.
[b] When atmospheric corrosion resistance is desired, Type 3 can be specified.
[c] For use with fasteners with diameter range given.
[d] See AISC *Specification* Section J3.2 for limitations on use of ASTM A449, A354 Gr. BC, and A354 Gr. BD.
[e] ASTM A572/A572M permits rod diameters up to 11 in., but practicality of threading should be confirmed before specifying.

Figure 3.12 (*continued*) Applicable ASTM Specifications for Various Types of Structural Fasteners

3.6.4 Steel for Welding

Steel used for welding is called *filler metal* because it essentially fills the gap between the base metal pieces it is joining. The most critical aspect of selecting filler metal, which actually corresponds to the welding electrode, is matching the electrode with the base metal. In all cases, the weld must not form the weak part of the joint. The American Welding Society provides the specification for appropriate matching of the base metal and electrodes in Table 5.4 of their standard ANSI/AWS D1.1/D1.1M-2020 and the AISC *Specification* includes some information in Chapter J. The most commonly used weld strength is 70 ksi. A discussion of welding processes and material matching is presented in Chapter 10.

3.6.5 Steel for Headed Stud Anchors

Steel headed stud anchors, more commonly known simply as shear studs, are mechanical fasteners welded to structural shapes and embedded in concrete that permit steel and concrete to work together. This is called *composite construction*. Because these studs are welded to the steel shape, their properties are specified jointly between AWS and ASTM. Shear studs are specified in AWS D1.1/D1.1M Clause 9, with material as required in Clause 9.2.6 which states that studs are to be made from cold drawn bar conforming to the requirements of ASTM A29. Type B is usual, and the corresponding mechanical requirements are stated in AWS D1.1/D1.1M Table 9.1 as $F_y = 51$ ksi and $F_u = 65$ ksi.

3.7 AVAILABILITY OF STRUCTURAL STEEL

Structural engineers normally use the list of shapes found in the AISC *Manual* as the basis for design. Unfortunately, all shapes are not equally available in the marketplace and the selection of shapes that are difficult to obtain could negatively impact the overall cost and schedule of a project. Some shapes are available from a wide variety of producers; for instance, a W10×30 could be obtained from five different mills as of April 2023. However, the largest shapes, such as the W44×335, are not produced by all mills. Also, several of the smaller M-shapes are not rolled by any mill. Shape availability data are maintained by the mills on the AISC Web site at www.aisc.org/steelavailability.

Another important source – increasingly the primary source – of steel is the steel service center. These organizations are warehouses throughout the country that obtain steel directly from the mills; many stock nearly the full range of shapes. Although the task of obtaining the steel needed for any given project falls to the steel fabricator, it is always beneficial to the engineer to have some knowledge of availability. AISC maintains a list of steel service centers at www.aisc.org which in April 2023 identified 17 AISC Member Steel Service Centers and the states they serve.

3.8 PROBLEMS

1. When was the first AISC *Specification* published and what was its purpose?

2. In addition to buildings, what other types of structures are included in the scope of the 2022 AISC *Specification*?

3. Sketch and label a typical stress-strain curve for steel subjected to a simple uniaxial tension test.

4. What is the value of the modulus of elasticity used for calculations according to the AISC *Specification*, and what does this value represent in relation to the graph of stress versus strain for steel?

5. What happens to a steel element when it is loaded beyond the elastic limit and then unloaded?

6. Describe the difference between the yield stress and ultimate stress of a steel element.

7. Sketch and label 10 different structural shape cross sections whose properties are given in the AISC *Manual*.

8. What are the nominal and actual depths of a W36×318 wide-flange member? What is the weight of this member per linear foot? (Hint: Use your AISC *Manual*.)

9. What are the nominal and actual depths of a W16×57 wide-flange member? What is the weight of this member per linear foot? (Hint: Use your AISC *Manual*.)

10. What are the nominal and actual depths of a W18×211 wide-flange member? What is the weight of this member per linear foot? (Hint: Use your AISC *Manual*.)

11. What are the nominal and actual depths of a W14×808 wide-flange member? Compare these to the nominal and actual depths of a W14×159. (Hint: Use your AISC *Manual*.)

12. What are the nominal and actual depths of a W12×305 wide-flange member? Compare these to the nominal and actual depths of a W12×14. (Hint: Use your AISC *Manual*.)

13. What are the nominal and actual depths of a W10×100 wide-flange member? Compare these to the nominal and actual depths of a W10×49. (Hint: Use your AISC *Manual*.)

14. What are the actual depth, flange width, and flange thickness of a W14×99? (Hint: Use your AISC *Manual*.)

15. What are the actual depth, flange width, and flange thickness of a W27×129? (Hint: Use your AISC *Manual*.)

16. What are the actual depth and flange width of an HP12×53? What are the thicknesses of the web and flange? Compare the HP12×53 to a W12×53 (Hint: Use your AISC *Manual*.)

17. What are the actual depth, average flange thickness, and web thickness of a C12×25? (Hint: Use your AISC *Manual*.)

18. What are the actual depth, average flange thickness, and web thickness of a C6×13? (Hint: Use your AISC *Manual*.)

19. What are the cross sectional area, leg dimensions and thickness of an L6×6×5/8? (Hint: Use your AISC *Manual*.)

20. What are the cross sectional area, leg dimensions and thickness of an L6×4×9/16? (Hint: Use your AISC *Manual*.)

21. What are the cross sectional area and weight per linear foot of an L4×3×1/2 member? (Hint: Use your AISC *Manual*.)

22. What are the actual depth, flange width, flange thickness, and stem thickness of a WT15×45? Compare to the properties for a W30×90. (Hint: Use your AISC *Manual*.)

23. What are the actual depth, flange width, flange thickness, and stem thickness of a WT13.5×42? Compare to the properties for a W27×84. (Hint: Use your AISC *Manual*.)

24. What are the outside dimensions of a rectangular HSS24×12×5/8? What are the nominal and design wall thicknesses? (Hint: Use your AISC *Manual*.)

25. What are the outside dimensions of a rectangular HSS10×6×1/2? What are the nominal and design wall thicknesses? (Hint: Use your AISC *Manual*.)

26. What is the outside diameter of a round HSS6.625×0.500? What are the nominal and design wall thicknesses? (Hint: Use your AISC *Manual*.)

27. What is the outside diameter of a round HSS10.000×0.375? What are the nominal and design wall thicknesses? (Hint: Use your AISC *Manual*.)

28. What is the distinction between a round HSS and a pipe?

29. What are the outside diameter and nominal and design wall thicknesses of a Pipe 10 xx-Strong? (Hint: Use your AISC *Manual*.)

30. What are the outside diameter and nominal and design wall thicknesses of a Pipe 6 Std.? (Hint: Use your AISC *Manual*.)

31. What is the difference between a rectangular bar section and a plate?

32. What are the three preferred types of steel for shapes in the construction industry?

33. What effects does the addition of carbon have on steel?

34. Name three elements that improve the corrosion resistance of steel.

35. What grade of steel is most commonly used today in the production of W-shapes, and what are its yield stress and tensile stress?

36. What grade of steel is most commonly used today in the production of round HSS? Rectangular HSS?

37. What are the differences between an A500 Grade C rectangular HSS and an A1085 rectangular HSS?

38. What grade of steel is preferred for the fabrication of I-shaped structural shapes other than W-shapes, and what are its yield stress and tensile stress? (Hint: See Figure 3.10.)

39. What grade of steel is typically used for high-strength bolts in construction?

40. What resources can be consulted to determine the availability of a particular steel structural shape?

Chapter 4

Tension Members

JMA DOME at Syracuse University
Photo courtesy of Geiger Engineers

4.1 INTRODUCTION

The most efficient way to carry a force in steel is through tension. Because tension forces result in a fairly uniform stress distribution in the member cross section, all of the material is able to work to its fullest capacity. The normal assumption that tensile forces are applied to a member through the centroid of the cross section means that other structural actions, such as buckling or bending, are not normally present to reduce the member's ability to carry load. Thus, tension members are perhaps the simplest to design and a good starting point for studying structural steel design.

Tension members are fairly common elements in building structures, although they are not found in every structure. The structural members considered in this chapter are those subjected to a concentric tensile force as their primary force. Secondary effects, such as load misalignment and the influence of connections, will be addressed; however, the interaction of tension and bending is saved for treatment in Chapter 8.

Table 4.1 lists the sections of the *Specification* and parts of the *Manual* discussed in this chapter.

4.2 TENSION MEMBERS IN STRUCTURES

A wide variety of tension members can be found in building structures. Among the more important are members of trusses, bracing members, hangers, and sag rods.

Tension members are found in trusses as chords, diagonals, and verticals. Figure 4.1 shows a typical simply supported truss, with the tension bottom chord and two of the tension diagonals indicated. Tension members used as bracing for structures are normally long and slender, as seen in Figure 4.2. Because these slender members are relatively flexible, they must be carefully designed and erected, particularly if there is any chance of load reversal, which would call on them to carry a compression load. Even the smallest compressive force in a member that has been designed as a tension-only member can cause significant strength or serviceability problems in the final structure.

Table 4.1 Sections of *Specification* and Parts of *Manual* Covered in this Chapter

	Specification
B3	Design Basis
B4.2	Design Wall Thickness for HSS
B4.3	Gross and Net Area Determination
D1	Slenderness Limitations
D2	Tensile Strength
D3	Effective Net Area
D4	Built-up Members
D5	Pin-Connected Members
D6	Eyebars
J3.3	Size and Use of Holes
J3.6	Maximum Spacing and Edge Distance
J3.7	Tensile and Shear Strength of Bolts and Threaded Parts
J4.1	Strength of Elements in Tension
J4.3	Block Shear Strength
J7	Bearing Strength
M2.5	Bolted Construction
	Manual
Part 1	Dimensions and Properties
Part 5	Design of Tension Members

Other examples of tension members are hangers that connect lower floors to some support above, as seen in Figure 4.3, and sag rods that support purlins in the roof structure or the girts in the walls of a steel-framed building. It is easy to see the importance of most tension members because they normally carry an obvious, direct load. For secondary members like sag rods, their failure can produce unsightly displacements in the walls, and could cause stability problems for the purlins or girts, but they are not likely to be found carrying significant direct loading.

4.3 CROSS-SECTIONAL SHAPES FOR TENSION MEMBERS

Tension members can be structural steel shapes, plates, or combinations of shapes and plates; eyebars and pin-connected plates; rods and bars; or wire rope and steel cables. Wire rope and steel cables are not covered by the *Specification* nor considered here, although they are important elements in the special structures where they occur.

Figure 4.1 A Simply Supported Truss with Tension Members Indicated
Photo courtesy Matt Melrose/LERA Consulting Structural Engineers

Figure 4.2 Tension Bracing Members

Figure 4.3 Tension Hangers

Eyebars are not in common use today but can be found in older applications, particularly in bridges, trusses and similar structures, as seen in Figure 4.4. Although rarely used, they are still covered in the *Specification*. The pin-connected plate shown in Figure 4.5 is actually a part of a connection. This configuration is used in a variety of applications, such as industrial structures, anchors for tension members, and connections in bridge girders.

Figure 4.4 Eyebars in an Historic Building Roof Structure

Figure 4.5 A Pin-Connected Member as part of a truss
Photo courtesy of Steel Structures Technology Center, Inc.

Several common shapes used for tension members are shown in Figure 4.6, and some typical built-up shapes are given in Figure 4.7. The solid round bar is frequently used, either as a threaded rod or welded to other members. The threaded end provides a simple connection to the structure, but the design must take into account the reduction in cross-sectional area caused by the threads. Rods with upset ends are occasionally used instead of the normal rods; the enlarged end permits threading without reducing the cross-sectional area below the main portion of the rod. The differences between these two types of rods can be seen in Figure 4.8.

Square, rectangular, and circular HSS have become more common as tension members over the past few years, largely due to their attractive appearance and ease of maintenance. However, the end connections may become complicated and expensive, depending on the particular application. HSS are especially useful for longer tension components, when slenderness and related serviceability considerations may be important.

Single angles, as shown in Figure 4.7a, are used extensively in towers, such as those supporting cellular telephone communications and high-voltage power lines. Double angles and double channels, as shown in Figure 4.7b and d, are probably the most popular tension members for planar trusses due to the fact that gusset plates can be conveniently placed in the space between the individual shapes. The end connections for these members are therefore straightforward to design and fabricate and allow for symmetry in the vertical plane. Star angles, as shown in Figure 4.7c are used when the gusset is thinner and there is a need to have more area engaged at the connection.

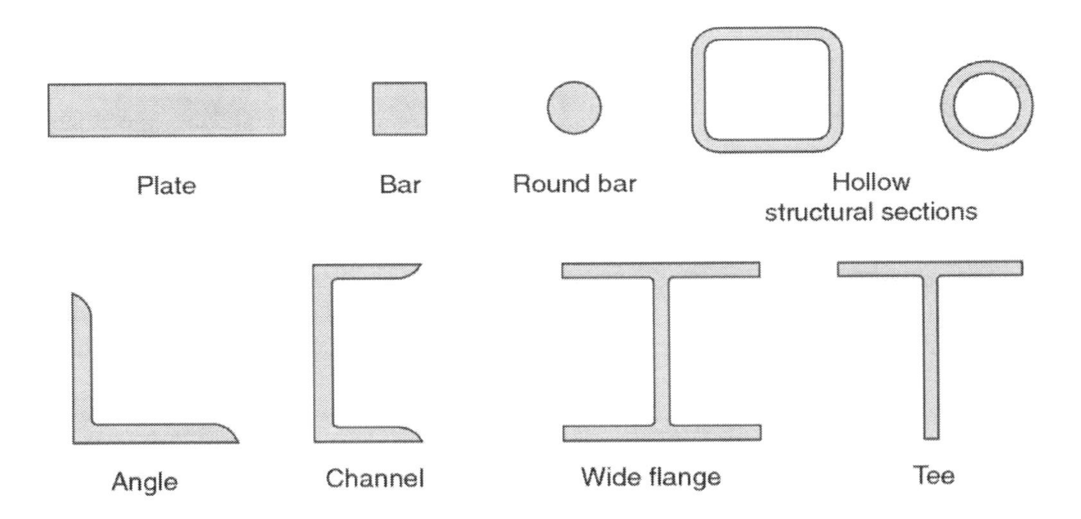

Figure 4.6 Common Shapes for Tension Members

Large tensile forces may require cross sections with an area dictating that the member be made from a wide-flange shape, a tee, double channels, or built-up shapes, such as those given in Figure 4.7e and f. Built-up cross sections were more common in the past, when rolled shapes were comparatively smaller in cross section and the cost of labor was lower; today larger rolled shapes are available and labor costs mean it is more economical to use them. Current structural applications of such elements are found in long-span roof trusses, bridge trusses, and bracing members in large industrial structures.

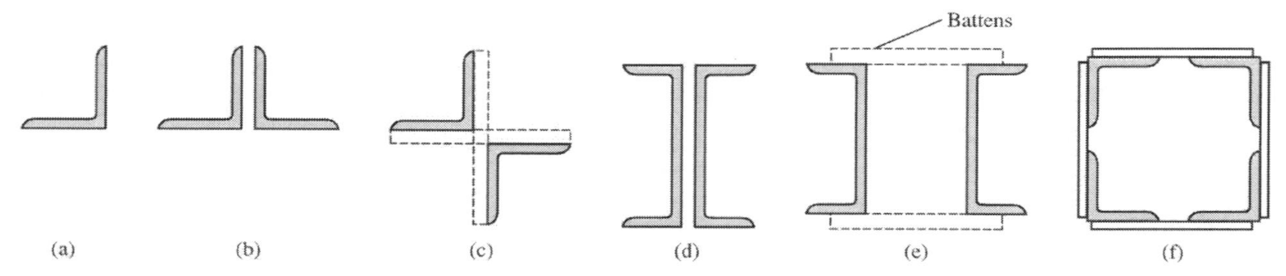

Figure 4.7 Typical Built-up Shapes Used as Tension Members

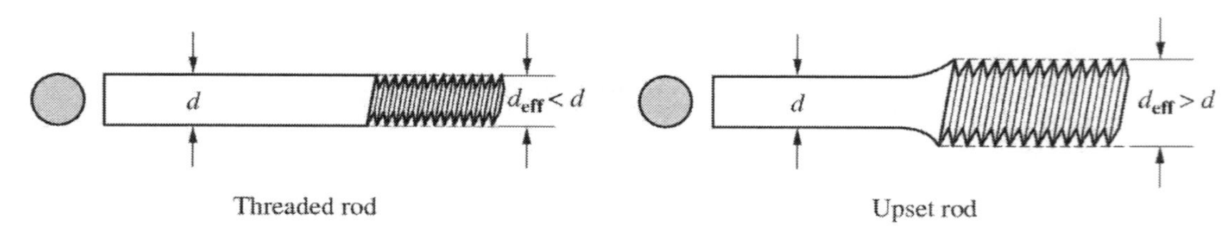

Figure 4.8 Threaded and Upset Rods

4.4 BEHAVIOR AND STRENGTH OF TENSION MEMBERS

Tension members are covered in Chapter D of the *Specification*. The first requirement given, Section D1, is not really a requirement at all but recognition that in the past there had been a limit on the slenderness, the ratio of length to radius of gyration, L/r, for tension members. That limit was really a recommendation since very slender members may be difficult to handle during construction, but it had no impact on the strength of tension members. Thus, the user note suggests that a limit on L/r, where L is the fabricated length of the member, of 300 would be reasonable.

Two possible limit states are defined in Section D2 for tension members: yielding and rupture. The controlling limit state depends on the ability of the member to undergo plastic deformation. Both of these failure modes represent limit states of strength that must be taken into account in the design of the tension member. The design basis for ASD and LRFD were presented in Sections 1.9 and 1.10, respectively. Equations B3-2 and B3-1 are repeated here to reinforce the relationship between the nominal strength, resistance factor, and safety factor presented throughout the *Specification*.

The requirement for ASD is

$$R_a \leq \frac{R_n}{\Omega} \qquad \text{(AISC B3-2)}$$

The requirement for LRFD is

$$R_u \leq \phi R_n \qquad \text{(AISC B3-1)}$$

4.4.1 Yielding

Yielding occurs when the uniformly distributed stress throughout the cross section reaches the yield stress over the length of the member. Although the member will continue to resist the load that caused yielding to occur, it will undergo excessive stretching, and this elongation will make the member unusable. The longer the member, the greater the elongation. Because the limit state of yielding on the gross section of the member is accompanied by this large deformation, it readily serves as a warning of any impending failure.

The yield limit state is defined as

$$P_n = F_y A_g \qquad \text{(AISC D2-1)}$$

where

P_n = nominal tensile yield strength
F_y = yield stress
A_g = gross area of the member

The design strength and allowable strength are to be determined using

$$\phi_t = 0.90 \text{ (LRFD)} \qquad \Omega_t = 1.67 \text{ (ASD)}$$

4.4.2 Rupture

Bolt holes in a member will cause stress concentrations to develop under the service load, as shown in Figure 4.9. Elastic theory shows that the stress concentration results in a peak stress approximately three times the average stress. As the peak stress reaches the yield stress, the member will continue to strain and load can continue to increase. With increasing load, the strain in the region of the hole increases into the strain hardening region, and the member ruptures once the stress in this area exceeds the ultimate strength. Although the material in the region of the holes yields initially, it does so over a very short length, resulting in a small total elongation. Thus, the material can reach its ultimate strength through strain hardening, without excessive elongation, and failure occurs through rupture. The limit state of rupture on the effective net area of the cross section is accompanied by small deformations from yielding, giving little or no warning of the impending sudden failure, and offering limited opportunities to take corrective action before the rupture.

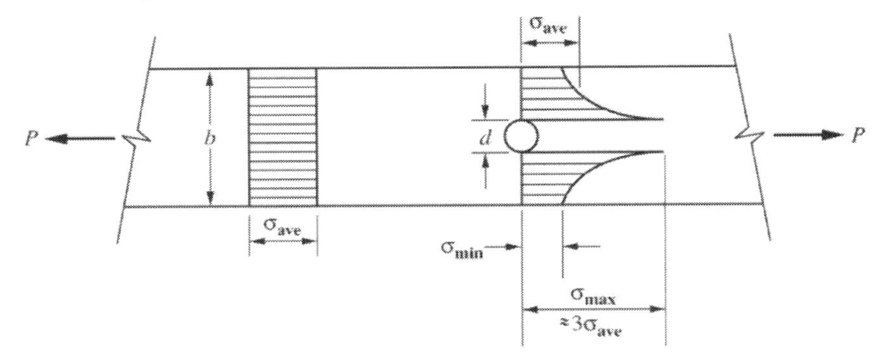

Figure 4.9 Stress Concentration Due to Bolt Hole in Member

The rupture limit state is defined as

$$P_n = F_u A_e \qquad \text{(AISC D2-2)}$$

where

P_n = nominal tensile rupture strength
F_u = ultimate stress
A_e = effective net area of the member

The design strength and allowable strength are to be determined using

$$\phi_t = 0.75 \text{ (LRFD)} \qquad \Omega_t = 2.00 \text{ (ASD)}$$

If the two limit states were to result in the same available strength, using the LRFD formulation,

$$0.75F_u A_e = 0.90F_y A_g \qquad (4.1)$$

or

$$A_e / A_g = 0.90F_y / 0.75F_u \qquad (4.2)$$

The limit state of yielding on the gross section governs when the right-hand side of Equation 4.1 is less than the left-hand side. Using Equation 4.2, yielding on the gross section governs if

$$A_e / A_g > 0.90F_y / 0.75F_u \qquad (4.3)$$

and rupture on the effective net section governs if

$$A_e / A_g < 0.90F_y / 0.75F_u \qquad (4.4)$$

Steel with a small (F_y/F_u) value, such as ASTM A36, with $0.9F_y/0.75F_u$ = 0.9(36)/(0.75(58)) = 0.74, will allow more of the cross section to be removed in the form of bolt holes before the rupture limit state governs than will steel with a higher (F_y/F_u) value, such as ASTM A992, with $0.9F_y/0.75F_u$ = 0.9(50)/(0.75(65)) = 0.92.

The comparisons discussed above are applicable only for normal bolted connections and their corresponding areas. The *Specification* equations are not intended to cover tension members with large cutouts. These require special design considerations, and are beyond the scope of this book because they are not common in most building structures.

Although welded connections do not normally require the removal of material from the cross section, the placement of the welds and the type of cross section may require a reduction from the gross area to determine the effective net area.

4.5 COMPUTATION OF AREAS

The design of tension members uses the following cross-sectional area definitions:

1. gross area, A_g
2. net area, A_n
3. effective net area, A_e

The criteria governing the computation of the various areas required for tension member analysis and design are given in Sections B4.3 and D3 of the *Specification*. They are discussed in further detail here.

4.5.1 Gross Area

The gross area of a member might also be thought of as the full cross-sectional area. A section is made perpendicular to the longitudinal axis of the element, along which the tensile force is acting, and the gross area, A_g, is the area of that cross section. No holes or other area reductions are considered when calculating the gross area.

In the case of plates, bars, and solid circular shapes, the value of A_g is found directly as the value of width times thickness, *bt*, for plates and bars, and as $\pi d^2/4$ for circular shapes, where d is the diameter. For structural steel shapes commonly used in construction, the *Manual* provides values for gross areas in Part 1. However, in lieu of using the tabulated values, A_g may be approximated as

$$A_g = \Sigma w_i t_i \tag{4.5}$$

where w_i and t_i are the width and thickness, respectively, of the rectangular cross-sectional element, i, of the shape. Equation 4.5 applies only to shapes that are composed of flat plate components, such as wide flanges. The calculation for hollow circular shapes is similarly straightforward. The gross area of HSS meeting the requirements of ASTM A500 is determined using 93 percent of the nominal wall thickness of the shape, as indicated in Section B4.2. Because HSS are consistently manufactured with a thickness at the low end of the tolerance limit, the values provided in the *Manual* are all based on this reduced thickness. For channels and S-shapes, both with sloping flanges, the inaccuracy of this approach means that the approach is rarely used. For all hot-rolled shapes, the area is easily obtained from the tables in Part 1 of the *Manual*.

The procedure for angles requires a slight modification. The angle may be treated as an equivalent flat plate, where the effective width is taken as the sum of the leg dimensions less the thickness, and the gross area is this effective width times the angle thickness.

An angle and its equivalent flat plate are shown in Figure 4.10.

4.5.2 Net Area

The net area is obtained by subtracting the area of any holes occurring at a particular section from the gross area at that section. Thus, holes associated with mechanical fasteners, such as bolts, and with welds, such as plug welds and slot welds, are considered. These fastening elements are normally used only to connect tension members to the adjacent parts of the structure, and the reduced areas therefore normally appear at member ends. However, if holes occur at any point along a tension member, their effect must be considered.

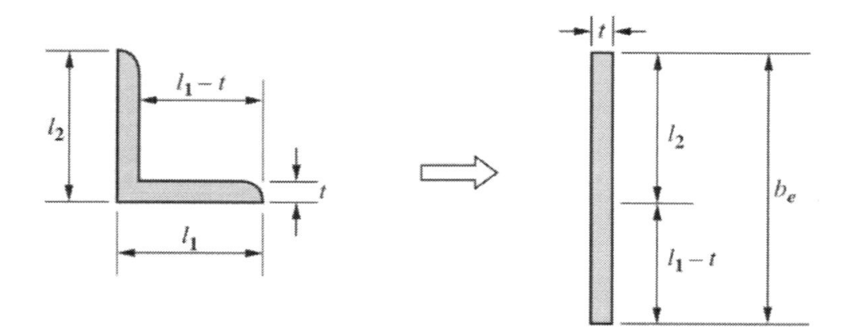

Figure 4.10 Angle and Its Equivalent Flat Plate

Plug and slot welds are made through holes in a member and are relatively uncommon in most structures in which the framing members are made from structural shapes and plates. Normal welded joints do not involve making holes in members. Thus, welds do not normally reduce the cross-sectional area of tension members, so the net area is equal to the gross area.

In the computation of net area for a tension member with bolted end connections, determining the size of the holes is important. The criteria for standard, oversize, and slotted holes are covered in *Specification* Section J3.3.

Standard Holes

Normal steel construction requires the specification of fastener size rather than hole size. The hole is then sized according to what is required to accommodate the fastener. The manner in which the hole is fabricated is also critical.

In recognizing the needs for fabrication and erection tolerances, standard bolt holes for bolts up to 7/8 in. in diameter are made 1/16 in. larger in diameter than the bolt to be inserted in the hole. For 1.0 in. diameter bolts and larger, the standard hole may be 1/8 in. larger in diameter than the bolt to be inserted in the hole. This increase for larger bolts is an option that some fabricators may deem unnecessary; however, since that may not be known by the designer ahead of time, the larger hole should be used in calculations. Thus, a 3/4 in. bolt requires a hole with a (3/4 + 1/16) = 13/16 in. diameter and a 1.0 in. diameter bolt requires a (1 + 1/8) = 1-1/8 in. diameter hole.

In the case of punched holes, the punching process may damage some of the material immediately adjacent to the hole. That material may not be considered fully effective in transmitting load and must also be deducted from the gross area along with the material that has actually been removed. Figure 4.11 shows an angle leg with a hole along with the punched out piece. As the punch is applied to the material, the edges around the hole are deformed, as seen in the taper of the punched-out piece. In discounting this region, the effective hole diameter, d_e, is increased by another 1/16 in., regardless of bolt diameter, according to *Specification* Section B4.3b. The deduction is conservatively applied to all holes, since one generally does not know how the holes will be produced.

Figure 4.11 Angle Leg with Punched Hole and Plug That Was Removed by Punching

The most common practice for producing holes is through punching. However, any method that meets the hole surface quality requirements of *Specification* Section M2.5 is permitted.

Because the decision to punch or drill a hole is a function of the steel fabricator's equipment capacity, for the design of tension members it is standard practice to deduct for holes with a diameter 1/8 in. greater than the specified bolt size up to 7/8 in. and 3/16 in. for 1.0 in. bolts and above.

The following examples demonstrate gross and net area calculations for several shapes.

EXAMPLE 4.1
Gross and Net Area

Goal: Determine the gross and net areas of a plate with a single line of holes.

Given: A single line of standard holes for 3/4 in. bolts is placed in a 6×1/2 plate, as shown in Figure 4.12a.

SOLUTION

Step 1: Determine the gross area at Section 1-1.

$$A_g = 6(1/2) = 3.0 \text{ in.}^2$$

Step 2: Determine the effective hole size for a 3/4 in. diameter bolt.

$$d_e = (3/4 + 1/16 + 1/16) = 7/8 \text{ in.}$$

Step 3: Determine the net area at section 2-2.

$$A_n = (b - d_e)t = (6.0 - 7/8)(1/2) = 2.56 \text{ in.}^2$$

EXAMPLE 4.2
Gross and Net Area

Goal: Determine the gross and net areas of a plate with a double line of holes.

Given: A double line of standard holes for 7/8 in. bolts are placed in a 10×3/4 plate, as shown in Figure 4.12b.

SOLUTION

Step 1: Determine the gross area at section 1-1.

$$A_g = 10(3/4) = 7.5 \text{ in.}^2$$

Step 2: Determine the effective hole size for a 7/8 in. diameter bolt.

$$d_e = (7/8 + 1/16 + 1/16) = 1.0 \text{ in.}$$

Step 3: Determine the net area at section 2-2, with $n = 2$ bolts.

$$A_n = (b - nd_e)t = (10.0 - 2(1.0))(3/4) = 6.00 \text{ in.}^2$$

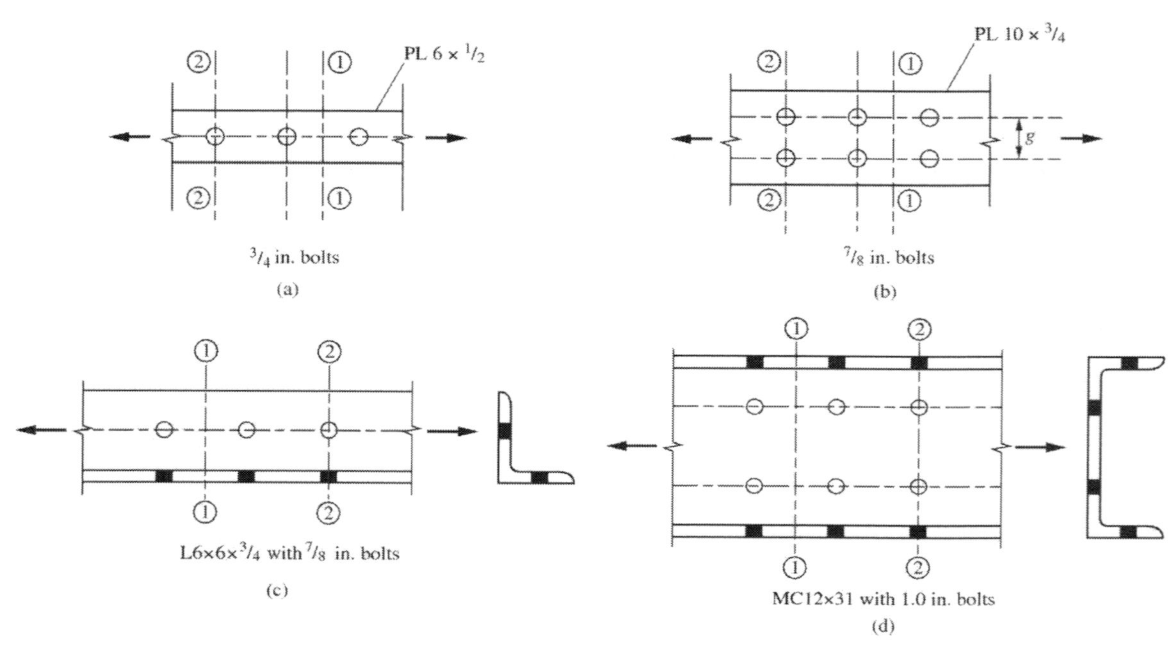

Figure 4.12 Plates and Shapes with Holes for Use with Examples 4.1 through 4.4

EXAMPLE 4.3 Gross and Net Area

Goal: Determine the gross and net areas of an angle with a single line of holes in each leg.

Given: A single line of standard holes for 7/8 in. bolts is placed on each leg of a 6×6×¾ angle as shown in Figure 4.12c.

SOLUTION

Step 1: Determine the gross area at section 1-1 from *Manual* Table 1-7.

$$A_g = 8.46 \text{ in.}^2$$

Step 2: Determine the effective hole size for a 7/8 in. diameter bolt.

$$d_e = (7/8 + 1/16 + 1/16) = 1.0 \text{ in.}$$

Step 3: Determine the net area at section 2-2, with $n = 2$ bolts.

$$A_n = A_g - nd_e t = 8.46 - 2(1.0)(3/4) = 6.96 \text{ in.}^2$$

EXAMPLE 4.4
Gross and Net Area

Goal: Determine the gross and net areas of a channel with multiple lines of holes.

Given: Four lines of standard holes for 1.0 in. bolts are placed in an MC12×31, as shown in Figure 4.12d. Two lines are in the web and one line is in each flange.

SOLUTION

Step 1: Determine the gross area at section 1-1 from *Manual* Table 1-6.

$$A_g = 9.12 \text{ in.}^2$$

Step 2: Determine the effective hole size for a 1.0 in. diameter bolt.

$$d_e = (1.0 + 1/8 + 1/16) = 1.19 \text{ in.}$$

Step 3: Determine the net area at section 2-2, with n_w = 2 bolts in the web and n_f = 2 bolts in the flanges. Determine the web and flange thicknesses from *Manual* Table 1-6.

$$t_w = 0.370 \text{ in. and } t_f = 0.700 \text{ in.}$$

$$\begin{aligned} A_n &= A_g - n_w d_e t_w - n_f d_e t_f \\ &= 9.12 - 2(1.19)(0.370) - 2(1.19)(0.700) \\ &= 6.57 \text{ in.}^2 \end{aligned}$$

Oversized and Slotted Holes

Section J3.3 of the *Specification* gives the required measurements for larger-than-standard or oversized holes, as well as for short-slotted and long-slotted holes. Figure 4.13 illustrates the criteria that apply for nominal bolt diameters of 5/8 in., 3/4 in. and 7/8 in.; refer to the *Specification* for data for other bolt sizes. These types of holes are used to facilitate the erection of the structure and, in some cases, to permit larger rotations or deformations to take place under loading.

Short Connecting Elements

Tension members within connections are usually short connecting elements such as links, flange plates, or gusset plates. When the member is short, and the net area and gross area are close to equal, there may not be sufficient length for the load to spread to a uniform distribution on the entire cross section. In this case, the area that is the first to yield may reach rupture at an early stage, and the rupture limit state would therefore be reached prematurely. This is an undesirable mode of failure, primarily because it is not ductile and could occur suddenly, with little or no warning. Section J4.1 indicates that the effective net area in these short connecting elements may be limited due to stress distribution as calculated by methods such as the Whitmore section, which is discussed in Chapter 11.

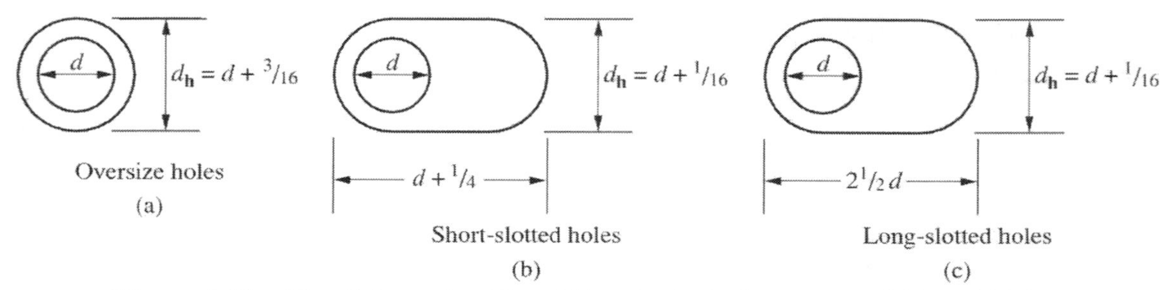

Figure 4.13 Size Criteria for 5/8 to 3/4 in. Bolts in Oversize and Slotted Holes

4.5.3 Influence of Hole Placement

The examples in Section 4.5.2 represent simple cases in which the net area is found in the section that produces the largest reduction in area, usually the section with the largest number of holes. However, hole placement does not always follow simple patterns where every section has the same number of holes. It is sometimes advantageous to use a pattern of staggered holes, such as those shown in Figure 4.14. Figure 4.14a shows an arrangement of staggered holes for a plate, and Figure 4.14b shows an example for an angle. When there are multiple holes, the center-to-center distance between adjacent holes in the direction parallel to the primary applied force is defined as the pitch, s. When there is more than one line of holes parallel to the line of force, the center-to-center distance between adjacent holes in the direction perpendicular to the primary applied force is the gage, g.

It is not clear from Figure 4.14 what the governing net section would be for either case. For the plate, sections 1-1 and 2-2 give identical A_n values, in which a deduction for one hole is taken for each line. Another possibility would be to follow a line that incorporates two holes, starting along line 1 and ending along line 2, as shown by the diagonal dashed line. The *Specification* refers to this line as a "chain" because it links individual holes. In this case, the area of two holes would be deducted from the gross cross section. However, this approach would be no different than if both holes were along the same straight line, line 1 or line 2. It seems reasonable in this situation that an approach that would deduct both holes would deduct too much, because these holes are staggered and are not along the same straight line. The correct solution should be somewhere between the result of deducting for one hole and that of deducting for two holes.

A simplified approach to address the interaction of staggered holes was adopted long ago by previous AISC *Specifications*. Although numerous studies have been conducted since this original simplification was introduced, none have proposed a significantly more accurate approach that is equally easy to implement.

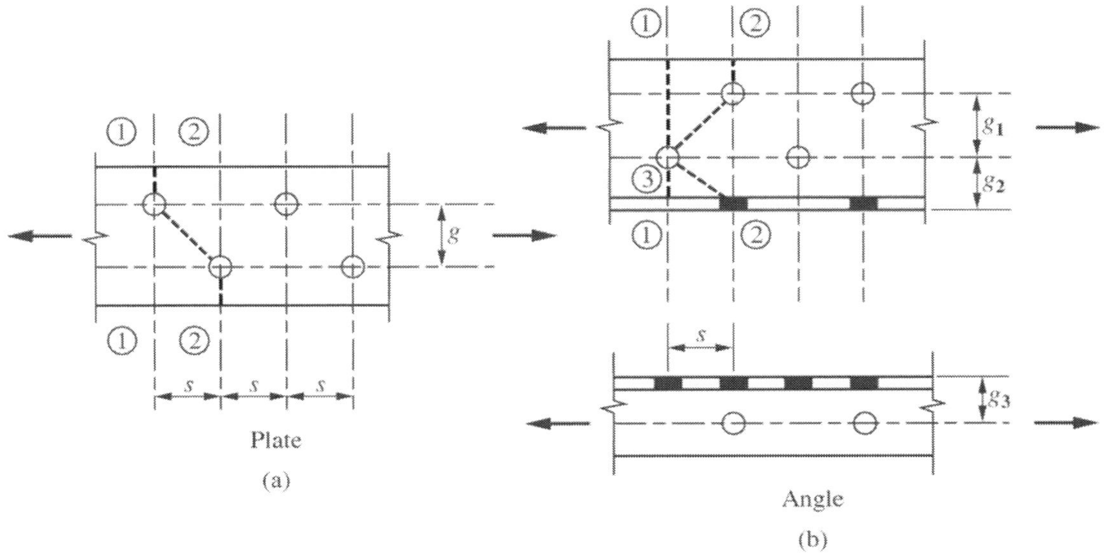

Figure 4.14 Staggered Hole Patterns in Plate and Angle

The *Specification* approach, given in Section B4.3b, requires that every potential failure line be assessed with the full area of each intersected hole deducted and something added back for the increased strength provided by the diagonal path. For every diagonal on a potential failure path, the quantity $s^2/4g$ is added back into the net width, b_n, to account for the overestimation of the required deduction when a full adjacent hole has been deducted. Examples 4.5 and 4.6 show the application of the staggered hole criterion in the middle of the member where the tension flows through the member. When holes are at the end of a member as part of a connection the flow of the tension force through the connectors should be considered.

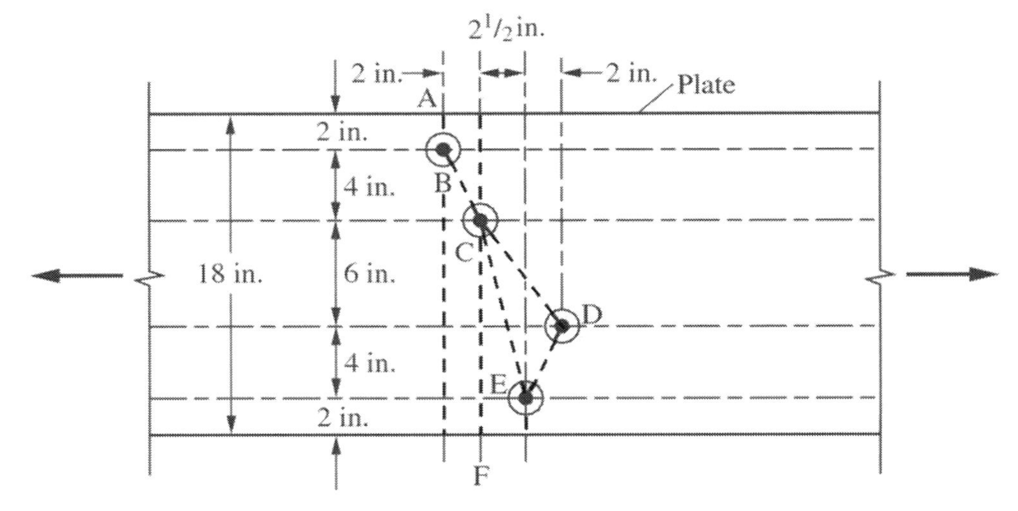

Figure 4.15 Hole Pattern for 18 in. Plate Used in Example 4.5

EXAMPLE 4.5
Net Width and Net Area of Plate

Goal: Determine the net width of a plate with staggered holes. Then determine the net area.

Given: The hole pattern for an 18×7/8 in. plate with holes for 3/4 in. bolts that is loaded in tension as shown in Figure 4.15. A and F represent the edges of the plate and B, C, D, and E represent the hole locations

SOLUTION

Step 1: Chain ABF (a straight line through one hole):
Deduct for one hole $(3/4+1/8)=-0.875$ in.

Step 2: Chain ABCF:
Deduct for two holes $2(3/4+1/8)=-1.75$ in.
For BC add $s^2/4g=(2.0)^2/(4(4.0))=+0.25$ in.
Total deduction $=-1.50$ in.

Step 3: Chain ACEF
Deduct for two holes $2(3/4+1/8)=-1.75$ in.
For CE add $s^2/4g=(2.5)^2/(4(10.0))=+0.16$ in.
Total deduction $=-1.59$ in.

Step 4: Chain ABCEF
Deduct for three holes $3(3/4+1/8)=-2.63$ in.
For BC add $s^2/4g=(2.0)^2/(4(4.0))=+0.25$ in.
For CE add $s^2/4g=(2.5)^2/(4(10.0))=+0.16$ in.
Total deduction $=-2.22$ in.

Step 5: Chain ABCDEF
Deduct for four holes $4(3/4+1/8)=-3.50$ in.
For BC add $s^2/4g=(2.0)^2/(4(4.0))=+0.25$ in.
For CD add $s^2/4g=(4.5)^2/(4(6.0))=+0.84$ in.
For DF add $s^2/4g=(2.0)^2/(4(4.0))=+0.25$ in.
Total deduction $=-2.16$ in.

Step 6: Deduct the largest quantity to obtain the least net width. For chain ABCEF, the net width is

$$b_n=18.0-2.22=15.8 \text{ in.}$$

Step 7: Use the least net width to determine the net area

$$A_n = b_n t = 15.8(7/8) = 13.8 \text{ in.}^2$$

Note: the hole pattern shown for this example is not common; it was selected for convenience to illustrate how rupture paths could vary.

EXAMPLE 4.6
Gross Area and Net Area

Goal: Determine the gross and net areas of an angle with staggered holes.

Given: An L6×4×1/2 with holes for 7/8 in. bolts are placed in the middle of the member as shown in Figure 4.16.

SOLUTION

Step 1: Determine the width of the equivalent flat plate representing the width of the angle.

$$w_e = l_1 + l_2 - t = 6 + 4 - 1/2 = 9.50 \text{ in.}$$

Step 2: Determine the gross area of the equivalent plate.

$$A_g = w_e t = 9.50(1/2) = 4.75 \text{ in.}^2$$

This is the same as found in *Manual* Table 1-7.

Step 3: Determine the gages for each bolt line.
The workable gages for the holes as found in Table 1-7A are shown in Figure 4.16. The gage between the holes closest to the heel of the angle in the two legs must be adjusted to account for the angle thickness. Thus,

$$(g + g_1 - t) = 2.50 + 2.25 - 0.50 = 4.25 \text{ in.}$$

Step 4: Determine the net area.
The governing net section will be section 2-2 or section 2-1-2. There is no need to consider section 1-1 because b_{n1} (section 1-1) will clearly be greater than b_{n2} (section 2-2).

For section 2-2, the net width is

$$b_{n2} = (9.5 - 2(7/8 + 1/8)) = 7.5 \text{ in.}$$

and the net area is

$$A_{n2} = 7.5(0.5) = 3.75 \text{ in.}^2$$

For section 2-1-2, this chain has two staggers of bolt holes, and both have the same pitch (s = 2.50 in). The gages are different, with one at 4.25 in. and the other at 2.50 in.

The net width for this chain is

$$b_{n3} = 9.5 - 3(7/8 + 1/8) + \left(\frac{2.50^2}{4(2.5)}\right) + \left(\frac{2.50^2}{4(4.25)}\right) = 7.49 \text{ in.}$$

and the net area is

$$A_{n3} = 7.49(0.5) = 3.75 \text{ in.}^2$$

Step 5: Select the least net area.

The lowest net area controls. In this case, the two chains yield the same net area,

$$A_n = A_{n2} = A_{n3} = 3.75 \text{ in.}^2$$

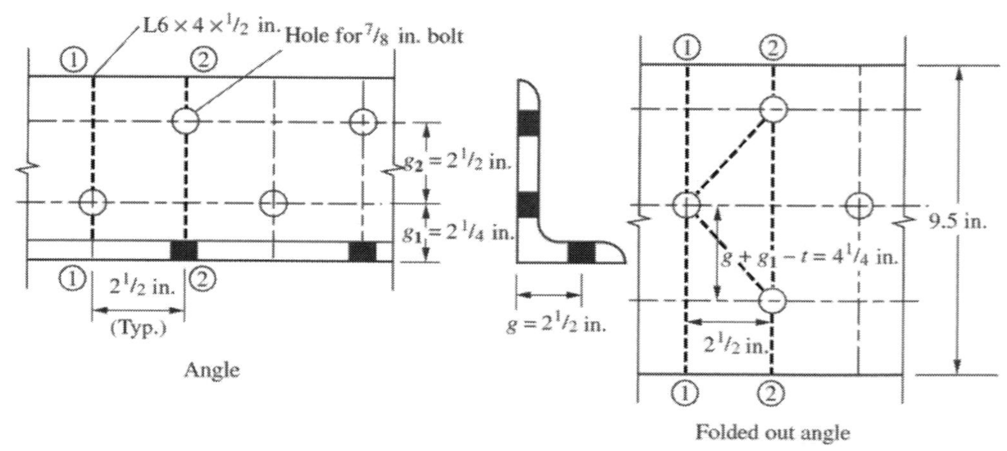

Figure 4.16 Hole Pattern for L6×4×1/2 Used in Example 4.6

4.5.4 Effective Net Area

When all elements of a tension member are attached to connecting elements, the entire member participates fully in transferring the load to the connection. However, when not all elements are attached to connecting elements, they cannot all participate fully. Figure 4.17 shows an angle with one leg attached to a connecting element and the other, the outstanding leg, unattached. To account for the inability of this unattached leg to transfer load, the net area used in calculating the rupture strength must be reduced to an effective net area, A_e.

This phenomenon occurs because the uniform stresses, occurring near the mid-length of the member at some distance from the connection, must be transferred through the more restricted area where the connection is located. The portion of the member area that is participating effectively in the transfer of force is smaller than the full net area. Thus, the net area is reduced to the effective net area.

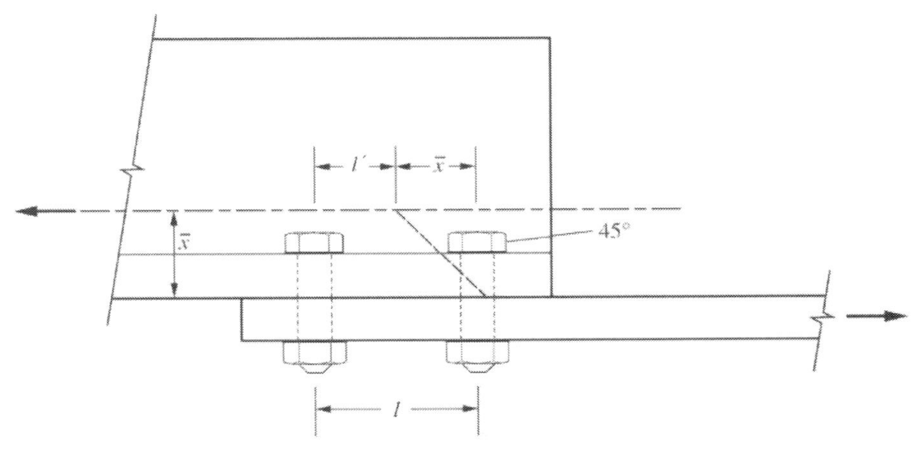

Figure 4.17 Conceptual Basis for Shear Lag Reduction Factor

This general behavior is called *shear lag*. Since its introduction into the *Specification*, it has been approximated through the use of the shear lag reduction factor, U, such that

$$A_e = UA_n \qquad \text{(AISC D3-1)}$$

Specification Table D3.1 provides values of the shear lag factor, U, for a wide variety of elements. For all tension members, except plates, HSS and those with longitudinal welds, when the tension load is transmitted to some but not all of the cross-sectional elements by bolts or longitudinal welds in combination with transverse welds, the effective length of the connection is reduced to $l' = l - \bar{x}$, where $\bar{x}$ is the distance from the attached face to the member centroid and l is the length of the connection, as shown in Figure 4.17. The reduction in net area is then taken in proportion to the reduction in effective length, l'/l. Thus, the reduction becomes the shear lag factor shown for Case 2:

$$U = \frac{l'}{l} = \frac{l - \bar{x}}{l} = 1 - \frac{\bar{x}}{l} \qquad (4.6)$$

Figure 4.18 shows the definition of connection length, l, for both a bolted and a welded connection. For a welded connection, the length of the connection, l, is the length of the weld. For a bolted connection, l is the center-to-center distance between the bolt holes at each end of the connection.

Table D3.1 of the *Specification* also provides simplified approaches to finding the shear lag factor when certain criteria are met.

Case 7 of Table D3.1: for W-, M-, S-, and HP-shapes or for tees cut from these shapes, the following apply:

Flange connected with three or more fasteners per line in the direction of loading where b_f is the flange width	$b_f \geq 2d/3$	$U = 0.90$
	$b_f < 2d/3$	$U = 0.85$

and d is the depth of the section or for tees, the depth of section from which the tee was cut

Web connected with four or more fasteners per line in the direction of loading	$U = 0.70$

Case 8 of Table D3.1: for single and double angles:

With four or more fasteners per line in the direction of loading	$U = 0.80$
With three fasteners per line in the direction of loading	$U = 0.60$

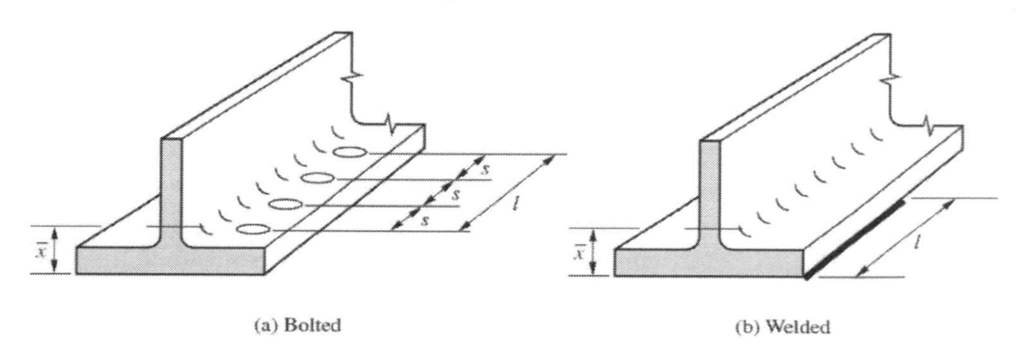

Figure 4.18 Definition of Connection Length, l, for Bolted and Welded Connection

If U is calculated using Case 7 or Case 8 as well as Case 2, the larger value is permitted to be used. For any other conditions, the table values should be used. However, for open cross sections (such as W-, M-, S-, C-, HP-, WT-, and ST-shapes, single angles, and double angles), there is no need to take U as less than the ratio of the gross area of the connected elements to the member gross area. This will also prove useful when there is only one row of connectors transverse to the member axis.

When tension members are connected with staggered bolts and not all elements of the member are connected, the effects of bolt stagger, shear lag and tension flow through the connection must be considered.

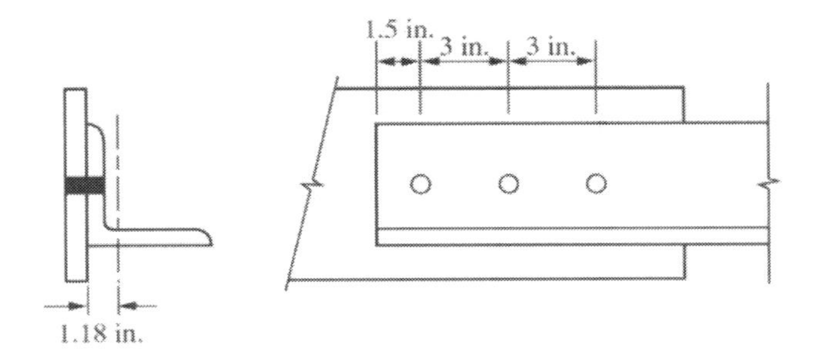

Figure 4.19 Single-Angle Tension Member for Example 4.7

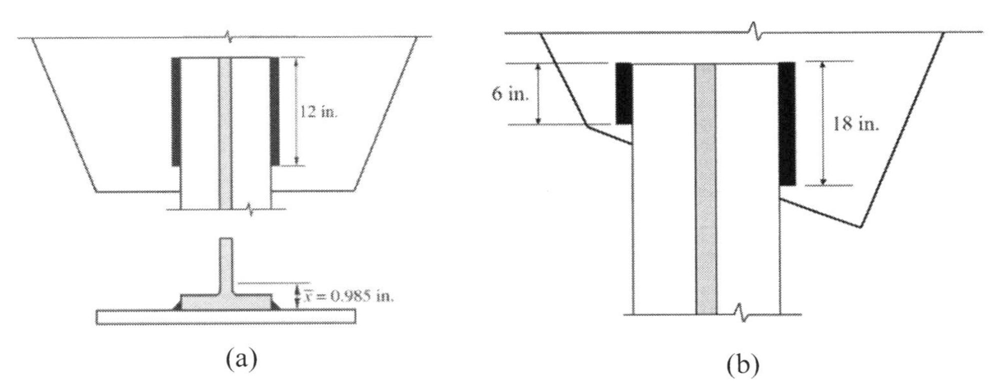

Figure 4.20 A WT Welded to a Gusset Plate for Examples 4.8 and 4.9

EXAMPLE 4.7
Tensile Strength of an Angle

Goal: Determine the design strength (LRFD) and the allowable strength (ASD) of an angle.

Given: Consider an L4×4×1/2 attached through one leg to a gusset plate with 3/4 in. bolts in standard holes as shown in Figure 4.19. Use A572 Gr. 50 steel.

SOLUTION

Step 1: Determine the gross and net areas needed for calculations.

$$A_g = 3.75 \text{ in.}^2 \text{ (from } \textit{Manual} \text{ Table 1-7)}$$

$$A_n = 3.75 - (3/4 + 1/8)(1/2) = 3.31 \text{ in.}^2$$

Step 2: Determine the shear lag factor and the effective net area.

First using Case 8 from Table D3.1, there are only three fasteners in a line for this connection, $U = 0.60$.

The minimum U for this open shape connected as shown is

$$U = \frac{\text{Gross area of connected leg}}{\text{Gross area of angle}} = \frac{4.0(0.5)}{3.75} = 0.533$$

Based on connection length, the Case 2 shear lag factor for a 6 in. connection length and an angle with $\bar{x}$ = 1.18 in. (from *Manual* Table 1-7) is

$$U = 1 - \frac{\bar{x}}{L} = 1 - \frac{1.18}{6} = 0.80$$

Thus, use $U = 0.80$, the largest permitted value, and

$$A_e = 0.80(3.31) = 2.65 \text{ in.}^2$$

For LRFD

Step 3: For the limit state of yielding,

$$P_n = F_y A_g = 50(3.75) = 188 \text{ kips}$$

$$\phi P_n = 0.90(188) = 169 \text{ kips}$$

Step 4: For the limit state of rupture,

$$P_n = F_u A_e = 65(2.65) = 172 \text{ kips}$$

$$\phi P_n = 0.75(172) = 129 \text{ kips}$$

Step 5: Determine the design strength. The limit state of rupture controls and the design strength is

$$\phi P_n = 129 \text{ kips}$$

For ASD

Step 3: For the limit state of yielding,

$$P_n = F_y A_g = 50(3.75) = 188 \text{ kips}$$

$$\frac{P_n}{\Omega_t} = \frac{188}{1.67} = 113 \text{ kips}$$

Step 4: For the limit state of rupture,

$$P_n = F_u A_e = 65(2.65) = 172 \text{ kips}$$

$$\frac{P_n}{\Omega_t} = \frac{172}{2.00} = 86.0 \text{ kips}$$

Step 5: Determine the allowable strength. The limit state of rupture controls and the allowable strength is

$$\frac{P_n}{\Omega_t} = 86.0 \text{ kips}$$

Note that the controlling limit state is the same for ASD and LRFD.

EXAMPLE 4.8
Tensile Strength of a Tee

Goal: Determine the design strength (LRFD) and the allowable strength (ASD) of a WT.

Given: Consider a WT6×32.5 attached to a gusset plate with equal length longitudinal welds as shown in Figure 4.20a. Use A992 steel.

SOLUTION

Step 1: Determine the gross and net areas:

$$A_g = 9.54 \text{ in.}^2 \quad \text{(from } \textit{Manual} \text{ Table 1-8)}$$

Because the force is transferred by welds only, there are no holes and

$$A_n = A_g$$

Step 2: Determine the shear lag factor and the effective net area.

The minimum U for this open shape connected as shown is

$$U = \frac{\text{Gross area of connected flange}}{\text{Gross area of WT}} = \frac{12.0(0.605)}{9.54} = 0.761$$

Based on connection length, the shear lag factor for a 12 in. connection length with longitudinal welds only and a tee with b_f = 12.0 and $\bar{x}$ = 0.985 in. (from *Manual* Table 1-8) is

$$U = \frac{3l^2}{3l^2 + w^2}\left(1 - \frac{\bar{x}}{l}\right) = \frac{3(12.0)^2}{3(12.0)^2 + 12.0^2}\left(1 - \frac{0.985}{12}\right) = 0.688$$

Thus, use $U = 0.761$ and

$$A_e = 0.761(9.54) = 7.26 \text{ in.}^2$$

For LRFD
Step 3 For the limit state of yielding,

$$P_n = F_y A_g = 50(9.54) = 477 \text{ kips}$$

$$\phi P_n = 0.90(477) = 429 \text{ kips}$$

Step 4: For the limit state of rupture,

$$P_n = F_u A_e = 65(7.26) = 472 \text{ kips}$$

$$\phi P_n = 0.75(472) = 354 \text{ kips}$$

Step 5: Therefore, the limit state of rupture controls and the design strength is

$$\phi P_n = 354 \text{ kips}$$

For ASD

Step 3: For the limit state of yielding,

$$P_n = 50(9.54) = 477 \text{ kips}$$

$$\frac{P_n}{\Omega_t} = \frac{477}{1.67} = 286 \text{ kips}$$

Step 4: For the limit state of rupture,

$$P_n = 65(7.26) = 472 \text{ kips}$$

$$\frac{P_n}{\Omega_t} = \frac{472}{2.0} = 236 \text{ kips}$$

Step 5: Therefore, the limit state of rupture controls and the allowable strength is

$$\frac{P_n}{\Omega_t} = 236 \text{ kips}$$

EXAMPLE 4.9
Tensile Strength of a Tee

Goal: Determine the design strength (LRFD) and the allowable strength (ASD) of a WT.

Given: Consider a WT6×32.5 attached to a gusset plate with unequal length longitudinal welds as shown in Figure 4.20b. Use A992 steel.

SOLUTION

Step 1: Determine the gross and net areas:

$$A_g = 9.54 \text{ in.}^2 \quad \text{(from } \textit{Manual} \text{ Table 1-8)}$$

Because the force is transferred by welds only,

$$A_n = A_g$$

Step 2: Determine the shear lag factor.

The minimum U for this open shape connected as shown is

$$U = \frac{\text{Gross area of connected flange}}{\text{Gross area of WT}} = \frac{12.0(0.605)}{9.54} = 0.761$$

Based on connection length,

$$l = \frac{l_1 + l_2}{2} = \frac{6+18}{2} = 12 \text{ in.}$$

the shear lag factor for a 12 in. connection length with longitudinal welds only and a tee with $b_f = 12.0$ and $\bar{x} = 0.985$ in. (from *Manual* Table 1-8) is

$$U = \frac{3l^2}{3l^2 + w^2}\left(1 - \frac{\bar{x}}{l}\right) = \frac{3(12)^2}{3(12)^2 + 12.0^2}\left(1 - \frac{0.985}{12}\right) = 0.688$$

Thus, use $U = 0.761$

Step 3: Determine the effective net area.

$$A_e = 0.761(9.54) = 7.26 \text{ in.}^2$$

For LRFD Step 4: The effective net area is the same as it was for Example 4.8, thus tension rupture will again control

$$\phi P_n = 354 \text{ kips}$$

For ASD Step 4: The effective net area is the same as it was for Example 4.8, thus tension rupture will again control

$$\frac{P_n}{\Omega_t} = 236 \text{ kips}$$

4.6 DESIGN OF TENSION MEMBERS

To design a structural steel tension member, the member size must be determined and then the appropriate limit states checked. The only additional issue to address is the slenderness of the member. For tension members, slenderness is defined as the member fabricated length divided by the least radius of gyration, L/r. The *Specification* has, in the past, placed a limit on the slenderness of tension members. However, since slenderness of a tension member is primarily a serviceability limit state, there is currently no specified limitation on tension member slenderness, as indicated in Section D1. The designer should exercise caution when selecting tension members with very high slenderness ratios, that is, those near the former limit of $L/r = 300$, because these members could easily be damaged during erection and might cause other problems due to their flexibility in the transverse direction.

Because the main task in tension member design is to determine the area of the member, the two limit states of yielding and rupture can be used to determine minimum gross and net areas such that

$$A_{g\,min} = \frac{P_u}{\phi_t F_y} \text{ (LRFD)} \quad \text{or} \quad A_{g\,min} = \frac{\Omega_t P_a}{F_y} \text{ (ASD)}$$

and

$$A_{e\,min} = \frac{P_u}{\phi_t F_u} \text{ (LRFD)} \quad \text{or} \quad A_{e\,min} = \frac{\Omega_t P_a}{F_u} \text{ (ASD)}$$

Because connection details are not normally known in the early stages of member selection, it may not be possible to determine the actual deductions necessary to obtain the exact effective net area of the member being designed. One approach would be to assume a fixed percentage deduction for the effective net area. The designer would decide the magnitude of this deduction and then confirm the adequacy of this deduction at the completion of the design. A second approach would be to use Cases 7 or 8, if applicable, since these are maximum reductions.

Part 5 of the *Manual* provides tables for tension member design that give the strength of tension members based on the limit states of yielding on the gross area and rupture on an effective net area equal to $0.75A_g$. This is a reasonable estimate of a value of A_e that is practical to achieve with typical end connections, not a minimum value or a prescriptive way to calculate A_e. If the actual effective net area differs from this assumed value, the designer can simply adjust the strength accordingly.

EXAMPLE 4.10a
Tension Member Design by LRFD

Goal: Select a double-angle tension member for use as a web member in a truss and determine the maximum area reduction that would be permitted for bolt holes and shear lag. Using *Manual* Table 5-8, check the available strength of the member.

Given: The member must carry a dead load of $P_D = 67.5$ kips and a live load of $P_L = 202.5$ kips. For the load combination $1.2P_D + 1.6P_L$, the LRFD required strength is $P_u = 405$ kips. Use equal leg angles of A572 Gr. 50 steel.

SOLUTION

Step 1: Determine the minimum required gross area based on the limit state of yielding where $\phi = 0.9$:

$$A_{g\ \min} = P_u/\phi F_y = 405/(0.9(50)) = 9.0 \text{ in.}^2$$

Step 2: Based on this minimum gross area, from *Manual* Table 1-15, select

2L6×6×7/16 with $A_g = 10.2$ in.^{2}

Step 3: Determine the minimum effective net area based on the limit state of rupture where $\phi = 0.75$:

$$A_{e\ \min} = P_u/\phi F_u = 405/(0.75(65)) = 8.31 \text{ in.}^2$$

Step 4: Thus, the combination of holes and shear lag may not reduce the area of this pair of angles by more than

$$A_e/A_g = 8.31/10.2 = 0.815$$

or a new double-angle section must be chosen.

Step 5: From Manual Table 5-8, the available strengths are given as

For yielding

$$\phi_t P_n = 459 \text{ kips}$$

For rupture, using $A_e = 0.75A_g$

$$\phi_t P_n = 373 \text{ kips}$$

Step 6: Since the available rupture strength is less than the required strength, determine the maximum permitted reduction for bolt holes and shear lag.

$$A_e / A_g = 405(0.75)/373 = 0.814$$

Which is what was already determined in Step 4

EXAMPLE 4.10b
Tension Member Design by ASD

Goal: Select a double-angle tension member for use as a web member in a truss and determine the maximum area reduction that would be permitted for bolt holes and shear lag.

Given: The member must carry a dead load of P_D = 67.5 kips and a live load of P_L = 202.5 kips. For the load combination $P_D + P_L$, the ASD required strength is P_a = 270 kips. Use equal leg angles of A572 Gr. 50 steel.

SOLUTION

Step 1: Determine the minimum required gross area based on the limit state of yielding where $\Omega = 1.67$:

$$A_{g\ \min} = P_a/(F_y/\Omega) = 270/(50/1.67) = 9.02 \text{ in.}^2$$

Step 2: Based on this minimum gross area, from *Manual* Table 1-15, select 2L6×6×7/16 with A_g = 10.2 in.^{2}

Step 3: Determine the minimum effective net area based on the limit state of rupture where $\Omega = 2.00$:

$$A_{e\ \min} = P_a/(F_u/\Omega) = 270/(65/2.00) = 8.31 \text{ in.}^2$$

Step 4: Thus, the combination of holes and shear lag may not reduce the area of this pair of angles by more than

$$A_e / A_g = 8.31/10.2 = 0.815$$

or a new double-angle section must be chosen.

Note that since the ratio P_L/P_D = 3.0, the required areas for the LRFD and ASD solutions are the same.

Step 5: From Manual Table 5-8, the available strengths are given as
For yielding

$$\phi_t P_n = 305 \text{ kips}$$

For rupture, using $A_e = 0.75A_g$

$$\phi_t P_n = 249 \text{ kips}$$

Step 6: Since the available rupture strength is less than the required strength, determine the maximum permitted reduction for bolt holes and shear lag.

$$A_e/A_g = 270(0.75)/249 = 0.813$$

Which is essentially what was already determined in Step 4

EXAMPLE 4.11a
Tension Member Design by LRFD

Goal: Select a WT9 for use as a tension member and determine the maximum area reduction that would be permitted for bolt holes and shear lag.

Given: The member must carry an LRFD required strength of $P_u = 818$ kips. Use A992 steel.

SOLUTION

Step 1: Determine the minimum required gross area based on the limit state of yielding where $\phi = 0.9$:

$$A_{g\ \min} = P_u/\phi F_y = 818/(0.9(50)) = 18.2 \text{ in.}^2$$

Step 2: Based on the minimum gross area, from *Manual* Table 1-8, select WT9×65 with $A_g = 19.2$ in.^{2}

Step 3: Determine the minimum effective net area needed to resist the applied force where $\phi = 0.75$:

$$A_{e\ \min} = P_u/\phi F_u = 818/(0.75(65)) = 16.8 \text{ in.}^2$$

Step 4: The combination of holes and shear lag may not reduce the area of this WT by more than

$$A_e/A_g = 16.8/19.2 = 0.875$$

or a new WT section must be chosen. If the $A_e/A_g \geq 0.75$, one could use *Manual* Table 5-3 to select a WT9×79 with $A_g = 23.3$ in.^{2} and a rupture strength, $\phi P_n = 848$ kips for $A_e = 0.75A_g$.

EXAMPLE 4.11b
Tension Member Design by ASD

Goal: Select a WT9 for use as a tension member and determine the maximum area reduction that would be permitted for bolt holes and shear lag.

Given: The member must carry an ASD required strength of $P_a = 545$ kips. Use A992 steel.

SOLUTION

Step 1: Determine the minimum required gross area based on the limit state of yielding where $\Omega = 1.67$:

$$A_{g\ \min} = P_a/(F_y/\Omega) = 545/(50/1.67) = 18.2 \text{ in.}^2$$

Step 2: Based on the minimum gross area, from *Manual* Table 1-8, select WT9×65 with $A_g = 19.2$ in.^{2}

Step 3: Determine the minimum effective net area needed to resist the applied force where $\Omega = 2.00$:

$$A_{e\ min} = P_a/(F_u/\Omega) = 545/(65/2.00) = 16.8 \text{ in.}^2$$

Step 4: The combination of holes and shear lag may not reduce the area of this WT by more than

$$A_e/A_g = 16.8/19.2 = 0.875$$

or a new WT section must be chosen. If the $A_e/A_g \geq 0.75$, one could use *Manual* Table 5-3 to select a WT9×79 with $A_g = 23.3$ in.^{2} and a rupture strength, $P_n/\Omega = 566$ kips for $A_e = 0.75A_g$.

4.7 BLOCK SHEAR

When a portion of a member tears out as shown in Figure 4.21a, a combination of tension and shear results as shown in Figure 4.21b, and the failure is known as a *block shear failure*. Even though this failure mode is primarily the result of a connection failure, it may possibly control the overall strength of a tension member. The resistance to tear-out is provided by a combination of shear on the plane parallel to the tension force and tension on the plane perpendicular to it.

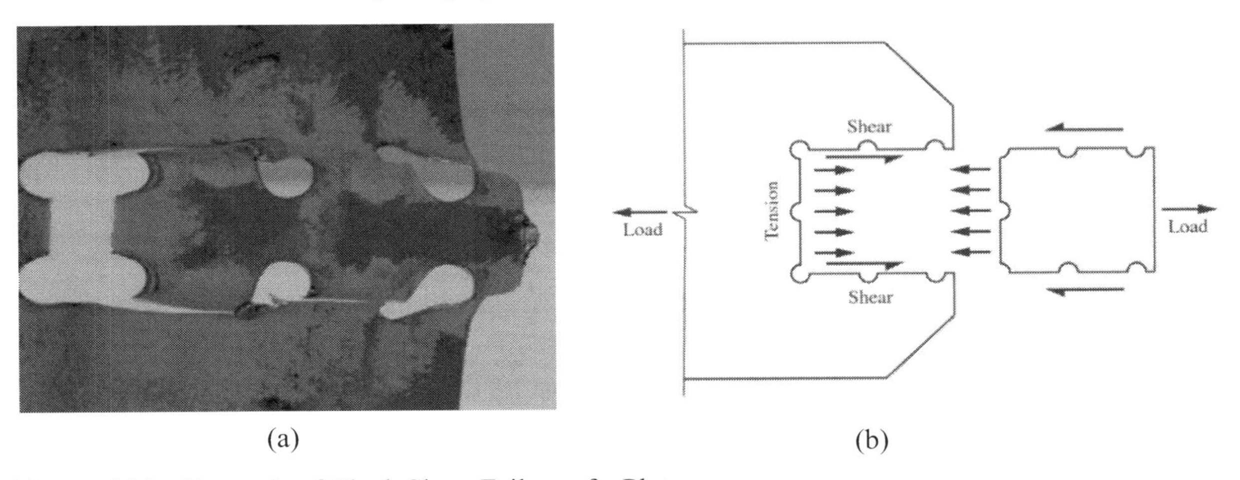

Figure 4.21 Example of Block Shear Failure of a Plate
Photo courtesy Robert Driver

Rupture will always be the controlling mode on the tension face of the failure block, due to the relatively short length of material that will be available to yield. The controlling limit states on the shear face will be either yielding or rupture, whichever has the lower strength. Unlike the situation for overall member strength, in which the yield and rupture limit states were found to have different resistance and safety factors, block shear uses the same values for both limit states. Thus, a simple comparison of nominal strengths is appropriate for determining the controlling limit state. Section J4.3 of the *Specification* gives the block shear strength as

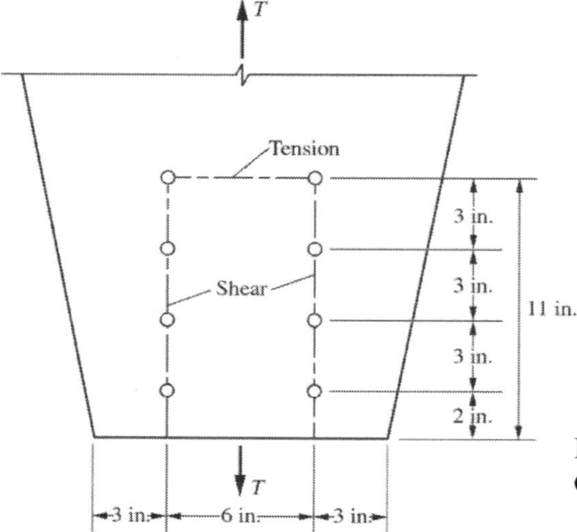

Figure 4.22 Block Shear Geometry for Example 4.12

$$R_n = 0.6F_uA_{nv} + U_{bs}F_uA_{nt} \leq 0.6F_yA_{gv} + U_{bs}F_uA_{nt} \qquad \text{(AISC J4-5)}$$

where

A_{gv} = gross area in shear
A_{nv} = net area in shear
A_{nt} = net area in tension
U_{bs} = 1.0 if the tension stress is uniform and 0.5 if the tension stress is not uniform. For tension members, the tensile stress is assumed to be uniform. Thus, U_{bs} = 1.0 will be used (U_{bs} = 0.5 is addressed in Chapter 10.)

The design and allowable strengths are determined using

$$\phi = 0.75 \text{ (LRFD)} \qquad \Omega = 2.0 \text{ (ASD)}$$

The block shear strength given by Equation J4-5 can be simplified by recognizing that it simply requires the combination of the minimum shear strength, yielding or rupture, with the tensile rupture strength. It is also important to note that the calculation of net area for both tension and shear is determined by deducting the area of the holes from the gross areas.

EXAMPLE 4.12
Gusset Plate Tension Strength

Goal: Determine whether the gusset plate has sufficient strength in block shear.

Given: The gusset plate shown in Figure 4.22 has a plate thickness of 1/2 in. The required strength for LRFD is P_u = 325 kips and for ASD is P_a = 220 kips. The steel is A572 Gr. 50 and the holes are punched for 7/8 in. Bolts. The block between the lines of bolts is more critical than the two blocks outside the lines of bolts because the width in tension outside the lines of bolts is larger than the width inside.

SOLUTION

Step 1: Determine the areas needed to perform the calculations.

$$A_{nt} = \left(6 - \left(7/8 + 1/8\right)\right)\left(1/2\right) = 2.50 \text{ in.}^2$$

$$A_{gv} = 2\left(11\right)\left(1/2\right) = 11.0 \text{ in.}^2$$

$$A_{nv} = 2\left(11.0 - 3.5\left(7/8 + 1/8\right)\right)\left(1/2\right) = 7.50 \text{ in.}^2$$

Step 2: Determine the nominal block shear strength:

Determine the tension rupture strength

$$F_u A_{nt} = 65\left(2.50\right) = 163 \text{ kips}$$

Consider shear yield and shear rupture and select the least nominal strength; thus,

$$0.6F_y A_{gv} = 0.6\left(50\right)\left(11.0\right) = 330 \text{ kips}$$

$$0.6F_u A_{nv} = 0.6\left(65\right)\left(7.50\right) = 293 \text{ kips}$$

Selecting the shear rupture term and combining it with the tension rupture term gives a block shear nominal strength, using U_{bs} = 1.0, of

$$R_n = 293 + 1.0\left(163\right) = 456 \text{ kips}$$

For LRFD Step 3: For LRFD, the design strength is

$$\phi R_n = 0.75\left(456\right) = 342 > 325 \text{ kips}$$

Because the design strength is greater than the required strength of 325 kips, the gusset plate is adequate to resist this force based on block shear.

For ASD Step 3: For ASD, the allowable strength is

$$R_n / \Omega = 456/2.00 = 228 > 220 \text{ kips}$$

Because the allowable strength is greater than the required strength of 220 kips, the gusset plate is adequate to resist this force based on block shear.

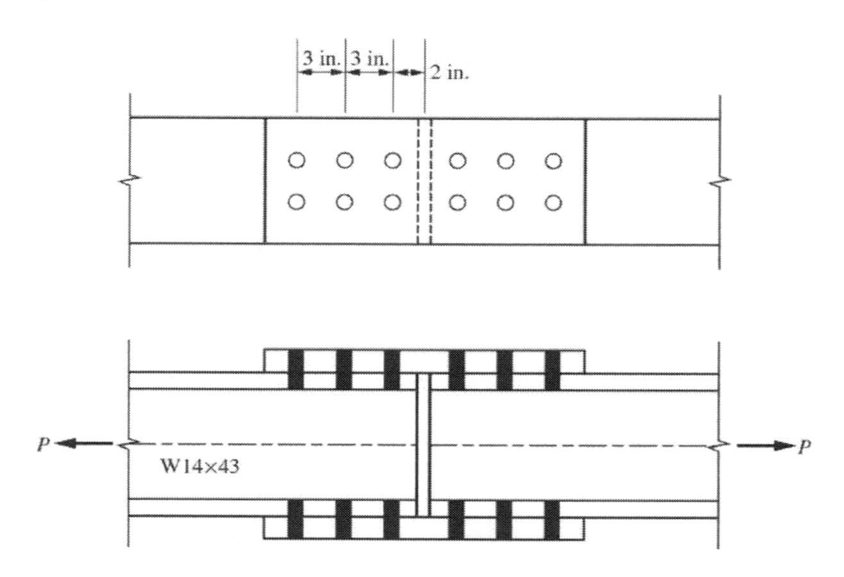

Figure 4.23 Spliced Tension Member for Example 4.13

EXAMPLE 4.13a ***Tension Strength of Spliced Members by LRFD***

Goal: Determine the design strength of a splice between two W-shapes.

Given: Two W14×43 A992 wide flanges are spliced by flange plates, as shown in Figure 4.23, with 7/8 in. diameter bolts arranged as shown in standard holes. The LRFD available strength of a group of six bolts in a flange is 227 kips. The plates will be selected such that they do not limit the member strength.

SOLUTION

Step 1: Determine the design strength for the limit state of tensile yielding of the W14×43.

From *Manual* Table 1-1, $A_g = 12.6 \text{ in.}^2$

Thus, $P_n = F_y A_g = 50(12.6) = 630 \text{ kips}$

$$\phi P_n = 0.90(630) = 567 \text{ kips}$$

Step 2: Determine the net area of the W14×43. The area to be deducted for each flange, with t_f = 0.530 in. and two bolt holes at the critical section, is

$$2(7/8 + 1/8)(0.530) = 1.06 \text{ in.}^2$$

Thus, with the deduction for two flanges from the gross area

$$A_n = 12.6 - 2(1.06) = 10.5 \text{ in.}^2$$

Step 3: Determine the shear lag factor. The W14×43 is treated as two tee sections, each a WT7×21.5. The $\bar{x}$ for each WT is found in *Manual* Table 1-8 as 1.31 in. With the length of the connection, $L = 2(3.0) = 6.0$ in.,

$$U = 1 - \frac{1.31}{6.0} = 0.782$$

Specification Table D3.1 provides that for this case, if $b_f \geq 2d/3$, a value of $U = 0.90$ may be used and if $b_f < 2d/3$ $U = 0.85$.
From *Manual* Table 1-1, b_f = 8.0 in. and d = 13.7 in. thus:

$$b_f = 8.0 < 2(13.7)/3 = 9.13 \text{ in.}$$

Therefore use $U = 0.85$

Step 4: Determine the design strength for the limit state of tensile rupture.

$$A_e = UA_n = 0.85(10.5) = 8.93 \text{ in.}^2$$

$$P_n = F_u A_e = 65(8.93) = 580 \text{ kips}$$

$$\phi P_n = 0.75(580) = 435 \text{ kips}$$

Step 5: Determine the design block shear strength of the flanges.

The block shear limit state must be checked for tear-out of the flanges, as shown in Figure 4.24. The calculations will be carried out for one block as shown in the figure and the total obtained by adding the results for all four flange sections.

Rupture on the tension plane:

$$F_u A_{nt} = 65(2.0 - (1/2)(7/8 + 1/8))(0.530) = 51.7 \text{ kips}$$

Yield on the shear plane:

$$0.6 F_y A_{gv} = 0.6(50)(8.00)(0.530) = 127 \text{ kips}$$

Rupture on the shear plane:

$$0.6 F_u A_{nv} = 0.6(65)(8.00 - 2.5(7/8 + 1/8))(0.530) = 114 \text{ kips}$$

Because shear rupture is less than shear yield, the design strength for a single block shear element is, with $U_{bs} = 1.0$,

$$R_n = (114 + 1.0(51.7)) = 166 \text{ kips}$$

$$\phi R_n = 0.75(166) = 125 \text{ kips}$$

and the block shear strength of the W14×43 is

$$\phi R_n = 4(125) = 500 \text{ kips}$$

Step 6: Compare the design strength for each limit state, assuming plates have been selected so as to not limit the capacity.

Bolt design shear strength (given, 2 flanges ×227 kips/(bolt group))	454 kips
Tensile yielding of member	567 kips
Tensile rupture of the member	435 kips
Block shear for the member	500 kips

Step 7: Tensile rupture of the member controls the design. Therefore, the design strength of the splice is

$$\phi R_n = 435 \text{ kips}$$

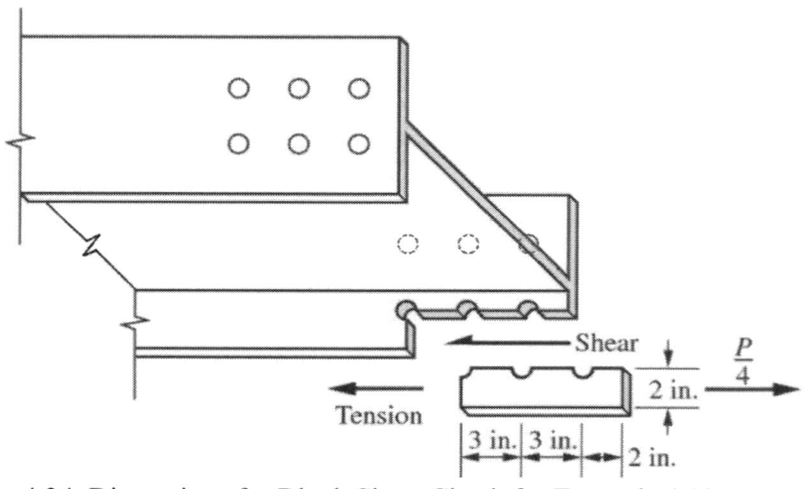

Figure 4.24 Dimensions for Block Shear Check for Example 4.13

EXAMPLE 4.13b
Tension Strength of Spliced Members by ASD

Goal: Determine the design strength of a splice between two W-shapes.

Given: Two W14×43 A992 wide flanges are spliced by flange plates, as shown in Figure 4.23, with 7/8 in. diameter bolts arranged as shown in standard holes. The ASD available strength of a group of six bolts in a flange is 151 kips. The plates will be selected such that they do not limit the member strength.

SOLUTION

Step 1: Determine the design strength for the limit state of tensile yielding of the W14×43.

From *Manual* Table 1-1, $A_g = 12.6 \text{ in.}^2$

$$P_n = F_y A_g = 50(12.6) = 630 \text{ kips}$$

$$P_n/\Omega = 630/1.67 = 377 \text{ kips}$$

Step 2: Determine the net area of the W14×43. The area to be deducted for each flange, with t_f = 0.530 in. and two bolt holes at the critical section, is

$$2(7/8+1/8)(0.530) = 1.06 \text{ in.}^2$$

Thus, with the deduction for two flanges from the gross area

$$A_n = 12.6 - 2(1.06) = 10.5 \text{ in.}^2$$

Step 3: Determine the shear lag factor. The W14×43 is treated as two tee sections, each a WT7×21.5. The $\bar{x}$ for each WT is found in *Manual* Table 1-8 as 1.31 in. With the length of the connection L = 2(3.0) = 6.0 in.,

$$U = 1 - \frac{1.31}{6.0} = 0.782$$

Specification Table D3.1 provides that for this case, if $b_f \geq 2d/3$, a value of U = 0.90 may be used and if $b_f < 2d/3$ U = 0.85.

From *Manual* Table 1-1, b_f = 8.0 in. and d = 13.7 in. thus:

$$b_f = 8.0 < 2(13.7)/3 = 9.13 \text{ in.}$$

Therefore use U = 0.85

Step 4: Determine the allowable strength for the limit state of tensile rupture.

$$A_e = UA_n = 0.85(10.5) = 8.93 \text{ in.}^2$$

$$P_n = F_u A_e = 65(8.93) = 580 \text{ kips}$$

$$P_n/\Omega = 580/2.00 = 290 \text{ kips}$$

Step 5: Determine the allowable block shear strength of the flanges.

The block shear limit state must be checked for tear-out of the flanges, as shown in Figure 4.24. The calculations will be carried out for one block as shown in the figure and the total obtained by adding the results for all four flange sections.

Rupture on the tension plane:

$$F_u A_{nt} = 65\left(2.0 - (1/2)(7/8+1/8)\right)(0.530) = 51.7 \text{ kips}$$

Yield on the shear plane:

$$0.6F_y A_{gv} = 0.6(50)(8.00)(0.530) = 127 \text{ kips}$$

Rupture on the shear plane:

$$0.6F_uA_{nv} = 0.6(65)(8.00 - 2.5(7/8 + 1/8))(0.530) = 114 \text{ kips}$$

Because shear rupture is less than shear yield, the allowable strength for a single block shear element is, with $U_{bs} = 1.0$,

$$R_n = (114 + 1.0(51.7)) = 166 \text{ kips}$$

$$R_n/\Omega = 166/2.00 = 83.0 \text{ kips}$$

and the block shear strength of the W14×43 is

$$R_n/\Omega = 4(83.0) = 332 \text{ kips}$$

Step 6: Compare the design strength for each limit state, assuming plates have been selected so as to not limit the capacity.

Bolt design shear strength (given, 2 flanges ×151 kips/(bolt group))	302 kips
Tensile yielding of member	377 kips
Tensile rupture of the member	290 kips
Block shear for the member	332 kips

Step 7: Tensile rupture of the member controls the design. Therefore, the design strength of the splice is

$$R_n/\Omega = 290 \text{ kips}$$

EXAMPLE 4.14a
Tension Strength of an Angle by LRFD

Goal: Determine the design strength of one of a pair of angles in a tension member.

Given: The truss diagonal member in Figure 4.25 consists of a pair of angles L4×3×3/8 that are loaded in tension. The bolts to be used are 3/4 in. and the steel is A36. The bolt design shear strength for one angle in this connection is 53.7 kips.

SOLUTION

Step 1: Determine the single angle design strength for the limit state of yielding. From *Manual* Table 1-7 $A_g = 2.49$ in^2.

$$P_n = F_yA_g = 36(2.49) = 89.6 \text{ kips}$$

$$\phi P_n = 0.90(89.6) = 80.6 \text{ kips}$$

Step 2: Determine the single angle design strength for the limit state of rupture.

$$A_n = 2.49 - (3/4 + 1/8)(3/8) = 2.16 \text{ in.}^2$$

The shear lag coefficient is

$$U = 1 - \frac{\bar{x}}{l} = 1 - \frac{0.775}{6.0} = 0.871$$

and the effective net area is

$$A_e = A_n U = 2.16(0.871) = 1.88 \text{ in.}^2$$

The nominal strength is

$$P_n = F_u A_e = 58(1.88) = 109 \text{ kips}$$

Therefore the design strength is

$$\phi P_n = 0.75(109) = 81.8 \text{ kips}$$

Step 3: Determine the single angle design strength in block shear.

Rupture on the tension plane:

$$F_u A_{nt} = 58\left(1.5 - (1/2)(3/4 + 1/8)\right)(3/8) = 23.1 \text{ kips}$$

Yield on the shear plane:

$$0.6 F_y A_{gv} = 0.6(36)(7.25)(3/8) = 58.7 \text{ kips}$$

Rupture on the shear plane:

$$0.6 F_u A_{nv} = 0.6(58)\left(7.25 - 2.5(3/4 + 1/8)\right)(3/8) = 66.1 \text{ kips}$$

Because shear yield is less than shear rupture, the design strength for a single block shear element is, with $U_{bs} = 1.0$,

$$R_n = \left(58.7 + 1.0(23.1)\right) = 81.8 \text{ kips}$$

$$\phi R_n = 0.75(81.8) = 61.4 \text{ kips}$$

Step 4: Compare the single angle design strength for each limit state.

Bolt design strength	53.7 kips
Yielding of the member	80.6 kips
Rupture of the member	81.8 kips
Block shear for the member	61.4 kips

Step 5: The bolt design strength controls the design. Therefore, the design strength of one angle is

$$\phi R_n = 53.7 \text{ kips}$$

The design strength for the pair of angles is

$$\phi R_n = 2(53.7) = 107 \text{ kips}$$

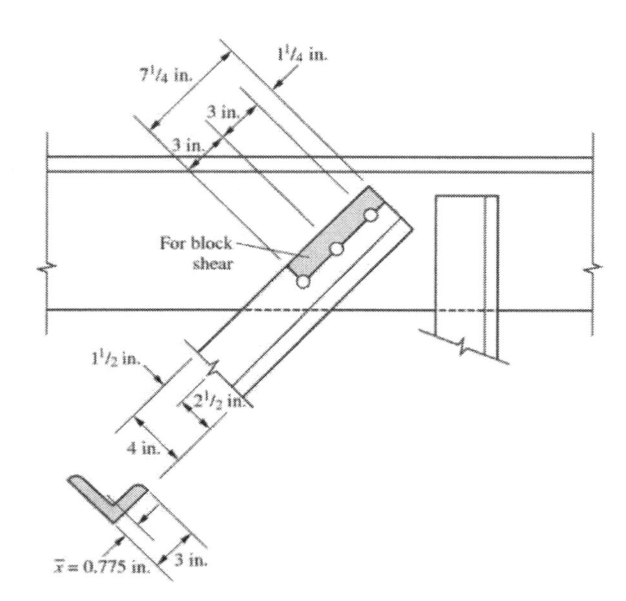

Figure 4.25 Truss Diagonal Member for Example 4.14

EXAMPLE 4.14b
Tension Strength of an Angle by ASD

Goal: Determine the allowable strength of one of a pair of angles in a tension member.

Given: The truss diagonal member in Figure 4.25 consists of a pair of angles L4×3×3/8 that are loaded in tension. The bolts to be used are 3/4 in. and the steel is A36. The bolt allowable shear strength for one angle in this connection is 35.7 kips.

SOLUTION

Step 1: Determine the single angle allowable strength for the limit state of yielding. From *Manual* Table 1-7 $A_g = 2.49$ in^2.

$$P_n = F_y A_g = 36(2.49) = 89.6 \text{ kips}$$

$$P_n/\Omega = 89.6/1.67 = 53.7 \text{ kips}$$

Step 2: Determine the single angle allowable strength for the limit state of rupture.

$$A_n = 2.49 - (3/4 + 1/8)(3/8) = 2.16 \text{ in.}^2$$

The shear lag coefficient is

$$U = 1 - \frac{\bar{x}}{l} = 1 - \frac{0.775}{6.0} = 0.871$$

and the effective net area is

$$A_e = A_n U = 2.16(0.871) = 1.88 \text{ in.}^2$$

The nominal strength is

$$P_n = F_u A_e = 58(1.88) = 109 \text{ kips}$$

Therefore the allowable strength is

$$P_n/\Omega = 109/2.00 = 54.5 \text{ kips}$$

Step 3: Determine the single angle allowable strength in block shear.

Rupture on the tension plane:

$$F_u A_{nt} = 58(1.5 - (1/2)(3/4 + 1/8))(3/8) = 23.1 \text{ kips}$$

Yield on the shear plane:

$$0.6F_y A_{gv} = 0.6(36)(7.25)(3/8) = 58.7 \text{ kips}$$

Rupture on the shear plane:

$$0.6F_u A_{nv} = 0.6(58)(7.25 - 2.5(3/4 + 1/8))(3/8) = 66.1 \text{ kips}$$

Because shear yield is less than shear rupture, the allowable strength for a single block shear element is, with U_{bs} = 1.0,

$$R_n = (58.7 + 1.0(23.1)) = 81.8 \text{ kips}$$

$$R_n/\Omega = 81.8/2.00 = 40.9 \text{ kips}$$

Step 4: Compare the single angle allowable strength for each limit state.

Bolt allowable strength	35.7 kips
Yielding of the member	53.7 kips
Rupture of the member	54.5 kips
Block shear for the member	40.9 kips

Step 5: The bolt allowable strength controls the design. Therefore, the allowable strength of one angle is

$$R_n/\Omega = 35.7 \text{ kips}$$

The design strength for the pair of angles is

$$R_n/\Omega = 2(35.7) = 71.4 \text{ kips}$$

4.8 PIN-CONNECTED MEMBERS

When a pin connection is to be made in a tension member, a hole is cut in both the member and the parts to which it is to be attached. A pin is inserted in the hole and a mechanical means is found to keep the elements together. This type of connection is the closest to a true frictionless pin as can be made. Figure 4.26 shows the end of a pin-connected member and the dimensions needed to determine its strength. These members are not particularly common in buildings; they are used mainly for special applications, such as hangers in suspension structures or connecting links in bridge structures.

Specification Section D5 identifies the limit states for which pin-connected members must be designed. These are (1) tensile rupture, (2) shear rupture, (3) bearing on the projected area of the pin, and (4) yielding on the gross section. The strength of the pin-connected tension member is taken as the lowest strength predicted by each of these limit states.

In addition, Section D5.2 provides dimensional requirements that ensure these are the only limit states that need to be evaluated. The requirements include locating the pin hole equidistant between the edges of the plate and the size of the gap between the pin and the hole when the pin is expected to provide for relative movement between the connected parts. It also establishes that the width, w, of the plate at the pin hole shall not be less than $2b_e + d$ and the minimum extension, a, beyond the bearing end, parallel to the axis of the member, shall not be less than $1.33b_e$.

For tension rupture

$$P_n = F_u\left(2tb_e\right) \qquad \text{(AISC D5-1)}$$

$$\phi_t = 0.75\text{(LRFD)} \qquad \Omega_t = 2.00\text{ (ASD)}$$

where t is the thickness of the plate and b_e is the effective width at either side of the hole, and $b_e = 2t + 0.63$ in inches, but not more than the actual distance, b, from the edge of the hole to the edge of the part, measured perpendicular to the direction of the force.

For shear rupture

$$P_n = 0.6C_r F_u A_{sf} \qquad \text{(AISC D5-2)}$$

$$\phi_{sf} = 0.75\text{(LRFD)} \qquad \Omega_{sf} = 2.00\text{ (ASD)}$$

where $A_{sf} = 2t(a + d/2)$, a is the shortest distance from the edge of the pin hole to the edge of the member measured parallel to the direction of the force, d is the pin diameter, and $C_r = 1.0$ when the pin hole diameter is 1/32 in. larger than the pin diameter or 0.95 when the pin hole diameter is between 1/32 and 1/16 in. larger than the pin diameter.

For bearing on the projected area of the pin, from Section J7,

$$P_n = 1.8F_y A_{pb} \qquad \text{(AISC J7-1)}$$

$$\phi_t = 0.75\text{ (LRFD)} \qquad \Omega_t = 2.00\text{ (ASD)}$$

where $A_{pb} = td$, the projected area of the pin.

For yielding,

$$P_n = F_y A_g \qquad \text{(AISC D2-1)}$$

$$\phi_t = 0.90\text{ (LRFD)} \qquad \Omega_t = 1.67\text{ (ASD)}$$

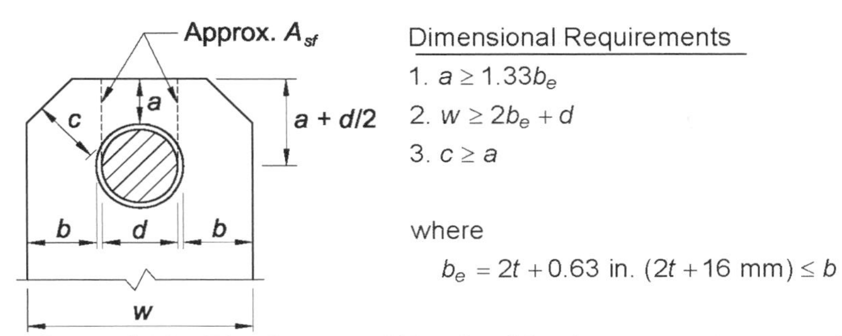

Figure 4.26 Pin-Connected Tension Member Copyright © American Institute of Steel Construction, Reprinted with Permission. All Rights Reserved

EXAMPLE 4.15a ***Pin-Connected Member Design by LRFD***

Goal: Design a pin-connected member using LRFD.

Given: A dead load of 30 kips and a live load of 70 kips are to be supported. The steel is A572 Gr. 50 with a yield stress of 50 ksi and an ultimate strength of 65 ksi. Use a 3/4 in. plate with a 4 in. pin.

SOLUTION

Step 1: Determine the required strength.

$$P_u = 1.2(30) + 1.6(70) = 148 \text{ kips}$$

Step 2: Determine the minimum required effective net width for the limit state of tensile rupture based on Equation D5-1,

$$(b_e)_{\min} = \frac{P_u}{\phi F_u 2t} = \frac{148}{0.75(65)(2)(0.750)} = 2.02 \text{ in.}$$

and the maximum b_e from Section D5.1 is

$$(b_e)_{\max} = 2t + 0.63 = 2(0.750) + 0.63 = 2.13 \text{ in.}$$

Therefore, try a 9.0 in. plate, which will give an actual distance to the edge of the plate equal to 2.5 in. which is greater than $(b_e)_{\min}$. Thus, $(b_e)_{\max}$ is used to calculate the rupture strength of the plate.

Step 3: Determine the design strength of this 9×3/4 plate for the limit state of tension rupture using Equation D5-1.

$$\phi P_n = \phi F_u(2tb_e) = 0.75(65)(2(0.750)(2.13)) = 156 > 148 \text{ kips}$$

Step 4: Determine the design strength for the limit state of shear rupture using Equation D5-2. For a 9 in. plate and a 4 in. pin, $a \geq b = 2.5$ in. The end distance, a, must also be greater than $1.33b_e$. Thus, take $a =$ 3.0 in. > 1.33(2.13) = 2.83 in. Use a pin hole 1/16 in. greater than the 4 in. pin diameter based on Section D5.2(b), thus $C_r = 0.95$.

$$A_{sf} = 2t(a + d/2) = 2(0.750)(3.0 + 4.0/2) = 7.50 \text{ in.}^2$$

$$\phi P_n = \phi 0.6 C_r F_u A_{sf} = 0.75(0.6(0.95)(65))(7.50) = 208 > 148 \text{ kips}$$

Step 5: Determine the design strength for the limit state of bearing on the projected area of the 4 in. pin using Equation J7-1.

$$A_{pb} = td = 0.750(4.0) = 3.0 \text{ in.}^2$$

$$\phi P_n = \phi 1.8 F_y A_{pb} = 0.75(1.8)(50)(3.0) = 203 > 148 \text{ kips}$$

Step 6: Determine the design strength for the limit state of tensile yielding on the gross area of the member using Equation D2-1.

$$\phi P_n = \phi F_y A_g = 0.9(50)(0.750)(9.00) = 304 > 148 \text{ kips}$$

Step 7: Since each limit state has design strength greater than the required strength, the 9×3/4 pin-connected member with a 4 in. pin in a 4-1/16 in. hole will be sufficient to carry the applied load.

EXAMPLE 4.15b ***Pin-Connected Member Design by ASD***

Goal: Design a pin-connected member using ASD.

Given: A dead load of 30 kips and a live load of 70 kips are to be supported. The steel is A572 Gr. 50 with a yield stress of 50 ksi and an ultimate strength of 65 ksi. Use a 3/4 in. plate with a 4 in. pin.

SOLUTION

Step 1: Determine the required strength.

$$P_a = 30 + 70 = 100 \text{ kips}$$

Step 2: Determine the minimum required effective net width for the limit state of tensile rupture based on Equation D5-1,

$$(b_e)_{\min} = \frac{P_a \Omega}{F_u 2t} = \frac{100(2.00)}{65(2)(0.750)} = 2.05 \text{ in.}$$

and the maximum b_e from Section D5.1 is

$$(b_e)_{\max} = 2t + 0.63 = 2(0.750) + 0.63 = 2.13 \text{ in.}$$

Therefore, try a 9.0 in. plate, which will give an actual distance to the edge of the plate equal to 2.5 in. which is greater than $(b_e)_{\min}$. Thus, $(b_e)_{\max}$ is used to calculate the rupture strength of the plate.

Step 3: Determine the allowable strength of this 9×3/4 plate for the limit state of tension rupture using Equation D5-1.

$$P_n/\Omega = F_u(2tb_e)/\Omega = 65(2(0.750)(2.13))/2.00 = 104 > 100 \text{ kips}$$

Step 4: Determine the allowable strength for the limit state of shear rupture using Equation D5-2. For a 9 in. plate and a 4 in. pin, $a \geq b = 2.5$ in. The end distance, a, must also be greater than $1.33b_e$. Thus, take $a =$ 3.0 in. > 1.33(2.13) = 2.83 in. Use a pin hole 1/16 in. greater than the 4 in. pin diameter, thus $C_r = 0.95$.

$$A_{sf} = 2t(a + d/2) = 2(0.750)(3.0 + 4.0/2) = 7.50 \text{ in.}^2$$

$$P_n/\Omega = 0.6C_rF_uA_{sf}/\Omega = 0.6(0.95)(65)(7.50)/2.00 = 139 > 100 \text{ kips}$$

Step 5: Determine the allowable strength for the limit state of bearing on the projected area of the 4 in. pin using Equation J7-1.

$$A_{pb} = td = 0.750(4.0) = 3.0 \text{ in.}^2$$

$$P_n/\Omega = 1.8F_yA_{pb}/\Omega = 1.8(50)(3.0)/2.00 = 135 > 100 \text{ kips}$$

Step 6: Determine the allowable strength for the limit state of tensile yielding on the gross area of the member using Equation D2-1.

$$P_n/\Omega = F_yA_g/\Omega = 50(0.750)(9.00)/1.67 = 202 > 100 \text{ kips}$$

Step 7: Since each limit state has allowable strength greater than the required strength, the 9×3/4 pin-connected member with a 4 in. pin in a 4-1/16 in. hole will be sufficient to carry the applied load.

4.9 EYEBARS AND RODS

Eyebar tension members are not commonly used in new construction but may be found in special applications where the objective is a design that has some historical context. Thus, the provisions for the design of these members are still found in Section D6 of the *Specification*. Historically, they were commonly used as tension members in trusses and as links forming the main tension member in suspension bridges. Eyebars are designed only for the limit state of yielding on the gross section because the dimensional requirements preclude the possibility of failure at any load below that level. Figure 4.27 shows a schematic of an eyebar and Figure 4.4 shows an eyebar in a building application.

Rods are commonly used for tension members in situations where the required tensile strength is small. These tension members are generally considered secondary members and include sag rods, hangers, and tie rods, although there are cases where rods play a significant role in carrying primary forces. Rods may also be used as part of the lateral bracing system in walls and roofs.

Although it is possible to connect rods by welding to the structure, threading and bolting is the most common means of connection. Rods can be threaded in two ways. Standard rods have threads that reduce the cross-sectional area through the removal of material. The upset rod has enlarged ends, with the threads reducing that area to something larger than the gross area of the rod. The strength of the rod depends on the manner in which the threads are applied.

A standard threaded rod is treated as a bolt for the determination of strength. Nominal strength is given in *Specification* Section J3.7 as $F_n = 0.75F_u$ over the area of the unthreaded body of the rod, which gives

$$P_n = 0.75F_u A_b \tag{4-7}$$

and for design,

$$\phi_t = 0.75 \text{ (LRFD)} \qquad \Omega_t = 2.00 \text{ (ASD)}$$

If the rod has a threaded upset end, the nominal strength can be taken as F_u, since the 0.75 factor is used to account for the approximate ratio of the effective tension area of the threaded portion of the rod to the actual rod area.

4.10 BUILT-UP TENSION MEMBERS

Section D4 of the *Specification* allows tension members that are fabricated from a combination of shapes and plates. Their strength is determined in the same way as the strength for single-shape tension members. However, the designer must remember that in bolted built-up members, bolts are usually placed along the member length to tie the various shapes together. These bolts result in holes along the member length, not just at the ends, so that rupture on the effective net section may become the controlling limit state at a location other than the member end. Welded built-up members will not normally experience rupture on the effective net area.

Perforated cover plates or tie plates can be used to tie the separate shapes together. Limitations on the spacing of these elements are also provided in Section D4, and requirements for the placement of bolts can be found in Section J3.6.

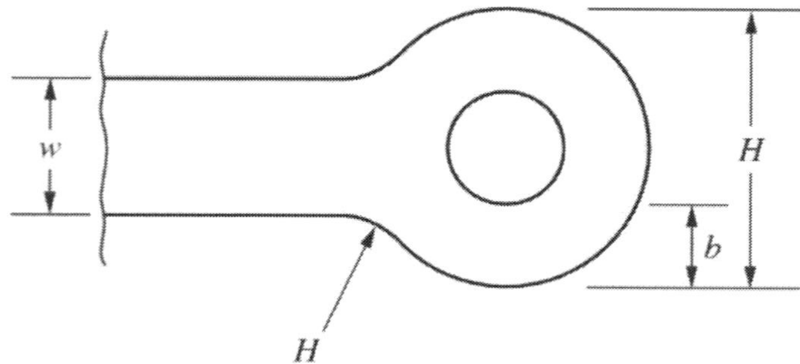

Figure 4.27 Eyebar Geometry

4.11 TRUSS MEMBERS

The most common tension members found in building structures are the tension web and chord members of trusses. Trusses are normally found as roof structures and as transfer structures within a building. Depending on the particular load patterns that a truss might experience, a truss member might be called upon to always resist tension or to resist tension in some cases and compression in others. In cases in which a member is required to carry both tension and compression, it will need to be sized accordingly. Because the compression strength of a member is normally significantly less than the tension strength of that same member, as will be seen in Chapter 5, compression may actually control the design.

The typical truss member can be composed of either single shapes or a combination of shapes. When composed of a combination of shapes, the requirements discussed in Section 4.10 must be included. Otherwise, truss members are designed the same way as any other tension member discussed in this chapter. Examples 4.10, 4.11, and 4.14 showed the application of the tension provisions to several truss tension members.

4.12 BRACING MEMBERS

As with truss members, members used to provide lateral load resistance for a building might also be called upon to carry tension under some conditions and compression under others. They too would need to be designed to resist both loads. However, it is sometimes more economical to provide twice as many tension members and to assume that if a tension member were called upon to resist compression, it would buckle and therefore carry no load. This would permit all bracing to be designed as tension-only members and almost certainly permit them to have a smaller cross section than if they were required to resist compression. An additional simplification that this assumption entails is the elimination of potential compression members from the analysis for member forces. This may result in the structure being a determinate structure rather than an indeterminate one, thus simplifying the analysis.

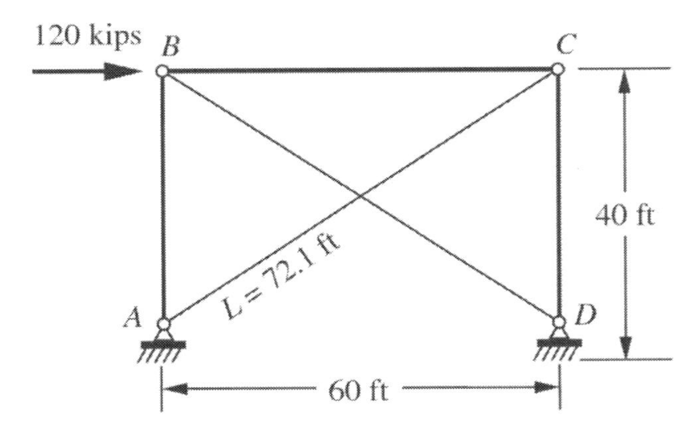

Figure 4.28 Braced Frame for Example 4.16

EXAMPLE 4.16a
Tension Member Design by LRFD

Goal: Select an appropriate tension member for use as a brace in an X-braced frame.

Given: The member is a tension-only diagonal as shown in Figure 4.28 and must resist a force resulting from the horizontal wind load, $W = 120$ kips. Use a pair of equal leg angles or a threaded rod of A36 steel.

SOLUTION

Step 1: Determine the force in the brace labeled AC if the brace labeled BD is assumed not to act since it would be in compression for this loading. The most critical load combination would be LRFD Load Combination 5, 0.9 Dead + 1.0 Wind. Using similar triangles,

$$P_u = 1.0(120)\left(\frac{72.1}{60.0}\right) = 144 \text{ kips}$$

Step 2: Determine the minimum required gross area based on the limit state of yielding.

$$A_{g\ \min} = 144/(0.90(36)) = 4.44 \text{ in.}^2$$

Step 3: Based on this minimum gross area, from *Manual* Table 1-15, select

2L2-1/2×2-1/2×1/2 with $A_g = 4.52$ in.^{2}

Step 4: Determine the minimum effective net area based on the limit state of rupture.

$$A_{e\ \min} = 144/(0.75(58)) = 3.31 \text{ in.}^2$$

Step 5: Thus, the combination of holes and shear lag may not reduce the area of this pair of angles by more than

$$A_e/A_g = 3.31/4.52 = 0.732$$

or another double-angle must be selected.

Step 6: Determine the slenderness ratio of the selected double angles and compare it with the recommended maximum of 300.

From *Manual* Table 1-15, the lower radius of gyration $r_x = 0.735$ in. and

$$\frac{L}{r_x} = \frac{72.1(12)}{0.735} = 1180 > 300$$

A check of *Manual* Table 1-15 shows that no pair of double angles will satisfy the slenderness limit of 300.

Step 7: Select a threaded rod to meet the strength requirement. The maximum slenderness limit does not apply to rods.

For a threaded rod, the only limit state to consider is tensile rupture. Based on Equation 4-7,

$$P_u \le \phi P_n = \phi(0.75F_u)A_b$$

Thus, the minimum area of the unthreaded portion of the rod is

$$A_b = \frac{P_u}{\phi(0.75F_u)} = \frac{144}{0.75(0.75)(58)} = 4.41 \text{ in.}^2$$

Therefore, select a 2-3/8 in. diameter threaded rod with $A = 4.43$ in.^{2}.

EXAMPLE 4.16b
Tension Member Design by ASD

Goal: Select an appropriate tension member for use as a brace in an X-braced frame.

Given: The member is a tension-only diagonal as shown in Figure 4.28 and must resist a force resulting from the horizontal wind load, $W = 120$ kips. Use a pair of equal leg angles or a threaded rod of A36 steel.

SOLUTION

Step 1: Determine the force in the brace labeled AC if the brace labeled BD is assumed not to act since it would be in compression for this loading. The most critical load combination would be ASD Load Combination 5, Dead + 0.6Wind. Using similar triangles,

$$P_a = 0.6(120)\left(\frac{72.1}{60.0}\right) = 86.5 \text{ kips}$$

Step 2: Determine the minimum required gross area based on the limit state of yielding.

$$A_{g\ \min} = 86.5/(36/1.67) = 4.01 \text{ in.}^2$$

Step 3: Based on this minimum gross area, from *Manual* Table 1-15, select 2L3-1/2×3-1/2×5/16 with $A_g = 4.20$ in.^{2}

Step 4: Determine the minimum effective net area based on the limit state of rupture.

$$A_{e\ \min} = 86.5/(58/2.00) = 2.98 \text{ in.}^2$$

Step 5: Thus, the combination of holes and shear lag may not reduce the area of this pair of angles by more than

$$A_e/A_g = 2.98/4.20 = 0.710$$

or another double-angle must be selected.

Step 6: Determine the slenderness ratio of the selected double angles and compare it with the recommended maximum of 300.

From *Manual* Table 1-15, the lower radius of gyration $r_x = 1.08$ in. and

$$\frac{L}{r_x} = \frac{72.1(12)}{1.08} = 801 > 300$$

A check of *Manual* Table 1-15 shows that no pair of double angles will satisfy the slenderness limit of 300.

Step 7: Select a threaded rod to meet the strength requirement. The maximum slenderness limit does not apply to rods.

For a threaded rod, the only limit state to consider is tensile rupture. Based on Equation 4-7,

$$P_a \leq P_n/\Omega = (0.75F_u)A_b/2.00$$

Thus, the minimum area of the unthreaded portion of the rod is

$$A_b = \frac{2.00P_a}{0.75F_u} = \frac{2.00(86.5)}{0.75(58)} = 3.98 \text{ in.}^2$$

Therefore, select a 2-3/8 in. diameter threaded rod with $A = 4.43$ in.^{2}.

4.13 PROBLEMS

1. Determine the gross and net areas for a 10×3/4 in. plate with a single line of standard holes for 7/8 in. bolts.

2. Determine the gross and net areas for an 8×1/2 in. plate with a single line of standard holes for 3/4 in. bolts.

3. Determine the gross and net areas for a 12×5/8 in. plate with a single line of standard holes for 1 in. bolts.

4. Determine the gross and net areas for a 12×1/2 in. plate with two lines of standard holes for 7/8 in. bolts.

5. Determine the gross and net areas for a 12×3/4 in. plate with two lines of standard holes for 3/4 in. bolts.

6. Determine the gross and net areas for a 16×1 in. plate with three lines of standard holes for 3/4 in. bolts.

7. Determine the gross and net areas for an L5×5×1/2 with two lines, one in each leg, of standard holes for 3/4 in. bolts.

8. Determine the gross and net areas for an L4×4×5/8 with two lines, one in each leg, of standard holes for 7/8 in. bolts.

9. Determine the gross and net areas for an L8×4×7/8 with three lines of standard holes, two in the 8 in. leg and one in the 4 in. leg, for 7/8 in. bolts.

10. Determine the gross and net areas for a WT8×18 with three lines of standard holes for 3/4 in. bolts. Each element of the WT will be attached to the connection.

11. Determine the gross and net areas for a WT12×34 with three lines of standard holes for 7/8 in. bolts. Each element of the WT will be attached to the connection.

12. Determine the gross and net areas for a WT7×66 with three lines of standard holes for 3/4 in. bolts. Each element of the WT will be attached to the connection.

13. Determine the gross and net areas for a WT9×53 with four lines of standard holes for 3/4 in. bolts. Each element of the WT will be attached to the connection (two lines of bolts in the stem).

14. Determine the gross and net areas for a C15×40 with five lines of standard holes for 7/8 in. bolts. Each flange will contain one line of bolts and the web will contain three lines of bolts.

15. Determine the gross and net areas for a C10×30 with four lines of standard holes for 3/4 in. bolts. Each flange will contain one line of bolts and the web will contain two lines of bolts.

16. Determine the gross and net areas for a C10×15.3 with two lines of standard holes in the web for 7/8 in. bolts.

17. Determine the net width for a 10×1/2 in. plate with 3/4 in. bolts placed in three lines as shown in the lap splice in Figure P4.17.

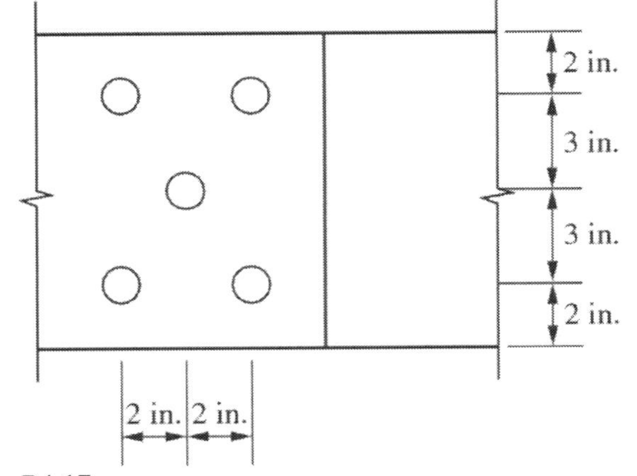

P4.17

18. Determine the net width for a 12×1/2 in. plate with 3/4 in. bolts placed in three lines as shown in Figure P4.18.

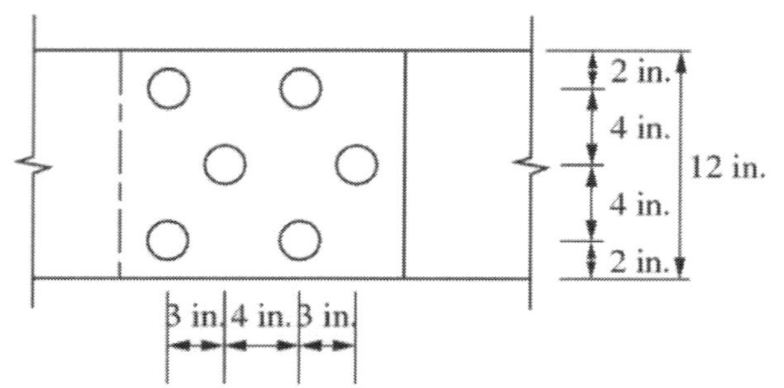

P4.18

19. Determine the net area for the L6×4×9/16 with 3/4 in. bolts shown in Figure P4.19.

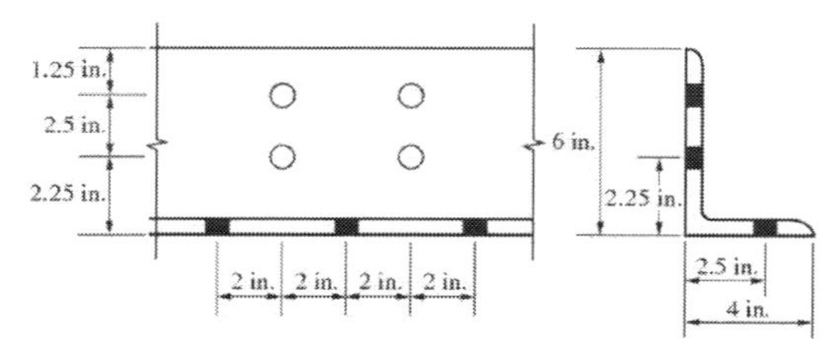

P4.19

20. Determine the net area for the L6×4×1/2 with 3/4 in. bolts shown in Figure P4.20.

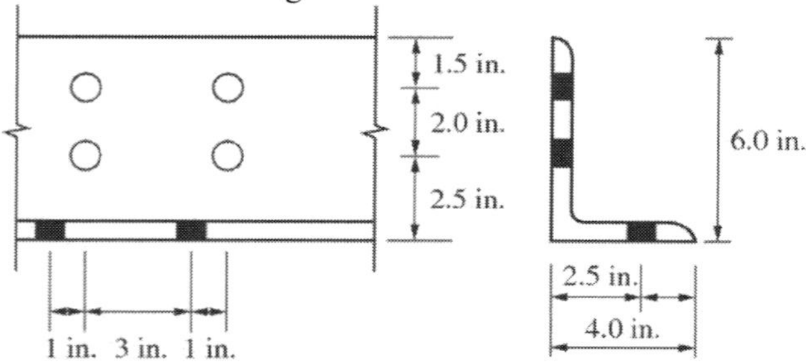

P4.20

21. Determine the gross and net areas for a double-angle tension member composed of two L4×4×1/2 shapes as shown in Figure P4.21 with holes for 3/4 in. bolts staggered in each leg.

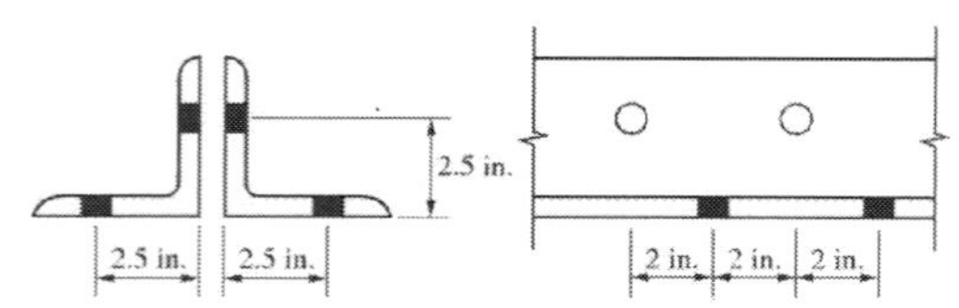

P4.21

22. Determine the gross and net areas for a double-angle tension member composed of two L6×6×9/16 shapes as shown in Figure P4.22 with holes for 3/4 in. bolts staggered in each leg.

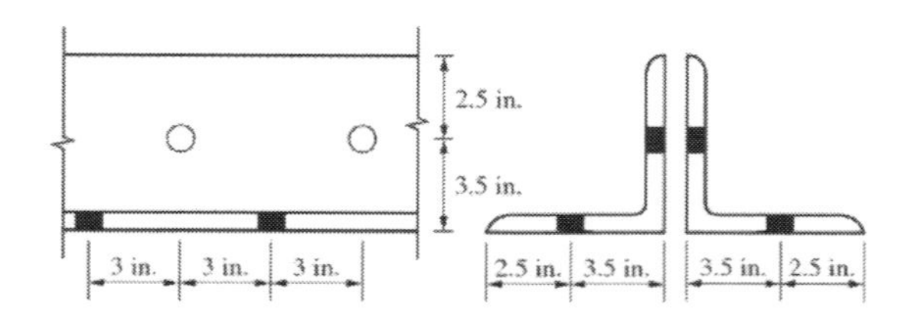

P4.22

23. For a WT8×50 attached through a flange to a 12×3/4 in. plate with six 3/4 in. bolts at a spacing of 3 in. and placed in two rows as shown in Figure P4.23, determine the shear lag factor (consider Case 2 and Case 7) and effective net area of the WT.

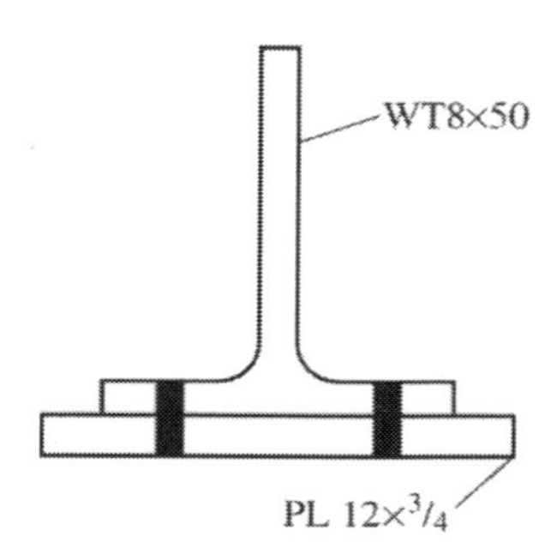

P4.23

24. For a WT7×34 attached through a flange to a 10×1 in. plate with eight 7/8 in. bolts at a spacing of 3 in., placed in two rows of four bolts as shown in Figure P4.24, determine the shear lag factor (consider Case 2 and Case 7) and effective net area of the WT.

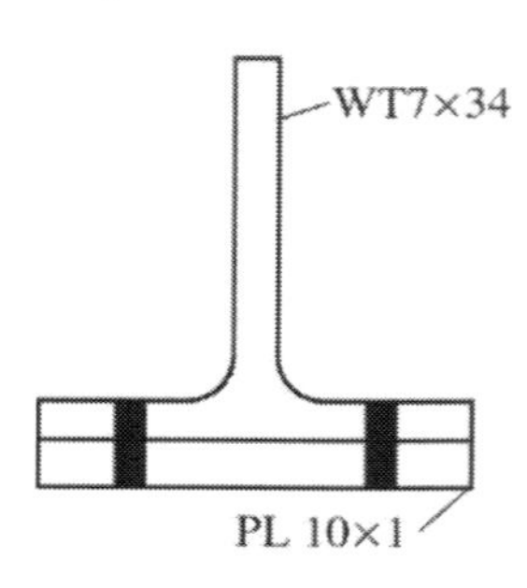

P4.24

25. A single L6×6×1 is used as a tension brace in a multi-story building. One leg of the angle is attached to a gusset plate with a single line of three 7/8 in. bolts at a spacing of 3 in. Determine the shear lag factor (consider Case 2 and Case 8) and effective net area.

26. A single L7×4×3/4 is used as a tension brace in a multi-story building. The 7 in. leg of the angle is attached to a gusset plate with a single line of four 7/8 in. bolts at a spacing of 3 in. Determine the shear lag factor (consider Case 2 and Case 8) and effective net area.

27. A single L8×6×1/2 is used as a tension brace in a multi-story building. The 8 in. leg of the angle is attached to a gusset plate with two lines of four 3/4 in. bolts at a spacing of 3 in. Determine the shear lag factor (consider Case 2 and Case 8) and effective net area.

28. The WT8×50 of Problem 23 is welded along the tips of the flange for a length of 12 in. on each flange and at the end, perpendicular to the load. Determine the shear lag factor (Case 2) and effective net area for the WT. If the transverse weld at the member end is not included, determine the shear lag factor and effective net area for the WT (be sure to select the correct case).

29. Determine the available strength of a 12×1/2 in. A36 plate connected to two 12 in. plates, as shown in Figure P4.29, with two lines of 3/4 in. bolts. Considering yielding and net section rupture, determine the (a) design strength by LRFD and (b) allowable strength by ASD.

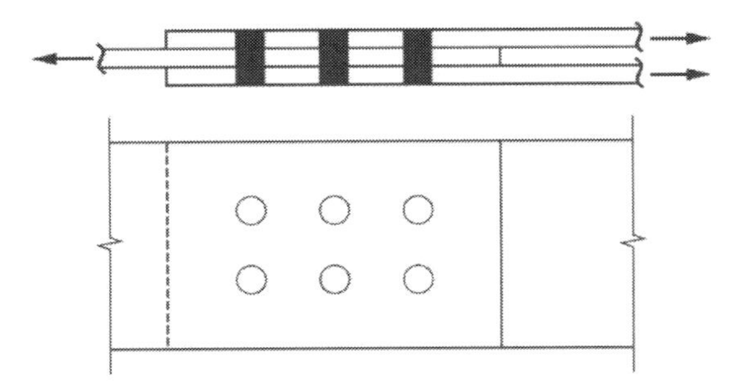

P4.29

30. Determine the available strength of an L6×4×3/4 attached through the long leg to a gusset plate with eight 7/8 in. bolts, in two lines of four bolts, at a 3 in. spacing. Use A572 Gr 50 steel. Considering yielding and net section rupture, determine the (a) design strength by LRFD and (b) allowable strength by ASD.

31. Determine the available strength of a C8×11.5 attached through the web to a gusset plate with ten 3/4 in. bolts, in two lines of five bolts, at a 3 in. spacing. Use A992 steel. Considering yielding and net section rupture, determine the (a) design strength by LRFD and (b) allowable strength by ASD.

32. Determine the available strength of a C12×30 attached through the web to a gusset plate with eight 7/8 in. bolts, in two lines of four bolts, at a 3 in. spacing. Use A992 steel. Considering yielding and net section rupture, determine the (a) design strength by LRFD and (b) allowable strength by ASD.

33. Determine the available strength of an 8×1/2 in. A572 Gr. 50 plate connected with three lines of 7/8 in. bolts. Considering yielding and net section rupture, determine the (a) design strength by LRFD and (b) allowable strength by ASD.

34. For the WT8×50, A992 steel addressed in Problem 23, considering yielding and net section rupture, determine the (a) design strength by LRFD and (b) allowable strength by ASD.

35. For the WT7×34, A992 steel addressed in Problem 24, considering yielding and net section rupture, determine the (a) design strength by LRFD and (b) allowable strength by ASD.

36. Determine the available strength of a WT7×19, A992 steel, with the flanges welded to a 1/2 in. gusset plate by a 10 in. weld along each side of the flange. Considering yielding and net section rupture, determine the (a) design strength by LRFD and (b) allowable strength by ASD.

37. Determine the available strength of a WT9×20, A992 steel, with the flanges welded to a 1/2 in. gusset plate by a 12 in. weld along each side of the flange. Considering yielding and net section rupture, determine the (a) design strength by LRFD and (b) allowable strength by ASD.

38. Determine the available strength of a WT9×20, A992 steel, with the flanges welded to a 1/2 in. gusset plate with unequal length longitudinal welds as shown in Figure 4.20b. Considering yielding and net section rupture, determine the (a) design strength by LRFD and (b) allowable strength by ASD.

39. Determine the available strength of a WT7×19, A992 steel, with the flanges welded to a 1/2 in. gusset plate with unequal length longitudinal welds, a 5 in. weld and an 11 in. weld. Considering yielding and net section rupture, determine the (a) design strength by LRFD and (b) allowable strength by ASD.

40. Design a 10 ft long, single-angle tension member to support a live load of 45 kips and a dead load of 15 kips (L/D = 3). The member is to be connected through one leg. Estimate three bolts in a single line. Use A992 steel and limit the slenderness ratio to 300. Consider yield and net section rupture, and design by (a) LRFD and (b) ASD.

41. Design a 10 ft long, single-angle tension member as in Problem 40 with the same total service load, 60 kips. Using a live load of 6.7 kips and a dead load of 53.3 kips (L/D = 0.126), (a) design by LRFD and (b) design by ASD.

42. Design a 10 ft long, single-angle tension member as in Problem 40 with the same service load using a live load of 50 kips and a dead load of 10 kips (L/D = 5). Design by (a) LRFD and (b) ASD.

43. Design a 33 ft long WT tension wind brace for a multi-story building to resist a wind force of 290 kips. Use A992 steel and 7/8 in. bolts connected to the flange only. Assume two lines of bolts with at least three bolts per line. Limit the slenderness ratio to 300. Consider yield and net section rupture, and design by (a) LRFD and (b) ASD.

44. Design a 20 ft long double angle tension wind brace for a multi-story building to resist a wind force of 100 kips. Use A992 steel and 3/4 in. bolts connected to the outstanding legs only. Assume two lines of bolts with at least three bolts per line. Limit the slenderness ratio to 300. Consider yield and net section rupture, and design by (a) LRFD and (b) ASD.

45. Design a W12 A992 tension member for a truss that will carry a dead load of 70 kips and a live load of 210 kips. The flanges will be bolted to the connecting plates with 7/8 in. bolts located so that four bolts will occur in any net section. Assume at least three bolts per line. Consider yield and net section rupture, and design by (a) LRFD and (b) ASD.

46. Design a W14 A992 tension member for a truss that will carry a dead load of 120 kips and a live load of 360 kips. The flanges will be bolted to the connecting plates with 7/8 in. bolts located so that four bolts will occur in any net section. Assume at least three bolts per line. Consider yield and net section rupture, and design by (a) LRFD and (b) ASD.

47. An L4×3×5/8 is attached to a gusset plate with three 3/4 in. bolts spaced at 3 in. with an end and edge distance of 1.5 in. as shown in Figure P4.47. Determine the available block shear strength of the A36 angle by (a) LRFD and (b) ASD.

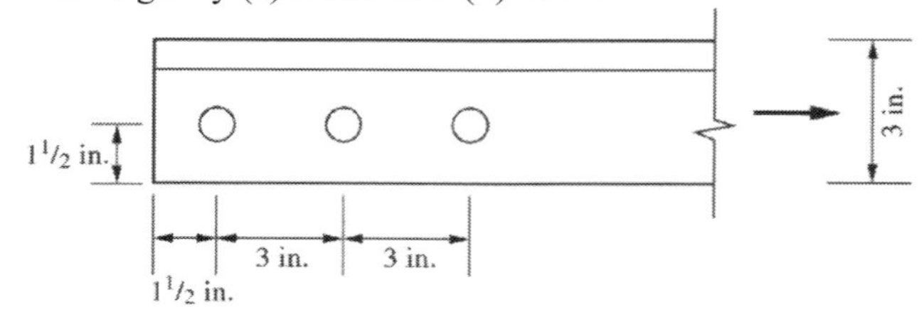

P4.47

48. A 7×5/8 in. A36 plate is shown in Figure P4.48, where the holes are for 3/4 in. bolts. Determine the available block shear strength by (a) LRFD and (b) ASD.

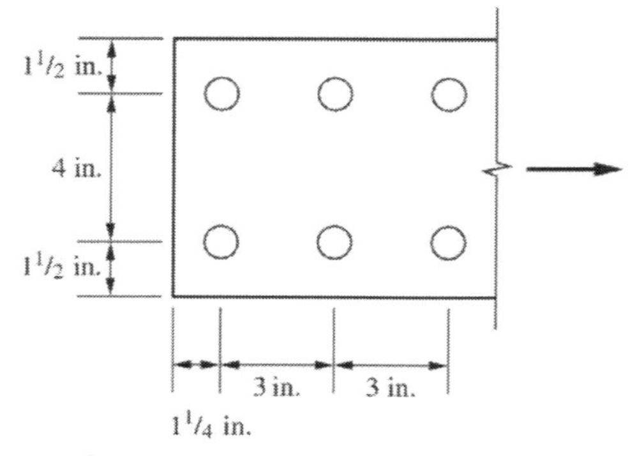

P4.48

49. Determine the available block shear strength for the A992 steel WT7×41, attached through the flange with eight 3/4 in. bolts as shown in Figure P4.49, by (a) LRFD and (b) ASD.

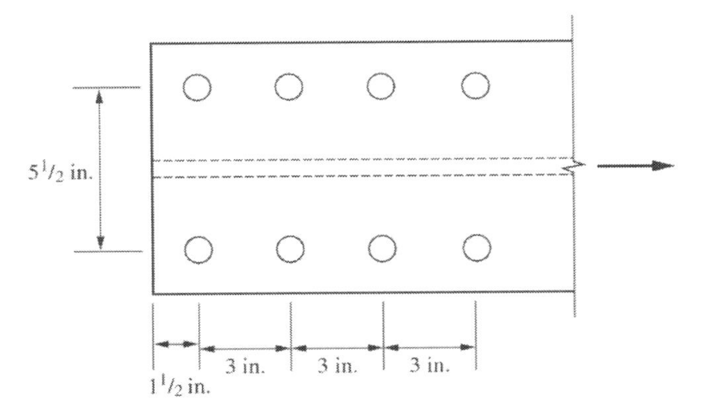

P4.49

50. Determine the available block shear strength for the A992 steel WT12×34, attached through the flange with eight 5/8 in. bolts as shown in Figure P4.49, by (a) LRFD and (b) ASD.

51. For the WT7x41 addressed in Problem 49, considering yielding, net section rupture, and block shear, determine the (a) design strength by LRFD and (b) allowable strength by ASD and indicate the controlling limit state.

51. For the WT12x34 addressed in Problem 50, considering yielding, net section rupture, and block shear, determine the (a) design strength by LRFD and (b) allowable strength by ASD and indicate the controlling limit state.

52. Two 3/4 in. A572 Grade 50 steel plates and six 3/4 in. bolts are used in a bolted lap splice (Figure P4.52). Consider yield, net section rupture, and block shear rupture. Determine the (a) design strength by LRFD and (b) the allowable strength by ASD and indicate the controlling limit state.

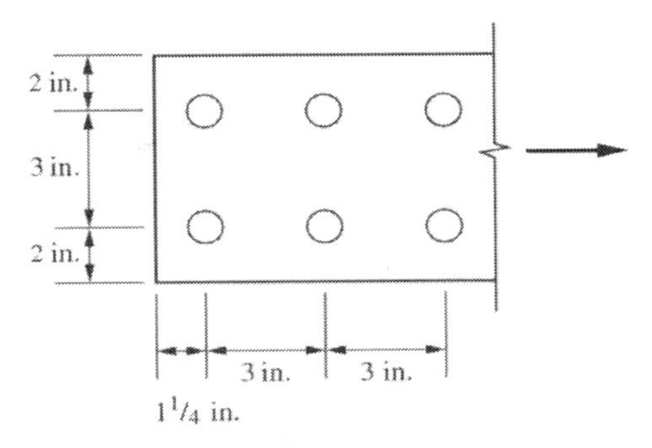

P4.52

53. Two 1.0 in. A36 steel plates and eight 7/8 in. bolts are used in a bolted lap splice (Figure P4.53). Consider yield, net section rupture, and block shear rupture. Determine the (a) design strength by LRFD and (b) the allowable strength by ASD and indicate the controlling limit state.

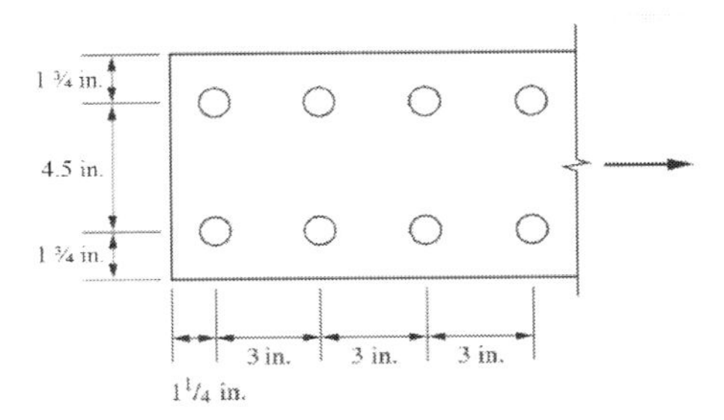

P4.53

54. A36 steel and 7/8 in. bolts are used in the bolted splice shown in Figure P4.54. Consider yield, net section rupture, and block shear rupture. Determine the (a) design strength by LRFD and (b) the allowable strength by ASD and indicate the controlling limit state.

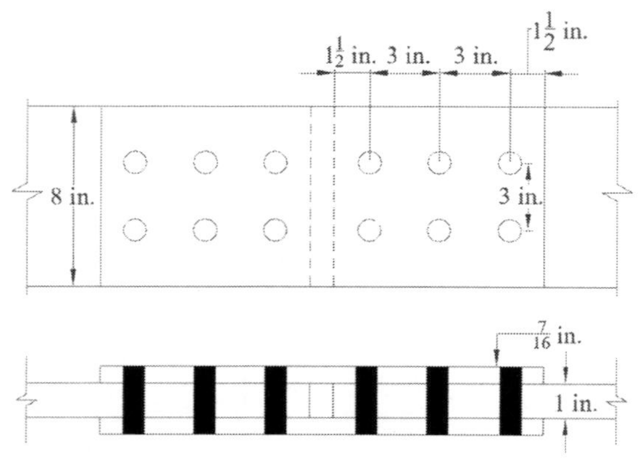

P4.54

55. A572 Grade 50 steel and 3/4 in. bolts are used in the bolted splice shown in Figure P4.54. Consider yield, net section rupture, and block shear rupture. Determine the (a) design strength by LRFD and (b) the allowable strength by ASD and indicate the controlling limit state.

56. Integrated Design Project. Lateral load resistance in the north-south direction is provided by a chevron braced frame as shown in Figure 1.24. Before the forces in these bracing members can be determined, the specified wind load must be determined. At this stage in the design, a simplified approach to wind load calculation might yield the following loads at each level.

Roof	54.0 kips
Fourth floor	102.0 kips
Third floor	93.0 kips
Second floor	88.0 kips
Total wind load	337.0 kips

It is permissible to treat the chevron braced frames as vertical trusses with pinned joints at all intersections. Using the two braced frames to share equally the given wind loading, analyze the vertical trusses for the wind in one direction only. It will be apparent that the windward brace is in tension and the leeward brace is in compression. When the wind blows in the other direction, the brace forces reverse.

Design all the braces as if the tension forces control. Although it is usually compression forces that control brace design, in this chapter we are focusing on tension. It would be useful to consider the use of several types of members, e.g., angles, channels, and W-shapes, to resist these forces.

Chapter 5

Compression Members

270 Park Avenue, New York, NY
Photo courtesy of Severud Associates
© DBOX for JPM Chase

5.1 COMPRESSION MEMBERS IN STRUCTURES

Compression members are structural elements subjected to axial forces that tend to push the ends of the members toward each other. The most common compression member in a building structure is a *column*. Columns are vertical members that support the horizontal elements of a roof or floor system. Several columns can be seen in Figure 5.1 as part of a building structure. They are the primary elements that provide the vertical space to form an occupiable volume. Other compression members are found in trusses as chord and web members and as bracing members in floors and walls. Other names often used to identify compression members are *struts* and *posts*. Throughout this chapter the terms *compression member* and *column* will be used interchangeably.

The compression members discussed in this chapter experience only axial forces. In real structures, additional load effects are often exerted on a compression member that would tend to combine bending with the axial force. These combined force members are called beam-columns and are discussed in Chapter 8. The majority of the provisions that apply to compression members are located in Chapter E of the *Specification*.

Table 5.1 lists the sections of the *Specification* and parts of the *Manual* discussed in this chapter.

5.2 CROSS-SECTIONAL SHAPES FOR COMPRESSION MEMBERS

Compression members carry axial forces, so the primary cross-sectional property of interest is the area. Thus, the simple relationship between force and stress,

$$f = \frac{P}{A} \tag{5.1}$$

Figure 5.1 Columns in a Multistory Building
Photo courtesy of Greg Grieco

is applicable. As long as this relationship dictates compression member strength, all cross sections with the same area will perform in the same way. In real structures, however, other factors influence the strength of the compression member, and the distribution of the area becomes important.

In building structures, the typical compression member is a column and the typical column is a rolled wide-flange member. Later discussions of compression member strength will show that the W-shape does not have the most efficient distribution of material for compression members. It does, however, provide a compression member that can easily be connected to other members of the system such as beams and other columns. This feature significantly influences its selection as an appropriate column cross section.

Figure 5.2 shows examples of rolled and built-up shapes that are used as compression members. Many of these are the same shapes used for the tension members discussed in Chapter 4. This is reasonable because the forces being considered in these two cases are both axial, although they act in the opposite direction. However, other factors that influence the strength of compression members will dictate additional criteria for the selection of the most efficient shapes for these members.

The tee and angle shown in Figure 5.2c and d are commonly used as chords and webs of trusses. In these applications, the geometry of the shapes helps simplify the connections between members. Angles are also used in pairs as built-up compression members, with the connecting element between the two angles as shown in Figure 5.2h. The channel can be found in trusses as a single element or combined with another channel as shown in Figure 5.2b, i, l, and m. Built-up columns can also be found using channels. The hollow structural sections (HSS) shown in Figure 5.2e, f, and g are commonly found as columns in buildings, particularly one-story structures where the connections to the shape can be simplified by carrying beams over the columns. The distribution of the material in these shapes is the most efficient for columns.

Table 5.1 Sections of *Specification* and Parts of *Manual* Covered in this Chapter

	Specification
B3	Design Basis
B4	Classification of Sections for Local Buckling
E1	General Provisions
E2	Effective Length
E3	Flexural Buckling of Members without Slender Elements
E4	Torsional and Flexural-Torsional Buckling of Single Angles and Members without Slender Elements
E5	Single-Angle Compression Members
E6	Built-up Members
E7	Members with Slender Elements
	Manual
Part 1	Dimensions and Properties
Part 4	Design of Compression Members
Part 6	Design of Members Subject to Combined Loading

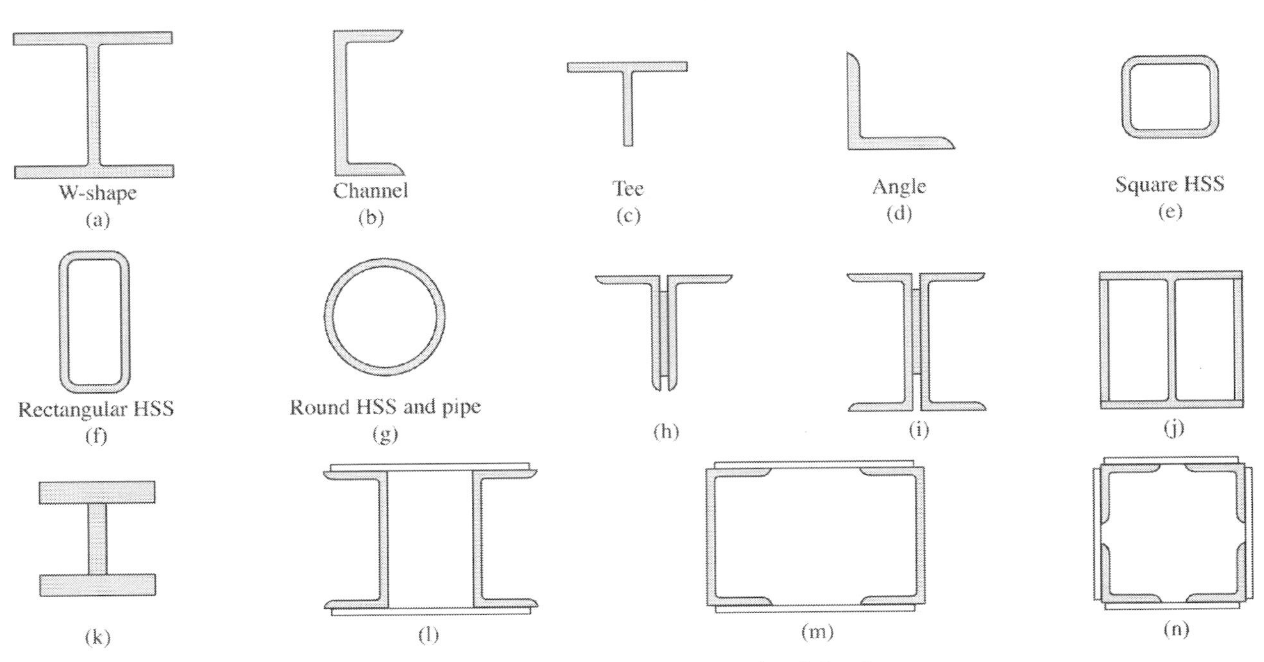

Figure 5.2 Rolled Shapes and Built-up Shapes for Compression Members

5.3 COMPRESSION MEMBER STRENGTH

If no other factors were to impact the strength of a compression member, the simple axial stress relationship given in Equation 5.1 could be used to describe member strength. Thus, the maximum force that a compression member could resist at yield would be

$$P_y = F_y A_g \tag{5.2}$$

where P_y is the yield load, sometimes called the squash load; F_y is the yield stress; and A_g is the gross area. This is the response that would be expected if a very short specimen, one whose length approximates its other two dimensions, were to be tested in compression. This type of column test specimen, shown in Figure 5.3a, is called a *stub column*. Because most compression members will have a length that greatly exceeds its other dimensions, length effects cannot be ignored. A more realistic column is shown in a test frame in Figure 5.3b.

5.3.1 Euler Column

To address the impact of length on compression member behavior, a simple model, as shown in Figure 5.4, is used. The Swiss mathematician Leonard Euler first presented this analysis in 1759. A number of assumptions are made in this column model: (1) the column ends are frictionless pins, (2) the column is perfectly straight, (3) the load is applied along the centroidal axis, and (4) the material behaves elastically. Based on these assumptions, this column model is usually called the *perfect column* or the *pure column*.

Figure 5.4a shows the perfect column with an applied load that will not cause any lateral displacement or yielding. In this arrangement, the load can be increased with no lateral displacement of the column. However, at a particular load, defined as the critical load or the buckling load, P_{cr}, the column will displace laterally as shown in Figure 5.4b. In this configuration, the dashed line represents the original position of the member, and the solid line represents the displaced position. Note that an axis system is presented in the figure, with the z-axis along the member length and the y-axis transverse to the member length. This places the x-axis perpendicular to the plane of the figure. The x- and y-axes correspond to the centroidal axes of the cross section.

A free body diagram of the lower portion of the column in its displaced position is shown in Figure 5.4c. If moments are taken about point C, equilibrium requires

$$M_z = P_{cr} y$$

From the principles of mechanics and using small displacement theory, the differential equation relating moment to curvature of the deflected member is given as

$$\frac{d^2 y}{dz^2} = -\frac{M_z}{EI_x}$$

(a)

(b)

Figure 5.3 Column Testing. (a) Stub column. (b) Long column
(a) Photo courtesy Prof. Dr. Mario Fontana, (b) Photo courtesy Mohammed Ali Morovat and Michael Engelhardt, University of Texas at Austin

Combining these two equations and rearranging the terms yields the differential equation of equilibrium,

$$\frac{d^2y}{dz^2} + \frac{P_{cr}}{EI_x} y = 0$$

If the coefficient of the second term is taken as $k^2 = P_{cr}/EI_x$, the differential equation for the column becomes

$$\frac{d^2y}{dz^2} + k^2 y = 0$$

which is a standard second-order linear ordinary differential equation. The solution to this equation is given by

$$y = A\sin kz + B\cos kz \tag{5.3}$$

where A and B are constants of integration. To further evaluate this equation, the boundary conditions must be applied. Because at $z = 0$, $y = 0$ and at $z = L$, $y = 0$, we find that

$$B = 0$$

and

$$A\sin kL = 0$$

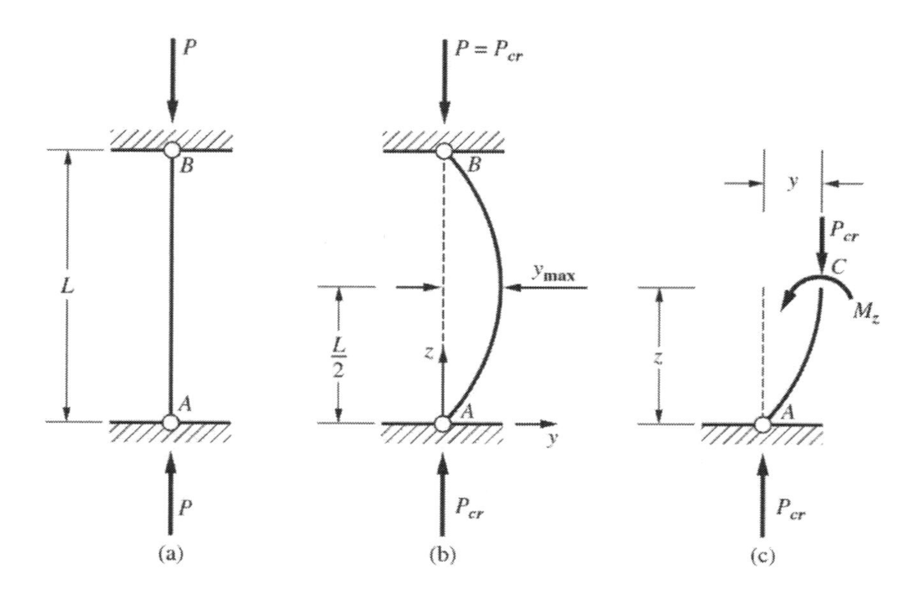

Figure 5.4 Stability Conditions for Elastic Columns

For Equation 5.3 to have a nontrivial solution, (sin kL) must equal zero. This requires that $kL = n\pi$, where n is any integer. Substituting for k and rearranging yields

$$P_{cr} = \frac{n^2\pi^2 EI_x}{L^2} \tag{5.4}$$

Because n can be taken as any integer, Equation 5.4 has a minimum when $n = 1$. This is called the *Euler buckling load* or the *critical buckling load* and is given as

$$P_{cr} = \frac{\pi^2 EI_x}{L^2} \tag{5.5}$$

If values for B and kL are substituted into Equation 5.3, the shape of the buckled column can be determined from

$$y = A\sin\left(n\pi\frac{z}{L}\right) \tag{5.6}$$

Because any value for A will satisfy Equation 5.6, a unique magnitude for the displacement cannot be determined; however, it is clear that the shape of the buckled column is a half sine curve when $n = 1$. This is shown in Figure 5.5a. For other values of n, different buckled shapes will result along with the higher critical buckling load. When $n > 1$, these shapes are referred to as higher mode shapes. Several cases are shown in Figure 5.5b, c, and d. In all cases, the basic shape is the sine curve. In order for these higher modes to occur, some type of physical restraint against buckling is required at the point where the buckled shape crosses the original, undeflected shape. This can be accomplished with the addition of braces, which is discussed later.

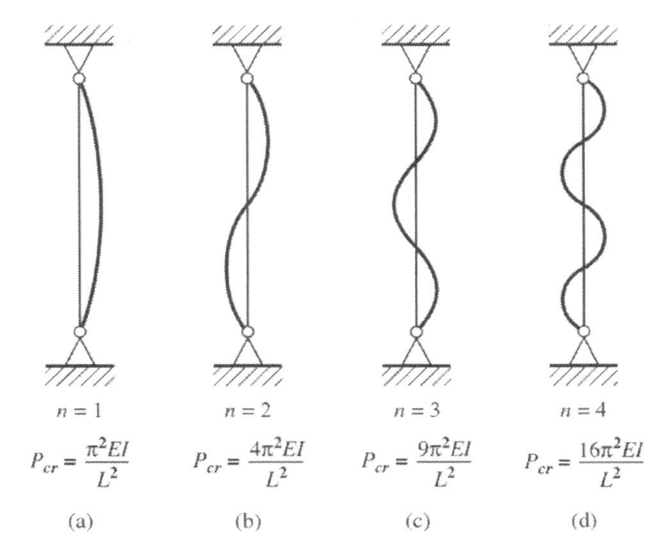

Figure 5.5 Shape of Buckled Columns

We now have two equations to predict the column strength: Equation 5.2, which does not address length; and Equation 5.5, which does. These two equations are plotted in Figure 5.6. Because the derivation of the Euler equation was based on elastic behavior and the column cannot carry more load than the yield load, there is an upper limit to the column strength.

If the length at which this limit occurs is taken as L_y, it can be determined by setting Equation 5.2 equal to Equation 5.5 and solving for length, giving

$$L_y = \pi\sqrt{\frac{EI_x}{F_y A_g}}$$

To simplify this equation, the radius of gyration, r, will be used, where

$$r = \sqrt{\frac{I}{A}}$$

Because the moment of inertia depends on the axis being considered, and A is the gross area of the section, which is independent of axis, r will depend on the buckling axis. In the derivation just developed, the axis of buckling for the column of Figure 5.4 was taken as the x-axis; thus,

$$L_y = \pi r_x\sqrt{\frac{E}{F_y}}$$

For this theoretical development, a column whose length is less than L_y would fail by yielding and could be called a short column, whereas a column with a length greater than L_y would fail by buckling and be called a long column.

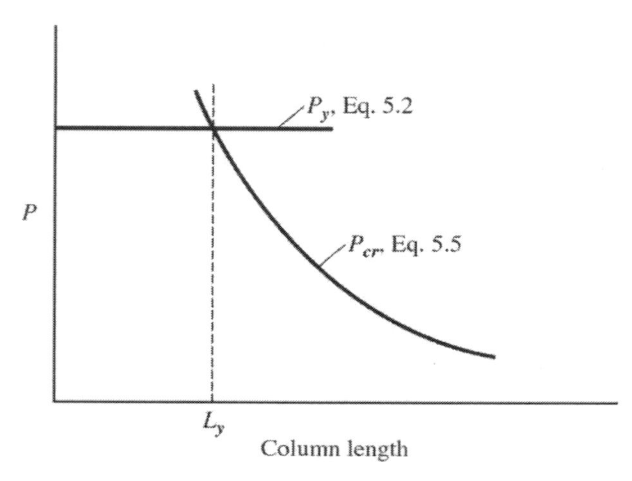

Figure 5.6 Column Strength Based on Length

It is also helpful to write Equation 5.5 in terms of stress. Dividing both sides by the area and substituting again for the radius of gyration yields

$$F_{cr} = \frac{\pi^2 E}{\left(\frac{L}{r}\right)^2} \tag{5.7}$$

In this equation, the radius of gyration is left unsubscripted so that it can be applied to whichever axis is determined to be the critical axis. A plot of stress versus L/r would be of the same shape as the plot of force versus L in Figure 5.6.

5.3.2 Other Boundary Conditions

Derivation of the buckling equations presented as Equations 5.5 and 5.7 included the boundary condition of frictionless pins at both ends. For perfect columns with other boundary conditions, the moment will not be zero at the ends, and this will result in a nonhomogeneous differential equation. Solving the resulting differential equation and applying the appropriate boundary conditions will lead to a buckling equation of a form similar to the previous equations. To generalize the buckling equation for other end conditions, the column length, L, is replaced by the column effective length, KL, where K is the effective length factor. Thus, the general buckling equations become

$$P_{cr} = \frac{\pi^2 EI}{(KL)^2} \tag{5.8}$$

and

$$F_{cr} = \frac{\pi^2 E}{\left(\frac{KL}{r}\right)^2} \tag{5.9}$$

Figure 5.7 depicts the original pin-ended column with several examples of columns showing the influence of different end conditions. All columns are shown with the lower support fixed against lateral translation. Three of the columns have upper ends that are also restrained from lateral translation, and three others have upper ends that are free to translate. The effective length can be visualized as the length between inflection points, where the curvature reverses. This result is similar to the original derivation when n was taken as some integer other than 1. It is most easily seen in Figure 5.7b and c but can also be seen in Figure 5.7d by visualizing the extended buckled shape above the column as shown in Figure 5.8. In all cases, the buckled curve is a segment of the sine curve. The most important thing to observe is that the column with fixed ends in Figure 5.7b has an effective length of $0.5L$, whereas the column in Figure 5.7a has an effective length of L. Thus, the fixed-end column will have four times the strength of the pin-ended column.

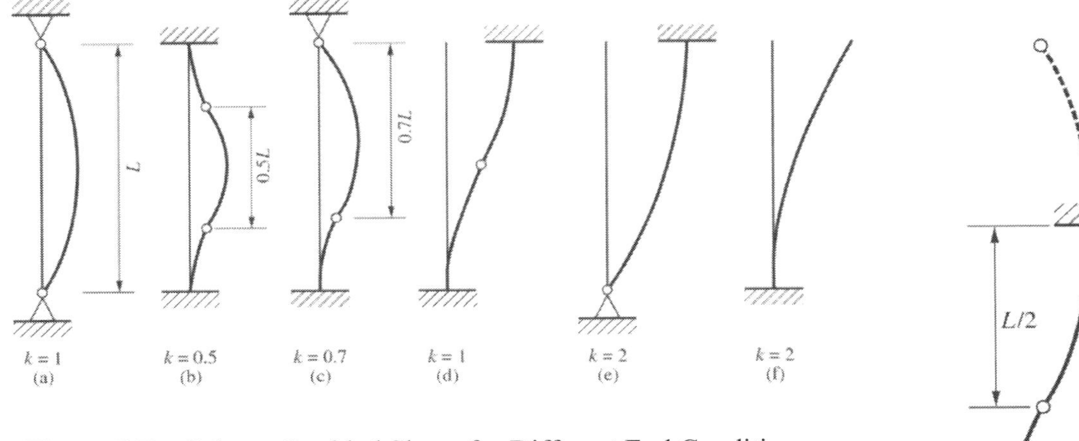

Figure 5.7 Column Buckled Shape for Different End Conditions

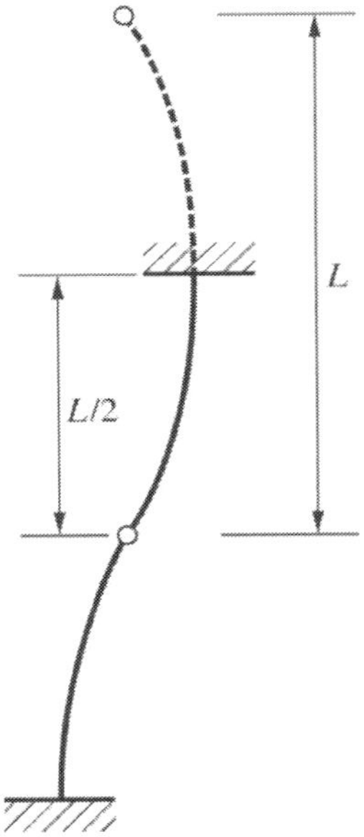

Figure 5.8 Extended Shape of Buckled Column from Figure 5.7d

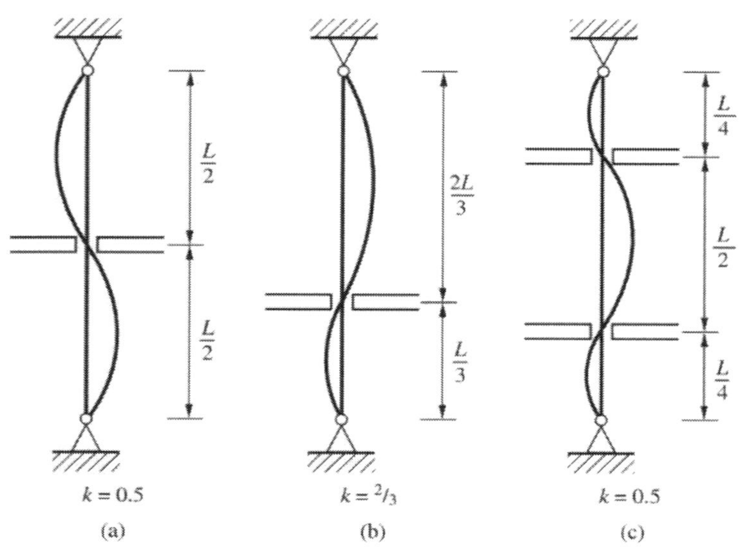

Figure 5.9 Buckled Shape for Columns with Intermediate Braces

5.3.3 Combination of Bracing and End Conditions

The influence of intermediate bracing on the effective length was touched upon in the discussion of the higher modes of buckling. In those cases, the buckling resulted in equal-length segments that reflected the mode number. Thus, a column with n = 2 had two equal segments, whereas a column with n = 3 buckled with three equal segments. If physical braces are used to provide buckling resistance to the column, the effective length will depend on the location of the braces. Figure 5.9 shows three columns with pinned ends and intermediate supports. The column in Figure 5.9a is the same as the column in Figure 5.5b. The effective length is $0.5L$, so $K = 0.5$. The column in Figure 5.9b shows lateral braces in an unsymmetrical arrangement with one segment equal to $L/3$ and the other to $2L/3$. Although the exact location of the inflection point would be slightly into the longer segment, normal practice is to take the longest unbraced length as the effective length; thus $KL = 2L/3$, so $K = 2/3$. The column in Figure 5.9c is braced at two locations. The longest unbraced length for this case gives an effective length $KL = 0.5L$ and a corresponding $K = 0.5$. A general rule can be stated that, when the column ends are pinned, the longest unbraced length is the effective length for buckling in that direction.

When other end conditions are present, these two influences must be combined. The columns of Figure 5.10 illustrate the influence of combinations of end supports and bracing on the column effective length. The end conditions would influence only the effective length of the end segment of the column. For the column in Figure 5.10a, the lower segment has $L = a$, and that segment would buckle with an effective length $KL = a$. The upper segment has $L = b$ but also has a fixed end. Thus, it would buckle with an effective length $KL = 0.7b$, obtained by combining the end conditions of Figure 5.7c with the length, b. Thus, the relationship between lengths a and b determine which end of the column dictates the overall column effective length. As an example, the column in Figure 5.10b shows that the lowest segment would set the column effective length at $0.35L$.

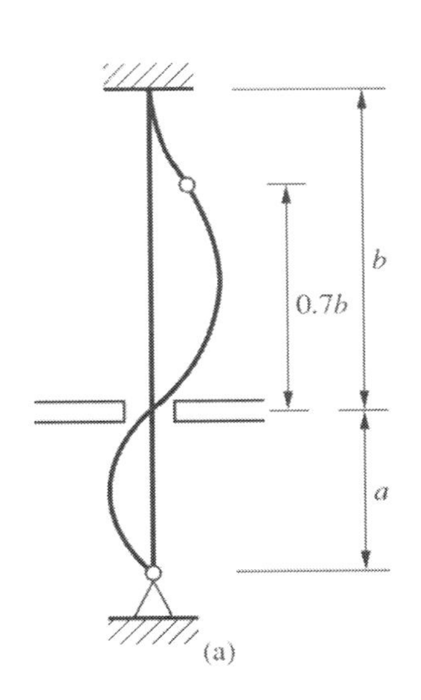

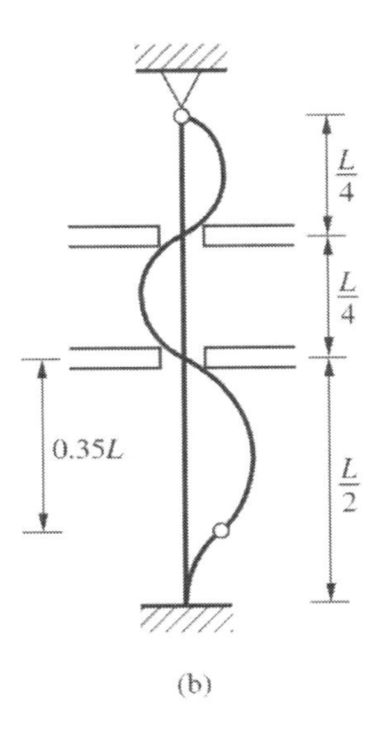

Figure 5.10 Buckled Shape for Columns with Different End Conditions and Intermediate Braces

EXAMPLE 5.1
Theoretical Column Strength

Goal: Determine the theoretical strength for a pin-ended column and whether it will first buckle or yield.

Given: A W10×33, A992, column with a length of 20 ft.

SOLUTION

Step 1: Determine the load that would cause buckling.

With no other information, it must be assumed that this column will buckle about its weak axis, if it buckles at all, because the effective length, $KL = 20$ ft for both axes.

From *Manual* Table 1-1, $I_y = 36.6$ in.^{4} and $A_g = 9.71$ in.^{2}. The load that would cause it to buckle is

$$P_{cr} = \frac{\pi^2 EI_y}{(KL)^2} = \frac{\pi^2(29{,}000)(36.6)}{(20(12))^2} = 182 \text{ kips}$$

Step 2: Determine the load that would cause yielding.

$$P_y = F_y A_g = 50(9.71) = 486 \text{ kips}$$

Step 3: Conclusion: Because $P_{cr} < P_y$, the theoretical column strength is

$$P = 182 \text{ kips}$$

and the column would buckle before it could reach its yield stress.

EXAMPLE 5.2
Critical Buckling Load

Goal: Determine the overall column length that, if exceeded, would theoretically cause the column to buckle elastically before yielding.

Given: A W8×31 column with fixed supports. Use steels with (a) $F_y = 40$ ksi and (b) $F_y = 100$ ksi.

SOLUTION

Step 1: From *Manual* Table 1-1, $I_y = 37.1$ in.^{4} and $A_g = 9.13$ in.^{2}.

Part a

Step 2: Determine the force that would cause the column to yield when $F_y =$ 40 ksi.

$$P_y = F_y A_g = 40(9.13) = 365 \text{ kips}$$

Step 3: To determine the length that would cause this same load to be the buckling load for the pinned-pinned case, set this force equal to the buckling load equation and determine the effective length from

$$365 \text{ kips} = \frac{\pi^2 EI_y}{(KL)^2} = \frac{\pi^2(29{,}000)(37.1)}{(KL)^2}$$

which gives

$$KL = \sqrt{\frac{\pi^2(29{,}000)(37.1)}{365}} = 171 \text{ in.}$$

So the effective length is

$$KL = \frac{171}{12} = 14.3 \text{ ft}$$

Step 4: From Figure 5.7b, a fixed-end column has an effective length equal to one-half the actual length, buckling will not occur if the actual length is less than or equal to:

$$L = KL/0.5 = 2(14.3) = 28.6 \text{ ft for a column with } F_y = 40 \text{ ksi}$$

Part b

Step 5: Determine the force that would cause the column to yield when $F_y =$ 100 ksi.

$$P_y = F_y A_g = 100(9.13) = 913 \text{ kips}$$

Part 6: To determine the effective length that would cause this same load to be the buckling load for the pinned-pinned case, set this force equal to the buckling force and determine the length from

$$913 \text{ kips} = \frac{\pi^2 EI_y}{(KL)^2} = \frac{\pi^2 (29{,}000)(37.1)}{(KL)^2}$$

which gives

$$KL = \sqrt{\frac{\pi^2 (29{,}000)(37.1)}{913}} = 108 \text{ in.}$$

So the effective length is

$$KL = \frac{108}{12} = 9.0 \text{ ft}$$

Step 7: From Figure 5.7b, a fixed-end column has an effective length equal to one-half the actual length, buckling will not occur if the actual length is less than or equal to:

$$L = KL/0.5 = 2(9.0) = 18.0 \text{ ft} \quad \text{for a column with } F_y = 100 \text{ ksi}$$

5.3.4 Real Column

Physical testing of specimens that effectively model columns found in real building structures, like that seen in Figure 5.3b, has shown that column strength was not as great as either the buckling load predicted by the Euler buckling equation or the squash load predicted by material yielding. This inability of the theory to predict actual behavior was recognized early, and numerous factors were found to be the cause. Three main factors influence column strength: material inelasticity, column initial out-of-straightness, and end conditions. The influence of column end conditions has already been discussed with respect to effective length determination. Material inelasticity and initial out-of-straightness, which also significantly impact real column strength, are discussed here.

Inelastic behavior of a column directly results from built-in or residual stresses in the cross section. These residual stresses are, in turn, the direct result of the manufacturing process. Steel is produced with heat, and heat is also necessary to form the steel into the shapes used in construction. Once the shape is fully formed, it is cooled. During this cooling process residual stresses are developed. Figure 5.11 shows a wide-flange cross section in various stages of cooling. Initially, as shown in Figure 5.11a, the tips of the flanges with the most surface area to give off heat begin to cool. This material contracts as it cools, eventually reaching the ambient temperature. At this point, the fibers in this part of the section reach what is expected to be their final length.

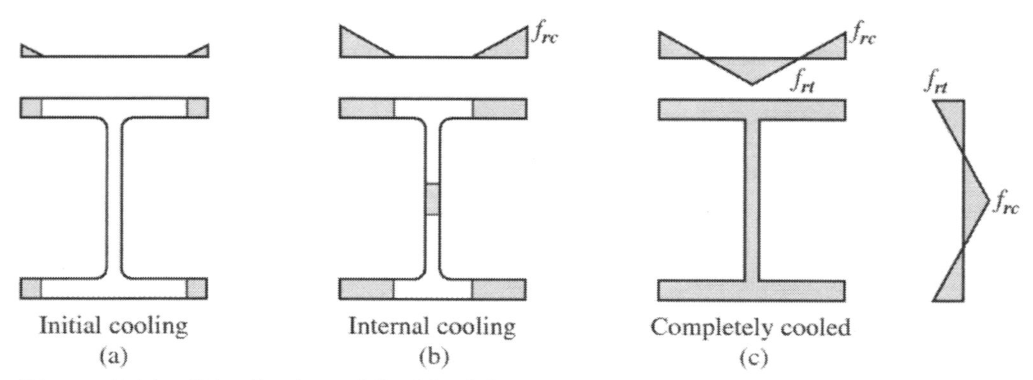

Figure 5.11 Distribution of Residual Stresses

As adjacent fibers cool, they too contract. In the process of contracting, these subsequently cooling fibers pull on the previously cooled fibers, placing the latter under some amount of compressive stress. Figure 5.11b shows a cross section with additional flange elements cooled. When the previously cooled portion of the cross section provides enough stiffness to restrain the contraction of the subsequently cooling material, a tensile stress is developed in the now-cooling material because it cannot contract as it would without this restraint. When completely cooled, as shown in Figure 5.11c, the tips of the flanges and the middle of the web are put into compression, and the flange-web juncture is put into tension. Thus, the first fibers to cool are in compression, whereas the last to cool are in tension.

Several different representations of the residual stress distribution have been suggested. One distribution is shown in Figure 5.11c. The magnitude of the maximum residual stress does not depend on the material yield strength but is a function of material thickness. In addition, the compressive residual stress is of critical interest in the consideration of compression members. The magnitude of this residual stress varies from 10 ksi to about 30 ksi, depending on the shape. The higher values are found in wide flanges with the thickest flange elements.

To understand the overall impact of these residual stresses on column behavior, a stub column will again be investigated. Figure 5.12 shows the stress-strain relation for a short column, one that will not buckle but exhibits the influence of residual stresses. As the column is loaded with an axial load, the member shortens and the corresponding strain and stress are developed, as if this were a perfectly elastic specimen. The response of a perfectly elastic, perfectly plastic column is shown by the dashed line in Figure 5.12. When the applied stress is added to a member with residual compressive stress, the stub column begins to shorten at a greater rate as the tips of the flange become stressed beyond the yield stress. This point is identified in Figure 5.12 as F_p, the proportional limit Thus, the stress-strain curve moves off the straight dashed line and follows the curved solid line. Continuing to add load to the column results in greater strain for a given stress, and the column eventually reaches the yield stress of the perfectly elastic material. Thus, the only difference between the behavior of the actual column and the usual test specimen used to determine the stress-strain relationship is that the real column behaves inelastically as those portions of its cross section with compressive residual stresses reach the material yield stress.

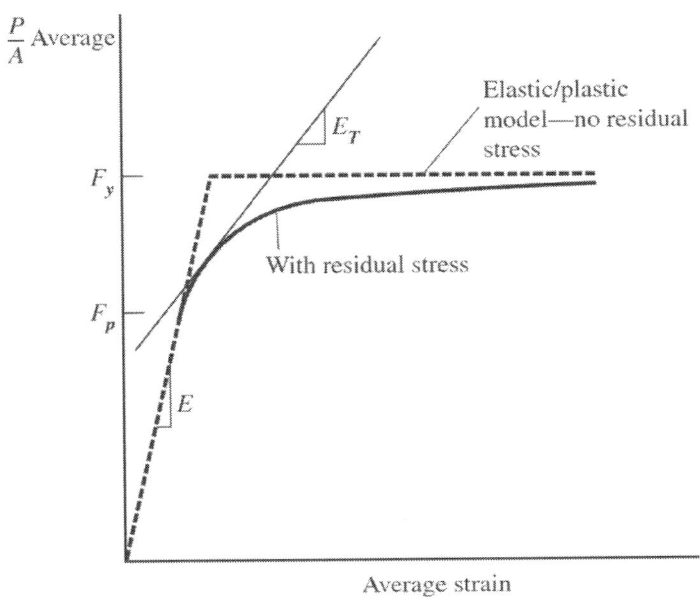

Figure 5.12 Stub Column Stress-Strain Diagrams with and without Residual Stress

If a new term, the *tangent modulus*, E_T, is defined as the slope of a tangent to the actual stress-strain curve at any point shown in Figure 5.12, an improved prediction of column buckling strength can be obtained by modifying the Euler buckling equation so that

$$P_{cr} = \frac{\pi^2 E_T I_x}{(KL)^2}$$

Thus, as the column is loaded beyond its elastic limit, E_T decreases, and the buckling strength does also. This partially accounts for the inability of the Euler buckling equation to accurately predict column strength.

Another factor to significantly impact column strength is the column initial out-of-straightness. Once again, the manufacturing process for steel shapes impacts the ability of the column to carry the predicted load. In this case, the problem is related to the fact that no structural steel member comes out of the production process perfectly straight. In the past, the AISC *Code of Standard Practice* had limited the initial out-of-straightness to 1/1000 of the length between points with lateral support. Although this appears to be a small variation from straightness, it still impacts column strength.

Figure 5.13a shows a perfectly elastic, pin-ended column with an initial out-of-straightness, δ. A comparison of this column diagram with that used to derive the Euler column, Figure 5.4, shows that the moment along the column length will be greater for this initially crooked column in its buckled position than it would have been for an initially straight column. Thus, the solution to the differential equation would be different. In addition, because the applied load works at an eccentricity from the column along its length, even before buckling, a moment is applied to the column that has not yet been accounted for. Figure 5.13b shows the load versus lateral displacement diagram for this initially crooked column compared to that of the initially straight column. This column not only exhibits greater lateral displacement, it also has a lower maximum strength.

When these two factors are combined, the Euler equation cannot properly describe column behavior on its own. Thus, the development of curves to predict column behavior has historically been a matter of curve-fitting the test data in an attempt to present a simple representation of column behavior.

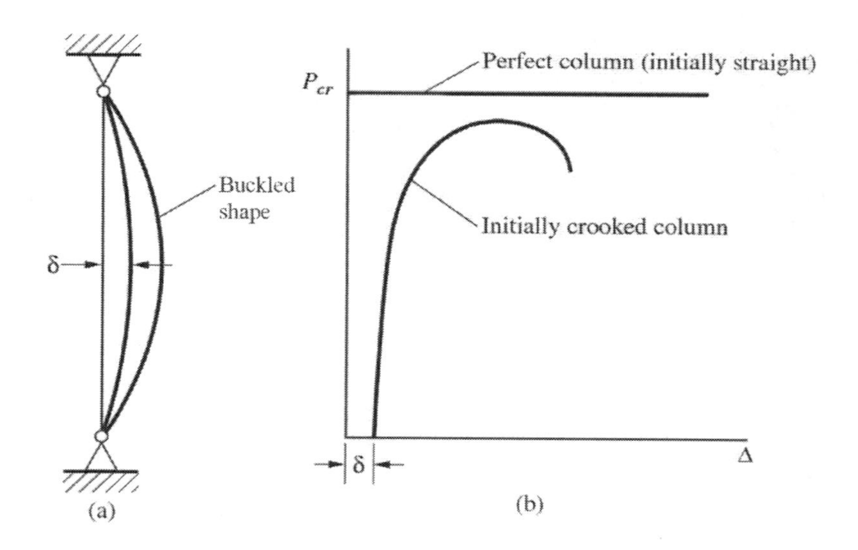

Figure 5.13 Influence of Initial Out-of-Straightness on Column Strength

5.3.5 AISC Provisions

The compression members discussed thus far have either yielding or overall column buckling as the controlling limit state. Figure 5.14 plots sample column test data compared to the Euler equation and the squash load. The Structural Stability Research Council proposed three equations to predict column behavior. To simplify column design, AISC selected a single curve described using two segments as their representation of column strength.

The design basis for ASD and LRFD were presented in Sections 1.9 and 1.10, respectively. The strength equations are repeated here in order to reinforce the relationship between the nominal strength, resistance factor, and safety factor presented throughout the *Specification*.

The requirement for ASD is

$$R_a \leq \frac{R_n}{\Omega} \qquad \text{(AISC B3.2)}$$

The requirement for LRFD is

$$R_u \leq \phi R_n \qquad \text{(AISC B3.1)}$$

As indicated earlier, the *Specification* provides the relationship to determine nominal strength and the corresponding resistance factor and safety factor for each limit state to be considered. The provisions for compression members with nonslender elements, i.e., no local buckling, are given in *Specification* Section E3. The nominal column strength for the limit state of flexural buckling of members with nonslender elements is

$$P_n = F_n A_g \qquad \text{(AISC E3-1)}$$

and

$$\phi_c = 0.9(\text{LRFD}) \qquad \Omega_c = 1.67(\text{ASD})$$

where A_g is the gross area of the section and F_n is the nominal flexural buckling stress. (The Euler column derivation in Section 5.3.1 addressed the limit state of flexural buckling.)

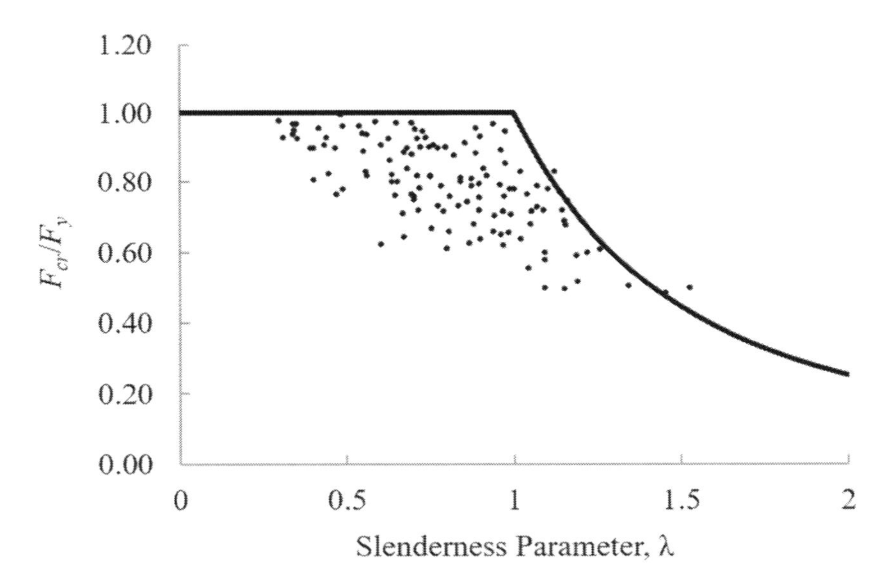

Figure 5.14 Sample Column Test Data Compared to Theoretical Column Strength

The *Specification* defines L_c as the effective length and shows it equal to *KL*. This is the same effective length factor, *K*, discussed earlier. To capture column behavior when inelastic buckling dominates column strength, that is, where residual stresses become important, the *Specification* provides that when

$$L_c / r \leq 4.71\sqrt{E / F_y} \text{ or } \frac{F_y}{F_e} \leq 2.25$$

$$F_n = \left[0.658^{\frac{F_y}{F_e}}\right] F_y \qquad \text{(AISC E3-2)}$$

To capture behavior when inelastic buckling is not a factor and initial crookedness is dominant, that is when

$$L_c / r > 4.71\sqrt{E / F_y} \text{ or } \frac{F_y}{F_e} > 2.25$$

$$F_n = 0.877 F_e \qquad \text{(AISC E3-3)}$$

where F_e is the elastic buckling stress; the Euler buckling stress previously presented as Equation 5.9 and restated here is

$$F_e = \frac{\pi^2 E}{\left(\frac{L_c}{r}\right)^2} \qquad \text{(AISC E3-4)}$$

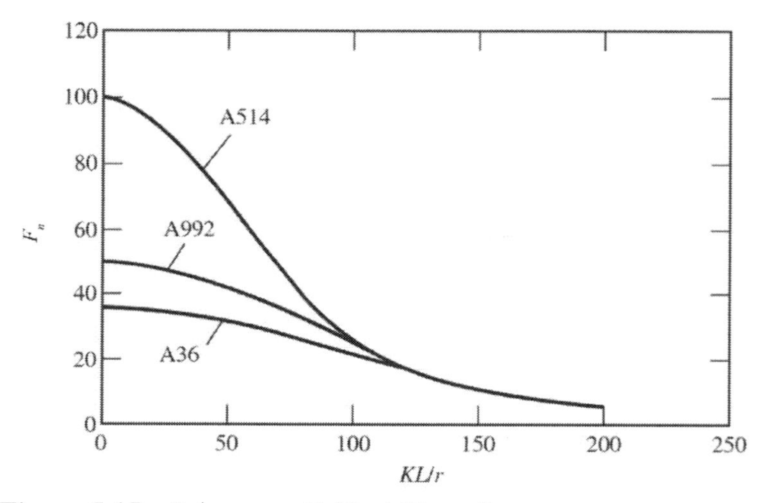

Figure 5.15 L_c/r versus Critical Strength

The nominal flexural buckling stresses for three different steels, A36, A992, and A514, versus the slenderness ratio, L_c/r, are shown in Figure 5.15. For very slender columns, the buckling stress is independent of the material yield. The division between elastic and inelastic behavior, Equations E3-2 and E3-3, corresponds to L_c/r values of 134, 113, and 80.2 for steels with a yield of 36, 50, and 100 ksi, respectively.

Early editions of the LRFD *Specification* defined the exponent of Equation E3-2 in a slightly different form that makes the presentation a bit simpler. If a new term is defined such that

$$\lambda_c^2 = \frac{F_y}{F_e} = \left(\frac{L_c}{\pi r}\right)^2 \frac{F_y}{E}$$

then the dividing point between elastic and inelastic behavior, where

$$\frac{L_c}{r} = 4.71\sqrt{\frac{E}{F_y}}$$

becomes

$$\lambda_c = \frac{L_c}{\pi r}\sqrt{\frac{F_y}{E}} = \frac{4.71}{\pi} = 1.5$$

By substituting $\lambda_c^2 = F_y/F_e$, the nominal flexural buckling stress for $\lambda_c \leq 1.5$ becomes

$$F_n = \left(0.658^{\lambda_c^2}\right)F_y \tag{5.10}$$

and for $\lambda_c > 1.5$,

$$F_n = \frac{0.877}{\lambda_c^2}F_y \tag{5.11}$$

A plot of the ratio of nominal flexural buckling stress to yield stress as a function of the slenderness parameter, λ_c, is given in Figure 5.16. Using this formulation, it is evident that regardless of the steel yield stress, the ratio of nominal flexural buckling stress to yield stress is the same when plotted against the slenderness parameter, λ_c. Table 5.2 provides these numerical values in a convenient, usable form.

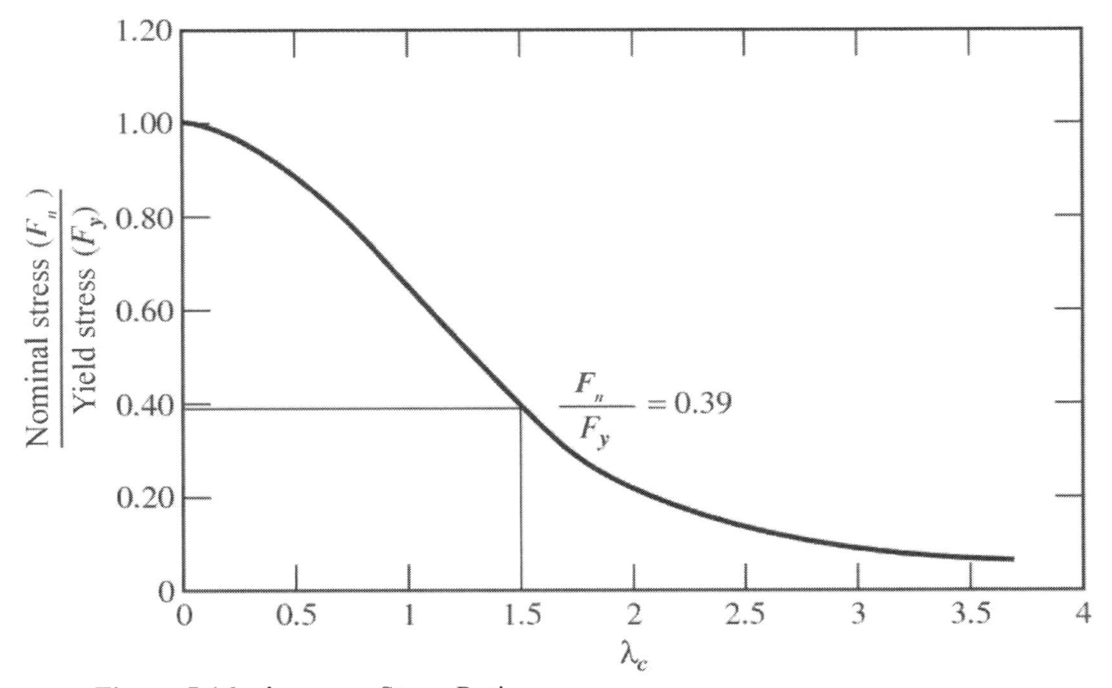

Figure 5.16 λ_c versus Stress Ratio

Earlier editions of the ASD and LRFD *Specifications* indicated that there should be an upper limit on the magnitude of the slenderness ratio at L_c/r = 200. The intent for this limit was to have the engineer recognize that for very slender columns, the nominal flexural buckling stress was so low as to make the column very inefficient. This limit has been removed in recent editions of the *Specification* because there are many factors influencing column strength that indicate that a very slender column might actually be acceptable. Section E2 simply informs the designer, through a User Note, that column slenderness should preferably be kept to something less than 200. It also points to the same recommendation that had been given for tension members, that the slenderness of the member, as fabricated, preferably should not exceed 300. *Manual* Table 4-14 gives the nominal flexural buckling stress for values of slenderness ratio, L_c/r, from 0 to 200 in increments of 1.0 for steels with six different yield stresses.

Table 5.2 Ratio of Nominal Stress to Yield Stress

λ_c	F_n/F_y	λ_c	F_n/F_y	λ_c	F_n/F_y	λ_c	F_n/F_y
0.00	1.000	0.95	0.685	1.90	0.243	2.85	0.108
0.05	0.999	1.00	0.658	1.95	0.231	2.90	0.104
0.10	0.996	1.05	0.630	2.00	0.219	2.95	0.101
0.15	0.991	1.10	0.603	2.05	0.209	3.00	0.0974
0.20	0.983	1.15	0.575	2.10	0.199	3.05	0.0943
0.25	0.974	1.20	0.547	2.15	0.190	3.10	0.0913
0.30	0.963	1.25	0.520	2.20	0.181	3.15	0.0884
0.35	0.950	1.30	0.493	2.25	0.173	3.20	0.0856
0.40	0.935	1.35	0.466	2.30	0.166	3.25	0.0830
0.45	0.919	1.40	0.440	2.35	0.159	3.30	0.0805
0.50	0.901	1.45	0.415	2.40	0.152	3.35	0.0781
0.55	0.881	1.50	0.390	2.45	0.146	3.40	0.0759
0.60	0.860	1.55	0.365	2.50	0.140	3.45	0.0737
0.65	0.838	1.60	0.343	2.55	0.135	3.50	0.0716
0.70	0.815	1.65	0.322	2.60	0.130	3.55	0.0696
0.75	0.790	1.70	0.303	2.65	0.125	3.60	0.0677
0.80	0.765	1.75	0.286	2.70	0.120	3.65	0.0658
0.85	0.739	1.80	0.271	2.75	0.116	3.70	0.0641
0.90	0.712	1.85	0.256	2.80	0.112	3.75	0.0624

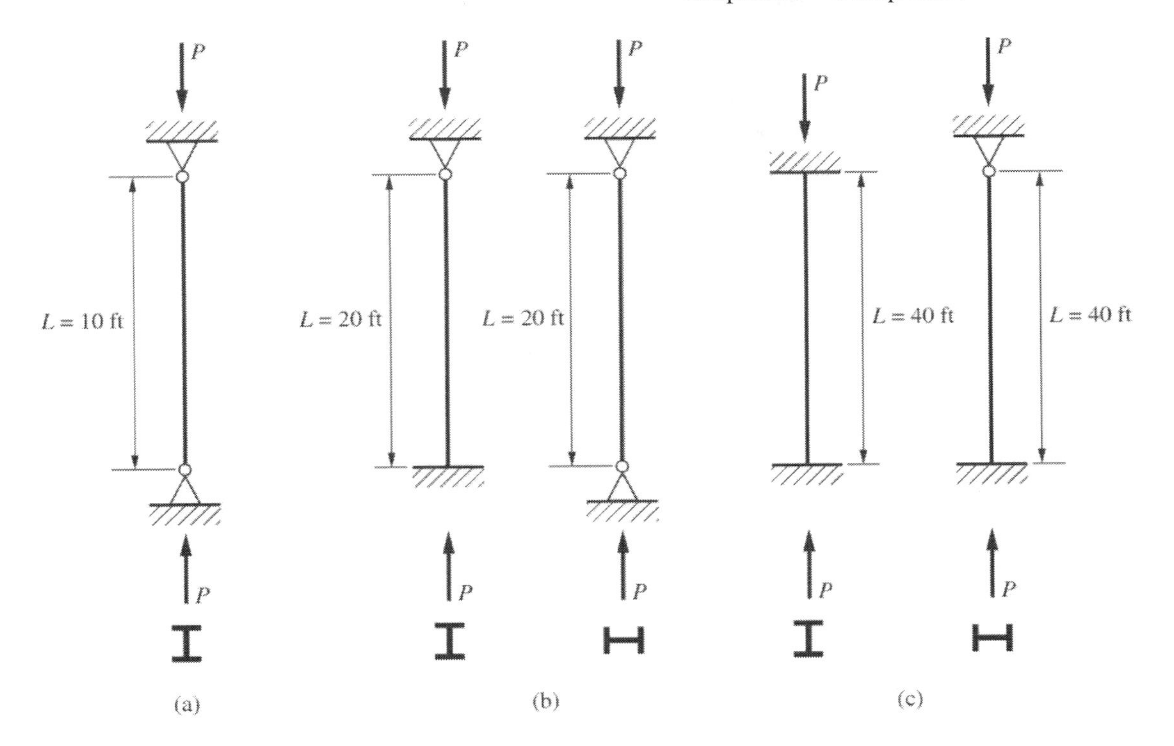

Figure 5.17 Columns for Examples 5.3, 5.4, and 5.5.

EXAMPLE 5.3
Column Strength by AISC Provisions

Goal: Determine the available column strength.

Given: A W12×79 pin-ended column with a length of 10.0 ft. as shown in Figure 5.17a. Use A992 steel.

SOLUTION

Step 1: From *Manual* Table 1-1, $r_x = 5.34$ in., $r_y = 3.05$ in., and $A = 23.2$ in.^{2}.

Step 2: Determine the controlling effective slenderness ratio.

Because the length is 10.0 ft and the column has pinned ends, $L_c = KL = 10.0$ ft for both the x-axis and y-axis. Thus,

$$\frac{L_{cx}}{r_x} = \frac{10.0(12)}{5.34} = 22.5$$

and

$$\frac{L_{cy}}{r_y} = \frac{10.0(12)}{3.05} = 39.3$$

Since

$$\frac{L_{cy}}{r_y} > \frac{L_{cx}}{r_x}$$

the y-axis controls.

Step 3: Determine which column strength equation to use. Since

$$\frac{L_c}{r} = 39.3 < 4.71\sqrt{\frac{E}{F_y}} = 4.71\sqrt{\frac{29,000}{50}} = 113$$

use Equation E3-2

Step 4: Determine the Euler buckling stress.

$$F_e = \frac{\pi^2(29,000)}{(39.3)^2} = 185 \text{ ksi}$$

Step 5: Determine the nominal stress from Equation E3-2.

$$F_n = 0.658^{\left(\frac{F_y}{F_e}\right)}F_y = 0.658^{\left(\frac{50}{185}\right)}(50) = 44.7 \text{ ksi}$$

Step 6: Determine the nominal strength.

$$P_n = 44.7(23.2) = 1040 \text{ kips}$$

For LRFD Step 7: Determine the design strength for LRFD.

$$\phi P_n = 0.9(1040) = 936 \text{ kips}$$

For ASD Step 7: Determine the allowable strength for ASD.

$$\frac{P_n}{\Omega} = \frac{1040}{1.67} = 623 \text{ kips}$$

EXAMPLE 5.4 ***Column Strength by AISC Provisions***

Goal: Determine the available column strength.

Given: A W10×49 column with a length of 20.0 ft, one end pinned and the other end fixed for the y-axis, and both ends pinned for the x-axis, as shown in Figure 5.17b. Use A992 steel.

SOLUTION

Step 1: From *Manual* Table 1-1, $r_x = 4.35$ in., $r_y = 2.54$ in., and $A = 14.4$ in.^{2}.

Step 2: Determine the effective length factors from Figure 5.7.

Comparing the columns shown in Figure 5.17b with those shown in Figure 5.7, the effective length factors are $K_y = 0.7$ and $K_x = 1.0$.

Step 3: Determine the x- and y-axis slenderness ratios.

$$\frac{L_{cx}}{r_x} = \frac{K_x L}{r_x} = \frac{1.0(20.0)(12)}{4.35} = 55.2$$

$$\frac{L_{cy}}{r_y} = \frac{K_y L}{r_y} = \frac{0.7(20.0)(12)}{2.54} = 66.1$$

Step 4: Using the larger slenderness ratio, determine which column strength equation to use. Since

$$\frac{L_c}{r} = 66.1 < 4.71\sqrt{\frac{E}{F_y}} = 4.71\sqrt{\frac{29{,}000}{50}} = 113,$$

use Equation E3-2

Step 5: Determine the Euler buckling stress.

$$F_e = \frac{\pi^2(29{,}000)}{(66.1)^2} = 65.5 \text{ ksi}$$

Step 6: Determine the nominal stress from Equation E3-2.

$$F_n = 0.658^{\left(\frac{F_y}{F_e}\right)} F_y = 0.658^{\left(\frac{50}{65.5}\right)}(50) = 36.3 \text{ ksi}$$

Step 7: Determine the nominal strength.

$$P_n = 36.3(14.4) = 523 \text{ kips}$$

For LRFD Step 8: Determine the design strength for LRFD.

$$\phi P_n = 0.9(523) = 471 \text{ kips}$$

For ASD Step 8: Determine the allowable strength for ASD.

$$P_n/\Omega = 523/1.67 = 313 \text{ kips}$$

EXAMPLE 5.5
Column Strength by AISC Provisions

Goal: Determine the available column strength.

Given: A W14×53 column with a length of 40.0 ft, both ends fixed for the y-axis, and one end pinned and one end fixed for the x-axis, as shown in Figure 5.17c. Use A992 steel.

SOLUTION

Step 1: From *Manual* Table 1-1, $r_x = 5.89$ in., $r_y = 1.92$ in., and $A = 15.6$ in.^{2}.

Step 2: Determine the effective length factors from Figure 5.7.

Comparing the columns shown in Figure 5.17c with those shown in Figure 5.7, the effective length factors are $K_y = 0.5$ and $K_x = 0.7$.

Step 3: Determine the x- and y-axis slenderness ratios.

$$\frac{L_{cx}}{r_x} = \frac{K_x L}{r_x} = \frac{0.7(40.0)(12)}{5.89} = 57.0$$

$$\frac{L_{cy}}{r_y} = \frac{K_y L}{r_y} = \frac{0.5(40.0)(12)}{1.92} = 125$$

Step 4: Using the larger slenderness ratio, determine which column strength equation to use. Since

$$\frac{L_c}{r} = 125 > 4.71\sqrt{\frac{E}{F_y}} = 4.71\sqrt{\frac{29{,}000}{50}} = 113$$

use Equation E3-3

Step 5: Determine the Euler buckling stress.

$$F_e = \frac{\pi^2(29{,}000)}{(125)^2} = 18.3 \text{ ksi}$$

Step 6: Determine the nominal stress from Equation E3-3.

$$F_n = 0.877F_e = 0.877(18.3) = 16.0 \text{ ksi}$$

Step 7: Determine the nominal strength.

$$P_n = 16.0(15.6) = 250 \text{ kips}$$

For LRFD Step 8: Determine the design strength for LRFD.

$$\phi P_n = 0.9(250) = 225 \text{ kips}$$

For ASD
Step 8 Determine the allowable strength for ASD.

$$P_n/\Omega = 250/1.67 = 150 \text{ kips}$$

5.4 ADDITIONAL LIMIT STATES FOR COMPRESSION

Two limit states for compression members were discussed in Section 5.3, yielding and flexural buckling. The strength equations provided in *Specification* Section E3 clearly show that the upper limit for column strength, F_yA_g, is reached only for the zero-length column. Thus, the provisions are presented in the *Specification* as applying to the limit state of flexural buckling only, even though they do consider yielding.

Singly symmetric, unsymmetric, and certain doubly symmetric members may also be limited by torsional buckling or flexural-torsional buckling. The strength provisions for these limit states are given in Section E4 of the *Specification* and are discussed here in Section 5.8.

For some column profiles, another limit state may actually control overall column strength. The individual elements of a column cross section may buckle locally at a stress below the stress that would cause the overall column to buckle. If this is the case, the column is said to be a *column with slender elements*. The impact of these slender elements on column strength is determined through the use of an effective area which is smaller than the actual area of the member. The additional provisions for these types of members are presented in Section 5.6.

5.5 LENGTH EFFECTS

The effective lengths that have been discussed were all related to fairly simple columns with easily defined end conditions and bracing locations. Once a column is recognized as being a part of a real structure, determining the effective length becomes more involved. Moreover, for more complex structures, it might be simpler to determine the buckling strength of the structure through analysis than through the use of the effective length factor, K. Using that analysis, the elastic buckling stress of the individual columns, F_e, can be determined. This can then be used directly in the column strength equations. However, for this book, column elastic buckling is determined through a calculation of effective length. This approach may incorporate some simplifications that would not be made in an actual buckling analysis and, depending on the approach used to determine K, may include assumptions of behavior that the actual structure may not satisfy.

A first attempt at incorporating some realistic aspects of structures is shown in Table C-A-7.1 of the Commentary and here in Figure 5.18. The columns shown in this figure are the same as those shown in Figure 5.7, and the same K-factors are shown and identified here as the theoretical K-values. What is new here is the presentation of recommended design values when ideal conditions are approximated. Most of these recommended values are based on the fact that perfectly rigid connections are difficult to obtain. Thus, for example, a fixed-end column (case a of Figure 5.18) would have a theoretical $K = 0.5$, but if the end connections were to actually rotate, even just a small amount, the effective length would increase. As the end rotation increases toward what would occur for a pin-ended column, K would approach 1.0. Thus, the recommended value of K is 0.65. A similar assessment of the other cases with a fixed end should lead to an understanding of the idea behind these recommended values, each being a bit higher than the theoretical value because actual column end conditions are unlikely to match the theoretical assumptions. In addition to the recommended values for K, this Commentary table shows the column end conditions differently than they have been shown historically and in this book. For instance, the upper support for Figure 5.18 column (c) is shown here with rollers so that it is clear that the support can transfer the load directly to the column. In Figure 5.7 column (a) this is shown with the same symbol as the lower support, with no attempt to show graphically that the load is transferred to the column. Although the distinction is critical in structural analysis, it has always been assumed for individual columns that the graphical distinction was not necessary

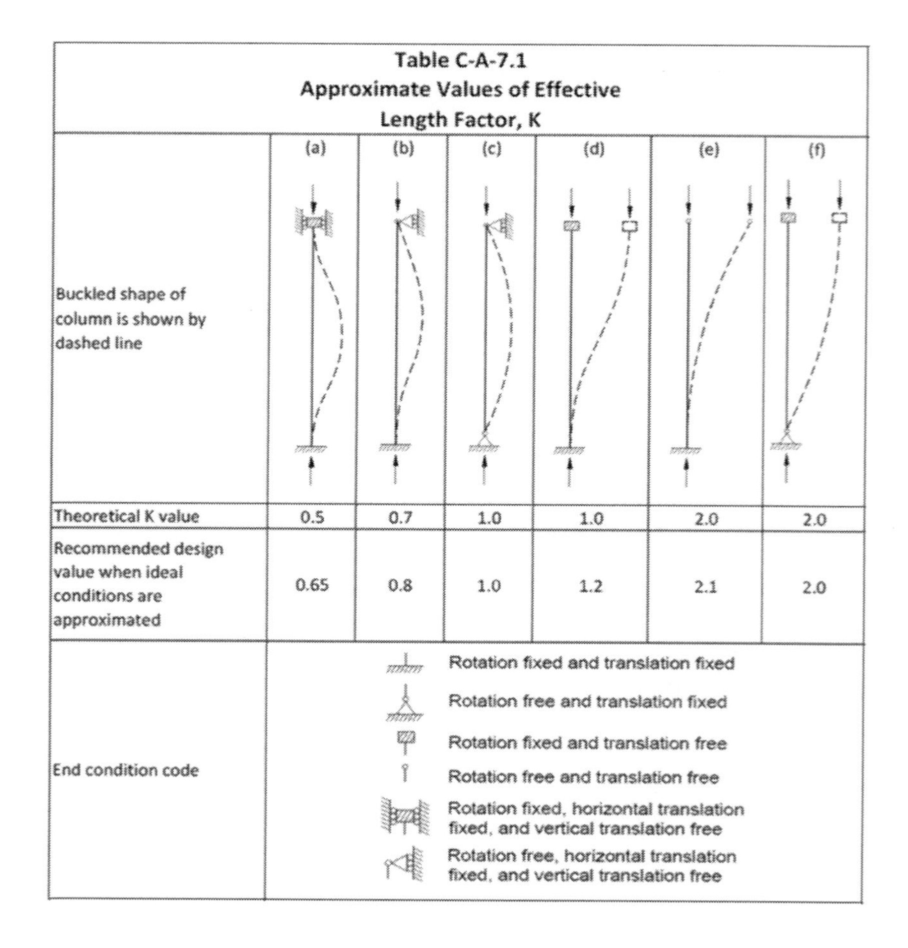

Table C-A-7.1
Approximate Values of Effective Length Factor, K

	(a)	(b)	(c)	(d)	(e)	(f)
Buckled shape of column is shown by dashed line						
Theoretical K value	0.5	0.7	1.0	1.0	2.0	2.0
Recommended design value when ideal conditions are approximated	0.65	0.8	1.0	1.2	2.1	2.0

End condition code:

- Rotation fixed and translation fixed
- Rotation free and translation fixed
- Rotation fixed and translation free
- Rotation free and translation free
- Rotation fixed, horizontal translation fixed, and vertical translation free
- Rotation free, horizontal translation fixed, and vertical translation free

Figure 5.18 Values of Effective Length Factor, K

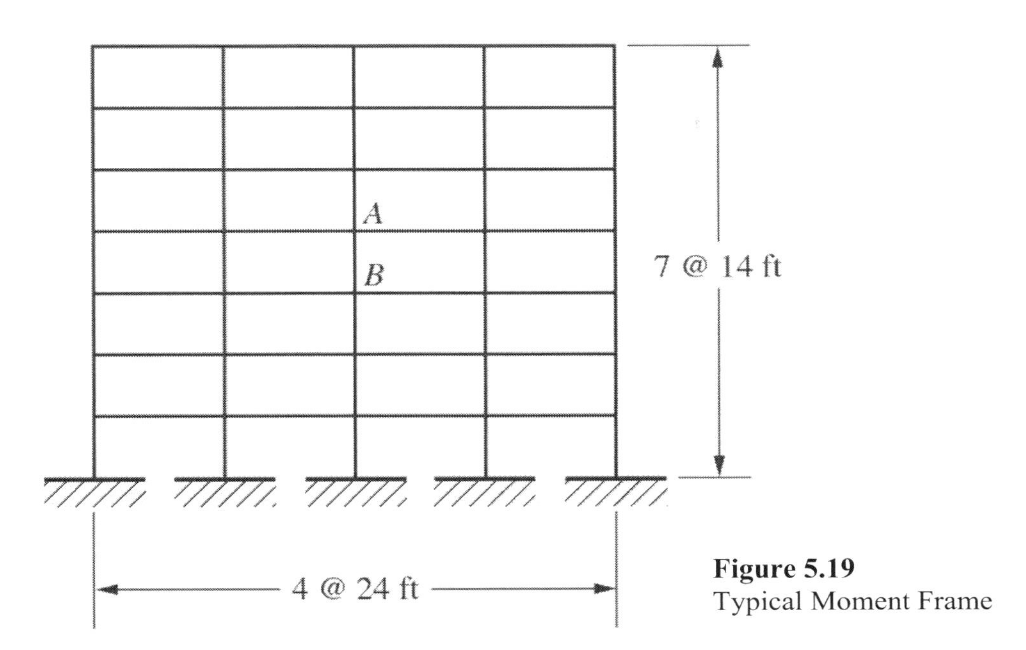

Figure 5.19
Typical Moment Frame

When a column is part of a frame, as shown in Figure 5.19, the stiffness of the members framing into the column impact the rotation that could occur at the column ends.

As with the rigid supports discussed for the columns in Figure 5.18, these end conditions permit the column end to rotate. The amount of this rotation is something between the zero rotation of a fixed support and the free rotation of a pin support. When the column under consideration is part of a frame where the ends of the column are not permitted to displace laterally relative to each other, the frame is called a *braced frame*, a *sidesway prevented frame*, or a *sidesway inhibited frame*—shown as cases a, b, and c in Figure 5.18. For a column in a braced frame, the possible K-factors range from 0.5 to 1.0. In frames of this type, K is often taken as 1.0, a conservative approximation that simplifies design. In fact, *Specification* Appendix 7, Section 7.2.3(a) says that in braced frames K shall be taken as 1.0 unless analysis shows that a lower value is appropriate.

When the column under consideration is in a frame in which the ends are permitted to move laterally, the frame is called a *moment frame*, an *unbraced frame*, a *sidesway permitted frame*, or a *sidesway uninhibited frame*— shown as cases d, e, and f in Figure 5.18. For the three cases shown there, the lowest value of K is 1.0. The other extreme case, not shown in Figure 5.18, is a pin-ended column in an unbraced frame. The effective length of this column would theoretically be infinite. Thus, K-values for columns in moment frames range from 1.0 to infinity.

The determination of reliable effective length factors and thus reliable effective lengths is a critical aspect of column design. Several approaches are presented in the literature, but the most commonly used approach is through the alignment charts presented in the Commentary to Appendix 7. The development of these charts is based on a set of assumptions that are often violated in real structures; nevertheless, the alignment charts are used extensively and often modified in an attempt to account for variations from these assumptions.

These assumptions, as given in the Commentary to Appendix 7, are:

1. Behavior is purely elastic.
2. All members have a constant cross section.
3. All joints are rigid.
4. For columns in frames with sidesway inhibited, rotations at opposite ends of the restraining beams are equal in magnitude and opposite in direction, producing single curvature bending.
5. For columns in frames with sidesway uninhibited, rotations at opposite ends of the restraining beams are equal in magnitude and direction, producing reverse curvature bending.
6. The stiffness parameter $L\sqrt{P/EI}$ of all columns is equal.
7. Joint restraint is distributed to the column above and below the joint in proportion to EI/L for the two columns.
8. All columns buckle simultaneously.
9. No significant axial compression force exists in the girders.
10. Shear deformations are neglected

Using these assumptions, the following equation can be obtained for columns in sidesway inhibited frames.

$$\frac{G_A G_B}{4}(\pi/K)^2+\left(\frac{G_A+G_B}{2}\right)\left(1-\frac{\pi/K}{\tan(\pi/K)}\right)+\frac{2\tan(\pi/2K)}{(\pi/K)}-1=0 \qquad \text{(AISC C-A-7-1)}$$

For sidesway uninhibited frames, the following equation is obtained.

$$\frac{G_A G_B(\pi/K)^2-36}{6(G_A+G_B)}-\frac{(\pi/K)}{\tan(\pi/K)}=0 \qquad \text{(AISC C-A-7-2)}$$

In Equations C-A-7-1 and C-A-7-2, the terms G_A and G_B relate to the relative stiffness of the columns and beams framing into the column at ends A and B, respectively, as given by

$$G=\frac{\Sigma(EI/L)_{col}}{\Sigma(EI/L)_g} \qquad \text{(AISC C-A-7-3)}$$

If the beams and columns behave elastically, as noted in assumption 1, this reduces to

$$G=\frac{\Sigma(I/L)_{col}}{\Sigma(I/L)_g} \qquad (5.12)$$

Equations C-A-7-1 and C-A-7-2 are transcendental equations that do not have a closed-form solution. With the computer methods readily available today, iterative solutions are easily obtained. However, that was not always the case, and a graphical solution was developed in the early 1960s that has become a standard approach for obtaining solutions. Such graphical solutions are called *nomographs* or *alignment charts*. Figure 5.20 shows the nomograph for sidesway inhibited frames, and Figure 5.21 gives the chart for sidesway uninhibited frames.

Since these alignment charts are based on the assumptions given previously, Section 7.2 Commentary provides several adjustments that may be made to model the actual structure more accurately. One of those adjustments is to account for column end conditions. A column end simply supported on a footing would have G theoretically equal to infinity. But, unless the connection is designed and constructed as a true pin, it is more reasonable to take $G = 10$ for practical design. Similarly, for a column end rigidly attached to a properly designed footing, G would theoretically be zero, but it is reasonable to take G = 1.0 which would account for a small amount of potential rotation.

Approximate solutions to Equations C-A-7-1 and C-A-7-2 have also been presented in design rules and the literature. The French have used the following equations in their design rules since 1966[1]. For sidesway inhibited,

$$K = \frac{3G_A G_B + 1.4(G_A + G_B) + 0.64}{3G_A G_B + 2(G_A + G_B) + 1.28} \tag{5.13}$$

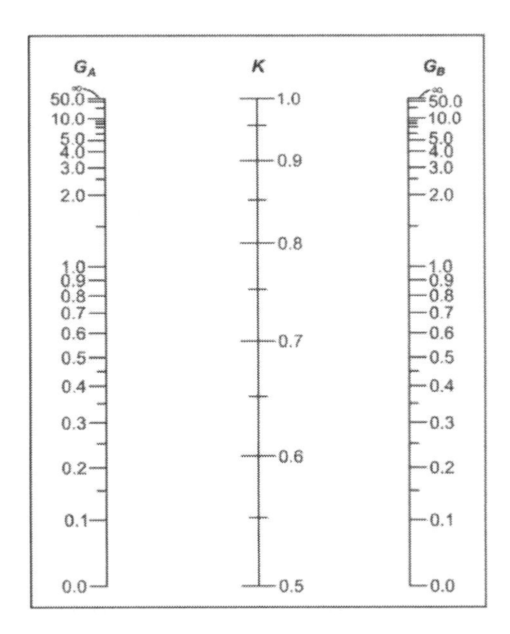

Figure 5.20 Alignment Chart for a Braced Frame (Sidesway Inhibited)

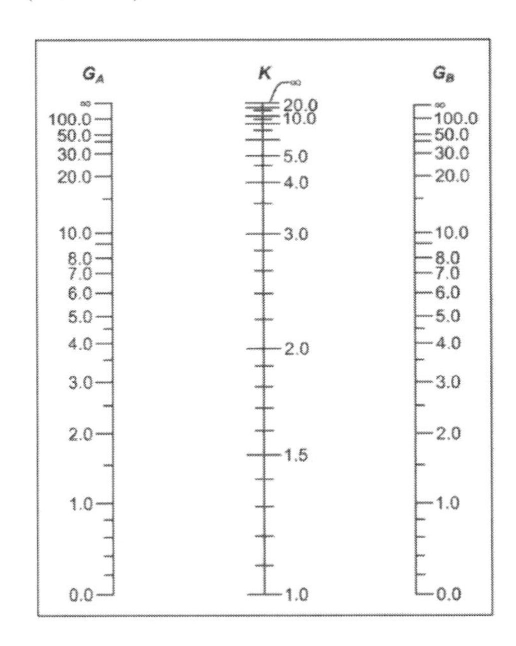

Figure 5.21 Alignment Chart for an Unbraced Frame (Sidesway Uninhibited)

---

[1] Dumonteil, P. (1992), "Simple Equations for Effective Length Factors," *Engineering Journal*, American Institute of Streel Construction, Vol. 29, No. 3, pp. 111-115.

For sidesway uninhibited,

$$K = \sqrt{\frac{1.6G_A G_B + 4(G_A + G_B) + 7.5}{G_A + G_B + 7.5}} \tag{5.14}$$

These approximate equations are said to be accurate within 2 percent. For design this should easily yield results as accurate as those obtained by reading a value from the alignment charts.

For the special case where $G_A = G_B$, even simpler equations can be expressed. For sidesway inhibited,

$$K = \frac{G + 0.4}{G + 0.8} \tag{5.15}$$

For sidesway uninhibited,

$$K = \sqrt{0.8G + 1.0} \tag{5.16}$$

Equations 5.15 and 5.16 might be particularly useful for preliminary design.

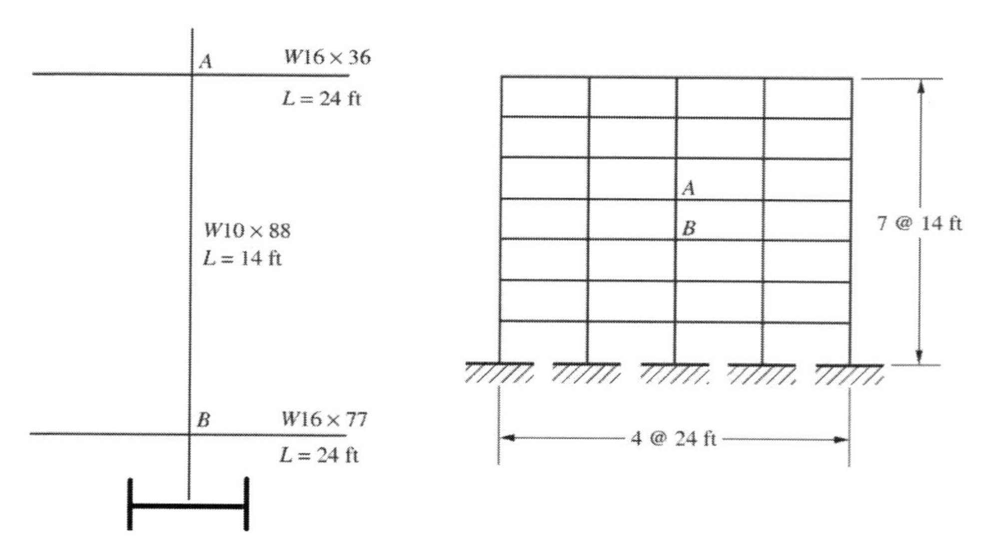

Figure 5.22 Multi-story Frame for Examples 5.6 and 5.7

EXAMPLE 5.6
Column Effective Length

Goal: Determine the column effective length using (a) the alignment chart and (b) Equation 5.14.

Given: The column AB in a moment frame is shown in Figure 5.22. Assume that the column has its web in the plane of the frame. The beams also have their webs in the plane of the frame and thus beams and columns are bending about their major, x-axis. It would be very unusual for a beam in a moment frame to have the primary bending moments about other than the x-axis. However, columns may be oriented for bending about either principal axis.

SOLUTION

Part a

Step 1: Determine member properties from *Manual* Table 1-1.

end A:

$$\text{W16×36;} \quad I_{gx} = 448 \text{ in.}^4$$
$$\text{W10×88;} \quad I_{cx} = 534 \text{ in.}^4$$

end B:

$$\text{W16×77;} \quad I_{gx} = 1110 \text{ in.}^4$$
$$\text{W10×88;} \quad I_{cx} = 534 \text{ in.}^4$$

Step 2: Determine the stiffness ratio at each end using Equation 5.12

$$G_A = \frac{2\left(\dfrac{534}{14}\right)}{2\left(\dfrac{448}{24}\right)} = 2.04$$

$$G_B = \frac{2\left(\dfrac{534}{14}\right)}{2\left(\dfrac{1110}{24}\right)} = 0.825$$

Step 3: Use the alignment chart shown in Figure 5.21 for a sidesway uninhibited frame. Enter G_A and G_B on the appropriate scales and construct a straight line between them, as shown in Figure 5.23. The intersection with the scale for K gives the effective length factor, in this case,

$$K = 1.42$$

Thus,

$$L_c = KL = 1.42(14.0) = 19.9 \text{ ft}$$

Part b

Step 4: Determine K using the stiffness ratios, G_A and G_B, determined in part (a) Step 2 and Equation 5.14.

$$K = \sqrt{\frac{1.6(2.04)(0.825) + 4(2.04 + 0.825) + 7.5}{2.04 + 0.825 + 7.5}} = 1.45$$

Thus,

$$L_c = KL = 1.45(14.0) = 20.3 \text{ ft}$$

Note that K determined graphically from the alignment chart and K calculated with Equation 5.14 are very close, as might be expected.

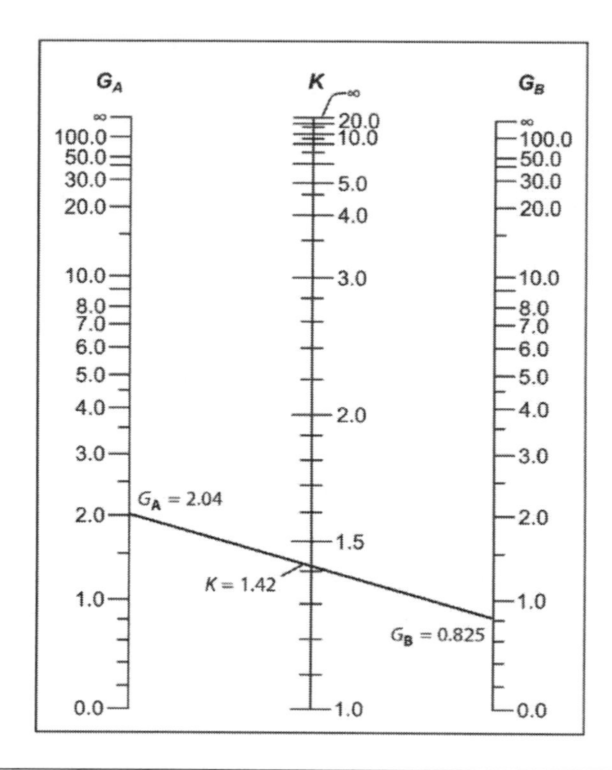

Figure 5.23 Alignment Chart for Example 5.6

EXAMPLE 5.7
Column Effective Length

Goal: Determine the column effective length for the column of Example 5.6 using the alignment chart if the column is bending about its weak axis.

Given: The column *AB* in a moment frame is shown in Figure 5.22. However, for this example assume that the column has its web perpendicular to the plane of the frame, thus it is bending about its minor or weak axis.

SOLUTION

Step 1: Find member properties from *Manual* Table 1-1.

end A:

$$W16\times36;\quad I_{gx} = 448 \text{ in.}^4$$
$$W10\times88;\quad I_{cy} = 179 \text{ in.}^4$$

end B:

$$W16\times77;\quad I_{gx} = 1110 \text{ in.}^4$$
$$W10\times88;\quad I_{cy} = 179 \text{ in.}^4$$

Step 2: Determine the stiffness ratio at each end using Equation 5.12.

$$G_A = \frac{2\left(\dfrac{179}{14}\right)}{2\left(\dfrac{448}{24}\right)} = 0.685$$

$$G_B = \frac{2\left(\frac{179}{14}\right)}{2\left(\frac{1110}{24}\right)} = 0.276$$

Step 3: Use the alignment chart shown in Figure 5.21 for a sidesway uninhibited frame. Enter the values for G_A and G_B on the appropriate scales and draw a straight line between them. The line's intersection with the scale for K gives the effective length factor—in this case,

$$K = 1.16$$

Thus,

$$L_c = KL = 1.16(14.0) = 16.2 \text{ ft}$$

Step 4: Note that the reduction in moment of inertia of the columns results in the beams providing more end restraint, reducing the effective length factor for the column and thus reducing the column effective length.

5.5.1 Effective Length for Inelastic Columns

The assumption of elastic behavior for all members of a frame is regularly violated. We have already seen the role that residual stresses play in determining column strength through inelastic behavior. Thus, it is useful to accommodate this inelastic behavior in the determination of K-factors. The assumption of elastic behavior is important in the calculation of G as the simplification is made to move from Equation C-A-7-3 to Equation 5.12. Returning to Equation C-A-7-3 and assuming that all columns framing into a joint have the same modulus of elasticity—which is equal to the tangent modulus, E_T shown in Figure 5.12—and that the beams behave elastically, the definition of G for inelastic behavior becomes

$$G_{inelastic} = \frac{E_T(\Sigma(I/L)_c)}{E(\Sigma(I/L)_g)} \tag{5.17}$$

If G for elastic behavior is taken as $G_{elastic}$, then $G_{inelastic}$ can be formulated as

$$G_{inelastic} = \left(\frac{E_T}{E}\right) G_{elastic} \tag{5.18}$$

Thus, including inelastic column behavior simply results in a modification of G. The ratio of tangent modulus to elastic modulus is always less than 1, so the assumption of elastic behavior for this application leads to a conservative estimate, as can be seen by entering the nomograph with lower G-values and determining the corresponding K-factor. Before a straightforward approach to including inelastic effects in the determination of effective length can be proposed, the relationship between the tangent modulus and the elastic modulus must be established.

The Commentary to Appendix 7 of the *Specification* indicates that $\tau_b = E_T/E$, as given in Chapter C for the direct analysis method, should be used to account for column inelasticity in the effective length method. Thus, if $\alpha P_r/P_{ns} \leq 0.5$

$$\tau_b = 1.0 \qquad \text{(AISC C2-2a)}$$

and if $\alpha P_r/P_{ns} > 0.5$

$$\tau_b = 4\left(\frac{\alpha P_r}{P_{ns}}\right)\left[1-\left(\frac{\alpha P_r}{P_{ns}}\right)\right] \qquad \text{(AISC C2-2b)}$$

where α = 1.0 for LRFD and α = 1.6 for ASD. P_{ns} is the cross section compression strength. For members without slender elements $P_{ns} = P_y$. For compression members with slender elements, $P_{ns} = F_yA_e$ which is addressed in Section 5.6. *Manual* Table 4-13 provides values for τ_b based on the required strength, P_r/A_g. The use of Table 4-13 assumes that the column is loaded to its full available strength. If it is not, the table provides a conservative assessment of the inelastic stiffness reduction factor and the effective length.

EXAMPLE 5.8
Inelastic Column Effective Length

Goal: Determine the inelastic column effective length using the alignment chart.

Given: Determine the inelastic effective length for the column in Example 5.6. The column has an LRFD required strength of P_u = 950 kips and an ASD required strength of P_a = 633 kips. Use Equation 5.14 in place of the alignment chart. The column is A992 steel.

SOLUTION

Step 1: From *Manual* Table 1-1, for a W10×88 A = 26.0 in.^{2}, and from Example 5.6, the elastic stiffness ratios are G_A = 2.04 and G_B = 0.825.

For LRFD

Step 2: Determine the required stress based on the required strength.

$$\frac{P_r}{A} = \frac{P_u}{A} = \frac{950}{26.0} = 36.5 \text{ ksi}$$

Step 3: Determine the stiffness reduction factor from *Manual* Table 4-13, interpolating between 36 and 37 ksi.

$$\tau_b = 0.788$$

Step 4: Determine the inelastic stiffness ratios by multiplying the elastic stiffness ratios by the stiffness reduction factor.

$$G_{iA} = 0.788(2.04) = 1.61$$

$$G_{iB} = 0.788(0.825) = 0.650$$

Step 5: Determine K from Equation 5.14.

$$K = \sqrt{\frac{1.6(1.61)(0.650) + 4(1.61 + 0.650) + 7.5}{1.61 + 0.650 + 7.5}} = 1.37$$

Thus,

$$L_c = KL = 1.37(14.0) = 19.2 \text{ ft}$$

Note that the effective length factor and thus the effective length is less than that determined in Example 5.6, as expected.

For ASD

Step 2: Determine the required stress based on the required strength.

$$\frac{P_r}{A} = \frac{P_a}{A} = \frac{633}{26.0} = 24.3 \text{ ksi}$$

Step 3: Determine the stiffness reduction factor from *Manual* Table 4-13, interpolating between 24 and 25 ksi.

$$\tau_b = 0.691$$

Step 4: Determine the inelastic stiffness ratios by multiplying the elastic stiffness ratios by the stiffness reduction factor.

$$G_{iA} = 0.691(2.04) = 1.41$$

$$G_{iB} = 0.691(0.825) = 0.570$$

Step 5: Determine K from Equation 5.14.

$$K = \sqrt{\frac{1.6(1.41)(0.570) + 4(1.41 + 0.570) + 7.5}{1.41 + 0.570 + 7.5}} = 1.33$$

Thus,

$$L_c = KL = 1.33(14.0) = 18.6 \text{ ft}$$

Note that the effective length factor and thus the effective length is less than that determined in Example 5.6, as expected.

5.5.2 Effective Length when Supporting Gravity Only Columns

Another condition that influences column buckling and thus the effective length factor is the existence of columns that carry only gravity load and contribute nothing to the lateral load resistance or stability of the structure. Figure 5.24a illustrates a simple structure of this type where the stability or lateral load resisting column is the flagpole column on the left, column A, and the gravity only column is the pin ended column on the right, column B. The load P is applied to column A and the load Q is applied to column B.

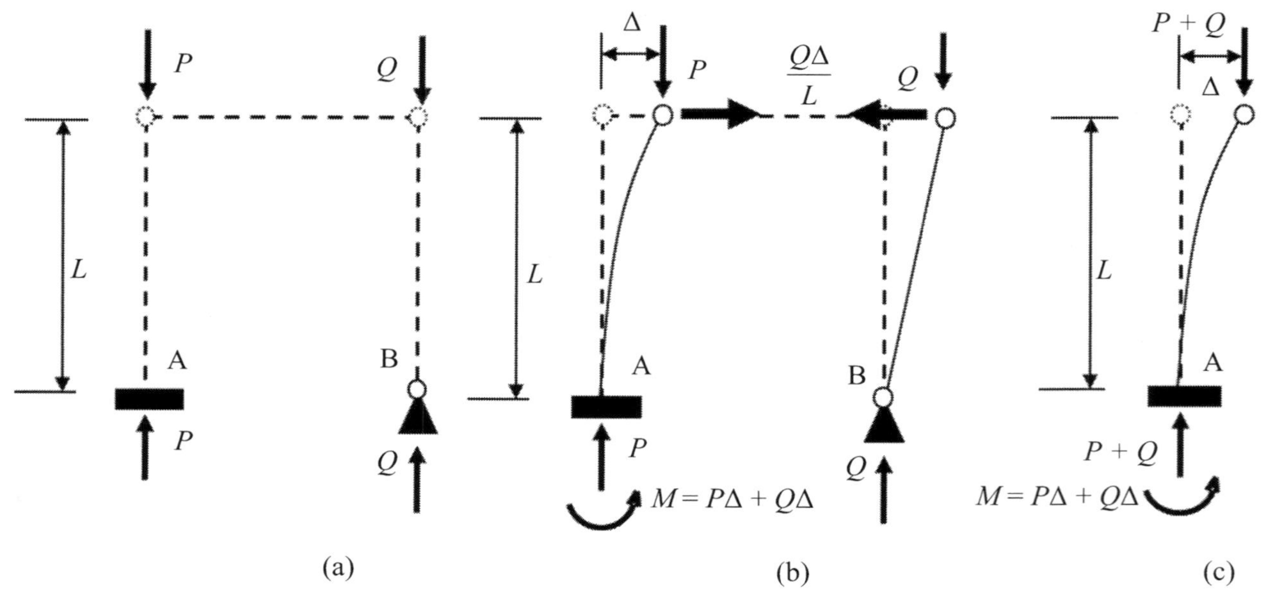

Figure 5.24 Flagpole Column Providing Lateral Restraint for a Gravity Only Column

If $Q = 0$, column A behaves as if column B did not exist since it just goes along for the ride. However, when Q is not zero, buckling of column A leads to lateral displacement, Δ, at the top of columns A and B. Thus, equilibrium requires a lateral force be exerted at the top of column B. This force must be resisted by column A as shown in Figure 5.24b. In the displaced position illustrated in Figure 5.24b, equilibrium of column A requires a resisting moment at the support of $M = P\Delta + Q\Delta$. Figure 5.24c shows column A with an applied load, $(P + Q)$, which in the displaced position produces a moment at the support of $M = P\Delta + Q\Delta$. Thus, the column in Figure 5.24c can be thought of as a representation of column A in Figure 5.24b with only slight error.

Since these two columns are considered equal, if column A can support the load $(P + Q)$, it should be adequate for column A to support its load, P, and the effect of the load Q on column B. Considering elastic buckling this can be stated as

$$(P + Q) = \frac{\pi^2 EI}{(K_o L)^2} \tag{5.19}$$

Where K_o is the K-factor for column A. For this example, the theoretical K-factor for the flagpole column is $K_o = 2$.

Another way to approach this problem would be to continue to consider that column A supports only the load P but use a K-factor that accounts for the influence of the load on the gravity only column, K_n. This can be stated as

$$P = \frac{\pi^2 EI}{\left(K_n L\right)^2} \tag{5.20}$$

Since these two equations represent the same structure, they can be solved for $\pi^2 EI/L^2$. Thus, from Equation 5.19 $\pi^2 EI/L^2 = K_o^2\left(P+Q\right)$ and from Equation 5.20 $\pi^2 EI/L^2 = K_n^2 P$. Setting these equal and solving for K_n yields

$$K_n = K_o\sqrt{\frac{P+Q}{P}} = K_o\sqrt{1+\frac{Q}{P}} \tag{5.21}$$

Thus, a column that supports a load P and also must provide stability for load Q on gravity only columns may be designed using this modified effective length factor K_n.

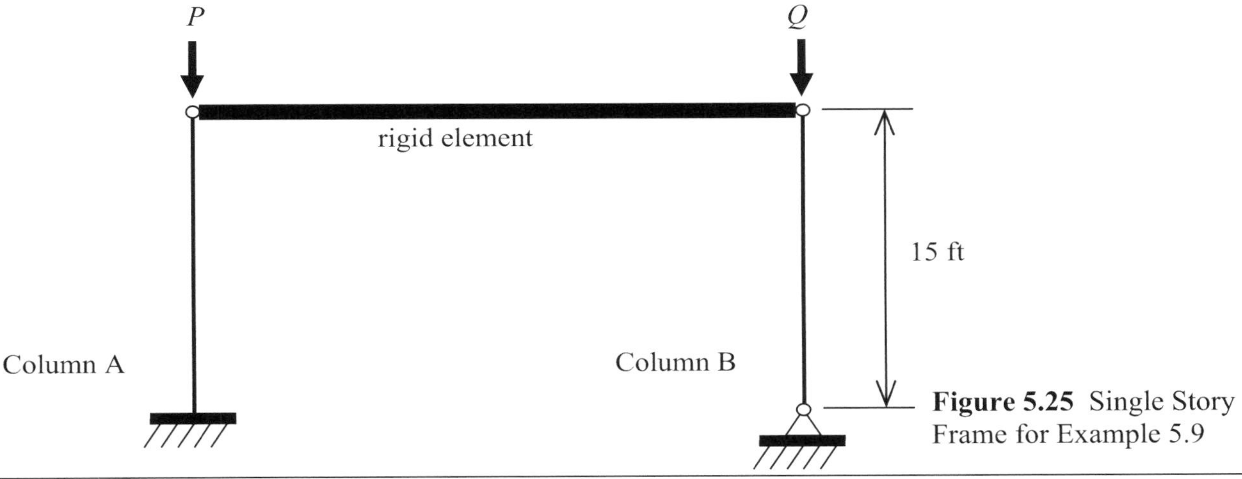

Figure 5.25 Single Story Frame for Example 5.9

EXAMPLE 5.9 ***Gravity Only Columns and Effective Length***

Goal: Determine the in-plane nominal strength of the column that is required to carry a concentrated load and provide lateral stability for gravity only columns. Also, determine the strength of the column if there is no load on the gravity only column.

Given: A W14×90 column shown in Figure 5.25 is to a) carry an applied load, P, and provide lateral restraint to a gravity only column carrying the load $2P$ and b) carry an applied load, P, with no load on the gravity only column. The W14×90 is oriented so the web is in the plane of the frame. Use A992 steel.

SOLUTION

Step 1: Determine the effective length factor for the W14×90 column without considering the gravity only column. Since this is a flagpole column, from Figure 5.7f, the theoretical K-factor is 2.0

Part a

Step 2: Using Equation 5.21, determine the modified effective length factor to account for the gravity only column load, $Q = 2P$. Thus

$$K_n = K_o\sqrt{1+\frac{Q}{P}} = 2.0\sqrt{1+\frac{2P}{P}} = 3.46$$

Step 3: From *Manual* Table 1-1

$$A = 26.5 \text{ in.}^2 \text{ and } r_x = 6.14$$

Step 4: Determine which column strength equation to use.
Since

$$\frac{K_n L}{r_x} = \frac{3.46(15(12))}{6.14} = 101 < 4.71\sqrt{\frac{29{,}000}{50}} = 113$$

use Equation E3-2

Step 5: Determine the Euler buckling stress

$$F_e = \frac{\pi^2(29{,}000)}{(101)^2} = 28.1 \text{ ksi}$$

Step 6: Determine the nominal stress from Equation E3-2

$$F_n = \left(0.658^{\left(\frac{F_y}{F_e}\right)}\right)F_y = \left(0.658^{\left(\frac{50}{28.1}\right)}\right)50 = 23.7 \text{ ksi}$$

Step 7: Determine the nominal strength

$$P_n = 23.7(26.5) = 628 \text{ kips}$$

Part b

Step 8: With no load on the gravity only column, $K_n = K_o = 2.0$. Determine which column strength equation to use.
Since

$$\frac{K_n L}{r_x} = \frac{2.0(15(12))}{6.14} = 58.6 < 4.71\sqrt{\frac{29{,}000}{50}} = 113$$

use Equation E3-2

Step 9: Determine the Euler buckling stress

$$F_e = \frac{\pi^2(29{,}000)}{(58.6)^2} = 83.3 \text{ ksi}$$

Step 10: Determine the nominal stress from Equation E3-2

$$F_n = \left(0.658^{\left(\frac{F_y}{F_e}\right)}\right)F_y = \left(0.658^{\left(\frac{50}{83.3}\right)}\right)50 = 38.9 \text{ ksi}$$

Step 11: Determine the nominal strength

$$P_n = 38.9(26.5) = 1030 \text{ kips}$$

5.6 SLENDER ELEMENTS IN COMPRESSION

As mentioned in Section 5.4, the columns discussed thus far are controlled by overall column buckling. For some shapes, another form of buckling may actually control column strength: local buckling of the elements that make up the column shape. Whether the shape is rolled or built up, it can be thought of as being composed of a group of interconnected plates. Depending on how these plates are supported by each other, they could buckle at a stress below the critical buckling stress of the overall column. This is *local buckling*, also called *plate buckling*, and is shown in Figure 5.26. Local buckling is described through a plate critical buckling equation similar to the Euler buckling equation for columns. The critical buckling stress for an axially loaded plate is

$$F_{cr} = \frac{k\pi^2 E}{12(1-\nu^2)\left(\frac{b}{t}\right)^2} \tag{5.22}$$

where k is a plate buckling coefficient that depends on the plate loading, edge conditions, and length-to-width ratio; ν is Poisson's ratio; and b/t is the ratio of the width perpendicular to the compression force to the thickness of the plate. The width-to-thickness ratio is called the *plate slenderness* and is similar in function to the column slenderness. This critical stress plotted as a function of width-to-thickness ratio is shown as the dashed curve in Figure 5.27.

As with overall column buckling, an inelastic transition exists between elastic buckling and element yielding. This transition is due to the existence of residual stresses and imperfections in the element, just as in the case of overall column buckling, and results in the inelastic portion of the curve shown in Figure 5.27. The point identified in the figure as F_p-λ_p indicates where the elastic curve and the inelastic curve become tangent. In addition, for plates with low b/t ratios, strain hardening plays a critical role in their behavior, indicated by λ_o, and plates with large b/t ratios have significant post-buckling strength as shown in the figure.

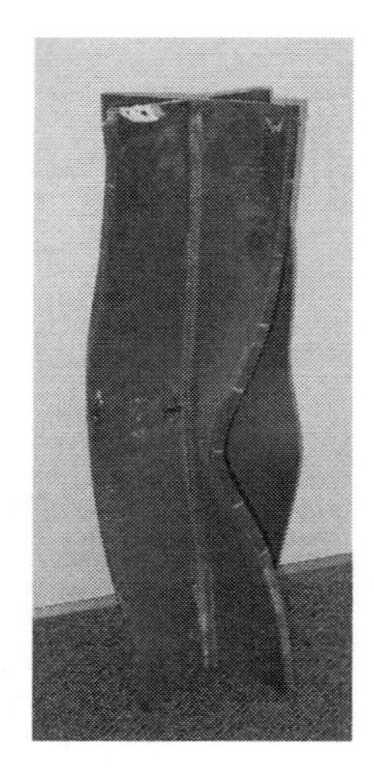

Figure 5.26 Column Tested to Failure through Local Buckling
Photo courtesy Perry Green

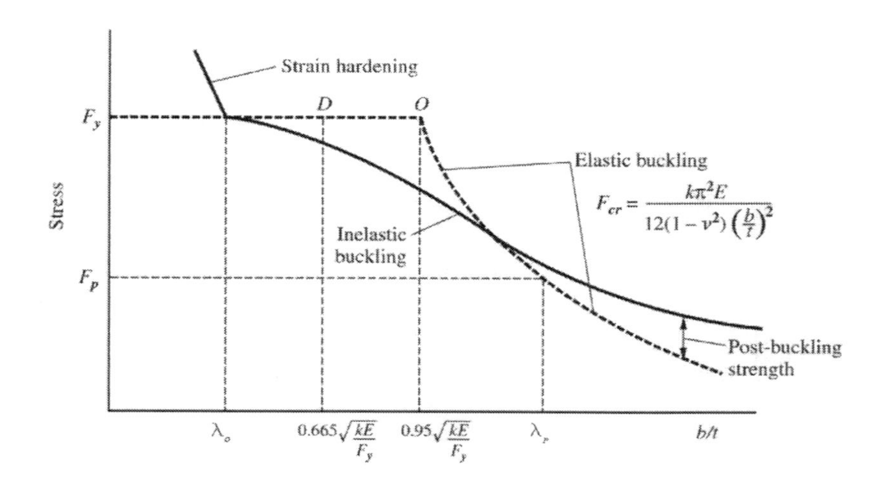

Figure 5.27 Plate Strength in Compression

To ensure that local buckling will not control column strength, the critical plate buckling stress for local buckling is limited to the critical buckling stress for overall column buckling. This approach results in a different maximum plate slenderness value for each corresponding column slenderness value. However, to alert the designer to the need for consideration of plate buckling, an initial check on element slenderness is made assuming that the stress in the plate has reached the yield stress. The development of the *Specification* provisions starts by finding a plate slenderness that sets the plate buckling stress equal to the column yield stress. Equation 5.22 then becomes

$$\frac{b}{t} = \sqrt{\frac{k\pi^2 E}{12(1-\nu^2)F_y}} \tag{5.23}$$

Taking $\nu = 0.3$, the standard value for steel, this plate slenderness becomes

$$\frac{b}{t} = 0.95\sqrt{\frac{kE}{F_y}} \tag{5.24}$$

which is shown as point O in Figure 5.27. This point is well above the inelastic buckling curve. In order to obtain a b/t that would bring the inelastic buckling stress closer to the yield stress, a somewhat arbitrary slenderness limit is taken as 0.7 times the limit that corresponds to the column yield stress, which gives

$$\frac{b}{t} = 0.665\sqrt{\frac{kE}{F_y}}$$

This is indicated as point D in Figure 5.27.

The remaining factor to be determined is the plate buckling coefficient, k. This factor is a function of the stress distribution, the edge support conditions and the aspect ratio of the plate. For a plate with uniform compression on opposite ends and simply supported on all four sides, the minimum value of k can be shown to be 4.0. For actual columns, with the variety of potential cross section shapes available, the determination of the value of k becomes much more complicated as the actual edge supports, stress distribution and aspect ratio vary.

The limiting width-to-thickness ratios are given in *Specification* Table B4.1a. These limits may be given as

$$\lambda_r = c_3\sqrt{\frac{E}{F_y}} \tag{5.25}$$

where c_3 is given in Table 5.4 for several elements in uniform compression taken from Table B4.1a of the *Specification*. The apparent plate buckling coefficient, k, used to obtain these values is also given in Table 5.4. For shapes with element slenderness exceeding these λ_r values plate buckling must be considered. As already shown, these limits are based on the assumption that the column is stressed to F_y. Since columns are rarely stressed to that level, it is very possible that what appears to be a slender element compression member based on Table B4.1a may not actually see its strength limited by the limit state of local buckling.

Table 5.4 Parameters for Consideration of Compression Member Local Buckling

Case*		λ	c_3	k	c_4	c_5
1	Flanges of rolled I-shaped sections, plates projecting from rolled I-shaped sections; outstanding legs of pairs of angles connected with continuous contact, flanges of channels, and flanges of tees	b/t	0.56	0.71	0.834	0.184
3	Legs of single angles, legs of double angles with separators, and all other unstiffened elements	b/t	0.45	0.46	0.671	0.148
4	Stems of tees	d/t	0.75	1.27	1.12	0.246
5	Webs of doubly symmetric rolled and built-up I-shaped sections and channels	h/t_w	1.49	5.0	1.95	0.351
6	Walls of rectangular HSS	b/t	1.40	4.43	1.93	0.386

* From Table B4.1a in the *Specification*.

For W-shapes, Case 1 in Table B4.1a, with F_y = 50 ksi, the flange slenderness limit is $\lambda_{rf} = 0.56\sqrt{E/F_y} = 13.5$, and all W-shapes have a flange slenderness less than this limit. For webs of these W-shapes, Case 5 in Table B4.1a, $\lambda_{rw} = 1.49\sqrt{E/F_y} = 35.9$, and many available W-shapes have a web slenderness that exceeds this limit and are classified as slender.

Design of slender element compression members according to the *Specification* follows the same requirements as those for compression members without slender elements, with one modification. To account for slender element behavior, the full area of the slender element cannot be used. Thus, a reduced effective area, A_e, is used in place of the gross area, A_g, to determine column strength. For columns with slender elements, Section E7 indicates that column strength is given by

$$P_n = F_n A_e \qquad \text{(AISC E7-1)}$$

which is to be used in place of Equation E3-1 but with the same nominal stress, F_n.

Once the designer is directed to the slender element provisions of Section E7 it is apparent that the nominal stress based on the controlling limit state must first be determined. Then, using that stress, the actual plate element slenderness at the transition from elastic to inelastic buckling can be determined. For the web of a rolled W-shape, Case 5 in Table B4.1a, the limiting width-to-thickness ratio becomes

$$\lambda_r = \frac{h}{t_w} = c_3\sqrt{\frac{E}{F_n}} = 1.49\sqrt{\frac{E}{F_n}}$$

If the width-to-thickness ratio of the web does not exceed this value, the usable width of the web is the actual width and no change in area is required. If the width-to-thickness ratio of the web does exceed this value, a reduced width must be determined through

$$b_e = b\left(1 - c_1\sqrt{\frac{F_{el}}{F_n}}\right)\sqrt{\frac{F_{el}}{F_n}} \qquad \text{(AISC E7-3)}$$

where the elastic plate buckling stress from Equation 5.22 is presented as

$$F_{el} = \left(c_2\frac{\lambda_r}{\lambda}\right)^2 F_y \qquad \text{(AISC E7-5)}$$

The constants c_1 and c_2 are given in *Specification* Table E7.1. If Equations 5.25 and E7-5 are substituted into Equation E7-3, the effective width becomes

$$b_e = c_2 c_3 t\sqrt{\frac{E}{F_n}}\left(1 - \frac{c_1 c_2 c_3}{(b/t)}\sqrt{\frac{E}{F_n}}\right) \qquad (5.26)$$

Equation 5.26 may be simplified by substituting $c_4 = c_2 c_3$ and $c_5 = c_1 c_2 c_3$ which yields

$$b_e = c_4 t\sqrt{\frac{E}{F_n}}\left(1 - \frac{c_5}{(b/t)}\sqrt{\frac{E}{F_n}}\right) \qquad (5.27)$$

Values for c_4 and c_5 are given in Table 5.4.

Continuing the consideration of W-shape webs by making the appropriate substitutions into Equation 5.27 yields

$$b_e = 1.95t\sqrt{\frac{E}{F_n}}\left[1 - \frac{0.351}{(b/t)}\sqrt{\frac{E}{F_n}}\right] \tag{5.28}$$

Once the effective width of a slender element is obtained, the corresponding effective area of the member can be determined. Since hot-rolled shapes have fillets at the junction of the plate elements, the best approach for determining the effective area is to use the gross area and deduct the appropriate ineffective element area. This will be illustrated in the example.

EXAMPLE 5.10
Strength of Column with Slender Elements

Goal: Determine the available strength of a compression member with a slender web.

Given: Use an A992 W16×26 as a column with $L_{cy} = 5.0$ ft.

Note: In *Manual* Table 1-1, this shape is identified with footnote c, indicating that it must be considered as a slender element member for compression. It is the most slender-web W-shape available and is not normally used as a column.

SOLUTION

Step 1: From *Manual* Table 1-1,

$$A = 7.68 \text{ in.}^2, \quad h/t_w = 56.8, \quad t_w = 0.250 \text{ in.}, r_y = 1.12 \text{ in.}$$

Step 2: Determine the web slenderness limit from *Specification* Table B4.1a, Case 5.

$$\lambda_{rw} = 1.49\sqrt{\frac{E}{F_y}} = 1.49\sqrt{\frac{29{,}000}{50}} = 35.9$$

Step 3: Check the slenderness of the web.

$$\frac{h}{t_w} = 56.8 > \lambda_{rw} = 35.9$$

Thus, the shape must be treated as one with a slender web. It has already been established that all W-shapes, with $F_y = 50$ ksi, have nonslender flanges, so that check will not be made here.

Step 4: Determine the Euler buckling stress, F_e, for $L_c = 5.0$ ft.

$$F_e = \frac{\pi^2(29{,}000)}{\left(\frac{5(12)}{1.12}\right)^2} = 99.7 \text{ ksi}$$

Step 5: Determine F_n

$$F_e = 99.7 \text{ ksi } > F_y/2.25 = 50/2.25 = 22.2 \text{ ksi}$$

Therefore, use Equation E3-2

$$F_n = 0.658^{\left(\frac{50}{99.7}\right)}(50) = 40.5 \text{ ksi}$$

Step 6: Check the slenderness of the web against the new limit using F_n in place of F_y.

$$\lambda_{rw} = 1.49\sqrt{\frac{E}{F_n}} = 1.49\sqrt{\frac{29{,}000}{40.5}} = 39.9$$

since

$$\frac{h}{t_w} = 56.8 > \lambda_{rw} = 39.9$$

the web is slender and the effective web width must be determined.

Step 7: Determine the effective width of the web using Equation 5.28

$$b_e = 1.95t\sqrt{\frac{E}{F_n}}\left[1 - \frac{0.351}{(b/t)}\sqrt{\frac{E}{F_n}}\right]$$

$$= 1.95(0.250)\sqrt{\frac{29{,}000}{40.5}}\left[1 - \frac{0.351}{56.8}\sqrt{\frac{29{,}000}{40.5}}\right] = 10.9 \text{ in.}$$

Step 8: Determine the actual web width.

The width of the web plate is given by h. However, a value of h is not specifically available in the *Manual*, so with $h/t_w = 56.8$ and $t_w = 0.250$, h can be determined as

$$h = (h/t_w)t_w = 56.8(0.250) = 14.2 \text{ in.}$$

Step 9: Determine the effective area.

Because $b_e < h$, use b_e to determine A_e. To properly account for the fillets at the web-flange junction, the area of the ineffective web is deducted from the gross area of the shape; thus,

$$A_e = A_g - (h - b_e)t_w = 7.68 - (14.2 - 10.9)(0.250) = 6.86 \text{ in.}^2$$

Step 10: Determine the nominal strength of the column.

$$P_n = F_n A_e = 40.5(6.86) = 278 \text{ kips}$$

For LRFD

Step 11: Determine the design strength for this slender web column with $L_c = 5.0$ ft.

$$\phi P_n = 0.9(278) = 250 \text{ kips}$$

For ASD

Step 11: Determine the design strength for this slender web column with L_c = 5.0 ft.

$$P_n/\Omega = 278/1.67 = 166 \text{ kips}$$

EXAMPLE 5.11
Strength of Column with Slender Elements

Goal: Determine the available strength of a compression member with a slender web.

Given: Use the W16×26 column from Example 5.10 but with L_{cy} = 15.0 ft.

Note: This shape has already been shown to have a slender web based on Table B4.1a

SOLUTION

Step 1: From *Manual* Table 1-1,

$A = 7.68$ in.^{2}, $h/t_w = 56.8$, $t_w = 0.250$ in., $r_y = 1.12$ in.

Step 2: Determine the web slenderness limit from *Specification* Table B4.1a, Case 5.

$$\lambda_{rw} = 1.49\sqrt{\frac{E}{F_y}} = 1.49\sqrt{\frac{29.000}{50}} = 35.9$$

Step 3: Check the slenderness of the web.

$$\frac{h}{t_w} = 56.8 > \lambda_{rw} = 35.9$$

Thus, the shape has a slender web. It has already been established that all W-shapes with F_y = 50 ksi have nonslender flanges.

Step 4: Determine the Euler buckling stress, F_e, for L_c = 15.0 ft.

$$F_e = \frac{\pi^2(29{,}000)}{\left(\frac{15(12)}{1.12}\right)^2} = 11.1 \text{ ksi}$$

Step 5: Determine F_n.

$$F_e = 11.1 \text{ ksi} < F_y/2.25 = 50/2.25 = 22.2 \text{ ksi}$$

Therefore, use Equation E3-3

$$F_n = 0.877(11.1) = 9.73 \text{ ksi}$$

Step 6: Check the slenderness of the web against the new limit using F_n in place of F_y.

$$\lambda_{rw} = 1.49\sqrt{\frac{E}{F_n}} = 1.49\sqrt{\frac{29,000}{9.73}} = 81.3$$

since

$$\frac{h}{t_w} = 56.8 < \lambda_{rw} = 81.3$$

the column will not be limited by local buckling of the web

Step 7: Determine the nominal strength of the column.

$$P_n = F_n A_g = 9.73(7.68) = 74.7 \text{ kips}$$

For LRFD Step 8: Determine the design strength for this slender web column with L_c = 15.0 ft.

$$\phi P_n = 0.9(74.7) = 67.2 \text{ kips}$$

For ASD Step 8: Determine the design strength for this slender web column with L_c = 15.0 ft.

$$P_n/\Omega = 74.7/1.67 = 44.7 \text{ kips}$$

Examples 5.10 and 5.11 illustrate that a W-shape that appears to be a slender element shape based on Table B4.1a may not actually be limited in strength because of that slender element, the limit state of local buckling. It can be shown that the W16×26 considered in these examples will have its strength limited by local buckling for columns with effective length, L_c, up to about 10.8 ft. Above that effective length, the slender elements will not impact overall column strength.

5.7 COLUMN DESIGN TABLES

A review of the AISC column equations, E3-2 and E3-3, shows that the only factor other than shape geometry and material strength that influences the determination of column strength is the slenderness ratio. Therefore, it is convenient to tabulate column strength as a function of slenderness. Part 4 of the *Manual* contains tables for W-shapes, HP-shapes, and HSS and several singly symmetric shapes. Figure 5.28 shows a sample of *Manual* Table 4-1a for several W14 sections with F_y = 50 ksi. As with all of the available strength tables in the *Manual*, both allowable strength (ASD) and design strength (LRFD) values are given. Tables 4-1b and 4-1c are provided for selected W-shapes that are commonly available with F_y = 65 and 70 ksi respectively.

The values in these column tables are based on the assumption that the column will buckle about its weak axis. For all W-shapes this is the y-axis, so the values in the tables are given in terms of the effective length with respect to the least radius of gyration, r_y. Their use is quite straightforward when the critical buckling length is about this axis. An approach that permits the use of these tables when the strong axis controls will be addressed following the example.

Table 4-1a (continued)

Available Strength in Axial Compression, kips

W-Shapes

F_y = 50 ksi — W14

Shape		W14×													
lb/ft		82		74		68		61		53		48		43[c]	
Design		P_n/Ω_c	$\phi_c P_n$	P_n/Ω_c	$\phi_c P_n$	P_n/Ω_c	$\phi_c P_n$	P_n/Ω_c	$\phi_c P_n$	P_n/Ω_c	$\phi_c P_n$	P_n/Ω_c	$\phi_c P_n$	P_n/Ω_c	$\phi_c P_n$
		ASD	LRFD	ASD	LRFD	ASD	LRFD	ASD	LRFD	ASD	LRFD	ASD	LRFD	ASD	LRFD
Effective length, L_c (ft), with respect to least radius of gyration, r_y	0	719	1080	653	981	599	900	536	805	467	702	422	634	374	562
	6	676	1020	614	922	562	845	503	756	421	633	380	572	339	510
	7	661	993	600	902	550	826	492	739	406	610	366	551	327	491
	8	644	968	585	879	536	805	479	720	389	585	351	527	312	470
	9	626	940	568	854	520	782	465	699	371	557	334	502	297	447
	10	606	910	550	827	503	756	450	676	351	528	316	475	281	422
	11	584	878	531	797	485	729	433	651	331	497	298	447	264	397
	12	562	844	510	767	466	701	416	626	310	465	279	419	247	371
	13	538	809	489	735	446	671	398	599	288	433	259	390	229	345
	14	514	772	467	701	426	640	380	571	267	401	240	360	212	318
	15	489	735	444	667	405	608	361	543	246	369	221	331	194	292
	16	464	697	421	633	384	577	342	514	225	338	202	303	177	267
	17	438	659	398	598	362	544	323	485	205	308	183	276	161	242
	18	413	620	375	563	341	512	304	456	185	278	166	249	145	218
	19	387	582	352	529	320	480	285	428	166	250	149	224	130	196
	20	362	545	329	495	299	449	266	399	150	226	134	202	117	177
	22	314	472	285	428	258	388	229	345	124	186	111	167	97.1	146
	24	267	402	243	365	219	330	195	293	104	157	93.2	140	81.6	123
	26	228	343	207	311	187	281	166	249	88.8	133	79.4	119	69.5	104
	28	197	295	179	268	161	242	143	215	76.6	115	68.5	103	59.9	90.1
	30	171	257	156	234	140	211	125	187	66.7	100	59.7	89.7	52.2	78.5
	32	150	226	137	205	123	185	110	165	58.6	88.1				
	34	133	200	121	182	109	164	97.0	146						
	36	119	179	108	162	97.5	147	86.5	130						
	38	107	160	96.9	146	87.5	131	77.7	117						
	40	96.3	145	87.5	131	79.0	119	70.1	105						
Available Strength Parameters for Concentrated Forces[i]															
P_{wo}, kips		123	185	104	155	90.6	136	77.5	116	77.1	116	67.4	101	56.9	85.4
P_{wi}, kip/in.		17.0	25.5	15.0	22.5	13.8	20.8	12.5	18.8	12.3	18.5	11.3	17.0	10.2	15.3
P_{wb}, kips		201	302	138	207	108	163	80.1	120	76.7	115	59.5	89.5	43.0	64.7
P_{fb}, kips		137	206	115	173	97.0	146	77.8	117	81.5	123	66.2	99.6	52.6	79.0
Properties															
L_p, ft		8.76		8.76		8.69		8.65		6.78		6.75		6.68	
L_r, ft		33.2		31.0		29.3		27.5		22.3		21.1		20.0	
A_g, in.^{2}		24.0		21.8		20.0		17.9		15.6		14.1		12.6	
I_x, in.^{4}		881		795		722		640		541		484		428	
I_y, in.^{4}		148		134		121		107		57.7		51.4		45.2	
r_y, in.		2.48		2.48		2.46		2.45		1.92		1.91		1.89	
r_x/r_y		2.44		2.44		2.44		2.44		3.07		3.06		3.08	
$P_{ex}L_c^2/10^4$, kip-in.^{2}		25200		22800		20700		18300		15500		13900		12300	
$P_{ey}L_c^2/10^4$, kip-in.^{2}		4240		3840		3460		3060		1650		1470		1290	

ASD	LRFD
Ω_c = 1.67	ϕ_c = 0.90

[c] Shape is slender for compression with F_y = 50 ksi; tabulated values have been adjusted accordingly.
[i] Flange local buckling, web local buckling, and web compression buckling are considered. Web local crippling, web sidesway buckling, and web panel zone shear are not addressed in this table.
Note: Heavy line indicates L_c/r_y equal to or greater than 200.

Figure 5.28 Available Strength in Axial Compression Copyright © American Institute of Steel Construction, Reprinted with Permission. All rights reserved.

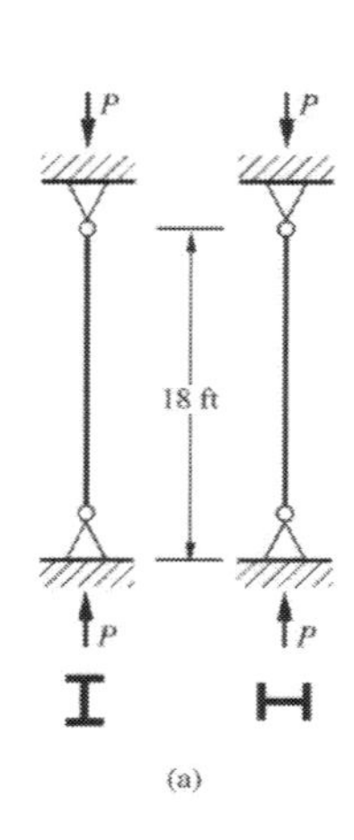

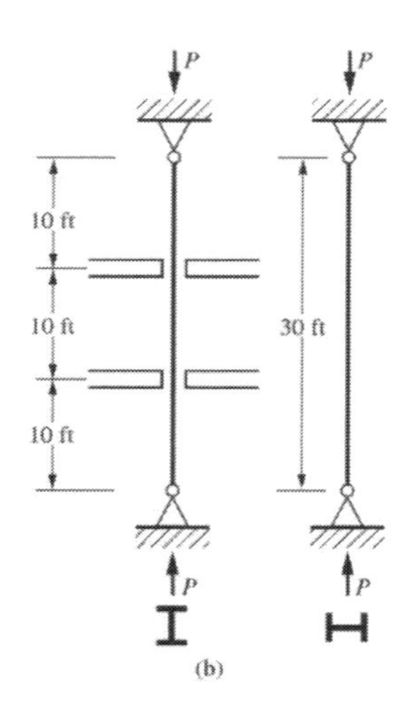

Figure 5.29 Columns for Examples 5.12 and 5.13

EXAMPLE 5.12a
Column Design by LRFD

Goal: Determine the least-weight section to carry the loads given using the limited selection available in Figure 5.28.

Given: The column is shown in Figure 5.29a. It must resist the following loads in the appropriate combinations: $P_D = 56$ kips, $P_L = 172$ kips, and $P_W = 176$ kips. Use A992 steel. Assume the live load comes from a distributed load less than 100 psf, so that the LRFD load factor on live load may be taken as 0.5 for load combination 4.

SOLUTION

Step 1: Determine the maximum required strength using the LRFD load combinations from Section 2.4.

1. $1.4P_D = 1.4(56) = 78.4$ kips

2. $1.2P_D + 1.6P_L = 1.2(56) + 1.6(172) = 342$ kips

4. $1.2P_D + 0.5P_L + 1.0P_W$
$= 1.2(56) + 0.5(172) + 1.0(176) = 329$ kips

6. $0.9P_D + 1.0P_W = 0.9\,(56) + 1.0(176) = 226$ kips

So the column must carry $P_u = 342$ kips.

Step 2: The column has the same effective length about the x- and y-axes, so enter the table in Figure 5.28 with $L_c = 18$ ft. Scanning across the table at $L_c = 18$ ft and checking the LRFD values, select the least-weight shape in this portion of the table that can support this load.

Select a W14×61 with a design compression strength

$$\phi P_n = 456 \text{ kips}$$

EXAMPLE 5.12b
Column Design by ASD

Goal: Determine the least-weight section to carry the loads given using the limited selection available in Figure 5.28.

Given: The column is shown in Figure 5.29a. It must resist the following loads in the appropriate combinations: $P_D = 56$ kips, $P_L = 172$ kips, and $P_W = 176$ kips. Use A992 steel.

SOLUTION

Step 1: Determine the maximum required strength using the ASD load combinations from Section 2.4.

1. $P_D = 56$ kips
2. $P_D + P_L = 56 + 172 = 228$ kips
5. $P_D + 0.60P_W = 56 + 0.6(176) = 162$ kips
6. $P_D + 0.75\ P_L + 0.75(0.6P_W)$
$= 56 + 0.75(172) + 0.75(0.6(176)) = 264$ kips
7. $0.6\ P_D + 0.6\ P_W = 0.6(56) + 0.6(176) = 139$ kips

So the column must carry $P_a = 264$ kips.

Step 2: The column has the same effective length about the x- and y-axes, so enter the table in Figure 5.28 with $L_c = 18$ ft. Scanning across the table at $L_c = 18$ ft and checking the ASD values, select the least-weight shape in this portion of the table that can support this load

Select a W14×61 with an allowable compression strength

$$P_n/\Omega = 304 \text{ kips}$$

If the largest slenderness ratio for a particular column happens to be for x-axis buckling, the tables may not be entered directly with the x-axis effective length because the table effective length is intended to be used in conjunction with the least radius of gyration. However, it is possible to determine a modified effective length that, when used in the table, will result in the correct column strength.

When the x-axis controls column strength, the slenderness ratio used in the column equations is L_{cx}/r_x. To use the column tables, an effective length, $(L_c)_{eff}$, must be determined that, when combined with r_y, gives the same slenderness ratio. So

$$\frac{(L_c)_{eff}}{r_y} = \frac{L_{cx}}{r_x}$$

Solving this equation for $(L_c)_{eff}$ yields

$$(L_c)_{eff} = \frac{L_{cx}}{(r_x / r_y)}$$

With this modified effective length, the tables can be entered, and a suitable column selected. There is one difficulty with this process, however. Until a column section is known, the value for r_x/r_y cannot be determined. To account for this, a quick scan of the column tables should be made to estimate r_x/r_y. Then, when a section is selected, the assumption can be verified and an adjustment made if necessary.

EXAMPLE 5.13
Column Design

Goal: Determine the least-weight section to carry the force given using the limited selection available through Figure 5.28. Design by LRFD and ASD.

Given: The column is shown in Figure 5.29b. Use the loading from Example 5.12.

SOLUTION

Step 1: Determine the effective length for each axis.

Bracing of the y-axis, shown in Figure 5.29b, yields $L_{cy} = 10.0$ ft. The unbraced x-axis has $L_{cx} = 30.0$ ft.

Step 2: Determine $(L_c)_{eff}$ for the x-axis.

Select a representative r_x/r_y from Figure 5.28. There are two general possibilities. Assume that the larger shapes might be needed to carry the load and try $r_x/r_y = 2.44$. Thus,

$$(L_c)_{eff} = \frac{L_{cx}}{(r_x/r_y)} = \frac{30.0}{2.44} = 12.3 \text{ ft}$$

Step 3: Determine the controlling effective length.

Because $(L_c)_{eff} = 12.3$ ft is greater than $L_{cy} = 10.0$ ft, enter the table with $L_c = 12.3$ ft and interpolate between 12 ft and 13 ft.

For LRFD

Step 4: From Example 5.12a the column must have a design strength greater than $P_u = 342$ kips with $L_c = 12.3$ ft. Try a W14×43, which happens to be the smallest column available with the limited selection available in Figure 5.28. This column has $r_x/r_y = 3.08$.

Step 5: Determine $(L_c)_{eff}$ with this new r_x/r_y. Thus,

$$(L_c)_{eff} = \frac{30.0}{3.08} = 9.74 \text{ ft}$$

Step 6: Determine the new controlling effective length.

Because $(L_c)_{eff} = 9.74$ ft is now less than $L_{cy} = 10.0$ ft, enter the table with 10.0 ft and note that the W14×43 has a design strength of 422 kips, which is greater than the required strength of 342 kips.

Step 7: Therefore, use the selected

W14×43

Note: The W14×43 is identified in the table by a footnote as slender for F_y = 50 ksi. This is not an issue for our design because the impact of any slender element has already been taken into account in generating the table as stated in the same footnote.

Using the full complement of tables available in the *Manual* results in a smaller W12 section having the ability to carry the given load.

For ASD

Step 4: From Example 5.12b the column must have an allowable strength greater than P_a = 264 kips with L_c = 12.3 ft. Try a W14×48. This column has $r_x/r_y = 3.06$.

Step 5: Determine $(L_c)_{eff}$ with this new r_x/r_y. Thus,

$$(L_c)_{eff} = \frac{30.0}{3.06} = 9.80 \text{ ft}$$

Step 6: Determine the new controlling effective length.

Because $(L_c)_{eff}$ = 9.80 ft is now less than L_{cy} = 10.0 ft, enter the table with 10.0 ft and see that the W14×43 has an allowable strength of 281 kips, which is greater than the required strength of 268 kips and r_x/r_y = 3.08 which is greater than that for the W14×48 so L_{cy} = 10.0 ft will still control.

Step 7: Therefore, use the selected

W14×43

Note: The W14×43 is identified in the table by a footnote as slender for F_y = 50 ksi. This is not an issue for our design because the impact of any slender element has already been taken into account in generating the table as stated in the same footnote.

Using the full complement of tables available in the *Manual* results in a smaller W12 section having the ability to carry the given load.

Table 4-1a in Part 4 of the *Manual* includes shapes from a W8×31 up to a W14×873. All of the shapes included are considered column shapes and have reasonably similar strengths about the *x*- and *y*-axes. That is, the shapes are close to being square and r_x/r_y is not extremely large, ranging from 1.59 to 3.08. Any of the other available W-shapes may be used for columns if desired, but it must be recognized that the relationship between the *x*- and *y*-axes is such that the *y*-axis will control unless significant bracing is provided. These shapes are generally considered beam shapes. Since beams are intended to be used to carry flexure, the relationship between the *x*- and *y*-axes is not as critical. For example, for a W16×26 with a length of 24 ft braced at the ends only for the *x*-axis and $r_x/r_y = 5.59$, the *y*-axis will control unless it is braced at least every 4.29 ft.

The W-shape column tables in Part 4 of the *Manual* for $F_y = 50$ ksi also exclude the smallest W-shapes in an attempt to direct the design engineer toward using shapes that are more appropriate when considering connections. That does not mean that these smaller shapes are not acceptable for use as columns. The tables in Part 6 of the *Manual*, which will be discussed in Chapter 8, can be used for the design of columns and they include all of the W-shapes.

EXAMPLE 5.14
Column Design

Goal: Determine the least-weight section to carry the force given using the small shapes provided in the W-shape tables in *Manual* Part 6. Design by LRFD and ASD.

Given: The A992 column has an effective length for both axes of 10 ft and must carry a concentrated dead load of 8 kips and a concentrated live load of 24 kips.

SOLUTION

For LRFD

Step 1: Determine the required strength for the load combination 1.2D + 1.6L.

$$P_u = 1.2(8.0) + 1.6(24.0) = 48.0 \text{ kips}$$

Step 2: Using *Manual* Table 6-2, select the lightest column to support this load.

Select the W4×13.

$$\phi P_n = 60.1 \text{ kips}$$

For ASD

Step 1: Determine the required strength for the load combination D + L.

$$P_a = 8.0 + 24.0 = 32.0 \text{ kips}$$

Step 2: Using *Manual* Table 6-2, select the lightest column to support this load.

Select the W4×13.

$$\frac{P_n}{\Omega} = 40.0 \text{ kips}$$

5.8 TORSIONAL BUCKLING AND FLEXURAL-TORSIONAL BUCKLING

Up to this point, the discussion has addressed the limit states of flexural buckling and local buckling. Two additional limit states for column behavior must be addressed: torsional buckling and flexural-torsional buckling. Doubly symmetric shapes normally fail through flexural buckling, as discussed earlier in this chapter, or through torsional buckling. Singly symmetric and unsymmetric shapes can fail through flexural, torsional, or flexural-torsional buckling. Because the shapes normally used for steel members are not well suited to resist torsion, except for closed HSS, it is usually desirable to avoid any torsional limit states through proper bracing of the column or by avoiding torsional loading.

If either of the torsional limit states must be evaluated, the applicable *Specification* provisions are found in Section E4, except for the special cases associated with single angles, which are found in Section E5. For doubly symmetric, singly symmetric and unsymmetric members braced so that they buckle torsionally about their shear center, specific elastic buckling stress equations are provided. For doubly symmetric members with bracing offset from the shear center, separate elastic buckling stress equations are given depending on if the bracing is offset from the strong or weak axis. For all these shapes, once the elastic buckling stress, F_e, is determined, it is then used in Equations E3-2 and E3-3 to determine the nominal compressive stress, F_n. The equations given in Section E4 defining the elastic buckling stress for the limit states of torsional and flexural-torsional buckling are also found in several other books with varying notation, including *Buckling Strength of Metal Structures.*[2] The equations of Section E3 are used to account for such factors as inelastic buckling, initial out-of-straightness, and residual stresses.

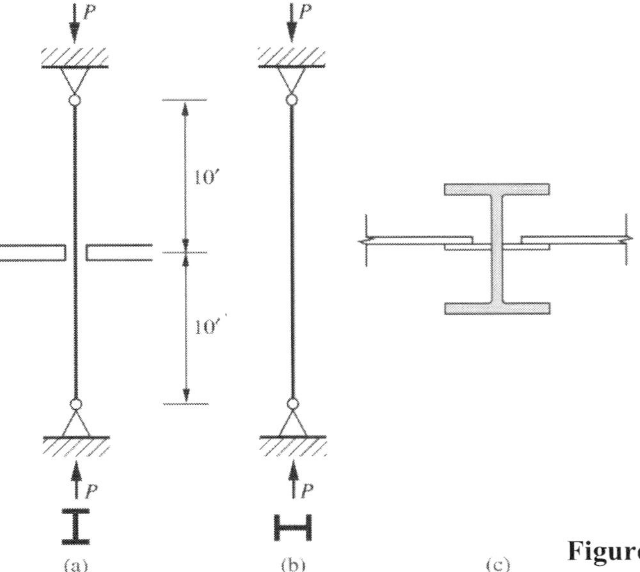

Figure 5.30 Column for Example 5.15

[2] Bleich, F. *Buckling Strength of Metal Structures*. New York: McGraw-Hill, 1952.

Because single-angle compression members are so common, the *Specification* provides a simplified approach for those members meeting a specific set of criteria. By limiting the way that load is applied to the ends of a single-angle compression member, an effective slenderness is established, which is then used in Equations E3-2 and E3-3 to determine the nominal compressive stress, F_n.

The limit state of torsional buckling is not normally considered in the design of W-shape columns when the y-axis is the controlling axis for flexural buckling. Torsional buckling generally does not govern, and when it does, the critical load differs very little from the strength determined from flexural buckling. For other member types, such as WT or double-angle compression members often used in trusses, torsional limit states are quite important.

An additional factor in determining strength based on these limit states is the torsional effective length. The Commentary recommends that, conservatively, the torsional effective length be taken as the column length and provides several other possibilities if greater accuracy is desired.

EXAMPLE 5.15
Strength of a W-Shape Column with Torsional Buckling

Goal: Determine the available strength of a W-shape column and consider torsional buckling.

Given: A W14×48 A992 column as shown in Figure 5.30 is braced laterally and torsionally at its ends. At mid-height it is braced to resist buckling about the y-axis, but it cannot resist torsional buckling based on the bracing shown in Figure 5.30c.

SOLUTION

Step 1: From *Manual* Table 1-1,

$$A_g = 14.1 \text{ in.}^2,\ I_x = 484 \text{ in.}^4,\ I_y = 51.4 \text{ in.}^4,\ r_x = 5.85 \text{ in.},\ r_y = 1.91 \text{ in.},\ C_w = 2240 \text{ in.}^6,\ J = 1.45 \text{ in.}^4,\ h/t_w = 33.6$$

Step 2: Determine the web slenderness limit.

$$\lambda_{rw} = 1.49\sqrt{\frac{E}{F_y}} = 1.49\sqrt{\frac{29{,}000}{50}} = 35.9$$

Step 3: Check slenderness of the web and flange.

$$\frac{h}{t_w} = 33.6 < \lambda_{rw} = 35.9$$

Therefore, the web is not slender. As previously discussed, all W-shapes with $F_y = 50$ ksi have nonslender flanges.

Step 4: Determine the nominal stress for y-axis buckling.

$$\left(\frac{L_c}{r}\right)_y = \frac{10(12)}{1.91} = 62.8$$

$$F_e = \frac{\pi^2 E}{\left(\frac{L_c}{r}\right)^2} = \frac{\pi^2 (29{,}000)}{(62.8)^2} = 72.6 \text{ ksi} > \frac{F_y}{2.25} = 22.2 \text{ ksi}$$

Therefore, use Equation E3-2:

$$F_n = (0.658)^{\left(\frac{50}{72.6}\right)}(50) = 37.5 \text{ ksi}$$

Step 5: Determine the nominal stress for x-axis buckling.

$$\left(\frac{L_c}{r}\right)_x = \frac{20(12)}{5.85} = 41.0$$

$$F_e = \frac{\pi^2 E}{\left(\frac{L_c}{r}\right)^2} = \frac{\pi^2 (29{,}000)}{(41.0)^2} = 170 \text{ ksi} > \frac{F_y}{2.25} = 22.2 \text{ ksi}$$

Therefore, use Equation E3-2:

$$F_n = (0.658)^{\left(\frac{50}{170}\right)}(50) = 44.2 \text{ ksi}$$

Step 6: Determine the nominal stress for z-axis buckling, or twisting about the shear center, using Section E4(a) Equation E4-2.

$$F_e = \left[\frac{\pi^2 E C_w}{(L_{cz})^2} + GJ\right]\frac{1}{I_x + I_y}$$

$$= \left[\frac{\pi^2 (29{,}000)(2240)}{(20(12))^2} + 11{,}200(1.45)\right]\frac{1}{484 + 51.4}$$

$$= 51.1 \text{ ksi} > \frac{F_y}{2.25} = 22.2 \text{ ksi}$$

Therefore use Equation E3-2:

$$F_n = (0.658)^{\left(\frac{50}{51.1}\right)}(50) = 33.2 \text{ ksi}$$

Step 7: Select the lowest nominal stress determined in Steps 4, 5, and 6.

$$F_n = 33.2 \text{ ksi}$$

Since the controlling nominal stress comes from Step 6, the strength of the column is controlled by torsional buckling.

Step 8: Determine the nominal strength of the column

$$P_n = 33.2(14.1) = 468 \text{ kips}$$

Note: Determination of F_n in steps 4, 5, and 6 could have been delayed until after the controlling, smallest, value of F_e had been determined and then F_n determined only once.

For LRFD Step 9: Determine the column design strength.

$$\phi P_n = 0.9(468) = 421 \text{ kips}$$

For ASD Step 9: Determine the column allowable strength.

$$\frac{P_n}{\Omega} = \frac{468}{1.67} = 280 \text{ kips}$$

EXAMPLE 5.16 ***Strength of a WT-Shape Compression Member***

Goal: Determine the available strength of a WT-shape compression member with consideration of flexural, torsional, and flexural-torsional buckling.

Given: A WT7×34 A992 column is 10.0 ft long and is braced laterally and torsionally at its ends only.

SOLUTION

Step 1: From *Manual* Table 1-8,

$A_g = 10.0$ in.^{2}, $I_x = 32.6$ in.^{4}, $I_y = 60.7$ in.^{4}, $r_x = 1.81$ in., $r_y = 2.46$ in., $t_f = 0.720$, $C_w = 3.21$ in.^{6}, $J = 1.50$ in.^{4}, $d/t_w = 16.9$, $\bar{y} = 1.29$ in., $b_f/2t_f = 6.97$

Step 2: Determine the flange and stem slenderness limits from Table B4.1 cases 1 and 4.

$$\lambda_{rf} = 0.56\sqrt{\frac{E}{F_y}} = 0.56\sqrt{\frac{29{,}000}{50}} = 13.5$$

$$\lambda_{rw} = 0.75\sqrt{\frac{E}{F_y}} = 0.75\sqrt{\frac{29{,}000}{50}} = 18.1$$

Step 3: Check slenderness of the flange and stem.

$$\frac{b_f}{2t_f} = 6.97 \le \lambda_{rf} = 13.5$$

$$\frac{d}{t_w} = 16.9 < \lambda_{rw} = 18.1$$

Therefore the WT has nonslender flange and stem.

Step 4: Determine the nominal strength for flexural buckling. Since $L_{cx} = L_{cy}$ and the x-axis has the smallest radius of gyration, flexural buckling will be controlled by the x-axis.

$$\left(\frac{L_c}{r}\right)_x = \frac{10(12)}{1.81} = 66.3 \le 4.71\sqrt{\frac{E}{F_y}} = 4.71\sqrt{\frac{29,000}{50}} = 113$$

$$F_{ex} = \frac{\pi^2 E}{\left(\frac{L_c}{r}\right)^2} = \frac{\pi^2(29,000)}{(66.3)^2} = 65.1 \text{ ksi} > \frac{F_y}{2.25} = 22.2 \text{ ksi}$$

Therefore, use Equation E3-2:

$$F_n = (0.658)^{\left(\frac{F_y}{F_e}\right)} F_y = (0.658)^{\left(\frac{50}{65.1}\right)}(50) = 36.3 \text{ ksi}$$

and

$$P_n = 36.3(10.0) = 363 \text{ kips}$$

Step 5: To determine flexural-torsional buckling, the elastic buckling stress for y-axis buckling is required.

$$\left(\frac{L_c}{r}\right)_y = \frac{10(12)}{2.46} = 48.8$$

$$F_{ey} = \frac{\pi^2 E}{\left(\frac{L_c}{r}\right)^2} = \frac{\pi^2(29,000)}{(48.8)^2} = 120 \text{ ksi}$$

Step 6: Determine the flexural-torsional elastic buckling stress for z-axis buckling using Equation E4-3. The shear center of a WT-shape is at the stem-flange intersection. Thus, the distance from the centroid to the shear center is

$$x_o = 0,\ y_o = \bar{y} - \frac{t_f}{2} = 1.29 - \frac{0.720}{2} = 0.930 \text{ in.}$$

and from Equation E4-9

$$r_o^2 = x_o^2 + y_o^2 + \frac{I_x + I_y}{A_g} = 0.0 + 0.930^2 + \frac{32.6 + 60.7}{10.0} = 10.2$$

and Equation E4-8

$$H = 1 - \frac{x_o^2 + y_o^2}{\bar{r}_o^2} = 1 - \frac{0 + 0.930^2}{10.2} 0.915$$

From the user note in Section E4, take $C_w = 0$ in Equation E4-7. Thus,

$$F_{ez} = \frac{GJ}{A_g \bar{r}_o^2} = \frac{11,200(1.50)}{10.0(10.2)} = 165 \text{ ksi}$$

Step 7: Determine the flexural-torsional elastic buckling stress for the singly symmetric member using Equation E4-3.

$$F_e = \left(\frac{F_{ey} + F_{ez}}{2H}\right)\left(1 - \sqrt{1 - \frac{4F_{ey}F_{ez}H}{(F_{ey} + F_{ez})^2}}\right)$$

$$= \left(\frac{120 + 165}{2(0.915)}\right)\left(1 - \sqrt{1 - \frac{4(120)(165)(0.915)}{(120 + 165)^2}}\right) = 105 \text{ ksi}$$

Step 8: Determine the nominal stress using the flexural-torsional elastic buckling stress

$$F_e = 105 \text{ ksi} > \frac{F_y}{2.25} = 22.2 \text{ ksi}$$

Therefore, using Equation E3-2

$$F_n = (0.658)^{\frac{50}{105}}(50) = 41.0 \text{ ksi}$$

Step 9: Determine the nominal strength of the compression member for the limit state of flexural-torsional buckling.

$$P_n = 41.0(10.0) = 410 \text{ kips}$$

For LRFD Step 10: Determine the compression member design strength. Since the nominal strength for flexural buckling about the *x*-axis is less than the flexural-torsional buckling strength,

$$\phi P_n = 0.9(363) = 327 \text{ kips}$$

For ASD Step 10: Determine the compession member allowable strength. Since the nominal strength for flexural buckling about the *x*-axis is less than the flexural-torsional buckling strength,

$$\frac{P_n}{\Omega} = \frac{363}{1.67} = 217 \text{ kips}$$

EXAMPLE 5.17 ***Strength of a W-Shape Column with Constrained-Axis Torsional Buckling***

Goal: Determine the available strength of a W-shape column and consider torsional buckling when the lateral bracing is offset from the shear center.

Given: A W14×48 A992 column as shown in Figure 5.30 and considered in Example 5.15 is braced laterally and torsionally at its ends. At mid-height it is braced to resist buckling about the *y*-axis. The y-axis bracing is moved from the shear center, as shown in Figure 5.30c, to the face of the flange. Thus constrained-axis torsional buckling must be assessed.

SOLUTION

Step 1: From *Manual* Table 1-1,

$$A_g = 14.1 \text{ in.}^2,\ I_x = 484 \text{ in.}^4,\ I_y = 51.4 \text{ in.}^4,\ r_x = 5.85 \text{ in.},\ r_y = 1.91 \text{ in.},$$
$$d = 13.8 \text{ in.},\ h_o = 13.2 \text{ in.},\ t_f = 0.595,\ C_w = 2240 \text{ in.}^6,\ J = 1.45 \text{ in.}^4,$$
$$h/t_w = 33.6$$

Step 2: Determine the web slenderness limit.

$$\lambda_{rw} = 1.49\sqrt{\frac{E}{F_y}} = 1.49\sqrt{\frac{29{,}000}{50}} = 35.9$$

Step 3: Check slenderness of the web and flange.

$$\frac{h}{t_w} = 33.6 < \lambda_{rw} = 35.9$$

Therefore, the web is not slender. As previously discussed, all W-shapes with $F_y = 50$ ksi have nonslender flanges.

Step 4: Determine the nominal stress for *y*-axis buckling.

$$\left(\frac{L_c}{r}\right)_y = \frac{10(12)}{1.91} = 62.8$$

$$F_e = \frac{\pi^2 E}{\left(\frac{L_c}{r}\right)^2} = \frac{\pi^2(29{,}000)}{(62.8)^2} = 72.6 \text{ ksi} > \frac{F_y}{2.25} = 22.2 \text{ ksi}$$

Therefore, use Equation E3-2:

$$F_n = (0.658)^{\left(\frac{50}{72.6}\right)}(50) = 37.5 \text{ ksi}$$

Step 5: Determine the nominal stress for *x*-axis buckling.

$$\left(\frac{L_c}{r}\right)_x = \frac{20(12)}{5.85} = 41.0$$

$$F_e = \frac{\pi^2 E}{\left(\frac{L_c}{r}\right)^2} = \frac{\pi^2 (29{,}000)}{(41.0)^2} = 170 \text{ ksi} > \frac{F_y}{2.25} = 22.2 \text{ ksi}$$

Therefore, use Equation E3-2:

$$F_n = (0.658)^{\left(\frac{50}{170}\right)}(50) = 44.2 \text{ ksi}$$

Step 6: Determine the nominal stress for z-axis buckling with the bracing offset along the minor axis using Section E4(d) Equation E4-10.

The bracing offset is $y_a = d/2 = 13.8/2 = 6.9$ in. and $x_a = 0$. Thus, from Equation E4-11,

$$r_o^2 = \left(r_x^2 + r_y^2 + y_a^2 + x_a^2\right) = \left(5.85^2 + 1.91^2 + 6.9^2 + 0\right) = 85.5 \text{ in.}^2$$

and

$$F_{ez} = \left[\frac{\pi^2 EI_y}{(L_{cz})^2}\left(\frac{h_o^2}{4} + y_a^2\right) + GJ\right]\frac{1}{A_g r_o^2}$$

$$= \left[\frac{\pi^2 (29{,}000)(51.4)}{(20(12))^2}\left(\frac{13.2^2}{4} + 6.9^2\right) + 11{,}200(1.45)\right]\frac{1}{14.1(85.5)}$$

$$= 32.8 \text{ ksi} > \frac{F_y}{2.25} = 22.2 \text{ ksi}$$

Therefore use Equation E3-2:

$$F_n = (0.658)^{\left(\frac{50}{32.8}\right)}(50) = 26.4 \text{ ksi}$$

Step 7: Select the lowest nominal stress determined in Steps 4, 5, and 6.

$$F_n = 26.4 \text{ ksi}$$

Since the controlling nominal stress comes from Step 6, the strength of the column is controlled by torsional buckling.

Step 8: Determine the nominal strength of the column

$$P_n = 26.4(14.1) = 372 \text{ kips}$$

Note: As was the case for Example 5.15, determination of F_n in steps 4, 5, and 6 could have been delayed until after the controlling, smallest, value of F_e had been determined and then F_n determined only once.

Also note that moving the lateral brace from the shear center, as in Example 5-15, to the face of the flange has reduced the strength of the column.

For LRFD Step 9: Determine the column design strength.

$$\phi P_n = 0.9(372) = 335 \text{ kips}$$

For ASD Step 9: Determine the column allowable strength.

$$\frac{P_n}{\Omega} = \frac{372}{1.67} = 223 \text{ kips}$$

5.9 SINGLE-ANGLE COMPRESSION MEMBERS

Single-angle compression members would be designed for flexural-torsional buckling according to the provisions in *Specification* Section E4 except for an exclusion for angles with $b/t \leq 0.71\sqrt{E/F_y}$. All hot rolled, A36 angles satisfy this exclusion limit so they need not be checked for flexural-torsional buckling. However, since the preferred material for angles is A572 Gr. 50, one must check to be sure that the provisions in *Specification* Section E4 are applicable.

Studies show that the compressive strength of single angles can be reasonably predicted using the compression member equations of *Specification* Section E3 if a modified effective length is used and the member satisfies the following limiting criteria as found in *Specification* Section E5.

1. Members are loaded at their ends in compression through the same one leg.
2. Members are attached by either welding or a connection containing a minimum of two bolts.
3. There are no intermediate transverse loads.
4. L_c/r as determined in this section does not exceed 200.
5. For unequal leg angles, the ratio of the long leg width to short leg width is less than 1.7.

Two cases are given for these provisions: (1) angles that are individual members or web members of planar trusses, and (2) angles that are web members in box or space trusses. This distinction is intended to reflect the difference in restraint provided by the elements to which the compression members are attached.

The first set of equations is for angles that

1. are individual members or web members of planar trusses.
2. are equal-leg angles or unequal-leg angles connected through the longer leg.
3. have adjacent web members attached to the same side of a gusset plate or truss chord.

Buckling is assumed to occur about the geometric axis parallel to the attached leg. Since this may be either the x- or y-axis, the *Specification* uses the subscript a and then defines r_a as the radius of gyration about the axis parallel to the attached leg.

If $\dfrac{L}{r_a} \leq 80$,

$$\frac{L_c}{r} = 72 + 0.75\frac{L}{r_a} \qquad \text{(AISC E5-1)}$$

and if $\dfrac{L}{r_a} > 80$

$$\frac{L_c}{r} = 32 + 1.25\frac{L}{r_a} \qquad \text{(AISC E5-2)}$$

These effective lengths must be modified if the unequal-leg angles are attached through the shorter legs. The provisions of *Specification* Section E5 should be reviewed for these angles as well as for similar angles in box or space trusses.

EXAMPLE 5.18
Strength of Single-Angle Compression Member

Goal: Determine the available strength of a 10.0 ft single-angle compression member using A572 Gr. 50 steel and the provisions of *Specification* Section E5.

Given: A 4×4×1/2 angle is a web member in a planar truss. It is attached by two bolts at each end through the same leg.

SOLUTION

Step 1: Check angle leg slenderness,

$$\frac{b}{t} = \frac{4}{0.5} = 8.0 < 0.71\sqrt{\frac{E}{F_y}} = 0.71\sqrt{\frac{29{,}000}{50}} = 17.1$$

Therefore, it is permissible to use the provisions of *Specification* Section E5 and, since

$$\frac{b}{t} = \frac{4}{0.5} = 8.0 < 0.45\sqrt{\frac{E}{F_y}} = 0.45\sqrt{\frac{29{,}000}{50}} = 10.8$$

The provisions of Section E7 do not apply.

Step 2: From *Manual* Table 1-7,

$A = 3.75$ in.^{2} and $r_x = 1.21$.

Step 3: Determine the slenderness ratio for the axis parallel to the connected leg, $r_a = r_x$.

$$\frac{L}{r_a} = \frac{L}{r_x} = \frac{10.0(12)}{1.21} = 99.2$$

Step 4: Determine which equation will give the effective slenderness ratio.

Because

$$\frac{L}{r_a} = 99.2 > 80$$

use Equation E5-2.

Step 5: Determine the effective slenderness ratio from Equation E5.2.

$$\frac{L_c}{r} = 32 + 1.25(99.2) = 156 < 200$$

Step 6: Determine which column strength equation to use.

Because

$$\frac{L_c}{r} = 156 > 4.71\sqrt{\frac{29{,}000}{50}} = 113$$

use Equation E3-3.

Step 7: Determine the Euler buckling stress.

$$F_e = \frac{\pi^2 E}{\left(\frac{L_c}{r}\right)^2} = \frac{\pi^2(29{,}000)}{(156)^2} = 11.8 \text{ ksi}$$

Step 8: Determine the nominal stress from Equation E3-3.

$$F_n = 0.877F_e = 0.877(11.8) = 10.3 \text{ ksi}$$

Step 9: Determine the nominal strength.

$$P_n = F_{cr}A = 10.3(3.75) = 38.6 \text{ kips}$$

For LRFD Step 10: Determine the design strength.

$$\phi P_n = 0.9(38.6) = 34.7 \text{ kips}$$

For ASD Step 10: Determine the allowable strength.

$$P_n/\Omega = 38.6/1.67 = 23.1 \text{ kips}$$

For single angle compression members that do not meet the criteria set forth in Section E5 for use of the modified slenderness ratio equations, the provisions of Sections E3 or E7 must be followed. The provisions in Section E4 for torsional or flexural-torsional buckling do not need to be followed for hot-rolled angles that meet the leg slenderness exclusion of $b/t \le 0.71\sqrt{E/F_y}$. Thus, for these members, the strength for flexural buckling about the principal axes must be assessed.

EXAMPLE 5.19
Strength of Single-Angle Compression Member

Goal: Determine the available strength of a 10.0 ft single-angle compression member using A572 Gr. 50 steel.

Given: A 4×4×1/2 angle is a web member in a planar truss. It is attached by single bolts at each end through the same leg.

SOLUTION

Step 1: From *Manual* Table 1-7,

$$A = 3.75\text{ in.}^2 \text{ and } r_x = r_y = 1.21,\ r_z = 0.776\text{ in.}^2.$$

and from Example 5.17, the angle is not a slender element member and the provisions of Section E7 do not apply. Additionally, since

$$\frac{b}{t} = \frac{4}{0.5} = 8.0 < 0.71\sqrt{\frac{E}{F_y}} = 0.71\sqrt{\frac{29{,}000}{50}} = 17.1$$

the provisions of Section E4 need not be checked.

Step 2: Determine the slenderness ratio for the minor (weak) principal axis.

$$\frac{L_c}{r_z} = \frac{10.0(12)}{0.776} = 155$$

Step 3: Determine which column strength equation to use.

Because

$$\frac{L_c}{r_z} = 155 > 4.71\sqrt{\frac{29{,}000}{50}} = 113$$

Use Equation E3-3.

Step 4: Determine the Euler buckling stress.

$$F_e = \frac{\pi^2 E}{\left(\frac{L_c}{r}\right)^2} = \frac{\pi^2(29{,}000)}{(155)^2} = 11.9\text{ ksi}$$

Step 5: Determine the nominal stress from Equation E3-3.

$$F_{cr} = 0.877F_e = 0.877(11.9) = 10.4\text{ ksi}$$

Step 6: Determine the nominal strength.

$$P_n = F_{cr}A = 10.4(3.75) = 39.0 \text{ kips}$$

For LRFD
Step 7: Determine the design strength.

$$\phi P_n = 0.9(39.0) = 35.1 \text{ kips}$$

For ASD
Step 7: Determine the allowable strength.

$$P_n/\Omega = 39.0/1.67 = 23.4 \text{ kips}$$

5.10 BUILT-UP MEMBERS

Members composed of more than one shape are called *built-up members*. Several of these were illustrated in Figure 5.2h through n. Built-up compression members composed of two shapes are covered in *Specification* Section E6. Compressive strength is addressed by establishing the slenderness ratio and referring to *Specification* Section E3, E4, or E7 as appropriate.

If a built-up section buckles so that the fasteners between the shapes are not stressed in shear but simply "go along for the ride," the only requirement is that the slenderness ratio of the shape between fasteners be no greater than 0.75 times the controlling slenderness ratio of the built-up shape. If overall buckling would put the fasteners into shear, then the controlling slenderness ratio will be somewhat greater than the slenderness ratio of the built-up shape. This modified slenderness ratio is used to account for the effect of shearing deformations through the connectors. Thus, the effective slenderness ratio for a built-up member with snug-tight connectors will be greater than the same member with pre-tensioned or welded connectors. In addition, the spacing of the intermediate connectors will influence the modified slenderness ratio.

For intermediate connectors that are bolted snug-tight, the modified slenderness ratio is always greater than the slenderness ratio of the built-up member acting as a unit since there will always be some shearing deformation in the connectors. It is specified as

$$\left(\frac{L_c}{r}\right)_m = \sqrt{\left(\frac{L_c}{r}\right)_o^2 + \left(\frac{a}{r_i}\right)^2} \qquad \text{(AISC E6-1)}$$

If the intermediate connectors are welded or pre-tensioned bolted, the shearing deformation in the connectors is significantly less than for snug-tight connectors and the modified slenderness ratio may be equal to the slenderness ratio of the built-up member acting as a unit. For this case, the modified slenderness ratio is specified as, when $\frac{a}{r_i} \leq 40$,

$$\left(\frac{L_c}{r}\right)_m = \left(\frac{L_c}{r}\right)_o \qquad \text{(AISC E6-2a)}$$

and when $\frac{a}{r_i} > 40$,

$$\left(\frac{L_c}{r}\right)_m = \sqrt{\left(\frac{L_c}{r}\right)_o^2 + \left(\frac{K_i a}{r_i}\right)^2} \qquad \text{(AISC E6-2b)}$$

where

$\left(\frac{L_c}{r}\right)_o$ = column slenderness of built-up member acting as a unit

K_i = 0.5 for angles back-to-back

= 0.75 for channels back-to-back

= 0.86 for all other shapes

a = distance between connectors

r_i = minimum radius of gyration of individual component

The remaining provisions in *Specification* Section E6 address dimensions and detailing requirements. These provisions are based on judgment and experience and are provided to ensure that the built-up member behaves in a way consistent with the strength provisions already discussed. The ends of built-up compression members must be either welded or pre-tensioned bolted in order to ensure that the member can work together as a unit. Even the smallest amount of slip in the end connections could mean that the built-up member is unable to carry any more load than the components individually. Along the length of built-up members, the longitudinal spacing of connectors must be sufficient to provide for transfer of the required shear force in the buckled member. The Commentary of the *Specification* gives guidance on how to determine the magnitude of the forces in the connectors. A built-up compression member with connectors spaced so that the slenderness ratio of the shape between fasteners is no greater than 0.75 times the controlling slenderness ratio of the built-up shape will not automatically satisfy this strength requirement.

The *Manual* provides tables of properties for double angles, double channels, and I-shapes with cap channels in Part 1 and tables of compressive strength for double-angle compression members in Part 4.

EXAMPLE 5.20 ***Strength of a Built-up Double-Angle Compression Member***

Goal: Determine the available strength of a 10.0 ft double-angle compression member using A572 Gr. 50 steel.

Given: Two 5×3×5/16 angles, long legs back-to-back with a 3/8 in. gap are used as a chord member in a planar truss. The angles are welded at each end to a gusset plate and along the length at two intermediate points with a spacing of 40 in.

SOLUTION

Step 1: From *Manual* Table 1-15 for double angles

$A = 4.82 \text{ in.}^2$, $r_x = 1.61$ in., $r_y = 1.21$ in., $\bar{r}_o = 2.52$ in. and $H = 0.640$ in.

From *Manual* Table 1-7 for single angles

$r_z = 0.649$ in. and $J = 0.0832 \text{ in.}^4$

Check leg slenderness

$$\frac{b}{t} = \frac{5}{0.3125} = 16.0 < 0.71\sqrt{\frac{E}{F_y}} = 0.71\sqrt{\frac{29{,}000}{50}} = 17.1$$

Thus, flexural-torsional buckling of the individual angles need not be considered.

For local buckling,

$$\frac{b}{t} = \frac{5}{0.3125} = 16.0 > 0.45\sqrt{\frac{E}{F_y}} = 0.45\sqrt{\frac{29{,}000}{50}} = 10.8$$

Therefore, local buckling must be considered.

Step 2: Determine the slenderness ratio for each axis if the member works as a unit.

$$\frac{L}{r_x} = \frac{10.0(12)}{1.61} = 74.5$$

$$\frac{L}{r_y} = \frac{10.0(12)}{1.21} = 99.2$$

Step 3: Determine the effective slenderness ratio for buckling about the y-axis, the axis that will put the connectors in shear. Since the intermediate connectors are spaced at 40 in.

$$\frac{a}{r_i} = \frac{a}{r_z} = \frac{40}{0.649} = 61.6 > 40$$

Therefore use Equation E6.2b

$$\left(\frac{L_c}{r}\right)_m = \sqrt{\left(\frac{L_c}{r}\right)_o^2 + \left(\frac{K_i a}{r_i}\right)^2} = \sqrt{99.2^2 + \left(0.5(61.6)\right)^2} = 104$$

Step 4: Check the maximum permitted slenderness ratio between connectors

$$\frac{a}{r_i} = 61.6 < 0.75(104) = 78.0$$

Step 5: Determine the elastic buckling stress for flexural buckling using the modified slenderness ratio

$$F_{ey} = \frac{\pi^2 E}{\left(\frac{L_c}{r}\right)_m^2} = \frac{\pi^2(29{,}000)}{(104)^2} = 26.5 \text{ ksi}$$

Step 6: Determine the elastic buckling stress for torsional buckling using Equation E4-7 with $C_w = 0$ based on the user note.

$$F_{ez} = \frac{GJ}{A_g \bar{r}_o^2} = \frac{11{,}200(2(0.0832))}{4.82(2.52)^2} = 60.9 \text{ ksi}$$

Step 7: Determine the elastic buckling stress for flexural-torsional buckling using Equation E4-3.

$$F_e = \left(\frac{F_{ey} + F_{ez}}{2H}\right)\left(1 - \sqrt{1 - \frac{4F_{ey}F_{ez}H}{(F_{ey} + F_{ez})^2}}\right)$$

$$= \left(\frac{26.5 + 60.9}{2(0.640)}\right)\left[1 - \sqrt{1 - \frac{4(26.5)(60.9)(0.640)}{(26.5 + 60.9)^2}}\right] = 22.0 \text{ ksi}$$

Step 8: Determine the nominal stress.

Since the elastic buckling stress for flexural-torsional buckling is less than that for flexural buckling, use that to determine the critical stress.

$$\frac{F_y}{F_e} = \frac{50}{22.0} = 2.27 > 2.25$$

Therefore, use Equation E3-3

$$F_n = 0.877F_e = 0.877(22.0) = 19.3 \text{ ksi}$$

Step 9: Determine if the local buckling must be included.

For the short leg $b/t = 3.0/0.3125 = 9.6$

and for the long leg $b/t = 5.0/0.3125 = 16.0$

From Table B4.1a case 3

$$\lambda_r = 0.45\sqrt{\frac{E}{F_n}} = 0.45\sqrt{\frac{29,000}{19.3}} = 17.4$$

Since

$$b/t = 16.0 < \lambda_r = 17.4$$

the legs are not slender.

Step 10: Determine the nominal strength.

$$P_n = F_n A_g = 19.3(4.82) = 93.0 \text{ kips}$$

For LRFD Step 11: Determine the design strength.

$$\phi P_n = 0.9(93.0) = 83.7 \text{ kips}$$

For ASD Step 11: Determine the allowable strength.

$$P_n/\Omega = 93.0/1.67 = 55.7 \text{ kips}$$

5.11 COLUMN BASE PLATES

When columns are supported on material other than steel, such as concrete or masonry, it is necessary to distribute their load over an area significantly larger than the gross area of the column. In these situations, a column base plate similar to that shown in Figure 5.31 is used.

Column base plates may be attached to the column in the shop, as shown in Figure 5.31a, or shipped separately to the site and attached in the field. Columns are normally welded to the plate but may be attached with angles when large plates must be shipped separately. In either case, the selection of the dimensions and thickness of the plate follows the same rules.

Column base plates are normally attached to a footing or pier with anchor rods, and the space between the plate and the support is filled with a non-shrink grout. A leveling plate, leveling nuts, or shims (as shown in Figure 5.31b) are used to level the column base plate. In cases where the column supports an axial compression only, anchor rods are not designed to resist a specific force. However, all column base plates must be anchored with a minimum of four anchor rods according to the Occupational Safety and Health Administration (OSHA) regulations in *OSHA 29 CFR 1926 Subpart R Safety Standards for Steel Erection*. Figure 5.32 illustrates a column with base plate in plan (Figure 5.32a) and elevation (Figure 5.32b), including four anchor rods.

(a)

(b)

Figure 5.31 Example of a W-Shape Column and Base Plate
Photos courtesy Douglas Steel Fabricating Corporation

Column base plates are normally attached to a footing or pier with anchor rods, and the space between the plate and the support is filled with a non-shrink grout. A leveling plate, leveling nuts, or shims (as shown in Figure 5.31b) are used to level the column base plate. In cases where the column supports an axial compression only, anchor rods are not designed to resist a specific force. However, all column base plates must be anchored with a minimum of four anchor rods according to the Occupational Safety and Health Administration (OSHA) regulations in *OSHA 29 CFR 1926 Subpart R Safety Standards for Steel Erection*. Figure 5.32 illustrates a column with base plate in plan (Figure 5.32a) and elevation (Figure 5.32b), including four anchor rods.

To determine the area of bearing that is required, the strength of the material upon which the base plate is bearing must be evaluated. For concrete, Section J8 of the *Specification* gives provisions identical to those given in the concrete code, ACI 318. When the bearing plate is covering the full area of the concrete support, the nominal bearing strength is

$$P_n = P_p = 0.85 f_c' A_1 \qquad \text{(AISC J8-1)}$$

where f_c' is the specified concrete compressive strength and A_1 is the area of the plate and concrete. If the plate does not cover all of the concrete, there will be an increase in strength due to the spread of the load as it progresses down through the concrete. In this case the nominal bearing strength is given as

$$P_n = P_p = 0.85 f_c' A_1 \sqrt{A_2 / A_1} \leq 1.7 f_c' A_1 \qquad \text{(AISC J8-2)}$$

Here A_2 is the maximum area of concrete with the same shape as the bearing plate. The limit on the right side of the equation imposes a maximum ratio of areas of 4:1. If the supporting element is designed based on the bearing strength of the soil, it will be relatively easy to determine the extent to which the base plate covers the concrete foundation or pier. In all cases, $\phi = 0.65$ and $\Omega = 2.31$.

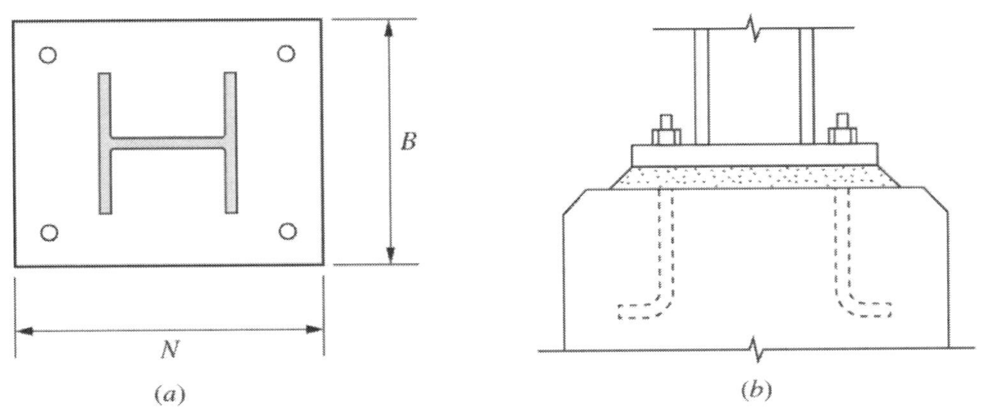

Figure 5.32 Column and Base Plate Section and Plan Including Anchor Rods

The thickness of a column base plate is a function of the bending strength of the plate. Since bending has not yet been covered in this book, this topic is deferred to Section 11.11. Those wishing to address base plate design further should proceed to Section 11.11 and to the example problems given there as well as AISC Design Guides 1 and 10.

5.12 PROBLEMS

1. Determine the theoretical buckling strength (Euler buckling load) for a W8×48 A992 column with an effective length of 20 ft. Will the theoretical column buckle or yield at this length?

2. Determine the theoretical buckling strength (Euler buckling load) for a W16×77 A36 column with an effective length of 12 ft. Will the theoretical column buckle or yield at this length?

3. Determine the theoretical buckling strength (Euler buckling load) for a W24×370 A992 column with an effective length of 20 ft. Will the theoretical column buckle or yield at this length?

4. Determine the theoretical buckling strength (Euler buckling load) for an HSS 10×5×3/8 A500 Grade C column with an effective length of 20 ft. Will the theoretical column buckle or yield at this length?

5. For a W12×72 A992 column, determine the effective length at which the theoretical buckling strength (Euler buckling load) will equal the yield strength.

6. For a W6×25 A992 column, determine the effective length at which the theoretical buckling strength (Euler buckling load) will equal the yield strength.

7. For an HP8×36 A572 Grade 50 column, determine the effective length at which the theoretical buckling strength (Euler buckling load) will equal the yield strength.

8. A W14×132 column has an effective length for *y*-axis buckling equal to 24 ft. Determine the effective length for the *x*-axis that will provide the same theoretical buckling strength (Euler buckling load).

9. A W14×53 column has an effective length for *x*-axis buckling equal to 20 ft. Determine the effective length for the *y*-axis that will provide the same theoretical buckling strength (Euler buckling load).

10. An HSS12×6×1/2 column has an effective length for *x*-axis buckling equal to 16 ft. Determine the effective length for the *y*-axis that will provide the same theoretical buckling strength (Euler buckling load).

11. A W14×132 A992 column has an effective length of 36 ft about both axes. Determine the available compressive strength for the column. Determine the (a) design strength by LRFD and (b) allowable strength by ASD. Is this an elastic or inelastic buckling condition?

12. Determine the available compressive strength for a W12×210 A992 column with an effective length about both axes of 40 ft. Determine the (a) design strength by LRFD and (b) allowable strength by ASD. Is this an elastic or inelastic buckling condition?

13. A W6×15 A992 column has an effective length of 8 ft about both axes. Determine the available compressive strength for the column. Determine the (a) design strength by LRFD and (b) allowable strength by ASD. Is this an elastic or inelastic buckling condition?

14. Determine the available compressive strength for an M10×7.5 A572 Gr 50 column with an effective length about both axes of 7 ft. Determine the (a) design strength by LRFD and (b) allowable strength by ASD. Is this an elastic or inelastic buckling condition?

15. A W14×211 A992 column has an effective length of 40 ft about both axes. Determine the available compressive strength for the column. Determine the (a) design strength by LRFD and (b) allowable strength by ASD. Is this an elastic or inelastic buckling condition?

16. Determine the available compressive strength for a W12×72 A992 column when the effective length is 20 ft about the *y*-axis and 40 ft about the *x*-axis. Determine the (a) design strength by LRFD and (b) allowable strength by ASD. Is this an elastic or inelastic buckling condition? Describe a common condition where the effective length is different about the different axes.

17. A W8×24 A992 column has an effective length of 12.5 ft about the *y*-axis and 28 ft about the *x*-axis. Determine the available compressive strength and indicate whether this is due to elastic or inelastic buckling. Determine the (a) design strength by LRFD and (b) allowable strength by ASD.

18. Determine the available compressive strength for an HSS 5×5×3/8 A500 Grade C column where the effective length is 10 ft about the *y*-axis and 15 ft about the *x*-axis. Determine the (a) design strength by LRFD and (b) allowable strength by ASD. Is this an elastic or inelastic buckling condition?

19. A round HSS 16.000×0.375 A500 Grade C column has an effective length of 20 ft. Determine the available compressive strength and indicate whether this is due to elastic or inelastic buckling. Determine the (a) design strength by LRFD and (b) allowable strength by ASD.

20. A W8×40 is used as a 12 ft column in a braced frame with W16×26 beams at the top and bottom as shown in Figure P5.20. The columns above and below are also 12 ft W8×40s. The beams provide moment restraint at each column end. Determine the effective length using the alignment chart and the available compressive strength, and the (a) design strength by LRFD and (b) allowable strength by ASD. Assume that the columns are oriented for (i) buckling about the weak axis and (ii) buckling about the strong axis. All steel is A992.

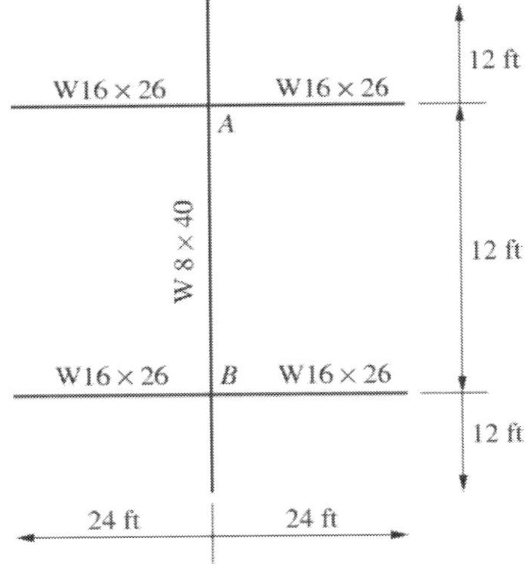

P5.20

21. If the structure described in Problem 20 is an unbraced frame, determine the effective length and compressive strength as requested in Problem 20.

22. A W12×136 column is shown in Figure P5.22 with end conditions that approximate ideal conditions. Using the recommended approximate values from Commentary Table C-A-7.1, determine the effective lengths for the *y*-axis and the *x*-axis. Which effective length will control the column strength?

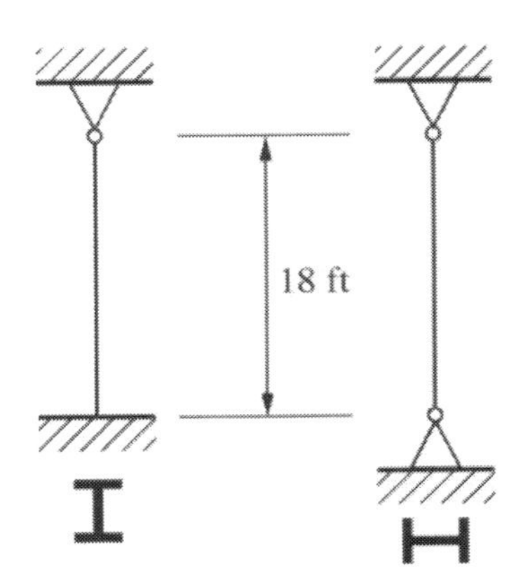

P5.22

23. A W12×96 column is shown in Figure P5.23 with end conditions that approximate ideal conditions. Using the recommended approximate values from Commentary Table C-A-7.1, determine the effective lengths for the *y*-axis and the *x*-axis. Which effective length will control the column strength?

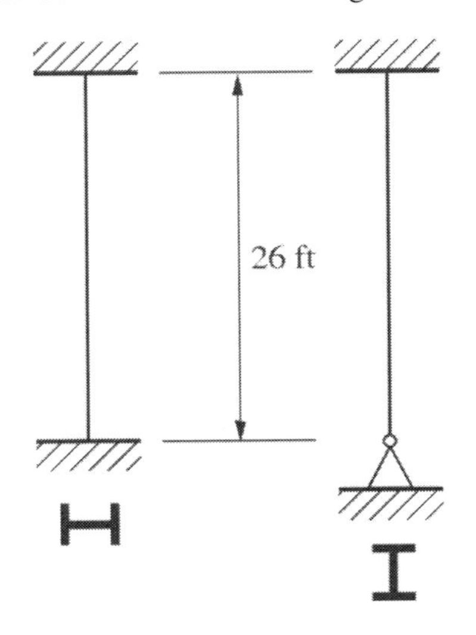

P5.23

24. A W10×54 column with an effective length of 30 ft for both axes is called upon to carry a compressive dead load of 80 kips and a compressive live load of 100 kips. Determine whether the column will support the load by (a) LRFD and (b) ASD. Evaluate the strength for (i) $F_y = 50$ ksi and (ii) $F_y = 70$ ksi.

25. A W14×257 A992 column in a building has effective lengths of 16 ft for both axes. Determine whether the column will carry a compressive dead load of 800 kips and a compressive live load of 1100 kips by (a) LRFD and (b) ASD.

26. An A992 W12×53 is used as a column in a building with an effective length for each axis of 15 ft. Determine whether the column will carry a compressive dead load of 85 kips and a compressive live load of 255 kips by (a) LRFD and (b) ASD.

27. An A992 W8×58 is used in a structure to support a dead load of 60 kips and a live load of 100 kips. The column has an effective length of 22 ft. Determine whether the column will support the load by (a) LRFD and (b) ASD.

28. An A992 W10×54 is used as a column in a building with an effective length of 30 ft. Determine whether the column will carry a compressive dead load of 24 kips and a compressive live load of 72 kips by (a) LRFD and (b) ASD.

29. An A992 W16×77 is used as a column in a building to support a dead load of 130 kips and a live load of 200 kips. The column effective length is 20 ft for the *y*-axis and 30 ft for the *x*-axis. Determine whether the column will support the load by (a) LRFD and (b) ASD.

30. An A992 W21×111 is used as a column in a building to support a dead load of 120 kips and a live load of 300 kips. The column effective length is 22 ft for the *y*-axis and 33 ft for the *x*-axis. Determine whether the column will support the load by (a) LRFD and (b) ASD.

31. An A992 W24×146 is used as a column in a building to support a dead load of 245 kips and a live load of 500 kips. The column has an effective length about the *y*-axis of 18 ft and an effective length about the *x*-axis of 36 ft. Determine whether the column will support the load by (a) LRFD and (b) ASD.

32. An A500 Gr. C HSS7×7×1/2 is used as a column to support a dead load of 175 kips and a live load of 100 kips. The column has an effective length of 10 ft. Determine whether the column will support the load by (a) LRFD and (b) ASD.

33. For the W10×77 column with bracing and end conditions shown in Figure P5.33, determine the theoretical effective length for each axis and identify the axis that will limit the column strength.

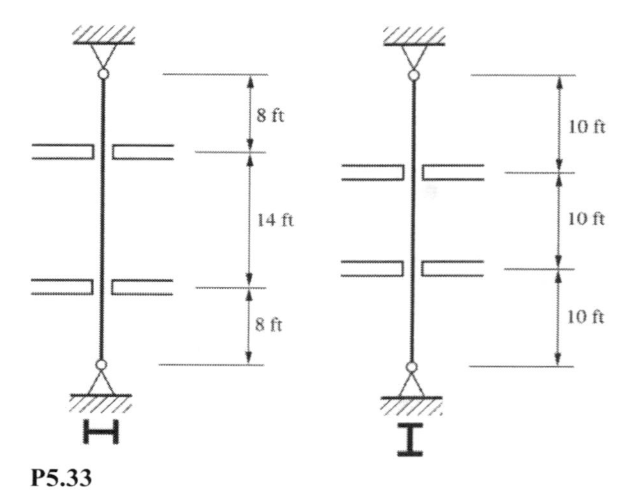

P5.33

34. For the W8×24 column with bracing and end conditions shown in Figure P5.34, determine the theoretical effective length for each axis and identify the axis that will limit the column strength.

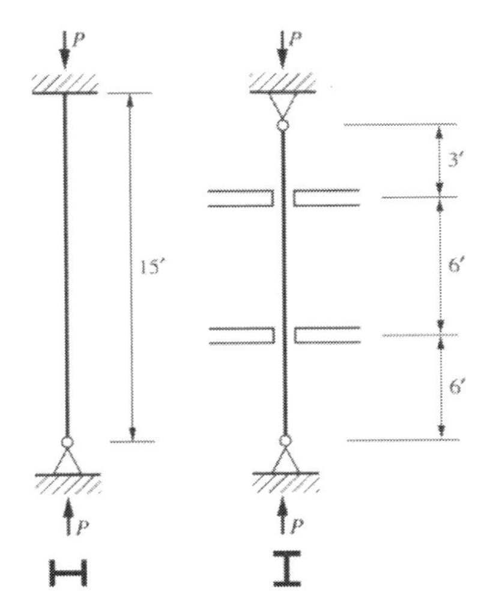

P5.34

35. A W10×100 column with end conditions and bracing is shown in Figure P5.35. Determine the least theoretical bracing and its location about the y-axis, in order that the y-axis not control the strength of the column.

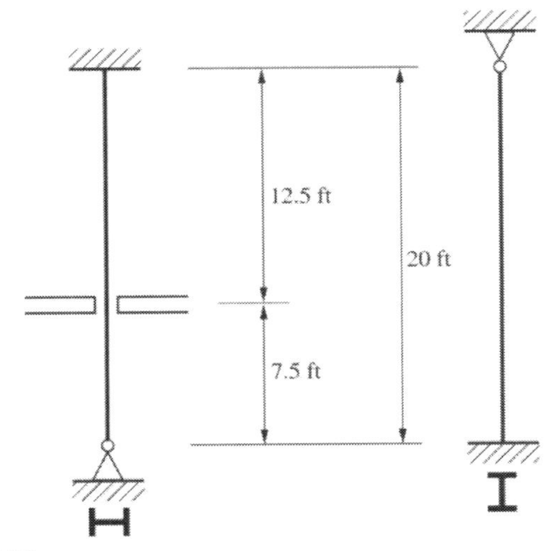

P5.35

36. A W14×176 column with end conditions and bracing is shown in Figure P5.36. Determine the least theoretical bracing and its location about the y-axis, in order that the y-axis not control the strength of the column.

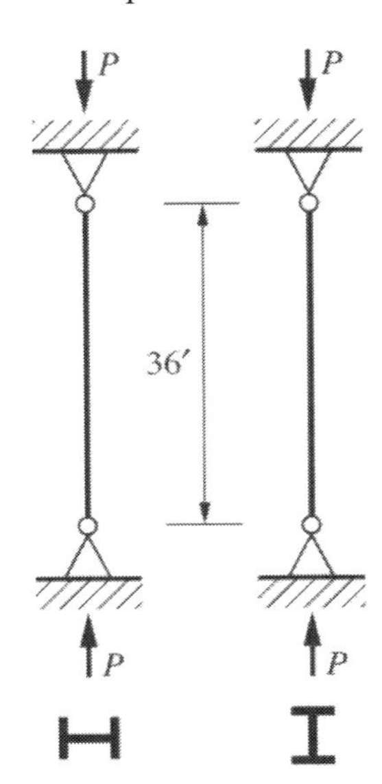

P5.36

37. A W12×72 column is an exterior 2nd story column (gridline A) with strong axis buckling in the plane of the frame in an unbraced multi-story frame. The column below is also a W12×72. Beams and dimensions are as shown in Figure P5.37. Determine the effective length for this condition and the corresponding compressive strength by (a) LRFD and (b) ASD. All steel is A992.

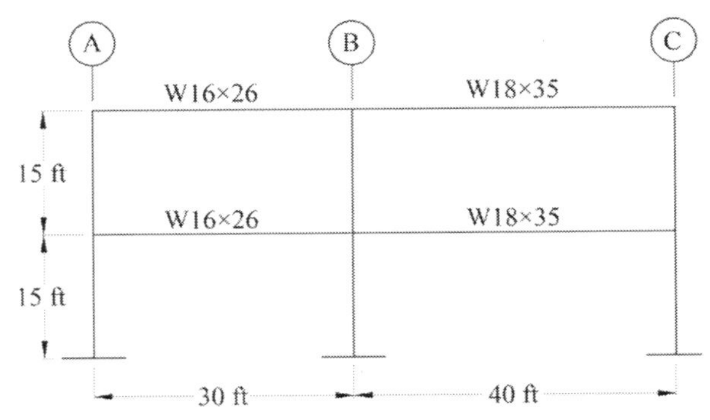

P5.37

38. A W12×72 column is an exterior 2nd story column (at gridline C) with strong axis buckling in the plane of the frame in an unbraced multi-story frame. The column below is also a W12×72. Beams and dimensions are as shown in Figure P5.37. Determine the effective length for this condition and the corresponding compressive strength by (a) LRFD and (b) ASD. All steel is A992.

39. A W12×72 column is an exterior 1st story column (at gridline C) with strong axis buckling in the plane of the frame in an unbraced multi-story frame. The column above is also a W12×72. Beams and dimensions are as shown in Figure P5.37. Using the AISC Commentary recommended stiffness, *G*, for the base of a fixed base column, determine the effective length for this condition and the corresponding compressive strength by (a) LRFD and (b) ASD. All steel is A992.

40. A W12×72 column is an interior 2nd story column (at gridline B) with strong axis buckling in the plane of the frame in an unbraced multi-story frame. The column below is also a W12×72. Beams and dimensions are as shown in Figure P5.37. Determine the effective length for this condition and the corresponding compressive strength by (a) LRFD and (b) ASD. All steel is A992.

41. A W12×72 column is an interior 1st story column (at gridline B) with strong axis buckling in the plane of the frame in an unbraced multi-story frame. The column above is also a W12×72. Beams and dimensions are as shown in Figure P5.37. Using the AISC Commentary recommended stiffness, *G*, for the base of a fixed base column, determine the effective length for this condition and the corresponding compressive strength by (a) LRFD and (b) ASD. All steel is A992.

42. Repeat Problem 39 if the column support were given as a pin.

43. Repeat Problem 41 if the column support were given as a pin.

44. A W12×50 column is an interior column with strong axis buckling in the plane of the frame in an unbraced multi-story frame. The columns above and below are also W12×50. The beams framing in at the top are W16×31 and those at the bottom are W16×40. The columns are 14 ft and the beam span is 25 ft. The column carries a dead load of 75 kips and a live load of 150 kips. Determine the inelastic effective length for this condition and the corresponding compressive strength by (a) LRFD and (b) ASD. All steel is A992.

45. Select the least-weight W12 A992 column to carry a live load of 130 kips and a dead load of 100 kips with an effective length about both axes of 15 ft by (a) LRFD and (b) ASD.

46. Select the least-weight W14 A992 column to carry a dead load of 200 kips and a live load of 600 kips if the effective length about both axes is 22 ft by (a) LRFD and (b) ASD.

47. Select the least-weight W10 A992 column to carry a dead load of 80 kips and a live load of 280 kips with an effective length about both axes of 15 ft by (a) LRFD and (b) ASD.

48. Select the least-weight W8 A992 column to carry a dead load of 20 kips and a live load of 50 kips with an effective length about both axes of 25 ft by (a) LRFD and (b) ASD.

49. Select the least-weight W6 A992 column to carry a dead load of 12 kips and a live load of 36 kips with an effective length about both axes of 8 ft by (a) LRFD and (b) ASD.

50. Select the least-weight W8 A992 column to carry a dead load of 13 kips and a live load of 39 kips with an effective length about both axes of 14 ft by (a) LRFD and (b) ASD.

51. A column with pin ends for both axes must be selected to carry a compressive dead load of 95 kips and a compressive live load of 285 kips. The column is 14 ft long and is in a braced frame. Select the lightest-weight W12 to support this load by (a) LRFD and (b) ASD.

52. If the column in Problem 48 had an effective length of 32 ft, select the lightest-weight W12 to support this load by (a) LRFD and (b) ASD.

53. A W14 A992 column must support a dead load of 80 kips and a live load of 300 kips. The column is 22 ft long and has end conditions that approximate the ideal conditions of a fixed support at one end and a pin support at the other. Select the lightest-weight W14 to support this load by (a) LRFD and (b) ASD.

54. Select the least-weight W8 A992 column to support a dead load of 170 kips with an effective length of 16 ft by (a) LRFD and (b) ASD.

55. A column with an effective length of 21 ft must support a dead load of 120 kips, a live load of 175 kips, and a wind load of 84 kips. Select the lightest W14 A992 member to support the load by (a) LRFD and (b) ASD.

56. A W14×99 A992 column is 20 ft long, pinned at each end, and braced at mid-height to prevent lateral movement for buckling about the *y*-axis. However, the *y*-axis bracing is not adequate to resist torsion. Considering flexural and torsional buckling, determine the nominal strength of this compression member.

57. An A36 single-angle compression web member of a truss is 10 ft long and attached to gusset plates through the same leg at each end with a minimum of two bolts. The member must carry a dead load of 8 kips and a live load of 10 kips. Select the least-weight equal leg angle to carry this load by (a) LRFD and (b) ASD.

58. If the compression web member of Problem 57 were loaded concentrically, determine the least-weight single angle to carry the load by (a) LRFD and (b) ASD.

59. Determine the web and flange width-to-thickness ratios and determine if a W14×43 A913 Gr 70 compression member requires consideration as a slender element member.

60. Determine the web and flange width-to-thickness ratios and determine if a W8×10 A992 compression member requires consideration as a slender element member.

61. A W16×36 A992 compression member has a slender web when used in uniform compression. Determine the available strength by (a) LRFD and (b) ASD when the effective length is (i) 6 ft and (ii) 12 ft.

62. The W14×43 is the only A992 column shown in the *Manual* column tables that has a slender web. Determine the available strength for this column if the effective length is 8 ft and show whether the slender web impacts that strength by (a) LRFD and (b) ASD.

63. Determine the available strength of a WT6×25 A992 steel compression chord of a truss with effective length L_c = 14 ft. Consider the member braced laterally and torsionally at its ends only. Determine by (a) LRFD and (b) ASD.

64. Determine the available strength of a WT12×125 A992 steel column with effective length L_c = 18 ft. Consider the member braced laterally and torsionally at its ends only. Determine by (a) LRFD and (b) ASD.

65. Determine the available strength of a C9×20 A992 steel compression chord of a truss with effective length, L_c = 16 ft. Consider the member braced laterally and torsionally at its ends only. Determine by (a) LRFD and (b) ASD.

66. Determine the available strength of an A572 Gr 50, 20 ft long, 2L6×4×5/8 double-angle compression member in a planar truss. The angles are long legs back-to-back with a 3/8 in. gap. The angles are welded at each end to a gusset plate and along the length at two intermediate points with a spacing of 80 in. Determine the (a) design strength by LRFD and (b) allowable strength by ASD. Compare the results to that found in *Manual* Table 4-9.

67. Determine the available strength of an A572 Gr 50, 15 ft long, 2L6×3-1/2×1/2 double-angle compression member in a planar truss. The angles are long legs back-to-back with a 3/8 in. gap. The angles are connected with pretensioned bolts at each end to a gusset plate and along the length at two intermediate points with snug-tight bolts at a spacing of 60 in. Determine the (a) design strength by LRFD and (b) allowable strength by ASD. Compare the results to that found in *Manual* Table 4-9 and explain why they differ.

68. Determine the available strength of an A572 Gr 50, 18 ft long, 2L3×3×1/2 double-angle compression member in a planar truss. The angles have a back-to-back gap of 3/8 in. The angles are welded at each end to a gusset plate and along the length at two intermediate points with a spacing of 72 in. Determine the (a) design strength by LRFD and (b) allowable strength by ASD. Compare the results to that found in *Manual* Table 4-9.

69. Redo Problem 68 if the intermediate points are connected with snug-tight bolts and compare the results to Problem 68.

Multi-Chapter Problems

70. Using the framing plan shown in Figure P5.70 (presented earlier as Figure 2.9), design the columns marked 4 and 5. This is the same structure used in Section 2.5, where load calculations with live load reductions were discussed. Those calculations can be reused here. Load case 2 for dead plus live load is to be considered. The building is an office building with a nominal live load of 50 pounds per square foot (psf) and a calculated dead load of 70 psf.

4: Interior column D-2 regardless of deck span direction
5: Exterior column D-4 regardless of deck span direction

Design for column length L = 14 ft and K = 1.0 using (a) LRFD and (b) ASD.

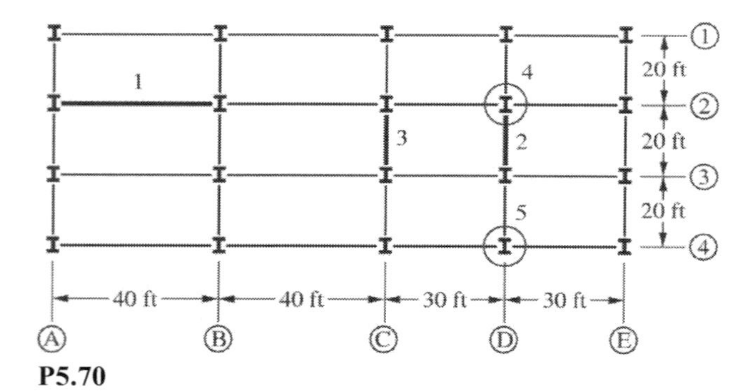

P5.70

71. The framing plan shown in Figure P5.71 is the same as the one shown in Figure P2.24 for an 18-story office building. It must support a floor and roof dead load of 80 psf and a floor live load of 50 psf and a roof live load of 30 psf. In all cases, the decking spans in a direction from line A toward line E. Determine the required axial strength for the columns and design the columns as required below for (a) design by LRFD and (b) design by ASD. The required axial load strengths were determined in Problem 24 of Chapter 2. Use a story height of 13.5 ft in a braced frame so that K = 1.0.

iv: The column at the corner on lines 1 and E that supports eight floor levels plus the roof.

v: The column on the edge at the intersection of lines 4 and A that supports eight floor levels plus the roof.

vi: The interior column at the intersection of line 4 and the point between lines C and D that supports three floor levels plus the roof.

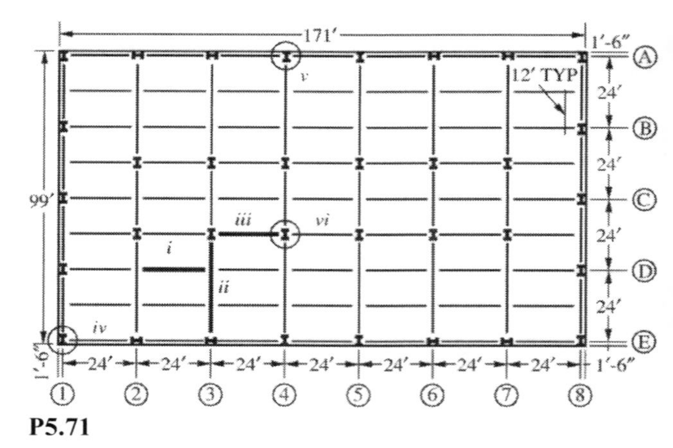

P5.71

72. Integrated Design Project. Using the gravity column loads determined in Chapter 2, design the gravity-only columns. Design columns as single-story members. (It is often more economical to use multi-story columns because of construction costs.) Select the final columns so that they are two-story from below grade to the second floor, two-story from the second floor to the fourth floor, and then use a single-story column to support the roof.

Design the columns in the braced frame for the gravity loads determined in Chapter 2 and the wind load determined in Chapter 4. Remember that the wind load must be considered to act in two directions, so use the largest compression forces from wind to combine with the gravity loads.

Using the wind load analysis from Chapter 4, design all the braces as compression members. Compare the tension design with the compression design and select the appropriate final members.

Chapter 6

Bending Members

One Vanderbilt Avenue, New York, NY
Photo courtesy of Severud Associates,

6.1 BENDING MEMBERS IN STRUCTURES

A *bending member* carries load applied normal to its longitudinal axis and transfers it to its support points through bending moments and shears. In building construction, the most common application of bending members is to provide support for floors or roofs. These beams can be either simple span or continuous, and normally transfer their load to other structural members such as columns, girders, and walls. Although the terms *beam* and *girder* are often used interchangeably because both are bending members, *beam* normally refers to a bending member that directly supports an applied load whereas *girder* usually refers to a bending member that supports a beam. The distinction is not important for design because the same criteria apply to all bending members.

The most commonly used shapes for bending members are the I-shaped cross sections and, of these, the W-shape is dominant. However, there are numerous situations where other shapes are used as bending members. L-shapes are commonly used as lintels over openings, T-shapes are found as chords of trusses that may be called upon to resist bending along with axial forces, and C-shapes may coexist with W-shapes in floor systems.

In addition to the use of the standard shapes, engineers often find it necessary to develop their own shapes by combining shapes and/or plates. Several examples of these built-up shapes are shown in Figure 6.1. Although the use of these built-up shapes is permitted by the *Specification*, they may not be economical, except for particular applications where the added labor costs associated with fabrication can be justified. Otherwise, standard rolled (I-shapes) and manufactured (HSS) shapes should be considered. The complexity that results from the wide variety of possible shapes is the reason so many separate provisions are included in Chapter F of the *Specification*. Shapes made up of plates, like shown in Figure 6.1(a) and (b), are generally referred to as plate girders and they will be addressed here in Chapter 7.

The most common and economical bending members are those that can reach the full material yield strength without being limited by buckling of any cross-sectional elements. These members are referred to as compact members and are addressed first.

Table 6.1 lists the sections of the *Specification* and parts of the *Manual* that are discussed in this chapter.

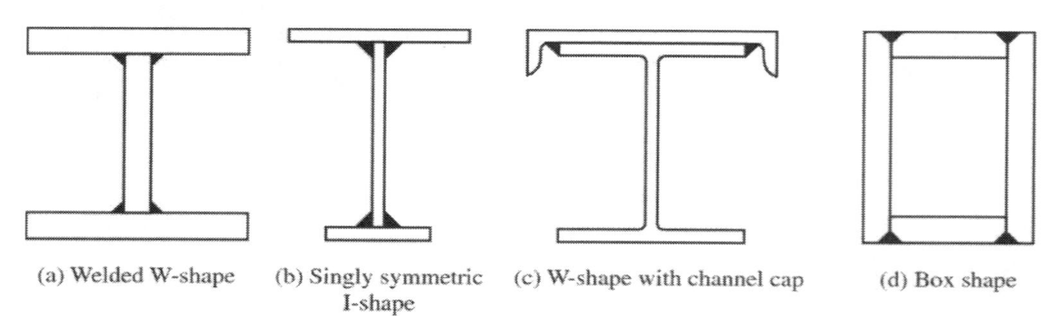

Figure 6.1 Built-up Beams

6.2 STRENGTH OF BEAMS

As load is applied to a bending member, resulting in a bending moment, stresses are developed in the cross section. For a load at or below the nominal level, the load magnitude established in the building code, it is reasonable to expect the entire beam cross section to behave elastically. The stresses and strains are distributed as shown in Figure 6.2a. This elastic behavior occurs whenever the material's behavior corresponds to the initial straight-line portion of the stress-strain curve of Figure 3.5.

From the basic principles of strength of materials, the relationship between the applied moment and resulting stresses is given by the familiar flexure formula:

$$f_y = \frac{My}{I} \tag{6.1}$$

where

M = any applied moment that stresses the section in the elastic range
y = distance from the neutral axis to the point where the stress is to be determined
I = moment of inertia
f_y = resulting bending stress at location y

Table 6.1 Sections of *Specification* and Parts of *Manual* Covered in This Chapter

	Specification
B3	Design Basis
B4	Classification of Sections for Local Buckling
F1	General Provisions
F2	Doubly Symmetric Compact I-Shaped Members and Channels Bent about Their Major Axis
F3	Doubly Symmetric I-Shaped Members with Compact Webs and Noncompact or Slender Flanges Bent about Their Major Axis
F6	I-Shaped Members and Channels Bent about Their Minor Axis
F9	Tees and Double Angles Loaded in the Plane of Symmetry
F10	Single Angles
Chapter G	Design of Members for Shear
H1	Doubly and Singly Symmetric Members Subject to Flexure and Axial Force
J10	Flanges and Webs with Concentrated Forces
Chapter L	Design for Serviceability
Appendix 1	Design by Advanced Analysis
	Manual
Part 1	Dimensions and Properties
Part 3	Design of Flexural Members

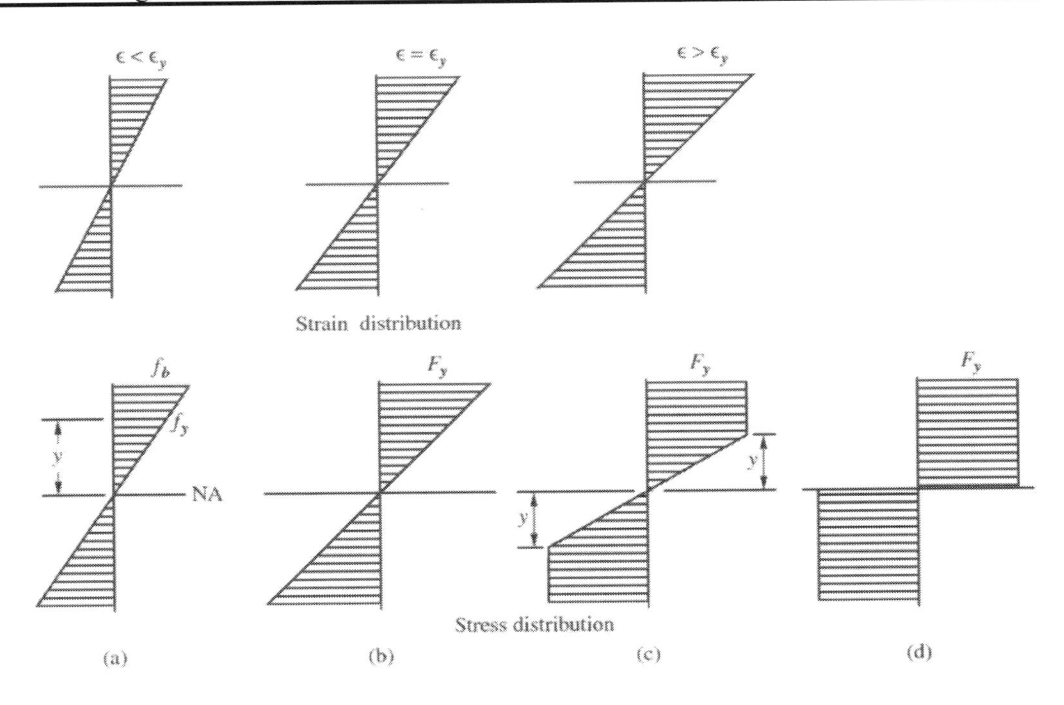

Figure 6.2 Cross-Sectional Bending Stresses and Strains: (a) elastic; (b) yield; (c) partial plastic; (d) plastic

Normally the stress at the extreme fiber, that is, the fiber most distant from the neutral axis, is of interest because the greatest stress occurs at this point. The distance from the neutral axis to the extreme fiber may be taken as c and the flexure formula becomes

$$f_b = \frac{Mc}{I} = \frac{M}{S} \tag{6.2}$$

where

S = section modulus
f_b = extreme fiber bending stress

The moment that causes the extreme fiber to reach the yield stress, F_y, is called the *yield moment*, M_y. The corresponding stress and strain diagrams are shown in Figure 6.2b. If the load is increased beyond the yield moment, the strain in the extreme fiber increases; however, the stress remains at F_y because these fibers are behaving as depicted by the plateau on the stress-strain diagram, shown previously in Figures 3.5 and 3.6. The stress at some points on the cross section closer to the neutral axis also reach the yield stress while those even closer, from the neutral axis to the location dimensioned as y in Figure 6.2c, remain elastic.
As the moment continues to increase, the portion of the cross section experiencing the yield stress continues to increase until the entire section experiences the yield stress, as shown in Figure 6.2d. Note that there is no corresponding strain diagram given for this loading condition since the strain would be unlimited over the entire cross section. The moment that causes this stress distribution is called the *plastic moment*, M_p. Since the strain is unlimited, the cross section undergoes unlimited rotation, and the condition is referred to as a plastic hinge. A cross section that is capable of attaining this stress distribution and corresponding moment is referred to as a compact section. At every stage of loading, equilibrium of the cross section requires, at all times, that the total internal tension force be equal to the total internal compression force. The basic principles of strength of materials are addressed in numerous texts, such as *Mechanics of Materials.*[1]

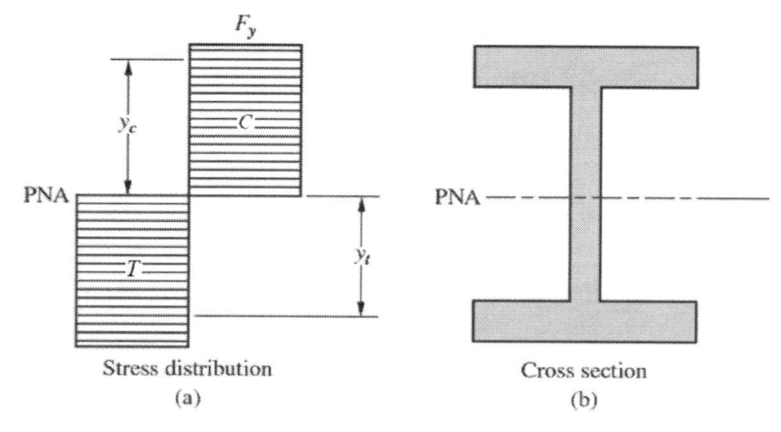

Figure 6.3 Equilibrium in a Doubly Symmetric Wide-Flange Shape

[1] Philpot, Timothy. *Mechanics of Materials: An Integrated Learning System*. Wiley, 2013.

For the doubly symmetric wide flange shape shown in Figure 6.3, equilibrium for the plastic moment occurs when the portion of the shape above tne elastic neutral axis is stressed to the yield stress in compression while the portion below the elastic neutral axis is stressed to the yield stress in tension. For shapes that are not symmetric about the axis being considered, the area above the elastic neutral axis is not equal to the area below the elastic neutral axis. Thus, a new axis that gives equal areas in tension and compression must be defined. This new axis dividing the section into two equal areas is the *plastic neutral axis* (PNA). For symmetric shapes the elastic and plastic neutral axes coincide, as was the case for the wide flange. For shapes that are not symmetric about the axis being considered, these neutral axes are at different locations.

Because equilibrium means that the tension and compression forces are equal and opposite, they form a force couple. Although moments can be taken about any reference point for a force couple, it is common practice to take moments about the PNA. The moment that corresponds to this fully yielded stress distribution, M_p, is determined from

$$M_p = F_y(A_c\, y_c) + F_y(A_t\, y_t) \tag{6.3}$$

where A_t and A_c are the equal tension and compression areas, respectively, and y_c and y_t are the distances from the centroid of the area to the PNA for the tension and compression areas, respectively. Equation 6.3 may be simplified to

$$M_p = F_y\left(\frac{A}{2}\right)(y_c + y_t) \tag{6.4}$$

The two terms multiplied by the yield stress are functions only of the geometry of the cross section and are normally combined and called the *plastic section modulus*, Z. Thus, the plastic moment is given as

$$M_p = F_y Z \tag{6.5}$$

The plastic section modulus is tabulated for all available shapes in Part 1 of the *Manual*.

Chapter F of the *Specification* contains the provisions for design of flexural members subject to bending. For a given beam to attain its full plastic moment strength, it must be compact and satisfy the lateral support criteria established in Section F2. If these criteria are not met, the strength is defined as something less than M_p. The criteria to be satisfied are defined by two limit states in addition to yielding: local buckling (a noncompact shape) and lateral torsional buckling. These limit states and their impact on beam strength are discussed in Sections 6.4 and 6.5.

EXAMPLE 6.1
Plastic Moment Strength for a Symmetric Shape

Goal: Determine the plastic moment strength of a W-shape using the model of three rectangular plates and compare the calculated plastic section modulus to the value determined from the *Manual*.

Given: A W24×192 is modeled as shown in Figure 6.4. Assume $F_y = 50$ ksi.

SOLUTION

Step 1: Determine the location of the plastic neutral axis.

Because the shape is symmetric, the plastic neutral axis is located on the axis of symmetry.

Step 2: Determine the plastic section modulus as the sum of the moments of the areas about the plastic neutral axis.

$$Z = \frac{A}{2}(y_c + y_t) = 2\left(A_f y_f + \frac{A_w}{2} y_w\right)$$

$$Z = 2\left[13.0(1.46)\left(\frac{22.58}{2} + \frac{1.46}{2}\right) + \frac{22.58(0.810)}{2}\left(\frac{22.58}{4}\right)\right] = 560 \text{ in.}^3$$

Step 3: Determine the plastic moment strength as the plastic section modulus times the yield stress using Equation 6.5

$$M_p = F_y Z = 50(560) = 28{,}000 \text{ in.-kips}$$

or

$$M_p = 28{,}000/12 = 2330 \text{ ft-kips}$$

Step 4: Compare the calculated plastic section modulus value with that from *Manual* Table 1-1.

From the table, $Z_x = 559$ in.^{3}.

This shows that the impact of the simplification in using rectangular plates and ignoring the fillets at the flange-web junction is small.

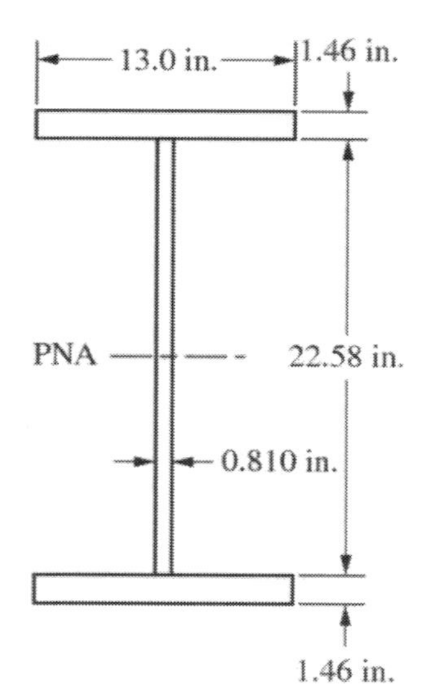

Figure 6.4 W24×192 Model for Example 6.1

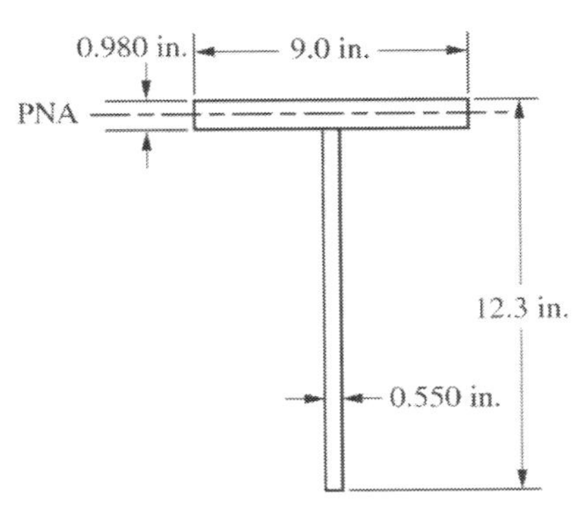

Figure 6.5 T-Beam Model for Example 6.2

EXAMPLE 6.2
Plastic Section Modulus for a Nonsymmetric Shape

SOLUTION

Goal: Locate the plastic neutral axis and determine the plastic section modulus for a WT.

Given: A WT12×51.5 modeled as two plates is shown in Figure 6.5. Assume that $F_y = 50$ ksi.

Step 1: Determine the area of the T-shape.

$$A_{flange} = 9.00(0.980) = 8.82 \text{ in.}^2$$

$$A_{stem} = 0.550(12.3 - 0.980) = 6.23 \text{ in.}^2$$

$$A_{total} = 8.82 + 6.23 = 15.1 \text{ in.}^2$$

Step 2: Determine one-half of the area, because one-half of the area must be above the plastic neutral axis and one-half must be below.

$$\frac{A_{total}}{2} = \frac{15.1}{2} = 7.55 \text{ in.}^2$$

Step 3: Determine whether the plastic neutral axis is in the flange or the stem. Because half of the area is less than the area of the flange, the plastic neutral axis is in the flange and

$$y_p = \frac{7.55}{9.0} = 0.839 \text{ in.}$$

with the plastic neutral axis measured from the top of the flange.

Step 4: Determine the plastic section modulus as the sum of the moments of the areas about the plastic neutral axis.

$$Z = 7.55\left(\frac{0.839}{2}\right) + (8.82 - 7.55)\left(\frac{0.980 - 0.839}{2}\right)$$

$$+ 6.23\left(0.980 - 0.839 + \frac{12.3 - 0.980}{2}\right)$$

$$= 3.17 + 0.0895 + 36.1 = 39.4 \text{ in.}^3$$

Step 5: Compare these values with the values in *Manual* Table 1.8.

$$y_p = 0.841 \text{ in.} \quad \text{and} \quad Z = 39.2 \text{ in.}^3$$

This shows that the impact of the simplification in using rectangular plates and ignoring the fillets at the flange-web junction is small.

Another useful term is the shape factor defined as the ratio of the plastic section modulus to the elastic section modulus, shape factor $= Z/S$. Values for the shape factor are not given in the *Manual* but are readily determined and their significance will be discussed later.

6.3 DESIGN OF COMPACT LATERALLY SUPPORTED WIDE-FLANGE BEAMS

Section F1(b) of the *Specification* requires that all beams be restrained at the supports against twist about their longitudinal axis. For a compact beam to attain its plastic moment strength, it must also be laterally supported at some specified length along its compression flange. This may be accomplished through the attachment of intermediate beams or by attaching decking at regular intervals along the length of the member, as shown in Figure 6.6. When this is the case the beam is said to have full lateral support. Section 6.2 showed that the nominal strength of a compact member with full lateral support—that is, in which no buckling limit states control—is determined by the limit state of yielding. For this limit state, *Specification* Section F2 provides that

$$M_n = M_p = F_y Z \qquad \text{(AISC F2-1)}$$

Specification Section F1 also indicates that for all flexural limit states, design strength and allowable strength are to be determined using

$$\phi = 0.90 \;(\text{LRFD}) \quad \Omega = 1.67 \;(\text{ASD})$$

Figure 6.6 Lateral Support of Beams using Steel Deck

The design basis from Sections B3.1 and B3.2, discussed in Chapter 1, are repeated here. The requirement for ASD is

$$R_a \le \frac{R_n}{\Omega} \qquad \text{(AISC B3-2)}$$

The requirement for LRFD is

$$R_u \le \phi R_n \qquad \text{(AISC B3-1)}$$

Figure 6.7 shows a portion of *Manual* Table 3-2, which lists W-shapes in order of major axis plastic section modulus. Shapes are grouped so the least weight member with the largest plastic section modulus is given in bold at the top of each group. This facilitates selection of the least weight W-shape for beams controlled by the limit state of yielding. Since A992 is the preferred specification for W-shapes, the tabulated strength values in Table 3-2 are based on F_y = 50 ksi. This figure will be used in the following example to aid in the design of a compact, laterally supported beam.

EXAMPLE 6.3a
Beam Design by LRFD

Goal: Select the least-weight wide flange member for the conditions given.

Given: An A992 beam, simply supported at both ends, spans 20 ft and is loaded at midspan with a dead load of 8.0 kips and a live load of 24.0 kips, as shown in Figure 6.8, in addition to its self-weight. Assume full lateral support and a compact section.

SOLUTION

Step 1: Determine the required strength using the governing LRFD load combination from Section 2.4.

$$P_u = 1.2P_D + 1.6P_L = 1.2(8.0) + 1.6(24.0) = 48.0 \text{ kips}$$

$$M_u = \frac{P_u L}{4} = \frac{48(20)}{4} = 240 \text{ ft-kips}$$

Step 2: Determine the required plastic section modulus. For a compact, fully braced section,

$$M_n = M_p = F_y Z$$

Thus, because *Specification* Section B3.1 provides that the required moment be less than or equal to the available moment, $M_u \le \phi M_n = \phi F_y Z$, and

$$Z_{req} = \frac{M_u}{\phi F_y} = \frac{240(12)}{0.90(50)} = 64.0 \text{ in.}^3$$

Z_x

Table 3-2 (continued)
W-Shapes
Selection by Z_x

$F_y = 50$ ksi

Shape	Z_x	M_{px}/Ω_b	$\phi_b M_{px}$	M_{rx}/Ω_b	$\phi_b M_{rx}$	BF/Ω_b	$\phi_b BF$	L_p	L_r	I_x	V_{nx}/Ω_v	$\phi_v V_{nx}$
		kip-ft	kip-ft	kip-ft	kip-ft	kips	kips				kips	kips
	in.^{3}	ASD	LRFD	ASD	LRFD	ASD	LRFD	ft	ft	in.^{4}	ASD	LRFD
W18×35	**66.5**	**166**	**249**	**101**	**151**	**8.14**	**12.3**	**4.31**	**12.3**	**510**	**106**	**159**
W12×45	64.2	160	241	101	151	3.80	5.80	6.89	22.4	348	81.1	122
W16×36	64.0	160	240	98.7	148	6.24	9.36	5.37	15.2	448	93.8	141
W14×38	61.5	153	231	95.4	143	5.37	8.20	5.47	16.2	385	87.4	131
W10×49	60.4	151	227	95.4	143	2.46	3.71	8.97	31.6	272	68.0	102
W8×58	59.8	149	224	90.8	137	1.70	2.55	7.42	41.6	228	89.3	134
W12×40	57.0	142	214	89.9	135	3.66	5.54	6.85	21.1	307	70.2	105
W10×45	54.9	137	206	85.8	129	2.59	3.89	7.10	26.9	248	70.7	106
W14×34	**54.6**	**136**	**205**	**84.9**	**128**	**5.01**	**7.55**	**5.40**	**15.6**	**340**	**79.8**	**120**
W16×31	**54.0**	**135**	**203**	**82.4**	**124**	**6.86**	**10.3**	**4.13**	**11.8**	**375**	**87.5**	**131**
W12×35	51.2	128	192	79.6	120	4.34	6.45	5.44	16.6	285	75.0	113
W8×48	49.0	122	184	75.4	113	1.67	2.55	7.35	35.2	184	68.0	102
W14×30	**47.3**	**118**	**177**	**73.4**	**110**	**4.63**	**6.95**	**5.26**	**14.9**	**291**	**74.5**	**112**
W10×39	46.8	117	176	73.5	111	2.53	3.78	6.99	24.2	209	62.5	93.7
W16×26[v]	**44.2**	**110**	**166**	**67.1**	**101**	**5.93**	**8.98**	**3.96**	**11.2**	**301**	**70.5**	**106**
W12×30	43.1	108	162	67.4	101	3.97	5.96	5.37	15.6	238	64.0	95.9
W14×26	**40.2**	**100**	**151**	**61.7**	**92.7**	**5.33**	**8.11**	**3.81**	**11.0**	**245**	**70.9**	**106**
W8×40	39.8	99.3	149	62.0	93.2	1.64	2.46	7.21	29.9	146	59.4	89.1
W10×33	38.8	96.8	146	61.1	91.9	2.39	3.62	6.85	21.8	171	56.4	84.7
W12×26	**37.2**	**92.8**	**140**	**58.3**	**87.7**	**3.61**	**5.46**	**5.33**	**14.9**	**204**	**56.1**	**84.2**
W10×30	36.6	91.3	137	56.6	85.1	3.08	4.61	4.84	16.1	170	63.0	94.5
W8×35	34.7	86.6	130	54.5	81.9	1.62	2.43	7.17	27.0	127	50.3	75.5
W14×22	**33.2**	**82.8**	**125**	**50.6**	**76.1**	**4.78**	**7.27**	**3.67**	**10.4**	**199**	**63.0**	**94.5**
W10×26	31.3	78.1	117	48.7	73.2	2.91	4.34	4.80	14.9	144	53.6	80.3
W8×31[f]	30.4	75.8	114	48.0	72.2	1.58	2.37	7.18	24.8	110	45.6	68.4
W12×22	**29.3**	**73.1**	**110**	**44.4**	**66.7**	**4.68**	**7.06**	**3.00**	**9.13**	**156**	**64.0**	**95.9**
W8×28	27.2	67.9	102	42.4	63.8	1.67	2.50	5.72	21.0	98.0	45.9	68.9
W10×22	**26.0**	**64.9**	**97.5**	**40.5**	**60.9**	**2.68**	**4.02**	**4.70**	**13.8**	**118**	**49.0**	**73.4**
W12×19	**24.7**	**61.6**	**92.6**	**37.2**	**55.9**	**4.27**	**6.43**	**2.90**	**8.61**	**130**	**57.3**	**86.0**
W8×24	23.1	57.6	86.6	36.5	54.9	1.60	2.40	5.69	18.9	82.7	38.9	58.3
W10×19	**21.6**	**53.9**	**81.0**	**32.8**	**49.4**	**3.18**	**4.76**	**3.09**	**9.73**	**96.3**	**51.0**	**76.5**
W8×21	20.4	50.9	76.5	31.8	47.8	1.85	2.77	4.45	14.8	75.3	41.4	62.1

ASD	LRFD
$\Omega_b = 1.67$	$\phi_b = 0.90$
$\Omega_v = 1.50$	$\phi_v = 1.00$

[f] Shape exceeds compact limit for flexure with $F_y = 50$ ksi; tabulated values have been adjusted accordingly.

[v] Shape does not meet the h/t_w limit for shear in AISC *Specification* Section G2.1(a) with $F_y = 50$ ksi; therefore, $\phi_v = 0.90$ and $\Omega_v = 1.67$.

Figure 6.7 W-Shapes: Selection by Z_x

Step 3: Using the required plastic section modulus, select the minimum-weight W-shape from the table shown in Figure 6.7 which facilitates selection of the least weight W-shape. Start at the bottom of the Z_x column and move up until a shape in bold with at least $Z_x = 64.0$ in.^{3} is found.

Select W18×35 with Z=66.5 in.^{3}

This is the least weight W-shape, based on the section weight, which provides the required plastic section modulus.

Step 4: An alternative approach, also using Figure 6.7, would be to enter the table with the required moment, $M_u = 240$ ft-kips, and proceed up the ϕM_n column. The same section will be selected with this approach.

Step 5: Determine the additional required strength based on the actual weight of the chosen beam. The beam weighs 35 lb/ft, which gives an additional moment of

$$M_{u(self\text{-}weight)} = 1.2\left(\frac{0.035(20)^2}{8}\right) = 1.2(1.75) = 2.10 \text{ ft-kips}$$

Step 6: Since the maximum moment for both the applied load and the self-weight occur at the same point, combine this moment with the moment due to superimposed load to determine the new required strength.

$$M_u = 240 + 2.10 = 242 \text{ ft-kips}$$

Step 7: Determine the new required plastic section modulus.

$$Z_{req} = \frac{M_u}{\phi F_y} = \frac{(242)(12)}{0.90(50)} = 64.5 \text{ in.}^3$$

Step 8: Make the final selection. This required plastic section modulus is less than that provided by the W18×35 already chosen. Therefore, select the

W18×35

Note that this is a compact section.

EXAMPLE 6.3b
Beam Design By ASD

Goal: Select the least-weight wide flange member for the conditions given.

Given: An A992 beam, simply supported at both ends, spans 20 ft and is loaded at mid-span with a dead load of 8.0 kips and a live load of

24.0 kips, as shown in Figure 6.8, in addition to its self-weight. Assume full lateral support and a compact section.

SOLUTION

Step 1: Determine the required strength using the ASD load combinations from Section 2.4.

$$P_a = P_D + P_L = (8.0) + (24.0) = 32.0 \text{ kips}$$

$$M_a = \frac{P_a L}{4} = \frac{32.0(20)}{4} = 160 \text{ ft-kips}$$

Step 2: Determine the required plastic section modulus. For a compact, fully braced section,

$$M_n = M_p = F_y Z$$

Thus, because *Specification* Section B3.2 provides that the required moment be less than or equal to the available moment, $M_a \le M_n/\Omega = F_y Z/\Omega$, and

$$Z_{req} = \frac{M_a}{F_y/\Omega} = \frac{160(12)}{(50/1.67)} = \frac{160(12)}{30} = 64.0 \text{ in.}^3$$

Step 3: Using the required plastic section modulus, select the minimum-weight W-shape from the table shown in Figure 6.7 which facilitates selection of the lest weight W-shape. . Start at the bottom of the Z_x column and move up until a shape in bold with at least Z_x = 64.0 in.^{3} is found.

Select W18×35 with Z = 66.5 in.^{3}

This is the least weight W-shape, based on section weight, which provides the required plastic section modulus.

Step 4: An alternative approach, again using Figure 6.7, would be to enter the table with the required moment, M_a = 160 ft-kips, and proceed up the M_n/Ω column. The same section will be selected with this approach.

Step 5: Determine the additional required strength based on the actual weight of the chosen beam. The beam weighs 35 lb/ft, which gives an additional moment of

$$M_{a(self\text{-}weight)} = \frac{0.035(20)^2}{8} = 1.75 \text{ ft-kips}$$

Step 6: Since the maximum moment for both the applied load and the self-weight occur at the same point, combine this moment with the moment due to superimposed load to determine the new required strength.

$$M_a = 160 + 1.75 = 162 \text{ ft-kips}$$

Step 7: Determine the new required plastic section modulus.

$$Z_{req} = \frac{M_a}{F_y/\Omega} = \frac{(162)(12)}{30} = 64.8 \text{ in.}^3$$

Step 8: Make the final selection. This required plastic section modulus is less than that provided by the W18 × 35 already chosen. Therefore, select the

W18×35

Note that this is a compact section

Note that since the live-to-dead load ratio in Example 6.3 is 3, before inclusion of the self-weight, the required plastic section modulus in Examples 6.3a and 6.3b are the same. Once the self-weight is added to the problem, the live-to-dead load ratio is no longer 3 and the required plastic section modulus is seen to be slightly different between the LRFD and ASD examples.

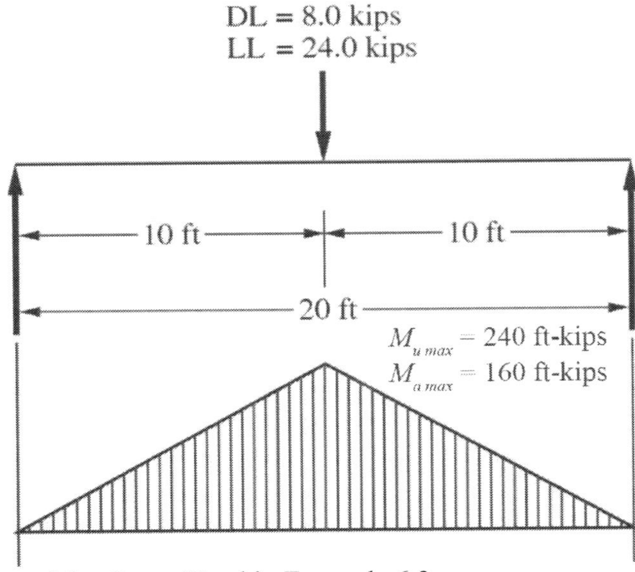

Figure 6.8 Beam Used in Example 6.3

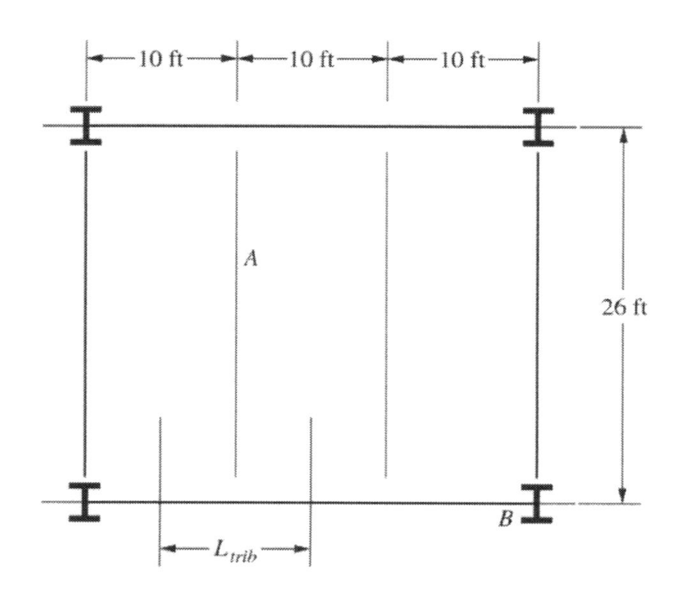

Figure 6.9 Framing Plan for Example 6.4.

EXAMPLE 6.4a ***Beam Design by LRFD***

Goal: Design a W-shape floor beam for the intermediate beam marked *A* on the floor plan shown in Figure 6.9.

Given: The beam is loaded uniformly from the floor with a live load of 60 pounds per square foot (psf) and a dead load of 80 psf, in addition to its self-weight. The beam will have full lateral support provided by the floor deck, and a compact section will be selected. Use A992 steel.

SOLUTION

Step 1: Determine the required load and moment.

$$w_u = (1.2w_D + 1.6w_L)L_{trib} = (1.2(60) + 1.6(80))(10) = 2000\text{ lb/ft}$$

$$M_u = \frac{w_u L^2}{8} = \frac{2.0\text{ kips/ft }(26\text{ ft})^2}{8} = 169\text{ ft-kips}$$

Step 2: Determine the required plastic section modulus.

For a compact, fully braced beam, $M_n = M_p = F_y Z$. Section B3.1 of the *Specification* requires that

$$M_u \le \phi M_n = \phi F_y Z$$

Therefore

$$Z_{req} = \frac{M_u}{\phi F_y} = \frac{(169)(12)}{0.90(50)} = 45.1\text{ in.}^3$$

Step 3: Using the table in Figure 6.7, select the least weight W-shape based on least weight.

$$W14\times30 \text{ with } Z = 47.3 \text{ in.}^3$$

Step 4: Determine the additional required strength based on the actual weight of the chosen beam. The beam weighs 30 lb/ft, which gives an additional moment of

$$M_{u(self\text{-}weight)} = 1.2\left(\frac{0.030(26)^2}{8}\right) = 1.2(2.54) = 3.04 \text{ ft-kips}$$

Step 5: Since the maximum moment for both the applied load and the self-weight occur at the same point, combine this moment with the moment due to superimposed load to determine the new required strength.

$$M_u = 169 + 3.04 = 172 \text{ ft-kips}$$

Step 6: Determine the new required plastic section modulus.

$$Z_{req} = \frac{M_u}{\phi F_y} = \frac{(172)(12)}{0.90(50)} = 45.9 \text{ in.}^3$$

Step 7: Make the final selection. This required plastic section modulus is less than that provided by the W14×30 already chosen. Therefore, select the

$$W14\times30$$

Step 8: As shown in Example 6.3, an alternative approach is to use the required moment, M_u = 172 ft-kips, and enter the ϕM_n column to determine the same W-shape.

EXAMPLE 6.4b
Beam Design by ASD

Goal: Design a W-shape floor beam for the intermediate beam marked *A* on the floor plan shown in Figure 6.9.

Given: The beam is loaded uniformly from the floor with a live load of 60 psf and a dead load of 80 psf, in addition to its self-weight. The beam will have full lateral support provided by the floor deck, and a compact section will be selected. Use A992 steel.

SOLUTION

Step 1: Determine the required load and moment.

$$w_a = (w_D + w_L)L_{trib} = (60 + 80)(10) = 1400 \text{ lb/ft}$$

$$M_a = \frac{w_a L^2}{8} = \frac{1.40 \text{ kips/ft}(26 \text{ ft})^2}{8} = 118 \text{ ft-kips}$$

Step 2: Determine the required plastic section modulus.

For a compact, fully braced beam, $M_n = M_p = F_y Z$. Section B3.2 of the *Specification* requires that

$$M_a \leq \frac{M_n}{\Omega} = \frac{F_y Z}{\Omega}$$

Therefore,

$$Z_{req} = \frac{M_a}{F_y/\Omega} = \frac{(118)(12)}{30} = 47.2 \text{ in.}^3$$

Step 3: Using the table in Figure 6.7, select the least weight W-shape based on least weight.

$$\text{W14×30 with } Z = 47.3 \text{ in.}^3$$

Step 4: Determine the additional required strength based on the actual weight of the chosen beam. The beam weighs 30 lb/ft, which gives an additional moment of

$$M_{a(self\text{-}weight)} = \frac{0.030(26)^2}{8} = 2.54 \text{ ft-kips}$$

Step 5: Since the maximum moment for both the applied load and the self-weight occur at the same point, combine this moment with the moment due to superimposed load to determine the new required strength.

$$M_a = 118 + 2.54 = 121 \text{ ft-kips}$$

Step 6: Determine the new required plastic section modulus.

$$Z_{req} = \frac{M_a}{F_y/\Omega} = \frac{(121)(12)}{30} = 48.4 \text{ in.}^3$$

Step 7: Make the final selection. This required plastic section modulus is more than that provided by the W14×30. Therefore, select the next heavier shape from the table

W16×31

By inspection, this additional 1 lb/ft is acceptable

Step 8: As shown in Example 6.3, an alternative approach is to use the required moment, $M_a = 121$ ft-kips, and enter the M_n/Ω column to determine the same W-shape.

An alternative approach for including beam self-weight is to start with a preliminary estimate of the beam weight. That way it will already be included when the first trial shape is selected. Then the assumed weight is compared to the weight of the selected shape. If the assumed weight is greater than the selected weight, the beam is adequate. If there is a significant discrepancy, either over or under, a second trial shape can be selected.

6.4 DESIGN OF COMPACT LATERALLY UNSUPPORTED WIDE-FLANGE BEAMS

6.4.1 Lateral-Torsional Buckling

The compression region of a bending member cross section has a tendency to buckle similarly to how a pure compression member buckles. The major difference is that the bending tension region helps to resist that buckling. The upper half of the wide flange member in bending acts as a T in pure compression. This T is fully braced about its horizontal axis by the web so it will not buckle in that direction, but it can be unbraced for some distance for buckling about its vertical axis. Thus, it will have a tendency to buckle laterally. Because the tension region also tends to restrain the lateral buckling, the shape actually buckles in a combined lateral and torsional mode. The beam midspan deflects in the plane down and buckles laterally, causing it to twist. This is illustrated in Figure 6.10 by a beam in a laboratory test that has failed due to lateral-torsional buckling.

Figure 6.10 Example of Lateral-Torsional Buckling of a Beam in a Test Frame
Photo courtesy of Donald W. White

The beam appears to have a tendency to fall over on its weak axis. In order to resist this tendency, *Specification* Sections B3.4 and F1(b) require that all bending members be restrained at their support points against rotation about their longitudinal axis. If the beam has sufficient lateral and/or torsional support along its length, a compact cross section can develop the yield stress before buckling. This is the case that was discussed in Sections 6.2 and 6.3. If the beam tends to buckle before the yield stress is reached, the nominal moment strength is less than the plastic moment.

To ensure that a beam cross section can develop its full plastic moment strength without lateral-torsional buckling, *Specification* Section F2.2 limits the slenderness to

$$\frac{L_b}{r_y} \leq 1.76\sqrt{\frac{E}{F_y}} \tag{6.6}$$

where

L_b = unbraced length of the compression flange
r_y = radius of gyration for the shape about the y-axis

The practical application of this limitation is to use the unbraced length alone, rather than in combination with the radius of gyration in the form of a slenderness ratio. This results in the requirement for attaining the full plastic moment strength given the *Specification* as $L_b \leq L_p$ where

$$L_p = 1.76 r_y \sqrt{\frac{E}{F_y}} \qquad \text{(AISC F2-5)}$$

Thus, L_p is the maximum unbraced length that would permit the shape to reach its plastic moment strength. This value is tabulated for each W-shape and can be found in *Manual* Table 3-2, a portion of which was shown in Figure 6.7, and several other locations in the *Manual*.

When the unbraced length of a beam exceeds L_p, its strength is reduced due to the tendency of the member to buckle laterally at a load level below what would cause the plastic moment to be reached. For simply supported doubly symmetric members, a closed-form solution is well established in the literature. The *Guide to Stability Design Criteria for Metal Structures*[2] shows the development of the *Specification* equations based on the closed-form solution.

The elastic lateral-torsional buckling (LTB) strength of a W-shape is given in *Specification* Section F2.2 as

$$M_n = F_{cr} S_x \leq M_p \qquad \text{(AISC F2-3)}$$

where

$$F_{cr} = \frac{C_b \pi^2 E}{\left(\dfrac{L_b}{r_{ts}}\right)^2}\sqrt{1 + 0.078\frac{Jc}{S_x h_o}\left(\frac{L_b}{r_{ts}}\right)^2} \qquad \text{(AISC F2-4)}$$

---

[2] Ziemian, R. D. *Guide to Stability Design Criteria for Metal Structures*, 6th ed. New York: John Wiley & Sons, 2010.

which can be combined to give the nominal moment strength for elastic lateral-torsional buckling as

$$M_n = \frac{C_b \pi^2 E S_x}{\left(\frac{L_b}{r_{ts}}\right)^2} \sqrt{1 + 0.078 \frac{Jc}{S_x h_o} \left(\frac{L_b}{r_{ts}}\right)^2} \tag{6.7}$$

where the terms in the equations are as defined in Section F2.2. A beam buckles elastically if the actual stress in the member at buckling does not exceed F_y at any point. Because all hot rolled shapes have built-in residual stresses as discussed for columns in Section 5.3.4, there is a practical limit to the usefulness of this elastic LTB equation. The *Specification* sets the level of the residual stress at $0.3F_y$ so that only $0.7F_y$ is available to resist a bending moment from applied load while still behaving elastically. This limit results in the maximum elastic moment, $M_{rLTB} = 0.7F_y S_x$. If Equation F2-4 is set equal to $0.7F_y$, a limiting unbraced length, L_r, beyond which the member buckles elastically may be determined. This limit as provided in *Specification* Section F2 is

$$L_r = 1.95 r_{ts} \frac{E}{0.7F_y} \sqrt{\frac{Jc}{S_x h_o} + \sqrt{\left(\frac{Jc}{S_x h_o}\right)^2 + 6.76\left(\frac{0.7F_y}{E}\right)^2}} \qquad \text{(AISC F2-6)}$$

With an unbraced length between L_p and L_r, the beam behaves inelastically. In this range, the nominal moment, M_n, is reasonably well predicted by a straight-line equation when compared to the results of laboratory tests. The *Specification* equation for the nominal moment strength is

$$M_n = C_b \left[M_p - \left(M_p - 0.7F_y S_x\right)\left(\frac{L_b - L_p}{L_r - L_p}\right)\right] \qquad \text{(AISC F2-2)}$$

Equation F2-2 represents interpolation between the end points of a straight line. These end points correspond to the plastic moment, M_p given by Equation F2-1, at an unbraced length of L_p and an elastic moment, $M_{rLTB} = 0.7F_y S_x$, at an unbraced length L_r. Although the determination of F_{cr} and L_r from Equations F2-4 and F2-6 may look somewhat daunting, the *Manual* has extensive tables that permit their determination with little effort, and the equations also can be readily automated using spreadsheet calculations.

The complete picture of the nominal moment strength of a beam as a function of unbraced length is presented in Figure 6.11, where the curve segments are labeled according to the appropriate strength equations. Curves similar to these are available in *Manual* Table 3-10 for each W-shape and Table 3-11 for C- and MC-shapes. For all of these curves, the moment gradient factor, C_b, is taken as 1.0. Examples of these curves are given in Figure 6.12 where solid circles represent the end of the plastic moment region and open circles represent the end of the inelastic straight line. Solid lines indicate that there are no equal weight or lighter shapes with a higher capacity while dashed lines indicate otherwise.

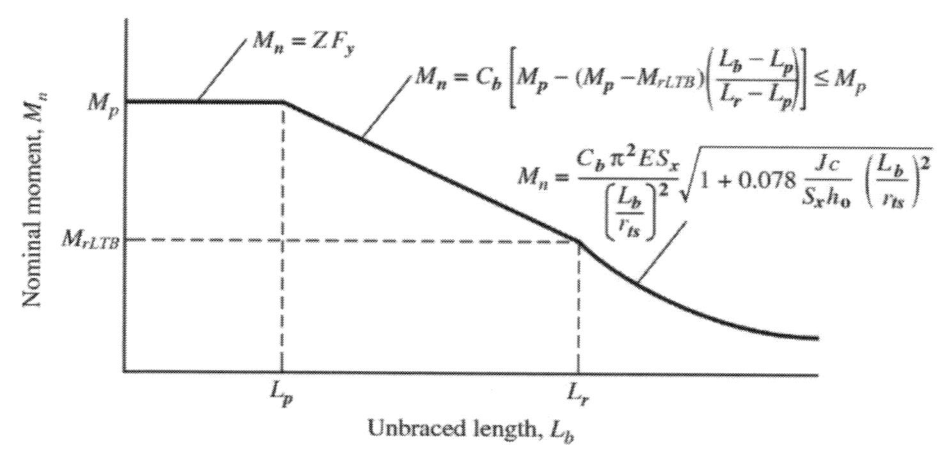

Figure 6.11 Lateral-Torsional Buckling

EXAMPLE 6.5a
Beam Design by LRFD

Goal: Select a W-shape beam considering the unbraced length of the compression flange using *Manual* Table 3-10.

Given: The beam spans 30.0 ft and has a required strength, including a self-weight estimate of 60 lb/ft, of 282 ft-kips. It will have lateral support at its ends and at midspan. Use A992 steel.

SOLUTION

Step 1: Determine the required moment strength and unbraced length. From the given information,

$$M_u = 282 \text{ ft-kips}$$

$$L_b = 15.0 \text{ ft}$$

Step 2: Select the lightest-weight W-shape from the portion of *Manual* Table 3-10 given in Figure 6.12.

Enter Table 3-10 at a design strength of ϕM_n = 282 ft-kips and an unbraced length of L_b = 15.0 ft. The first solid line to the upper right indicates the lightest-weight shape with the most strength sufficient to carry this moment at this unbraced length.

Step 3: Therefore, select a

W21×55

Since the beam weight is less than that assumed for the problem, this is an acceptable selection.

Note that there is another shape at the same weight, a W18×55, that has sufficient strength to carry this moment but has less overall moment strength than the one selected.

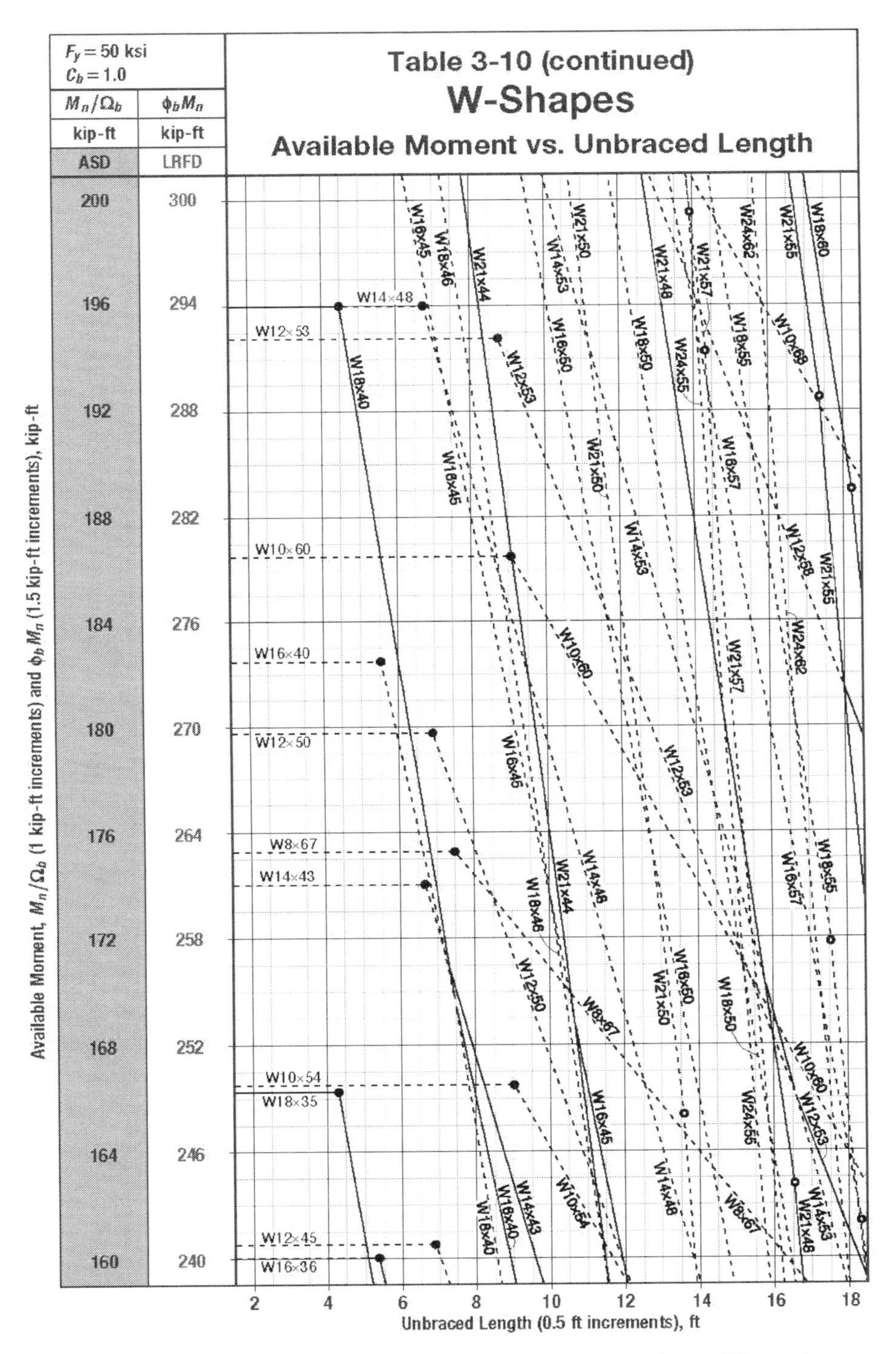

Figure 6.12 W-Shapes: Available Moment versus Unbraced Length

EXAMPLE 6.5b
Beam Design by ASD

Goal: Select a W-shape beam considering the unbraced length of the compression flange using *Manual* Table 3-10.

Given: The beam spans 30.0 ft and has a required strength, including a self-weight estimate of 60 lb/ft, of 188 ft-kips. It will have lateral support at its ends and at midspan. Use A992 steel.

SOLUTION

Step 1: Determine the required moment strength and unbraced length. From the given information,

$$M_a = 188 \text{ ft-kips}$$

$$L_b = 15.0 \text{ ft}$$

Step 2: Select the lightest-weight W-shape from the portion of *Manual* Table 3-10 given in Figure 6.12.

Enter Table 3-10 at an allowable strength of $M_n/\Omega = 188$ ft-kips, and an unbraced length of $L_b = 15$ ft. The first solid line to the upper right indicates the lightest-weight shape with the most strength sufficient to carry this moment at this unbraced length.

Step 3: Therefore, select a

W21×55

Since the beam weight is less than that assumed for the problem, this is an acceptable selection.

Note that there is another shape at the same weight, a W18×55, that has sufficient strength to carry this moment but has less overall moment strength than the one selected.

When M_n is to be determined through a calculation, an additional simplification can be applied to the straight-line portion of the curve. From Equation F2-2, the ratio

$$\left(\frac{M_p - 0.7F_yS_x}{L_r - L_p}\right) = \left(\frac{M_p - M_{rLTB}}{L_r - L_p}\right)$$

is a constant for each beam shape and is defined as the variable BF when written in terms of nominal strength. This constant is tabulated in *Manual* Table 3-2, and shown in Figure 6.7, as ϕBF and BF/Ω for LRFD and ASD, respectively. Thus, for nominal strength, with C_b again taken as 1.0, Equation F2-2 can be rewritten as

$$M_n = M_p - BF(L_b - L_p) \tag{6.8a}$$

and for LRFD as

$$\phi M_n = \phi M_p - \phi BF(L_b - L_p) \tag{6.8b}$$

and for ASD as

$$\frac{M_n}{\Omega} = \frac{M_p}{\Omega} - \frac{BF}{\Omega}(L_b - L_p) \tag{6.8c}$$

Further inspection of *Manual* Table 3-2 shows that two additional moment values are given, $\phi_b M_{rx}$ and M_{rx}/Ω_b. These values are the LRFD and ASD values for $M_{rLTB} = 0.7F_y S_x$.

6.4.2 Moment Gradient

The nominal strength of a beam limited by lateral-torsional buckling as defined when C_b = 1.0 assumes that the moment is uniform across the entire length of the beam as shown in Figure 6.13a. For the limit state of lateral-torsional buckling this is the most severe loading case possible, because it stresses the entire length of the beam to its maximum level, just as for a column. For any other loading pattern, and resulting moment diagram, the compressive force in the beam would vary with the moment diagram. Thus, the reduced stresses along the member length would result in a reduced tendency for lateral-torsional buckling and an increase in strength. The variation in moment over a particular unbraced segment of the beam is called the *moment gradient,* which describes how the moment varies along a specific length.

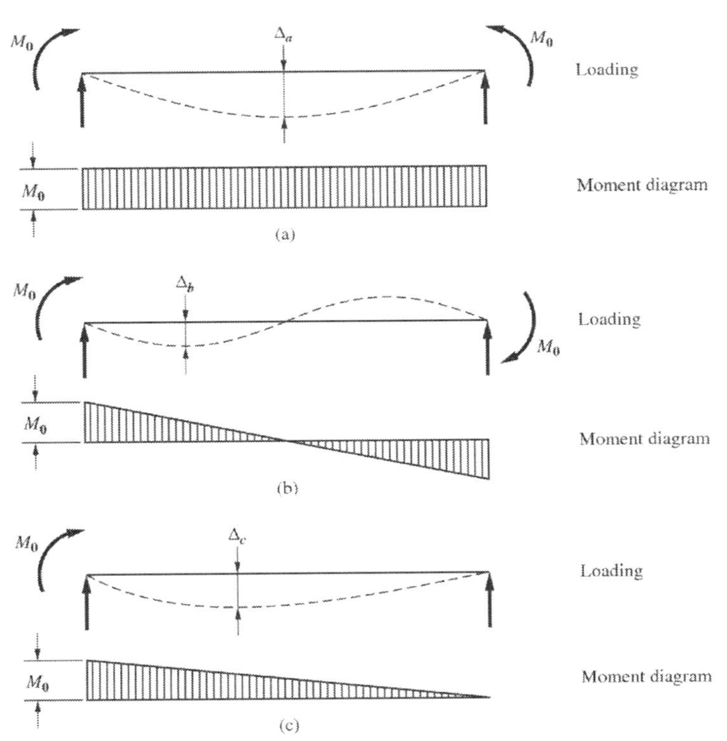

Figure 6.13 Resistance to the Maximum Moment under Three Different Loading Conditions

For the normal case of loading that produces a moment diagram that is not constant, the nominal moment strength calculated through Equations F2-2, F2-3, and F2-4 will be increased to account for the moment gradient when C_b is not taken as 1.0. The lateral-torsional buckling modification factor, C_b, accounts for nonuniform moment diagrams over the unsupported length. There have been many proposed equations for the determination of C_b in the literature. The *Guide to Stability Design Criteria for Metal Structures* (see footnote 2) provides a good summary of these proposals. The equation presented in *Specification* Section F1 is

$$C_b = \frac{12.5M_{\max}}{2.5M_{\max} + 3M_A + 4M_B + 3M_C} \tag{F1-1}$$

where

M_{max} = absolute value of maximum moment in the unbraced segment
M_A = absolute value of moment at quarter point of the unbraced segment
M_B = absolute value of moment at centerline of the unbraced segment
M_C = absolute value of moment at three-quarters point of the unbraced segment

C_b = 1.0 for a uniform moment and can be conservatively taken as 1.0 for other cases. In doing so, however, the designer may be sacrificing significant economy. Figure 6.14 provides examples of loading conditions, bracing locations, and the corresponding C_b values found in *Manual* Table 3-1.

The effect of the moment gradient factor, C_b, is to alter the nominal moment–unbraced length relationship by a constant, as shown in Figure 6.15. The shaded area shows the increase in moment capacity as a result of the use of C_b. Regardless of how small the unbraced length might be, the nominal moment strength of the member can never exceed the plastic moment strength. Thus, the upper portion of the curve in Figure 6.15 is terminated at M_p.

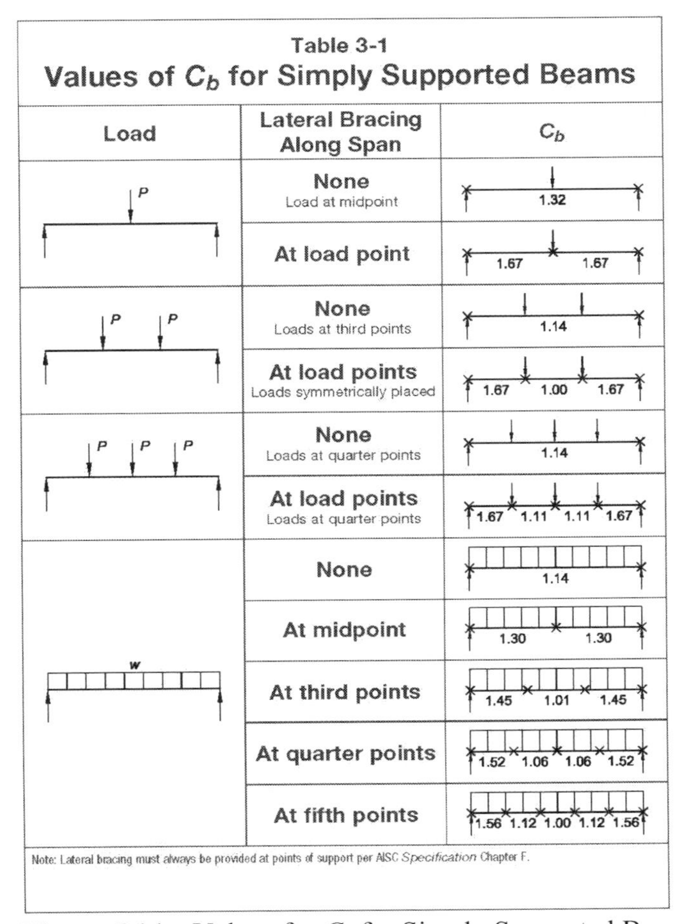

Table 3-1
Values of C_b for Simply Supported Beams

Load	Lateral Bracing Along Span	C_b
P (load at midpoint)	None Load at midpoint	1.32
	At load point	1.67 1.67
P P (loads at third points)	None Loads at third points	1.14
	At load points Loads symmetrically placed	1.67 1.00 1.67
P P P (loads at quarter points)	None Loads at quarter points	1.14
	At load points Loads at quarter points	1.67 1.11 1.11 1.67
w (uniform load)	None	1.14
	At midpoint	1.30 1.30
	At third points	1.45 1.01 1.45
	At quarter points	1.52 1.06 1.06 1.52
	At fifth points	1.56 1.12 1.00 1.12 1.56

Note: Lateral bracing must always be provided at points of support per AISC *Specification* Chapter F.

Figure 6.14 Values for C_b for Simply Supported Beams

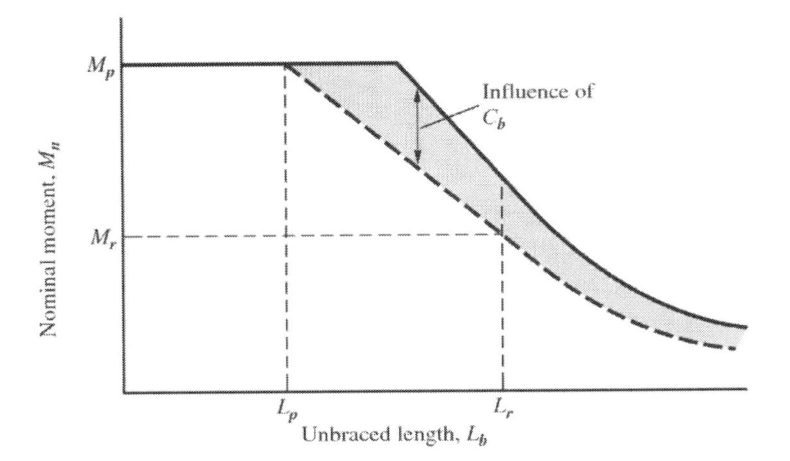

Figure 6.15 Effect of Moment Gradient

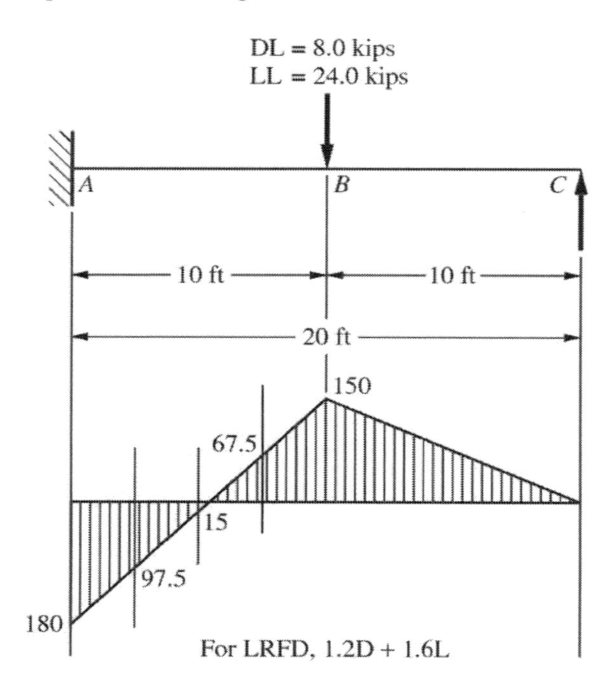

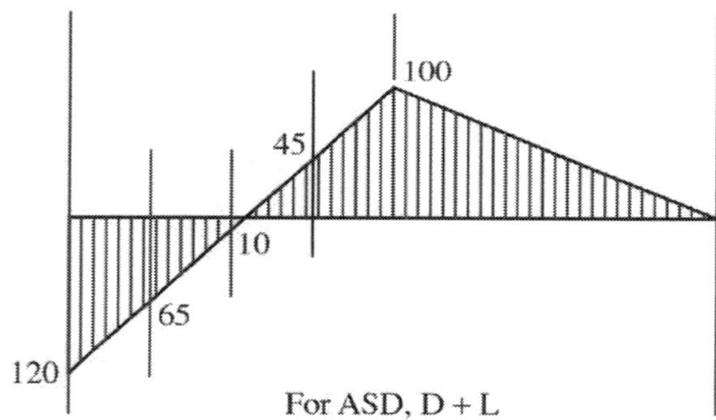

Figure 6.16 Beam Used in Example 6.6

EXAMPLE 6.6a ***Beam Strength and Design by LRFD Considering Moment Gradient***

Goal: Determine whether the W14×34 beam shown in Figure 6.16 will carry the given load. Ignore the self-weight of the beam. Consider the moment gradient (a) $C_b = 1.0$ and (b) C_b from Equation F1-1, and then (c) determine the least-weight section to carry the load using the correct C_b.

Given: Figure 6.16 shows a beam that is fixed at one support and pinned at the other. The beam has a concentrated dead load of 8 kips and a concentrated live load of 24 kips at midspan. A lateral brace is located at each support and at the load point.

SOLUTION

Step 1: Determine the required strength. For the load combination of 1.2D + 1.6L,

$$P_u = 1.2(8.0) + 1.6(24.0) = 48.0 \text{ kips}$$

Step 2: Determine the maximum moment from an elastic analysis of the indeterminate beam. At the fixed end the moment is shown in Figure 6.16 as

$$M_u = 180 \text{ ft-kips}$$

Step 3: Determine the values needed from *Manual* Table 3-2 in order to use Equation 6.8b. For a W14×34,

$$Z = 54.6 \text{ in.}^3,\ L_p = 5.40 \text{ ft},\ L_r = 15.6 \text{ ft},$$
$$\phi M_p = 205 \text{ ft-kips, and } \phi BF = 7.55 \text{ kips}$$

Part a $C_b = 1.0$

Step 4: Determine the design moment strength for lateral bracing of the compression flange at the supports and the load, $L_b = 10$ ft.

Because

$$L_b = 10 \text{ ft} > L_p$$
$$< L_r$$

$$\phi M_n = C_b\left(\phi M_p - \phi BF(L_b - L_p)\right) \le \phi M_p$$
$$\phi M_n = 1.0\left(205 - 7.55(10.0 - 5.40)\right) = 170 \text{ ft-kips} < 205 \text{ ft-kips}$$

As an alternative approach, *Manual* Table 3-10 can be entered with an unbraced length of 10 ft and the design strength of the W14×34 determined to be 170 ft-kips.

Therefore, since

$$\phi M_n = 170 \text{ ft-kips} < M_u = 180 \text{ ft-kips}$$

the W14×34 beam will not work if $C_b = 1.0$

Part b Use the Calculated Value of C_b

Step 5: Determine the correct C_b for the two unbraced segments of the beam.

For the unbraced segment BC, the first beam in Figure 6.14 can be used to obtain $C_b = 1.67$ since the moment gradient for the segment from the load to the support in Figure 6.14 is the same as the moment gradient for segment BC of the beam being considered here. This C_b corresponds to the maximum moment of 150 ft-kips at point B on the beam. The W14×34 can resist this moment without consideration of C_b, as shown in part (a) Step 4 above.

For the unbraced segment AB, Figure 6.14 cannot be used and C_b must be calculated. Using Equation F1-1 and the moment values given in Figure 6.16 at the quarter points of the segment AB,

$$C_b = \frac{12.5(180)}{2.5(180)+3(97.5)+4(15.0)+3(67.5)} = 2.24$$

Step 6: Determine the design moment strength using the calculated value of C_b and the design moment strength determined from part (a), with Equation 6.8b amplified by C_b and limited to ϕM_p.

$$\begin{aligned}\phi M_n &= C_b\left(\phi M_p - \phi BF(L_b - L_p)\right) \le \phi M_p \\ &= 2.24\left(205 - 7.55(10.0 - 5.40)\right) \\ &= 2.24(170) = 381 \text{ ft-kips} > \phi M_p = 205 \text{ ft-kips}\end{aligned}$$

Therefore, the limiting strength of the beam is

$$\phi M_n = 205 \text{ ft-kips} > 180 \text{ ft-kips}$$

So the W14×43 is adequate for bending.

Note that including the self-weight of the beam is not likely to add more to the required strength than that already available with the 205 ft-kips provided. However, it still must be checked. It has been ignored here for simplicity since this is an indeterminate beam and the inclusion of self-weight potentially changes the magnitude and location of the maximum moments.

Part c Considering that C_b = 2.24 and 1.67 for the two segments, a smaller section can be tried.

Step 7: Assuming the beam can be treated as fully braced as a result of including the influence of the moment gradient, C_b not equal to 1.0, select from *Manual* Table 3-2 a W16×31 which is the lightest W-shape to carry the given moment, M_u = 180 ft-kips.

Determine the values needed to evaluate the shape from *Manual* Table 3-2.

$$\phi M_p = 203 \text{ ft-kips},\ L_p = 4.13 \text{ ft},\ L_r = 11.8 \text{ ft},\ \phi BF = 10.3 \text{ kips}$$

Step 8: First consider unbraced segment BC. As determined in Step 5 above, C_b = 1.67. Because L_b = 10 ft > L_p = 4.13 ft, and < L_r = 11.8 ft, use Equation 6.8b with C_b.

$$\phi M_n = C_b\left(\phi M_p - \phi BF(L_b - L_p)\right) \leq \phi M_p$$
$$= 1.67\left(203 - 10.3(10.0 - 4.13)\right)$$
$$= 1.67(143) = 239 \text{ ft-kips } > \phi M_p = 203 \text{ ft-kips}$$

Note that for this unbraced segment, without the use of C_b,

$$\phi M_n = 143 \text{ ft-kips } < 150 \text{ ft-kips}$$

Thus, with C_b included

$$\phi M_n = 239 \text{ ft-kips} > \phi M_p = 203 \text{ ft-kips}.$$

Therefore the design strength is

$$\phi M_n = 203 \text{ ft-kips} > 150 \text{ ft-kips}$$

Step 9: For unbraced segment AB, again $L_b = 10$ ft $> L_p = 4.13$ ft, and $< L_r = 11.8$ ft, and again use Equation 6.8b with $C_b = 2.24$ as determined in Step 5.

$$\phi M_n = C_b\left(\phi M_p - \phi BF(L_b - L_p)\right)) \leq \phi M_p$$
$$= 2.24\left(203 - 10.3(10.0 - 4.13)\right) = 2.24(143)$$
$$= 320 \text{ ft-kips } > \phi M_p = 203 \text{ ft-kips}$$

where

$$\phi M_n = 320 \text{ ft-kips} > \phi M_p = 203 \text{ ft-kips}.$$

Step 10: Thus, the design strength is

$$\phi M_n = 203 \text{ ft-kips} > 180 \text{ ft-kips}$$

So the W16×31 will also work.

As in step 6, self-weight is not likely to exceed available strength.

EXAMPLE 6.6b
Beam Strength and Design by ASD Considering Moment Gradient

Goal: Determine whether the W14×34 beam shown in Figure 6.16 will carry the given load. Ignore the self-weight of the beam. Consider the moment gradient (a) $C_b = 1.0$ and (b) C_b from Equation F1-1, and then (c) determine the least-weight section to carry the load using the correct C_b.

Given: Figure 6.16 shows a beam that is fixed at one support and pinned at the other. The beam has a concentrated dead load of 8 kips and a concentrated live load of 24 kips at midspan. A lateral brace is located at each support and at the load point.

SOLUTION

Step 1: Determine the required strength. For the load combination of D + L,

$$P_a = 8.0 + 24.0 = 32.0 \text{ kips}$$

Step 2: Determine the maximum moment from an elastic analysis of the indeterminate beam. At the fixed end the moment is shown in Figure 6.16 as

$$M_a = 120 \text{ ft-kips}$$

Step 3: Determine the needed values from *Manual* Table 3-2 in order to use Equation 6.8b. For a W14×34,

$$Z = 54.6 \text{ in.}^3,\ L_p = 5.40 \text{ ft},\ L_r = 15.6 \text{ ft},$$
$$M_p/\Omega = 136 \text{ ft-kips, and } BF/\Omega = 5.01 \text{kips}$$

Part a $C_b = 1.0$

Step 4: Determine the allowable moment strength for lateral bracing of the compression flange at the supports and the load, $L_b = 10$ ft.

Because

$$L_b = 10 \text{ ft} > L_p$$
$$< L_r$$

$$M_n/\Omega = \left(M_p/\Omega - (BF/\Omega)(L_b - L_p)\right) \le M_p/\Omega$$
$$= \left(136 - 5.01(10.0 - 5.40)\right) = 113 \text{ ft-kips} < 136 \text{ ft-kips}$$

As an alternative approach, *Manual* Table 3-10 can be entered with an unbraced length of 10 ft and the design strength of the W14×34 determined to be 113 ft-kips.

Therefore, since

$$M_n/\Omega = 113 \text{ ft-kips } < M_a = 120 \text{ ft-kips}$$

the W14×34 beam will not work if $C_b = 1.0$

Part b Use the Calculated Value of C_b

Step 5: Determine the correct C_b for the two unbraced segments of the beam.

For the unbraced segment *BC*, the first beam in Figure 6.14 can be used to obtain $C_b = 1.67$ since the moment gradient for the segment from the load to the support in Figure 6.14 is the same as the moment gradient for segment *BC* of the beam being considered here. This C_b corresponds to the maximum moment of 100 ft-kips at point *B* on the beam. The W14×34 can resist this moment without consideration of C_b, as shown in part (a) Step 4 above.

For the unbraced segment AB, Figure 6.14 cannot be used and C_b must be calculated. Using Equation F1-1 and the moment values given in Figure 6.16 at the quarter points of the segment AB,

$$C_b = \frac{12.5(120)}{2.5(120)+3(65.0)+4(10.0)+3(45.0)} = 2.24$$

Step 6: Determine the allowable moment strength using the calculated value of C_b and the allowable moment strength determined from part (a), with Equation 6.8b amplified by C_b and limited to M_p/Ω.

$$M_n/\Omega = C_b\left(M_p/\Omega - (BF/\Omega)(L_b - L_p)\right) \le M_p/\Omega$$
$$= 2.24\left(136 - 5.01(10.0 - 5.40)\right)$$
$$= 2.24(113) = 253 \text{ ft-kips} > M_p/\Omega = 136 \text{ ft-kips}$$

Therefore, the limiting strength of the beam is

$$M_n/\Omega = 136 \text{ ft-kips} > 120 \text{ ft-kips}$$

So the W14×43 is adequate for bending.

Note that including the self-weight of the beam is not likely to add more to the required strength than that already available with the 136 ft-kips provided. However, it still must be checked. It has been ignored here for simplicity since this is an indeterminate beam and the inclusion of self-weight potentially changes the magnitude and location of the maximum moments.

Part c Considering that $C_b = 2.24$ and 1.67 for the two segments, a smaller section can be tried.

Step 7: Assuming the beam can be treated as fully braced as a result of including the influence of the moment gradient, C_b not equal to 1.0, select from *Manual* Table 3-2 a W16×31 which is the lightest W-shape to carry the given moment, $M_a = 120$ ft-kips.

Determine the values needed to evaluate the shape from *Manual* Table 3-2.

$$M_p/\Omega = 135 \text{ ft-kips},\ L_p = 4.13 \text{ ft},\ L_r = 11.8 \text{ ft},\ BF/\Omega = 6.86 \text{ kips}$$

Step 8: First consider unbraced segment BC. As determined in Step 5 above, $C_b = 1.67$. Because $L_b = 10$ ft $> L_p = 4.13$ ft, and $< L_r = 11.8$ ft, use Equation 6.8b with C_b.

$$M_n/\Omega = C_b\left(M_p/\Omega - (BF/\Omega)(L_b - L_p)\right) \le M_p/\Omega$$
$$= 1.67\left(135 - 6.86(10.0 - 4.13)\right) = 1.67(94.7)$$
$$= 158 \text{ ft-kips} > M_p/\Omega = 135 \text{ ft-kips}$$

Note that for this unbraced segment, without the use of C_b,

$$M_n/\Omega = 94.7 \text{ ft-kips} < 100 \text{ ft-kips}$$

Thus, with C_b included

$$M_n/\Omega = 158 \text{ ft-kips} > M_p/\Omega = 135 \text{ ft-kips}.$$

Therefore the allowable strength is

$$M_n/\Omega = 135 \text{ ft-kips} > 120 \text{ ft-kips}$$

Step 9: For unbraced segment AB, again L_b = 10 ft > L_p = 4.13 ft, and < L_r = 11.8 ft, and again use Equation 6.8b with C_b= 2.24 as determined in Step 5.

$$M_n/\Omega = C_b\left(M_p/\Omega - (BF/\Omega)(L_b - L_p)\right) \le M_p/\Omega$$
$$= 2.24\left(135 - 6.86(10.0 - 4.13)\right) = 2.24(94.7)$$
$$= 212 \text{ ft-kips} > M_p/\Omega = 135 \text{ ft-kips}$$

where

$$M_n/\Omega = 212 \text{ ft-kips} > M_p/\Omega = 135 \text{ ft-kips}.$$

Step 10: Thus, the allowable strength is

$$M_n/\Omega = 135 \text{ ft-kips} > 120 \text{ ft-kips}$$

So the W16×31 will also work.

As in step 6, self-weight is not likely to exceed available strength.

Manual Table 3-10 was shown in Example 6.5 to be an efficient aid in design considering unbraced length. When design is to consider moment gradient as well as unbraced length, the charts in Table 3-10 can be equally useful. Since the curves in Table 3-10 are based on C_b = 1.0, some modifications to the required moment strength must be made before entering the table. Starting with Equation F2-2 and dividing both sides by C_b gives M_n/C_b equal to the nominal strength based on C_b = 1.0. Thus, if Table 3-10 is entered with M_n/C_b a member can be selected. Of course, the limit on Equation F2-2 must still be addressed as will be shown in Example 6.7.

EXAMPLE 6.7a
Beam Design by LRFD

Goal: Select a W-shape beam and consider the unbraced length of the compression flange and the moment gradient, using *Manual* Table 3-10.

Given The beam spans 30.0 ft and has a required strength of 282 ft-kips, including an estimated beam self-weight of 50 lb/ft, as in Example 6.5a. It is loaded with concentrated loads at the midspan and will have lateral support at its ends and at midspan. Use A992 steel.

SOLUTION

Step 1: Determine the required moment strength and unbraced length. From the given information,

$$M_u = 282 \text{ ft-kips}$$

$$L_b = 15.0 \text{ ft}$$

Step 2: Determine the appropriate C_b from Figure 6.14.

For a beam with a concentrated load at midspan and lateral supports at the load and at the ends, $C_b = 1.67$.

Step 3: Determine the moment to use for entering *Manual* Table 3-10.

To account for the increased strength due to the moment gradient, the required moment is divided by C_b, and that value is used to enter the table along with the given unbraced length.

Thus, enter the table at the design moment strength

$$\frac{\phi M_n}{C_b} = \frac{282}{1.67} = 169 \text{ ft-kips}$$

and unbraced length $L_b = 15.0$ ft.

Step 4: The first solid line above and to the right indicates the lightest-weight shape sufficient to carry this moment at this unbraced length. Since this is not found in Figure 6.12, use the actual *Manual* Table 3-10 and select a W16×40.

Step 5: Determine the maximum strength of the W16×40. Remember that in no case can ϕM_n be greater than ϕM_p.

From *Manual* Table 3-2,

$$\phi M_p = 274 \text{ ft-kips}$$

Since this is less than the required strength, $M_u = 282$ ft-kips, this W-shape will not be adequate even with the use of C_b.

Step 6: Select a shape with $\phi M_p \geq 282$ ft-kips from *Manual* Table 3-2. Thus, select

$$W18\times40,\ \phi M_p = 294 \text{ ft-kips}.$$

Step 7: Confirm in *Manual* Table 3-10 that the line, either solid or dashed, for the W18×40 is above and to the right of the intersection of $\phi M_n = 169$ ft-kips and L_b = 15 ft. Since this is not the case, the W18×40 is not adequate.

Step 8: By trial and error, checking the shapes that are up and to the right of this point in Table 3-10, select

W16×45

With $\phi M_p = 309 \geq 282$ ft-kips

Since the beam weight is less than that assumed for the problem, this is an acceptable selection.

EXAMPLE 6.7b
Beam Design by ASD

Goal: Select a W-shape beam and consider the unbraced length of the compression flange and the moment gradient, using *Manual* Table 3-10.

Given: The beam spans 30.0 ft and has a required strength of 188 ft-kips, including an estimated beam self-weight of 50 lb/ft, as in Example 6.5b. It is loaded with concentrated loads at the midspan and will have lateral support at its ends and at midspan. Use A992 steel.

SOLUTION

Step 1: Determine the required moment strength and unbraced length. From the given information,

$$M_a = 188 \text{ ft-kips}$$

$$L_b = 15.0 \text{ ft}$$

Step 2: Determine the appropriate C_b from Figure 6.14.

For a beam with a concentrated load at midspan and lateral supports at the load and at the ends, $C_b = 1.67$.

Step 3: Determine the moment to use for entering Manual Table 3-10.

To account for the increased strength due to the moment gradient, the required moment is divided by C_b, and that value is used to enter the table along with the given unbraced length.

Thus, enter the table at allowable moment strength

$$(M_n/\Omega)/C_b = 188/1.67 = 113 \text{ ft-kips}$$

and unbraced length $L_b = 15.0$ ft.

Step 4: The first solid line above and to the right indicates the least-weight shape sufficient to carry this moment at this unbraced length. Since this is not found in Figure 6.12, use the actual *Manual* Table 3-10 and select a W16×40.

Step 5: Determine the maximum strength of the W16×40. Remember that in no case can M_n/Ω be greater than M_p/Ω

From *Manual* Table 3-2,

$$M_p/\Omega = 182 \text{ ft-kips}$$

Since this is less than the required strength, $M_a = 188$ ft-kips, this W-shape will not be adequate even with the use of C_b.

Step 6: Select a shape with $M_p/\Omega \geq 188$ ft-kips from *Manual* Table 3-2. Thus, select a

$$\text{W18×40 with } M_p/\Omega = 196 \text{ ft-kips}$$

Step 7: Confirm in *Manual* Table 3-10 that the line, either solid or dashed, for the W18×40 is above and to the right of the intersection of $M_n/\Omega = 113$ ft-kips and $L_b = 15$ ft. Since this is not the case, the W18×40 is not adequate.

Step 8: By trial and error, checking the shapes that are up and top the right of this point in Table 3-10, select

W16×45

With $M_p/\Omega = 205 \geq 188$ ft-kips

Since the beam weight is less than that assumed for the problem, this is an acceptable selection.

6.5 DESIGN OF NONCOMPACT BEAMS

6.5.1 Local Buckling

Local buckling occurs when a compression element of a cross section buckles under load before it reaches the yield stress. Because this buckling occurs at a stress lower than the yield stress, the shape is not capable of reaching the plastic moment. Thus, the shape is not compact and the nominal strength of the member is something less than M_p. Buckling of the flange and web elements, and lateral-torsional buckling of the section, do not occur in isolation, so it is difficult to illustrate them individually. Figure 6.17 primarily illustrates local buckling of the compression flange of a wide-flange beam during loading in an experimental test. These failures occur when the flange or web are slender, and they can be predicted through the use of the plate buckling equation discussed in Chapter 5. The projecting flange of a wide-flange member is considered an "unstiffened" element because the web supports only one edge, while the other edge is unsupported and free to rotate. The wide-flange web is connected at both its edges to the flanges, so it is considered a "stiffened" element.

Figure 6.17 Example of Flange Local Buckling
Photo courtesy of Donald W. White

Cases 10 and 15 in Table B4.1b of the *Specification* provides the limiting width-to-thickness ratios, λ_p, for the flange and web, respectively, to ensure that the full plastic moment strength can be reached. When both the flange and web have slenderness ratios (b/t_f and h/t_w, respectively) less than or equal to the λ_p values given in the table, the shapes are called *compact shapes*. If either element exceeds this value, the shape cannot be called compact, and the nominal strength must be reduced. These latter shapes are discussed here.

For the flange of a W-shape to be compact (case 10 in Table B4.1b), its width-to-thickness ratio must satisfy the following limit:

$$\lambda_f = \frac{b}{t} \le \lambda_{pf} = 0.38\sqrt{\frac{E}{F_y}} \tag{6.9}$$

where $b/t = b_f/2t_f$. For the web to be compact (case 15 in Table B4.1b), the limiting ratio is

$$\lambda_w = \frac{h}{t_w} \le \lambda_{pw} = 3.76\sqrt{\frac{E}{F_y}} \tag{6.10}$$

Using the common A992 steel with $F_y = 50$ ksi, these limits become:
for a compact flange

$$\frac{b_f}{2t_f} \leq \lambda_{pf} = 9.15$$

and for a compact web

$$\frac{h}{t_w} \leq \lambda_{pw} = 90.6$$

A comparison of these limits with the data given in *Manual* Table 1-1 shows that the majority of the W-shapes have compact flanges for F_y = 50 ksi and all have compact webs.

Case	Description of Element	Width Thickness Ratio	Limiting Width-Thickness Ratios		Example
			λ_p (compact)	λ_r (noncompact)	
10	Flanges of rolled I-shaped sections, channels and tees	b/t	$0.38\sqrt{E/F_y}$	$1.0\sqrt{E/F_y}$	
11	Flanges of doubly and singly symmetric I-shaped built-up sections	b/t	$0.38\sqrt{E/F_y}$	$0.95\sqrt{k_c E/F_L}$ [a], [b]	
15	Webs of doubly symmetric I-shaped sections and channels	h/t_w	$3.76\sqrt{E/F_y}$	$5.70\sqrt{E/F_y}$	
19	Webs of rectangular HSS and boxes	h/t	$2.42\sqrt{E/F_y}$	$5.70\sqrt{E/F_y}$	

Figure 6.18 Definition of Element Slenderness taken from *Specification* Table B4.1

Figure 6.18 illustrates these dimensions for several commonly used sections along with the slenderness limits (width-to-thickness ratios) as found in *Specification* Table B4.1b. Other shapes can also be found in Table B4.1b.

6.5.2 Flange Local Buckling

The full range of nominal moment strength, M_n, of a cross section can be expressed as a function of flange slenderness, $\lambda_f = \lambda = b/t$. This relationship is illustrated in Figure 6.19. The three regions in the figure identify three types of behavior. The first region represents plastic behavior, in which the shape is capable of attaining its full plastic moment strength. This strength was discussed in Section 6.2. Shapes that fall into this region are called *compact*. The behavior exhibited in the middle region is inelastic, and shapes that fit this category are called *noncompact*. Shapes that fall into the last region exhibit elastic buckling and are called *slender shapes*. The provisions for doubly symmetric I-shaped members that exhibit these last two forms of behavior are given in *Specification* Section F3.

For I-shaped sections, the dividing line between compact and noncompact flanges was given in Equation 6.9. The division between noncompact and slender flange sections is a function of the residual stresses present in the hot rolled member. As was the case with lateral-torsional buckling, the *Specification* assumes that elastic behavior continues up to the point where the elastic moment, $M_{rFLB} = 0.7F_yS_x$. As with lateral-torsional buckling, this limit is a function of the residual stress set at a value of $0.3F_y$. This corresponds to a flange slenderness, as found in *Specification* Table B4.1b, of

$$\lambda_{rf} = 1.0\sqrt{\frac{E}{F_y}} \qquad (6.11)$$

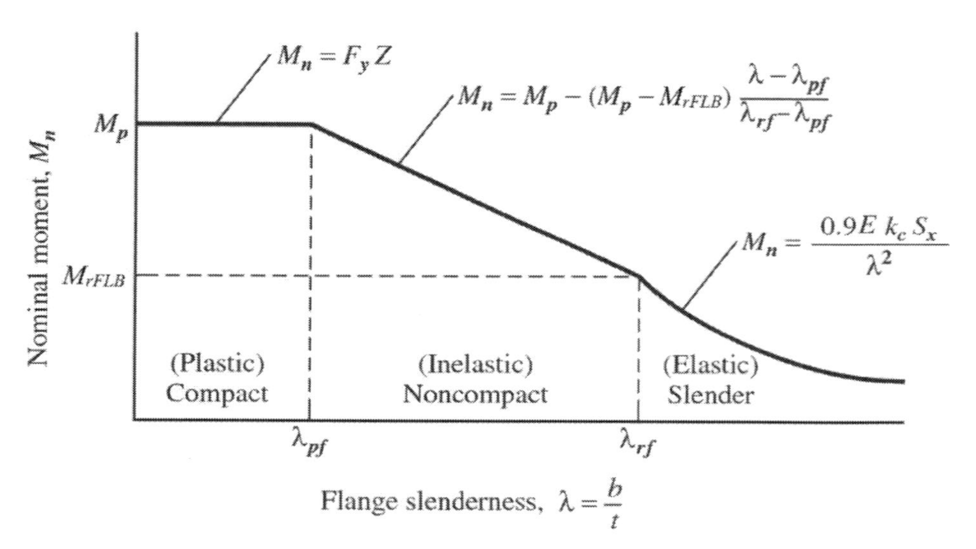

Figure 6.19 Flange Local Buckling Strength.

The strength at the junction of compact behavior and noncompact behavior is the plastic moment strength

$$M_n = M_p = F_y Z$$

At the junction of noncompact and slender behavior, the moment is an elastic moment defined as

$$M_{rFLB} = 0.7F_yS_x$$

The strength for noncompact shapes is represented by a straight line between these points. Thus,

$$M_n = \left[M_p - \left(M_p - M_{rFLB} \right) \left(\frac{\lambda - \lambda_{pf}}{\lambda_{rf} - \lambda_{pf}} \right) \right] \tag{6.12}$$

or as given in the *Specification*

$$M_n = \left[M_p - \left(M_p - 0.7 F_y S_x \right) \left(\frac{\lambda - \lambda_{pf}}{\lambda_{rf} - \lambda_{pf}} \right) \right] \qquad \text{(AISC F3-1)}$$

For A992 steel with $F_y = 50$ ksi, Equation 6.11 provides an upper limit to the noncompact flange of

$$\lambda_{rf} = 1.0\sqrt{\frac{E}{F_y}} = 1.0\sqrt{\frac{29{,}000}{50}} = 24.1$$

A review of *Manual* Table 1-1 for $b_f / 2t_f$ shows that there are no W-shapes with flanges that exceed this limit. Thus, all wide-flange shapes have either compact or noncompact flanges. A further review of the tables shows that only 10 W-shapes have noncompact flanges for $F_y = 50$ ksi.

6.5.3 Web Local Buckling

A comparison of the slenderness criteria for web local buckling given in Equation 6.10 with the data available in *Manual* Table 1-1 for h/t_w indicates that all W-shapes have compact webs for $F_y = 50$ ksi. Thus there is no need to address web local buckling for W-shapes. Noncompact and slender webs, however, are addressed for built-up members in Chapter 7, where plate girders are treated.

EXAMPLE 6.8
Bending Strength of Noncompact Beam

Goal: Determine the nominal moment strength and then the design moment strength (LRFD) and allowable moment strength (ASD) for a W-shape with noncompact flange.

Given: A simply supported W6×15 spans 10 ft. It is braced at the ends and at the midspan ($L_b = 5$ ft). The steel is A992.

SOLUTION

Step 1: Determine the nominal strength.

Check the limits for flange local buckling. For the flange, from *Manual* Table 1-1,

$$\frac{b_f}{2t_f} = 11.5 > \lambda_{pf} = 0.38\sqrt{\frac{E}{F_y}} = 0.38\sqrt{\frac{29{,}000}{50}} = 9.15$$

Therefore, the flange is not compact. Checking for a slender flange, even though our previous review of the *Manual* data indicated that no W-shapes exceeded this requirement,

$$\frac{b_f}{2t_f} = 11.5 < \lambda_{rf} = 1.0\sqrt{\frac{E}{F_y}} = 1.0\sqrt{\frac{29{,}000}{50}} = 24.1$$

Because $\lambda_{pf} < b_f/2t_f < \lambda_{rf}$, the shape has a noncompact flange.

Step 2: Check the limit states for web local buckling.

For the web, from *Manual* Table 1-1,

$$\frac{h}{t_w} = 21.6 < \lambda_{pw} = 3.76\sqrt{\frac{E}{F_y}} = 3.76\sqrt{\frac{29{,}000}{50}} = 90.6$$

So the web is compact, as expected from our earlier evaluation of all W-shapes.

Step 3: Because the shape is noncompact (flange), determine the nominal moment strength by Equation F3-1. From *Manual* Table 1-1 Z_x = 10.8 in.^{3} and S_x= 9.72 in.^{3} and

$$M_p = F_y Z_x = 50(10.8) = 540 \text{ in.-kips}$$

Thus, for flange local buckling

$$M_n = \left[M_p - (M_p - 0.7F_y S_x)\left(\frac{\lambda - \lambda_{pf}}{\lambda_{rf} - \lambda_{pf}}\right)\right]$$

$$= \left[540 - (540 - 0.7(50)9.72)\left(\frac{11.5 - 9.15}{24.1 - 9.15}\right)\right] = 509 \text{ in.-kips}$$

or

$$M_n = \frac{509 \text{ in.-kip}}{12 \text{ in./ft}} = 42.4 \text{ ft-kips}$$

Step 4: Check for the limit state of lateral-torsional buckling.

For this shape, from *Manual* Table 1-1 r_y = 1.45 in. and

$$L_p = 1.76 r_y \sqrt{E/F_y} = 1.76(1.45)\sqrt{29{,}000/50} = 61.5 \text{ in.}$$

Thus, L_p = 5.13 ft, which is greater than L_b = 5.0 ft, so the beam is adequately braced to resist the plastic moment. For lateral-torsional buckling,

$$M_n = M_p = F_y Z = 50(10.8) = 540 \text{ in.-kips}$$

or

$$M_n = 540/12 = 45.0 \text{ ft-kips}$$

Step 5: Because the moment based on flange local buckling, 42.4 ft-kips, is less than the moment based on lateral-torsional buckling, 45.0 ft-kips, local buckling controls and

$$M_n = 42.4 \text{ ft-kips}$$

For LRFD Step 6: Determine the design moment for LRFD.

$$\phi M_n = 0.9(42.4) = 38.2 \text{ ft-kips}$$

For ASD Step 6: Determine the allowable moment for ASD

$$M_n/\Omega = 42.4/1.67 = 25.4 \text{ ft-kips}$$

EXAMPLE 6.9
Bending Strength of a Noncompact Beam

Goal: Determine the (a) design moment strength (LRFD) and (b) allowable moment strength (ASD) for a W-shape with noncompact flange using Table 3-2.

Given: A simply supported W14×90 spans 15 ft and carries a uniformly distributed load. It is braced only at the supports. The steel is A992. Note in Table 3-2 that this W-shape is identified with a superscript f indicating in a footnote that the flange is not compact.

SOLUTION

Step 1: Check the unbraced length to determine if lateral-torsional buckling is a limit state that must be considered with L_b = 15.0 ft. From Manual Table 3-2,

$$L_p = 15.1 > L_b = 15.0 \text{ ft}$$

Thus, lateral-torsional buckling does not influence the beam strength.

Step 2: Knowing that the shape is not compact, check the limits for flange local buckling.

From *Manual* Table 1-1,

$$\frac{b_f}{2t_f} = 10.2 > \lambda_{pf} = 0.38\sqrt{\frac{E}{F_y}} = 0.38\sqrt{\frac{29{,}000}{50}} = 9.15$$

Therefore, the flange is not compact. We know that no W-shapes have slender flanges but we need that limit. Thus,

$$\lambda_{rf} = 1.0\sqrt{E/F_y} = 1.0\sqrt{29{,}000/50} = 24.1$$

Step 3: Determine the nominal moment strength using Equation F3-1. From *Manual* Table 1-1 Z_x = 157 in.^{3} and S_x = 143 in.^{3} and

$$M_p = F_y Z_x = 50(157) = 7850 \text{ in.-kips}$$

and

$$M_n = \left[M_p - \left(M_p - 0.7F_y S_x\right)\left(\frac{\lambda - \lambda_{pf}}{\lambda_{rf} - \lambda_{pf}}\right)\right]$$

$$= \left[7850 - \left(7850 - 0.7(50)(143)\right)\left(\frac{10.2 - 9.15}{24.1 - 9.15}\right)\right] = 7650 \text{ in.-kips}$$

Thus,

$$M_n = \frac{7650}{12} = 638 \text{ ft-kips}$$

For LRFD Step 4: Determine the design strength and compare it to $\phi_b M_{px}$ from Table 3-2

$$\phi M_n = 0.9(638) = 574 \text{ ft-kips}$$

which is identical to the value from Table 3-2

For ASD Step 4: Determine the allowable strength and compare it to M_{px}/Ω_b from Table 3-2

$$M_n/1.67 = 638/1.67 = 382 \text{ ft-kips}$$

which is identical to the value from Table 3-2

Step 5: Conclusion: Manual Table 3-2 includes consideration of the flange compactness for noncompact flange W-shapes.

6.6 DESIGN OF BEAMS FOR WEAK AXIS BENDING

Up to this point, I-shaped beams have been assumed to be bending about an axis parallel to their flanges, called the x-axis. A quick scan of the shape property tables in the *Manual* shows that the section modulus and plastic section modulus about the x-axis are larger than the corresponding values about the other orthogonal axis, the y-axis. Thus, bending about the x-axis is called *strong axis* or *major axis bending*, whereas bending about the y-axis is called *weak axis* or *minor axis bending*. Although beams are not normally oriented for bending about this weak axis, a situation may arise when it is necessary to determine the strength of a beam in this orientation.

Design of I-shaped beams for weak axis bending is relatively easy. Section F6 of the *Specification* applies to I-shaped members and channels bent about their minor axis. Two limit states are identified: yielding and flange local buckling. The flange and web referred to here are the same elements as for the case when the shape is bending about its major axis. Thus, the limits on flange slenderness in Case 13 are the same as discussed earlier for Case 10. Since the web is at the plastic neutral axis, there is no requirement to check web local buckling. For those few W-shapes with noncompact flanges, an equation similar to that used previously for noncompact flanges is required.

For the limit state of yielding,

$$M_n = M_p = F_y Z_y \leq 1.6 F_y S_y \qquad \text{(AISC F6-1)}$$

An I-shaped member bending about its weak axis has properties close to those of a rectangle. For the rectangle, the ratio of the plastic moment to the elastic yield moment, called the shape factor, equals 1.5. The addition of the web alters the elastic section modulus and plastic section modulus so that the shape factor for these weak axis bending members exceeds 1.5. To ensure an appropriate level of rotational capacity at the plastic limit state, the shape factor for weak axis bending is limited to 1.6. All but seven W-shapes meet this limitation.

Although I-shaped members are not often called upon to carry moment about the y-axis as pure bending members, they are called upon to participate in combined bending as discussed in Section 6.12 and combined with axial load as discussed in Chapter 8.

EXAMPLE 6.10
Weak Axis Bending Strength of Beam

Goal: Determine the nominal moment strength and then the design moment strength (LRFD) and the allowable moment strength (ASD) for bending about the y-axis.

Given: A simply supported W10×30, A992 steel, is loaded to cause bending about its weak axis.

SOLUTION

Step 1: Determine the nominal strength.

Check the limits for flange local buckling. For the flange, from *Manual* Table 1-1,

$$\frac{b_f}{2t_f} = 5.70 < \lambda_{pf} = 0.38\sqrt{\frac{E}{F_y}} = 0.38\sqrt{\frac{29{,}000}{50}} = 9.15$$

Therefore, the flange is compact.

Step 2: For the limit state of yielding, from Equation F6-1 with Z_y = 8.8 in.^{3} and S_y = 5.75 in.^{3} from *Manual* Table 1-1

$$M_n = F_y Z_y = 50(8.84) = 442 \text{ in.-kips}$$

$$\leq 1.6 F_y S_y = 1.6(50)(5.75) = 460 \text{ in.-kips}$$

Thus the strength is controlled by the plastic moment and

$$M_n = \frac{442}{12} = 36.8 \text{ ft-kips}$$

For LRFD
Step 3: For LRFD, determine the design moment.

$$\phi M_n = 0.9(36.8) = 33.1 \text{ ft-kips}$$

For ASD Step 3: For ASD, determine the allowable moment.

$$M_n/\Omega = 36.8/1.67 = 22.0 \text{ ft-kips}$$

EXAMPLE 6.11
Weak Axis Bending Strength of Beam

Goal: Determine the nominal moment strength, design moment strength (LRFD), and allowable moment strength (ASD) for bending about the *y*-axis.

Given: A simply supported W40×392, A992 steel, is loaded to cause bending about its weak axis.

SOLUTION

Step 1: Determine the nominal moment strength.

Check the limits for flange local buckling. For the flange, from *Manual* Table 1-1,

$$\frac{b_f}{2t_f} = 2.45 < \lambda_{pf} = 0.38\sqrt{\frac{E}{F_y}} = 0.38\sqrt{\frac{29{,}000}{50}} = 9.15$$

Therefore, the flange is compact.

Step 2: For the limit state of yielding, from Equation F6-1 with $Z_y = 212 \text{ in.}^3$ and $S_y = 130 \text{ in.}^3$ from *Manual* Table 1-1

$$M_n = F_y Z_y = 50(212) = 10{,}600 \text{ in.-kips}$$

$$\leq 1.6F_y S_y = 1.6(50)(130) = 10{,}400 \text{ in.-kips}$$

Thus the strength is controlled by the upper limit and

$$M_n = \frac{10{,}400}{12} = 867 \text{ ft-kips}$$

For LRFD Step 3: For LRFD, determine the design moment.

$$\phi M_n = 0.9(867) = 780 \text{ ft-kips}$$

For ASD Step 3: For ASD, determine the allowable moment.

$$M_n/\Omega = 867/1.67 = 519 \text{ ft-kips}$$

6.7 DESIGN OF BEAMS FOR SHEAR

Chapter G of the *Specification* establishes the requirements for beam shear. Although shear failures are uncommon with rolled sections, the strength must still be confirmed. A beam can fail in shear by yielding or buckling. Beam webs also need to be checked for shear rupture on the net area of the web when bolt holes are present. Shear rupture is addressed in the discussion of connections in Chapter 10.

The nominal shear yielding strength is based on the von Mises criterion,[3] which states that for an unreinforced beam web that is stocky enough not to fail by buckling, the shear strength can be taken as $F_y/\sqrt{3} = 0.58F_y$. The *Specification* rounds this stress to $0.6F_y$ and provides, in Section G2, the shear strength as

$$V_n = 0.6F_y A_w C_{v1} \qquad \text{(AISC G2-1)}$$

where A_w is the area of the web, taken as the total depth times the web thickness.

The web shear coefficient, C_{v1}, is used to account for shear web buckling. Thus, if the web is capable of reaching yield, $C_{v1} = 1.0$. To ensure that the beam web is capable of reaching yield before buckling, the *Specification* sets the limit on web slenderness at

$$\frac{h}{t_w} \leq 1.10\sqrt{\frac{k_v E}{F_y}}$$

where $k_v = 5.34$ for unstiffened webs. All current ASTM A6 rolled I-shaped members have webs that meet the criteria for $k_v = 5.34$, and all A992 W-shapes meet the criteria for web yielding, $C_{v1} = 1.0$.

Thus, the nominal shear strength of all rolled W-shapes, using Equation G2-1 with $C_{v1} = 1.0$, can be taken as

$$V_n = 0.6F_y A_w \qquad (6.13)$$

Determining the shear design strength or allowable strength is complicated by a variation in resistance and safety factors. To keep the beam shear strength provisions the same in the 2022 *Specification* as in the allowable stress specifications prior to 2005, the resistance and safety factors for a particular set of rolled I-shapes was liberalized. Thus, for webs of rolled I-shapes with $h/t_w \leq 2.24\sqrt{E/F_y}$,

$$\phi = 1.00 \ \text{(LRFD)} \quad \Omega = 1.50 \ \text{(ASD)}$$

For all other shapes,

$$\phi = 0.90 \ \text{(LRFD)} \quad \Omega = 1.67 \ \text{(ASD)}$$

---

[3] Mendelson, A. *Plasticity: Theory and Application*, New York: Macmillan, 1968.

EXAMPLE 6.12 ***Shear Strength of Beam***

Goal: Determine the nominal shear strength, design shear strength (LRFD), and allowable shear strength (ASD) when the beam is bending about the x-axis and compare to the shear strength given in *Manual* Table 3-2.

Given: A W16×31, A992 steel member is loaded to cause shear in its web.

SOLUTION

Step 1: Determine the nominal strength.

Check the limits for yielding of the web. From the user note at the end of Section G2.1 we know that all W-shapes will yield in shear. Thus,

$$C_{v1} = 1.0$$

and from Manual Table 1-1

$$d = 15.9 \text{ in.},\ t_w = 0.275 \text{ in., and } h/t_w = 51.6$$

Step 2: For the limit state of yielding,

$$V_n = 0.6F_y A_w C_{v1}$$
$$= 0.6(50)(15.9)(0.275)(1.0) = 131 \text{ kips}$$

Step 3: Determine the resistance factor and the safety factor.

$$\frac{h}{t_w} = 51.6 < 2.24\sqrt{\frac{E}{F_y}} = 2.24\sqrt{\frac{29{,}000}{50}} = 53.9$$

Therefore,

$$\phi_v = 1.00 \quad \text{and} \quad \Omega_v = 1.50$$

For LRFD Step 4: For LRFD, determine the design shear strength.

$$\phi V_n = 1.0(131) = 131 \text{ kips}$$

From *Manual* Table 3-2 $\phi V_n = 131$ kips. Thus the answers are the same.

For ASD Step 4: For ASD, determine the allowable shear strength.

$$V_n/\Omega = 131/1.50 = 87.3 \text{ kips}$$

From *Manual* Table 3-2 $V_n/\Omega = 87.5$ kips. Thus, the results are close. The difference is due to round off.

EXAMPLE 6.13 ***Shear Strength of Beam***

Goal: Determine the nominal shear strength, design shear strength (LRFD), and allowable shear strength (ASD) when the beam is bending about the x-axis and compare to the shear strength given in *Manual* Table 3-2.

Given: A W16×26, A992 steel member is loaded to cause shear in its web.

SOLUTION

Step 1: Determine the nominal strength.

Check the limits for yielding of the web. From the user note at the end of Section G2.1 we know that all W-shapes will yield in shear. Thus,

$$C_{v1} = 1.0$$

and from Manual Table 1-1

$$d = 15.7 \text{ in.}, t_w = 0.250 \text{ in., and } h/t_w = 56.8$$

Step 2: For the limit state of yielding,

$$V_n = 0.6F_yA_wC_{v1}$$
$$= 0.6(50)(15.7)(0.250)(1.0) = 118 \text{ kips}$$

Step 3: Determine the resistance factor and the safety factor.

$$\frac{h}{t_w} = 56.8 > 2.24\sqrt{\frac{E}{F_y}} = 2.24\sqrt{\frac{29,000}{50}} = 53.9$$

Therefore,

$$\phi_v = 0.90 \quad \text{and} \quad \Omega_v = 1.67$$

For LRFD Step 4: For LRFD, determine the design shear strength.

$$\phi V_n = 0.90(118) = 106 \text{ kips}$$

From *Manual* Table 3-2 $\phi V_n = 106 \text{ kips}$. Thus, the answers are the same

For ASD Step 4: For ASD, determine the allowable shear strength.

$$V_n/\Omega = 118/1.67 = 70.7 \text{ kips}$$

From *Manual* Table 3-2 $V_n/\Omega = 70.5 \text{ kips}$. Thus, the results are close. The difference is due to round off.

6.8 CONTINUOUS BEAMS

Beams that span over more than two supports are called *continuous beams*. Unlike simple beams, continuous beams are indeterminate and must be analyzed by applying more than the three basic equations of equilibrium. Although indeterminate analysis is not within the scope of this book, a few aspects should be addressed, even if only briefly.

The *Manual* includes shears, moments, and deflections for several continuous beams with various uniform load patterns in *Manual* Table 3-23. These results come from an elastic indeterminate analysis and can be used for the design of any beams that fit the support and loading conditions.

It has long been known that material ductility permits steel members to redistribute load. When one section of a member becomes overloaded, it can redistribute a portion of its load to a less loaded section. This redistribution can be accounted for through an analysis method called plastic analysis or through a number of more modern methods capable of modeling the real behavior of the members. These methods may collectively be called *advanced analysis* and may be used in structural steel design through the provisions of Appendix 1 of the *Specification*. This appendix also permits use of the simplified plastic analysis approach for continuous beams through Appendix Section 1.3.

To allow the designer to take advantage of some of the redistribution that is accounted for in plastic analysis, Appendix 8, Section 8.2 gives provisions for approximate inelastic moment redistribution in beams. Design of beams and girders that are compact and have sufficiently braced compression flanges may take advantage of this simplified redistribution approach. The compact criteria are those already discussed, whereas the unbraced length criteria are different. To use the simplified redistribution, for doubly symmetric I-shaped beams the unbraced length of the compression flange, L_b, must be less than that given as

$$L_m = \left[0.12 + 0.076\left(\frac{M_1}{M_2}\right)\right]\left(\frac{E}{F_y}\right) r_y \qquad \text{(AISC A-8-9)}$$

where M_1 is the smaller and M_2 is the larger moment at the ends of the unbraced length. The moment ratio is positive when the moments cause reverse curvature and negative when they cause single curvature.

When these criteria are satisfied, the beam can be proportioned for 0.9 times the negative moments at points of support. This redistribution is permitted only for gravity-loading cases and moments determined through an elastic analysis. When this reduction in negative moment is used, the positive moment must be increased to maintain equilibrium. This can be accomplished simply by adding to the maximum positive moment 0.1 times the average original negative moments.

EXAMPLE 6.14a
Continuous Beam Design by LRFD

Goal: Select a compact, fully braced section for use as a continuous beam.

Given: The beam is continuous over three spans of 30 ft each. It supports a live load of 2.5 kip/ft and a dead load of 1.8 kip/ft. Use A992 steel.

SOLUTION

Step 1: Determine the required strength.

The design load is $w_u = 1.2(1.8) + 1.6(2.5) = 6.16$ kip/ ft.

From the beam shear, moment, and deflection diagrams in *Manual* Table 3-23, case 39, the critical span is the exterior span, where the negative moment is

$$-M_{BA} = 0.100 w_u l^2 = 0.100(6.16)(30)^2 = 554 \text{ ft-kips}$$

and the positive moment is

$$+M_{BA} = 0.0800 w_u l^2 = 0.0800(6.16)(30)^2 = 444 \text{ ft-kips}$$

Step 2: Consider redistribution of moments according to *Specification* Appendix 8, Section 8.2.

A design could be carried out for a maximum moment of 554 ft-kips, but with redistribution this moment may be reduced to

$$M_{BA} = 0.9(554) = 499 \text{ ft-kips}$$

provided that the positive moment is increased by the average negative moment reduction. Thus,

$$M_{AB} = 444 + \frac{0 + 0.1(554)}{2} = 472 \text{ ft-kips}$$

Step 3: Determine the required plastic section modulus.

Even with the increase in positive moment, the negative moment is still the maximum moment, so for a moment of 499 ft-kips,

$$Z_{req} = \frac{499(12)}{0.9(50)} = 133 \text{ in.}^3$$

Step 4: Select the least-weight W-shape from *Manual* Table 3-2.

W24×55 with $Z = 134$ in.^{3}

EXAMPLE 6.14b
Continuous Beam Design by ASD

Goal: Select a compact, fully braced section for use as a continuous beam.

Given: The beam must be continuous over three spans of 30 ft each. It must support a live load of 2.5 kip/ft and a dead load of 1.8 kip/ft. Use A992 steel.

SOLUTION

Step 1: Determine the required strength.

The design load is $w_a = (1.8) + (2.5) = 4.3$ kip/ft.

From the beam shear, moment, and deflection diagrams in *Manual* Table 3-23, case 39, the critical span is the exterior span, where the negative moment is

$$-M_{BA} = 0.100 w_a l^2 = 0.100(4.3)(30)^2 = 387 \text{ ft-kips}$$

and the positive moment is

$$+M_{BA} = 0.0800 w_a l^2 = 0.0800(4.3)(30)^2 = 310 \text{ ft-kips}$$

Step 2: Consider redistribution of moments according to *Specification* Appendix 8, Section 8.2.

A design could be carried out for a maximum moment of 387 ft-kips, but with redistribution this moment may be reduced to

$$M_{BA} = 0.9(387) = 348 \text{ ft-kips}$$

provided that the positive moment is increased by the average negative moment reduction. Thus,

$$M_{AB} = 310 + \frac{0 + 0.1(387)}{2} = 329 \text{ ft-kips}$$

Step 3: Determine the required plastic section modulus.

Even with this increase in the positive moment, the negative moment is still the maximum moment, so for a moment of 348 ft-kips

$$Z_{req} = \frac{348(12)}{(50/1.67)} = 139 \text{ in.}^3$$

Step 4: Select the least-weight W-shape from *Manual* Table 3-2.

W21×62 with $Z = 144$ in.^{3}

Note that in the preceding example, the reduction in negative moment leads to selection of a smaller W-shape than would have resulted from design for the original negative moment.

6.9 PLASTIC ANALYSIS AND DESIGN OF CONTINUOUS BEAMS

Up to this point, it has been assumed that the plastic moment strength of a bending member could be compared to the maximum moment on a beam resulting from an elastic analysis to satisfy the strength requirements of the *Specification*. This is accurate for determinate members in which the occurrence of the plastic moment at the single point of maximum moment results in the development of a single plastic hinge, which would lead to member collapse. However, for indeterminate structures, such as continuous beams, more than one plastic hinge must form before the beam would actually collapse, and this provides some additional capacity that an elastic analysis cannot capture. The formation of plastic hinges in the appropriate locations causes a collapse, and the geometry of this collapse is called a *failure* or *collapse mechanism*. This is the approach referred to as *plastic analysis*, permitted by Appendix 1, Section 1.3 of the *Specification* for use with LRFD only. The *Specification* only addresses beam design by plastic analysis. Use beyond beams, such as for frames, is up to the designer; proper consideration of second-order effects on stability is required. Thus, only the determination of plastic collapse mechanisms for beams is considered here.

The formation of a beam failure mechanism may best be understood by following the load history of a fixed-end beam with a uniformly distributed load. The beam and moment diagrams that result from an elastic indeterminate analysis are given in Figure 6.20a. The largest moments occur at the fixed ends and are given by $wL^2/12$. If the load on the beam is increased, the beam behaves elastically until the moments on the ends equal the plastic moment strength of the member, as shown in Figure 6.20b. Because the application of additional load causes the member to rotate at its ends while maintaining the plastic moment, these points behave as pins. These pins are called *plastic hinges*. In this case, the load is designated as w_1. The member can continue to accept load beyond this w_1, functioning as a simple beam, until a third plastic hinge forms at the beam midspan. The formation of this third hinge makes the beam unstable, thus forming the collapse mechanism. The mechanism and corresponding moment diagram are given in Figure 6.20c.

For the collapse mechanism just described, equilibrium requires that the simple beam moment, $w_uL^2/8$, equal twice the plastic moment; thus,

$$M_p = w_u L^2/16 \tag{6.14}$$

Had this beam been designed based on an elastic analysis, it would have required a moment capacity greater than or equal to $w_uL^2/12$. Using a plastic analysis, a smaller plastic moment strength, equal to $w_uL^2/16$, must be provided for in the design. Thus, in this case of an indeterminate beam, plastic analysis has the potential to result in a smaller member being required to carry this same load.

An additional advantage to the use of plastic analysis for indeterminate beams is the simplicity of the analysis. By observation, regardless of the overall geometry of the continuous beam, each segment between supports can be evaluated independently of the other segments. This means that any beam segment, continuous at each end and loaded with a uniformly distributed load, exhibits the same collapse mechanism. Thus, the relation between the applied load and the plastic moment will be as given in Equation 6.14. Plastic analysis results for additional loading and beam configurations are given in Figure 6.21. Additional examples, as well as the development of these relations through application of energy principles, can be found in several books including, *Applied Plastic Design in Steel*.[4]

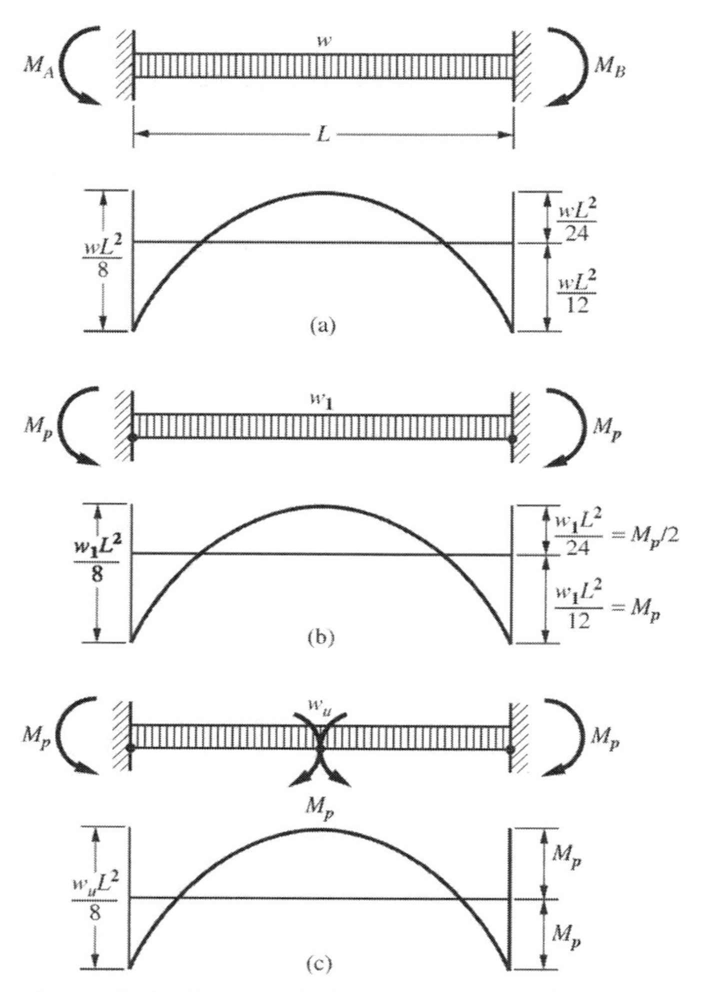

Figure 6.20 Beam and Moment Diagrams for the Development of a Plastic Mechanism

To ensure that a given beam cross section can undergo the necessary rotation at each plastic hinge, the *Specification* requires that $F_y \leq 65$ ksi, that the section be compact, and that the compression flange be braced such that the unbraced length in the area of the hinge is less than L_{pd} given as Equation A-1-5 in *Specification* Appendix 1, Section 1.3. If these limits are not satisfied, the member design must be based on an elastic analysis.

[4] Disque, R. O. *Applied Design in Steel*. New York: Van Nostrand Reinhold, 1971.

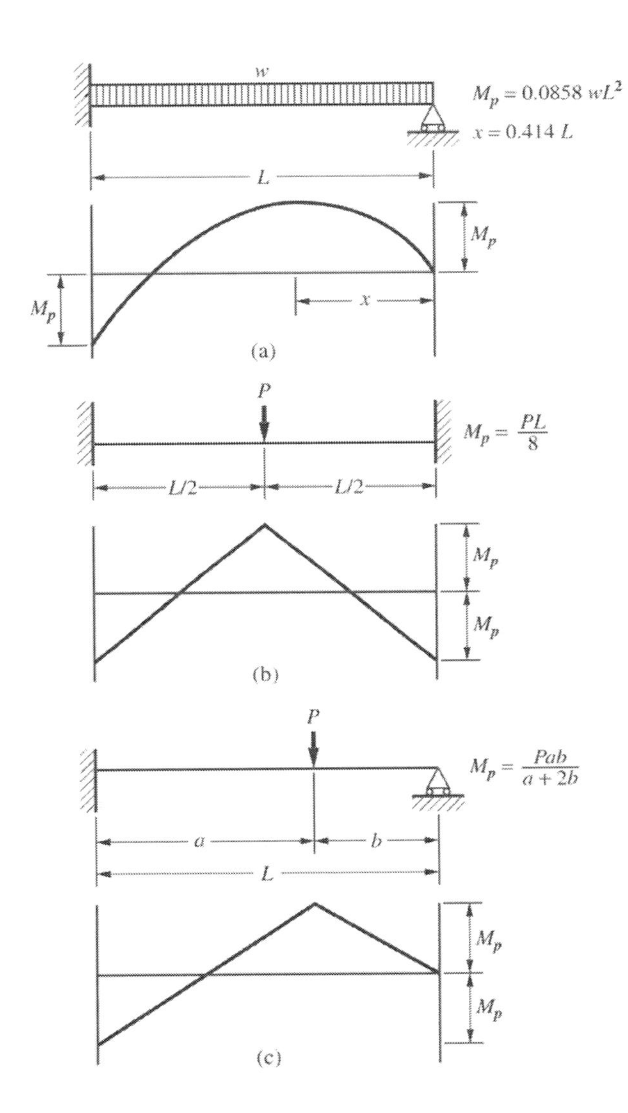

Figure 6.21 Loading and Beam Configurations Resulting from Plastic Analysis

---

EXAMPLE 6.15
Beam Design using Plastic Analysis (LRFD only)

Goal: Design a beam using plastic analysis and A992 steel. Plastic analysis is applicable only for LRFD load combinations.

Given: A beam is simply supported at one end and fixed at the other, similar to that shown in Figure 6.16 for Example 6.6. It spans 20 ft and is loaded at its midspan with a dead load of 16.0 kips and a live load of 48.0 kips. It is assumed that the final section will be compact and adequately braced.

SOLUTION

Step 1: Determine the required strength.

$$P_u = 1.2(16.0) + 1.6(48.0) = 96.0 \text{ kips}$$

Using the plastic analysis results from Figure 6.21c, with $P = P_u$,

$$\phi M_{p\ req} = \frac{Pab}{(a+2b)} = \frac{96.0(10.0)(10.0)}{10.0+2(10.0)} = 320 \text{ ft-kips}$$

Step 2: Select the required W-shape from *Manual* Table 3-2.

$$\text{W}21\times 44, \quad \phi M_p = 358 \text{ ft-kips}$$

Step 3: Check the initial assumptions on compactness and lateral bracing.

A check of the compact flange and web criteria shows that this shape is compact. With continuous bracing, the beam can be designed through plastic analysis.

Thus, select a

$$\text{W}21\times 44$$

6.10 T-SHAPED MEMBERS IN BENDING

T-shaped members are normally cut from I-shaped members by splitting the I-shape down the longitudinal axis. Thus, they are often referred to as split tees. When they are made from W-shapes they are called WTs. Similarly, when made from M-shapes as MTs and S-shapes as STs. Provisions for beams formed by combining a pair of angles to form a T and for beams made from a split I-shape and loaded in the axis of symmetry are both found in Section F9 of the *Specification*. These singly symmetric members may be loaded with the stem in tension or compression. Four limit states must be considered in the design of these T-shaped members: yielding, lateral-torsional buckling, flange local buckling, and stem local buckling. The *Specification* combines the provisions for tees and double angles, and those for stem or web legs in tension or compression. This section addresses tees while double angles will be treated after the discussion of single angle bending members.

6.10.1 Yielding of Tees

For the limit state of yielding

$$M_n = M_p \qquad \text{(AISC F9-1)}$$

and M_p is limited, depending on the orientation of the section. For the stem in tension

$$M_p = F_y Z_x \le 1.6 M_y \qquad \text{(AISC F9-2)}$$

and for the stem in compression

$$M_p = M_y \qquad \text{(AISC F9-4)}$$

where

$$M_y = F_y S_x . \qquad \text{(AISC F9-3)}$$

The limit on Equation F9-2 is necessary to ensure that the member is capable of rotating sufficiently to attain the plastic moment strength without the extreme fibers of the shape reaching into the strain-hardening region. This is the same limit that was discussed for I-shaped members bending about their weak axis and can be thought of as limiting the section shape factor. Limiting the plastic moment to the elastic moment, as in Equation F9-4 when the stem is in compression, is a conservative assumption that reflects the limited knowledge available to predict what moment the tee in this orientation could actually attain.

6.10.2 Lateral-Torsional Buckling of Tees

Lateral-torsional buckling also must account for the orientation of the shape. For stems in tension, the limiting unbraced length for the limit state of yielding is the same as it was for doubly symmetric I-shapes. As given in Section F9 it is

$$L_p = 1.76 r_y \sqrt{\frac{E}{F_y}} \qquad \text{(AISC F9-8)}$$

The limiting unbraced length for inelastic lateral-torsional buckling is

$$L_r = 1.95\left(\frac{E}{F_y}\right)\frac{\sqrt{I_y J}}{S_x}\sqrt{2.36\left(\frac{F_y}{E}\right)\frac{dS_x}{J}+1} \qquad \text{(AISC F9-9)}$$

If the unbraced length is less than L_p, the limit state of lateral-torsional buckling does not apply.

If the unbraced length is between L_p and L_r, the nominal strength, when the stem is in tension, is given by a straight line as

$$M_n = M_p - \left(M_p - M_y\right)\left[\frac{L_b - L_p}{L_r - L_p}\right] \qquad \text{(AISC F9-6)}$$

It should be noted that this is similar to the straight-line equations used for other types of members but when $L_b = L_r$, M_n is set to M_y.

When the unbraced length is greater than L_r and the stem is in tension, the nominal strength is given by

$$M_n = M_{cr} \qquad \text{(AISC F9-7)}$$

and

$$M_{cr} = \frac{1.95E}{L_b}\sqrt{I_y J}\left(B + \sqrt{1 + B^2}\right) \qquad \text{(AISC F9-10)}$$

where

$$B = 2.3\left(\frac{d}{L_b}\right)\sqrt{\frac{I_y}{J}} \quad \text{(AISC F9-11)}$$

For the stem in compression at any point along the span, whenever the unbraced length is greater than L_p

$$M_n = M_{cr} \leq M_y \quad \text{(AISC F9-13)}$$

and again

$$M_{cr} = \frac{1.95E}{L_b}\sqrt{I_y J}\left(B + \sqrt{1 + B^2}\right) \quad \text{(AISC F9-10)}$$

but the sign on B is changed to negative. Thus,

$$B = -2.3\left(\frac{d}{L_b}\right)\sqrt{\frac{I_y}{J}} \quad \text{(AISC F9-12)}$$

6.10.3 Flange Local Buckling of Tees

The limit state of flange local buckling for tees reflects the same behavior as for the I-shapes from which they are cut. The limiting width-to-thickness ratios are the same as discussed earlier, and the nominal strength equation is the same, except that it is limited to $1.6M_y$. For compact flanges, the limit state of flange local buckling does not apply. For noncompact flanges, $\lambda_p < \lambda = b_f/2t_f \leq \lambda_r$,

$$M_n = \left[M_p - \left(M_p - 0.7F_y S_{xc}\right)\left(\frac{\lambda - \lambda_{pf}}{\lambda_{rf} - \lambda_{pf}}\right)\right] \leq 1.6M_y \quad \text{(AISC F9-14)}$$

and for slender flanges, $\lambda_r < \lambda$,

$$M_n = \frac{0.7ES_{xc}}{\left(\frac{b_f}{2t_f}\right)^2} \quad \text{(AISC F9-15)}$$

S_{xc} is the section modulus referred to the compression flange. If the stem is in compression, this limit state does not apply.

6.10.4 Stem Local Buckling of Tees

When the stem is in flexural compression, the nominal strength for the limit state of stem local buckling is given by

$$M_n = F_{cr} S_x \quad \text{(AISC F9-16)}$$

S_x is the elastic section modulus and F_{cr} is the critical stress, which is dependent on the stem slenderness, d/t_w. Thus, when

$$\frac{d}{t_w} \leq 0.84\sqrt{\frac{E}{F_y}}$$

the stem will not buckle and

$$F_{cr} = F_y \qquad \text{(AISC F9-17)}$$

When

$$0.84\sqrt{\frac{E}{F_y}} < \frac{d}{t_w} \leq 1.52\sqrt{\frac{E}{F_y}}$$

$$F_{cr} = \left[1.43 - 0.515\frac{d}{t_w}\sqrt{\frac{F_y}{E}}\right]F_y \qquad \text{(AISC F9-18)}$$

and when $\dfrac{d}{t_w} > 1.52\sqrt{\dfrac{E}{F_y}}$

$$F_{cr} = \frac{1.52E}{\left(\dfrac{d}{t_w}\right)^2} \qquad \text{(AISC F9-19)}$$

If the stem is in tension, this limit state does not apply.

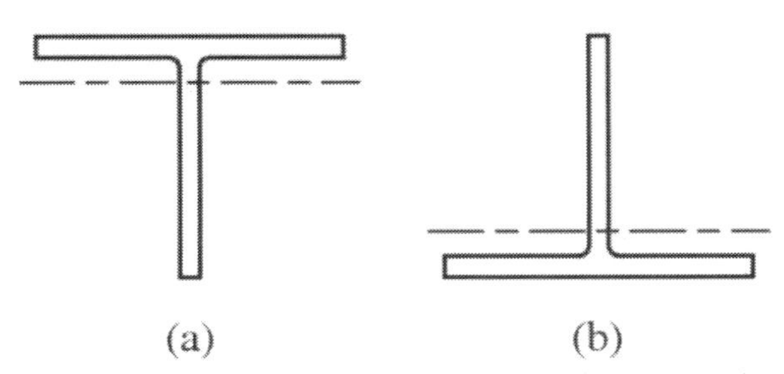

Figure 6.22 T-Beam Orientation for Example 6.16

EXAMPLE 6.16
Bending Strength of WT-Shape

Goal: Determine the nominal moment strength for the given WT member if the stem is in (a) tension (Figure 6.22a) and (b) compression (Figure 6.22b).

Given: An A992 WT9×17.5 is used as a beam to support gravity loads and has lateral support provided at 5 ft intervals.

SOLUTION

Step 1: Determine the section properties for the WT-shape from *Manual* Table 1-8.

$Z_x = 11.2$ in.^{3}, $S_x = 6.21$ in.^{3}, $d = 8.85$ in., $t_w = 0.300$ in.,
$I_y = 7.67$ in.^{4}, $J = 0.252$ in.^{4}, $r_y = 1.22$ in., $b_f/2t_f = 7.06$

Part a Determine the nominal moment strength for the stem in tension. The WT is oriented as shown in Figure 6.22a.

Step 2: Determine the nominal moment strength for the limit state of yielding using Equation F9-2.

$$M_y = F_y S_x = 50(6.21) = 311 \text{ in.-kips}$$

$$M_p = F_y Z_x \le 1.6M_y$$

$$= 50(11.2) = 560 \text{ in.-kips}$$

$$\le 1.6(311) = 498 \text{ in.-kips}$$

Thus,

$$M_n = 498 \text{ in.-kips for the limit state of yielding}$$

Step 3: Determine the nominal moment strength for the limit state of lateral-torsional buckling. From Equation F9-8

$$L_p = 1.76r_y\sqrt{\frac{E}{F_y}} = 1.76(1.22)\sqrt{\frac{29{,}000}{50}} = 51.7 \text{ in.} \Rightarrow 4.31 \text{ ft}$$

and Equation F9-9

$$L_r = 1.95\left(\frac{E}{F_y}\right)\frac{\sqrt{I_yJ}}{S_x}\sqrt{2.36\left(\frac{F_y}{E}\right)\frac{dS_x}{J}+1}$$

$$= 1.95\left(\frac{29{,}000}{50}\right)\frac{\sqrt{7.67(0.252)}}{6.21}\sqrt{2.36\left(\frac{50}{29{,}000}\right)\frac{8.85(6.21)}{0.252}+1}$$

$$= 348 \text{ in.} \Rightarrow 29.0 \text{ ft}$$

Thus, since $L_p < L_b = 5.0 \text{ ft} < L_r$ the beam will buckle inelastically and using Equation F9-6

$$M_n = M_p - (M_p - M_y)\left[\frac{L_b - L_p}{L_r - L_p}\right]$$

$$= 498 - (498 - 311)\left[\frac{5.0 - 4.31}{29.0 - 4.31}\right] = 493 \text{ in.-kips}$$

Step 4: Consider the limit state of flange local buckling.

$$\frac{b_f}{2t_f} = 7.06 < \lambda_{pf} = 0.38\sqrt{\frac{E}{F_y}} = 0.38\sqrt{\frac{29{,}000}{50}} = 9.15$$

Thus, the flange is compact and the limit state of flange local buckling does not apply

Step 5: The controlling limit state for the WT with the stem in tension is the smaller strength given by the limit states of yielding and lateral-torsional buckling. Thus, for the limit state of lateral-torsional buckling

$$M_n = 493/12 = 41.1 \text{ ft-kips}$$

Part b Determine the nominal moment strength for the stem in compression. The WT is oriented as shown in Figure 6.22b.

Step 6: Determine the nominal moment strength for the limit state of yielding using Equation F9-4

M_p is limited to M_y so that, from Step 2,

$$M_n = M_p = M_y = 311 \text{ in.-kips}$$

Step 7: Determine the nominal moment strength for the limit state of lateral-torsional buckling for this orientation with $L_b = 5.0$ ft $> L_p = 4.31$ ft.

Determine B from Equation F9-11 using the negative sign since the stem is in tension. Thus,

$$B = -2.3\left(\frac{d}{L_b}\right)\sqrt{\frac{I_y}{J}} = -2.3\left(\frac{8.85}{5(12)}\right)\sqrt{\frac{7.67}{0.252}} = -1.87$$

and from Equation F9-10

$$M_{cr} = \frac{1.95E}{L_b}\sqrt{I_y J}\left(B + \sqrt{1 + B^2}\right)$$

$$M_n = M_{cr} = \frac{1.95(29{,}000)}{5.0(12.0)}\sqrt{7.67(0.252)}\left[-1.87 + \sqrt{1 + (-1.87)^2}\right]$$

$$= 328 \text{ in.-kips}$$

Step 8: Determine the nominal moment strength for the limit state of stem local buckling.

The stem slenderness is

$$\frac{d}{t_w} = \frac{8.85}{0.300} = 29.5 > 0.84\sqrt{\frac{E}{F_y}} = 0.84\sqrt{\frac{29{,}000}{50}} = 20.2$$

$$< 1.52\sqrt{\frac{E}{F_y}} = 1.52\sqrt{\frac{29{,}000}{50}} = 36.6$$

Therefore, from Equation F9-18

$$F_{cr} = \left[1.43 - 0.515\frac{d}{t_w}\sqrt{\frac{F_y}{E}}\right]F_y$$

$$= \left[1.43 - 0.515\frac{8.85}{0.300}\sqrt{\frac{50}{29,000}}\right](50) = 40.0 \text{ ksi}$$

and

$$M_n = F_{cr}S_x = 40.0(6.21) = 248 \text{ in.-kips}$$

Step 9: Determine the controlling limit state strength for the WT with the stem in compression. The nominal moment strength is the smallest of the limit states checked, yielding, lateral-torsional buckling, and stem local buckling. Thus, for the limit state of stem local buckling

$$M_n = 248/12 = 20.7 \text{ ft-kips}$$

Note: This example shows that using a WT-shape with the stem in compression significantly compromises the strength of the member. Even so, beams with this orientation are often easier to construct for applications such as lintels in masonry walls.

6.11 SINGLE-ANGLE BENDING MEMBERS

When single angles are used as bending members, they can be bending about one of the geometric axes, parallel to the legs, or about one of the principal axes. They are often used as lintels over openings in masonry walls, where they are bending about the geometric axes. Unfortunately, this most useful orientation of the single-angle bending member is also the most complex orientation for the determination of strength. Figures 6.23a and b show a single angle oriented for bending about the geometric axis, and Figures 6.23c and d show the angle oriented for bending about the minor and major principal axes, respectively.

For the doubly symmetric I-shapes and singly symmetric tees discussed to this point, the shear center is located on an axis of symmetry. Typically, loading is applied symmetrically through the shear center. The WT-shapes considered in Section 6.10 were only discussed for loading on the axis of symmetry. For single angles this is not the case. An angle bending about a geometric axis and loaded through the shear center, an axis through the center of the leg thickness, will experience single axis bending. Unfortunately, it is quite common for angles to be loaded through a point other than the shear center, in which case the angle will experience both bending and torsion. The treatment of torsion is beyond the scope of this book although it is addressed in *Specification* Chapter H.

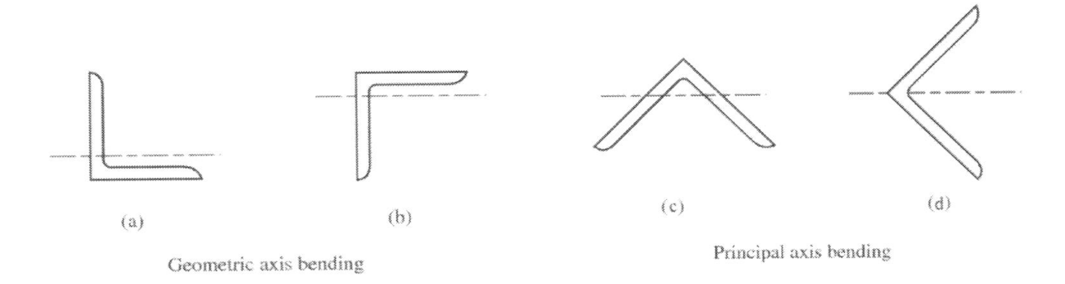

Figure 6.23 Single-Angle Bending about Geometric Axis and Principal Axis

Specification Section F10 gives the provisions for single-angle bending members. The limit states to be checked for these members are yielding, lateral-torsional buckling, and leg local buckling.

6.11.1 Yielding

The ratio of the plastic section modulus to the elastic section modulus, shape factor, for angles can easily become quite large. Thus, in order to be sure that the angle is not strained into the strain–hardening region, the nominal moment for the limit state of yielding is taken as

$$M_n = 1.5\, M_y \qquad \text{(AISC F10-1)}$$

where $M_y = F_y S$, and S is taken as the lowest section modulus about the axis of bending. This applies for either bending about a principal axis or a geometric axis.

6.11.2 Leg Local Buckling

Legs of angles in compression have the same tendency to buckle as other compression elements. *Specification* Table B4-1b defines the limiting slenderness, b/t, in case 12, as

$$\lambda_p = 0.54\sqrt{\frac{E}{F_y}}$$

and

$$\lambda_r = 0.91\sqrt{\frac{E}{F_y}}$$

The bending strength of a single-angle member as a function of leg slenderness is shown in Figure 6.24. In the region of noncompact behavior, $\lambda_p < b/t \le \lambda_r$, the nominal moment strength is given by the straight-line as

$$M_n = F_y S_c\left(2.43 - 1.72\left(\frac{b}{t}\right)\sqrt{\frac{F_y}{E}}\right) \qquad \text{(AISC F10-6)}$$

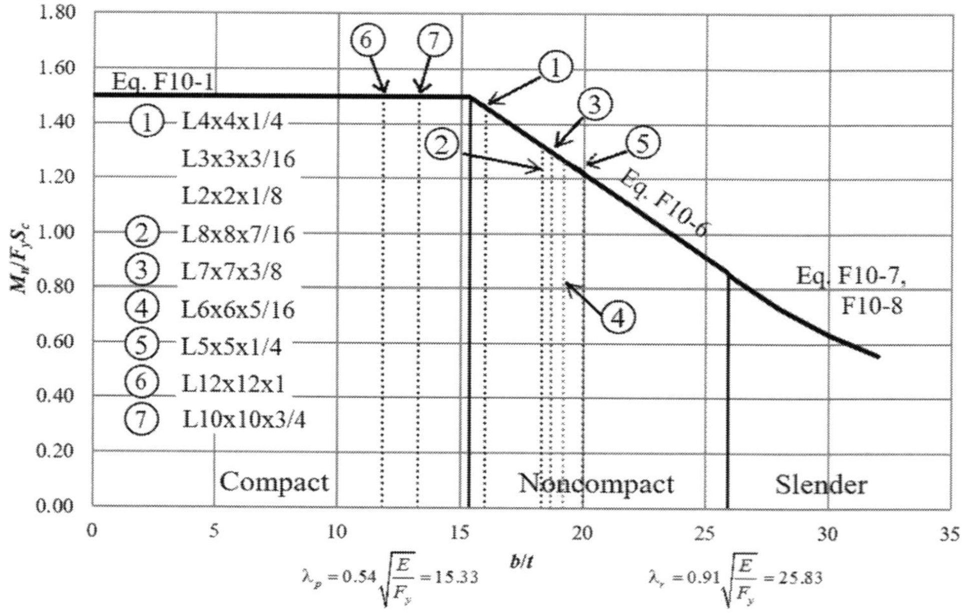

Figure 6.24 Strength of an A36 Single Angle as a Function of Leg Slenderness

and the nominal moment strength for the region of slender behavior is given as a combination of Equations F10-7 and F10-8 as

$$M_n = \frac{0.71ES_c}{\left(\frac{b}{t}\right)^2} \qquad \text{(AISC F10-7, AISC 10-8)}$$

where S_c is the elastic section modulus to the toe in compression, relative to the axis of bending, either geometric or principal axis. For bending about one of the geometric axes for an equal-leg angle with no lateral-torsional restraint, S_c must be taken as 0.8 times the geometric axis section modulus to reflect the effect of actual bending stresses about an inclined axis that result due to the absence of lateral-torsional restraint. The largest width-to-thickness ratios for each leg width of angles given in *Manual* Table 1-7 are shown in Figure 6.24. The angle leg with the largest width-to-thickness ratio is $b/t = 5/0.25 = 20$. It can be seen from the figure that none of these angles will be considered slender for F_y = 36 ksi.

6.11.3 Lateral-Torsional Buckling

The limit state of lateral-torsional buckling is a function of the axis of bending and whether the toe of the angle is at a maximum stress in tension or compression. For bending about the minor principal axis and for angles with full lateral support, the limit state of lateral-torsional buckling does not apply. Thus, provisions are given only for bending about the major principal axis and for the special case of bending about the geometric axis of equal leg angles.

For all cases of lateral-torsional buckling, the nominal moment strength is a function of the elastic lateral-torsional buckling moment, M_{cr}, which is determined for major principal axis bending or geometric axis bending of equal leg angles such that for $M_y/M_{cr} \le 1.0$

$$M_n = \left(1.92 - 1.17\sqrt{\frac{M_y}{M_{cr}}}\right)M_y \le 1.5M_y \qquad \text{(AISC F10-2)}$$

and for $M_y/M_{cr} > 1.0$

$$M_n = \left(0.92 - \frac{0.71M_{cr}}{M_y}\right)M_{cr} \qquad \text{(AISC F10-3)}$$

The nominal bending strength divided by the yield moment for the limit state of lateral-torsional buckling is given in Figure 6.25 as a function of the yield moment divided by the elastic lateral-torsional buckling moment. It can be seen that as the yield moment divided by the elastic lateral-torsional buckling moment increases, the nominal bending strength of the member divided by the yield moment decreases.

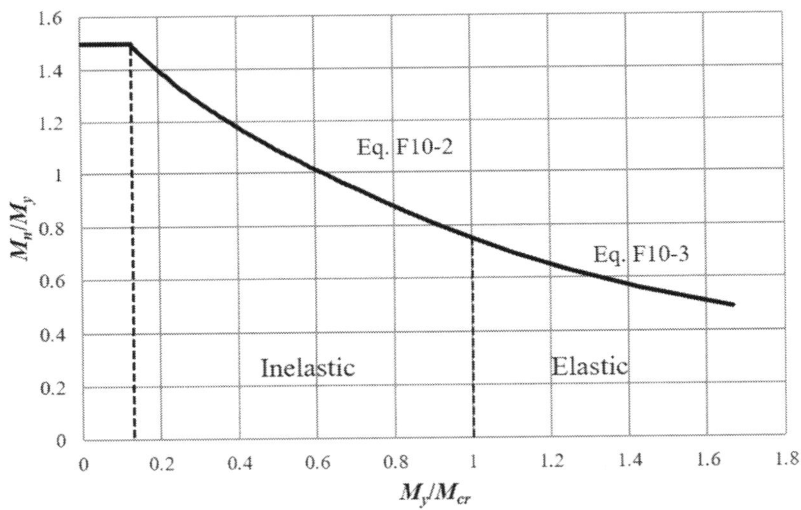

Figure 6.25 Strength of a Single Angle as a Function of Lateral-Torsional Buckling

For bending about the major principal axis, Figure 6.23d, the elastic lateral-torsional buckling moment is

$$M_{cr} = \frac{9EAr_z tC_b}{8L_b}\left(\sqrt{1+\left(4.4\frac{\beta_w r_z}{L_b t}\right)^2} + 4.4\frac{\beta_w r_z}{L_b t}\right) \qquad \text{(AISC F10-4)}$$

The variables in Equation F10-4 are the same as previously defined except for the new variable, β_w, which is a section property for single angles bent about their major principal axis. It is zero for equal leg angles, positive for angles with their short leg in compression and negative for angles with their long leg in compression. Values of β_w are given in Commentary Table C-F10.1. Note that β_w is independent of angle thickness. All unequal leg angles must be evaluated for lateral-torsional buckling about the major principal axis, even if the actual applied load is bending the angle about one of its geometric axes. Thus, for this condition, the applied geometric axis bending moment must be converted to moments about both principal axes.

For geometric axis bending the strength of equal leg angles may be determined for conditions of no lateral-torsional restraint or for lateral-torsional restraint at the point of maximum moment only. For any other condition of lateral-torsional restraint, they must be evaluated for bending about the principal axes.

When there is no lateral restraint, the yield moment M_y, is taken as $0.80F_yS$ where S is the geometric section modulus. The elastic lateral-torsional buckling moment for loading that puts the maximum stress as compression at the toe is

$$M_{cr} = \frac{0.58Eb^4tC_b}{L_b^2}\left(\sqrt{1+0.88\left(\frac{L_bt}{b^2}\right)^2}-1\right) \quad \text{(AISC F10-5a)}$$

and when the maximum stress is tension at the toe

$$M_{cr} = \frac{0.58Eb^4tC_b}{L_b^2}\left(\sqrt{1+0.88\left(\frac{L_bt}{b^2}\right)^2}+1\right) \quad \text{(AISC F10-5b)}$$

Note that when the toe is at a maximum stress in tension, the elastic lateral-torsional buckling moment is greater than when the maximum stress is compression at the toe. This means that the nominal moment strength for lateral-torsional buckling will be greater when the maximum stress is tension at the toe.

For the case of lateral-torsional restraint at the point of maximum moment, M_{cr} is taken as 1.25 times the value determined from Equations F10-5 and M_y is taken as $1.0F_yS$.

EXAMPLE 6.17
Bending Strength of a Single Angle

Goal: Determine the available bending strength of an equal-leg angle loaded about the geometric axis through the shear center.

Given: A 6×6×5/16 A572 Gr. 50 angle is required to span 8.0 ft on a simple span with lateral supports at the ends only. The vertical leg is up so that it is stressed in compression due to the uniform gravity load.

SOLUTION

Step 1: Determine the required properties from *Manual* Table 1-7

$$S_x = S_y = 2.95 \text{ in.}^3 \quad b = 6.0 \text{ in.} \quad t = 0.3125 \text{ in.}$$

Step 2: Determine the nominal moment strength for the limit state of yielding. Use Equation F10-1.

$$M_n = 1.5F_yS_x = 1.5(50(2.95)) = 221 \text{ in.-kips}$$

Step 3: Determine the nominal moment strength for the limit state of lateral-torsional buckling, first determine the elastic lateral-torsional buckling moment. For an equal-leg angle with maximum compression in the toe (leg up and top in compression), and C_b = 1.14 for the uniform load, use Equation F10-5a

$$M_{cr} = \frac{0.58Eb^4tC_b}{L_b^2}\left(\sqrt{1+0.88\left(\frac{L_bt}{b^2}\right)^2}-1\right)$$

$$= \frac{0.58(29{,}000)(6.0)^4(0.3125)(1.14)}{(8.0(12))^2}\left(\sqrt{1+0.88\left(\frac{8.0(12)(0.3125)}{(6.0)^2}\right)^2}-1\right)$$

$$= 227 \text{ in.-kips}$$

Step 4: Determine the yield moment to be used when there is no lateral-torsional restraint. This accounts for the presence of some torsion by using $0.8S_x$. Thus,

$$M_y = F_y(0.8S_x) = 50(0.8(2.95)) = 118 \text{ in.-kips}$$

Step 5: Since $M_y/M_{cr} = 118/227 = 0.520 < 1.0$, use Equation F10-2 to determine the nominal moment strength for the limit state of lateral-torsional buckling.

$$M_n = \left(1.92 - 1.17\sqrt{\frac{M_y}{M_{cr}}}\right)M_y \le 1.5M_y$$

$$= \left(1.92 - 1.17\sqrt{\frac{118}{227}}\right)118$$

$$= 127 \text{ in.-kips} \le 1.5M_y = 1.5(118) = 177 \text{ in.-kips}$$

Thus,

$$M_n = 127 \text{ in.-kips}$$

Step 6: Check the leg slenderness to determine if the strength limit state of leg local buckling must be determined. For the 6×5/16 angle, $b/t = 6.0/0.3125 = 19.2$. From Specification Table B4.1b, the slenderness limits are

$$\lambda_p = 0.54\sqrt{\frac{E}{F_y}} = 0.54\sqrt{\frac{29{,}000}{50}} = 13.0$$

$$\lambda_r = 0.91\sqrt{\frac{E}{F_y}} = 0.91\sqrt{\frac{29{,}000}{50}} = 21.9$$

Therefore, the angle is noncompact so the strength for the limit state of leg local buckling must be determined.

Step 7: As with lateral-torsional buckling, for the limit state of leg local buckling, the reduced section modulus must be used when bending is about the geometric axis. Thus, using Equation F10-6

$$M_n = F_y S_c\left(2.43 - 1.72\left(\frac{b}{t}\right)\sqrt{\frac{F_y}{E}}\right)$$

$$= 50(0.8(2.95))\left(2.43 - 1.72\left(\frac{6.0}{0.3125}\right)\sqrt{\frac{50}{29{,}000}}\right)$$

$$= 125 \text{ in.-kips}$$

Step 8: The nominal bending strength is the lowest value based on the limit states of yielding, lateral-torsional buckling, and leg local buckling. Thus, leg local buckling controls and

$$M_n = 125 \text{ in.-kips}$$

For LRFD Step 9: Determine the design strength

$$\phi M_n = 0.9(125) = 113 \text{ in.-kips}$$

For ASD Step 9: Determine the allowable strength

$$\frac{M_n}{\Omega} = \frac{125}{1.67} = 74.9 \text{ in.-kips}$$

6.12 DOUBLE ANGLE MEMBERS IN BENDING

T-shaped bending members were discussed in Section 6.10 and it was pointed out that tees and double angles are both treated in *Specification* Section F9. The same four limit states considered for tees must be considered for double angle members: yielding, lateral-torsional buckling, flange local buckling, and stem local buckling. As with tees, double angles must be loaded in the plane of symmetry. Again, these limit states are treated differently if the web legs are in compression or tension. The following describes the application of *Specification* Section F9 to double angle bending members.

6.12.1 Yielding of Double Angles

For the limit state of yielding

$$M_n = M_p \quad \text{(AISC F9-1)}$$

and M_p is limited, depending on the orientation of the section. For the web legs in tension

$$M_p = F_y Z_x \leq 1.6 M_y \quad \text{(AISC F9-2)}$$

and for web legs in compression

$$M_p = 1.5 M_y \quad \text{(AISC F9-5)}$$

where

$$M_y = F_y S_x \quad \text{(AISC F9-3)}$$

The limit on Equation F9-2 is required for double angles just as it was for tees, to ensure that the member is capable of rotating sufficiently to attain the plastic moment strength without the extreme fibers of the shape reaching into the strain-hardening region. Limiting the plastic moment when web legs are in compression to $1.5M_y$ is slightly more restrictive than the limit in Equation F9-2 but is significantly greater than it was for tees.

6.12.2 Lateral-Torsional Buckling of Double Angles

Lateral-torsional buckling also must account for the orientation of the shape. For web legs in tension, the provisions are the same as they were for tees. The limiting unbraced lengths are

$$L_p = 1.76 r_y \sqrt{\frac{E}{F_y}} \quad \text{(AISC F9-8)}$$

and

$$L_r = 1.95\left(\frac{E}{F_y}\right)\frac{\sqrt{I_y J}}{S_x}\sqrt{2.36\left(\frac{F_y}{E}\right)\frac{dS_x}{J}+1} \quad \text{(AISC F9-9)}$$

As usual, if the unbraced length is less than L_p, the limit state of lateral-torsional buckling does not apply.

If the unbraced length is between L_p and L_r, the nominal strength, when the web legs are in tension, is given by the same straight-line that was used for tees,

$$M_n = M_p - \left(M_p - M_y\right)\left[\frac{L_b - L_p}{L_r - L_p}\right] \quad \text{(AISC F9-6)}$$

When the unbraced length is greater than L_r and the stem is in tension, the nominal strength is given by

$$M_n = M_{cr} \quad \text{(AISC F9-7)}$$

and

$$M_{cr} = \frac{1.95E}{L_b}\sqrt{I_y J}\left(B + \sqrt{1+B^2}\right) \quad \text{(AISC F9-10)}$$

where

$$B = 2.3\left(\frac{d}{L_b}\right)\sqrt{\frac{I_y}{J}} \quad \text{(AISC F9-11)}$$

For web legs in compression at any point along the span, whenever the unbraced length is greater than L_p, the nominal moment strength is determined as it was for single angles, through Equations F10-2 and F10-3.

For $M_y/M_{cr} \leq 1.0$

$$M_n = \left(1.92 - 1.17\sqrt{\frac{M_y}{M_{cr}}}\right)M_y \leq 1.5M_y \quad \text{(AISC F10-2)}$$

and for $M_y/M_{cr} > 1.0$

$$M_n = \left(0.92 - \frac{0.71M_{cr}}{M_y}\right)M_{cr} \quad \text{(AISC F10-3)}$$

The elastic lateral-torsional buckling moment is determined as

$$M_{cr} = \frac{1.95E}{L_b}\sqrt{I_y J}\left(B + \sqrt{1+B^2}\right) \quad \text{(AISC F9-10)}$$

with

$$B = -2.3\left(\frac{d}{L_b}\right)\sqrt{\frac{I_y}{J}} \quad \text{(AISC F9-12)}$$

and

$$M_y = F_y S_x \quad \text{(AISC F9-3)}$$

6.12.3 Leg Local Buckling of Double Angles

Leg local buckling of double angles is treated the same as leg local buckling of single angles by reference from *Specification* Section F9 to Section F10. As discussed for single angles, the limiting slenderness, b/t, in case 12, is

$$\lambda_p = 0.54\sqrt{\frac{E}{F_y}}$$

and

$$\lambda_r = 0.91\sqrt{\frac{E}{F_y}}$$

For noncompact angles, $\lambda_p < b/t \le \lambda_r$, the nominal moment strength is given by

$$M_n = F_y S_c\left(2.43 - 1.72\left(\frac{b}{t}\right)\sqrt{\frac{F_y}{E}}\right) \qquad \text{(AISC F10-6)}$$

For slender angles the nominal moment strength is given as a combination of Equations F10-7 and F10-8 as

$$M_n = \frac{0.71ES_c}{\left(\frac{b}{t}\right)^2}$$

For both, S_c is the elastic section modulus of the double angle member to the toe in compression.

EXAMPLE 6.18
Bending Strength of a Double Angle

Goal: Determine the available bending strength of a double equal-leg angle loaded in the plane of symmetry.

Given: Two 6×6×5/16 A572 Gr. 50 angles are required to span 8.0 ft on a simple span with lateral supports at the ends only. The vertical legs are in contact and are oriented up so that they are stressed in compression due to the uniform gravity load.

SOLUTION

Step 1: Determine the required properties from *Manual* Table 1-7 for a single angle

$$S_x = 2.95 \text{ in.}^3 \quad I_x = 13.0 \text{ in.}^4 \quad b = 6.0 \text{ in.} \quad t = 0.3125 \text{ in.}$$

From Manual Table 1-15 for double angles with no separation, s,

$$r_y = 2.47 \text{ in. and } 2S_x = 2(2.95) = 5.90 \text{ in.}^3$$

Step 2: Determine the nominal moment strength for the limit state of yielding. Use Equation F9-5.

$$M_n = 1.5F_yS_x = 1.5(50(5.90)) = 443 \text{ in.-kips}$$

Step 3: Determine the nominal moment strength for the limit state of lateral-torsional buckling. First determine the unbraced length beyond which lateral-torsional buckling must be considered, L_p, based on Equation F9-8.

$$L_p = 1.76r_y\sqrt{\frac{E}{F_y}} = 1.76(2.47)\sqrt{\frac{29{,}000}{50}} = 105 \text{ in.}$$

Since L_b = 8.0 ft = 96.0 in.< 105 in. lateral torsional buckling does not need to be checked.

Step 6: Check the leg slenderness to determine if the strength limit state of leg local buckling must be determined using Section F10.3 by reference from Section F9.4. For two 6×6×5/16 angles with the back-to-back legs in flexural compression, $b/t = 6.0/0.3125 = 19.2$. From Specification Table B4.1b, the slenderness limits are

$$\lambda_p = 0.54\sqrt{\frac{E}{F_y}} = 0.54\sqrt{\frac{29{,}000}{50}} = 13.0$$

$$\lambda_r = 0.91\sqrt{\frac{E}{F_y}} = 0.91\sqrt{\frac{29{,}000}{50}} = 21.9$$

Therefore, the angles are noncompact just as for the single angle of Example 6.17 so the strength for the limit state of leg local buckling must be determined.

Step 7: Unlike for the single angle, the reduced section modulus is not used. Thus, using Equation F10-6

$$M_n = F_y S_c\left(2.43 - 1.72\left(\frac{b}{t}\right)\sqrt{\frac{F_y}{E}}\right)$$

$$= 50(5.90)\left(2.43 - 1.72\left(\frac{6.0}{0.3125}\right)\sqrt{\frac{50}{29{,}000}}\right) = 312 \text{ in.-kips}$$

Step 8: The nominal bending strength is the lowest value based on the limit states of yielding and leg local buckling. Thus, leg local buckling controls and

$$M_n = 312 \text{ in.-kips}$$

For LRFD Step 9: Determine the design strength

$$\phi M_n = 0.9(312) = 281 \text{ in.-kips}$$

For ASD Step 9: Determine the allowable strength

$$M_n/\Omega = 312/1.67 = 187 \text{ in.-kips}$$

The strength of the double angle beam in Example 6.18 is more than twice that of the single angle beam in Example 6.17, even with the same limit state, leg local buckling, controlling the strength. Because of symmetry about the plane of loading, the reduced section modulus that was used for the single angle is not used in determining the local buckling strength for the double angle member. In addition, lateral-torsional buckling did not even need to be checked for this double angle beam since the unbraced length was less than L_p, where it potentially could have controlled for the single angle beam. Clearly some of the modes of failure for a single angle beam and a double angle beam can be quite different.

6.13 MEMBERS IN BIAXIAL BENDING

Bending members are often called upon to resist forces that result in bending about two orthogonal axes. Examples of this member type are crane girders and roof purlins in industrial buildings. Regardless of the actual orientation of an applied moment, it is possible to break the moment into components about the two principal axes, as shown in Figure 6.26. Once this is accomplished, the ability of the section to resist the combined moments can be determined through the interaction equation.

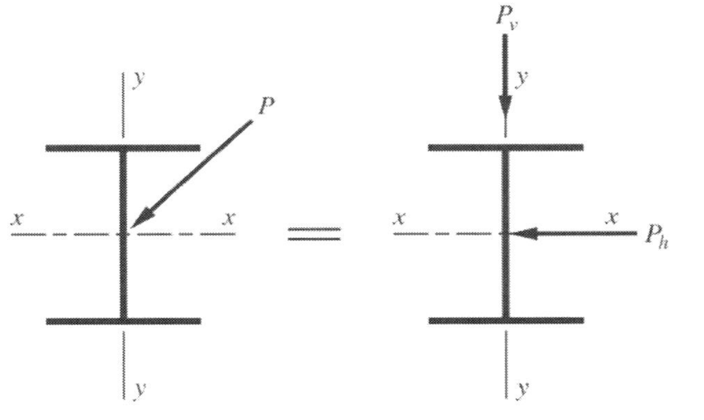

Figure 6.26 Biaxial Bending of I-Shaped Beam

Specification Chapter H addresses the interaction of forces. For the combination of moments, a simple linear interaction equation is used, as shown in Figure 6.27. This is taken from the equation provided in *Specification* Section H1 for combined axial load and moment. When the axial load is zero, Equation H1-1b reduces to

$$\frac{M_{rx}}{M_{cx}} + \frac{M_{ry}}{M_{cy}} \leq 1.0$$

where the moment terms relate to the x- and y-axes, the numerator is the required strength, and the denominator is the available strength, determined as though the member were bending about only one axis at a time. Thus, if the required x-axis moment is 79 percent of the x-axis strength, only 21 percent of the y-axis strength is available to resist moment. More attention is given to the use of interaction equations when axial load is combined with the bending moment in Chapter 8.

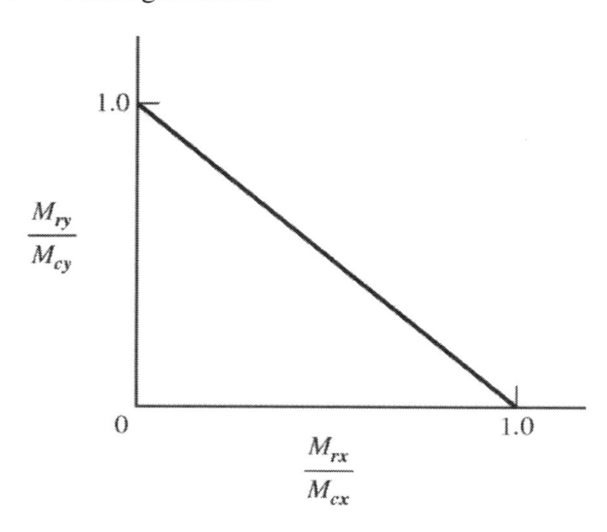

Figure 6.27 Simple Linear Interaction Diagram for Biaxial Bending

6.14 SERVICEABILITY CRITERIA FOR BEAMS

There are several serviceability considerations that the designer must address. A general set of provisions is found in *Specification* Chapter L. Although failure to satisfy these criteria may not impact the strength of the member or overall structure, it may lead to the first signs of difficulty for successful completion of a project. The specific criteria should be discussed in detail with the designer's client so that the quality of the final product is consistent with the expectations of the owner. Although excessive floor deflection may not be a safety issue, it may lead to cracking of finishes and an unacceptable appearance. It may also be an indicator that other serviceability issues are likely, such as vibration. Experience may indicate that an annoying amount of floor vibration may be present at first, but occupants become used to it with time. The client may be unwilling to deal with this period of dissatisfaction and insist that the system be designed to prevent vibration complaints. This must be known at the beginning of a project, not after the occupants move in and find the floor response objectionable. The engineer must be sure to identify these considerations for the owner so that the decisions made are appropriate to achieve the expected outcome.

Beams generally have two serviceability issues that must be addressed directly, deflection and vibration. With today's high strength steels, deflection criteria may often control designs for typical building spans and loads. A third issue is overall building drift, where the beams in moment frames play a significant role in the response.

6.14.1 Deflection

Deflection is the normal response of a beam to its imposed load. It is impossible to erect a beam with zero deflection under load, but the designer will be able to limit that deflection with proper attention to this serviceability limit state. Deflections must be addressed for a variety of loading cases. Deflection under dead load is critical because it impacts the construction process, including the amount of concrete fill needed to form a flat and level floor. Live load deflection is critical because it impacts the finishes of elements attached to the floor, such as ceilings and walls, and may be visible to the occupants. Experience has demonstrated that live load deflection is not a problem if it is limited to 1/360 of the span. Dead load deflection limitations are a function of the particular structural element and loading. AISC *Design Guide 3: Serviceability Design Considerations for Steel Buildings* covers deflection and other serviceability design criteria.

6.14.2 Vibration

Although vibration of floor systems is not a safety consideration, it can be a very annoying response and very difficult to correct after the building is occupied. The most common problem occurs with wide-open spaces with very little damping, such as an open office floor plan. To reduce the risk of annoyance, a general rule is to space the beams or joists sufficiently far apart so that the slab thickness is large enough to provide the needed stiffness and damping. AISC *Design Guide 11: Vibration of Steel-Framed Structural Systems Due to Human Activity* covers the design of steel-framed floor systems for human comfort.

6.14.3 Drift

Under lateral loading, a building will sway sideways. This lateral displacement is called *drift*. As with deflection and vibration, drift is usually not a safety consideration, but it can be annoying and have a negative impact on nonstructural elements, causing cracks in finishes. Beams and girders are important in reducing the drift for moment frames, and their final size might actually be determined by drift considerations. However, the impact of drift considerations on beams cannot be determined without also evaluating the other parts of the lateral load–resisting system. This serviceability limit state is treated in Chapter 8. Drift is also discussed in *Design Guide 3*.

6.14.4 Deflection Calculation

Because beam deflection is a serviceability consideration, calculations are carried out using the specific loads under which the serviceability considerations are to be checked. This can be live load, dead load, or some combination of loads, but normally does not include any load factors. Thus, regardless of whether a design is completed using LRFD or ASD, serviceability considerations are checked for the same loads. Numerous elastic analysis techniques are available to determine the maximum deflection of a given beam and loading. Some common loading conditions with their corresponding maximum deflections are shown in Figure 6.28. These and many others are given in *Manual* Table 3-23.

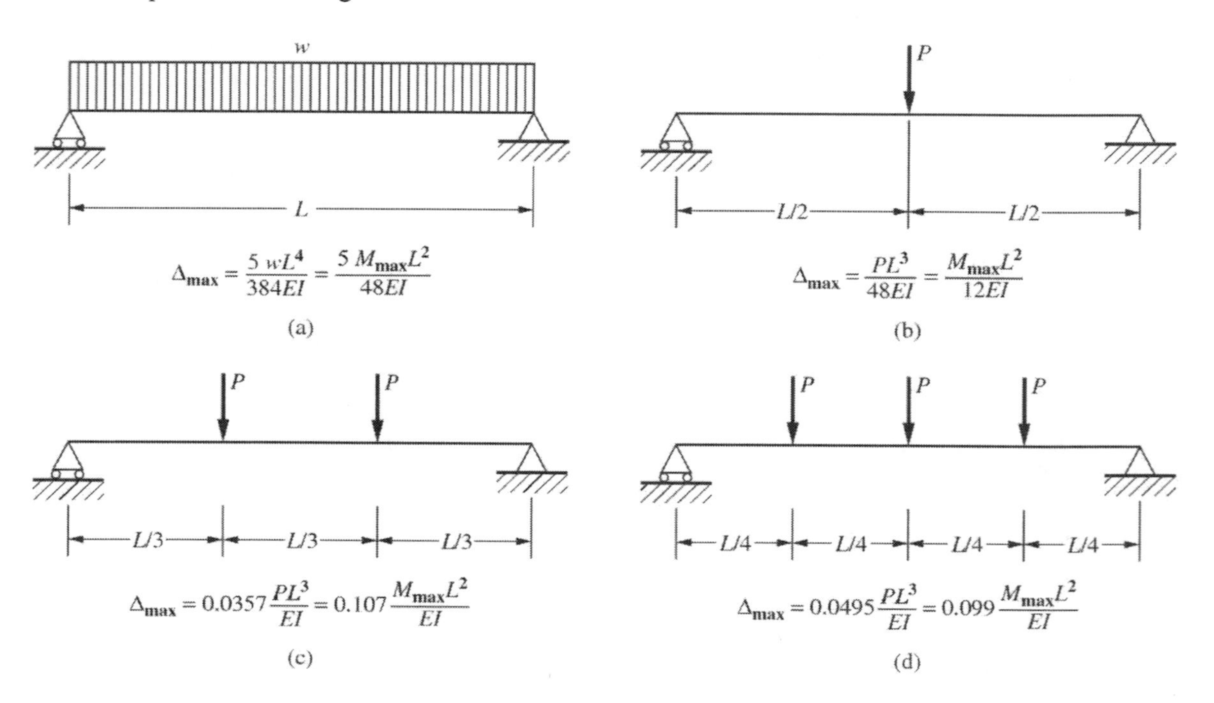

Figure 6.28 Some Common Loading Conditions with Their Corresponding Maximum Deflections

EXAMPLE 6.19 ***Live Load Deflection***

Goal: Check the live load deflection of a previously designed beam.

Given: Use the information from Example 6.3, where a W18×35 was selected. Limit the live load deflection for an acceptable design to 1/360 of the span.

SOLUTION

Step 1: Collect the required information from Example 6.3.

For the W18×35, $I = 510$ in.^{4}. The live load is 24 kips applied at the center of a 20 ft span.

Step 2: Determine the live load deflection. Using the deflection equation found in Figure 6.28 for case (b),

$$\Delta = \frac{PL^3}{48EI} = \frac{24(20.0)^3(12)^3}{48(29{,}000)(510)} = 0.467 \text{ in.}$$

Step 3: Compare the calculated deflection to the given limit.

The deflection limit is

$$\Delta_{max} = \frac{20.0(12)}{360} = 0.667 \text{ in.}$$

Because the calculated deflection is less than the deflection limit,

$$\Delta = 0.467 \text{ in.} < \Delta_{max} = 0.667 \text{ in.}$$

The deflection satisfies the set criteria

EXAMPLE 6.20
Beam Design through Deflection Limit

Goal: Select a W-shape to satisfy a live load deflection limit.

Given: Use the data from Example 6.19, except that the deflection limit is set to a more severe level of 1/1000 of the span. If the selected member does not meet the established criteria, select a W-shape that satisfies the limitation.

SOLUTION

Step 1: Check the new deflection limit.

$$\Delta_{max} = \frac{20.0(12)}{1000} = 0.240 \text{ in.}$$

From Example 6.19 we know already that the given beam deflects too much.

Step 2 Determine the minimum acceptable moment of inertia, I_{min}, necessary to ensure that the deflection does not exceed the given limit.

Rearranging the maximum deflection for a concentrated load at mid-span equation to solve for I_{min},

$$I_{min} = \frac{PL^3}{48E\,\Delta_{max}} = \frac{24(20.0)^3(12)^3}{48(29{,}000)(0.240)} = 993 \text{ in.}^4$$

Step 3: Select a beam with $I \geq 993$ in.^{4} and, from Example 6.3, one that satisfies the strength limit, $Z \geq 64.0$ in.^{3}.

From the moment of inertia tables, *Manual* Table 3-3, select a W21×55, $I = 1140$ in.^{4} and, from *Manual* Table 3-2, $Z = 126$ in.^{3}

This is the lightest-weight W-shape that will satisfy the required moment of inertia. Shapes are grouped in *Manual* Table 3-3 so the least weight member with the largest moment of inertia is given in bold at the top of each group.

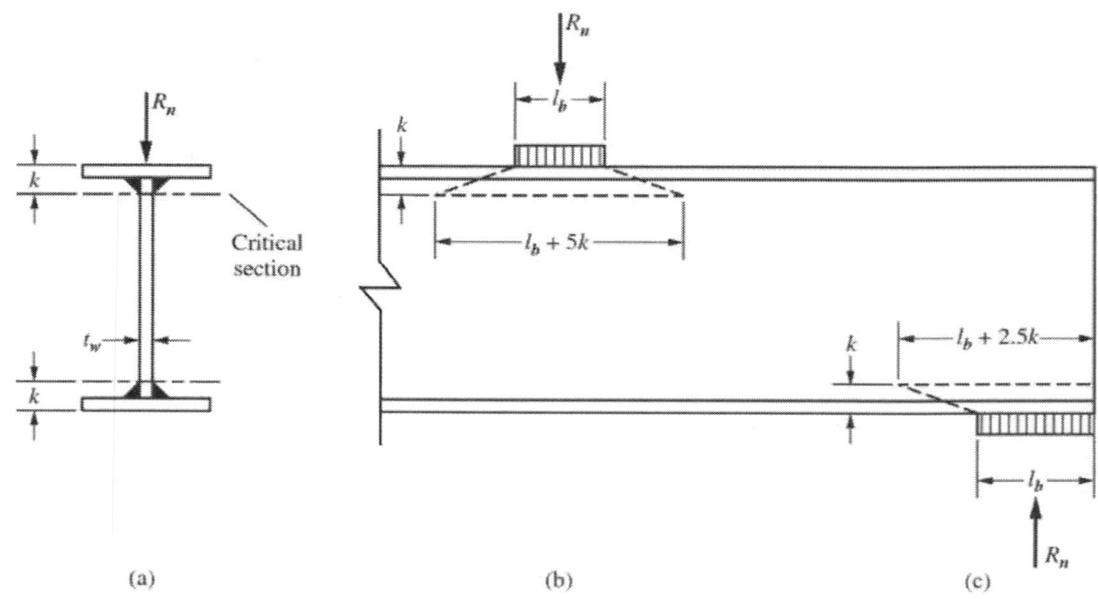

Figure 6.29 Single Concentrated Force Applied to a Beam

6.15 CONCENTRATED FORCES ON BEAMS

Before a beam can be called upon to carry a given load, that load must be transferred to the beam through some type of connection. In a similar manner, the beam reactions must be carried to the beam's supporting structure through some type of connection. Although the majority of beams are loaded through connections to their webs, some may be loaded by applying a concentrated force to the top flange, and some will have their reactions resisted by bearing on a supporting element. In these cases, a check must be made to establish that the beam web has sufficient strength to resist the applied forces.

Four limit states determine the load carrying strength of the web to resist these concentrated forces: web local yielding, web local crippling, web sidesway buckling, and flange local bending. These limit states are all described in Section J10 of the *Specification*. Although it is possible to select a beam with a web sufficiently thick that these limit states do not control, it is normally more economical to add bearing stiffeners under the concentrated loads to provide the necessary strength. For the limit states of web local yielding and web local crippling, it may be easiest to increase the length of bearing, which would eliminate the requirement for bearing stiffeners. However, if the web sidesway buckling limit state is exceeded, stiffeners are required. In applications where W-shapes are used, this limit state is particularly critical for continuous beams in the negative moment region where the bottom flange is the compression flange. The limit state of flange local bending is a concern only when tension loads are applied to beams. The design of stiffeners is covered in Section 7.4 of this book under the discussion of plate girders, because they are much more commonly found in that application. The limit states of web local yielding and web local crippling will be discussed here with an eye toward using bearing length as the controlling factor in resisting these limit states.

6.15.1 Web Local Yielding

When a single concentrated force is applied to a beam as shown in Figure 6.29, the force is assumed to be delivered to the beam over a length of bearing, l_b. It is then distributed through the flange and into the web. The narrowest portion of the web is the critical section. This occurs at the toe of the web-to-flange fillet, dimension k in Figure 6.29a. There are two variables listed in *Manual* Table 1-1 for this dimension, k_{des} and k_{det}. In this application k_{des} is to be used since it represents the lower bound on the actual k dimension found in the production of W-shapes. The distribution of the force along the web takes place at a slope of 1:2.5. Thus, when the critical section is reached, the force has been distributed over a length of l_b plus $2.5k$ in each direction. If the concentrated force is applied so that the force is distributed along the web in both directions, this distribution increases the bearing length by $5k$ as shown in Figure 6.29b. If the bearing is close to the end of the member, distribution takes place only in one direction, toward the midspan. The *Specification* defines "close to the member end" as being within the member depth from the end. Thus, the available length of the web is (l_b + $2.5k$), as shown in Figure 6.29c.

The nominal strength of the beam web when the concentrated force to be resisted is applied at a distance from the member end that is greater than the depth of the member, d, is

$$R_n = F_{yw} t_w \left(5k + l_b\right) \qquad \text{(AISC J10-2)}$$

When the concentrated force to be resisted is applied at a distance from the member end that is less than or equal to the depth of the member, d, the nominal strength is

$$R_n = F_{yw} t_w \left(2.5k + l_b\right) \qquad \text{(AISC J10-3)}$$

where

F_{yw} = yield stress of the web
l_b = length of bearing
k = k_{des}= distance from the outer face of the flange to the web toe of the fillet weld
t_w = web thickness

For web local yielding

$$\phi = 1.0 \text{ (LRFD)} \qquad \Omega = 1.50 \text{ (ASD)}$$

6.15.2 Web Local Crippling

The criteria for the limit state of web local crippling also depend on the location of the force with respect to the end of the member and the length of bearing, l_b. The *Specification* equations apply to both I-shaped members and HSS. When applied to I-shaped members, Q_f = 1.0.

When the concentrated compressive force is applied at a distance from the member end that is greater than or equal to $d/2$,

$$R_n = 0.80 t_w^2 \left[1 + 3\left(\frac{l_b}{d}\right)\left(\frac{t_w}{t_f}\right)^{1.5}\right] \sqrt{\frac{E F_{yw} t_f}{t_w}} Q_f \qquad \text{(AISC J10-4)}$$

When the force is applied at a distance less than $d/2$ and $l_b/d \leq 0.2$,

$$R_n = 0.40t_w^2\left[1+3\left(\frac{l_b}{d}\right)\left(\frac{t_w}{t_f}\right)^{1.5}\right]\sqrt{\frac{EF_{yw}t_f}{t_w}}Q_f \qquad \text{(AISC J10-5a)}$$

and when $l_b/d > 0.2$,

$$R_n = 0.40t_w^2\left[1+\left(\frac{4l_b}{d}-0.2\right)\left(\frac{t_w}{t_f}\right)^{1.5}\right]\sqrt{\frac{EF_{yw}t_f}{t_w}}Q_f \qquad \text{(AISC J10-5b)}$$

For web local crippling

$$\phi = 0.75 \text{ (LRFD)} \qquad \Omega = 2.0 \text{ (ASD)}$$

When the beam reaction is transferred to the supporting material, it is often necessary to use a beam bearing plate to spread the load over a sufficiently large area to avoid failure of the supporting material. These beam bearing plates are addressed in Section 11.11.

EXAMPLE 6.21a
Beam Bearing Strength by LRFD

Goal: Determine the required length of bearing for a W-shape to support (a) a midspan concentrated load and (b) concentrated reactions at the end of a simple span.

Given: A W18×35 A992 beam was selected in Example 6.3a as a simply supported member spanning 20 ft and carrying a midspan dead load of 8.0 kips and live load of 24.0 kips.

SOLUTION

Step 1: Determine section properties from *Manual* Table 1-1

$$d = 17.7 \text{ in.}, t_w = 0.300 \text{ in.}, t_f = 0.425 \text{ in.}, k_{des} = 0.827 \text{ in.}$$

Step 2: Determine the required bearing strength at the load point and at the reactions using the LRFD load combinations from Section 2.4.

$$P_u = 1.2P_D + 1.6P_L = 1.2(8.0) + 1.6(24.0) = 48.0 \text{ kips}$$

$$R_u = \frac{P_u}{2} = \frac{48}{2} = 24 \text{ kips}$$

Part a At the load

Step 3: For the limit state of yielding, determine the required length within the web to resist the applied load. The required length, based on Equation J10-2 where $N = (5k + l_b)$, is then

$$(5k + l_b) = \frac{P_n}{F_{yw}t_w} = \frac{P_u/\phi}{F_{yw}t_w} = \frac{48/1.0}{50(0.300)} = 3.20 \text{ in.}$$

Step 4: Determine the minimum length of bearing, l_b. Since this force is located so that distribution takes place in both directions,

$$l_b = 3.20 - 5(0.827) = -0.935 \text{ in.}$$

A negative bearing length in this calculation means that the distribution, 5k, is more than enough to carry the required load. Thus, the minimum bearing length for normal practice of 3.0 in. is adequate.

Step 5: For the limit state of web local crippling, determine the nominal strength if the bearing length is taken as a practical minimum of 3.0 in. From Equation J10-4,

$$R_n = 0.80t_w^2\left[1+3\left(\frac{l_b}{d}\right)\left(\frac{t_w}{t_f}\right)^{1.5}\right]\sqrt{\frac{EF_{yw}t_f}{t_w}}Q_f$$

$$= 0.80(0.300)^2\left[1+3\left(\frac{3.00}{17.7}\right)\left(\frac{0.300}{0.425}\right)^{1.5}\right]\sqrt{\frac{29{,}000(50)(0.425)}{0.300}}(1.0)$$

$$= 134 \text{ kips}$$

Step 6: Determine the design strength for web local crippling,

$$\phi R_n = 0.75(134) = 101 \text{ kips}$$

Step 7: For the final selection at mid-span, since

$$R_u = 48 \text{ kips} < \phi R_n = 101 \text{ kips}$$

a 3.0 in. bearing length will be sufficient.

Part b At the end reaction

Step 8: Since the end reactions are less than the concentrated load at mid-span, the 3.0 in. minimum bearing length will likely be sufficient for the end reactions also. Thus, check a 3.0 in. bearing.

$$R_n = F_{yw}t_w(2.5k + l_b) = 50(0.300)(2.5(0.827)+3.0) = 76.0 \text{ kips}$$

$$\phi R_n = 1.0(76.0) = 76.0 \text{ kips}$$

Step 9: Check the limit state of web local crippling. Since the reaction is at the end of the member at a distance less than $d/2$ and $l_b/d = 3.0/17.7 = 0.17 \le 0.2$ use Equation J10-5a.

$$R_n = 0.40t_w^2\left[1+3\left(\frac{l_b}{d}\right)\left(\frac{t_w}{t_f}\right)^{1.5}\right]\sqrt{\frac{EF_{yw}t_f}{t_w}}Q_f$$

$$= 0.40(0.300)^2\left[1+3\left(\frac{3.0}{17.7}\right)\left(\frac{0.300}{0.425}\right)^{1.5}\right]\sqrt{\frac{29{,}000(50)(0.425)}{0.300}}(1.0)$$

$$= 67.2 \text{ kips}$$

So,

$$\phi R_n = 0.75(67.2) = 50.4 \text{ kips}$$

Step 10: Determine the strength for the end reaction. Since web local crippling controls,

$$\phi R_n = 50.4 \geq 24.0 \text{ kips}$$

Step 11: Final selection.

Use a minimum bearing plate length along the beam web of 3.0 in. at the concentrated load and at the end reactions.

EXAMPLE 6.21b
Beam Bearing Strength by ASD

Goal: Determine the required length of bearing for a W-shape to support (a) a midspan concentrated load and (b) concentrated reactions at the end of a simple span.

Given: A W18×35 A992 beam was selected in Example 6.3b, as a simply supported member spanning 20 ft and carrying a midspan dead load of 8.0 kips and live load of 24.0 kips.

SOLUTION

Step 1: Determine section properties from *Manual* Table 1-1

$$d = 17.7 \text{ in.}, t_w = 0.300 \text{ in.}, t_f = 0.425 \text{ in.}, k_{des} = 0.827 \text{ in.}$$

Step 2: Determine the required bearing strength at the load point and at the reactions using the ASD load combinations from Section 2.4.

$$P_a = P_D + P_L = 8.0 + 24.0 = 32.0 \text{ kips}$$

$$R_a = \frac{P_a}{2} = \frac{32}{2} = 16 \text{ kips}$$

Part a At the load

Step 3: For the limit state of yielding, determine the required length within the web to resist the applied load. The required length, based on Equation J10-2 where $N = (5k + l_b)$, is then

$$(5k+l_b)=\frac{P_n}{F_{yw}t_w}=\frac{\Omega P_a}{F_y t_w}=\frac{1.5(32)}{50(0.300)}=3.20 \text{ in.}$$

Step 4: Determine the minimum length of bearing, l_b. Since this force is located so that distribution takes place in both directions,

$$l_b=3.20-5(0.827)=-0.935 \text{ in.}$$

A negative bearing length in this calculation means that the distribution, 5k, is more than enough to carry the required load. Thus, the minimum bearing length for normal practice of 3.0 in. is adequate.

Step 5: For the limit state of web local crippling, determine the nominal strength if the bearing length is taken as a practical minimum of 3.0 in. From Equation J10-4,

$$R_n=0.80t_w^2\left[1+3\left(\frac{l_b}{d}\right)\left(\frac{t_w}{t_f}\right)^{1.5}\right]\sqrt{\frac{EF_y t_f}{t_w}}Q_f$$

$$=0.80(0.300)^2\left[1+3\left(\frac{3.00}{17.7}\right)\left(\frac{0.300}{0.425}\right)^{1.5}\right]\sqrt{\frac{29{,}000(50)(0.425)}{0.300}}(1.0)$$

$$=134 \text{ kips}$$

Step 6: Determine the allowable strength for web local crippling,

$$\frac{R_n}{\Omega}=\frac{134}{2.00}=67.0 \text{ kips}$$

Step 7: For the final selection at mid-span, since

$$R_a=32 \text{ kips } < R_n/\Omega=67 \text{ kips}$$

a 3.0 in. bearing length will be sufficient.

Part b At the end reaction

Step 8: Since the end reactions are less than the concentrated load at mid-span, the 3.0 in. minimum bearing length will likely be sufficient for the end reactions also. Thus, check a 3.0 in. bearing.

$$R_n=F_{yw}t_w(2.5k+l_b)=50(0.300)(2.5(0.827)+3.0)=76.0 \text{ kips}$$

$$R_n/\Omega=76.0/1.5=50.7 \text{ kips}$$

Step 9: Check the limit state of web local crippling. Since the reaction is at the end of the member at a distance less than $d/2$ and $l_b/d=3.0/17.7=0.17\le 0.2$ use Equation J10-5a.

$$R_n = 0.40t_w^2\left[1+3\left(\frac{l_b}{d}\right)\left(\frac{t_w}{t_f}\right)^{1.5}\right]\sqrt{\frac{EF_{yw}t_f}{t_w}}Q_f$$

$$= 0.40(0.300)^2\left[1+3\left(\frac{3.0}{17.7}\right)\left(\frac{0.300}{0.425}\right)^{1.5}\right]\sqrt{\frac{29{,}000(50)(0.425)}{0.300}}(1.0)$$

$$= 67.2 \text{ kips}$$

So,

$$R_n/\Omega = 67.2/2.0 = 33.6 \text{ kips}$$

Step 10: Determine the strength for the end reaction. Since web local crippling controls,

$$R_n/\Omega = 33.6 > 16.0 \text{ kips}$$

Step 11: Final selection

Use a minimum bearing plate length along the beam web of 3.0 in. at the concentrated load and at the end reactions.

6.16 OPEN WEB STEEL JOISTS AND JOIST GIRDERS

The term *open web steel joist* refers to a building product made according to the design standards of the Steel Joist Institute (SJI). They are manufactured trusses regularly used for building floor and roof systems. An example of their use in a building is shown in Figure 6.30. As a building product, the structural engineer is not involved in actually designing the truss but rather selects the product from a table, much like selecting a particular W-shape. Although these members are made from steel, they are not designed according to any AISC standards but rather to the standards found in the SJI publication *Standard Specifications, Load Tables, and Weight Tables for Steel Joists and Joist Girders* (SJI 100-2020).

Some advantages of open web steel joists include the fact that they are lightweight and can easily span long distances, their open webs can easily accommodate passing mechanical systems through the structure, and in some applications, they are more economical than rolled steel shapes. Disadvantages include a lower load carrying capacity than rolled shapes, thus requiring much closer spacing; the inability to easily accommodate concentrated loads at points other than truss panel points; and potential vibration issues when they are used for floor systems. In all applications, the economic advantages or disadvantages of using an open web steel joist must be assessed for the specific conditions in question.

Four types of open web steel joists are defined in SJI standards: K-series, KCS-series, LH-series and DLH-series.

K-series joists are perhaps the most commonly used joists for floor and roof systems. They are available in depths from 10 in. to 30 in., and design tables are available covering spans up to 60 ft. The standard designation for K-series joists is 16K6, where the first number represents the depth, 16 in.; the letter indicates the series designation, K; and the third number represents place within the series, 6. The series designation identifies the details of manufacture of the truss, including the sizes of the elements that make up the truss. It is here that each manufacturer has the opportunity to design its own specific joist. For a 16K joist there are seven designations within the series: 2, 3, 4, 5, 6, 7, and 9. Each increase in designation indicates an increase in weight. For example, the seven designations of the 16K indicate joists that weigh approximately 5.5, 6.3, 7.0, 7.5, 8.1, 8.6, and 10.0 lb/ft. Weight is given as an approximate value since each manufacturer has the opportunity to produce joists with different cross section elements. The one constant, however, is that whatever joist is produced, it must have the strength indicated in the SJI standard. Figure 6.31 shows an example of the selection tables published by SJI. K-series joists are designed to carry a uniformly distributed load. Thus, the bending strength can vary along the span from a maximum at midspan to zero at the support, approximating a parabolic moment diagram, and the shear strength can vary from zero at midspan to a maximum at the support, approximating a triangular shear diagram. Design tables like those in Figure 6.31 give two numbers for each joist and span combination. Joist strength in terms of load per foot of span is given as the upper number, and load per foot of span to cause a deflection of 1/360 of the span is given as the lower number.

Figure 6.30 Application of Open Web Steel Joists
Photo courtesy Douglas Steel Fabricating Corporation

LRFD

LRFD K-SERIES ECONOMY TABLE - STANDARD UNITS

Joist Designation	14K6	18K5	20K5	22K5	24K4	24K5	16K6	26K5	18K6	20K6	22K6	24K6	16K7	26K6	18K7	20K7
Depth (In).	14	18	20	22	24	24	16	26	18	20	22	24	16	26	18	20
Approx. Wt. (lbs./ft)	7.7	7.7	7.7	7.7	7.8	7.9	8.1	8.1	8.4	8.4	8.5	8.5	8.6	8.6	8.9	8.9
Span (ft)																
14	825 550															
15	826 607															
16	825 467						825 550						825 550			
17	825 443						825 526						825 526			
18	825 408	825 550					825 490		825 550				825 490		825 550	
19	825 383	825 523	825 550				825 455		825 523	825 550			825 455		825 523	825 550
20	787 347	825 490	825 550				825 426		825 490	825 550			825 426		825 490	825 550
21	712 299	826 460	826 520	825 550			822 405		825 460	825 520	825 550		825 406		825 460	826 520
22	648 259	777 414	825 490	825 548			747 351		825 438	825 490	825 548		825 385		825 438	825 490
23	592 226	709 362	793 451	825 518	825 550	825 550	682 307		774 393	825 468	825 518	825 550	760 339		825 418	825 468
24	543 199	651 318	727 396	804 483	780 516	825 544	627 269		709 345	792 430	825 495	825 544	697 298		789 382	825 448
25	501 175	600 281	669 350	739 427	718 456	810 511	576 238	825 550	652 305	729 380	805 464	825 520	642 263	825 550	727 337	811 421
26	462 156	553 249	618 310	682 379	663 405	748 453	532 211	813 535	603 271	673 337	744 411	814 493	592 233	825 541	672 299	750 373
27	427 139	513 222	573 277	633 337	615 361	693 404	493 188	753 477	558 241	624 301	688 367	754 439	549 208	820 519	622 267	694 333
28	397 124	477 199	532 248	588 302	571 323	643 362	459 168	699 427	519 216	579 269	640 328	700 393	510 186	762 464	577 239	645 298
29		444 179	495 223	547 272	531 290	600 326	427 151	651 384	483 194	540 242	597 295	652 364	475 167	709 417	538 215	601 268
30		414 161	462 201	511 245	496 262	559 293	399 137	607 346	451 175	504 218	556 266	609 319	444 151	661 377	502 194	561 242
31		387 146	433 182	478 222	465 237	523 266	373 124	568 314	421 158	471 198	520 241	570 289	415 137	619 341	469 176	525 219
32		363 132	406 166	448 201	435 216	490 241	349 112	534 285	396 144	442 179	489 219	535 262	388 124	580 309	441 159	492 199
33		342 121	381 160	421 183	409 196	462 220		501 259	248 131	415 163	459 199	502 239		546 282	414 146	463 181
34		321 110	358 137	397 167	385 179	435 201		472 237	233 120	391 149	432 182	472 218		514 257	390 132	435 165
35		303 101	339 126	373 153	363 164	409 184		445 217	330 110	369 137	408 167	445 200		484 236	367 121	411 151
36		286 92	319 115	354 141	343 150	387 169		420 199	312 101	348 125	385 153	421 183		457 216	348 111	388 139
37			303 106	334 130	324 139	366 155		397 183		330 115	364 141	399 169		433 199		367 128
38			286 99	316 119	307 128	346 143		376 169		312 106	345 130	378 156		411 184		348 118
39			271 90	300 110	292 118	328 132		357 156		297 98	327 120	358 144		390 170		330 109
40			258 84	285 102	277 109	312 122		340 146		282 91	310 111	340 133		370 157		313 101
41				271 95	264 101	297 114		322 134			295 103	324 124		352 146		
42				259 86	262 94	283 106		307 125			282 96	309 116		336 136		
43				247 82	240 88	270 98		294 116			268 89	294 107		319 126		
44				235 78	229 82	258 92		280 108			256 83	280 100		306 118		
45					219 76	246 86		268 101				268 93		291 110		
46					208 71	235 80		256 95				256 87		279 103		
47					199 67	225 75		246 89				246 92		267 96		
48					192 63	216 70		235 83				235 77		256 90		
49								225 78						246 85		
50								216 73						235 80		
51								208 69						226 76		
52								199 65						217 71		

Figure 6.31 LRFD Economy Table for K-Series Open Web Steel Joists
Standard Specifications, Load Tables, and Weight Tables for Steel Joists and Joist Girders (2020), Courtesy of the Steel Joist Institute ©. Reprinted with Permission.

KCS-series joists are identified in the same fashion as the K-series joists. Thus, a 10KCS3 is a 10 in. deep KCS-series joist with a series designation of 3. The difference between K- and KCS-series joists is directly related to the moment and shear diagrams. KCS-series joists are designed for uniform moment over all interior panel points and a constant shear. These joists are particularly useful for supporting loads that combine uniformly distributed and concentrated forces. Unlike the K-series joists, the design tables for the KCS-series give only the moment and shear capacities.

LH-series joists are long-span joists. Their strengths are tabulated for spans from 25 ft to 96 ft. LH-series joists have depths from 18 in. to 48 in. A typical designation would be 28LH05. LH-series joists are used for floor and roof systems, and, like K-series joists, they are designed for a uniformly distributed load.

DLH-series joists are deep long-span joists intended primarily to support roof decks. These might be used for a roof joist that carries roof live and dead load plus a piece of mechanical equipment. DLH-series joists start at a depth of 52 in. and go up to 120 in. The design tables indicate spans up to 240 ft. The designation system is the same as for other joists; a 68DLH17 is a 68 in. deep DLH-series joist with a series designation of 17.

Another product that is designed according to the SJI standards is the *joist girder*. Joist girders are essentially pre-engineered trusses that are intended to support concentrated loads at the panel points. They are used to support open web steel joists that are evenly spaced and introduce the same load at each panel point. A possible designation for a joist girder is 44G8N12K. The first number indicates a 44 in. deep member and the G indicates a joist girder. The 8N indicates that there are eight joist spaces, which is the same as saying there are seven concentrated loads on the girder. The final number indicates the load in kips. If the load is for an LRFD design the last letter would be F, indicating a factored load, and the load magnitude would be the LRFD required strength.

EXAMPLE 6.22
Open Web Steel Joist Selection by LRFD

Goal: Select a K-series open web steel joist from the limited selection available in Figure 6.31 to satisfy strength and deflection criteria.

Given: Joists supporting a roof deck span 30 ft and are spaced 6.0 ft on center. The load is a uniformly distributed dead load of 20 psf and live load of 30 psf. Select the shallowest K-series open web steel joist available in Figure 6.31 to support this load. Limit live load deflection to 1/360 of the span.

SOLUTION

Step 1: Determine the required load in pounds per foot.

$$w_u = (1.2w_D + 1.6w_L)L_{trib} = (1.2(20) + 1.6(30))(6) = 432 \text{ lb/ft}$$

Step 2: Select a K-series joist for a span of 30 ft.

An 18K6 will support a factored load of 451 lb/ft. This is the upper number in the table. The other number in the table is 175 lb/ft. This is the serviceability load that will produce a deflection of 1/360 of the span.

Step 3: Check the deflection of the 18K6 for the serviceability live load of 30 psf.

For the 6.0 ft joist spacing, the joist carries a serviceability live load of 6(30) =180 lb/ft. Since this exceeds the load given in the table, the next larger joist should be selected if this deflection limit is to be satisfied.

Step 4: Check the 18K7.

For strength the capacity is

$$502 \text{ lb/ft} > 432 \text{ lb/ft}.$$

For deflection the capacity is

$$194 \text{ lb/ft} > 180 \text{ lb/ft}.$$

Step 5: Check to be sure the joist can carry its self-weight.

The joist weighs approximately 9 lb/ft. The factored dead load due to self-weight is

$$w_{self} = 1.2(9) = 10.8 \text{ lb/ft}$$

Thus, the total load the joist must carry is

$$w_u = 432 + 10.8 = 443 \text{ lb/ft} < 502 \text{ lb/ft}$$

Step 6: Final selection:

18K7

6.17 PROBLEMS

1. Determine the elastic section modulus and the plastic section modulus for a W40×199 modeled as three rectangles forming the flanges and the web. Compare the calculated values to those given in the *Manual*.

2. Determine the elastic section modulus and the plastic section modulus for a W36×330 modeled as three rectangles forming the flanges and the web. Compare the calculated values to those given in the *Manual*.

3. Determine the elastic section modulus and the plastic section modulus for a W33×130 modeled as three rectangles forming the flanges and the web. Compare the calculated values to those given in the *Manual*.

4. Determine the elastic section modulus and the plastic section modulus for a W24×55 modeled as three rectangles forming the flanges and the web. Compare the calculated values to those given in the *Manual*.

5. Determine the elastic section modulus and the plastic section modulus for a W18×35 modeled as three rectangles forming the flanges and the web. Compare the calculated values to those given in the *Manual*.

6. Determine the plastic section modulus for a W44×230 modeled as three rectangles forming the flanges and the web. Compare the calculated value to that given in the *Manual*.

7. Determine the plastic section modulus for a W27×84 modeled as three rectangles forming the flanges and the web. Compare the calculated value to that given in the *Manual*.

8. Determine the plastic section modulus for a W16×31 modeled as three rectangles forming the flanges and the web. Compare the calculated value to that given in the *Manual*.

9. Determine the elastic neutral axis, elastic section modulus, plastic neutral axis, and plastic section modulus for a WT15×45 modeled as two rectangles forming the flange and the stem. Compare the calculated values to those given in the *Manual*.

10. Determine the elastic neutral axis, elastic section modulus, plastic neutral axis, and plastic section modulus for a WT8×50 modeled as two rectangles forming the flange and the stem. Compare the calculated values to those given in the *Manual*.

11. Determine the elastic neutral axis, elastic section modulus, plastic neutral axis, and plastic section modulus for a WT5×22.5 modeled as two rectangles forming the flange and the stem. Compare the calculated values to those given in the *Manual*.

12. Determine the plastic section modulus for an HSS8×6×1/2 modeled as four rectangles forming the flanges and webs. Remember to use the design wall thickness for the plate thickness and ignore the corner radius. Compare the calculated value to that given in the *Manual*.

13. Determine the plastic section modulus for an HSS12×8×5/8 modeled as four rectangles forming the flanges and webs. Remember to use the design wall thickness for the plate thickness and ignore the corner radius. Compare the calculated value to that given in the *Manual*.

14. Determine the plastic section modulus for an HSS8×8×1/2 modeled as four rectangles forming the flanges and webs. Remember to use the design wall thickness for the plate thickness and ignore the corner radius. Compare the calculated value to that given in the *Manual*.

15. Determine the plastic section modulus for a round HSS10×0.500. Remember to use the design wall thickness in your calculations. Compare the calculated value to that given in the *Manual*.

16. Determine the elastic neutral axis, elastic section modulus, plastic neutral axis, and plastic section modulus for a C15×40 modeled as three rectangles forming the flanges and the web. Compare the calculated values to those given in the *Manual*.

17. Determine the elastic neutral axis, elastic section modulus, plastic neutral axis, and plastic section modulus for a C8×11.5 modeled as three rectangles forming the flanges and the web. Compare the calculated values to those given in the *Manual*.

18. Determine the elastic neutral axis, elastic section modulus, plastic neutral axis, and plastic section modulus for a MC13×50 modeled as three rectangles forming the flanges and the web. Compare the calculated values to those given in the *Manual*.

19. Determine the elastic neutral axis, elastic section modulus, plastic neutral axis, and plastic section modulus, all about the geometric axis, for a L4×4×1/2 modeled as two rectangles. Compare the calculated values to those given in the *Manual*.

20. Determine the elastic neutral axis, elastic section modulus, plastic neutral axis, and plastic section modulus, all about the geometric axis, for a L8×8×7/8 modeled as two rectangles. Compare the calculated values to those given in the *Manual*.

21. Determine the elastic neutral axis, elastic section modulus, plastic neutral axis, and plastic section modulus, all about the geometric axis, for a L3×3×1/4 modeled as two rectangles. Compare the calculated values to those given in the *Manual*.

22. A simply supported beam spans 25 ft and carries a uniformly distributed dead load of 0.6 kip/ft, including the beam self-weight, and a live load of 2.1 kip/ft. Determine the minimum required plastic section modulus and select the lightest-weight W-shape to carry the moment. Consider only the limit state of yielding and use A992 steel. Design by (a) LRFD and (b) ASD.

23. For flexure only, determine the lightest-weight W-shape to carry a uniform dead load of 1.2 kip/ft, including the beam self-weight, and a live load of 3.2 kip/ft on a simple span of 20 ft. Consider only the limit state of yielding and use A992 steel. Design by (a) LRFD and (b) ASD.

24. A beam is required to carry a uniform dead load of 0.80 kip/ft, including its self-weight, and a concentrated live load of 14 kips at the center of a 30 ft span. For bending only, determine the least-weight W-shape to carry the load. Consider only the limit state of yielding and use A992 steel. Design by (a) LRFD and (b) ASD.

25. A simply supported beam spans 32 ft and carries a uniformly distributed dead load of 1.8 kip/ft plus the beam self-weight and a live load of 5.4 kip/ft. Determine the minimum required plastic section modulus and select the lightest-weight W-shape to carry the moment. Consider only the limit state of yielding and use A992 steel. Design by (a) LRFD and (b) ASD.

26. For bending only, determine the lightest-weight W-shape to carry a uniform dead load of 4.5 kip/ft plus the beam self-weight and a live load of 3.5 kip/ft on a simple span of 24 ft. Consider only the limit state of yielding and use A992 steel. Design by (a) LRFD and (b) ASD.

27. A beam is required to carry a uniform dead load of 3.4 kip/ft plus its self-weight, and a concentrated dead load of 12 kips and a concentrated live load of 20 kips, both at the center of a 40 ft span. For bending only, determine the least-weight W-shape to carry the load. Consider only the limit state of yielding and use A992 steel. Design by (a) LRFD and (b) ASD.

28. Considering only bending, determine the lightest-weight W-shape to carry the following loads: a uniform dead load of 0.6 kip/ft plus self-weight, a concentrated dead load of 2.1 kips, and a concentrated live load of 6.4 kips, located at the center of a 16-ft span. Assume full lateral support and A992 steel. Design by (a) LRFD and (b) ASD.

29. Considering only bending, determine the lightest W-shape to carry a uniform dead load of 4.0 kip/ft plus the self-weight and a uniform live load of 2.3 kip/ft on a simple span of 10.0 ft. Assume full lateral support and A992 steel. Design by (a) LRFD and (b) ASD.

30. A 24 ft simple span laterally supported beam is required to carry a total uniformly distributed service load of 8.0 k/ft. Determine the lightest, A992, W-shape to carry this load, considering only flexure and the limit state of yielding, if the load is broken down as follows. Use LRFD.

- **a.** Live load = 1.0 k/ft; dead load = 7.0 kips/ft
- **b.** Live load = 3.0 k/ft; dead load = 5.0 kips/ft
- **c.** Live load = 5.0 k/ft; dead load = 3.0 kips/ft
- **d.** Live load = 7.0 k/ft; dead load = 1.0 kips/ft.

31. Repeat the designs specified in Problem 30 using ASD.

32. A 30 ft simply supported beam is loaded at the third points of the span with concentrated dead loads of 10.0 kips and live loads of 15.0 kips. Lateral supports are provided at the supports and at the load points. The self-weight of the beam will be ignored for this problem. Considering only flexure, determine the least-weight W-shape to carry the load. Use A992 steel and $C_b = 1.0$. Design by (a) LRFD and (b) ASD.

33. Determine the least-weight W-shape to support a concentrated dead load of 24 kips plus the beam self-weight and a concentrated live load of 15 kips on a 30 ft span. The concentrated loads are located at the midpoint of the span. Lateral supports are provided at the supports and at the load point. Use A992 steel and $C_b = 1.0$. Design for flexure by (a) LRFD and (b) ASD.

34. A 36 ft simply supported beam is loaded with concentrated loads 16 ft in from each support. On one end, the dead load is 8.0 kips and the live load is 18.0 kips. At the other end, the dead load is 4.0 kips and the live load is 9.0 kips. Include the self-weight of the beam in the design. Lateral supports are provided at the supports and the load points. Considering only bending, determine the least-weight W-shape to carry the load. Use A992 steel and $C_b = 1.0$. Design by (a) LRFD and (b) ASD.

35. An 18 ft simple span beam is loaded with a uniform dead load of 1.4 kip/ft, including the beam self-weight, and a uniform live load of 2.3 kip/ft. The lateral supports are located at the supports and at 6.0 ft intervals. Considering only bending, determine the least-weight W-shape to carry the load. Use A992 steel and $C_b = 1.0$. Design by (a) LRFD and (b) ASD.

36. A 40 ft simple span beam is loaded with a uniform dead load of 2.2 kip/ft plus the beam self-weight and a uniform live load of 3.6 kip/ft. The lateral supports are located at the supports and at the midpoint of the span. Considering only bending, determine the least-weight W-shape to carry the load. Use A992 steel and $C_b = 1.0$. Design by (a) LRFD and (b) ASD.

37. An A992 W18×60 is used on a 36 ft simple span to carry a uniformly distributed load. Using $C_b = 1.0$, determine the locations of lateral supports in order to provide just enough strength to carry (a) a design moment of 435 ft-kips and (b) an allowable moment of 290 ft-kips.

38. An A992 W16×31 is used on an 18 ft simple span to carry a uniformly distributed dead load of 0.95 kips/ft plus self-weight and a live load of 1.9 kips/ft. Using $C_b = 1.0$, determine the locations of lateral supports in order to provide just enough strength to carry this loading by (a) LRFD and (b) ASD.

39. Redesign the beam of Problem 33 using the correct value of C_b.

40. Redesign the beam of Problem 34 using the correct value of C_b.

41. Redesign the beam of Problem 35 using the correct value of C_b.

42. There are no tables in the *Manual* for the design of beams with F_y other than 50 ksi. Construct the moment strength vs. unbraced length curve, similar to those in *Manual* Table 3-10, for a W21×68 A572 Grade 65 steel. Show the curve up to an unbraced length of 40 ft.

43. Since the chart developed for Problem 42 was similar to those in *Manual* Table 3-10, it used $C_b = 1.0$. Redo Problem 42 with values of $C_b = 1.0$, 1.14, 1.32, and 1.67.

44. Plot the nominal moment strength vs. unbraced length for a W27×84 using A992 and A913 Grade 65 steel. Compare the nominal moment strengths for an unbraced length of 18 ft.

45. Plot the nominal moment strength vs. unbraced length for a W16×31 using A36 and A992 steel. Compare the nominal moment strengths for an unbraced length of 7.0 ft.

46. There are 4 W6 shapes given in *Manual* Table 1-1. Determine the flange width-to-thickness ratio for each shape and determine if the flange is compact, noncompact, or slender for A992 steel.

47. Determine the flange width-to-thickness ratio for an A992 W10×45 and determine if the shape has a compact, noncompact, or slender flange.

48. Determine the available flexural strength for a laterally supported A913 Gr 70 W8×24 on a simple span of 10 ft for (a) LRFD and (b) ASD.

49. Determine the available flexural strength for a laterally supported A913 Gr 70 W30×90 on a simple span of 40 ft for (a) LRFD and (b) ASD.

50. Determine the available flexural strength for a laterally supported A572 Gr 65 W24×104 on a simple span of 30 ft for (a) LRFD and (b) ASD.

51. Determine the available shear strength of a W24×62 A992 steel member by (a) LRFD and (b) ASD.

52. Determine the available shear strength of a W30×90 A992 steel member by (a) LRFD and (b) ASD.

53. Determine the available shear strength of a W27×129 A992 steel member by (a) LRFD and (b) ASD.

54. Determine the available shear strength of a W16×77 A992 steel member by (a) LRFD and (b) ASD.

55. Determine the available shear strength of a W24×162 A913 Gr 70 steel member by (a) LRFD and (b) ASD

56. A girder that carries a uniformly distributed dead load of 1.7 k/ft plus its self-weight and three concentrated live loads of 15 kips at the quarter points of the 36 ft span is to be sized. Using A992 steel, determine the lightest W-shape to carry the load with lateral supports provided at the supports and load points. Use the correct C_b, check shear, and limit live load deflection to 1/360 of span. Design by (a) LRFD and (b) ASD.

57. A girder that carries a uniformly distributed dead load of 1.5 k/ft plus its self-weight and two concentrated live loads of 15 kips at the third points of the 24 ft span is to be sized. Using A992 steel, determine the lightest W-shape to carry the load with lateral supports provided at the supports only. Use the correct C_b, check shear, and limit live load deflection to 1/240 of span. Design by (a) LRFD and (b) ASD.

58. A 32 ft simple span beam carries a uniform dead load of 2.4 k/ft plus its self-weight and a uniform live load of 3.0 k/ft. The beam is laterally supported at the supports only. Determine the minimum-weight W-shape to carry the load using A992 steel. Use the correct C_b, check shear, and limit live load deflection to 1/360 of span. Design by (a) LRFD and (b) ASD.

59. A 36 ft simple span beam carries a uniformly distributed dead load of 3.4 kip/ft plus its self-weight and a uniformly distributed live load of 2.4 kip/ft. Determine the least-weight W-shape to carry the load. Use the correct C_b, check shear, and limit live load deflection to 1/360 of the span. Use A992 steel and assume full lateral support. Design by (a) LRFD and (b) ASD.

60. A 20 ft simple span beam carries a uniformly distributed dead load of 1.5 kip/ft plus its self-weight and a uniformly distributed live load of 3.4 kip/ft. Determine the least-weight W-shape to carry the load. Use the correct C_b, check shear, and limit live load deflection to 1/360 of the span. Use A992 steel and assume full lateral support. Design by (a) LRFD and (b) ASD.

61. A simple span beam with a uniformly distributed dead load of 1.1 k/ft, including the self-weight, and concentrated dead loads of 3.4 kips and live loads of 6.0 kips at the third points of a 24-ft span, is to be designed with lateral supports at the third points and live load deflection limited to 1/360 of the span. Use the correct C_b, and be sure to check shear. Determine the least-weight W-shape to carry the loads. Use A992 steel. Design by (a) LRFD and (b) ASD.

62. A simple span beam with a uniformly distributed dead load of 1.0 k/ft, plus its self-weight, and concentrated dead loads of 2.5 kips and live loads of 8.0 kips at the quarter points of a 20-ft span, is to be designed with lateral supports at the supports and mid-span. The live load deflection must be limited to 1/240 of the span. Use the correct C_b and be sure to check shear. Determine the least-weight W-shape to carry the loads. Use A992 steel. Design by (a) LRFD and (b) ASD.

63. Determine the live load deflection for a W27×178 on a 40 ft span. The beam supports a uniformly distributed live load of 5.4 k/ft.

64. A W18×106 spans 30 ft with concentrated live loads of 10 kips at the third points and a single concentrated live load of 15 kips at mid-span. Determine the live load deflection.

65. A fixed-end beam on a 28 ft span is required to carry a total ultimate uniformly distributed load of 32.0 kips. Using plastic analysis and A992 steel, determine the design moment and select the lightest W-shape. Assume (a) full lateral support and (b) lateral support at the ends and center line.

66. A beam is fixed at one support and simply supported at the other. A concentrated ultimate load of 32.0 kips is applied at the center of the 40 ft span. Using plastic analysis and A992 steel, determine the lightest W-shape to carry the load when the nominal depth of the beam is limited to 18 in. Assume (a) full lateral support and (b) lateral supports at the ends and the load.

67. A fixed-end beam on a 40 ft span is required to carry a total ultimate uniformly distributed load of 72.5 kips. Using plastic analysis and A992 steel, determine the lightest-weight W-shape to carry the load. Assume full lateral support.

68. A three 30 ft span continuous beam is to be selected to carry a uniformly distributed dead load of 4.7 kip/ft, including its self-weight, and a uniformly distributed live load of 10.5 kip/ft. Be sure to check the shear strength of this beam. Use A992 steel and assume full lateral support. Design using moment redistribution by (a) LRFD and (b) ASD.

69. Determine the available minor axis bending strength of a W18×130 of A992 steel. Determine by (a) LRFD and (b) ASD.

70. Determine the available minor axis bending strength of a W10×12 of A992 steel. Determine by (a) LRFD and (b) ASD.

71. Determine the available minor axis bending strength of a W40×392 of A992 steel. Determine by (a) LRFD and (b) ASD.

72. Determine the available bending strength of a uniformly loaded L6×6×5/16, A572 Gr 50 steel, on an 8 ft span with full lateral support if the vertical leg is in compression. Determine by (a) LRFD and (b) ASD.

73. Determine the available bending strength of a uniformly loaded L6×6×5/16, A572 Gr 50 steel, on an 8 ft span with lateral support at the ends only with the vertical leg in compression. Determine by (a) LRFD and (b) ASD.

74. Determine the available bending strength of a uniformly loaded L6×6×5/16, A572 Gr 50 steel, on an 8 ft span with lateral support at the ends and at mid-span with the vertical leg in compression. Determine by (a) LRFD and (b) ASD.

75. Repeat Problem 73 if the vertical leg is in tension. Determine by (a) LRFD and (b) ASD.

76. Determine the available uniform load for an L6×4×7/8, A572 Gr 50 steel, spanning 8 ft and loaded for bending about the x-axis with the 6 in. leg in compression. The unbraced length is 8 ft. Determine by (a) LRFD and (b) ASD.

77. Determine the available bending strength of a WT8×25, A992 steel, if the stem is in compression and the unbraced length is 10 ft. Determine by (a) LRFD and (b) ASD.

78. Determine the available bending strength of a WT12×52, A992 steel, if the stem is in tension and the unbraced length is 8 ft. Determine by (a) LRFD and (b) ASD.

79. Determine the available bending strength of a 2L6×4×7/8, A572 Gr 50 steel, if the vertical legs are in tension and the unbraced length is 8 ft. Determine by (a) LRFD and (b) ASD.

80. Determine the available bending strength of a 2L6×4×7/8, A572 Gr 50 steel, if the vertical legs are in compression and the unbraced length is 8 ft. Determine by (a) LRFD and (b) ASD.

81. Determine the available bending strength of a 2L8×8×1, A572 Gr 50 steel, if the vertical legs are in tension and the unbraced length is 12 ft. Determine by (a) LRFD and (b) ASD.

82. Determine the available bending strength of a 2L8×8×1, A572 Gr 50 steel, if the vertical legs are in compression and the unbraced length is 12 ft. Determine by (a) LRFD and (b) ASD.

83. For the beam designed in Problem 34, determine the minimum required lengths of bearing (at the concentrated loads and at the supports) considering the limit state of web yielding and the limit state of web local crippling. Determine by (a) LRFD and (b) ASD.

84. For the girder designed in Problem 56, determine the minimum required lengths of bearing (at the concentrated loads and at the supports) considering the limit state of web yielding and the limit state of web local crippling. Determine by (a) LRFD and (b) ASD.

85. For the girder designed in Problem 57, determine the minimum required lengths of bearing (at the concentrated loads and at the supports) considering the limit state of web yielding and the limit state of web local crippling. Determine by (a) LRFD and (b) ASD.

86. For the beam designed in Problem 62, determine the minimum required lengths of bearing (at the concentrated loads and at the supports) considering the limit state of web yielding and the limit state of web local crippling. Determine by (a) LRFD and (b) ASD.

87. Obtain via the Internet a set of open web steel joist design tables from any manufacturer. Select the shallowest K-series joist to span 35 ft and support a roof system that carries a dead load of 25 psf and a live load of 35 psf. The joists will be spaced (i) 6.0 ft on center, (ii) 8.0 ft on center, and (iii) 10.0 ft on center. Limit live load deflection to span/360. Design by (a) LRFD and (b) ASD.

88. Obtain via the Internet a set of open web steel joist design tables from any manufacturer. Select the shallowest K-series joist to span 40 ft and support a roof system that carries a dead load of 30 psf and a live load of 40 psf. The joists will be spaced (i) 4.0 ft on center, (ii) 6.0 ft on center, and (iii) 8.0 ft on center. Limit live load deflection to span/360. Design by (a) LRFD and (b) ASD.

Multi-Chapter Questions

89. Using the framing plan shown in Figure P6.89 (presented earlier as Figure 2.9), design the beams marked 1, 2, and 3. This is the same structure used in Section 2.5, where load calculations with live load reductions were discussed. Those calculations can be reused here. Load case 2 for dead plus live load is to be considered. The building is an office building with a nominal live load of 50 pounds per square foot (psf) and a calculated dead load of 70 psf.

1: Girder AB on line 2-2 if the floor deck spans from line 1-1 to 2-2 to 3-3:

2: Floor beam 2-3 on line D-D if the floor deck spans from line C-C to D-D to E-E:

3: Floor beam 2-3 on line C-C if the floor deck spans from line B-B to C-C to D-D:

Design by (a) LRFD and (b) ASD.

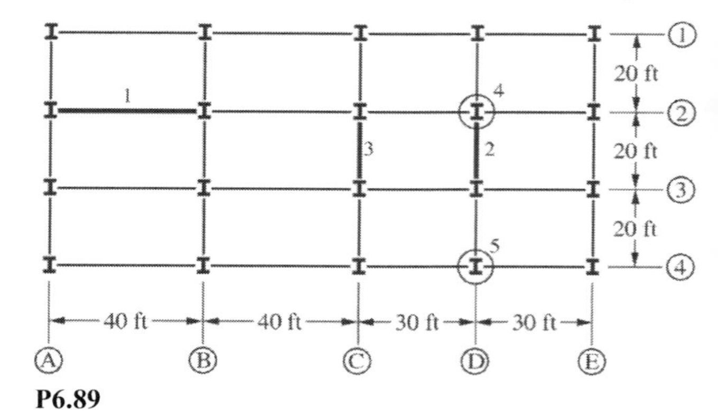

P6.89

90. The framing plan shown in Figure P6.90 is the same as that shown in Figure P2.24 for an 18-story office building. It must support a dead load of 80 psf and a live load of 50 psf. In all cases, the decking spans in the direction from line A toward line E. Determine the required moment and shear strength for the beams and design the beams as required below for (a) design by LRFD and (b) design by ASD. The required moment and shear strength were determined in Problem 20 of Chapter 2.

i. The beam between column lines 2 and 3 along line D.
ii. The girder on column line 3 between column line E and midway between lines D and C.
iii. The beam on the line between lines C and D and column lines 3 and 4.

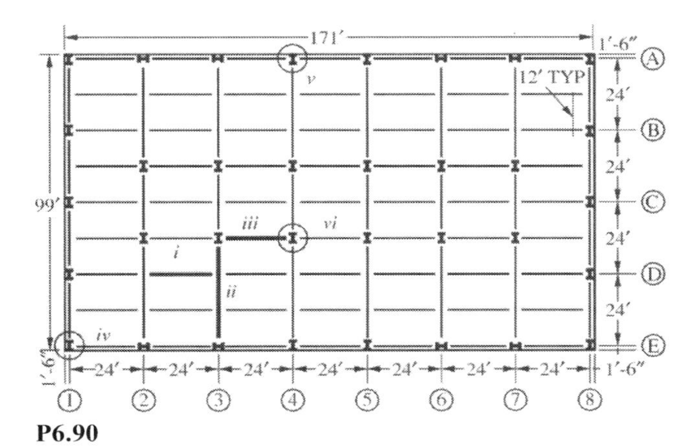

P6.90

91. Integrated Design Project. Design the beams in the gravity-only system for the loads determined in Chapter 2. Be sure to indicate all assumptions regarding lateral support.

Reconsider the proposed framing plans given in Figures 1.20 and 1.21. For the panel bounded by column lines A, C, 6, and 7, span the beams in the 30 ft direction and redesign. Compare the results with the designs done above.

For open web steel joists, manufacturer literature includes load tables. These may be found on the Internet. Select open web steel joists for the roof using the loading determined in Chapter 2. Use spacing different from that suggested in Figure 1.22 if that appears to be more appropriate.

Chapter 7

Plate Girders

One Vanderbilt Avenue, New York, NY.
Photo courtesy of Severud Associates

7.1 BACKGROUND

A plate girder is a bending member composed of individual steel plates. Although plate girders are normally the member of choice for situations in which the available rolled shapes are not deep or large enough to carry the intended load, there is no requirement that they will always be at the deep or large end of the spectrum of member sizes. Beams fabricated from individual steel plates and usually joined by fillet welds to meet a specific requirement are generally identified in the field as *plate girders*.

Plate girders are usually used in building structures for special situations such as those involving very long spans or very large loads. Perhaps their most common application is as a transfer girder, which is a bending member that supports a structure above and permits the column spacing to be changed below. They are also very common in industrial structures for use as crane girders and as support for large pieces of equipment. In commercial buildings, they are often used to span large open areas to meet particular architectural requirements; because of their normally greater depth and resulting stiffness, they tend to deflect less than other potential long span solutions. An example of a building application of the plate girder is shown in Figure 7.1.

The cross section of a typical plate girder is shown in Figure 7.2 and a large plate girder being transported on a truck is shown in Figure 7.3. Although it is possible to combine steel plates into numerous geometries, the plate girders addressed here are those formed from three plates, one for the web and two for the flanges. Because the web and flanges of the plate girder are fabricated from individual plates, they can be designed with the web and flange plates from the same grade of steel (a homogeneous plate girder) or from different grades of steel (a hybrid plate girder). For hybrid girders, the flanges are usually fabricated with a higher grade of steel than that used in the web. This takes advantage of the higher stresses that can be developed in the flanges, which are located a greater distance from the neutral axis than the web, resulting in a higher moment strength contribution. Hybrid girders are relatively common in bridge construction, though they are rarely used in buildings. In the past, hybrid girders have been included in the AISC *Specification*, but they are not specifically addressed in the current edition. Accordingly, they are not discussed in this book.

Figure 7.1 Application of a Plate Girder

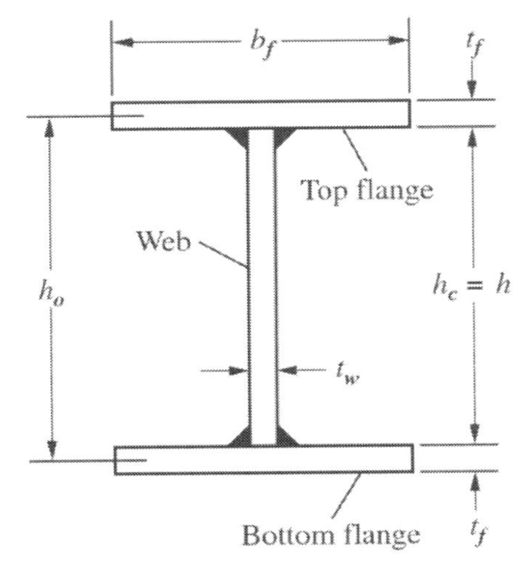

Figure 7.2 Typical Plate Girder Definitions

Figure 7.3 Plate Girder Being Transported
Photo courtesy LeJeune Steel Co.

Another type of plate girder is the singly symmetric girder, one with flanges that are not of the same size, as seen in Figure 7.4. Although singly symmetric plate girders are addressed in the AISC *Specification*, they are not particularly common in buildings and are not specifically addressed here. However, the principles for all of these plate girders are the same, and the careful application of the *Specification* provisions will lead to an economical and safe design for each of them.

Built-up doubly symmetric I-shaped members with compact webs are designed according to the same provisions as rolled I-shaped members presented in *Specification* Sections F2 and F3 and discussed in Chapter 6 of this book. The discussion of plate girders in this chapter addresses these built-up I-shapes with noncompact or slender webs.

Table 7.1 lists the sections of the *Specification* and parts of the *Manual* discussed in this chapter.

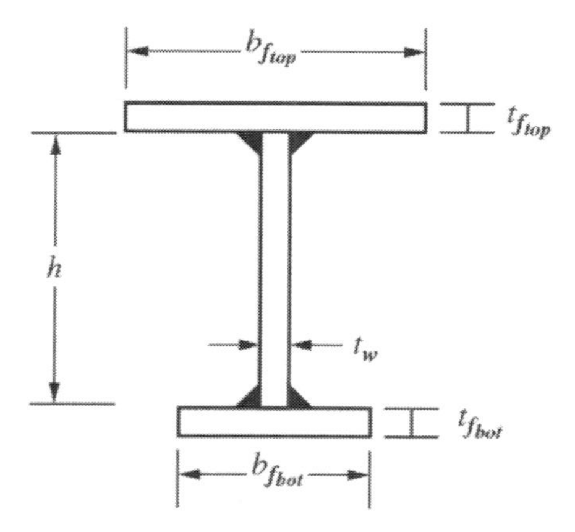

Figure 7.4 Singly Symmetric Plate Girder

Table 7.1 Sections of *Specification* and Parts of *Manual* Covered in This Chapter

	Specification
B4	Classification of Sections for Local Buckling
F4	Other I-Shaped Members with Compact or Noncompact Webs, Bent about Their Major Axis
F5	Doubly Symmetric and Singly Symmetric I-Shaped Members with Slender Webs Bent about Their Major Axis
G	Design of Members for Shear
J4.4	Strength of Elements in Compression
J10	Flanges and Webs with Concentrated Forces
	Manual
Part 3	Design of Flexural Members

7.2 HOMOGENEOUS PLATE GIRDERS IN BENDING

The behavior of plate girders will be addressed here by considering flexure and shear separately. In flexure, a plate girder is considered in this book as either noncompact or slender according to the proportions of the web. Flanges can be compact, noncompact, or slender, and the flange slenderness is treated the same as that discussed in Chapter 6. Thus, it is possible, for example, for a noncompact web plate girder to have a slender flange, potentially controlling the capacity of the member. The design rules for each type of girder, noncompact web or slender web, are considered separately.

With our discussion limited to doubly symmetric plate girders, the limit states that must be considered are compression flange yielding, compression flange local buckling, web local buckling, and lateral-torsional buckling. These are the same limit states considered for the rolled I-shapes in Chapter 6 with the addition of web local buckling. Remember that all W-shapes have compact webs so the web local buckling limit state was not considered for those bending members. The additional limit state of tension flange yielding found in *Specification* Sections F4 and F5 can be ignored, because the compression flange always controls over the tension flange in doubly symmetric homogeneous members. Plate girders with noncompact webs are addressed in *Specification* Section F4, and those with slender webs in Section F5. The nominal strength of plate girders, for all limit states, can be described as shown in Figure 7.5. As with the rolled I-shaped members discussed in Chapter 6, the behavior is plastic, inelastic, or elastic. Figure 7.5 shows that the plastic behavior corresponds to an area of the figure described as compact. Inelastic behavior corresponds to the area identified as noncompact, and elastic behavior corresponds to the area identified as slender.

Applying the same figure to the lateral-torsional buckling limit state, the fully braced region corresponds to plastic behavior, whereas the partially braced region corresponds to either inelastic or elastic buckling.

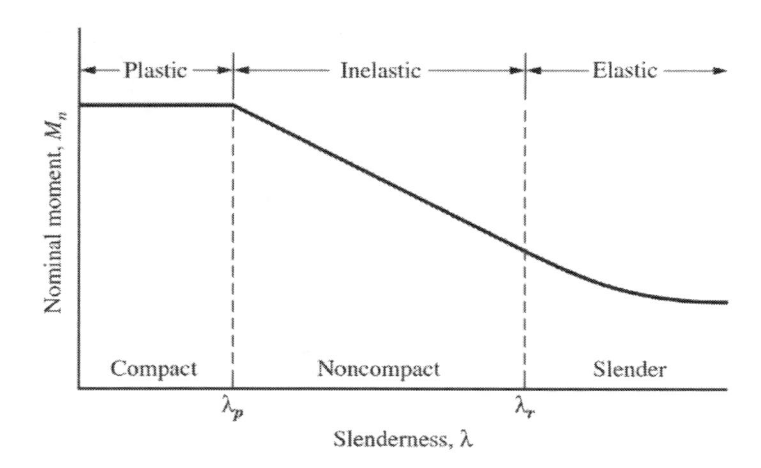

Figure 7.5 Plate Girder Nominal Flexural Strength

The flexural design strength (LRFD) and allowable strength (ASD) are determined just as they were for the flexural members discussed in Chapter 6 with

$$\phi = 0.9 \text{ (LRFD)} \qquad \Omega = 1.67 \text{ (ASD)}$$

The requirement for ASD is

$$R_a \leq \frac{R_n}{\Omega} \tag{AISC B3-2}$$

The requirement for LRFD is

$$R_u \leq \phi R_n \tag{AISC B3-1}$$

7.2.1 Noncompact Web Plate Girders

The influence of web slenderness on the strength of plate girders is not treated as a separate limit state to be assessed through its own set of requirements. Rather, web slenderness is treated as an effect on the flange yielding or flange local buckling strength and the lateral-torsional buckling strength as provided in *Specification* Section F4.

The slenderness parameter of the web is defined as $\lambda_w = h_c/t_w$, where h_c is twice the distance from the centroid to the inside face of the compression flange. For the doubly symmetric plate girder, case 15 in *Specification* Table B4.1b, this can be simplified to $\lambda_w = h/t_w$, where h is the clear distance between the flanges. Throughout this chapter, h_c in all *Specification* equations can be replaced by h. Similarly, S_{xc} can be replaced by S_x, and M_{yc} can be replaced by M_y. For a plate girder to be noncompact, the following requirements are set:

$$\lambda_{pw} < \lambda_w \leq \lambda_{rw}$$

where, according to Table B4.1b of the *Specification*,

$$\lambda_{pw} = 3.76\sqrt{\frac{E}{F_y}}$$

and

$$\lambda_{rw} = 5.70\sqrt{\frac{E}{F_y}}$$

The influence of the noncompact web is characterized through the web plastification factor, R_{pc}. This factor is used to assess the ability of the section to reach its full plastic capacity and modifies the flange yielding or local buckling and lateral torsional buckling limit states. The web plastification factor for a noncompact web is given in the *Specification* as

$$R_{pc} = \left[\frac{M_p}{M_{yc}} - \left(\frac{M_p}{M_{yc}} - 1\right)\left(\frac{\lambda - \lambda_{pw}}{\lambda_{rw} - \lambda_{pw}}\right)\right] \le \frac{M_p}{M_{yc}} \qquad \text{(AISC F4-9b)}$$

where

$$M_p = F_y Z_x \le 1.6 F_y S_x$$

and

$$M_{yc} = F_y S_{xc}$$

Equation F4-9b is shown in Figure 7.6 for two values of M_p/M_{yc}, one with a maximum value of 1.6 to account for the upper limit on M_p, and one arbitrarily taken at 1.2 as an example. The ratio M_p/M_{yc} is the shape factor that was discussed in Chapter 6. As was the case in that discussion, it must be limited to 1.6 in order to ensure that the necessary rotation can take place before strain hardening occurs as the section undergoes plastic deformation. The minimum R_{pc} is seen to be 1.0, regardless of M_p/M_{yc}. Thus, a conservative approach would be to take $R_{pc} = 1.0$. This is what the user note in Section F4 is implying when it says that Section F5 may conservatively be used for shapes that actually fall under the requirements of Section F4. Because a plate girder with a web that is only slightly noncompact would have significant additional strength reflected through the use of R_{pc}, and because the calculation of R_{pc} is not particularly difficult, there is perhaps no advantage to this simplification other than in preliminary design. Thus, Equation F4-9b will be used throughout this chapter, as appropriate.

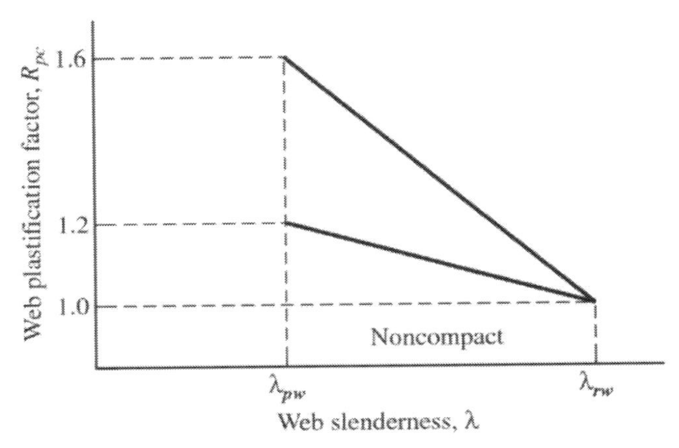

Figure 7.6 Web Plastification Factor

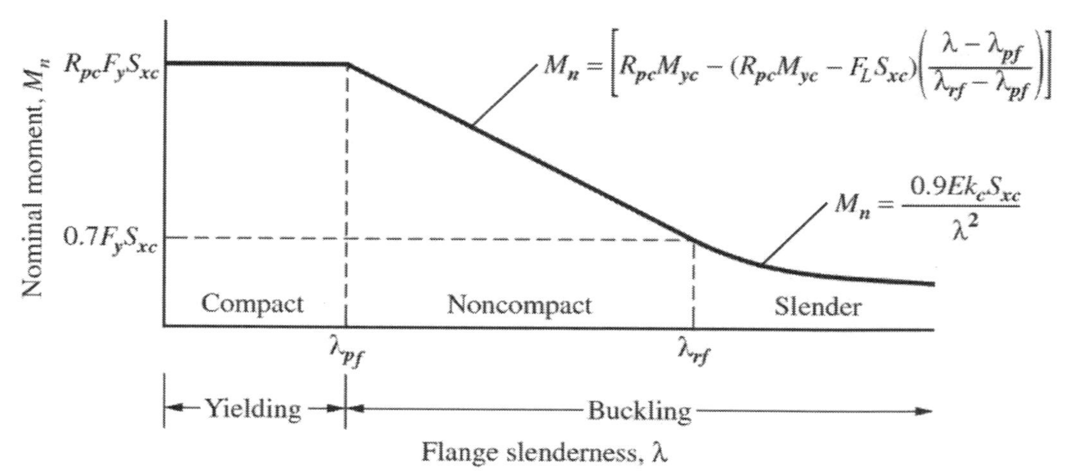

Figure 7.7 Nominal Flexural Strength Based on Flange Local Buckling

Flange Local Buckling

The requirements for compression flange local buckling are given in Section F4.3. The strength of a noncompact web plate girder as a function of flange slenderness is shown in Figure 7.7. For a compact flange girder, the impact of the noncompact web is to modify the strength by the factor R_{pc}, as given in Equation F4-1.

$$M_n = R_{pc}M_{yc} \tag{AISC F4-1}$$

At the lower limit for a noncompact web, when the web is just barely noncompact, $\lambda_w = \lambda_{pw}$, then

$$R_{pc} = \frac{M_p}{M_{yc}} \tag{7.1}$$

And the nominal strength for a compact flange girder is

$$M_n = \frac{M_p}{M_{yc}} M_{yc} = M_p \tag{7.2}$$

At the upper limit for a noncompact web, that is the web is very close to being slender, $\lambda_w = \lambda_{pr}$, R_{pc} = 1.0 and the nominal strength for a compact flange girder becomes

$$M_n = 1.0M_{yc} = F_yS_{xc} \tag{7.3}$$

For a noncompact flange girder, at the juncture between the noncompact and slender flange, $\lambda_f = \lambda_{rf}$, the nominal moment strength is given as

$$M_n = F_LS_{xc} = 0.7F_yS_{xc} \tag{7.4}$$

where the use of $F_L = 0.7F_y$, for doubly symmetric members, accounts for the stress above which the member inelastic buckling applies. This is the same as the residual stress assumed to have occurred in the hot rolled shapes, even though this is a welded shape.

The moment strength for a noncompact flange plate girder is found through the linear interpolation between the end points, as shown in Figure 7.7, and is given by Equation F4-13.

$$M_n = R_{pc}M_{yc} - (R_{pc}M_{yc} - F_L S_{xc})\left(\frac{\lambda - \lambda_{pf}}{\lambda_{rf} - \lambda_{pf}}\right) \qquad \text{(AISC F4-13)}$$

The influence of the noncompact web is diminished as the flange becomes more and more noncompact. Once the flange becomes slender, R_{pc} is no longer needed, and flange local buckling controls the strength of the girder.

For a slender flange girder, the behavior is an elastic buckling phenomenon depicted in Figure 7.7 and given as

$$M_n = \frac{0.9Ek_c S_{xc}}{\lambda^2} \qquad \text{(AISC F4-14)}$$

where $\lambda = b_f/2t_f$ and the plate buckling factor is

$$k_c = \frac{4}{\sqrt{h/t_w}} \qquad \text{(7-5)}$$

This plate buckling factor must be taken not less than 0.35 nor greater than 0.76 for the purpose of calculation, even though it may actually be outside that range.

Lateral-Torsional Buckling

Lateral-torsional buckling for noncompact web girders is addressed in Section F4.2. Lateral-torsional buckling behavior for noncompact web plate girders is, in principle, the same as for rolled beams. However, the equations in the *Specification* are slightly altered, and the web plastification factor must be included. As when considering the noncompact flange, the influence of the noncompact web is diminished as the lateral-torsional buckling response becomes more dominant. Figure 7.8 shows the strength of a noncompact web plate girder when considering lateral-torsional buckling.

For a plate girder with lateral supports at a spacing no greater than L_p, Equation F4-1, which includes the influence of R_{pc}, again defines the girder strength. This girder would be considered to have full lateral support. The definition of this limiting unbraced length is slightly different than it was for a compact web member. This difference is slight, but has been used in the *Specification*, because it gives more accurate results when used for singly symmetric girders. Thus, for noncompact web girders

$$L_p = 1.1r_t\sqrt{\frac{E}{F_y}} \qquad \text{(AISC F4-7)}$$

For I-shapes with a rectangular compression flange, the effective radius of gyration for lateral-torsional buckling, r_t, is the radius of gyration of the compression flange plus one third the compression portion of the web, given as

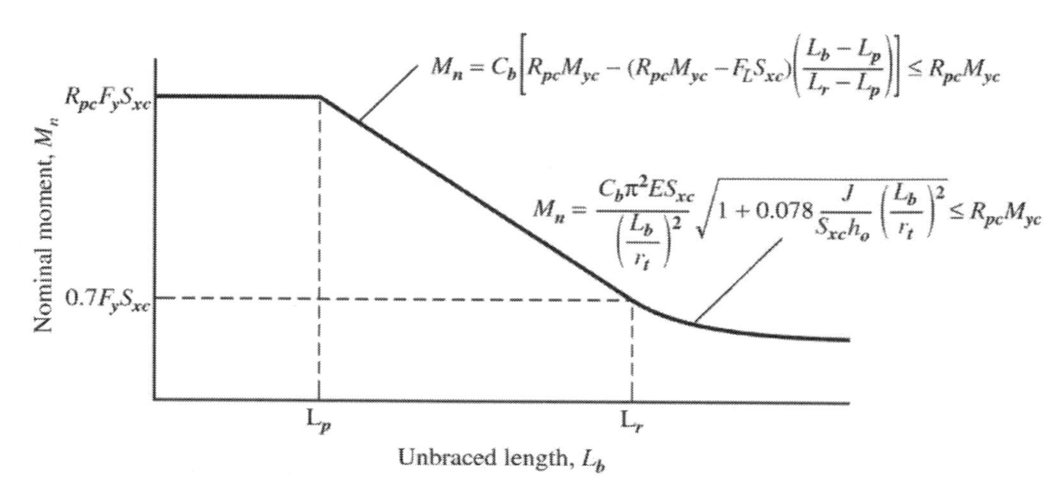

Figure 7.8 Nominal Flexural Strength Based on Unbraced Length

$$r_t = \frac{b_{fc}}{\sqrt{12\left(1+\frac{1}{6}a_w\right)}} \qquad \text{(AISC F4-11)}$$

where

$$a_w = \frac{h_c t_w}{b_{fc} t_{fc}} \qquad \text{(AISC F4-12)}$$

The strength of a section undergoing elastic lateral-torsional buckling, $L_b > L_r$, can be obtained through plate buckling theory. The *Specification* uses a combination of Equations F4-3 and F4-5 to determine the nominal moment strength as

$$M_n = \frac{C_b\pi^2 ES_{xc}}{\left(\frac{L_b}{r_t}\right)^2}\sqrt{1+0.078\frac{J}{S_{xc}h_o}\left(\frac{L_b}{r_t}\right)^2} \le R_{pc}M_{yc} \qquad (7.6)$$

This is essentially the same equation as was used for compact web girders as given through Equations F2-3 and F2-4 but with r_t replacing r_{ts}.

Because residual stresses occur in plate girders just as residual stresses do for rolled shapes, elastic buckling cannot occur if the applied stress pushes the actual stress on the shape beyond the yield stress. With the built-in stress taken as $0.3F_y$, the available elastic stress is again taken as $0.7F_y$. Thus, Equation 7.4 again gives the limiting strength, this time for elastic lateral-torsional buckling. Using this strength with Equation 7.6, the unbraced length that defines the limit of elastic lateral-torsional buckling is obtained as

$$L_r = 1.95r_t\frac{E}{F_L}\sqrt{\frac{J}{S_{xc}h_o}+\sqrt{\left(\frac{J}{S_{xc}h_o}\right)^2+6.76\left(\frac{F_L}{E}\right)^2}} \qquad \text{(AISC F4-8)}$$

If Equation F4-8 is modified to reflect only the doubly symmetric girders considered in this chapter, and F_L is taken as $0.7F_y$, it becomes

$$L_r = 1.95 r_t \frac{E}{0.7F_y}\sqrt{\frac{J}{S_x h_o} + \sqrt{\left(\frac{J}{S_x h_o}\right)^2 + 6.76\left(\frac{0.7F_y}{E}\right)^2}} \tag{7.7}$$

The lateral-torsional buckling strength when the member has an unbraced length between L_p and L_r is given by the same straight-line equation used previously. This time, however, it accounts for the noncompact web by including R_{pc} at the upper limit; thus, with $F_L = 0.7F_y$ and $S_{xc} = S_x$, Equation F4-2 becomes

$$M_n = C_b\left[R_{pc}M_{yc} - (R_{pc}M_{yc} - 0.7F_y S_x)\left(\frac{L_b - L_p}{L_r - L_p}\right)\right] \le R_{pc}M_{yc} \tag{7.8}$$

7.2.2 Slender Web Plate Girders

Slender web plate girders are covered in *Specification* Section F5. They are those built-up members with web slenderness, $\lambda_w = h_c/t_w$, exceeding the limit

$$\lambda_{rw} = 5.70\sqrt{\frac{E}{F_y}}$$

as given in Table B4.1b, case 15.

As was the case for noncompact web members, slender web members are evaluated by assessing how the web slenderness influences the other applicable limit states. The impact of the slender web on member strength is characterized through the bending strength reduction factor, R_{pg}. The bending strength reduction factor is given by Equation F5-6 and is shown in Figure 7.9.

$$R_{pg} = 1 - \frac{a_w}{1200 + 300a_w}\left(\frac{h_c}{t_w} - 5.7\sqrt{\frac{E}{F_y}}\right) \le 1.0 \qquad \text{(AISC F5-6)}$$

where a_w is as defined by Equation F4-12 but in this application is limited to a value no greater than 10.

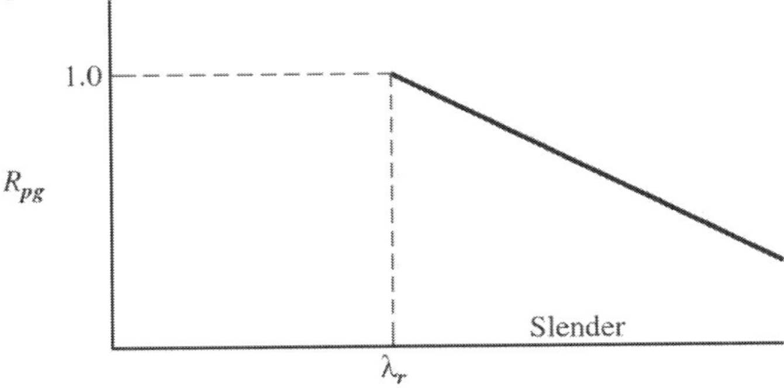

Figure 7.9 Bending Strength Reduction Factor

Specification Section F13.2 places limits on the proportions of members that can be designed according to its provisions. The web-to-flange area ratio, a_w, is limited to 10 to prevent the designer from using these provisions for members that are essentially webs with flanges that are little more than small stiffeners. In addition, the web slenderness is limited so that $h/t_w \leq 260$. This ensures that the girder is not so slender that the stated provisions do not properly reflect its behavior.

The bending strength reduction factor reduces the strength of the girder uniformly in all ranges of flange local buckling and lateral-torsional buckling. Thus, it could simply be held as a final reduction, as shown in the *Specification*, or could be used within the primary equations as shown below. Because it is uniformly applied to all of the limit states, the influence of a slender web on flexural strength is easily visualized.

Flange Local Buckling

The strength of a plate girder as a function of flange slenderness is shown in Figure 7.10. The slenderness parameters for the flange are defined in Table B4.1b and are the same as those used in Section 7.2.1. For a compact flange plate girder, $\lambda_f \leq \lambda_{pf}$,

$$M_n = R_{pg} F_y S_{xc} \qquad \text{(AISC F5-1)}$$

Because R_{pg} will not exceed 1.0, the maximum bending strength of the slender web girder is limited to the yield moment.

At the juncture between the noncompact and slender flange, $\lambda_f = \lambda_{rf}$, the strength is limited to elastic behavior, after accounting for built-in stresses. Thus

$$M_n = R_{pg}\left(0.7 F_y S_{xc}\right) \qquad (7.9)$$

The nominal moment strength for the slender web–noncompact flange plate girder is given by the linear transition which is a combination of Equations F5-7 and F5-8 as

$$M_n = R_{pg} S_{xc}\left[F_y - (0.3F_y)\left(\frac{\lambda - \lambda_{pf}}{\lambda_{rf} - \lambda_{pf}}\right)\right] \qquad (7.10)$$

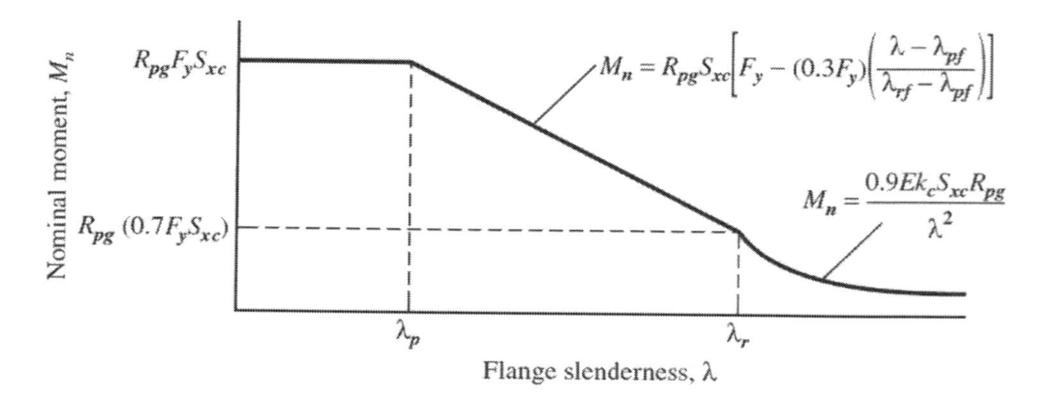

Figure 7.10 Nominal Flexural Strength Based on Flange Slenderness

For the slender web–slender flange member, the strength is the same as it was for the noncompact web–slender flange member and given in Equation F4-14, except for the use of R_{pg}. Thus, the combination of Equations F5-7 and F5-9 gives

$$M_n = \frac{0.9Ek_cS_{xc}R_{pg}}{\lambda^2} \tag{7.11}$$

The plate buckling factor, k_c, is as previously defined.

Lateral-Torsional Buckling

Lateral-torsional buckling for the slender web girder appears quite similar to that for the noncompact web girder and is shown in Figure 7.11. However, some differences must be noted. For a member to be considered as having full lateral support, its unbraced length is limited to a spacing not greater than L_p, where

$$L_p = 1.1r_t\sqrt{\frac{E}{F_y}} \qquad \text{(AISC F4-7)}$$

This is the same limit used for noncompact web girders but is different than that used for compact web members.

The elastic lateral-torsional buckling strength of the slender web girder is given in the *Specification* as a combination of Equations F5-1 and F5-4 as

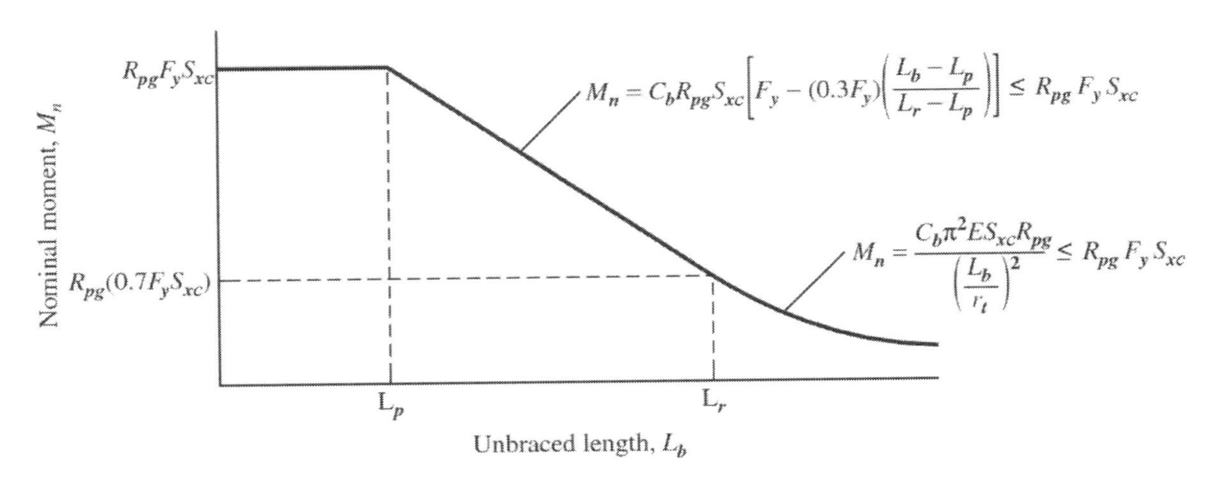

Figure 7.11 Nominal Flexural Strength Based on Unbraced Length

$$M_n = \frac{C_b \pi^2 E S_{xc} R_{pg}}{\left(\frac{L_b}{r_t}\right)^2} \leq R_{pg} F_y S_{xc} \tag{7.12}$$

When this strength is set equal to the corresponding strength limit given by Equation 7.9, the limiting unbraced length, L_r, becomes

$$L_r = \pi r_t \sqrt{\frac{E}{0.7F_y}} \qquad \text{(AISC F5-5)}$$

This limit for elastic lateral-torsional buckling is not the same as was used for the noncompact web girder. Thus, L_r is different for each of the three types of plate girder: compact web, noncompact web, and slender web.

For the inelastic lateral-torsional buckling region, the strength is given by a linear equation similar to those used previously, with the addition of the R_{pg} multiplier; thus, the combination of Equations F5-2 and F5-3 becomes

$$M_n = C_b R_{pg} S_{xc} \left[F_y - (0.3F_y) \left(\frac{L_b - L_p}{L_r - L_p} \right) \right] \leq R_{pg} F_y S_{xc} \tag{7.13}$$

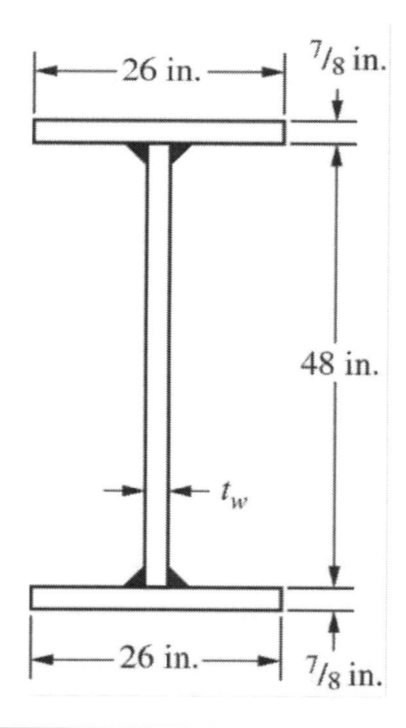

Figure 7.12 Plate Girder for Example 7.1

EXAMPLE 7.1
Plate Girder Flexural Strength

Goal: Determine the available flexural strength for two different plate girder designs, a web thickness of (a) 3/8 in. (noncompact) and (b) 1/4 in. (slender).

Given: The cross section of a homogeneous A36 plate girder is shown in

Figure 7.12. The span is 120 ft and the unbraced length of the compression flange is 20 ft. Assume $C_b = 1.0$.

SOLUTION

Step 1: Determine the section properties for both plate girders.

(a) 3/8 in. web
$t_w = 0.375$ in.
$A = 63.5$ in.^{2}
$I_x = 30{,}600$ in.^{4}
$I_y = 2560$ in.^{4}
$S_x = 1230$ in.^{3}
$Z_x = 1330$ in.^{3}

(b) 1/4 in. web
$t_w = 0.25$ in.
$A = 57.5$ in.^{2}
$I_x = 29{,}500$ in.^{4}
$I_y = 2560$ in.^{4}
$S_x = 1190$ in.^{3}
$Z_x = 1260$ in.^{3}
$r_y = 6.67$ in.

Part (a) For the plate girder with a 3/8 in. web plate

Step 2: Check the web slenderness in order to determine which sections of the *Specification* must be followed.

$$\lambda_w = \frac{h_w}{t_w} = \frac{48}{0.375} = 128 > \lambda_{pw} = 3.76\sqrt{\frac{E}{F_y}} = 3.76\sqrt{\frac{29{,}000}{36}} = 107$$

$$128 < \lambda_{rw} = 5.70\sqrt{\frac{E}{F_y}} = 5.70\sqrt{\frac{29{,}000}{36}} = 162$$

Thus, this is a noncompact web girder, and the provisions of Section F4 must be followed. The web plastification factor must be determined.

Step 3: Determine the shape factor and calculate R_{pc}.

$$\frac{M_p}{M_{yc}} = \frac{Z_x}{S_x} = \frac{1330}{1230} = 1.08 < 1.6$$

Therefore, use 1.08 in the calculation of R_{pc}, in Equation F4-9b.

$$R_{pc} = \left[\frac{M_p}{M_{yc}} - \left(\frac{M_p}{M_{yc}} - 1\right)\left(\frac{\lambda - \lambda_{pw}}{\lambda_{rw} - \lambda_{pw}}\right)\right] \le \frac{M_p}{M_{yc}}$$

$$= \left[1.08 - (1.08 - 1.0)\left(\frac{128 - 107}{162 - 107}\right)\right] = 1.05 \le 1.08$$

Step 4: Determine the nominal bending strength for the limit state of compression flange yielding.

$$M_n = R_{pc}F_yS_{xc}$$

$$= 1.05(36)(1230) = 46{,}500 \text{ in.-kips}$$

$$= \frac{46{,}500}{12} = 3880 \text{ ft-kips}$$

Step 5: Check the unbraced length for lateral torsional buckling, with L_b = 20 ft or 240 in. First determine the effective radius of gyration.

$$a_w = \frac{h_c t_w}{b_{fc} t_{fc}} = \frac{48(0.375)}{26.0(0.875)} = 0.791$$

and from Equation F4-11

$$r_t = \frac{b_{fc}}{\sqrt{12\left(1+\frac{a_w}{6}\right)}} = \frac{26.0}{\sqrt{12\left(1+\frac{0.791}{6}\right)}} = 7.05 \text{ in.}$$

From Equation F4-7

$$L_p = 1.1 r_t \sqrt{\frac{E}{F_y}}$$

$$= 1.1(7.05)\sqrt{\frac{29{,}000}{36}} = 220 \text{ in.}$$

$$= \frac{220}{12} = 18.3 \text{ ft}$$

And from Equation F4-8

$$L_r = 1.95 r_t \left(\frac{E}{F_L}\right)\sqrt{\frac{J}{S_{xc} h_o} + \sqrt{\left(\frac{J}{S_{xc} h_o}\right)^2 + 6.76\left(\frac{F_L}{E}\right)^2}}$$

$$= 1.95(7.05)\left(\frac{29{,}000}{0.7(36)}\right) \times$$

$$\sqrt{\frac{12.5}{1230(48.9)} + \sqrt{\left(\frac{12.5}{1230(48.9)}\right)^2 + 6.76\left(\frac{0.7(36)}{29{,}000}\right)^2}}$$

$$= 787 \text{ in.} = \frac{787}{12} = 65.6 \text{ ft}$$

where

$$J = \sum \frac{1}{3} bt^3 = 2\left(\frac{26(0.875)^3}{3}\right) + \frac{48(0.375)^3}{3} = 12.5 \text{ in.}^4$$

Step 6: Determine the nominal strength based on the limit state of lateral-torsional buckling.

Because the unbraced length is between L_p and L_r, the straight-line equation, Equation 7.8, is used; thus

$$M_n = C_b\left[R_{pc}M_{yc} - (R_{pc}M_{yc} - 0.7F_yS_x)\left(\frac{L_b - L_p}{L_r - L_p}\right)\right]$$

$$\le R_{pc}M_{yc}$$

$$= 1.0\left[3880 - \left(3880 - \frac{0.7(36)(1230)}{12}\right)\left(\frac{20.0 - 18.3}{65.6 - 18.3}\right)\right]$$

$$= 3830 \text{ ft-kips}$$

Step 7: Check for the limit state of compression flange local buckling.

$$\lambda_f = \frac{b_f}{2t_f} = \frac{26}{2(0.875)} = 14.9$$

and the limits are

$$\lambda_{pf} = 0.38\sqrt{\frac{E}{F_y}} = 0.38\sqrt{\frac{29{,}000}{36}} = 10.8$$

and, with $k_c = \dfrac{4}{\sqrt{48/0.375}} = 0.354 > 0.35$,

$$\lambda_{rf} = 0.95\sqrt{\frac{k_cE}{F_L}} = 0.95\sqrt{\frac{0.354(29{,}000)}{0.7(36)}} = 19.2$$

Step 8: Determine the nominal moment strength for the limit state of flange local buckling.

Because $\lambda_{pf} < \lambda_f < \lambda_{rf}$, the shape has a noncompact flange, so that, from Equation F4-13,

$$M_n = R_{pc}M_{yc} - (R_{pc}M_{yc} - F_LS_{xc})\left(\frac{\lambda - \lambda_{pf}}{\lambda_{rf} - \lambda_{pf}}\right)$$

$$= 3880 - \left(3880 - \frac{0.7(36)(1230)}{12}\right)\left(\frac{14.9 - 10.8}{19.2 - 10.8}\right) = 3250 \text{ ft-kips}$$

Step 9: Determine the lowest nominal moment for the limit states checked. For compression flange local buckling,

$$M_n = 3250 \text{ ft-kips}$$

For LRFD Step 10:

$$\phi M_n = 0.9(3250) = 2930 \text{ ft-kips}$$

For ASD

Step 10:
$$\frac{M_n}{\Omega} = \frac{3250}{1.67} = 1950 \text{ ft-kips}$$

Part (b) For the plate girder with a 1/4 in. web plate

Step 11: Check the web slenderness in order to determine which sections of the *Specification* must be followed. From step 2, $\lambda_{rw} = 162$ and

$$\lambda_w = \frac{h_c}{t_w} = \frac{48}{0.25} = 192 > \lambda_{rw} = 162$$

Thus, this is a slender web plate girder and the provisions of Section F5 must be followed.

Step 12: Determine the bending strength reduction factor, Equation F5-6.

$$a_w = \frac{h_c t_w}{b_{fc} t_{fc}} = \frac{48.0(0.25)}{26.0(0.875)} = 0.527 < 10$$

$$R_{pg} = 1 - \frac{a_w}{1200 + 300a_w}\left(\frac{h_c}{t_w} - 5.7\sqrt{\frac{E}{F_y}}\right) \leq 1.0$$

$$= 1 - \frac{0.527}{1200 + 300(0.527)}\left(\frac{48.0}{0.250} - 5.7\sqrt{\frac{29,000}{36}}\right) = 0.988 \leq 1.0$$

Step 13: Determine the nominal moment strength for the limit state of yielding. From Equation F5-1

$$M_n = \frac{R_{pg} F_y S_x}{12} = \frac{0.988(36)(1190)}{12} = 3530 \text{ ft-kips}$$

Step 14: Check the unbraced length for the limit state of lateral torsional buckling with $L_b = 20$ ft. The effective radius of gyration is

$$r_t = \frac{b_{fc}}{\sqrt{12\left(1 + \frac{a_w}{6}\right)}} = \frac{26.0}{\sqrt{12\left(1 + \frac{0.527}{6}\right)}} = 7.20 \text{ in.}$$

and from Equation F4-7

$$L_p = 1.1 r_t \sqrt{\frac{E}{F_y}}$$

$$= 1.1(7.20)\sqrt{\frac{29,000}{36}} = 225 \text{ in.}$$

$$= \frac{225}{12} = 18.8 \text{ ft}$$

From Equation F5-5

$$L_r = \pi r_t \sqrt{\frac{E}{0.7F_y}}$$

$$= \pi(7.20)\sqrt{\frac{29{,}000}{0.7(36)}} = 767 \text{ in.}$$

$$= \frac{767}{12} = 63.9 \text{ ft}$$

Step 15: Determine the nominal moment strength for the limit state of lateral torsional buckling.

Because the unbraced length is between L_p and L_r, the nominal moment strength for lateral-torsional buckling is given by Equation 7.13. Thus,

$$M_n = C_b R_{pg} S_{xc}\left[F_y - (0.3F_y)\left(\frac{L_b - L_p}{L_r - L_p}\right)\right] \leq R_{pg} F_y S_{xc}$$

$$= 1.0(0.988)(1190)\left[36 - 0.3(36)\left(\frac{20.0 - 18.8}{63.9 - 18.8}\right)\right]\left(\frac{1}{12}\right)$$

$$= 3500 \text{ ft-kips} < R_{pg} F_y S_{xc} = 3530 \text{ ft-kips}$$

Step 16: Check compression flange local buckling.

For compression flange local buckling, the flange slenderness is the same as it was in Part (a) of this problem, $\lambda_f = 14.9$, and the compact flange limit is also the same, $\lambda_{pf} = 10.8$. However, the limiting flange slenderness for the noncompact flange is different, because it is influenced by the web thickness through k_c. For the 1/4 in. web plate,

$$k_c = \frac{4}{\sqrt{h/t_w}}$$

$$= \frac{4}{\sqrt{48/0.250}} = 0.289 < 0.35$$

therefore,

$$k_c = 0.350$$

and

$$\lambda_{rf} = 0.95\sqrt{\frac{k_c E}{0.7F_y}} = 0.95\sqrt{\frac{0.350(29{,}000)}{0.7(36)}} = 19.1$$

Step 17: Determine the nominal moment strength for the limit state of flange local buckling, Equation 7.10.

$$M_n = R_{pg} S_{xc} \left[F_y - (0.3F_y) \left(\frac{\lambda - \lambda_{pf}}{\lambda_{rf} - \lambda_{pf}} \right) \right]$$

$$= 0.988(1190)\left[36 - 0.3(36)\left(\frac{14.9 - 10.8}{19.1 - 10.8}\right)\right]\left(\frac{1}{12}\right)$$

$$= 3000 \text{ ft-kips}$$

Step 18: Determine the lowest nominal moment for the limit states checked.

For compression flange local buckling

$$M_n = 3000 \text{ ft-kips}$$

For LRFD Step 19:

$$\phi M_n = 0.9(3000) = 2700 \text{ ft-kips}$$

For ASD Step 19:

$$\frac{M_n}{\Omega} = \frac{3000}{1.67} = 1800 \text{ ft-kips}$$

7.2.3 Compact Web Plate Girders

Doubly symmetric compact web flexural members are addressed in AISC *Specification* Section F3 and discussed in Chapter 6. In that presentation, the limit state of flange local buckling was limited to a discussion of compact and noncompact flanges. Discussion of slender flanges was delayed to this chapter since there are no hot rolled members with slender flanges.

The limiting flange width-to-thickness ratio is the same as for all other plate girders, case 11 in Table B4.1b. The nominal moment strength for elastic flange local buckling is given by

$$M_n = \frac{0.9Ek_c S_x}{\lambda^2} \qquad \text{(AISC F3-2)}$$

All variables in this equation are as defined in earlier sections. The limit states of yielding and lateral-torsional buckling are treated as they were presented in Chapter 6.

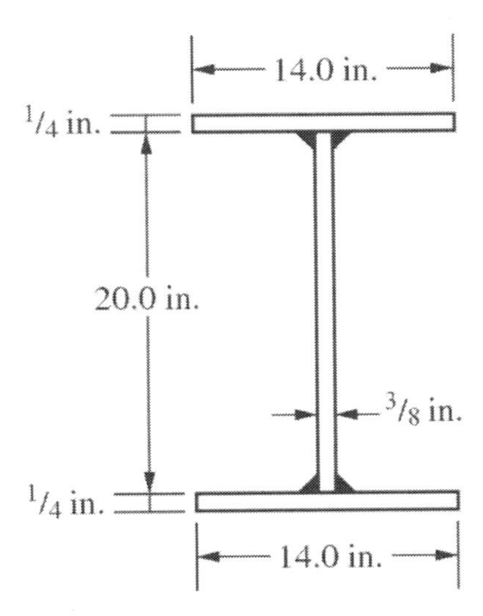

Figure 7.13 Plate Girder for Example 7.2

EXAMPLE 7.2
Plate Girder Flexural Strength

Goal: Determine the available moment strength for an A572 Gr. 50 I-shaped built-up member.

Given: The girder is shown in Figure 7.13. It has lateral supports for the compression flange at 8 ft intervals. $S_x = 94.4$ in.^{3} $r_y = 2.81$ in.

SOLUTION

Step 1: Check the web slenderness in order to determine which sections of the *Specification* must be followed. From Table B4.1b case 15,

$$\lambda_w = \frac{h_c}{t_w} = \frac{20}{0.375} = 53.3 < \lambda_{pw} = 3.76\sqrt{\frac{E}{F_y}} = 3.76\sqrt{\frac{29{,}000}{50}} = 90.6$$

Thus the girder has a compact web and should be designed in accordance with Section F3. These are the provisions that were discussed in Chapter 6.

Step 2: Check the unbraced length limits for lateral-torsional buckling with $L_b = 8.0$ ft.

From Equation F2-5

$$L_p = 1.76 r_y\sqrt{\frac{E}{F_y}} = 1.76(2.81)\sqrt{\frac{29{,}000}{50}} = 119 \text{ in.}$$

$$= \frac{119}{12} = 9.92 \text{ ft} > L_b = 8 \text{ ft}$$

Thus, lateral torsional buckling is not a factor.

Step 3: Check the slenderness limits for flange local buckling.

$$\lambda_f = \frac{b}{t} = \frac{b_f}{2t_f} = \frac{14.0}{2(0.250)} = 28.0$$

The limiting slenderness from Table B4.1b case 11 is

$$\lambda_{pf} = 0.38\sqrt{\frac{E}{F_y}} = 0.38\sqrt{\frac{29{,}000}{50}} = 9.15$$

and with

$$k_c = \frac{4}{\sqrt{h/t_w}} = \frac{4}{\sqrt{20.0/0.375}} = 0.548$$

$$\lambda_{rf} = 0.95\sqrt{\frac{k_c E}{0.7F_y}} = 0.95\sqrt{\frac{0.548\ (29{,}000)}{0.7(50)}} = 20.2$$

Since $\lambda_f = 28.0$ is greater than $\lambda_{rf} = 20.2$, the flange is slender.

Step 4: Determine the nominal strength for the limit state of flange local buckling.

Thus, for flange local buckling, using Equation F3-2,

$$M_n = \frac{0.9Ek_cS_x}{\lambda_f^2}$$

$$= \frac{0.9\ (29{,}000)(0.548)(94.4)}{(28.0)^2\ (12)} = 144 \text{ ft-kip}$$

Step 5: Determine the lowest nominal moment for the limit states checked.

Since web local buckling and lateral torsional buckling are not factors for this member, the only limit state considered is flange local buckling. This is the controlling limit state. Thus

$$M_n = 144 \text{ ft-kips}$$

For LRFD Step 6:

$$\phi M_n = 0.9(144) = 130 \text{ ft-kips}$$

For ASD Step 6:

$$\frac{M_n}{\Omega} = \frac{144}{1.67} = 86.2 \text{ ft-kips}$$

7.2.4 Proportioning Plate Girders

Design of a doubly symmetric I-shaped plate girder is similar to design of a W-shape flexural member, except that the possible solutions are almost limitless. Unlike design of a W-shape, there are no tables of most economical sections, like that found in *Manual* Table 3-2. Although older editions of the Manual, such as the 2nd ed. LRFD *Manual*, included properties for a selection of built-up I-shaped sections, more recent manuals have not included those tables.

The first step in proportioning a plate girder is likely the determination of the depth of the member. Several factors will impact the depth selected such as architectural considerations, available headroom, and support conditions. Another limiting factor is transportation. Generally, the maximum depth of a plate girder is 10 ft, based on transportation requirements. Typically, plate girders have a depth-to-span ratio of from 1/8 to 1/15 with an average of from 1/10 to 1/12. Deeper girders would be used for more highly loaded girders while shallower girders would be used for more lightly loaded applications.

Once a trial depth is selected, an approximate flange area can be determined by dividing the required moment by the trial depth times the yield stress. This assumes that all the moment strength is supplied by the flanges and that the flanges are compact. Clearly, the web will contribute to moment strength and the flange is not required to be compact. However, this is a starting point. Typically, the girder flange width to girder depth varies between 0.2 and 0.3. Thus, with the flange area determined, its width and thickness can be selected.

The thickness of the web influences both flexural strength, due to its influence on moment strength as already discussed, and shear strength, which will be discussed in the following sections. In order to be considered a plate girder, by our definition, the web must be either noncompact or slender. Thus, for an A36 girder, the maximum web thickness would be approximately $h/106$ while the maximum for an A572 Gr 50 girder would be $h/91$. A practical minimum web thickness is 5/16 in. However, for very deep plate girders, a web that is very thin may make for difficulties in fabrication. In addition, shear strength requirements will impact web plate thickness.

The available thickness of the plates used in plate girders will be restricted by the requirements previously presented in Table 3.2. Plates up to 3/8 in. thickness are available in 1/16 in. increments, from 3/8 in. to 1 in. they are available in 1/8 in. increments, and above 1 in. they are available in 1/4 in. increments. Additionally, plate widths are generally kept to 2 in. increments.

A plate girder sized according to these recommendations will give the designer a starting point. Limit states such as lateral-torsional buckling, flange local buckling, and the shear limit states yet to be discussed will all have an impact on the final strength of the plate girder.

7.3 HOMOGENEOUS PLATE GIRDERS IN SHEAR

Shear is an important factor in the behavior and design of plate girders because the webs have the potential to be relatively thin. Two design procedures are available for shear design of plate girders. One accounts for the post-buckling strength available through tension field action, whereas the other includes post-buckling strength of the web using the rotated stress field theory. Transverse stiffeners can be used to increase web shear strength but are not required unless the shear strength of the web is less than the required strength. Sometimes, a thicker web may be required even if stiffeners are used.

The limit states for shear are web yielding and web buckling. If tension field action is not considered, post-buckling strength is determined using a model that accounts for web stress redistribution in a member with or without transverse stiffeners. These provisions are covered in *Specification* Section G2.1. Under certain circumstances it is possible to take advantage of the post-buckling strength of the girder web through an approach called tension field action. Research has demonstrated that a plate girder with transverse stiffeners and a thin web can act similar to a Pratt truss once the web buckles, thus providing additional post-buckling strength. The buckled web of a plate girder is shown in Figure 7.14, and the truss model behavior is illustrated in Figure 7.15, where the buckled panel of the girder simulates the tension diagonal of the truss and the stiffener represents the vertical web member. The designer must decide whether to use this tension field action or to design without tension field action. One anomaly with the provisions for shear occurs when the tension field action approach actually produces less strength than the approach without tension field action. This is the result of the *Specification* using two different theoretical models to predict shear strength. It is an acceptable situation that simply must be checked when using tension field action if the maximum shear strength is desired.

It will be seen that web yielding controls the maximum strength of the girder web. If the size of the girder web permits web yielding, there will be no advantage to considering stiffeners, with or without tension field action.

Available shear strength for webs of plate girders, regardless of web width-to-thickness ratio, is determined using $\phi = 0.9$ and $\Omega = 1.67$.

Figure 7.14 Web Buckling Showing Tension Field Action
Photo courtesy Donald White

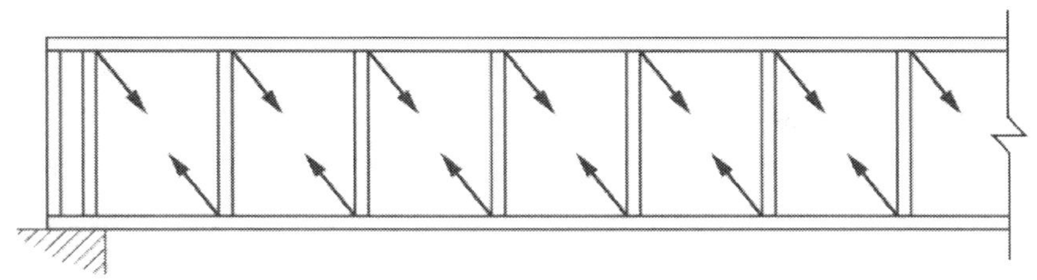

Figure 7.15 Plate Girder Showing Tension Field Action

7.3.1 Shear Strength

The primary provisions for shear strength of a plate girder are given in *Specification* Section G2.1(b). The nominal shear strength is a function of the slenderness of the web defined as $\lambda_{wv} = h/t_w$. The single limit used to define the ranges of behavior is

$$\lambda_{wvp} = 1.10\sqrt{\frac{k_v E}{F_y}}$$

The web plate buckling coefficient, k_v, for unstiffened webs of I-shaped members that meet the proportioning criteria of the *Specification*—that is, $\lambda_{wv} < 260$—is taken as $k_v = 5.34$. For stiffened webs, where stiffeners are spaced at a distance a, as shown in Figure 7.16,

$$k_v = 5 + \frac{5}{(a/h)^2} \qquad \text{(AISC G2-5)}$$

but is taken as 5.34 when $a/h > 3.0$.

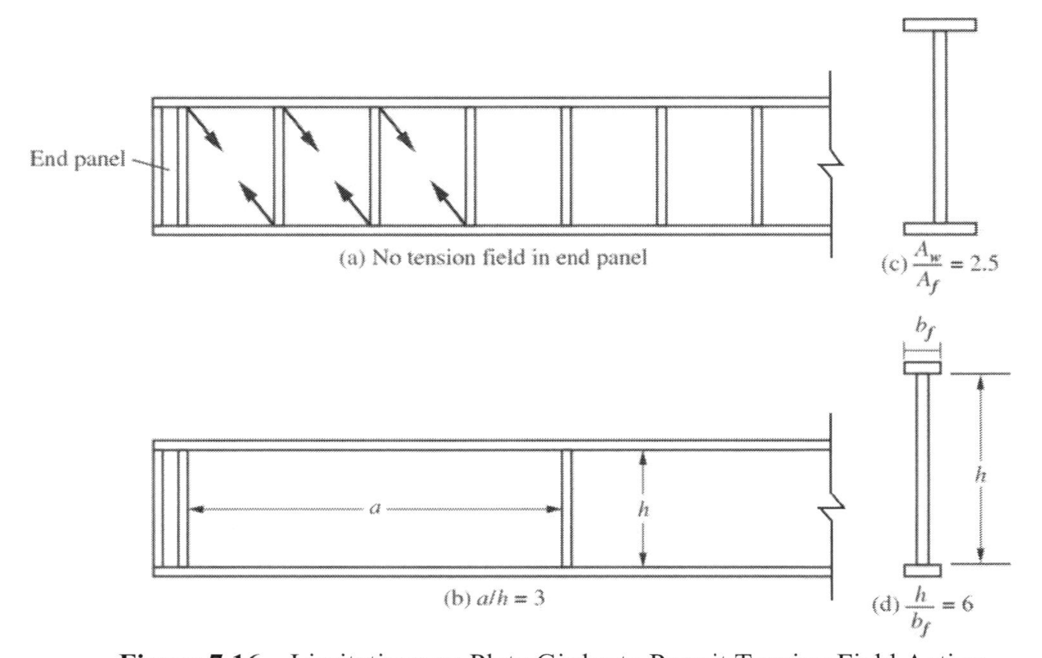

Figure 7.16 Limitations on Plate Girder to Permit Tension Field Action

The nominal shear strength of a girder is given by

$$V_n = 0.6F_yA_wC_{v1} \quad \text{(AISC G2-1)}$$

where

A_w = the overall depth times the web thickness
C_{v1}, the web shear coefficient, is a function of web shear slenderness

For $\lambda_{wv} \leq 1.10\sqrt{k_vE/F_y}$,

$$C_{v1} = 1.0 \quad \text{(AISC G2-3)}$$

For $\lambda_{wv} > 1.10\sqrt{k_vE/F_y}$,

$$C_{v1} = \frac{1.10\sqrt{k_vE/F_y}}{h/t_w} \quad \text{(AISC G2-4)}$$

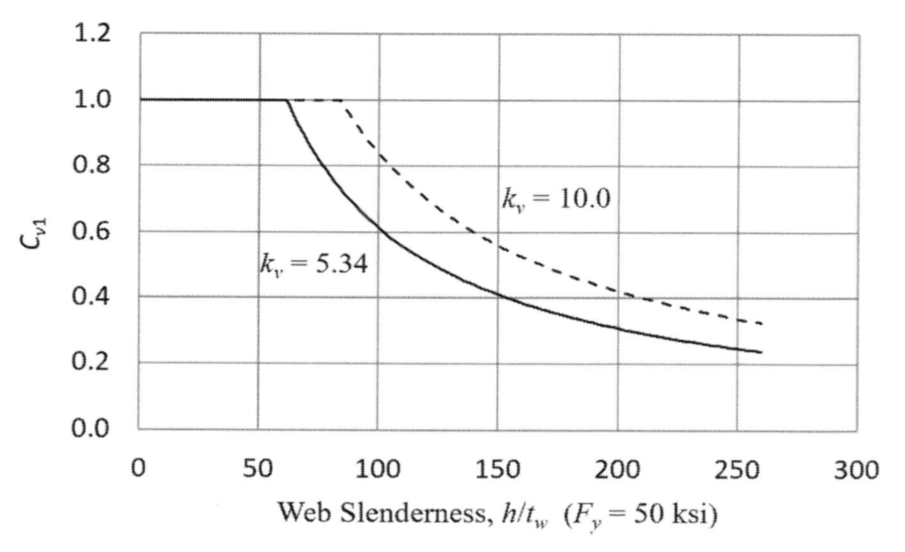

Figure 7.17 Web Shear Coefficient as a Function of Web Shear Slenderness

The web shear coefficient is shown in Figure 7.17 for two cases of the web plate buckling coefficient, $k_v = 5.34$ and $k_v = 10.0$. For a web with $C_{v1} = 1.0$ the web reaches its yield strength. For a web with shear slenderness such that C_{v1} is less than 1.0, the web buckles. Comparing the two curves in Figure 7.17 shows the impact of adding stiffeners to a girder; a girder with no stiffeners, $k_v = 5.34$; and a girder with stiffeners spaced so that the panels are square, $a/h = 1.0$ and $k_v = 10.0$.

7.3.2 Tension Field Action

Although the *Specification* does not require that tension field action be considered, a designer may take advantage of tension field action when stiffeners are present. The impact of tension field action is generally to increase the web shear strength, with the nominal shear strength determined as a combination of web buckling strength and web post-buckling strength. Both of these strength components are functions of stiffener spacing.

Tension field action is addressed in *Specification* Section G2.2 for interior panels and G2.3 for end panels. To include tension field action in the strength calculation, the plate girder must meet certain limitations. Figure 7.16 illustrates these four limitations as described here.

End Panels

Figure 7.16a shows a plate girder with a potentially buckled web. The diagonal tension that is developed in the web brings two orthogonal components of force to the flange-stiffener intersection. The vertical stiffener resists the vertical component, and the flange resists the horizontal component, just as for the Pratt truss. The end panel has no adjacent panel to help resist the horizontal component, so this last panel must resist the shear force either through beam shear without consideration of tension field action or through a modified form of tension field action. Thus, there are two approaches for determining the shear strength with tension field action, one for interior panels as given in Section G2.2 and one for end panels as given in Section G2.3. The provisions in G2.3 are new to this edition of the *Specification* and were developed through research as discussed in the Commentary. Every stiffened plate girder has end panels that must be addressed specifically. Often, this results in narrower panels at the ends of tension field girders. The requirements for use of tension field action in end panels are not addressed in this book so all end panels will be assumed to be controlled by the provisions of G2.1.

Proportions of Panels

The *Specification* limits the proportions of stiffened panels such that tension field action may not be considered if

$$\frac{a}{h} > 3$$

Figure 7.16b shows a portion of a stiffened plate girder with stiffeners placed at the limit of $a/h = 3$. The panel is quite elongated, and its effectiveness at resisting vertical forces is significantly reduced in comparison to that of a panel with a smaller aspect ratio, such as that shown in Figure 7.16a.

Proportions of Web to Flange

Section G2.2(b) provides two equations for shear strength of interior panels with tension field action, depending on the relative proportions of the web and flange. The first, when applied to doubly symmetric plate girders, is the ratio of web area to flange area not exceeding 2.5 and web height to flange width not exceeding 6. If these limits are exceeded, the flanges are not sufficient to fully resist the developed diagonal tension forces. Figures 7.16c and d show plate girders that meet these limits. In this case, the nominal shear strength is given as

$$V_n = 0.6F_y A_w \left(C_{v2} + \frac{1 - C_{v2}}{1.15\sqrt{1 + (a/h)^2}} \right) \qquad \text{(AISC G2-7)}$$

If these limits are not met, the nominal shear strength is given as

$$V_n = 0.6F_y A_w \left(C_{v2} + \frac{1 - C_{v2}}{1.15\left(a/h + \sqrt{1 + (a/h)^2}\right)} \right) \qquad \text{(AISC G2-8)}$$

The web shear buckling coefficient, C_{v2}, is a function of the web width-to-thickness ratio. It is based on a model of shear buckling that takes no post-buckling strength into account, so it is somewhat different from the coefficient, C_{v1}, given in *Specification* Section G2.1 when tension field action is not considered.

When $h/t_w \le \lambda_{wvp} = 1.10\sqrt{k_v E/F_y}$

$$C_{v2} = 1.0 \qquad \text{(AISC G2-9)}$$

when $\lambda_{vp} = 1.10\sqrt{k_v E/F_y} < h/t_w \le \lambda_{wvr} = 1.37\sqrt{k_v E/F_y}$

$$C_{v2} = \frac{1.10\sqrt{k_v E/F_y}}{h/t_w} \qquad \text{(AISC G2-10)}$$

and when $h/t_w > \lambda_{wvr} = 1.37\sqrt{k_v E/F_y}$

$$C_{v2} = \frac{1.51 k_v E}{(h/t_w)^2 F_y} \qquad \text{(AISC G2-11)}$$

Figure 7.18 shows the web shear strength for interior panels in terms of $V_n/(0.6F_y A_w)$ for a girder with and without tension field action. These girders meet the more stringent web to flange proportion limits that permit the use of Equation G2-7. Strength is given as a function of web shear slenderness for three panel sizes. Equation G2-7 can be rewritten to show that the strength due to tension field action is simply the combination of the pre-buckling strength and the post-buckling strength as

$$V_n = 0.6F_y A_w C_{v2} + 0.6F_y A_w \left[\frac{1 - C_{v2}}{1.15\sqrt{1 + (a/h)^2}} \right]$$

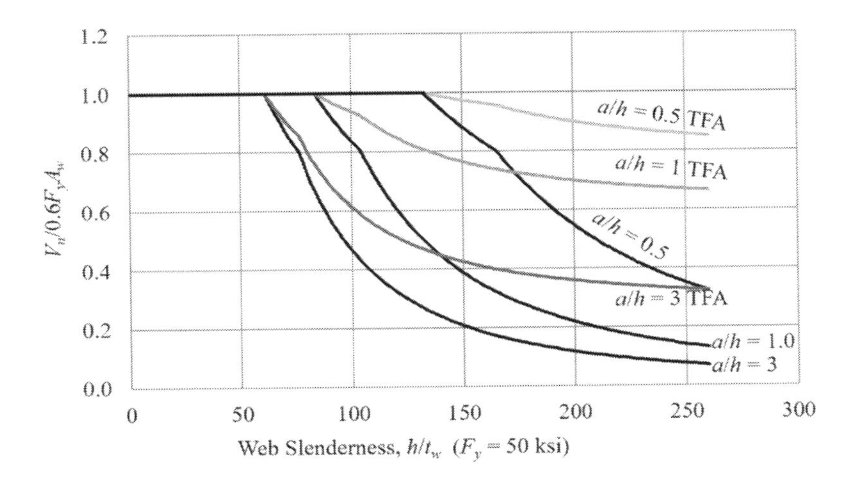

Figure 7.18 Web Shear Strength for Interior Panel of Girder with and without Tension Field Action

The pre-buckling strength can be seen in Figure 7.18 as the darker curve, the strength of the girder without tension field action. The addition of the post-buckling strength using the tension field action model shifts the curves up for each particular a/h shown in the figure as the lighter curves.

If the end panels in a tension field plate girder are not designed according to Section G2.3, they must be especially rigid in order for the remainder of the web to properly function as a Pratt truss; thus the stiffener spacing for the panel next to the support must be less than that within the span, and shear in the end panel must conform to the rules for a girder without tension field action as given in Section G2.1.

Design aids for plate girders in shear are found in Part 3 of the *Manual*, Tables 3-16 for F_y = 36 ksi and 3-17 for F_y = 50 ksi, for girders without tension field action and for interior panels with tension field action.

7.4 STIFFENERS FOR PLATE GIRDERS

When stiffeners are required for a plate girder, they can be either intermediate stiffeners or bearing stiffeners. The purpose of intermediate stiffeners is to increase girder shear strength, either by controlling the buckling strength of the girder web or by permitting additional post-buckling strength to be achieved. These stiffeners are distributed along the girder length and result in panel sizes with aspect ratios, a/h, that impact girder shear strength. Bearing stiffeners usually occur at the locations of concentrated loads or reactions. They permit the transfer of concentrated forces that could not already be transferred through direct bearing on the girder web.

7.4.1 Intermediate Stiffeners

The *Specification* requirements for intermediate stiffeners are prescriptive in nature. There are no forces for which these stiffeners must be sized; they are simply sized to meet the specific limitations provided in Section G2.4. There are two requirements for stiffener sizing. First is a width-to-thickness limit and second is a minimum moment of inertia.

For transverse stiffeners, the width-to-thickness ratio, $(b/t)_{st}$, is limited such that

$$\left(\frac{b}{t}\right)_{st} \le 0.56\sqrt{\frac{E}{F_{yst}}} \qquad \text{(AISC G2-16)}$$

where F_{yst} is the yield strength of the stiffener. Since intermediate stiffeners are not designed for strength, it is not uncommon for the stiffener and the web to have different strengths when a higher strength material is used for the web.

Transverse stiffeners used to develop the available web shear strength must have a moment of inertia, I_{st}, about an axis in the web center for stiffener pairs or about the face in contact with the web plate for single stiffeners as shown in Figure 7.19, such that the minimum moment of inertia based on consideration of no post-buckling strength is

$$I_{st2} = \left[\frac{2.5}{(a/h)^2} - 2\right] b_p t_w^3 \ge 0.5 b_p t_w^3 \qquad \text{(AISC G2-19)}$$

where b_p is the smaller of a and h.

The minimum moment of inertia required when the full post-buckling strength or tension field action strength is included is defined in the *Specification* as

$$I_{st1} = \frac{h^4 \rho_{st}^{1.3}}{40}\left(\frac{F_{yw}}{E}\right)^{1.5} \qquad \text{(AISC G2-18)}$$

where

ρ_{st} = the larger of F_{yw}/F_{yst} and 1.0
F_{yw} = yield stress of the web

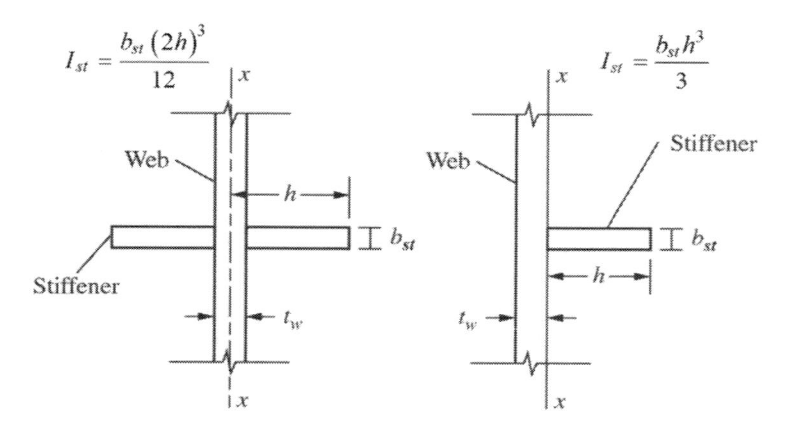

Figure 7.19 Web Stiffener Minimum Moment of Inertia

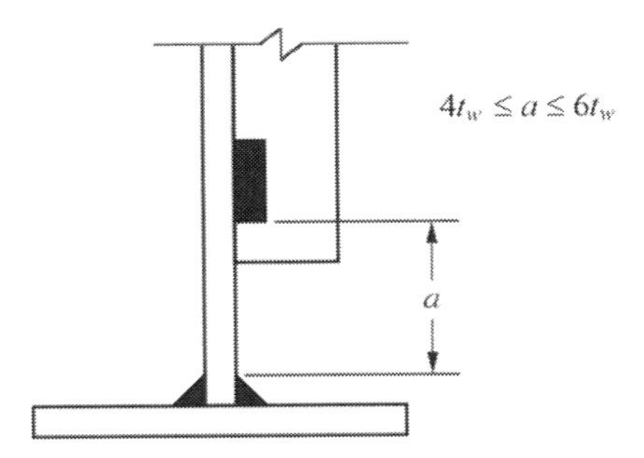

Figure 7.20 Detailing Requirement for Intermediate Stiffeners

If the required shear strength is less than the full available shear strength, a linear interpolation, as indicated by Equation G2-17, may be used to determine the actual required stiffener moment of inertia. Thus,

$$I_{st} \geq I_{st2} + \left(I_{st1} - I_{st2}\right)\rho_w \qquad \text{(AISC G2-17)}$$

which accounts for the amount of post-buckling strength, either with or without tension field action, included in the shear strength calculation, where

I_{st} = moment of inertia of the transverse stiffener
I_{st1} = minimum moment of inertia of the transverse stiffener required for the development of the full shear post-buckling strength, V_{c1}
I_{st2} = minimum moment of inertia of the transverse stiffener required to develop buckling strength without considering any post-buckling strength, V_{c2}
V_r = required shear strength in the panel being considered
V_{c1} = available post-buckling shear strength as defined in Section G2.1 or G2.2
V_{c2} = available shear strength when considering no post-buckling strength, using C_{v2} in Equation G2-1 in place of C_{v1}.
ρ_w = the maximum shear ratio within the panels on each side of the transverse stiffener,

$$\left[\frac{V_r - V_{c2}}{V_{c1} - V_{c2}}\right] \geq 0$$

In addition, the *Specification* provides detailing requirements for intermediate stiffeners. They can be stopped short of the tension flange and, when used in pairs, do not need to be attached to the compression flange. The weld by which they are attached to the web is to be terminated between four and six times the web thickness from the near toe to the web-flange weld, as shown in Figure 7.20, but there is no specific requirement for sizing that weld. Normally it would be sized based on the plate thickness. When single stiffeners are used, they must be attached to the compression flange, if it consists of a rectangular plate, to resist any uplift tendency due to torsion in the flange. Because intermediate stiffeners provide a convenient mechanism to transfer bracing forces to the girder, these stiffeners also must be connected to the compression flange and must be capable of transmitting one percent of the total flange force.

7.4.2 Bearing Stiffeners

Bearing stiffeners are required when the strength of the girder web is not sufficient to resist the concentrated forces exerted on it. Although bearing stiffeners can be required for rolled I-shaped members and were discussed in Section 6.14, they are much more likely to be required for plate girders, particularly at the girder supports. *Specification* Section J10 addresses the appropriate limit states. Normally the forces to be resisted are compressive in nature. For those cases, the limit states of web local yielding, web local crippling, and web sidesway buckling must be checked. When the applied load is tensile, web local yielding and flange local bending must be considered. If the strength of the web is insufficient to resist the applied force, bearing stiffeners can be used.

The relationship between available strength and nominal strength varies for each limit state associated with web strength. Thus, either design strengths or allowable strengths, not nominal strengths, must be compared to determine the minimum web strength. The appropriate resistance factors and safety factors are given with the following discussion of the corresponding limit states.

Web Local Yielding

When a single concentrated force, tension or compression, is applied to a girder, as shown in Figure 7.21, the force is assumed to be delivered to the girder over a length of bearing, l_b, and is then distributed through the flange and into the web. For a plate girder, the web just below the web-flange weld, dimensioned as k in Figure 7.21a, is the critical location. For plate girders, this dimension, k, is taken as the thickness of the flange plus the dimension of the weld. As was the case for the W-shape, the distribution takes place along a line with a slope of 1:2.5. Thus, when the critical section is reached, the force has been distributed over a length of l_b plus $2.5k$ in each direction. If the concentrated force is applied so that the force distributes along the web in both directions, this distribution increases the bearing length by $5k$, as shown in Figure 7.21b. If the bearing is close to the end of the member, distribution takes place only in one direction, toward the midspan. The *Specification* defines "close to the member end" as being within the member depth from the end. Thus, the available length of the web is $(l_b + 2.5k)$, as shown in Figure 7.21c.

The nominal strength of the girder web when the concentrated force to be resisted is applied at a distance from the member end that is greater than the depth of the member, d, is

$$R_n = F_{yw}t_w\left(5k + l_b\right) \qquad \text{(AISC J10-2)}$$

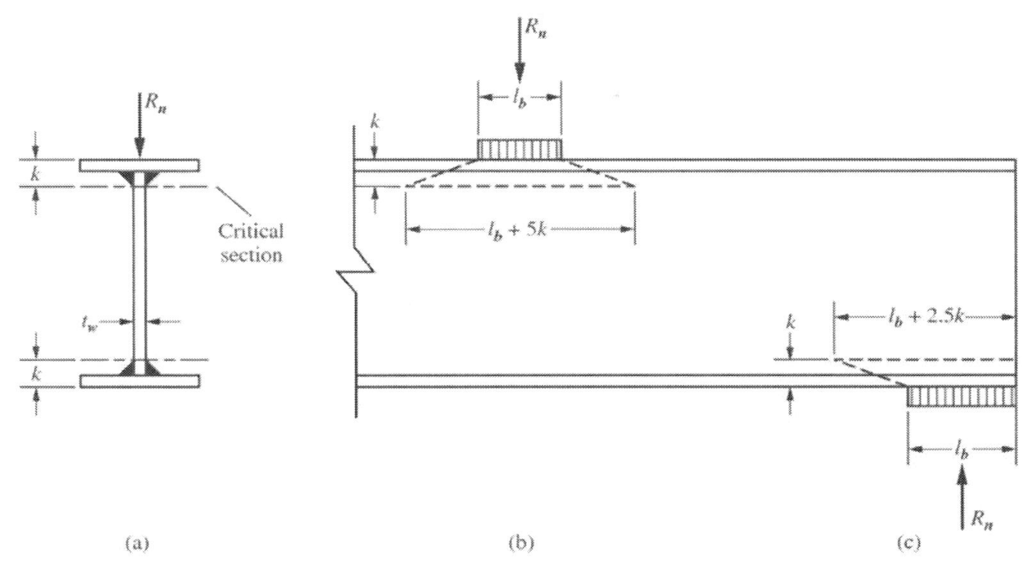

Figure 7.21 Single Concentrated Force Applied to Beam

When the concentrated force to be resisted is applied at a distance from the member end that is less than or equal to the depth of the member, *d*, the nominal strength is

$$R_n = F_{yw} t_w \left(2.5k + l_b\right) \qquad \text{(AISC J10-3)}$$

where

- F_{yw} = yield stress of the web
- l_b = length of bearing
- k = distance from the outer face of the flange to the web toe of the fillet weld
- t_w = web thickness

For web local yielding,

$$\phi = 1.0 \text{ (LRFD)} \qquad \Omega = 1.50 \text{ (ASD)}$$

Web Local Crippling

The criteria for the limit state of web local crippling of a plate girder is the same as it was for rolled W-shapes. It again depends on the location of the force with respect to the end of the girder and Q_f =1.0 for non HSS members.

When the concentrated compressive force is applied at a distance from the member end that is greater than or equal to $d/2$,

$$R_n = 0.80 t_w^2 \left[1 + 3\left(\frac{l_b}{d}\right)\left(\frac{t_w}{t_f}\right)^{1.5}\right]\sqrt{\frac{EF_{yw}t_f}{t_w}}Q_f \qquad \text{(AISC J10-4)}$$

When the force is applied at a distance less than $d/2$ and $l_b/d \leq 0.2$,

$$R_n = 0.40t_w^2\left[1+3\left(\frac{l_b}{d}\right)\left(\frac{t_w}{t_f}\right)^{1.5}\right]\sqrt{\frac{EF_{yw}t_f}{t_w}}Q_f \qquad \text{(AISC J10-5a)}$$

and when $l_b/d > 0.2$,

$$R_n = 0.40t_w^2\left[1+\left(\frac{4l_b}{d}-0.2\right)\left(\frac{t_w}{t_f}\right)^{1.5}\right]\sqrt{\frac{EF_{yw}t_f}{t_w}}Q_f \qquad \text{(AISC J10-5b)}$$

For web local crippling,

$$\phi = 0.75 \text{ (LRFD)} \qquad \Omega = 2.0 \text{ (ASD)}$$

Web Sidesway Buckling

The web of a plate girder is generally a noncompact or slender element, as has already been discussed. If the tension and compression flanges of the girder are not prevented from displacing laterally with respect to each other at the point of a compressive load, web sidesway buckling must be assessed. Two provisions are given for web sidesway buckling: (1) if the compression flange is restrained against rotation, such as when it is attached to a slab, and (2) if it is not.

When the compression flange is restrained against rotation and the ratio of web slenderness to lateral buckling slenderness, $(h/t_w)/(L_b/b_f) \leq 2.3$, the nominal strength is given as

$$R_n = \frac{C_r t_w^3 t_f}{h^2}\left[1+0.4\left(\frac{h/t_w}{L_b/b_f}\right)^3\right] \qquad \text{(AISC J10-6)}$$

If $(h/t_w)/(L_b/b_f) > 2.3$, the limit state of web sidesway buckling does not apply.

When the compression flange is not restrained against rotation and the ratio of web slenderness to lateral buckling slenderness, $(h/t_w)/(L_b/b_f) \leq 1.7$, the nominal strength is given as

$$R_n = \frac{C_r t_w^3 t_f}{h^2}\left[0.4\left(\frac{h/t_w}{L_b/b_f}\right)^3\right] \qquad \text{(AISC J10-7)}$$

and if $(h/t_w)/(L_b/b_f) > 1.7$, the limit state of web sidesway buckling does not apply.

In the above equations,

L_b = largest laterally unbraced length along either flange at the point of load
C_r = 960,000 ksi when, at the location of the force, $M_u < M_y$ or $1.5M_a < M_y$;
480,000 ksi when, at the location of the force, $M_u \geq M_y$ or $1.5M_a \geq M_y$

For web sidesway buckling,

$$\phi = 0.85 \text{ (LRFD)} \qquad \Omega = 1.76 \text{ (ASD)}$$

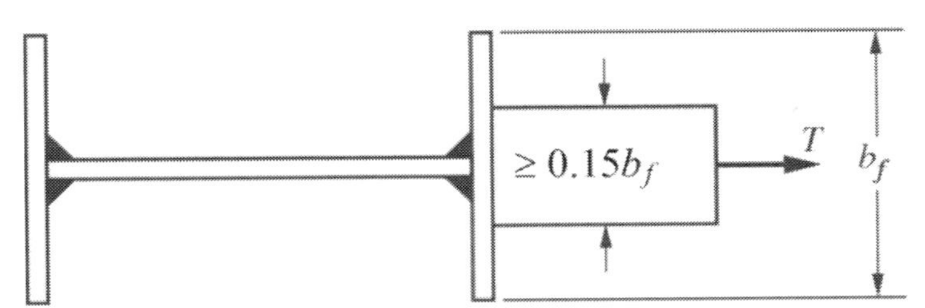

Figure 7.22 Flange Local Bending for an Applied Tension Load

Flange Local Bending

This limit state applies when a single tensile concentrated force is applied to the flange and the length of loading across the member flange is greater than or equal to $0.15b_f$, as shown in Figure 7.22. The nominal strength is

$$R_n = 6.25F_{yf}t_f^2 \qquad \text{(AISC J10-1)}$$

If the force is applied at a distance less than $10t_f$ from the member end, the strength must be reduced by 50 percent.

For flange local bending,

$$\phi = 0.9 \text{ (LRFD)} \qquad \Omega = 1.67 \text{ (ASD)}$$

7.4.3 Bearing Stiffener Design

Once the appropriate limit states are checked, a decision is made regarding the need for bearing stiffeners. Although it is possible to select a web plate that would not require bearing stiffeners, this is not usually the most economical approach, even though the addition of stiffeners is a high-labor, and thus high-cost, activity. When bearing stiffeners are to be sized, Section J10.8 of the *Specification* requires that they be sized according to the provisions for tension members or compression members, as appropriate.

Stiffeners designed to resist tensile forces must be designed according to the requirements of Section J4.1 which refer to Chapter D for the difference between the required strength and the minimum available limit state strength. The stiffener must be welded to the flange and web, and these welds must be sized to resist the force being transferred to the stiffeners.

Stiffeners required to resist compressive forces must be designed according to the provisions of Chapter E, except for stiffeners with $KL/r \leq 25$, which may be designed with $F_n = F_y$, according to Section J4.4, for the difference between the required strength and the minimum available limit state strength. These stiffeners must either bear on or be welded to the loaded flange and welded to the web, and these welds must be sized to resist the force being transferred to the stiffeners.

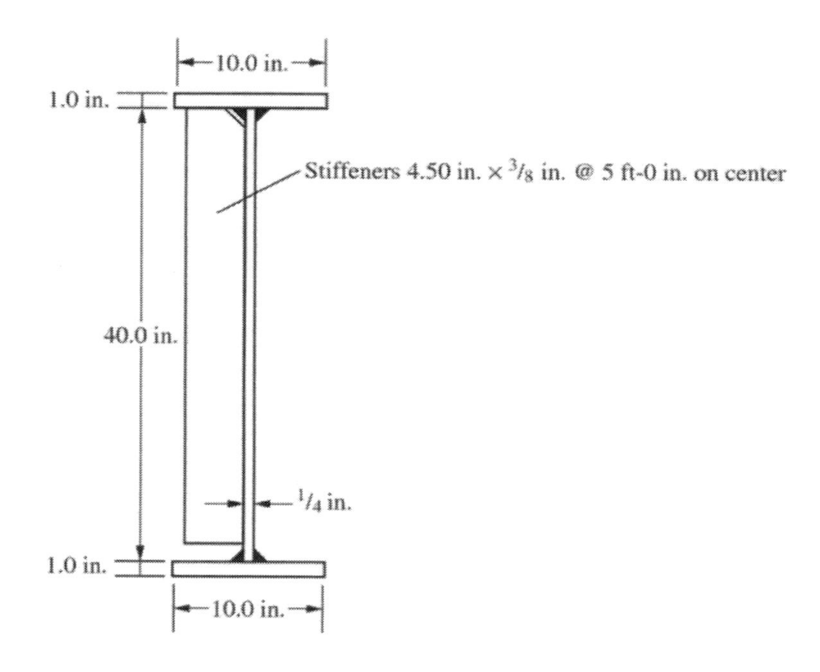

Figure 7.23 Plate Girder for Example 7.3

EXAMPLE 7.3
Plate Girder Flexural and Shear Strength

Goal: Determine the available moment and shear strength using tension field action in an interior panel and check the stiffener to see that it meets the criteria.

Given: A built-up member is shown in Figure 7.23. The beam is continuously laterally braced. Use A572 Gr. 50 for the member plates and A36 for the stiffeners. $S_x = 464$ in.^{3}.

SOLUTION

Step 1: Check the web slenderness to determine which section of the *Specification* must be used. Using Table B4.1b case 15

$$\lambda_w = \frac{h}{t_w} = \frac{40}{0.25} = 160 > \lambda_{rw} = 5.7\sqrt{\frac{29{,}000}{50}} = 137$$

Thus, this is a slender web plate girder, and the provisions of Section F5 must be applied.

Step 2: Determine the bending strength reduction factor.

The ratio of web area to flange area is given as

$$a_w = \frac{40.0(0.250)}{10.0(1.00)} = 1.00$$

And from Equation F5-6

$$R_{pg} = 1 - \frac{1.00}{1200 + 300(1.00)}\left(\frac{40.0}{0.250} - 5.7\sqrt{\frac{29{,}000}{50}}\right) = 0.985$$

Step 3: Determine the nominal moment strength for the limit state of yielding.

$$M_n = R_{pg}F_yS_{xc} = \frac{0.985(50)(464)}{12} = 1900 \text{ ft-kips}$$

Step 4: Determine the nominal moment strength for the limit state of lateral-torsional buckling.

The compression flange is fully braced, so this limit state does not apply.

Step 5: Determine the nominal moment strength for the limit state of flange local buckling.

The flange slenderness is, from Table B4.1b case 11

$$\lambda_f = \frac{b}{t} = \frac{b_f}{2t_f} = \frac{10.0}{2(1.0)} = 5.0 < \lambda_{pf} = 0.38\sqrt{\frac{29{,}000}{50}} = 9.15$$

Therefore, the flange is compact, and there is no reduction in strength.

Step 6: Determine the lowest nominal moment strength for the limit states considered, yielding, lateral-torsional buckling and flange local buckling. For the limit state of yielding,

$$M_n = 1900 \text{ ft-kips}$$

For LRFD Step 7: Determine the design moment strength.

$$\phi M_n = 0.9(1900) = 1710 \text{ ft-kips}$$

For ASD Step 7: Determine the allowable moment strength.

$$M_n/\Omega = 1900/1.67 = 1140 \text{ ft-kips}$$

Step 8: Determine the shear strength with tension field action. First check the girder proportions against the prescriptive requirements of Section G2.2.

Proportions of panel:

$$\frac{a}{h} = \frac{60.0}{40.0} = 1.50 < 3$$

Proportions of web to flange:

$$\frac{A_w}{A_f} = \frac{42.0(0.250)}{10.0(1.00)} = 1.05 < 2.5$$

Proportions of web height to flange width:

$$\frac{h}{b_f} = \frac{40.0}{10.0} = 4.00 < 6.0$$

Because the criteria have been satisfied, the tension field action provisions of Section G2.2(b)(1) may be used for all but the end panel.

Step 9: Check the web shear slenderness limits for determination of the shear strength coefficient. The web shear buckling coefficient from Equation G2-5 is

$$k_v = 5 + \frac{5}{(a/h)^2} = 5 + \frac{5}{(60.0/40.0)^2} = 7.22$$

Thus the limits for determining C_{v2} are

$$\lambda_{wvp} = 1.1\sqrt{\frac{k_v E}{F_y}} = 1.1\sqrt{\frac{7.22(29{,}000)}{50}} = 71.2$$

and

$$\lambda_{wvr} = 1.37\sqrt{\frac{k_v E}{F_y}} = 1.37\sqrt{\frac{7.22\ (29{,}000)}{50}} = 88.7$$

Because $\lambda_{wv} = h/t_w = 40.0/0.250 = 160$ is greater than λ_{wvp} and λ_{wvr}, the web plate will buckle elastically.

Step 10: Determine the shear strength coefficient from Equation G2-11.

$$C_{v2} = \frac{1.51(7.22)(29{,}000)}{(160)^2(50)} = 0.247$$

The nominal shear strength with tension field action from Equation G2-7 is

$$V_n = 0.6(50)(42.0)(0.250)\left((0.247 + \left[\frac{1 - 0.247}{1.15\sqrt{1 + (1.50)^2}} \right] \right) = 192 \text{ kips}$$

For LRFD Step 11: Determine the design shear strength.

$$\phi V_n = 0.9(192) = 173 \text{ kips}$$

For ASD Step 11: Determine the allowable shear strength.

$$V_n/\Omega = 192/1.67 = 115 \text{ kips}$$

Step 12: Check the intermediate stiffener size for meeting the criteria.

Check the width-to-thickness ratio of the stiffener from Equation G2-16

$$\left(\frac{b}{t}\right)_{st} = \frac{4.5}{0.375} = 12.0 \leq 0.56\sqrt{\frac{29,000}{36}} = 15.9$$

So the stiffener satisfies this requirement

For the single plate stiffener,

$$I_{st} = \frac{b_{st}h^3}{3} = \frac{0.375(4.50)^3}{3} = 11.4 \text{ in.}^4$$

Step 13: Check the minimum stiffener moment of inertia when tension field action is considered.

Since the required shear strength is not given, the stiffeners will be checked to see that they are capable of providing the full tension field strength of the web using Equation G2-18.

$$I_{st1} = \frac{h^4\rho_{st}^{1.3}}{40}\left(\frac{F_{yw}}{E}\right)^{1.5} = \frac{(40.0)^4\left(\frac{50}{36}\right)^{1.3}}{40}\left(\frac{50}{29,000}\right)^{1.5} = 7.02 \text{ in.}^4$$

and

$$I_{st1} = 7.02 < 11.4 \text{ in.}^4$$

Thus, these stiffeners are adequate to permit full use of the available shear strength with tension field action, as given in Step 11.

7.5 PROBLEMS

1. Determine the available moment strength of an A36 plate girder with a 50×1/2 in. web plate and equal flange plates of 12×1 in. Assume there is full lateral support. Determine by (a) LRFD and (b) ASD.

2. Determine the available moment strength of an A36 plate girder with a 60×1/2 in. web plate and equal flange plates of 12×1 in. Assume there is full lateral support. Determine by (a) LRFD and (b) ASD.

3. Determine the available moment strength of an A572 Gr. 50 plate girder with an 80×1/2 in. web plate and equal flange plates of 16×1/2 in. Assume there is full lateral support. Determine by (a) LRFD and (b) ASD.

4. Determine the available moment strength of an A572 Gr. 50 plate girder with a web plate of 40×1/2 in. and equal flange plates of 10×1 in. Assume there is full lateral support. Determine by (a) LRFD and (b) ASD.

5. Determine the available moment strength of an A572 Gr. 50 plate girder with a web plate of 60×1/2 in. and equal flange plates of 10×3/4 in. Assume there is full lateral support. Determine by (a) LRFD and (b) ASD.

6. Determine the available moment strength of an A36 plate girder with a web plate of 75×3/8 in. and equal flange plates of 14×1-1/4 in. Assume there is full lateral support. Determine by (a) LRFD and (b) ASD.

7. For a plate girder spanning 80 ft, carrying a uniformly distributed load, with lateral supports at the supports and midspan, determine the nominal moment strength. Use the correct C_b. The girder is an A36 member with a web plate of 45×3/4 in. and flange plates of 9×3/4 in.

8. For a plate girder spanning 80 ft carrying a concentrated load at midspan, with lateral supports at the supports and midspan, determine the nominal moment strength using the correct C_b. The girder is an A36 member with a web plate of 50×3/8 in. and flange plates of 9×3/4 in.

9. For a plate girder spanning 80 ft with concentrated loads at the third points and lateral supports at the supports and loads, determine the nominal moment strength using the correct C_b. The girder is an A36 member with a web plate of 65×3/8 in. and flange plates of 9×3/4 in.

10. For the plate girder of Problem 1 supporting a uniform load, determine the available moment strength if the girder spans 120 ft and has lateral supports at the ends and the third points of the span. Use the correct C_b. Determine by (a) LRFD and (b) ASD.

11. For the plate girder of Problem 3, determine the available moment strength if the girder spans 90 ft and has lateral supports at the ends and the quarter points of the span. Assume the girder is loaded with equal concentrated loads at the quarter points and use the correct C_b. Determine by (a) LRFD and (b) ASD.

12. For the plate girder of Problem 4 supporting a uniform load, determine the available moment strength if the girder spans 70 ft and has lateral supports at the ends only. Use the correct C_b. Determine by (a) LRFD and (b) ASD.

13. For the plate girder of Problem 6, determine the available moment strength if the girder spans 160 ft and has lateral supports at the ends and the third points of the span. Assume the girder supports equal concentrated loads at the third pints. Use the correct C_b. Determine by (a) LRFD and (b) ASD.

14. Determine the available moment strength of an A36 plate girder with a 50×1/2 in. web plate and equal flange plates of 12×1/2 in. Assume there is full lateral support. Determine by (a) LRFD and (b) ASD.

15. Determine the available moment strength of an A36 plate girder with a 60×1/2 in. web plate and equal flange plates of 12×1/2 in. Assume there is full lateral support. Determine by (a) LRFD and (b) ASD.

16. Determine the available moment strength of an A36 plate girder with an 85×1/2 in. web plate and equal flange plates of 12×1/2 in. Assume there is full lateral support. Determine by (a) LRFD and (b) ASD.

17. Determine the available moment strength of an A572 Gr. 50 plate girder with a web plate of 40×1/2 in. and equal flange plates of 12×1/2 in. Assume there is full lateral support. Determine by (a) LRFD and (b) ASD.

18. Determine the available moment strength of an A572 Gr. 50 plate girder with a web plate of 70×1/2 in. and equal flange plates of 12×1/2 in. Assume there is full lateral support. Determine by (a) LRFD and (b) ASD.

19. Determine the available moment strength of an A572 Gr. 50 plate girder with a web plate of 45×3/4 in. and flange plates of 14×1/2 in. Assume there is full lateral support. Determine by (a) LRFD and (b) ASD.

20. Determine the available moment strength for the girder from Example 7.1 for the now more commonly available A572 Gr 50 steel.

21. Determine the available shear strength of an A572 Gr. 50 plate girder without transverse stiffeners. The web plate is 100×3/4 in. and the flange plates are 15×1-1/2 in. Determine by (a) LRFD and (b) ASD.

22. Determine the available shear strength of the girder in Problem 1 if there are no intermediate stiffeners. Determine by (a) LRFD and (b) ASD.

23. Determine the available shear strength of the girder in Problem 2 if there are no intermediate stiffeners. Determine by (a) LRFD and (b) ASD.

24. Determine the available shear strength of the girder in Problem 3 if there are no intermediate stiffeners. Determine by (a) LRFD and (b) ASD.

25. Determine the available shear strength, without tension field action, of the girder in Problem 4 if there are intermediate stiffeners every 40 in. Determine by (a) LRFD and (b) ASD.

26. Determine the available shear strength, without tension field action, of the girder in Problem 5 if there are intermediate stiffeners every 80 in. Determine by (a) LRFD and (b) ASD.

27. Determine the available shear strength of the girder, without tension field action, in Problem 6 if there are intermediate stiffeners every 50 in. Determine by (a) LRFD and (b) ASD.

28. Repeat Problem 26 using tension field action. Determine by (a) LRFD and (b) ASD.

28. Repeat Problem 27 using tension field action. Determine by (a) LRFD and (b) ASD.

29. Determine the available shear strength of an A572 Gr. 50 plate girder with transverse stiffeners spaced every 100 in. The web plate is 100×3/4 in. and the flange plates are 15×1-1/2 in. Determine by (a) LRFD and (b) ASD.

30. Determine the available moment and available shear strength of an A36 plate girder on a 100 ft span with stiffeners at the ends and 20 in. from the ends, along with intermediate stiffeners spaced at 40 in. on center within the span. Assume there is full lateral support. The girder has a web plate of 40×1/4 in. and flange plates of 8×3/4 in. Determine by (a) LRFD and (b) ASD.

31. Consider the plate girder from Problem 30 and the intermediate stiffener between a 20 in. end panel and an adjacent 40 in. panel. Does a pair of A36 3×1/4 in. stiffener plates satisfy all intermediate transverse stiffener requirements for this location? Evaluate by (a) LRFD and (b) ASD.

32. Consider the plate girder from Problem 30 and the intermediate stiffener between a 20 in. end panel and an adjacent 40 in. panel. Does a single A36 3-1/2×3/8 in. stiffener plate satisfy all intermediate transverse stiffener requirements for this location? Evaluate by (a) LRFD and (b) ASD.

33. A girder is to be designed to span 125 ft and carry a uniformly distributed live load of 2.0 kip/ft and a uniformly distributed dead load of 1.9 kip/ft including the girder weight. Assume that the compression flange has full lateral support, and determine the web and flange plates necessary if the web plate is limited to a depth of 50 in. Determine any stiffener requirements for a girder without tension field action. Use A572 Gr. 50 steel. Design by (a) LRFD and (b) ASD.

34. Design a girder for the conditions of Problem 33 if the compression flange is braced at 1/3 points.

35. A girder is to be designed to span 100 ft and carry a uniformly distributed live load of 3.0 kip/ft and a uniformly distributed dead load of 3.6 kip/ft including the girder weight. The compression flange has full lateral support. Determine the web and flange plates necessary if the web plate is limited to a depth of 60 in. Determine any stiffener requirements for a girder without tension field action. Use A572 Gr. 50 steel. Design by (a) LRFD and (b) ASD.

36. Design a girder for the conditions of Problem 35 if the compression flange is braced at 1/3 points.

37. A girder is to be designed to span 150 ft and carry a uniformly distributed live load of 1.0 kip/ft, a uniformly distributed dead load of 2.4 kip/ft including the girder weight, and concentrated live loads of 50 kips at the third points of the span. The compression flange has lateral support at the load points. Determine the web and flange plates necessary if the web plate is limited to a depth of 75 in. Determine any stiffener requirements for a girder without tension field action. Use A572 Gr. 50 steel. Design by (a) LRFD and (b) ASD.

38. For the girder designed in Problem 37, determine the stiffener requirements if tension field action is used.

Chapter 8

Beam-Columns and Frame Behavior

The Pavilion at the Hospital of the University of Pennsylvania, Philadelphia, PA
Photo courtesy of HDR

8.1 INTRODUCTION

Beam-columns are members subjected to axial forces and bending moments simultaneously; thus their behavior falls somewhere between that of an axially loaded column and that of a beam under pure bending. It is thus possible to consider the beam or axially loaded member as special cases of the beam-column. Practical applications of the beam-column are numerous. They occur as chord members in trusses, as elements of rigidly connected frameworks, and as members of pin-connected structures with transverse or eccentric loads. It is not always possible to look at a member and determine whether it is a beam-column or not; some knowledge of the actual forces being carried by the member is required to categorize it as a beam-column. However, many structural members are subjected to these combined forces, and the beam-column is a very common element in building structures.

The manner in which the combined loads are transferred to a particular beam-column significantly impacts the ability of the member to resist those loads. Starting with the axially loaded column, bending moments can occur from various sources. Lateral load can be applied directly to the member, as is the case for a truss top chord or a column supporting the lateral load from a wall. Alternatively, the axial force can be applied at some eccentricity from the centroid of the column as a result of the specific connections. In addition, the member can receive end moments from its connection to other members of the structure, such as in a rigid frame. In all cases, the relation of the beam-column to the other elements of the structure is important in determining both the applied forces and the strength of the member.

To understand the behavior of beam-columns, it is common practice to look at the response predicted by an interaction equation. The response of a beam-column to an axial load P, major axis moment M_x, and minor axis moment M_y is presented on the three-dimensional diagram shown in Figure 8.1. Each axis in this diagram represents the capacity of the member when it is subjected to loading of one type only, whereas the curves represent the combination of two types of loading. The surface formed by connecting the three curves represents the interaction of axial load and biaxial bending. This interaction surface is of interest to the designer.

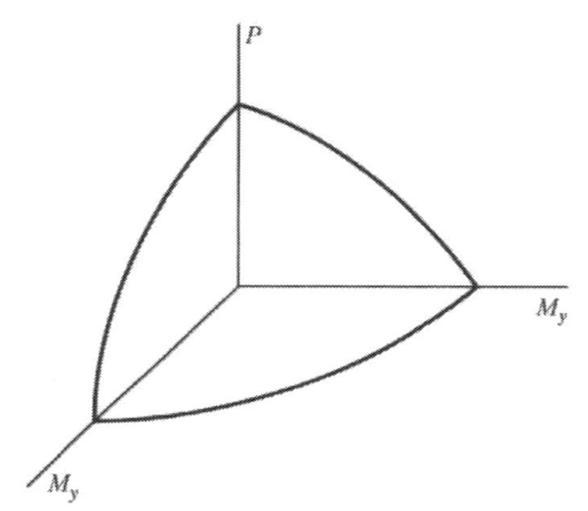

Figure 8.1 Ultimate Interaction Surface for a Stocky Beam-Column.

The end points of the curves shown in Figure 8.1 depend on the strength of the beam-columns as described for compression members (Chapter 5) and bending members (Chapter 6). The shape of the curves between these end points depends on the properties of the particular member as well as the properties of other members of the structure.

Table 8.1 lists the sections of the *Specification* and parts of the *Manual* discussed in this chapter.

8.2 SECOND-ORDER EFFECTS

The single most complicating factor in the analysis and design of a beam-column is what are known as *second-order effects*. Second-order effects are the changes in member forces and moments as the direct result of structural deformations. Because the commonly used elastic methods of structural analysis assume that all deformations are small, and because the equations of equilibrium are written using the undeformed configuration of the structure, these methods are not able to capture the additional second-order effects that occur in real structures without adjustment. The results of that type of analysis are called *first-order effects*—that is, first-order forces, first-order moments, and first-order displacements. To account for the influence of the deformations, an additional analysis must be performed. The results of this additional analysis are referred to as the second-order effects.

Table 8.1 Sections of *Specification* and Parts of *Manual* Covered in This Chapter

	Specification
B3	Design Basis
C	Design for Stability
H	Design of Members for Combined Forces and Torsion
Appendix 6	Stability Bracing for Columns and Beams
Appendix 7	Alternative Methods of Design for Stability
Appendix 8	Approximate Second-Order Analysis
	Manual
Part 1	Dimensions and Properties
Part 3	Design of Flexural Members
Part 4	Design of Compression Members
Part 6	Design of Members Subject to Combined Loading

Several approaches are available for including second-order effects in an analysis. A complete second-order inelastic analysis would take into account the actual deformation of the structure and the resulting forces, as well as the sequence of loading and the behavior of the structure after any of its components are stressed beyond the elastic limit. This approach to analysis is generally more complex than is necessary for normal design. A similar approach that includes the actual deformations but that does not include inelastic behavior is usually sufficient.

An approach that is consistent with normal design office practice and with how beam-columns have been handled for many years uses a first-order elastic analysis and amplification factors to approximate the second-order effects. This approach applies these amplification factors as multipliers to the results of the first-order analysis to obtain the second-order effects.

Two different deflection components that could occur in a beam-column influence the moments in that beam-column. The first, illustrated in Figure 8.2a, is the deflection along the length of the member that results from the moment along the member. In this case, the member ends must remain in their original position relative to each other; thus, no sway is considered. The moment created by the load, P, acting at an eccentricity δ_2 from the deformed member, is superimposed on the moment resulting from the applied end moments. Because the magnitude of this additional moment depends on the properties of the member itself, this is called the *member effect*.

When the beam-column is part of a structure that is permitted to sway, the displacements of the overall structure also influence the moments in the member. For a beam-column that is permitted to sway an amount Δ_2, as shown in Figure 8.2b, the additional moment is given by $P\Delta_2$. Because the lateral displacement of a given member is a function of the properties of all of the members in a given story, this is called the *structure effect*.

To understand the magnitude of the potential increase in moments on a column due to second-order effects, two simple calculations will be carried out. The first is for a 20 ft long column similar to that shown in Figure 8.2a. A W12×96 member is used to carry an axial load $P_u = 400$ kips and equal end moments of $M_u = 200$ ft-kips bending the member in single curvature. A first-order analysis yields an axial force in the column of

400 kips and a bending moment at every point along the column length of 200 ft-kips. The maximum deflection of the member at mid height due to the moment is

$$\delta = \frac{M_u L^2}{8EI} = \frac{200(20)^2(1728)}{8(29{,}000)(833)} = 0.715 \text{ in.}$$

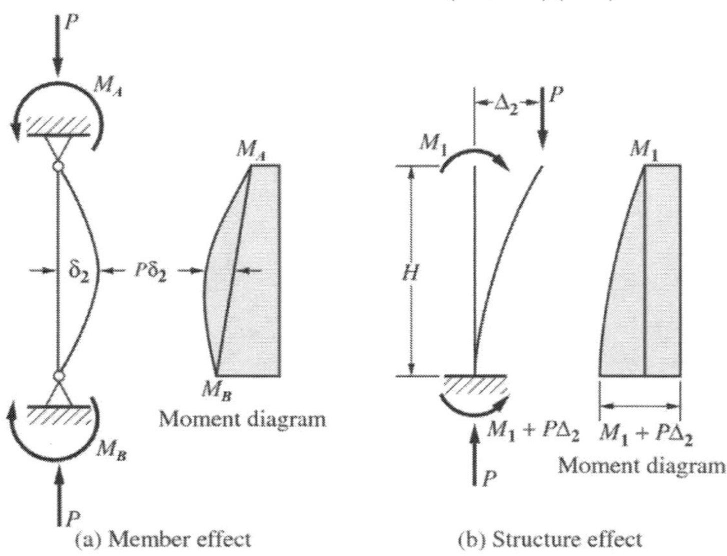

Figure 8.2 Displacements for Second-Order Effects

With the occurrence of this deflection, the applied load of 400 kips is now at an eccentricity from the member in its displaced position. Thus, an additional moment is induced into the member equal to

$$M_{\text{additional}} = \frac{400(0.715)}{12} = 23.8 \text{ ft-kips}$$

The addition of this additional moment to the original internal moment of 200 ft-kips yields the second-order moment,

$$M_{2nd} = 200 + 23.8 = 224 \text{ ft-kips}$$

Thus, there is an amplification of the moment by 224/200 = 1.12. If this were the final case, second-order analysis would be fairly simple. Unfortunately, the additional moment just determined also causes additional deflection, which, in turn, causes additional moment. This process continues until equilibrium is reached. The process is an iterative one, and is nonlinear.

A second example is a column similar to that shown in Figure 8.2b. The same W12×96 member is used, and the axial force is again P_u = 400 kips. In this case, the column is a cantilever with a moment of M_u = 200 ft-kips applied at the top. This moment will cause a horizontal deflection at the top of the column of

$$\Delta = \frac{M_u L^2}{2EI} = \frac{200(20)^2(1728)}{2(29{,}000)(833)} = 2.86 \text{ in.}$$

In this displaced position, the 400 kip load is now at an eccentricity from the fixed support, which induces an additional moment

$$M_{\text{additional}} = \frac{400(2.86)}{12} = 95.3 \text{ ft-kips}$$

The addition of this additional moment to the original support moment of 200 ft-kips yields the second-order moment

$$M_{2nd} = 200 + 95.3 = 295 \text{ ft-kips}$$

which is an increase of 1.48 times the first-order moment. Again, this is not the end of the required calculations; this additional moment causes additional deflections and additional moments.

Both of these second-order effects are significant in real structures and must be accounted for in the design of beam-columns according to Section C1 of the *Specification*. Procedures for incorporating these effects will be addressed once an overall approach to beam-column design is established.

8.3 INTERACTION PRINCIPLES

The interaction of axial load and bending within the elastic response range of a beam-column can be investigated through the straightforward techniques of *superposition*. This is the approach normally considered in elementary strength of materials in which the normal stress due to an axial force is added to the normal stress due to a bending moment.

Although the superposition of individual stress effects is both simple and correct for elastic stresses, there are significant limitations when applying this approach to the limit states of real structures. These include:

1. Superposition of stress is correct only for behavior within the elastic range, and only for similar stress types.
2. Superposition of strain can only be extended into the inelastic range when deformations are small.
3. Superposition cannot account for member deformations or stability effects such as local buckling.
4. Superposition cannot account for structural deflections and system stability.

With these limitations in mind, it is desirable to develop interaction equations that will reflect the true limit states behavior of beam-columns. Any limit state interaction equation must reflect the following characteristics:

Axial Load

1. Maximum column strength
2. Individual column slenderness

Bending Moment

1. Lateral support conditions
2. Sidesway conditions
3. Member second-order effects
4. Structure second-order effects
5. Moment variation along the member

The resulting equations must also provide a close correlation with test results and theoretical analyses for beam-columns, including the two limiting cases of pure bending and pure compression.

Application of the resulting interaction equations can be regarded as a process of determining available axial strength in the presence of a given bending moment or determining the available moment strength in the presence of a given axial load. An applied bending moment consumes a portion of the column strength, leaving a reduced axial load strength. When the two actions are added together, the resulting total load must not exceed the total column strength. Conversely, the axial load can be regarded as consuming a fraction of the moment strength. This fraction, plus the applied moments, must not exceed the maximum beam strength.

8.4 INTERACTION EQUATIONS

A simple form of the three-dimensional interaction equation is

$$\frac{P_r}{P_c} + \frac{M_{rx}}{M_{cx}} + \frac{M_{ry}}{M_{cy}} \leq 1.0 \tag{8.1}$$

where the terms with the subscript r represent the required strength and those with the subscript c represent the available strength.

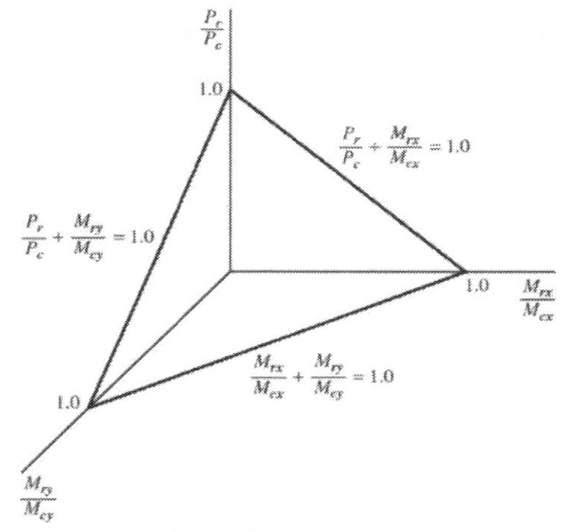

Figure 8.3 Simplified Interaction Surface

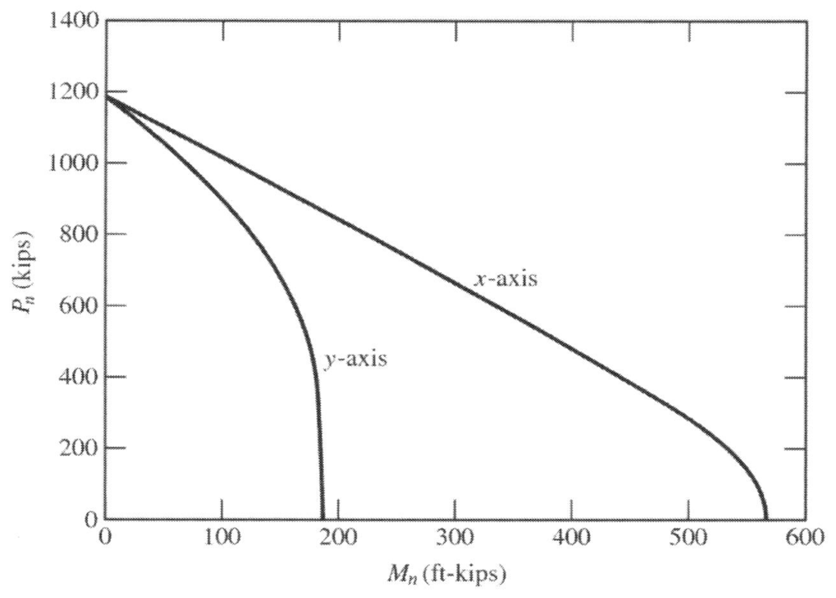

Figure 8.4a Interaction Diagram for Stub W14×82 Column.

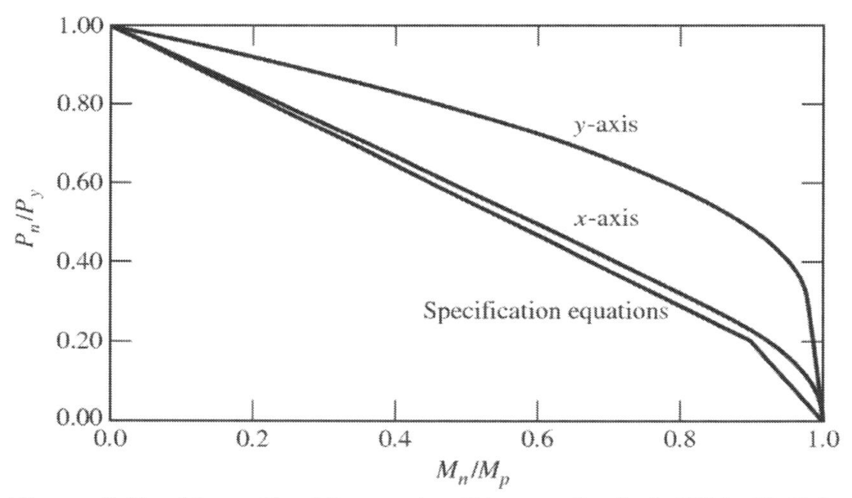

Figure 8.4b Normalized Interaction Diagram for Stub W14×82 Column.

This interaction equation is plotted in Figure 8.3. The figure shows that this results in a straight-line representation of the interaction between any two of the load components. The horizontal plane of Figure 8.3 represents the interaction of moments in the two principal axis directions without axial load, called biaxial bending, whereas the vertical planes along each axis represent the interaction of axial compression plus either major or minor axis bending. It should also be apparent that the three-dimensional aspect is represented by a plane surface with intercepts given by the straight lines on the three coordinate planes.

The interaction equations in Chapter H of the *Specification* result from fitting interaction equations that are similar to the form of Equation 8.1 to a set of data developed from an analysis of forces and moments for various plastic stress distributions on a stub column. Figure 8.4a shows the actual analysis results for a W14×82 stub column. Figure 8.4b shows the same data plotted as functions of the normalized axial strength, P_y, and flexural strength, M_p. In both cases, the influence of length on the axial or flexural strength is not included. Using curves of this type, developed for a wide variety of steel beam-column shapes, two equations were developed that are conservative and accurate for x-axis bending. When applied to y-axis bending, they are significantly more conservative; however, simplicity of design and the infrequent use of weak axis bending justify this extra level of conservatism.

An additional modification to these equations is required to account for length effects. Rather than normalizing the curves on the yield load and the plastic moment as was done in Figure 8.4b, the equations were developed around the nominal strength of the column and the nominal strength of the beam. The resulting equations are Equations H1-1a and H1-1b in the *Specification* and are plotted in Figure 8.5.

The equations shown here consider bending about both principal axes, whereas the plot in Figure 8.5 is for single-axis bending.

For $\dfrac{P_r}{P_c} \geq 0.2$,

$$\frac{P_r}{P_c} + \frac{8}{9}\left(\frac{M_{rx}}{M_{cx}} + \frac{M_{ry}}{M_{cy}}\right) \leq 1.0 \qquad \text{(AISC H1-1a)}$$

For $\dfrac{P_r}{P_c} < 0.2$,

$$\frac{P_r}{2P_c} + \left(\frac{M_{rx}}{M_{cx}} + \frac{M_{ry}}{M_{cy}}\right) \leq 1.0 \qquad \text{(AISC H1-1b)}$$

where

P_r = required compressive strength, kips
P_c = available compressive strength, kips
M_r = required flexural strength, ft-kips
M_c = available flexural strength, ft-kips
x = subscript relating symbol to strong axis bending
y = subscript relating symbol to weak axis bending

It is important to note that

1. The available column strength, P_c, is based on the axis of the column with the largest slenderness ratio. This is not necessarily the axis about which bending takes place.
2. The available bending strength, M_c, is based on the bending strength of the beam without axial load, including the influence of all the beam limit states.
3. The required compressive strength, P_r, is the force in the member, including second-order effects.

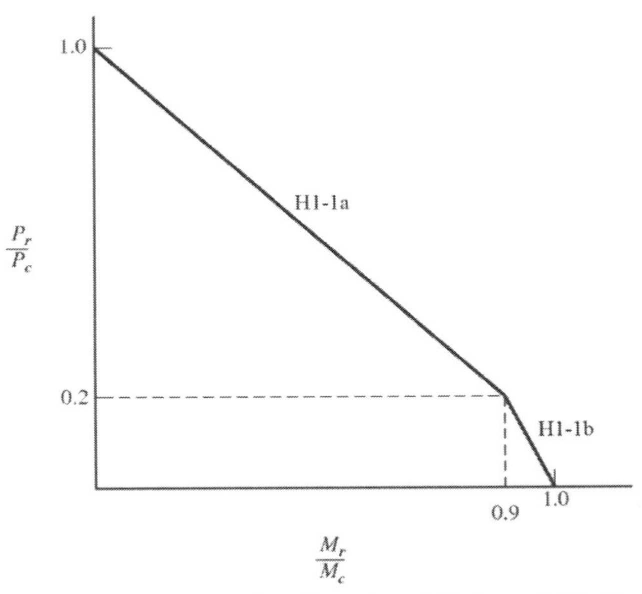

Figure 8.5 Interaction Equations H1-1a and H1-1b.

4. The required flexural strength, M_r, is the bending moment in the member, including second-order effects.

Second-order forces and moments can be determined through a second-order analysis or by a modification of the results of a first-order analysis using amplification factors as mentioned earlier. These amplification factors will be discussed as they relate to braced frames (Section 8.5) and moment frames (Section 8.6).

Additional provisions are available for cases where the axial strength limit state is out-of-plane buckling and the flexural strength limit state is lateral-torsional buckling for bending in plane. Equations H1-1a and b are conservative for this situation, but an additional approach is available. *Specification* Section H1.3 provides that (1) for the limit state of in-plane instability, Equations H1-1a and H1-1b should be used where the compressive strength is determined for buckling in the plane of bending and $M_{cx} = M_p$, and (2) for the limit states of out-of-plane buckling and lateral-torsional buckling

$$\frac{P_r}{P_{cy}}\left(1.5-0.5\frac{P_r}{P_{cy}}\right)+\left(\frac{M_{rx}}{C_b M_{cx}}\right)^2 \leq 1.0 \qquad \text{(AISC H1-3)}$$

where

P_{cy} = available compressive strength out of the plane of bending
M_{cx} = available lateral-torsional buckling strength for strong axis bending with $C_b = 1.0$
C_b = lateral-torsional buckling factor discussed in Chapter 6

If there is significant biaxial bending, meaning that the required-moment-to-available-moment ratio for y-axis bending is greater than or equal to 0.05, then this option is not available. Although this optional approach can provide a more economical solution in some cases, it is not used in the examples or problems in this book.

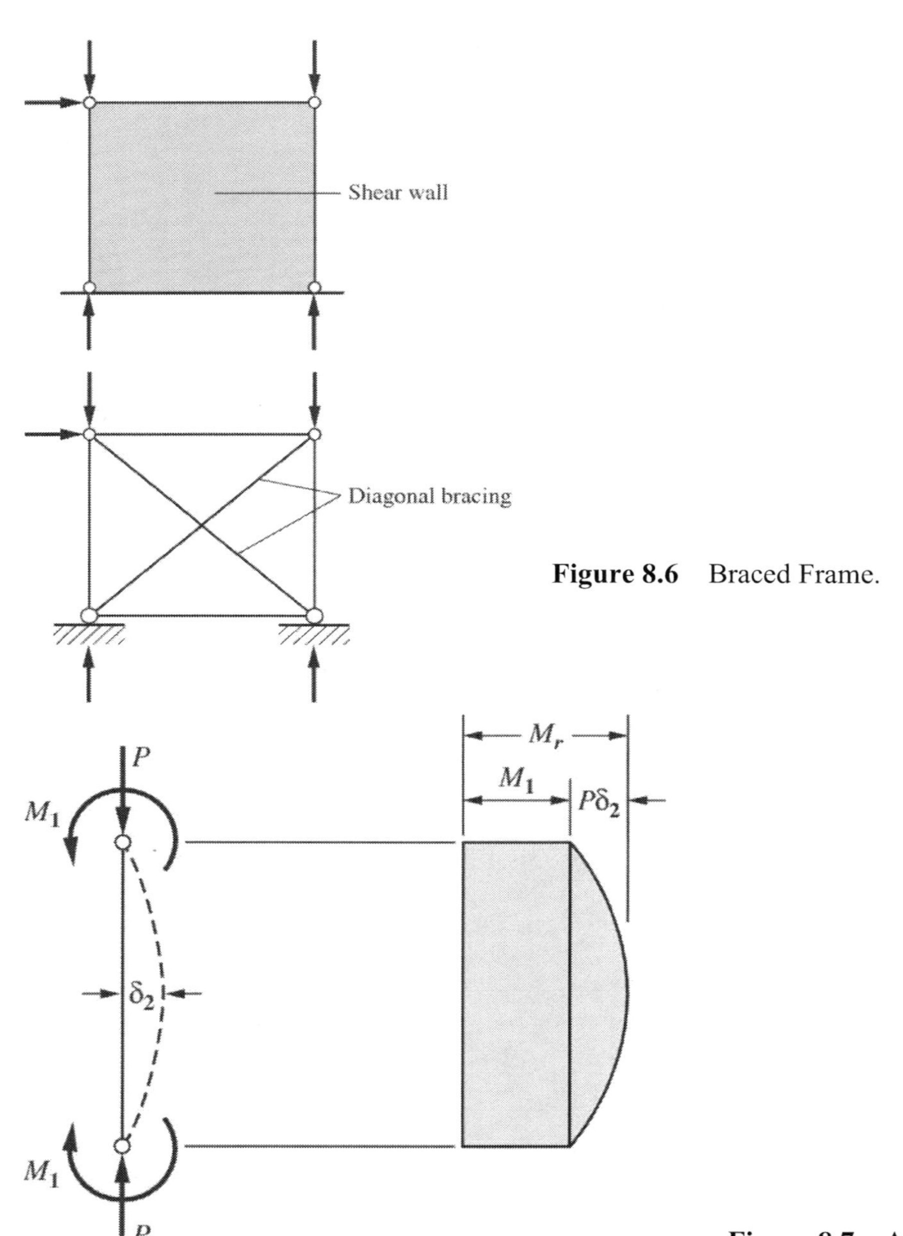

Figure 8.6 Braced Frame.

Figure 8.7 An Axially Loaded Column with Equal and Opposite End Moments.

8.5 BRACED FRAMES

A frame is considered to be braced if a positive system—that is, an actual system such as a shear wall (masonry, concrete, steel, or other material) or diagonal steel member—as illustrated in Figure 8.6, serves to resist the lateral loads, stabilize the frame under gravity loads, and resist lateral displacements. In these cases, columns are considered braced against lateral translation and the in-plane K-factor can be taken as 1.0, according to Appendix Section 7.2.3(a), unless a rational analysis indicates that a lower value is appropriate. This is the type of column that was discussed in Chapter 6. Later in this chapter the requirements for bracing to ensure that a structure can be considered a braced frame, as found in Appendix 6, are discussed.

If the column in a braced frame is rigidly connected to a girder, bending moments result from the application of the gravity loads to the girder. These moments can be determined through a first-order elastic analysis. The additional second-order moments resulting from the displacement along the column length can be determined through the application of an amplification factor.

The full derivation of the amplification factor has been presented by various authors.[1,2] Although this derivation is quite complex, a somewhat simplified derivation is presented here to help establish the background. An axially loaded column with equal and opposite end moments is shown in Figure 8.7a. This is the same column that was discussed in Section 8.2. The resulting moment diagram is shown in Figure 8.7b where the moments from both the end moments and the secondary effects are given.

The maximum moment occurring at the mid-height of the column, M_r, is shown to be

$$M_r = M_1 + P\delta_2$$

The amplification factor is defined as

$$AF = \frac{M_r}{M_1} = \frac{M_1 + P\delta_2}{M_1}$$

Rearranging terms yields

$$AF = \frac{1}{1 - \dfrac{P\delta_2}{M_1 + P\delta_2}}$$

Two simplifying assumptions will be made. The first is based on the assumption that δ is sufficiently small that

$$\frac{\delta_2}{M_1 + P\delta_2} \approx \frac{\delta_1}{M_1}$$

---

[1]Galambos, T. V., *Structural Members and Frames*. Englewood Cliffs, NJ: Prentice Hall, Inc., 1968.
[2]Johnson, B. G., Ed., *Guide to Stability Design Criteria for Metal Structures*, 3rd ed., SSRC, New York: Wiley, 1976.

and the second, using the beam deflection, $\delta_1 = M_1 L^2/8EI$, assumes that

$$\frac{M_1}{\delta_1} = \frac{8EI}{L^2} \approx \frac{\pi^2 EI}{L^2} = P_e$$

Because these simplifying assumptions are in error in opposite directions, they tend to be offsetting. This results in a fairly accurate prediction of the amplification. Thus,

$$AF = \frac{1}{1 - P/P_e} \tag{8.2}$$

A comparison between the actual amplification and that given by Equation 8.2 is shown in Figure 8.8.

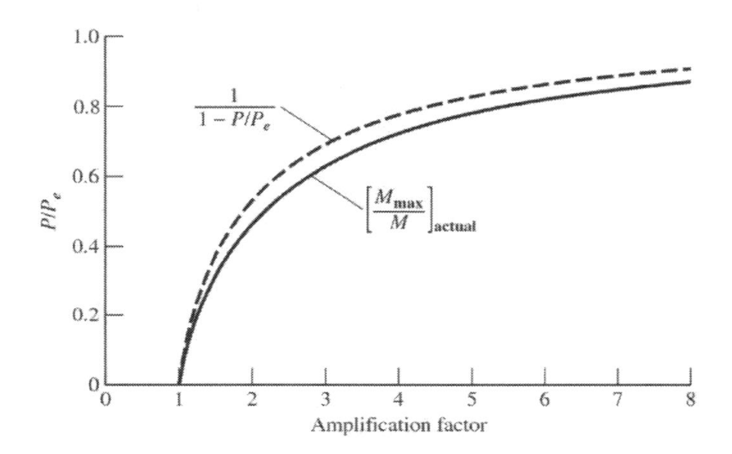

Figure 8.8 Amplified Moment: Exact and Approximate.

The discussion so far has assumed that the moments at each end of the column are equal and opposite, and that the resulting moment diagram is uniform. This is the most severe loading case for a beam-column braced against translation. If the moment is not uniformly distributed, the displacement along the member is less than previously considered and the resulting amplified moment is less than indicated. It has been customary in design practice to use the case of uniform moment as a base and to provide for other moment distributions by converting them to an equivalent uniform moment through the use of an additional factor, C_m.

Numerous studies have shown that a reasonably accurate correction results for beam-columns braced against translation and not subject to transverse loading between their supports, if the moment is reduced through its multiplication by C_m, where

$$C_m = 0.6 - 0.4(M_1/M_2) \qquad \text{(AISC A-8-4)}$$

M_1/M_2 is the ratio of the smaller to larger moments at the ends of the member unbraced length in the plane of bending. M_1/M_2 is positive when the member is bent in reverse curvature and negative when bent in single curvature.

For beam-columns in braced frames where the member is subjected to transverse loading between supports, C_m may be taken from Commentary Table C-A-8.1, or conservatively taken as 1.0.

The combination of the amplification factor, AF, and the equivalent moment factor, C_m, accounts for the total member secondary effects. This combined factor is given as B_1 in Appendix 8 of the *Specification* as

$$B_1 = \frac{C_m}{1 - \frac{\alpha P_r}{P_{e1}}} \geq 1.0 \qquad \text{(AISC A-8-3)}$$

where

α = 1.6 for ASD and 1.0 for LRFD to account for the nonlinear behavior of the structure at its ultimate strength

P_r = required strength, which may be taken as the first-order required strength, $P_{nt} + P_{lt}$, when used in moment frames

P_{e1} = Euler buckling load for the column in the plane of bending with an effective length factor, $K = 1.0$

Thus, the value of M_r in Equations H1-1a and H1-1b is taken as

$$M_r = B_1 M_{nt}$$

where M_{nt} is the maximum moment on the beam-column. The subscript nt indicates that for this case, the column does not undergo any lateral translation of its ends. It is possible for C_m to be less than 1.0 and for Equation A-8-3 to give an amplification factor less than 1.0. This indicates that the combination of the $P\delta$ effects and the nonuniform moment results in a moment less than the maximum moment on the beam-column from a first-order analysis. In this case, the amplification factor $B_1 = 1.0$.

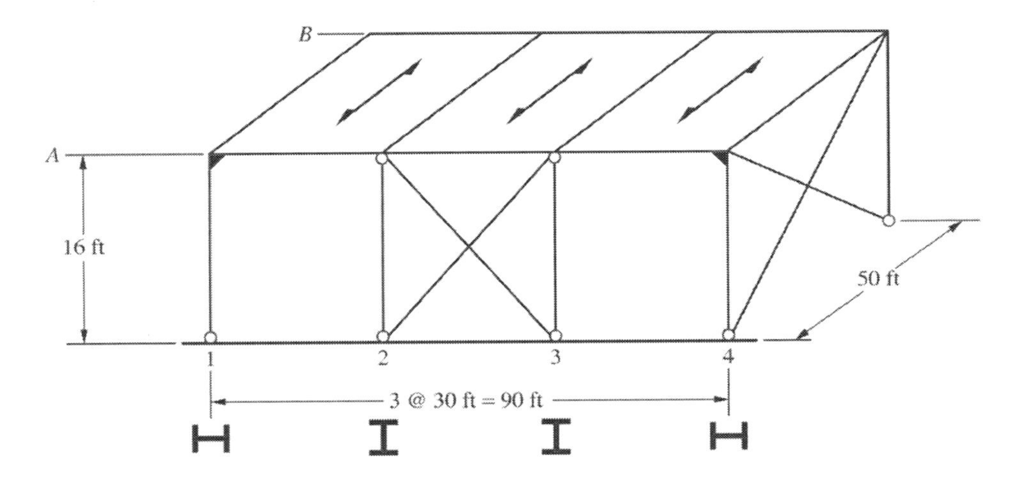

Figure 8.9a Three-Dimensional Braced Frame for a Single-Story Structure.

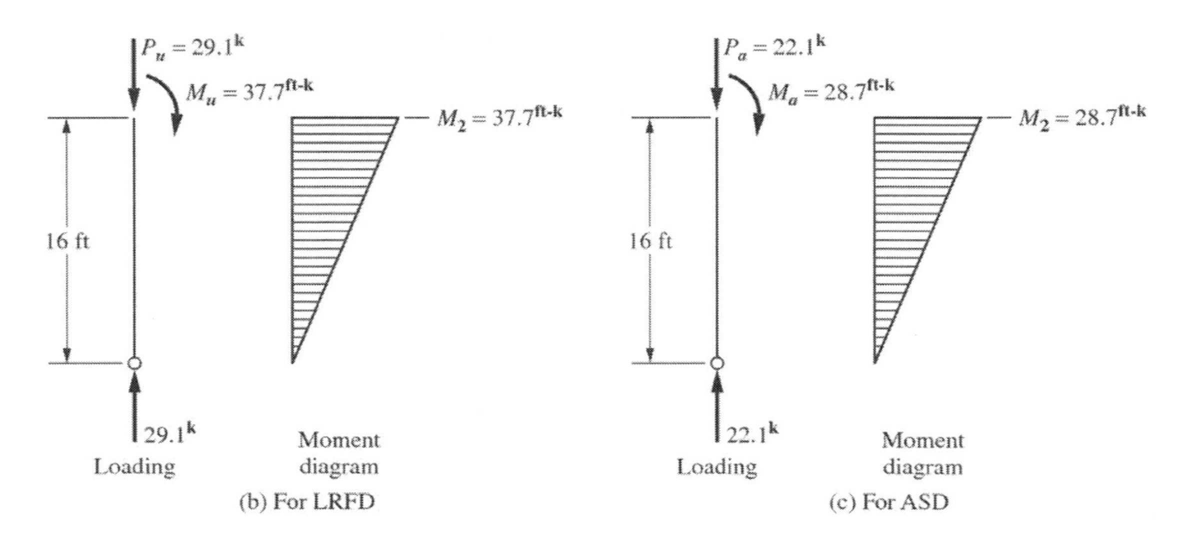

Figure 8.9b, c Three-Dimensional Braced Frame for a Single-Story Structure.

EXAMPLE 8.1a ***Braced Frame Column Design for Combined Axial and Bending by LRFD***

Goal: Design column A1 in Figure 8.9 for the given loads using the LRFD provisions and the second-order amplification factor provided in Appendix 8 of the *Specification*.

Given: The three-dimensional braced frame for a single-story structure is given in Figure 8.9. Rigid connections are provided at the roof level for columns A1, B1, A4, and B4. All other column connections are pinned. Dead Load = 50 psf, Snow Load = 20 psf, Roof Live Load = 10 psf, and Wind Load = 20 psf horizontal. Use A992 steel. Assume that the X-bracing is sufficiently stiffer than the rigid frames to resist all lateral load.

SOLUTION

Step 1: Determine the appropriate load combinations. From ASCE 7, Section 2.3, the following two combinations are considered.

ASCE 7 load combination 3

$$1.2D + 1.6(L_r \text{ or } S \text{ or } R) + (0.5L \text{ or } 0.5W)$$

ASCE 7 load combination 4

$$1.2D + 1.0W + 0.5L + 0.5(L_r \text{ or } S \text{ or } R)$$

Step 2: Determine the factored roof gravity loads for each load combination. For load combination 3

$$1.2(50) + 1.6(20) = 92 \text{ psf}$$

and for load combination 4

$$1.2(50) + 0.5(20) = 70 \text{ psf}$$

Because column A1 does not participate in the lateral load resistance, the worst case loading will use the uniformly distributed roof load of 92 psf.

Step 3: Carry out a preliminary first-order analysis. Because the structure is indeterminate, a number of approaches can be taken. If an arbitrary 6:1 ratio of moment of inertia for beams to columns is assumed, a moment distribution analysis yields the moment and force given in Figure 8.9b. Thus, the column will be designed to carry
$P_u = 29.1$ kips and $M_u = 37.7$ ft-kips

Step 4: Select a trial size for column A1 and determine its compressive strength and bending strength.

Try W10×33. (Section 8.8 addresses trial section selection.)

From *Manual* Table 1-1

$$A = 9.71 \text{ in.}^2,\ r_x = 4.19 \text{ in.},\ r_y = 1.94 \text{ in.},\ I_x = 171 \text{ in.}^4,\ r_x/r_y = 2.16$$

The column is oriented so that bending is about the x-axis of the column. It is braced against sidesway by the diagonal braces in panel A2–A3 and is pinned at the bottom and rigidly connected at the top in the plane of bending. The column is also braced out of the plane of bending by the brace in panel A1–B1. Because this column is part of a braced frame, $K = 1.0$ can be used. Although the *Specification* permits the use of a lower K-factor if justified by analysis, this is not recommended because it would likely require significantly more stiffness in the braced panel.

From *Manual* Table 4-1a, for y-axis buckling

$$\phi P_n = 214 \text{ kips for } L_c = 16.0 \text{ ft}$$

From *Manual* Table 3-10

$$\phi M_n = 113 \text{ ft-kips for } L_b = 16.0 \text{ ft}$$

Step 5: Check the W10×33 for combined axial load and bending in-plane.

For an unbraced length of 16 ft, the Euler load is

$$P_{e1} = \frac{\pi^2 EI}{L_{c1}^2} = \frac{\pi^2 (29{,}000)(171)}{(16.0(12))^2} = 1330 \text{ kips}$$

The column is bent in single curvature between bracing points, the end points, and the moment at the base is zero, so $M_1/M_2 = 0.0$. Thus

$$C_m = 0.6 - 0.4(0.0) = 0.6$$

Therefore, the amplification factor, with $\alpha = 1.0$, becomes

$$B_1 = \frac{C_m}{1 - \frac{\alpha P_r}{P_{e1}}} = \frac{0.6}{1 - \frac{1.0(29.1)}{1330}} = 0.613 \leq 1.0$$

The *Specification* requires that B_1 not be less than 1.0. Therefore, taking $B_1 = 1.0$,

$$M_{rx} = B_1(M_x) = 1.0(37.7) = 37.7 \text{ ft-kips}$$

To determine which equation to use, calculate

$$\frac{P_u}{\phi P_n} = \frac{29.1}{214} = 0.136 < 0.2$$

Therefore, use Equation H1-1b

$$\frac{P_u}{2\phi P_n} + \frac{M_u}{\phi M_n} \leq 1.0$$

$$0.5(0.136) + \frac{37.7}{113} = 0.402 < 1.0$$

Thus, the W10×33 will easily carry the given loads.

The solution to Equation H1-1b indicates that there is a fairly wide extra margin of safety. It would be appropriate to consider a smaller column for a more economical design.

EXAMPLE 8.1b
Braced Frame Column Design for Combined Axial and Bending by ASD

Goal: Design column A1 in Figure 8.9 for the given loads using the ASD provisions and the second-order amplification factor provided in Appendix 8 of the *Specification.*

Given: The three-dimensional braced frame for a single-story structure is given in Figure 8.9. Rigid connections are provided at the roof level for columns A1, B1, A4, and B4. All other column connections are pinned. Dead Load = 50 psf, Snow Load = 20 psf, Roof Live Load = 10 psf, and Wind Load = 20 psf horizontal. Use A992 steel. Assume that the X-bracing is sufficiently stiffer than the rigid frames to resist all lateral load.

SOLUTION

Step 1: Determine the appropriate load combinations. From ASCE 7, Section 2.4, the following two combinations are considered.

ASCE 7 load combination 3

$$D + (L_r \text{ or } S \text{ or } R)$$

ASCE 7 load combination 6

$$D + 0.75(0.6W) + 0.75(L_r \text{ or } S \text{ or } R)$$

Step 2: Determine the factored roof gravity loads for each load combination. For load combination 3

$$50 + 20 = 70 \text{ psf}$$

and for load combination 6

$$50 + 0.75(20) = 65 \text{ psf}$$

Because column A1 does not participate in the lateral load resistance, the worst case loading will use the uniformly distributed roof load of 70 psf.

Step 3: Carry out a preliminary first-order analysis. Because the structure is indeterminate, a number of approaches can be taken. If an arbitrary 6:1 ratio of moment of inertia for beams to columns is assumed, a moment distribution analysis yields the moment and force given in Figure 8.9c. Thus, the column will be designed to carry

$$P_a = 22.1 \text{ kips and } M_a = 28.7 \text{ ft-kips}$$

Step 4: Select a trial size for column A1 and determine its compressive strength and bending strength.

Try W10×33. (Section 8.8 addresses trial section selection.)

From *Manual* Table 1-1

$$A = 9.71 \text{ in.}^2,\ r_x = 4.19 \text{ in.},\ r_y = 1.94 \text{ in.},\ I_x = 171 \text{ in.}^4,\ r_x/r_y = 2.16$$

The column is oriented so that bending is about the x-axis of the column. It is braced against sidesway by the diagonal braces in panel A2–A3 and is pinned at the bottom and rigidly connected at the top in the plane of bending. The column is also braced out of the plane of bending by the brace in panel A1–B1. Because this column is part of a braced frame, $K = 1.0$ can be used. Although the *Specification* permits the use of a lower K-factor if justified by analysis, this is not recommended because it would likely require significantly more stiffness in the braced panel.

From *Manual* Table 4-1a for y-axis buckling

$$P_n/\Omega = 142 \text{ kips for } L_c = 16.0 \text{ ft}$$

From *Manual* Table 3-10

$$M_n/\Omega = 74.9 \text{ ft-kips for } L_b = 16.0 \text{ ft}$$

Step 5: Check the W10×33 for combined axial load and bending in-plane.

For an unbraced length of 16 ft, the Euler load is

$$P_{e1} = \frac{\pi^2 EI}{L_{c1}^2} = \frac{\pi^2(29{,}000)(171)}{(16.0(12))^2} = 1330 \text{ kips}$$

The column is bent in single curvature between bracing points, the end points, and the moment at the base is zero, so $M_1/M_2 = 0.0$. Thus

$$C_m = 0.6 - 0.4(0.0) = 0.6$$

Therefore, the amplification factor, with $\alpha = 1.6$. becomes

$$B_1 = \frac{C_m}{1 - \frac{\alpha P_r}{P_{e1}}} = \frac{0.6}{1 - \frac{1.6(22.1)}{1330}} = 0.616 \le 1.0$$

The *Specification* requires that B_1 not be less than 1.0. Therefore, taking $B_1 = 1.0$,

$$M_{rx} = B_1(M_x) = 1.0(28.7) = 28.7 \text{ ft-kips}$$

To determine which equation to use, calculate

$$\frac{P_u}{P_n/\Omega} = \frac{22.1}{142} = 0.156 < 0.2$$

Therefore, use Equation H1-1b

$$\frac{P_a}{2P_n/\Omega} + \frac{M_a}{M_n/\Omega} \le 1.0$$

$$0.5(0.156) + \frac{28.7}{74.9} = 0.461 < 1.0$$

Thus, the W10×33 will easily carry the given loads.

The solution to Equation H1-1b indicates that there is a fairly wide extra margin of safety. It would be appropriate to consider a smaller column for a more economical design.

8.6 Moment Frames

A moment frame depends on the stiffness of the beams and columns that make up the frame for stability under gravity loads and under combined gravity and lateral loads. Unlike braced frames, there is no external structure to lean against for stability. Columns in moment frames are subjected to both axial load and moment and experience lateral translation.

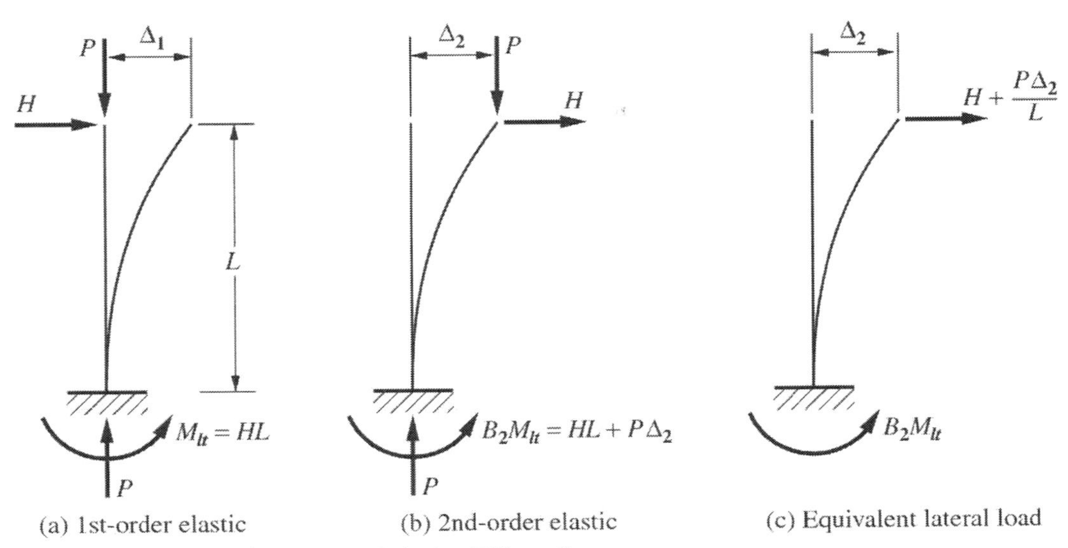

Figure 8.10 Structure Second-Order Effect: Sway.

The same interaction equations, Equations H1-1a and H1-1b, are used to design beam-columns in moment frames as were previously used for braced frames. However, in addition to the member second-order effects discussed in Section 8.5, there is the additional second-order effect that results from the sway or lateral displacement of the frame.

Figure 8.10 shows a cantilever or flag pole column under the action of an axial load and a lateral load. Figure 8.10a is the column as viewed for a first-order elastic analysis where equilibrium requires a moment at the bottom, $M_{lt} = HL$. The deflection that results at the top of the column, Δ_1, is the elastic deflection of a cantilever, so

$$\Delta_1 = \frac{HL^3}{3EI} \tag{8.3}$$

A second-order analysis yields the forces and displacements as shown in Figure 8.10b. The displacement, Δ_2, is the total displacement, including second-order effects, and the moment, including second-order effects, is

$$B_2 M_{lt} = HL + P\Delta_2 \tag{8.4}$$

An equivalent lateral load can be determined that results in the same moment at the bottom of the column as in the second-order analysis. This load is $H + P\Delta_2/L$ and is shown in Figure 8.10c.

It may be assumed, with only slight error, that the displacements at the top of the column for the cases in Figures 8.10b and c are the same. Thus, using the equivalent lateral load

$$\Delta_2 = \frac{(H + P\Delta_2/L)L^3}{3EI} = \frac{HL^3}{3EI}\left(1 + \frac{P\Delta_2}{HL}\right) = \Delta_1\left(1 + \frac{P\Delta_2}{HL}\right) \tag{8.5}$$

Equation 8.5 can now be solved for Δ_2, where

$$\Delta_2 = \frac{\Delta_1}{1 - \dfrac{P\Delta_1}{HL}} \tag{8.6}$$

and the result substituted into Equation 8.4. Solving the resulting equation for the amplification factor, B_2, and simplifying yields

$$B_2 = \frac{\Delta_2}{\Delta_1} = \frac{1}{1 - \dfrac{P\Delta_1}{HL}} \tag{8.7}$$

Considering that the typical beam-column will be part of some larger structure, this equation must be modified to include the effect of the multistory and multibay characteristics of the actual structure. This is easily accomplished by summing the total gravity load on the columns in the story and the total lateral load in the story. Thus, Equation 8.7 becomes

$$B_2 = \frac{1}{1 - \dfrac{\Sigma P\Delta_1}{\Sigma HL}} \tag{8.8}$$

This amplification factor is essentially that given in Appendix 8 of the *Specification* as Equation A-8-6, when combined with Equation A-8-7

$$B_2 = \frac{1}{1 - \dfrac{\alpha P_{story}}{P_{e\,story}}} = \frac{1}{1 - \dfrac{\alpha P_{story}\Delta_H}{R_M HL}} \geq 1.0 \tag{AISC A-8-6}$$

where

P_{story} = total gravity load on the story

$P_{e\,story}$ = measure of lateral strength of the structure $= R_M \dfrac{HL}{\Delta_H}$ (AISC A-8-7)

Δ_H = story drift from a first-order analysis due to the lateral load, H

α = 1.0 for LRFD and 1.6 for ASD to account for the nonlinear behavior of the structure at its ultimate strength

$$R_M = 1 - 0.15\left(\frac{P_{mf}}{P_{story}}\right) \tag{AISC A-8-8}$$

P_{mf} = the total vertical load in columns that are part of the lateral load resisting system

The variable R_M accounts for the influence of the member effect on the sidesway displacement that could not be accounted for in the simplified derivation above. If all the columns are moment frame columns, P_{mf}/P_{story} = 1.0 and R_M = 0.85. For braced frames, P_{mf} = 0 and R_M = 1.0. For frames with a combination of columns resisting lateral load through bending and gravity only or *leaning columns*, the value of R_M will be between these limits.

It is often desirable to limit the lateral displacement, or drift, of a structure during the design phase. ASCE 7 Appendix C Commentary provides some general guidance. This limit can be defined using a drift index, which is the story drift divided by the story height, Δ_H/L. The design then proceeds by selecting members so that the final structure performs as desired. This is similar to beam design, where deflection is the serviceability criterion. Because a limit on the drift index can be established without knowing member sizes, it can be used in Equation A-8-6; thus an analysis with assumed member sizes is unnecessary.

With this amplification for sidesway, the moment, M_r, to be used in Equations H1-1a and H1-1b, can be evaluated. M_r must include both the member and structure second-order effects. Thus, a first-order analysis without sidesway is carried out, yielding moments, M_{nt}, that is without translation, to be amplified by B_1. Next, a first-order analysis including lateral loads and permitting translation is carried out. This yields moments, M_{lt}, with translation, to be amplified by B_2. The resulting second-order moment is

$$M_r = B_1 M_{nt} + B_2 M_{lt} \qquad \text{(AISC A8-1)}$$

where

B_1 is given by Equation A-8-3

B_2 is given by Equation A-8-6

M_{nt} = first-order moments when the structure is not permitted to translate laterally

M_{lt} = first-order moments that result from just the lateral translation

M_{lt} could include moments that result from unsymmetrical frame properties or loading as well as from lateral loads. In most real structures, however, moments resulting from this lack of symmetry are usually small and are thus often ignored.

The second-order force is

$$P_r = P_{nt} + B_2 P_{lt} \qquad \text{(AISC A-8-2)}$$

The sum of P_{nt} and P_{lt} for the entire structure will equal the total gravity load on the structure, since the sum of P_{lt} will be zero. For the individual column, however, it is important to amplify the portion of the individual column force that comes from the lateral load.

For situations where there is no lateral load on the structure, it may be necessary to incorporate a minimum lateral load in order to capture the second-order effects of the gravity loads. This is covered in Section 8.7 where the three methods provided in the *Specification* for treating stability analysis and design are discussed.

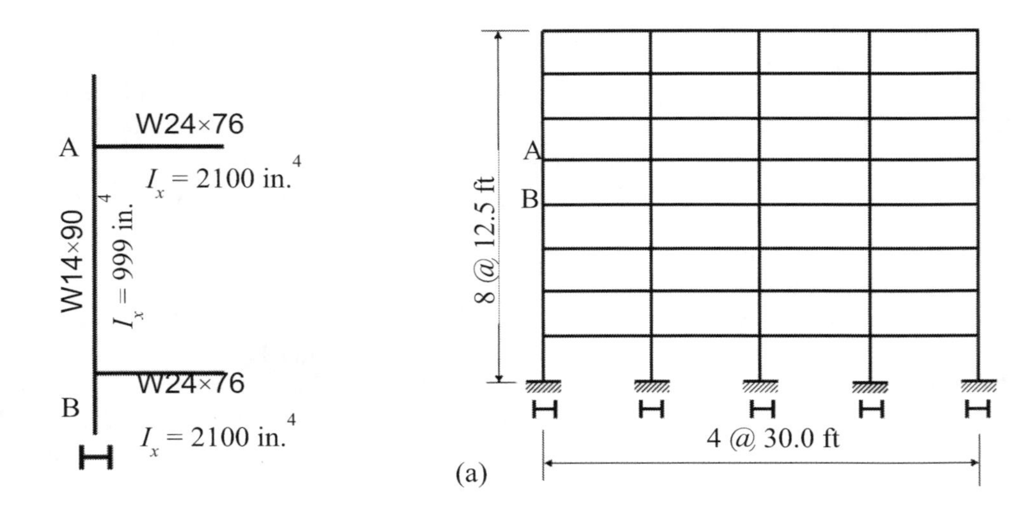

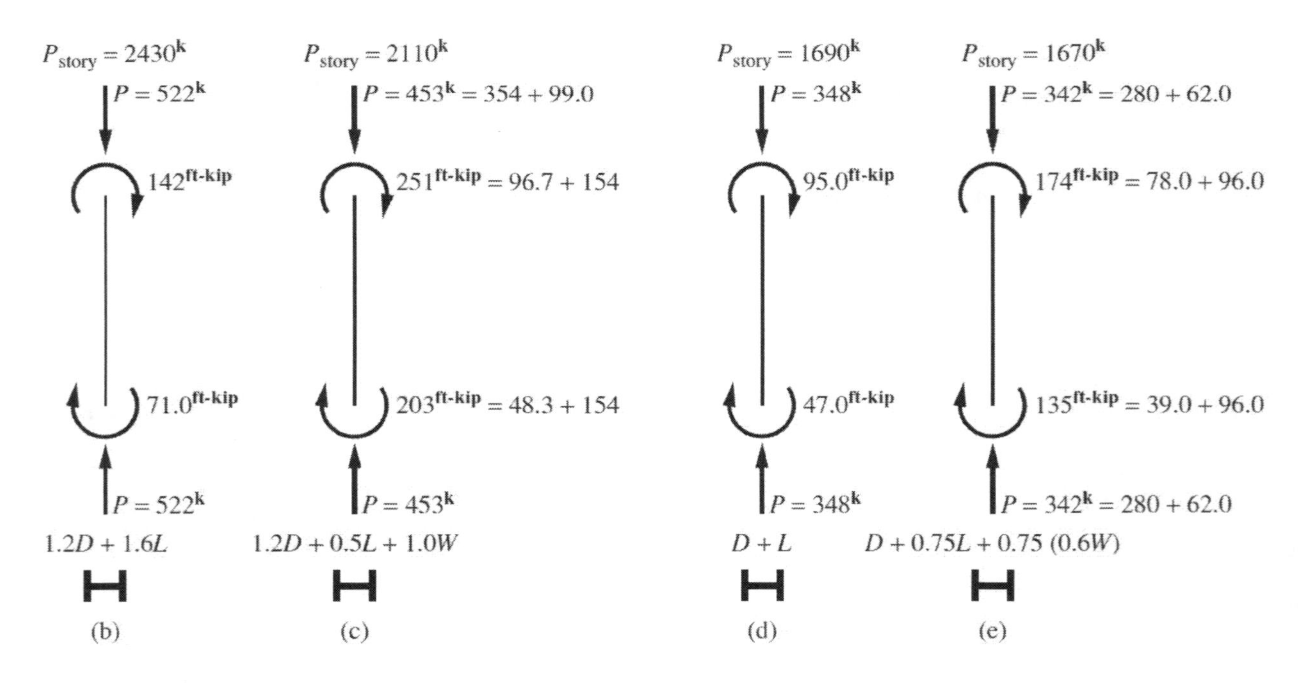

Figure 8.11 Exterior Column from an Intermediate Level of a Multistory Rigid Frame (Example 8.2).

EXAMPLE 8.2a
Moment Frame Strength Check for Combined Compression and Bending by LRFD

Goal: Using the LRFD provisions, determine whether the W14×90, A992 column shown in Figure 8.11 is adequate to carry the imposed loading.

Given: An exterior column from an intermediate level of a multi-story moment frame is shown in Figure 8.11. The column is part of a braced frame out of the plane of the figure. Figure 8.11a shows the elevation of the frame with the member to be checked labeled AB. The same column section will be used for the level above and below the column AB. A first-order analysis of the frame for gravity loads plus the minimum lateral load (the minimum lateral load will be discussed in Section 8.7) results in the forces shown in Figure 8.11b, whereas the results for gravity plus wind are shown in Figure 8.11c. Assume that the frame drift under service loads is limited to height/300 for a story shear, $H = 148$ kips.

SOLUTION

Step 1: Determine the column effective length factor in the plane of bending.

Using the effective length alignment chart introduced in Chapter 5 and given in Commentary Figure C-A-7.2, determine the effective length for buckling in the plane of the moment frame. At each joint there are two columns and one beam framing in. Thus,

$$G_A = G_B = \frac{\Sigma(I/L)_c}{\Sigma(I/L)_g} = \frac{2\left(\frac{999}{12.5}\right)}{\left(\frac{2100}{30.0}\right)} = 2.28$$

Thus, from Figure 5.21, $K = 1.66$.

Step 2: Determine the controlling effective length.

With $r_x/r_y = 1.66$ for the W14 × 90,

$$(L_{cx})_{eff} = (KL)_{eff} = \frac{(KL)_x}{r_x/r_y} = \frac{1.66(12.5)}{1.66} = 12.5 \text{ ft}$$

$$L_{cy} = KL_y = 1.0(12.5) = 12.5 \text{ ft}$$

Step 3: Since the effective length about each axis is 12.5 ft, determine the column design axial strength using $L_c = 12.5$.

From the column tables, *Manual* Table 4-1a, by interpolation for $L_c =$ 12.5 ft, $\phi P_n = 1060$ kips

Step 4: Determine the first-order moments and forces for the loading combination that includes wind, 1.2D +0.5L + 1.0W.

The column end moments given in Figure 8.11c are a combination of moments resulting from a nonsway gravity load analysis and a wind analysis:

Moment for end A:

$$M_{nt} = 96.7 \text{ ft-kips}$$
$$M_{lt} = 154 \text{ ft-kips}$$

Moment for end B:

$$M_{nt} = 48.3 \text{ ft-kips}$$
$$M_{lt} = 154 \text{ ft-kips}$$

Compression:

$$P_{nt} = 354 \text{ kips}$$
$$P_{lt} = 99.0 \text{ kips}$$

Step 5: Determine the second-order moments by amplifying the first-order moments.

No-translation amplification: The no-translation moments must be amplified by B_1. From Figure 8.11c it is seen that the end moments bend the column in reverse curvature:

$$\frac{M_1}{M_2} = \frac{48.3}{96.7} = 0.50$$

$$C_m = 0.6 - 0.4\left(M_1/M_2\right) = 0.6 - 0.4(0.50) = 0.4$$

$$P_{e1} = \frac{\pi^2 E}{L_{c1}^2} = \frac{\pi^2(29{,}000)(999)}{(1.0(12.5)(12))^2} = 12{,}700 \text{ kips}$$

Thus, with $\alpha = 1.0$ for LRFD and $P_r = 354 + 99 = 453$ kips, Equation A-8-3 yields

$$B_1 = \frac{C_m}{1 - \frac{\alpha P_r}{P_{e1}}} = \frac{0.4}{1 - \frac{(1.0)(453)}{12{,}700}} = 0.415 < 1.0$$

Therefore, $B_1 = 1.0$.

Translation amplification: The translation forces and moments must be amplified by B_2. The design drift limit of height/300 and Equation A-8-6 are used to determine B_2.

The total service lateral load on this story is given as

$$H = 148 \text{ kips}$$

Additional given information is that the total gravity load for this load combination in Figure 8.11c is

$$P_{story} = 2110 \text{ kips}$$

The drift limit under the service lateral load of 148 kips is

$$\Delta_H = L/300 = 12.5(12)/300 = 0.50 \text{ in.}$$

Remember that in the calculation of B_2, H can be taken as any convenient magnitude, as long as Δ_H is the corresponding displacement. This is because it is the ratio of H to Δ_H that is used in the determination of P_e story.

Thus, with $\alpha = 1.0$ for LRFD and $R_M = 0.85$ assuming all columns are moment frame columns, Equation A-8-7 gives

$$P_{e\ story} = \frac{R_M HL}{\Delta_H} = \frac{0.85(148)(12.5)(12)}{0.50} = 37{,}700 \text{ kips}$$

and Equation A-8-6 gives

$$B_2 = \frac{1}{1 - \dfrac{\alpha P_{story}}{P_{e\ story}}} = \frac{1}{1 - \left(\dfrac{(1.0)2110}{37{,}700}\right)} = 1.06 > 1.0$$

Thus, the second-order compressive force and moment are

$$P_r = P_{nt} + B_2(P_{lt}) = 354 + 1.06(99) = 459 \text{ kips}$$

$$M_r = B_1(M_{nt}) + B_2(M_{lt}) = 1.0(96.7) + 1.06(154) = 260 \text{ ft-kips}$$

These represent the required strength for this load combination.

Step 6: Determine whether the W14×90 will provide the required strength based on the appropriate interaction equation.

The unbraced length of the compression flange for pure bending is 12.5 ft, which is less than L_p = 15.1 ft for this section, taking into account that its flange is noncompact. Thus, from *Manual* Table 3-2, the design moment strength of the section is

$$\phi M_n = 574 \text{ ft-kips}$$

Determine the appropriate interaction equation. From Step 3, ϕP_n = 1060 kips;

$$\frac{P_u}{\phi P_n} = \frac{459}{1060} = 0.433 > 0.2$$

so use Equation H1-1a, which yields

$$\frac{P_u}{\phi P_n}+\frac{8}{9}\left(\frac{M_u}{\phi M_n}\right)\leq 1.0$$

$$0.433+\frac{8}{9}\left(\frac{260}{574}\right)=0.836<1.0$$

Thus,

the W14×90 is adequate for this load combination.

Step 7: Check the section for the gravity-only load combination, $1.2D + 1.6L$.

Because this is a gravity-only load combination, *Specification* Appendix Section 7.2.2, by reference to Section C2.2b, requires that the analysis include a minimum lateral load of 0.002 times the gravity load. This will be further discussed in Section 8.7. For this load combination, the total story gravity load must also be known and is given in Figure 8.11b as P_{story} = 2430 kips. Thus, for this frame the minimum lateral load is $0.002P_{story} = 0.002(2430) = 4.86$ kips at this level.

The forces and moments given in Figure 8.11b include the effects of this minimum lateral load. The magnitude of the lateral translation effect is small in this case. Since both the moment due to the minimum lateral load and the amplification factor, B_2, are expected to be small, the forces and moments used for this check will be assumed to come from a no-translation case, with little error. If the minimum lateral load would produce large moments or the amplification factor, B_2, calculated in Step 5, were large, this would not be a good assumption. Therefore, at end A, M_{nt} = 142 ft-kips, at end B M_{nt} = 71.0 ft-kips, and P_{nt} = 522 kips.

A quick review of the determination of B_1 from the first part of this solution shows that the only change is in the magnitude of the axial force and the member end moments; thus

$$\frac{M_1}{M_2}=\frac{71.0}{142}=0.50$$

$$C_m=0.6-0.4\left(M_1/M_2\right)=0.6-0.4(0.50)=0.4$$

$$P_{e1}=\frac{\pi^2 EI}{L_{c1}^2}=\frac{\pi^2(29{,}000)(999)}{(1.0(12.5)(12))^2}=12{,}700 \text{ kips}$$

Thus, with $\alpha = 1.0$ for LRFD,

$$B_1 = \frac{C_m}{1 - \dfrac{\alpha P_r}{P_{e1}}} = \frac{0.4}{1 - \dfrac{(1.0)(522)}{12,700}} = 0.417 < 1.0$$

Note that B_1 is again 1.0.

With the assumption that there is no lateral translation, $M_{lt} = 0.0$ and B_2 is unnecessary, thus

$$P_r = 522 \text{ kips}, \quad M_r = 1.0(142) = 142 \text{ ft-kips}$$

Again, using Equation H1-1a,

$$\frac{P_u}{\phi P_n} + \frac{8}{8}\left(\frac{M_u}{\phi M_n}\right) \le 1.0$$

$$\frac{522}{1060} + \frac{8}{9}\left(\frac{142}{574}\right) = 0.712 < 1.0$$

Thus,

the W14×90 is adequate for both load combinations.

EXAMPLE 8.2b
Moment Frame Strength Check for Combined Compression and Bending by ASD

Goal: Using the ASD provisions, determine whether the W14×90, A992 column shown in Figure 8.11 is adequate to carry the imposed loading.

Given: An exterior column from an intermediate level of a multi-story moment frame is shown in Figure 8.11. The column is part of a braced frame out of the plane of the figure. Figure 8.11a shows the elevation of the frame with the member to be checked labeled AB. The same column section will be used for the level above and below the column AB. A first-order analysis of the frame for gravity loads plus the minimum lateral load (the minimum lateral load will be discussed in Section 8.7) results in the forces shown in Figure 8.11d, whereas the results for gravity plus wind are shown in Figure 8.11e. Assume that the frame drift under service loads is limited to height/300 for a story shear, $H = 148$ kips.

SOLUTION

Step 1: Determine the column effective length factor in the plane of bending.

Using the effective length alignment chart introduced in Chapter 5 and given in Commentary Figure C-A-7.2, determine the effective length for buckling in the plane of the moment frame. At each joint there are two columns and one beam framing in. Thus,

$$G_A = G_B = \frac{\Sigma(I/L)_c}{\Sigma(I/L)_g} = \frac{2\left(\dfrac{999}{12.5}\right)}{\left(\dfrac{2100}{30.0}\right)} = 2.28$$

Thus, from Figure 5.20, $K = 1.66$.

Step 2: Determine the controlling effective length.

With $r_x/r_y = 1.66$ for the W14 × 90,

$$(L_{cx})_{eff} = (KL)_{eff} = \frac{(KL)_x}{r_x/r_y} = \frac{1.66(12.5)}{1.66} = 12.5 \text{ ft}$$

$$L_{cy} = KL_y = 1.0(12.5) = 12.5 \text{ ft}$$

Step 3: Since the effective length about each axis is 12.5 ft, determine the column allowable axial strength using $L_c = 12.5$.

From the column tables, *Manual* Table 4-1a, for $L_c = 12.5$ ft,

$$P_n/\Omega = 703 \text{ kips}$$

Step 4: Determine the first-order moments and forces for the loading combination that includes wind, D +0.75L + 0.75(0.6W).

The column end moments given in Figure 8.11e are a combination of moments resulting from a nonsway gravity load analysis and a wind analysis:

Moment for end *A*:

$$M_{nt} = 78.0 \text{ ft-kips}$$
$$M_{lt} = 96.0 \text{ ft-kips}$$

Moment for end *B*:

$$M_{nt} = 39.0 \text{ ft-kips}$$
$$M_{lt} = 96.0 \text{ ft-kips}$$

Compression:

$$P_{nt} = 280 \text{ kips}$$
$$P_{lt} = 62.0 \text{ kips}$$

Step 5: Determine the second-order moments by amplifying the first-order moments.

No-translation amplification: The no-translation moments must be amplified by B_1. From Figure 8.11e it is seen that the end moments bend the column in reverse curvature:

$$\frac{M_1}{M_2} = \frac{39.0}{78.0} = 0.50$$

$$C_m = 0.6 - 0.4(M_1/M_2) = 0.6 - 0.4(0.50) = 0.4$$

$$P_{e1} = \frac{\pi^2 E}{L_{c1}^2} = \frac{\pi^2 (29{,}000)(999)}{(1.0(12.5)(12))^2} = 12{,}700 \text{ kips}$$

Thus, with $\alpha = 1.6$ for ASD and $P_r = 280 + 62 = 342$ kips, Equation A-8-3 yields

$$B_1 = \frac{C_m}{1 - \frac{\alpha P_r}{P_{e1}}} = \frac{0.4}{1 - \frac{(1.6)(342)}{12,700}} = 0.418 < 1.0$$

Therefore, $B_1 = 1.0$.

Translation amplification: The translation forces and moments must be amplified by B_2. The design drift limit of height/300 and Equation A-8-6 are used to determine B_2.

The total service lateral load on this story is given as

$$H = 148 \text{ kips}$$

Additional given information is that the total gravity load for this load combination in Figure 8.11e is

$$P_{story} = 1670 \text{ kips}$$

The drift limit under the service lateral load of 148 kips is

$$\Delta_H = L/300 = 12.5(12)/300 = 0.50 \text{ in.}$$

Remember that in the calculation of B_2, H can be taken as any convenient magnitude, as long as Δ_H is the corresponding displacement. This is because it is the ratio of H to Δ_H that is used in the determination of $P_{e\,story}$.

Thus, with $\alpha = 1.6$ for ASD and $R_M = 0.85$ assuming all columns are moment frame columns, Equation A-8-7 gives

$$P_{e\,story} = \frac{R_M HL}{\Delta_H} = \frac{0.85(148)(12.5)(12)}{0.50} = 37{,}700 \text{ kips}$$

and Equation A-8-6 gives

$$B_2 = \frac{1}{1 - \frac{\alpha P_{story}}{P_{e\,story}}} = \frac{1}{1 - \left(\frac{(1.6)1670}{37,700}\right)} = 1.08 > 1.0$$

Thus, the second-order compressive force and moment are

$$P_r = P_{nt} + B_2(P_{lt}) = 280 + 1.08(62.0) = 347 \text{ kips}$$

$$M_r = B_1(M_{nt}) + B_2(M_{lt}) = 1.0(78.0) + 1.08(96.0) = 182 \text{ ft-kips}$$

These represent the required strength for this load combination.

Step 6: Determine whether this shape will provide the required strength based on the appropriate interaction equation.

The unbraced length of the compression flange for pure bending is 12.5 ft, which is less than L_p = 15.1 ft for this section, taking into account that its flange is noncompact. Thus, from *Manual* Table 3-2, the allowable moment strength of the section is

$$M_n/\Omega = 382 \text{ ft-kips}$$

Determine the appropriate interaction equation. From Step 3, ϕP_n = 1060 kips;

$$\frac{P_a}{P_n/\Omega} = \frac{347}{703} = 0.494 > 0.2$$

so use Equation H1-1a, which yields

$$\frac{P_a}{P_n/\Omega} + \frac{8}{9}\left(\frac{M_a}{M_n/\Omega}\right) \leq 1.0$$

$$0.494 + \frac{8}{9}\left(\frac{182}{382}\right) = 0.918 < 1.0$$

Thus,

the W14×90 is adequate for this load combination.

Step 7: Check the section for the gravity-only load combination, $D + L$.

Because this is a gravity-only load combination, *Specification* Appendix Section 7.2.2, by reference to Section C2.2b, requires that the analysis include a minimum lateral load of 0.002 times the gravity load. This will be further discussed in Section 8.7. For this load combination, the total story gravity load must also be known and is given in Figure 8.11d as P_{story} = 1690 kips. Thus, for this frame the minimum lateral load is $0.002P_{story} = 0.002(1690) = 3.38$ kips at this level.

The forces and moments given in Figure 8.11d include the effects of this minimum lateral load. The magnitude of the lateral translation effect is small in this case. Since both the moment due to the minimum lateral load and the amplification factor, B_2, are expected to be small, the forces and moments used for this check will be assumed to come from a no-translation case, with little error. If the minimum lateral load would produce large moments or the amplification factor, B_2, calculated in Step 5, were large, this would not be a good assumption. Therefore, at end A, M_{nt} = 95.0 ft-kips, at end B M_{nt} = 47.0 ft-kips, and P_{nt} = 348 kips.

A quick review of the determination of B_1 from the first part of this solution shows that the only change is in the magnitude of the axial force and the member end moments; thus

$$\frac{M_1}{M_2} = \frac{47.0}{95.0} = 0.50$$

$$C_m = 0.6 - 0.4(M_1/M_2) = 0.6 - 0.4(0.50) = 0.4$$

$$P_{e1} = \frac{\pi^2 EI}{L_{c1}^2} = \frac{\pi^2(29{,}000)(999)}{(1.0(12.5)(12))^2} = 12{,}700 \text{ kips}$$

Thus, with $\alpha = 1.6$ for ASD,

$$B_1 = \frac{C_m}{1 - \dfrac{\alpha P_r}{P_{e1}}} = \frac{0.4}{1 - \dfrac{(1.6)(348)}{12{,}700}} = 0.418 < 1.0$$

Note that B_1 is again 1.0.

With the assumption that there is no lateral translation,

M_{lt} = 0.0 and B_2 is unnecessary,

thus

$$P_r = 348 \text{ kips}, \quad M_r = 1.0(95.0) = 95.0 \text{ ft-kips}$$

Again using Equation H1-1a,

$$\frac{P_a}{P_n/\Omega} + \frac{8}{8}\left(\frac{M_a}{M_n/\Omega}\right) \le 1.0$$

$$\frac{348}{703} + \frac{8}{9}\left(\frac{95.0}{382}\right) = 0.716 < 1.0$$

Thus,

the W14×90 is adequate for both load combinations.

The moments in the beams and the beam-column connections must also be amplified for the critical case to account for the second-order effects. This is done by considering equilibrium of the beam-column joint. The amplified moments in the column above and below the joint are added together and this sum distributed to the beams which frame into the joint according to their stiffnesses. These moments then establish the connection design moments.

8.7 SPECIFICATION PROVISIONS FOR STABILITY ANALYSIS AND DESIGN

Up to this point, the discussion of the interaction of compression and bending has concentrated on the development of the interaction equations and one approach to incorporate second-order effects. The *Specification* actually provides three approaches to deal with these two closely linked issues. The most direct approach is to use a general second-order analysis in conjunction with the Direct Analysis Method described in Chapter C.

A general second-order analysis yields forces and moments that can be used directly in the interaction equations of Chapter H without the need to resort to amplification factors as just described. The disadvantage to this approach is that, since the extremely useful principal of superposition cannot be used (since the structural response is nonlinear), a complete nonlinear analysis must be carried out for each load combination. A discussion of general, or rigorous, methods of second-order analysis is beyond the scope of this book. Thus, in the remainder of this book, if second-order effects have not already been included in the analysis results given, the amplified first-order analysis approach will be used to obtain the required second-order forces and moments.

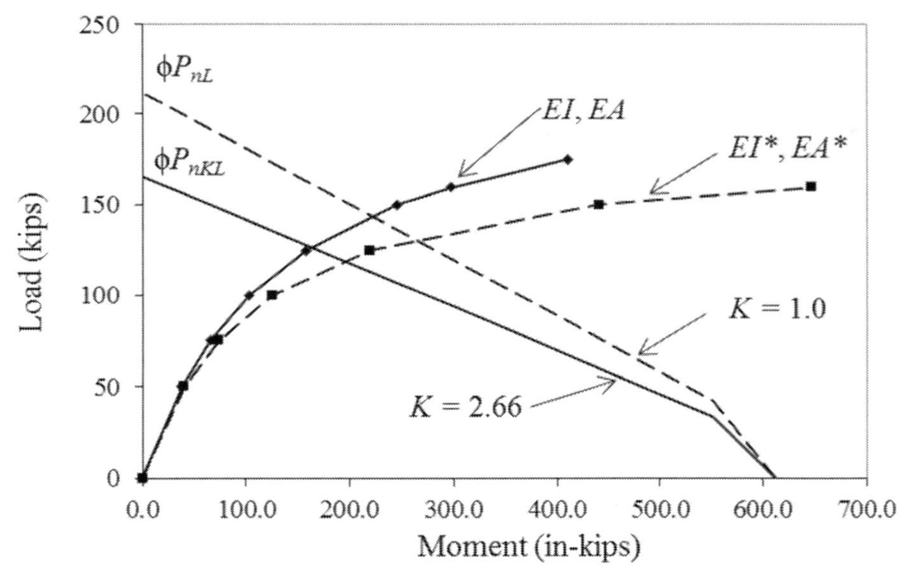

Figure 8.12 Comparison of the Effective Length Method and the Direct Analysis Method

8.7.1 Direct Analysis Method

The advantage of the direct analysis method of *Specification* Chapter C is that for the design of compression members, the effective length factor is taken as 1.0. For braced frames, K can always be taken as 1.0 based on Section 7.2.3(a), so the direct analysis method is particularly useful for moment frames where K is not normally 1.0. The

Specification requires that the stiffness of all elements contributing to the lateral load resistance of the structure be reduced in the analysis. Thus, rather than using EA for the axial stiffness of the members, the modified stiffness, $EA^* = 0.8EA$, is used. Similarly for flexural stiffness, the modified stiffness $EI^* = 0.8\tau_b EI$, where τ_b accounts for the influence of residual stresses on second-order effects, is used. It should be remembered that the influence of residual stresses on the strength of compression and bending members was already discussed in Chapters 5 and 6. The use of τ_b in this instance is to capture the influence of those same residual stresses on displacements and thus on second-order effects. This is the same τ_b used with the alignment charts in the determination of the effective length in Chapter 5.

Figure 8.12 shows a comparison between the effective length method already presented and the direct analysis method for a simple structure. Equations H1-1a and H1-1b are plotted for the effective length method and labeled with $K = 2.66$. This indicates that the compressive strength of the member has been determined using $K = 2.66$. The nonlinear load-moment curve is identified with EI and EA to indicate that the nominal stiffnesses are used to determine this behavior. The intersection of these two curves indicates that this load and moment combination satisfy the interaction equation. Equations H1-1a and H1-1b are also plotted for the direct analysis method and identified with $K = 1.0$. This indicates that the compressive strength of the member has been determined with $K = 1.0$. Note that regardless of which approach is selected to determine the compressive strength, the flexural strength is the same for both methods. The nonlinear load-moment curve is identified with EI^* and EA^* to indicate that the reduced stiffnesses were used to determine this behavior. The intersection of these two curves indicates that this load and moment combination satisfy the interaction equation. Next, note that the load magnitude for both of these intersections is nearly the same. Thus, the load that satisfies the interaction equation is the same regardless of which method is used. Since the direct analysis method did not require the determination of K, it is a significantly simpler method than the effective length method.

Another consideration that has only briefly been mentioned to this point is the requirement in Section C1 that the influence of geometric imperfections be considered. As with residual stresses, the influence of geometric imperfection on the strength of compression members has already been addressed through the *Specification* column strength equations. The requirement here is to consider the influence of out-of-plumbness on the stability of the structure. This may be accomplished by modeling the structure in its out-of-plumb condition or through the use of notional loads to simulate the out-of-plumbness. These notional loads will be discussed later in this section. It should be noted that this is not a requirement of the direct analysis method alone but a general requirement for determining required strength.

In addition to the direct analysis method, two other design methods are given in the *Specification*. They are found in Appendix 7. The limitations on when these methods can be applied are based on satisfying specific requirements of the direct analysis method, which then permit several simplifications to be used. The limitations applicable to each method are discussed along with the discussion of those methods.

8.7.2 Effective Length Method

Appendix 7.2 provides the requirements for the effective length method. This is the approach already described earlier in this chapter for braced and moment frames. It is valid as long as the structure supports gravity loads primarily through vertical columns, walls or frames and the ratio of second-order deflection to first-order deflection, Δ_2/Δ_1, is equal to or less than 1.5. Another way to state this second requirement is to remember that $\Delta_2/\Delta_1 = B_2$, so the effective length method is valid as long as $B_2 \leq 1.5$. Although this check was not made in Example 8.2, it can now be seen that it was acceptable to use the effective length method in that example, since for both LRFD and ASD, $B_2 \leq 1.5$. A special case occurs when $B_2 \leq 1.1$. In this case, columns in moment frames can be designed using $K = 1.0$. The effective length method is essentially the same method used in past practice with the addition of the requirement of a minimum lateral load to be applied in gravity-only load combinations. This is the notional load discussed above to account for initial out-of-plumbness. It is the same as the minimum lateral load used in Example 8.2 and will be discussed later in this section.

8.7.3 First-Order Analysis Method

A third method is given in the Appendix 7.3, the first-order analysis method. This approach permits design without direct consideration of second-order effects except through the application of additional notional lateral loads that account for structure out-of-plumbness and second-order effects. This is possible because of the limits placed on the implementation of this method. As with the effective length method, the structure must support gravity loads primarily through vertical columns, walls or frames and the ratio of the second-order drift to first-order drift must be less than or equal to 1.5. Additionally, the required axial strength in members that are essentially horizontal, such as beams, must be limited to

$$\alpha P_r \leq 0.08 P_e \qquad \text{(AISC A-7-1)}$$

where $P_e = \pi^2 EI/L^2$, the Euler buckling strength discussed in Chapter 5 and all compression members that participate in lateral load resistance must behave elastically according to

$$\alpha P_r \leq 0.5 P_{ns} \qquad \text{(AISC A-7-2)}$$

With the foregoing limitations and the application of the notional load given by

$$N_i = 2.1\alpha\left(\Delta/L\right)Y_i \geq 0.0042 Y_i \qquad \text{(AISC A-7-3)}$$

where Y_i equals the total gravity load applied at story i, compression members may be designed using $K = 1.0$.

8.7.4 Geometric Imperfections

Now, consider in more detail the requirement to consider geometric imperfections. *Specification* Section C1 requires that geometric imperfections be considered in the analysis and design of structures. There are two types of geometric imperfections that must be considered: initial out-of-straightness and initial out-of-plumbness. The strength equations of *Specification* Chapter E already account for initial out-of-straightness. To consider out-of-plumbness, the structure may be modeled in the out-of-plumb position or notional loads, as given in *Specification* Section C2.2b, may be used. The AISC *Code of Standard Practice* permits columns to be built with an out-of-plumbness tolerance of height/500. Figure 8.13a shows the upper two stories of a column that is out-of-plumb, $L/500 = 0.002L$. For level 1, with a load P_1 applied, a horizontal force of $0.002P_1$ is required for equilibrium as shown. The next story down, level 2, is also out-of-plumb by the same amount and the load from above is added to the load introduced at that level so that the column must carry $P_1 + P_2$. Since this load is also applied at the eccentricity of $0.002L$, equilibrium requires that the column be restrained by a force of $0002(P_1 + P_2)$ as shown for level 2. When these two columns are put together, it can be seen in Figure 8.13b that the horizontal force at the intersection becomes $0.002\ P_2$. Thus, Figure 8.13b shows for this two story out-of-plumb column, the horizontal forces that are required to keep it in equilibrium. The same effect could be accomplished if the column was modeled plumb and these restraining forces from Figure 8.13b were applied as loads as shown in Figure 8.13c.

Thus, the out-of-plumbness can be modeled with a lateral load equal to 0.2 percent of the gravity load introduced at each level of the structure. This lateral load is called a notional load and is taken as $N_i = 0.002\alpha Y_i$ where Y_i is the total gravity load on story i and α is 1.0 for LRFD and 1.6 for ASD, as discussed earlier. The *Specification* includes α as described here, but when the amplified first-order analysis method is used to obtain second-order effects, this is not necessary, since α is included in the B_1 and B_2 calculations. This notional load is the same as the "minimum lateral load" used in Example 8.2.

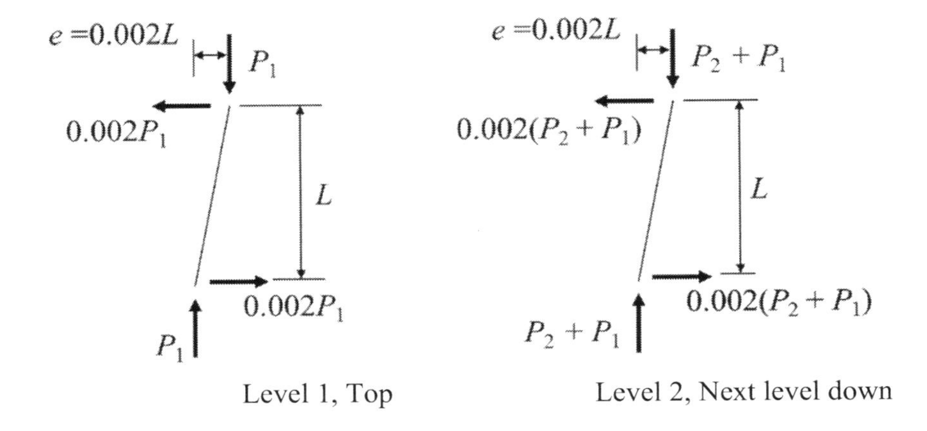

Figure 8.13a Notional Load Model for Geometric Imperfections.

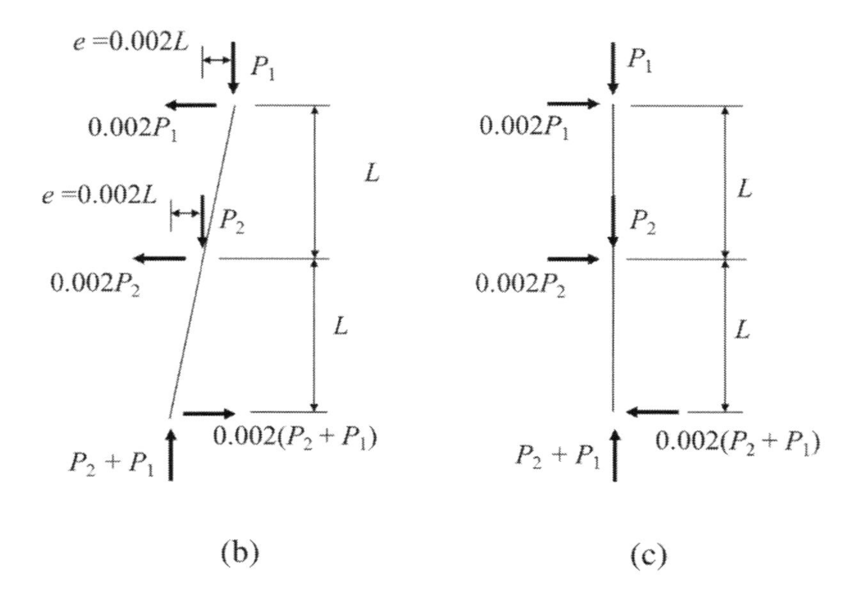

Figure 8.13b, c Notional Load Model for Geometric Imperfections.

8.7.5 Comparison of Methods

The three methods of frame stability analysis just described will be compared using a simple determinate structure. Figure 8.14 shows a one-bay unbraced frame with an LRFD gravity plus lateral load combination. Column A is a flagpole column and provides all of the lateral load resistance while column B is a gravity only column. Gravity only columns will be discussed in more depth in Section 8.10. Column A is a W14×90 bending about its strong axis, column B can be any size sufficient to support the gravity load since it does not contribute to the lateral load resistance, and the beam is assumed to be a rigid element.

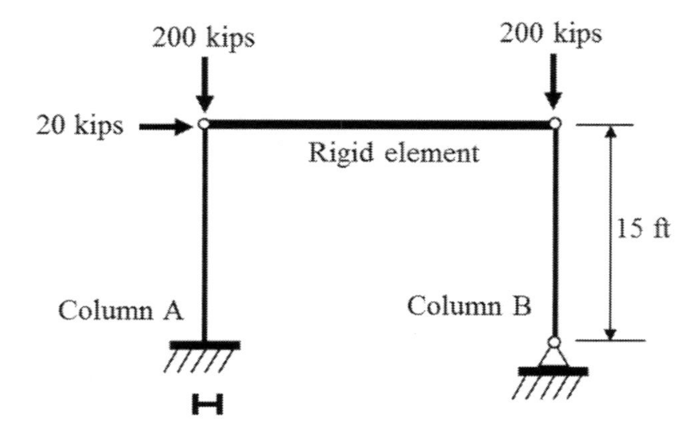

Figure 8.14 One-bay Unbraced Frame for Comparison of Analysis Methods

Effective Length Method: First the effective length method of Appendix Section 7.2 will be used along with the $B_1 - B_2$ amplification for second-order effects. If the structure is prevented from swaying, the *nt* analysis produces, for column A, $P_{nt} = 200$ kips and $M_{nt} = 0$ ft-kips. The lateral translation analysis produces, for column A, $P_{lt} = 0$ kips and $M_{nt} = 300$ ft-kips. Since there is no moment in the *nt* analysis, there are no P-δ effects (member effects) and no need to determine B_1. To assess the P-Δ effects (sway effects), B_2 will be determined. The 20 kip lateral load produces a drift calculated as for a cantilevered beam,

$$\Delta = \frac{HL^3}{3EI} = \frac{20(15(12))^3}{3(29{,}000)(999)} = 1.34 \text{ in.}$$

The total gravity load on the structure is 400 kips. Half of this load is on the lateral load resisting column A and half is on the gravity only column. Thus,

$$R_M = 1 - 0.15\left(\frac{P_{mf}}{P_{tory}}\right) = 1 - 0.15\left(\frac{200}{400}\right) = 0.925$$

and

$$P_{e\ story} = R_M \frac{HL}{\Delta_H} = 0.925\frac{20(15(12))}{1.34} = 2490 \text{ kips}$$

Thus,

$$B_2 = \frac{1}{1 - \dfrac{\alpha P_{story}}{P_{e\ story}}} = \frac{1}{1 - \dfrac{1.0(400)}{2490}} = 1.19$$

Next, consider the limitations on use of the effective length method. This structure supports gravity loads through vertical columns, so it meets the first limitation in Section 7.2.1. The second limitation requires that the second-order amplification, B_2, be less than or equal to 1.5. Since $B_2 = 1.19 \le 1.5$ the effective length method may be used for this frame.

The required strength, including second-order effects, is found through Equations A-8-1 and A-8-2.

$$M_r = B_1 M_{nt} + B_2 M_{lt} = 0 + 1.19(300) = 357 \text{ ft-kips}$$

and

$$P_r = P_{nt} + B_2 P_{lt} = 200 + 1.19(0) = 200 \text{ kips}$$

The next step in the effective length method is determination of the effective length factor. The effective length factor for the flagpole column alone is $K_x = 2.0$. Inelastic buckling need not be considered since it will not change the effective length factor for a cantilever column. However, as discussed in Chapter 5, the inclusion of the gravity only column with load will increase the effective length of column A. Using the approach presented in Chapter 5, with the load on the moment frame column, $P_{mf} = 200$ kips and the load on the gravity only column, $P_{\text{grav only}} = 200$ kips, the effective length factor is

$$K_x^* = K_x\sqrt{1 + P_{grav\ only}/P_{mf}} = 2.0\sqrt{1 + 200/200} = 2.83$$

Assuming that the frame is braced out of the plane of the frame, $K_y = 1.0$.

The available flexural strength of the W14×90 column can be determined from *Manual* Table 3-2 for an unbraced length of the compression flange $L_b = 15$ ft < $L_p = 15.1$ ft, $\phi M_n = 574\text{ ft-kips}$. The available compressive strength can be determined from *Manual* Table 4-1a. The controlling effective length is for x-axis buckling, thus $\left(L_{cx}\right)_{eff} = 2.83(15)/1.66 = 25.6\text{ ft}$ and $\phi P_n = 720\text{ kips}$. With the required strength and available strength determined, the interaction equation can be checked.

First determine which interaction equation should be used. Since $P_r/\phi P_n = 200/720 = 0.278 > 0.2$ use Equation H1-1a, thus

$$\frac{P_r}{P_c} + \frac{8}{9}\left(\frac{M_r}{M_c}\right) = \frac{200}{720} + \frac{8}{9}\left(\frac{357}{574}\right) = 0.278 + 0.553 = 0.831 < 1.0$$

So the W14×90 is shown to be adequate by the effective length method.

First-Order Analysis Method: The first-order analysis method of Appendix Section 7.3 may be used for those structures that meet the limitations of Section 7.3.1. These limitations are the same as for the effective length method with the addition of the requirement that the beams have a very low compression demand and columns behave elastically such that

$$\alpha P_r \le 0.5 P_{ns} \qquad \text{(AISC A-7-2)}$$

The beam in our example does not carry and axial load since all the lateral load is carried by bending of Column A. Since the W14×90 column does not have slender elements for compression, $P_{ns} = P_y = F_y A_g = 50(26.5) = 1330\text{ kips}$ and for the frame of Figure 8.14 $\alpha P_r = 200\text{ kips}$. Thus $200 \le 0.5(1330) = 666$ kips and the first-order analysis method may be used.

The required strength for the first-order analysis method is determined from a first-order analysis that includes a notional load defined by Equation A-7-3 added to the lateral load in all load combinations. This notional load accounts for both the initial out-of-plumbness of the structure and second-order effects. Thus,

$$N_i = 2.1\alpha\left(\Delta/L\right)Y_i \ge 0.0042Y_i \qquad \text{(AISC A-7-3)}$$

For our structure, Δ = 1.34 in. as before and Y_i = 400 kips so the notional load for this load combination is

$$N_i = 2.1(1.0)\left(1.34/\left(15(12)\right)\right)(400) = 6.25\text{ kips} \ge 0.0042(400) = 1.68\text{ kips}$$

Thus, the lateral load in the analysis will be increased from 20 kips to 26.3 kips. The results of the first-order analysis for the determinate structure are $P_u = 200\text{ kips}$ and

$M_u = 26.3(15) = 395$ ft-kips. Although this is called the first-order analysis method, it does require that the moment be amplified by B_1 found using

$$B_1 = \frac{C_m}{1-\alpha P_r/P_{e1}} \geq 1.0 \qquad \text{(AISC A-8-3)}$$

This amplification addresses the member effect and is influenced by the buckling strength of the column as a pin ended column in a no sway condition, P_{e1}, and the equivalent uniform moment factor, C_m. Thus,

$$C_m = 0.6 - 0.4(M_1/M_2) = 0.6 - 0.4(0/395) = 0.6$$

and with $EI^* = EI$,

$$P_{e1} = \frac{\pi^2 EI^*}{L_{c1}^2} = \frac{\pi^2(29{,}000)(999)}{\left(15(12)\right)^2} = 8830 \text{ kips}$$

which gives

$$B_1 = \frac{0.6}{1-200/8830} = 0.614 < 1.0$$

Therefore, there is no amplification needed so $P_r = P_u = 200$ kips and $M_r = M_u = 395$ ft-kips.

The available moment strength of the W14×90 column determined previously from Table 6-1 is unchanged, thus $\phi M_n = 574$ ft-kips. The controlling effective length is for y-axis buckling, thus $L_{cy} = 15.0$ ft and $\phi P_n = 1000$ kips. With the required strength and available strength determined, the interaction equation can be checked.

First determine which interaction equation should be used. Since

$$P_r/\phi P_n = 200/1000 = 0.20 \leq 0.2$$

use Equation H1-1a, thus

$$\frac{P_r}{P_c} + \frac{8}{9}\left(\frac{M_r}{M_c}\right) = \frac{200}{1000} + \frac{8}{9}\left(\frac{395}{574}\right) = 0.200 + 0.612 = 0.812 < 1.0$$

So the W14×90 is shown to be adequate by the first order analysis method.

Direct Analysis Method: The third method to be considered is the direct analysis method of Chapter C. There are no limitations on the use of the direct analysis method like there are on the effective length or first-order analysis methods and second-order effects and initial out-of-plumbness must be accounted for as they were for the effective length method. The only new requirement is that the stiffness of all members that contribute to the lateral load resistance be reduced in the analysis to $EI^* = 0.8\tau_b EI$ and $EA^* = 0.8EA$. It is this stiffness reduction that permits the use of an effective length factor equal to one when using the direct analysis method. From the discussion of the effective length method, it was seen that B_2 was less than 1.5 when using the unreduced stiffness thus the notional load to account for out-of-plumbness does not need to be added to the lateral load. Thus, from a first order analysis of the determinate structure, $P_u = 200$ kips and

$M_u = 20.0(15) = 300$ ft-kips. Initially assume that $\tau_b = 1.0$ so that the flexural stiffness of column A will be taken as $EI^* = 0.8EI$. Thus the 20 kip lateral load produces a drift calculated as for a cantilevered beam,

$$\Delta = \frac{HL^3}{3EI^*} = \frac{20(15(12))^3}{3(0.8)(29{,}000)(999)} = 1.68 \text{ in.}$$

The total gravity load on the structure is 400 kips. Half of this load is on the lateral load resisting column A and half is on the gravity only column. Thus again,

$$R_M = 1 - 0.15\left(\frac{P_{mf}}{P_{tory}}\right) = 1 - 0.15\left(\frac{200}{400}\right) = 0.925$$

and

$$P_{e\ story} = R_M \frac{HL}{\Delta_H} = 0.925 \frac{20(15(12))}{1.68} = 1980 \text{ kips}$$

Thus,

$$B_2 = \frac{1}{1 - \dfrac{\alpha P_{story}}{P_{e\ story}}} = \frac{1}{1 - \dfrac{1.0(400)}{1980}} = 1.25$$

Note that the drift increased from what was calculated for the effective length method and therefore the second-order amplification increased.

The required strength, including second-order effects is found through Equations A-8-1 and A-8-2.

$$M_r = B_1 M_{nt} + B_2 M_{lt} = 0 + 1.25(300) = 375 \text{ ft-kips}$$

and

$$P_r = P_{nt} + B_2 P_{lt} = 200 + 1.25(0) = 200 \text{ kips}$$

The available moment strength of the W14×90 column determined previously from Table 3-2 is unchanged, thus $\phi M_n = 574$ ft-kips. The controlling effective length is for y-axis buckling, thus, from Table 4-1a, $L_{cy} = 15.0$ ft and $\phi P_n = 1000$ kips. With the required strength and available strength determined, the interaction equation can be checked.

First determine which interaction equation should be used. Since $P_r/\phi P_n = 200/1000 = 0.20 \leq 0.2$ use Equation H1-1a, thus

$$\frac{P_r}{P_c} + \frac{8}{9}\left(\frac{M_r}{M_c}\right) = \frac{200}{1000} + \frac{8}{9}\left(\frac{375}{574}\right) = 0.200 + 0.581 = 0.781 < 1.0$$

Since the ratio of required strength to available strength is less than 0.5, a more detailed look at determination of τ_b is unnecessary. So, the W14×90 is adequate by the direct analysis method. Note that based on the results of the interaction equation, this approach is less conservative than the other two methods with the interaction equation result being the lowest of the three approaches. Since the only new requirement of the direct analysis method is to use a reduced stiffness in calculating second-order effects and this permits the use of an effective length factor in the lateral load resisting direction of one, this is clearly the simplest and most direct method available.

Three methods of analysis are available and all three have their place in design. It is up to the user to determine when to use each approach most efficiently.

EXAMPLE 8.3a ***Direct Analysis Method for Column Design by LRFD***

Goal: Using the LRFD provisions and the results from a second-order direct analysis, determine if a W14×132, A992 member is adequate to carry the given loads and moments.

Given: The column has a length of 16 ft and is braced at the ends only. The results of the second-order direct analysis are P_u = 800 kips, M_{ux} = 300 ft-kips, and M_{uy} = 76 ft-kips.

SOLUTION

Step 1: Determine the required strength.

Since the given results are from a second-order analysis, there is no need to amplify forces and moments; thus

$$P_r = P_u = 800 \text{ kips}, \quad M_{rx} = M_{ux} = 300 \text{ ft-kips}, \quad M_{ry} = M_{uy} = 76 \text{ ft-kips}$$

Step 2: Determine the available compressive strength of the column.

Since the given results are from a direct analysis, K = 1.0; thus, from *Manual* Table 4-1a with L_c = 16.0 ft,

$$\phi P_n = 1440 \text{ kips}$$

Step 3: Determine the available strength for bending about the x-axis.

With an unbraced length L_b = 16 ft, from *Manual* Table 3-2,

$$\phi M_p = 878 \text{ ft-kips}, \quad L_p = 13.3 \text{ ft}, \quad \phi BF = 7.74 \text{ kips}$$

and

$$\phi M_{nx} = \phi M_p - \phi BF\left(L_b - L_p\right) = 878 - 7.74(16.0 - 13.3) = 857 \text{ ft-kips}$$

Step 4: Determine the available strength for bending about the y-axis.

From *Manual* Table 3-4,

$$\phi M_{ny} = 424 \text{ ft-kips}$$

Step 5: Check the W14×132 for combined axial load and bending. To determine which equation to use, check

$$\frac{P_r}{\phi P_n} = \frac{800}{1440} = 0.556 \geq 0.2$$

Therefore, use Equation H1-1a.

$$\frac{P_r}{\phi P_n} + \frac{M_{rx}}{\phi M_{nx}} + \frac{M_{ry}}{\phi M_{ny}} \leq 1.0$$

$$0.556 + \frac{300}{857} + \frac{76}{424} = 1.09 > 1.0$$

Thus,

the W14×132 will not carry the given load.

EXAMPLE 8.3b *Direct Analysis Method for Column Design by ASD*

Goal: Using the ASD provisions and the results from a second-order direct analysis, determine if a W14×132, A992 member is adequate to carry the given loads and moments.

Given: The column has a length of 16 ft and is braced at the ends only. The results of the second-order direct analysis are P_a = 530 kips, M_{ax} = 200 ft-kips, and M_{ay} = 52 ft-kips.

SOLUTION

Step 1: Determine the required strength.

Since the given results are from a second-order analysis, there is no need to amplify forces and moments. Thus,

$$P_r = P_a = 530 \text{ kips}, \quad M_{rx} = M_{ax} = 200 \text{ ft-kips}, \quad M_{ry} = M_{ay} = 52 \text{ ft-kips}$$

Step 2: Determine the available compressive strength of the column.

Since the given results are from a direct analysis, K = 1.0. Thus, from *Manual* Table 4-1 with L_c = 16.0 ft,

$$\frac{P_n}{\Omega} = 960 \text{ kips}$$

Step 3: Determine the available strength for bending about the x-axis.

With an unbraced length L_b = 16 ft, from *Manual* Table 3-2,

$$\frac{M_p}{\Omega} = 584 \text{ ft-kips}, \quad L_p = 13.3 \text{ ft}, \quad \frac{BF}{\Omega} = 5.15 \text{ kips}$$

And

$$\frac{M_{nx}}{\Omega} = \frac{M_p}{\Omega} - \frac{BF}{\Omega}(L_b - L_p) = 584 - 5.15(16.0 - 13.3) = 570 \text{ ft-kips}$$

Step 4: Determine the available strength for bending about the y-axis.
From *Manual* Table 3-4

$$\frac{M_{ny}}{\Omega} = 282 \text{ ft-kips}$$

Step 5: Check the W14×132 for combined axial load and bending. To determine which equation to use, check

$$\frac{P_r}{P_n/\Omega} = \frac{530}{960} = 0.552 \geq 0.2$$

Therefore, use Equation H1-1a.

$$\frac{P_r}{P_n/\Omega} + \frac{M_{rx}}{M_{nx}/\Omega} + \frac{M_{ry}}{M_{ny}/\Omega} \leq 1.0$$

$$0.552 + \frac{200}{570} + \frac{52}{282} = 1.09 > 1.0$$

Thus,

the W14×132 will not carry the given load.

8.8 INITIAL BEAM-COLUMN SELECTION

Beam-column design is a trial-and-error process that requires that the beam-column section be known before any of the critical parameters can be determined for use in the appropriate interaction equations. There are numerous approaches to determining a preliminary beam-column size. Each incorporates its own level of sophistication and results in its own level of accuracy. Regardless of the approach used to select the trial section, one factor remains—the trial section must ultimately satisfy the appropriate interaction equation.

To establish a simple, yet useful, approach to selecting a trial section, Equation H1-1a is modified by multiplying each term by P_c which yields

$$P_r + \frac{8}{9}\frac{M_{rx}P_c}{M_{cx}} + \frac{8}{9}\frac{M_{ry}P_c}{M_{cy}} \leq P_c \tag{8.9}$$

Then multiplying the third term by M_{cx}/M_{cx}, letting

$$m = \frac{8P_c}{9M_{cx}}$$

and

$$U = \frac{M_{cx}}{M_{cy}}$$

and substituting into Equation 8.9 yields

$$P_r + mM_{rx} + mUM_{ry} \leq P_c \tag{8.10}$$

Because Equation 8.10 calls for the comparison of the left side of the equation to the column strength, P_c, Equation 8.10 can be thought of as an effective axial load: thus

$$P_{eff} = P_r + mM_{rx} + mUM_{ry} \leq P_c \tag{8.11}$$

The accuracy used in the evaluation of m and U dictates the accuracy with which Equation 8.11 represents the strength of the column being selected. Because at this point in a design the actual column section is not known, exact values of m and U cannot be determined.

Past editions of the AISC *Manual* have presented numerous approaches to the evaluation of these multipliers. A simpler approach, however, is more useful for preliminary design. If the influence of the length — that is, all buckling influence on P_c and M_{cx} — is neglected, the ratio, P_c/M_{cx}, becomes A/Z_x, and $m = 8A/9Z_x$. Evaluation of this m for all W6 to W44 shapes, with the inclusion of a units correction factor of 12, results in the average m values given in Table 8.2. If the relationship between the area, A, and the plastic section modulus, Z_x, is established using an approximate internal moment arm of $0.89d$, where d is the nominal depth of the member in inches, then $m = 24/d$. This value is also presented in Table 8.2. This new m is close enough to the average m that it may be readily used for preliminary design and does not require checking any table to determine a recommended value.

When bending occurs about the y-axis, U must also be evaluated. A review of the same W6 to W44 shapes results in the average U values given in Table 8.2. However, a review of the actual U values shows that there is significant variation within each nominal depth group. The smaller weight sections have a U greater that the average. Thus, to have a simpler approach, the values of U given in the table can be used for the first trial.

Table 8.2 Simplified Bending Factors

Shape	m_{avg}	$m = 24/d$	U_{avg}	U
W6	4.41	4.00	3.00	3
W8	3.25	3.00	3.17	3
W10	2.62	2.40	3.62	3
W12	2.08	2.00	3.47	3
W14	1.71	1.71	2.81	3
W16	1.69	1.50	5.35	5
W18	1.45	1.33	4.78	5
W21	1.29	1.43	5.61	5
W24	1.12	1.00	5.54	5
W27	0.98	0.89	5.31	5
W30	0.91	0.80	6.09	6
W33	0.82	0.73	6.13	6
W36	0.75	0.67	6.18	6
W40	0.71	0.60	7.23	6
W44	0.65	0.55	6.93	6

More accurate evaluations of these multipliers, including length effects, have been conducted[1]. It is up to the designer to determine if there is a need for this additional accuracy in a preliminary design. No matter how the initial trial section is determined, the actual *Specification* provisions must be satisfied which may require selection of a different section.

EXAMPLE 8.4a
Initial Trial Section Selection by LRFD

Goal: Determine the initial trial section for a column.

Given: The loadings of Figure 8.11c are to be used. Assume the column is a W14 and use A992 steel. Also, use the simplified values of Table 8.2, $m = 24/d$.

SOLUTION

Step 1: Obtain the required strength from Figure 8.11c. Use the first-order analysis results.

$$P_u = 453 \text{ kips}$$

$$M_u = 251 \text{ ft-kips}$$

Step 2: Determine the effective load by combining the axial force and the bending moment.

For a W14, $m = 1.71$, so

$$P_{eff} = 453 + 1.71(251) = 882 \text{ kips}$$

Step 3: Select a trial column size to carry the required force, P_{eff}.

Using an effective length $L_c = 12.5$ ft, from *Manual* Table 4-1a, the lightest W14 to carry this load is

$$\text{W14} \times 90 \text{ with } \phi P_n = 1060 \text{ kips}$$

Example 8.2a showed that this column adequately carries the imposed load. Because the approach used here is expected to be conservative, it would be appropriate to consider the next smaller selection, a W14×82, and check it against the appropriate interaction equations.

EXAMPLE 8.4b
Initial Trial Section Selection by ASD

Goal: Determine the initial trial section for a column.

Given: The loadings of Figure 8.11e are to be used. Assume the column is a W14, and use A992 steel. Also, use the simplified values of Table 8.2, $m = 24/d$.

[1] Reynolds, M. and Uang, C.-M., Updated Equivalent Axial Load Method for Design of Steel Beam-Columns, *Engineering Journal*, AISC. Vol. 56, No. 3, 2019.

SOLUTION

Step 1: Obtain the required strength from Figure 8.11e. Use the first-order analysis results.

$$P_a = 342 \text{ kips}$$
$$M_a = 174 \text{ ft-kips}$$

Step 2: Determine the effective load by combining the axial force and the bending moment.

For a W14, $m = 1.71$; thus

$$P_{eff} = 342 + 1.71(174) = 640 \text{ kips}$$

Step 3: Select a trial column size to carry the required force, P_{eff}.

Using an effective length L_c = 12.5 ft, from *Manual* Table 4-1a, the lightest W14 to carry this load is

$$\text{W14} \times 90 \text{ with } P_n/\Omega = 703 \text{ kips}$$

Example 8.2b showed that this column adequately carries the imposed load. Because the approach used here is expected to be conservative, it would be appropriate to consider the next smaller selection, a W14×82, and check it against the appropriate interaction equations.

Every column section selected must be checked through the appropriate interaction equations for the second-order forces and moments. Thus, the process for the initial selection should be quick and reasonable. The experienced designer will rapidly learn to rely on that experience rather than these simplified approaches.

8.9 BEAM-COLUMN DESIGN USING MANUAL PART 6

Manual Part 6, Design of Members Subject to Combined Loading contains Table 6-1 which includes the axial and flexural strength for all W-shapes. Although these tables are presented here as they relate to combined loading, they can also be used for compression only, bending only, tension only and shear. There is no information found in Table 6-1 that is not already included in other Parts of the *Manual* already discussed. The advantage for combined loading is that all of the available strength values needed are found in one location.

Figure 8.15 is a portion of *Manual* Table 6-1. It shows that the compressive strength for a given section is a function of the effective length about the weak axis of the member. The effective length is tabulated in the center of the table with the compressive strengths shown on the left portion of the table. This portion of the table is used in exactly the same way as the column tables in Part 4 of the *Manual*. The strong axis bending strength is a function of the unbraced length of the compression flange of the beam. Previously, this information was available only through the beam curves in Part 3 of the

Manual. In Table 6-1 it is tabulated on the right portion of the table with the same column of lengths now defined as the unbraced length of the compression flange. Weak axis bending is not a function of length, so only one value is given for each shape. Although not used for beam-columns, when tension is combined with bending, the table also provides tension yield and rupture strength.

EXAMPLE 8.5a
Combined Strength Check Using Manual Part 6 and LRFD

Goal: Check the strength of a beam-column using *Manual* Part 6 and compare to the results of Example 8.2a.

Given: It has already been shown that the W14×90 column of Example 8.2a is adequate by LRFD. Use the required strength values given in Example 8.2a and recheck this shape using the values found in Figure 8.15 or *Manual* Table 6-1.

SOLUTION

Step 1: Determine the values needed from *Manual* Table 6-1 (Figure 8.15). The column is required to carry a compressive force with an effective length about the y-axis of 12.5 ft and an x-axis moment with an unbraced length of 12.5 ft. Thus, from Figure 8.15,

$$\phi P_n = 1060 \text{ kips}$$

$$\phi M_n = 574 \text{ ft-kips}$$

Step 2: Determine which interaction equation to use.

$$\frac{P_r}{\phi P_n} = \frac{459}{1060} = 0.433 > 0.2$$

Therefore, use Equation H1-1a.

$$\frac{200}{1060} + \frac{8}{9}\left(\frac{260}{574}\right) = 0.433 + 0.403 = 0.836 < 1.0$$

Therefore, as previously determined in Example 8.2a, the shape is adequate for this column and this load combination. The results from *Manual* Table 6-1 are exactly the same as those determined from *Manual* Table 4-1a for compression and *Manual* Table 3-2 for bending.

EXAMPLE 8.5b
Combined Strength Check Using Part 6 and ASD

Goal: Check the strength of a beam-column using *Manual* Part 6 and compare to the results of Example 8.2b.

Given: It has already been shown that the W14×90 column of Example 8.2b is adequate by ASD. Use the required strength values given in Example 8.2b and recheck this shape using the values found in Figure 8.15 or *Manual* Table 6-1.

SOLUTION

Step 1: Determine the values needed from *Manual* Table 6-1 (Figure 8.15). The column is required to carry a compressive force with an effective length about the y-axis of 12.5 ft and an x-axis moment with an unbraced length of 12.5 ft. Thus, from Figure 8.15,

$$P_n/\Omega = 703 \text{ kips}$$

$$M_n/\Omega = 382 \text{ ft-kips}$$

Step 2: Determine which interaction equation to use.

$$\frac{P_r}{P_n/\Omega} = \frac{347}{703} = 0.494 > 0.2$$

Thus, use Equation H1-1a.

$$\frac{347}{703} + \frac{8}{9}\left(\frac{182}{382}\right) = 0.494 + 0.424 = 0.918 < 1.0$$

Thus, as previously determined in Example 8.2b, the shape is adequate for this column and this load combination. The results from *Manual* Tables 6-1 are exactly the same as those determined from *Manual* Table 4-1a for compression and *Manual* Table 3-2 for bending.

8.10 COMBINED SIMPLE AND MOMENT FRAMES

The practical design of steel structures often results in frames that combine segments of rigidly connected elements with segments that are pin connected as was the case in the example frame used in Section 8.7.5. If these structures rely on the moment frame to resist lateral load and to provide the overall stability of the structure, the rigidly connected columns are called upon to carry more load than appears to be directly applied to them. In these combined simple and moment frames, the simple columns "lean" on the moment frames in order to maintain their stability and thus are often called *leaning columns*. They are also called *gravity only columns,* which is a more appropriate term since they participate only in carrying gravity loads. These columns can be designed with an effective length factor, K = 1.0, regardless of the approach to analysis that has been taken. Because these gravity only columns have no lateral stability of their own, the moment frame columns must be designed to provide the lateral stability for the full frame. Although this combination of framing types makes design of a structure more complicated, it can also be economically advantageous, because the combination can reduce the number of moment connections for the full structure and thereby reduce overall cost.

Table 6-1 (continued)
Available Strength for Members Subject to Axial, Shear, Flexural, or Combined Forces
W-Shapes

W14

F_y = 50 ksi
F_u = 65 ksi

Shape	W14x							W14x					
lb/ft	109		99		90			109		99[f]		90[f]	
	P_n/Ω_c	$\phi_c P_n$	P_n/Ω_c	$\phi_c P_n$	P_n/Ω_c	$\phi_c P_n$		M_{nx}/Ω_b	$\phi_b M_{nx}$	M_{nx}/Ω_b	$\phi_b M_{nx}$	M_{nx}/Ω_b	$\phi_b M_{nx}$
Design	Available Compressive Strength, kips							Available Flexural Strength, kip-ft					
	ASD	LRFD	ASD	LRFD	ASD	LRFD	Effective Length, L_c, ft, with respect to least radius of gyration, r_y, or unbraced length, L_b, ft, for X-X axis bending	ASD	LRFD	ASD	LRFD	ASD	LRFD
	958	1440	871	1310	793	1190	0	479	720	430	646	382	574
	932	1400	848	1270	772	1160	6	479	720	430	646	382	574
	923	1390	839	1260	764	1150	7	479	720	430	646	382	574
	913	1370	830	1250	755	1140	8	479	720	430	646	382	574
	901	1350	819	1230	745	1120	9	479	720	430	646	382	574
	888	1340	807	1210	735	1100	10	479	720	430	646	382	574
	874	1310	794	1190	723	1090	11	479	720	430	646	382	574
	859	1290	780	1170	710	1070	12	479	720	430	646	382	574
	843	1270	766	1150	697	1050	13	479	720	430	646	382	574
	826	1240	750	1130	682	1030	14	475	714	427	642	382	574
	808	1210	733	1100	667	1000	15	470	706	422	635	382	574
	789	1190	716	1080	652	979	16	465	699	417	627	378	568
	770	1160	698	1050	635	955	17	460	691	413	620	373	560
	750	1130	680	1020	618	929	18	455	684	408	613	368	553
	729	1100	661	994	601	903	19	450	676	403	605	363	546
	708	1060	642	964	583	877	20	445	669	398	598	358	539
	664	998	602	904	547	822	22	435	654	388	583	349	524
	620	931	561	843	509	766	24	425	639	378	569	339	510
	574	863	519	781	472	709	26	415	623	369	554	329	495
	529	796	478	719	434	653	28	405	608	359	539	320	481
	485	729	438	658	397	597	30	395	593	349	524	310	466
	441	663	398	598	361	543	32	385	578	339	510	300	452
	399	600	360	541	326	490	34	375	563	329	495	291	437
	359	539	323	485	292	439	36	365	548	320	480	281	423
	322	484	290	435	262	394	38	355	533	310	466	271	408
	290	437	261	393	237	356	40	345	518	300	451	262	394
	263	396	237	356	215	323	42	335	503	290	436	252	379
	240	361	216	325	196	294	44	325	488	280	422	239	359
	220	330	198	297	179	269	46	315	473	269	404	226	339
	202	303	181	273	164	247	48	305	458	255	384	214	322
	186	279	167	251	151	228	50	291	438	243	365	204	306

Strength

Available Strength in Tensile Yielding, kips					
P_n/Ω_t	$\phi_t P_n$	P_n/Ω_t	$\phi_t P_n$	P_n/Ω_t	$\phi_t P_n$
958	1440	871	1310	793	1190
Available Strength in Tensile Rupture, kips					
P_n/Ω_t	$\phi_t P_n$	P_n/Ω_t	$\phi_t P_n$	P_n/Ω_t	$\phi_t P_n$
780	1170	709	1060	647	970
Available Strength in Major Axis Shear, kips					
V_{nx}/Ω_v	$\phi_v V_{nx}$	V_{nx}/Ω_v	$\phi_v V_{nx}$	V_{nx}/Ω_v	$\phi_v V_{nx}$
150	225	138	207	123	185
Available Strength in Flexure about Y-Y Axis, kip-ft					
M_{ny}/Ω_b	$\phi_b M_{ny}$	M_{ny}/Ω_b	$\phi_b M_{ny}$	M_{ny}/Ω_b	$\phi_b M_{ny}$
231	348	207	311	181	273

Properties

Limiting Unbraced Lengths, ft					
L_p	L_r	L_p	L_r	L_p	L_r
13.2	48.5	13.5	45.3	15.1	42.5
Area, in.^{2}					
32.0		29.1		26.5	
Moment of Inertia, in.^{4}					
I_x	I_y	I_x	I_y	I_x	I_y
1240	447	1110	402	999	362
r_y, in.					
3.73		3.71		3.70	
r_x/r_y					
1.67		1.66		1.66	

[f] Shape exceeds the compact limit for flexure with F_y = 50 ksi.

Figure 8.15 Combined Axial and Bending Strength for W-Shapes.

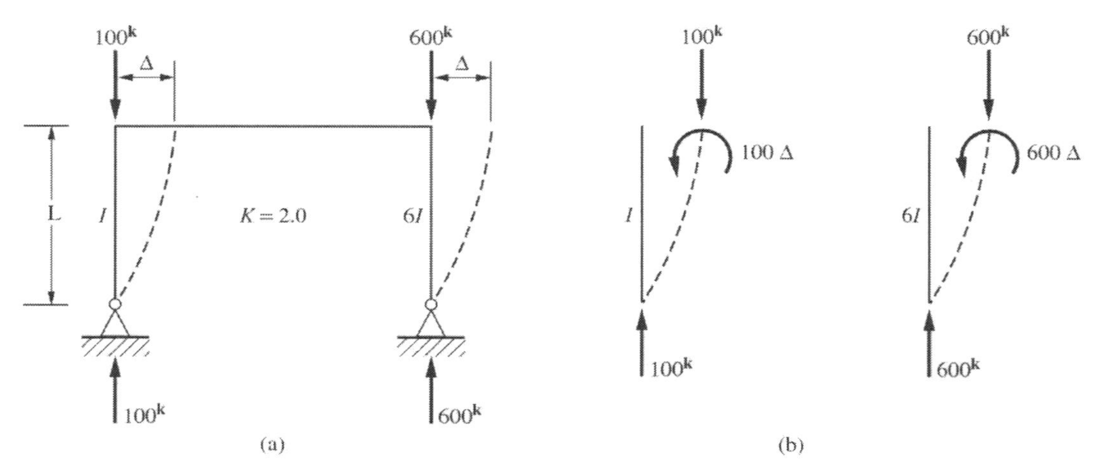

Figure 8.16 Pinned Base Unbraced Frame.

The two-column frame shown in Figure 8.16a is a moment frame with pinned-base columns and a rigidly connected beam. The column sizes are selected so that, under the loads shown, they buckle simultaneously in a sidesway mode, because their load is directly proportional to the stiffness of the members. Equilibrium in the displaced position is shown in Figure 8.16b. The lateral displacement of the frame, Δ, results in a moment at the top of each column equal to the load applied on the column times the displacement, as shown. These are the second-order effects discussed in Section 8.6. The total load on the frame is 700 kips, and the total $P\Delta$ moment is 700Δ, divided between the two columns based on the load that each carry.

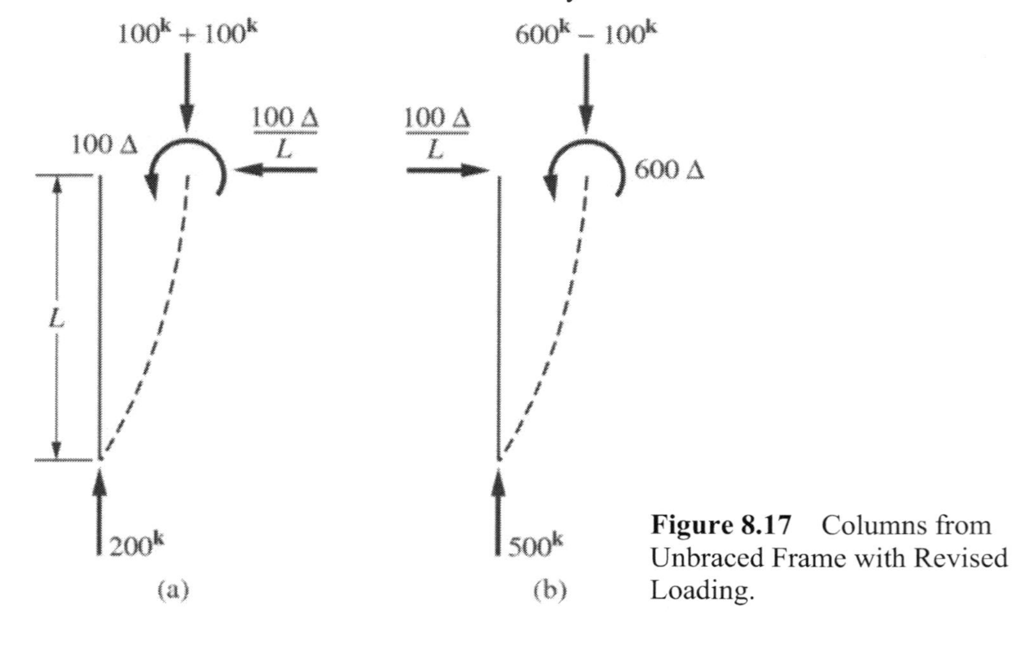

Figure 8.17 Columns from Unbraced Frame with Revised Loading.

If the load on the right-hand column is reduced to 500 kips, the column does not buckle sideways, because the moment at the top is now less than 600Δ. To reach the buckling condition, a horizontal force must be applied at the top of the column, as shown in Figure 8.17b. This force can result only from action on the left column that is transmitted through the beam. Equilibrium of the left column, shown in Figure 8.17a, requires that an additional column load of 100 kips be applied to that column in order for the load on the frame to be in equilibrium in this displaced position. The total frame capacity is still 700 kips and the total second-order moment is still 700Δ.

The maximum load that an individual column can resist is limited to that permitted for the column in a braced frame for which $K = 1.0$. In this example case, the left column could resist 400 kips and the right column 2400 kips. This is an increase of four times the load originally on the column, because the effective length factor for each column would be reduced from 2.0 to 1.0. The additional capacity of the left column is only with respect to the bending axis. The column would have the same capacity about the other axis as it did prior to reducing the load on the right column.

The ability of one column to carry increased load when another column in the frame is called upon to carry less than its critical load for lateral buckling is an important characteristic. This allows a pin-ended column to lean on a moment frame column, provided that the total gravity load on the frame can be carried by the rigid frame.

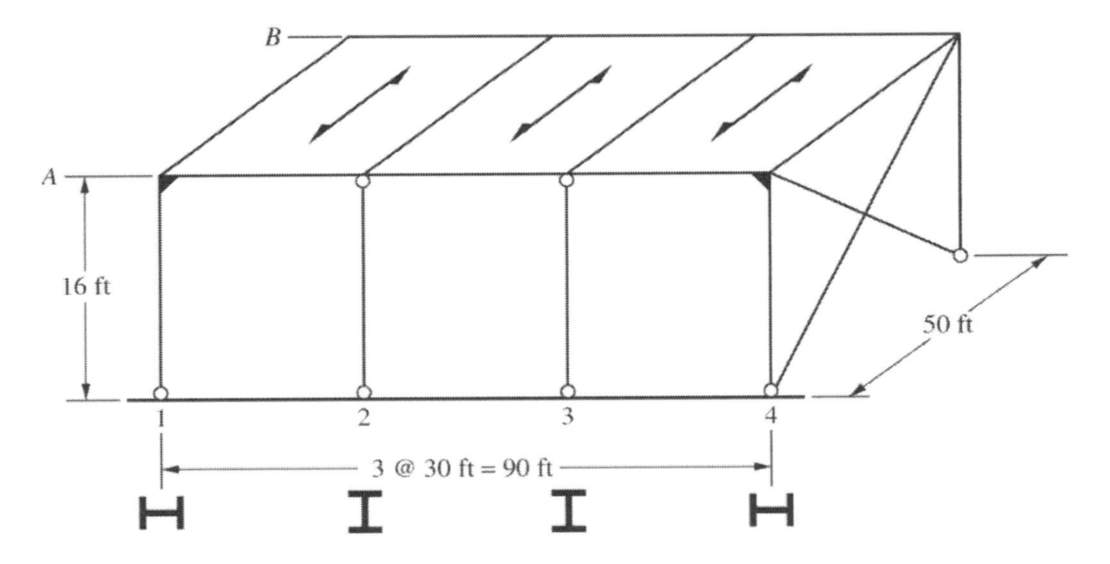

Figure 8.18 Frame Used in Example 8.6.

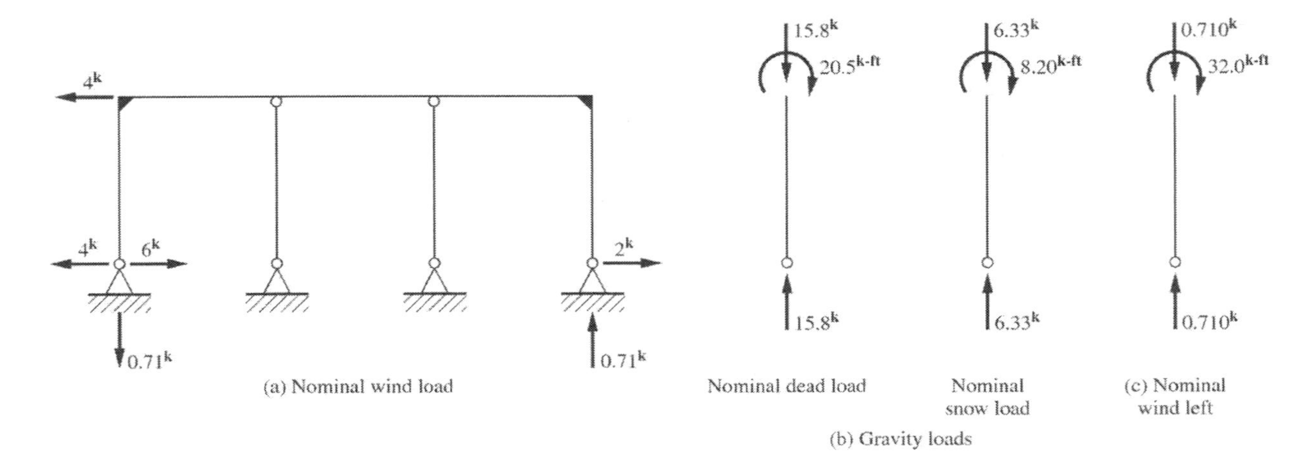

Figure 8.19 Nominal Wind Load, Snow Load, and Dead Load (Example 8.6).

EXAMPLE 8.6a ***Moment Frame Strength and Stability by LRFD***

Goal: Determine whether the structure shown in Figure 8.18 has sufficient strength and stability to carry the imposed loads.

Given: The frame shown in Figures 8.18 and 8.19 is similar to that in Example 8.1, except that the in-plane stability and lateral load resistance is provided by the moment frame action at the four corners. The exterior columns are W8×40, and the roof girder is assumed to be rigid. Out-of-plane stability and lateral load resistance is provided by X-bracing along column lines 1 and 4.

The loading is the same as that for Example 8.1: Dead Load = 50 psf, Snow Load = 20 psf, Roof Live Load = 10 psf, and Wind Load = 20 psf horizontal. Use A992 steel.

SOLUTION

Step 1: The analysis of the frame for gravity loads as given for Example 8.1 will be used. Because different load combinations may be critical, however, the analysis results for nominal Snow and nominal Dead Load are given in Figure 8.19b. The analysis results for nominal Wind Load acting to the left are given in Figure 8.19c.

Step 2: Determine the first-order forces and moments for the column on lines A-1.

For ASCE 7 load combination 3:

$$P_u = 1.2(15.8) + 1.6(6.33) + 0.5(0.710) = 29.1 + 0.355 = 29.5 \text{ kips}$$
$$M_u = 1.2(20.5) + 1.6(8.20) + 0.5(32.0) = 37.7 + 16.0 = 53.7 \text{ ft-kips}$$

For ASCE 7 load combination 4:

$P_u = 1.2(15.8) + 0.5(6.33) + 1.0(0.710) = 22.1 + 0.710 = 22.8$ kips

$M_u = 1.2(20.5) + 0.5(8.20) + 1.0(32.0) = 28.7 + 32.0 = 60.7$ ft-kips

Step 3: Determine the total story gravity load acting on one frame.

Dead = 0.05 ksf (90 ft)(50 ft)/2 frames = 113 kips

Snow = 0.02 ksf (90 ft)(50 ft)/2 frames = 45.0 kips

Step 4: Determine the second-order forces and moments for load combination 3. Gravity loads will be assumed to yield the no-translation effects, and wind load to yield the lateral translation effects.

From Step 2,

$P_{nt} = 29.1$ kips, $P_{lt} = 0.355$ kips, $M_{nt} = 37.7$ ft-kips, $M_{lt} = 16.0$ ft-kips

For the W8×40,

$A = 11.7$ in.^{2}, $I_x = 146$ in.^{4}, $r_x = 3.53$ in., $r_x/r_y = 1.73$

In the plane of the frame,

$$C_m = 0.6 = 0.4(M_1/M_2) = 0.6 - 0.4\left(\frac{0}{37.7}\right) = 0.6$$

$$P_{e1} = \frac{\pi^2 EI_x}{L_{c1}^2} = \frac{\pi^2(29{,}000)(146)}{(16.0(12))^2} = 1130 \text{ kips}$$

and with $P_r = P_{nt} + P_{lt} = 29.1 + 0.355 = 29.5$

$$B_1 = \frac{C_m}{1 - \dfrac{\alpha P_r}{P_{e1}}} = \frac{0.6}{1 - \dfrac{29.5}{1130}} = 0.616 < 1.0$$

Therefore, use $B_1 = 1.0$.

To determine the sway amplification, the total gravity load on the frame for this load combination from Step 3 is

$P_{story} = 1.2(113) + 1.6(45.0) = 208$ kips

A serviceability drift index of 0.003 is maintained under the actual wind loads. Therefore, $H = 4.0$ kips, and $\Delta/L = 0.003$ is used to determine the sway amplification factor. If this limit is not met at the completion of the design, the second-order effects must be recalculated.

The sway amplification is given by

$$B_2 = \frac{1}{1 - \left(\dfrac{\alpha P_{story}}{P_{e\ story}}\right)} > 1.0 \qquad \text{(AISC A-8-6)}$$

and

$$P_{e\ story} = R_M \frac{HL}{\Delta_H} \qquad \text{(AISC A-8-7)}$$

Since one third of the load is on the moment frame corner columns, Equation A-8-8 gives

$$R_M = 1 - 0.15\left(\frac{1}{3}\right) = 0.95$$

Thus, with $\alpha = 1.0$ for LRFD, Equation A-8-6 becomes

$$B_2 = \frac{1}{1 - \left(\dfrac{\alpha P_{story}}{R_M H}\left(\dfrac{\Delta}{L}\right)\right)} = \frac{1}{1 - \dfrac{1.0(208)}{0.95(4.0)}(0.003)} = 1.20$$

Thus, the second-order force and moment are

$M_r = 1.0(37.7) + 1.20(16.0) = 56.9$ ft-kips

$P_r = 29.1 + 1.20(0.355) = 29.5$ kips

Step 5: Determine whether the column satisfies the interaction equation.

Because the roof beam is assumed to be rigid in this example, use the recommended design value of $K = 2.0$ from Figure 5.17 case *f* in the plane of the frame, $L_{cx} = 2.0(16.0) = 32.0$ ft. Out of the plane of the frame, this is a braced frame where $K = 1.0$; thus, $L_{cy} = 16.0$ ft.

Determine the critical buckling axis.

$$(L_{cx})_{eff} = \frac{L_{cx}}{r_x / r_y} = \frac{32.0}{1.73} = 18.5 \text{ ft} > L_{cy} = 16.0 \text{ ft}$$

Thus, from *Manual* Table 6-1, using $L_c = (L_{cx})_{eff} = 18.5$ ft,

$$\phi P_n = 222 \text{ kips}$$

and from *Manual* Table 6-1 with an unbraced length of $L_b = 16$ ft

$$\phi M_{nx} = 128 \text{ ft-kips}$$

Determine the appropriate interaction equation to use.

$$\frac{P_r}{\phi P_n} = \frac{29.5}{222} = 0.133 < 0.2$$

Therefore, use Equation H1-1b.

$$\frac{P_u}{2\phi P_n} + \frac{M_u}{\phi M_n} \leq 1.0$$

$$\frac{29.5}{2(222)} + \frac{56.9}{128} = 0.511 < 1.0$$

Thus, the column is adequate for this load combination.

Step 6: Determine the first-order forces and moments for load combination 4 with the same assumption as to translation and no-translation effects. From Step 2.

$P_{nt} = 22.1$ kips, $P_{lt} = 0.710$ kips, $M_{nt} = 28.7$ ft-kips, $M_{lt} = 32.0$ ft-kips

Step 7: Determine the second-order forces and moments.

In the plane of the frame, as in Step 4,

$$C_m = 0.6 - 0.4(M_1/M_2) = 0.6 - 0.4\left(\frac{0}{28.7}\right) = 0.6$$

$$P_{e1} = \frac{\pi^2 EI_x}{L_{c1}^2} = \frac{\pi^2(29,\ 000)(146)}{(16(12))^2} = 1130 \text{ kips}$$

and

$$B_1 = \frac{C_m}{1 - \dfrac{\alpha P_r}{P_{e1}}} = \frac{0.6}{1 - \dfrac{22.8}{1130}} = 0.612 < 1.0$$

Therefore, use $B_1 = 1.0$.

To determine the sway amplification, the total gravity load on the frame is

$$P_{story} = 1.2(113) + 0.5(45.0) = 158 \text{ kips}$$

Again, a serviceability drift index of 0.003 is maintained under the actual wind loads. Therefore, $H = 4.0$ kips, and $\Delta/L = 0.003$ is used to determine the sway amplification factor. As before, $R_M = 0.95$ so

$$B_2 = \frac{1}{1 - \left(\dfrac{\alpha P_{story}}{R_M H}\left(\dfrac{\Delta}{L}\right)\right)} = \frac{1}{1 - \dfrac{1.0(158)}{0.95(4.0)}(0.003)} = 1.14$$

Thus, the second-order force and moment are

$$M_r = 1.0(28.7) + 1.14(32.0) = 65.2 \text{ ft-kips}$$
$$P_r = 22.1 + 1.14(0.710) = 22.9 \text{ kips}$$

Step 8: Determine whether the column satisfies the interaction equation.

Using the same strength values found in Step 5, determine the appropriate interaction equation.

$$\frac{P_r}{\phi P_n} = \frac{22.9}{222} = 0.103 < 0.2$$

Therefore, use Equation H1-1b.

$$\frac{P_u}{2\phi P_n} + \frac{M_u}{\phi M_n} \leq 1.0$$

$$\frac{22.9}{2(222)} + \frac{65.2}{128} = 0.561 < 1.0$$

Thus, the column is adequate for this load combination also.

Step 9: The W8×40 is shown to be adequate for gravity and wind loads in combination. Now, check to see that these columns have sufficient capacity to brace the interior pinned columns for load combination 3, which will put the greatest load on the gravity only columns.

Step 10: For stability in the plane of the frame, using the Yura approach discussed in Section 8.10, the total load on the structure is to be resisted by the four corner columns: thus

Dead Load = 0.05 ksf (50 ft)(90 ft)/4 columns = 56.3 kips/column
Snow Load = 0.02 ksf (50 ft)(90 ft)/4 columns = 22.5 kips/column

Thus, for load combination 3

$$P_u = 1.2(56.3) + 1.6(22.5) + 0.5(0.710) = 104 \text{ kips}$$
$$M_u = 1.2(20.5) + 1.6(8.20) + 0.5(32.0) = 53.7 \text{ ft-kips}$$

Step 11: Determine the second-order amplification.

As before, for the length $L_x = 16.0$ ft, $P_{e1} = 1130$ kips, and $C_m = 0.6$, the second-order amplification for member effect is

$$B_1 = \frac{C_m}{1 - \dfrac{\alpha P_r}{P_{e1}}} = \frac{0.6}{1 - \dfrac{104}{1130}} = 0.66 < 1.0$$

Therefore, use $B_1 = 1.0$ and $P_r = P_u = 104$ kips.

Sway amplification will be the same as determined in step 4, since the gravity load is the same; thus $B_2 = 1.20$. Therefore

$$M_r = M_u = 1.0(37.7) + 1.20(16.0) = 56.9 \text{ ft-kips}.$$

Step 12: Check the corner columns for interaction under these forces and moments.

As determined in Step 5 for in-plane buckling,

$$\phi P_{nx} = 222 \text{ kips}$$

$$\phi M_{nx} = 128 \text{ ft-kips}$$

Checking for the appropriate interaction equation,

$$\frac{P_u}{\phi P_n} = \frac{104}{222} = 0.468 > 0.2$$

Thus, use Equation H1-1a.

$$\frac{P_u}{\phi P_n} + \frac{8}{9}\left(\frac{M_{ux}}{\phi M_{nx}}\right) \leq 1.0$$

$$\frac{104}{222} + \frac{8}{9}\left(\frac{56.9}{128}\right) = 0.864 < 1.0$$

Thus, the W8×40 is adequate for both strength under combined load and stability for supporting the gravity only columns.

EXAMPLE 8.6b
Moment Frame Strength and Stability by ASD

Goal: Determine whether the structure shown in Figure 8.18 has sufficient strength and stability to carry the imposed loads.

Given: The frame shown in Figures 8.18 and 8.19 is similar to that in Example 8.1 except that the in-plane stability and lateral load resistance is provided by the rigid frame action at the four corners. The exterior columns are W8×40, and the roof girder is assumed to be rigid. Out-of-plane stability and lateral load resistance is provided by X-bracing along column lines 1 and 4.

The loading is the same as that for Example 8.1: Dead Load = 50 psf, Snow Load = 20 psf, Roof Live Load = 10 psf, and Wind Load = 20 psf horizontal. Use A992 steel.

SOLUTION

Step 1: The analysis of the frame for gravity loads as given for Example 8.1 will be used. Because different load combinations may be critical, however, the analysis results for nominal Snow and nominal Dead Load are given in Figure 8.19b. The analysis results for nominal Wind Load acting to the left are given in Figure 8.19c.

Step 2: Determine the first-order forces and moments for the column on lines A-1.

For ASCE 7 load combination 3:

$$P_a = (15.8) + (6.33) = 22.1 \text{ kips}$$

$M_a = (20.5) + (8.20) = 28.7$ ft-kips

For ASCE 7 load combination 6:

$P_a = (15.8) + 0.75(6.33) + 0.75(0.6(0.710)) = 20.9$ kips

$M_a = (20.5) + 0.75(8.20) + 0.75(0.6(32.0)) = 41.1$ ft-kips

Step 3: Determine the total story gravity load acting on one frame.

Dead = 0.05 ksf (90 ft)(50 ft)/2 frames = 113 kips

Snow = 0.02 ksf (90 ft)(50 ft)/2 frames = 45.0 kips

Step 4: Determine the second-order forces and moments for load combination 3. Gravity loads will be assumed to yield the no-translation effects. With no wind load, there will be no lateral translation effects; thus

From Step 2:

$P_{nt} = 22.1$ kips, $P_{lt} = 0$ kips, $M_{nt} = 28.7$ ft-kips, $M_{lt} = 0$ ft-kips

For the W8×40:

$A = 11.7$ in.^{2}, $I_x = 146$ in.^{4}, $r_x = 3.53$ in., $r_x/r_y = 1.73$

In the plane of the frame:

$$C_m = 0.6 - 0.4(M_1/M_2) = 0.6 - 0.4\left(\frac{0}{28.7}\right) = 0.6$$

$$P_{e1} = \frac{\pi^2 EI_x}{L_{c1}^2} = \frac{\pi^2(29{,}000)(146)}{(16.0(12))^2} = 1130 \text{ kips}$$

and

$$B_1 = \frac{C_m}{1 - \dfrac{\alpha P_r}{P_{e1}}} = \frac{0.6}{1 - \dfrac{1.6(22.1)}{1130}} = 0.619 < 1.0$$

Therefore, use $B_1 = 1.0$.

To determine the sway amplification, even though there are no lateral translation forces or moments for this combination, the total gravity load on the frame for this load combination from Step 3 is $P_{story} = (113) + (45.0) = 158$ kips

A serviceability drift index of 0.003 is maintained under the actual wind loads. Therefore, $H = 4.0$ kips, and $\Delta/L = 0.003$ is used to determine the sway amplification factor. If this limit is not met at the completion of the design, the second-order effects must be recalculated.

The sway amplification is given by

$$B_2 = \frac{1}{1 - \left(\dfrac{\alpha P_{story}}{P_{e\ story}}\right)} > 1.0 \qquad \text{(AISC A-8-6)}$$

and

$$P_{e\ story} = R_M \frac{HL}{\Delta_H} \qquad \text{(AISC A-8-7)}$$

Since one third of the load is on the moment frame corner columns, Equation A-8-8 gives

$$R_M = 1 - 0.15\left(\frac{1}{3}\right) = 0.95$$

Thus, with $\alpha = 1.6$ for ASD, Equation A-8-6 becomes

$$B_2 = \frac{1}{1 - \left(\dfrac{\alpha P_{story}}{R_M H}\left(\dfrac{\Delta}{L}\right)\right)} = \frac{1}{1 - \dfrac{1.6(158)}{0.95(4.0)}(0.003)} = 1.25$$

Thus, the second-order force and moment are

$M_r = 1.0(28.7) + 1.25(0) = 28.7$ ft-kips

$P_r = 22.1 + 1.25(0) = 22.1$ kips

Step 5: Determine whether the column satisfies the interaction equation.

Because the roof beam is assumed to be rigid in this example, use the recommended design value of $K = 2.0$ from Figure 5.17 case *f* in the plane of the frame, $L_{cx} = 2(16.0) = 32.0$ ft. Out of the plane of the frame, this is a braced frame where $K = 1.0$; thus, $L_{cy} = 16.0$ ft.

Determining the critical buckling axis.

$$\left(L_{cx}\right)_{eff} = \frac{L_{cx}}{r_x/r_y} = \frac{32}{1.73} = 18.5 \text{ ft} > L_{cy} = 16.0 \text{ ft}$$

Thus, from *Manual* Table 6-1, using $L_c = (L_{cx})_{eff} = 18.5$ ft,

$P_n/\Omega = 148$ kips

and from *Manual* Table 6-1, with an unbraced length of $L_b = 16$ ft,

$M_{nx}/\Omega = 84.9$ ft-kips

Determine the appropriate interaction equation to use.

$$\frac{P_r}{P_n/\Omega}=\frac{22.1}{148}=0.149<0.2$$

Therefore, use Equation H1-1b.

$$\frac{P_a}{2(P_n/\Omega)}+\frac{M_a}{(M_n/\Omega)}\leq 1.0$$

$$\frac{22.1}{2(148)}+\frac{28.7}{84.9}=0.413<1.0$$

Thus, the column is adequate for this load combination.

Step 6: Determine the first-order forces and moments for load combination 6. Gravity loads will be assumed to yield the no-translation effects, and wind load will yield the lateral translation effects. From Step 2.

P_{nt} = 20.5 kips, P_{lt} = 0.320 kips, M_{nt} = 26.7 ft-kips, M_{lt} = 14.4 ft-kips

Step 7: Determine the second-order forces and moments.

In the plane of the frame, as in Step 4,

$$C_m=0.6-0.4(M_1/M_2)0.6-0.4\left(\frac{0}{26.7}\right)=0.6$$

$$P_{e1}=\frac{\pi^2 EI_x}{L_{c1}^2}=\frac{\pi^2(29{,}000)(146)}{(16(12))^2}=1130\text{ kips}$$

and

$$B_1=\frac{C_m}{1-\dfrac{\alpha P_r}{P_{e1}}}=\frac{0.6}{1-\dfrac{1.6(20.8)}{1130}}=0.618<1.0$$

Therefore, use B_1 = 1.0.

To determine the sway amplification, the total gravity load on the frame is

$$P_{story}=(113)+0.75(45.0)=147\text{ kips}$$

Again, a serviceability drift index of 0.003 is maintained under the actual wind loads. Therefore, H = 4.0 kips and Δ/L = 0.003 is used to determine the sway amplification factor. As before, R_M = 0.95 so

$$B_2=\frac{1}{1-\left(\dfrac{\alpha P_{story}}{R_M H}\left(\dfrac{\Delta}{L}\right)\right)}=\frac{1}{1-\dfrac{1.6(147)}{0.95(4.0)}(0.003)}=1.23$$

Thus, the second-order force and moment are

$$M_r=1.0(26.7)+1.23(14.4)=44.4\text{ ft-kips}$$

and, adding in the lateral load effect amplified by B_2,

$$P_r = 20.5 + 1.23(0.320) = 20.9 \text{ kips}$$

Step 8: Determine whether the column satisfies the interaction equation.

Using the same values found in Step 5, determine the appropriate interaction equation.

$$\frac{P_r}{P_n/\Omega} = \frac{20.9}{148} = 0.141 < 0.2$$

Therefore, use Equation H1-1b.

$$\frac{P_a}{2(P_n/\Omega)} + \frac{M_a}{(M_n/\Omega)} \leq 1.0$$

$$\frac{20.9}{2(148)} + \frac{44.4}{84.9} = 0.594 < 1.0$$

Thus, the column is adequate for this load combination also.

Step 9: The W8×40 is shown to be adequate for gravity and wind loads in combination. Now, check to see that these columns have sufficient capacity to brace the interior pinned columns for gravity load only. This load combination puts the greatest load in the gravity only columns.

Step 10: For stability in the plane of the frame, using the Yura approach discussed in Section 8.10, the total load on the structure is to be resisted by the four corner columns: thus

Dead Load = 0.05 ksf (50 ft)(90 ft)/4 columns = 56.3 kips
Snow Load = 0.02 ksf (50 ft)(90 ft)/4 columns = 22.5 kips

Thus, for load combination 3,

$$P_a = (56.3) + (22.5) = 78.8 \text{ kips}$$
$$M_a = (20.5) + (8.20) = 28.7 \text{ ft-kips}$$

Step 11: Determine the second-order amplification

As before, for the length $L_x = 16.0$ ft, $P_{e1} = 1130$ kips and $C_m = 0.6$, the second-order amplification for member effect is

$$B_1 = \frac{C_m}{1 - \dfrac{\alpha P_r}{P_1}} = \frac{0.6}{1 - \dfrac{1.6(78.8)}{1130}} = 0.675 < 1.0$$

Therefore, use $B_1 = 1.0$ and $P_r = P_a = 78.8$ kips

Sway amplification will be the same as determined in step 4, since the gravity load is the same; thus, $B_2 = 1.25$. Therefore

$$M_r = 1.0(28.7) + 1.25(0) = 28.7 \text{ ft-kips}$$

Step 12: Check the corner columns for interaction under this force and moment.

As determined in Step 5 for in-plane buckling,

$$P_n/\Omega = 148 \text{ kips}$$
$$M_{nx}/\Omega = 84.9 \text{ ft-kips}$$

Checking for the appropriate interaction equation,

$$\frac{P_u}{P_n/\Omega} = \frac{78.8}{148} = 0.532 > 0.2$$

Thus, use Equation H1-1a.

$$\frac{P_a}{P_n/\Omega} + \frac{8}{9}\left(\frac{M_{ax}}{M_{nx}/\Omega}\right) \leq .0$$

$$\frac{78.8}{148} + \frac{8}{9}\left(\frac{28.7}{84.9}\right) = 0.833 < 1.0$$

Thus, the W8×40 is adequate for both strength under combined load and stability for supporting the gravity only columns.

8.11 PARTIALLY RESTRAINED FRAMES

The beams and columns in the frames considered up to this point have all been connected with moment-resisting fully restrained (FR) connections or simple pinned connections. These latter simple connections are defined in *Specification* Section B3.4a. Partially restrained (PR) connections, defined in *Specification* Section B3.4b along with FR connections, have historically been referred to as *semirigid connections*. When these PR connections are included as the connecting elements in a structural frame, they influence both the strength and stability of the structure.

Before considering the partially restrained frame, it will be helpful to look at the partially restrained beam. The relationship between the end moment and end rotation for a symmetric, uniformly loaded prismatic beam can be obtained from the well-known slope deflection equation as

$$M = -2\frac{EI\theta}{L} + \frac{WL}{12} \tag{8.12}$$

This equation is plotted in Figure 8.20a and labeled as the *beam line*.

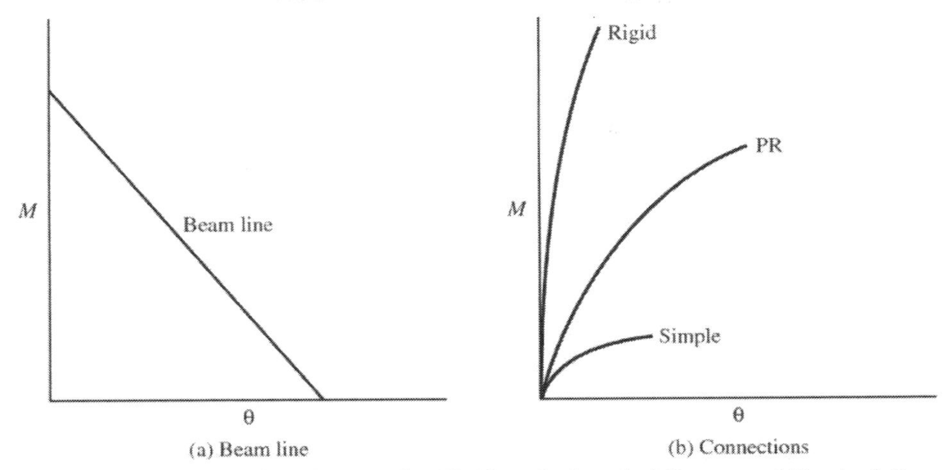

Figure 8.20 Moment Rotation Curves for Uniformly Loaded Beam and Typical Connections.

All PR connections exhibit some rotation as a result of an applied moment. The moment-rotation characteristics of these connections are the key to determining the type of connection and thus the behavior of the structure. Moment-rotation curves for three generic connections are shown in Figure 8.20b and are labeled rigid, simple, and PR. Numerous research studies have been conducted in an effort to identify the moment-rotation curves for real connections. Two compilations of these curves have been published.[5,6]

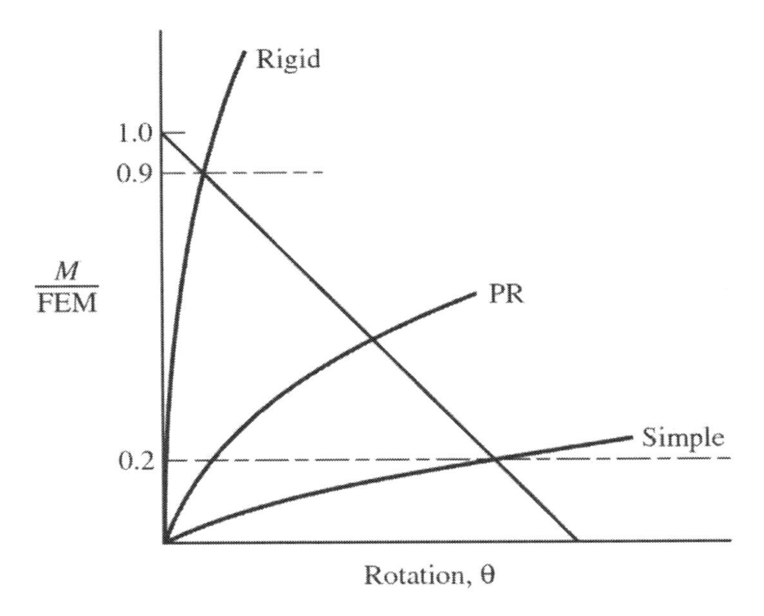

Figure 8.21 Beam Line and Connection Curves.

[5]Goverdhan, A. V., *A Collection of Experimental Moment Rotation Curves and Evaluation of Prediction Equations for Semi-Rigid Connections*, Master of Science Thesis, Vanderbilt University, Nashville, TN, 1983.
[6]Kishi, N., and Chen, W. F., *Data Base of Steel Beam-to-Column Connections*, CE-STR-86-26, West Lafayette, IN: Purdue University, School of Engineering, 1986.

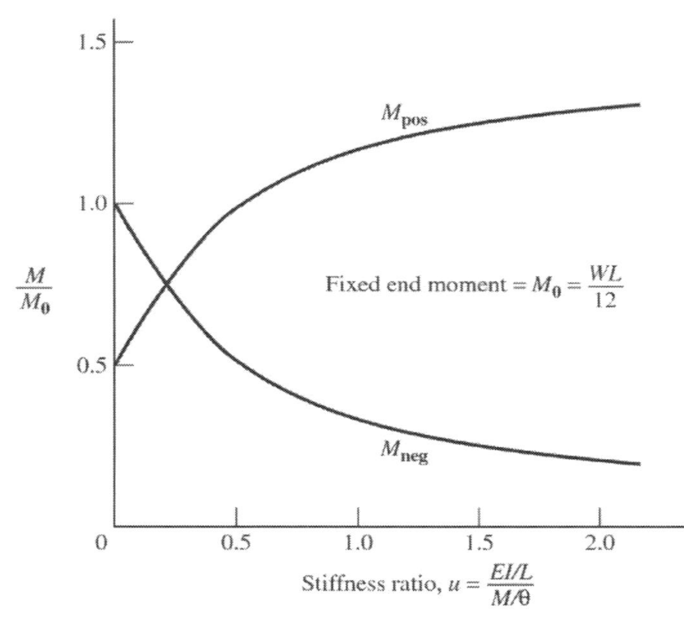

Figure 8.22 Influence of the PR Connection on the Maximum Positive and Negative Moments of a Beam.

The relationship between the moment-rotation characteristics of a connection and a beam can be seen by plotting the *beam line* and *connection curve* together, as shown in Figure 8.21. Equilibrium is obtained when the beam line and the connection curve intersect. Normal engineering practice treats connections capable of resisting at least 90 percent of the fixed-end moment as rigid and those capable of resisting no more than 20 percent of the fixed-end moment as simple. All connections that exhibit an ability to resist moment between these limits must be treated as partially restrained connections, accounting for their true moment-rotation characteristics.

The influence of the PR connection on the maximum positive and negative moments on the beam is seen in Figure 8.22. Here, the ratio of positive or negative moment to the fixed-end moment is plotted against the ratio of beam stiffness, EI/L, to a linear connection stiffness, M/θ. The moment for which the beam must be designed ranges from 0.75 times the fixed-end moment to 1.5 times the fixed-end moment, depending on the stiffness of the connection.

When PR connections are used to connect beams and columns to form PR frames, the analysis becomes much more complex. The results of numerous studies dealing with this issue have been reported. Although some practical designs have been carried out, widespread practical design of PR frames is not common. In addition to the problems associated with modeling a particular connection, the question of loading sequence arises. Because real, partially restrained connections behave nonlinearly, the sequence of applied loads influences the structural response. The approach to load application may have more significance than the accuracy of the connection model used in the analysis.

Although a complete, theoretical analysis of a partially restrained frame may currently be beyond the scope of normal engineering practice, a simplified approach exists that is not only well within the scope of practice, but also commonly carried out in everyday design and has been for over half a century. This approach can be referred to as

Flexible Moment Connections. It has historically been called *Type 2 with Wind*. The Flexible Moment Connection approach relies heavily on the nonlinear moment-rotation behavior of the PR connection although the actual curve is not used. In addition, it relies on a phenomenon called *shake-down*, which shows that the connection, although exhibiting nonlinear behavior initially, behaves linearly after a limited number of applications of lateral load.[7]

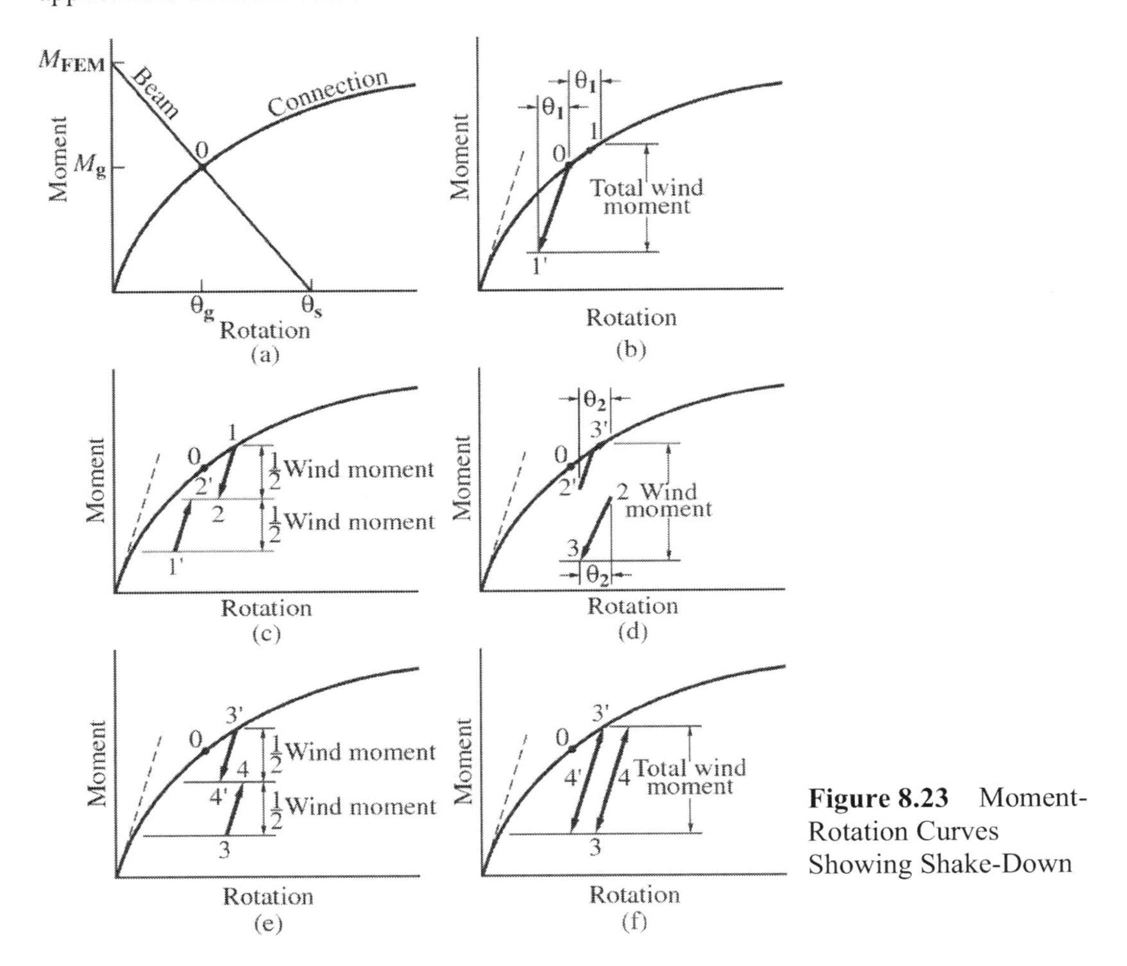

Figure 8.23 Moment-Rotation Curves Showing Shake-Down

The moment-rotation curve for a typical PR connection is shown in Figure 8.23a along with the beam line for a uniformly loaded beam. The point labeled 0 represents equilibrium for the applied gravity loads. The application of wind load produces moments at the beam ends that add to the gravity moment at the leeward end of the beam and subtract from the windward end. Because moment at the windward end is being removed, the connection behaves elastically with a stiffness close to the original connection stiffness, whereas at the leeward end, the connection continues to move along

[7]Geschwindner, L. F., and Disque, R. O., "Flexible Moment Connections for Unbraced Frames Subject to Lateral Forces—A Return to Simplicity." *Engineering Journal*, AISC, Vol. 42, No. 2, 2005, pp 99–112.

the nonlinear connection curve. Points labeled 1 and 1′ in Figure 8.23b represent equilibrium under the first application of wind to the frame.

When the wind load is removed, the connection moves from points 1 and 1′ to points 2 and 2′, as shown in Figure 8.23c. The next application of a wind load that is larger than the first and in the opposite direction will see the connection behavior move to points 3 and 3′. Note that on the windward side, the magnitude of this applied wind moment dictates whether the connection behaves linearly or follows the nonlinear curve, as shown in Figure 8.23d. Removal of this wind load causes the connection on one end to unload and on the other end to load, both linearly as shown in Figure 8.23e. Any further application of wind load, less than the maximum already applied, will see the connection behave linearly. In addition, the maximum moment on the connection is still close to that applied originally from the gravity load. Thus, the condition described in Figure 8.23f shows that shake-down has taken place and the connection now behaves linearly for both loading and unloading.

The design procedure used to account for this shake-down is straight forward. All beams are designed as simple beams using the appropriate load combinations. This assures that the beams are adequate, regardless of the actual connection stiffness, as was seen in Figure 8.22. Wind load moments are determined through a modified portal analysis where the leeward column is assumed not to participate in the lateral load resistance. Connections are sized to resist the resulting moments, again for the appropriate load combinations. In addition, it is particularly important to provide connections that have sufficient ductility to accommodate the large rotations that will occur, without overloading the bolts or welds under combined gravity and wind.

Columns must be designed to provide frame stability under gravity loads as well as gravity plus wind. The columns may be designed using the approach that was presented for columns in moment frames, but with two essential differences from the conventional rigid frame design:

1. Because the gravity load is likely to load the connection to its plastic moment capacity, the column can be restrained only by a girder on one side and this girder will act as if it is pinned at its far end. Therefore, in computing the girder stiffness rotation factor, I_g/L_g, for use in the effective length alignment chart, only one girder should be considered, and the girder length should be doubled.

2. One of the external columns, the leeward column for the wind loading case, cannot participate in frame stability, because it will be attached to a connection that is at its plastic moment capacity. The stability of the frame may be assured, however, by designing the remaining columns to support the total frame load.

For the exterior column, the moment in the beam to column joint is equal to the capacity of the connection. It is sufficiently accurate to assume that this moment is distributed one-half to the upper column and one-half to the lower column. For interior columns, the greatest realistically possible difference in moments resulting from the girders framing into the column should be distributed equally to the columns above and below the joint.

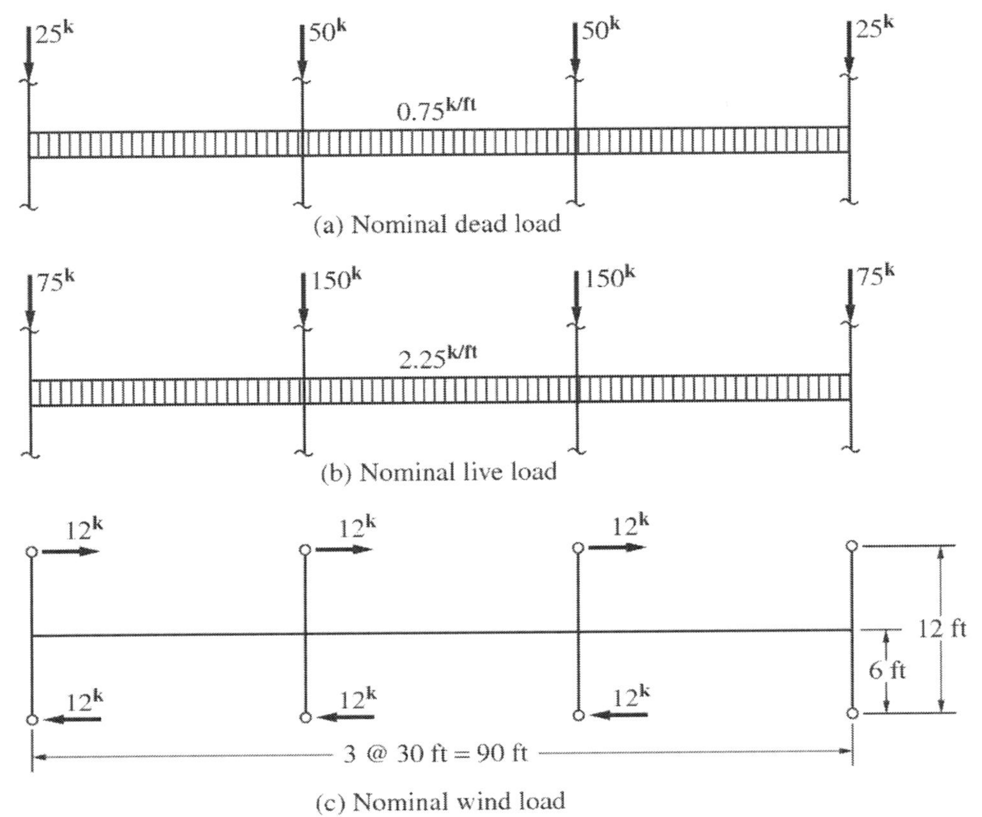

Figure 8.24 Intermediate Story of a Three-Story Building (Example 8.7).

EXAMPLE 8.7a ***Column Design with Flexible Wind Connections by LRFD***

Goal: Select girders and columns for a building with flexible wind connections.

Given: An intermediate story of a three-story building is given in Figure 8.24. Story height is 12 ft. The frame is braced in the direction normal to that shown. Use the LRFD provisions and A992 steel.

SOLUTION

Step 1: Determine the required forces and moments for the load combination

$$12.D + 0.5L + 1.0W.$$

The loads shown in Figure 8.24 are the code-specified nominal loads. The required forces are calculated using tributary areas as follows.

Gravity loads on exterior columns.

1.2D = 1.2(25 kips + 0.75 kips/ft (15 ft)) = 43.5 kips
0.5L = 0.5 (75 kips + 2.25 kips/ft (15 ft)) = 54.4 kips
Total =97.9 kips

Gravity loads on interior columns

1.2D = 1.2 (50 kips + 0.75 kips/ft (30 ft)) = 87.0 kips
0.5L = 0.5 (150 kips + 2.25 kips/ft (30 ft)) = 109 kips
Total= 196 kips

Gravity load on girders for the worst case 1.2D + 1.6L:

1.2D = 1.2 (0.75 kips/ft (30 ft)) = 27.0 kips
1.6L = 1.6 (2.25 kips/ft (30 ft)) = 108 kips
Total = 135 kips

Step 2: Design the girder for the simple beam moment assuming full lateral support using *Manual* Table 3-2 or 6-1.

$$M_u = WL/8 = 135(30.0)/8 = 506 \text{ ft-kips}$$

Therefore use

W21×62 (ϕM_n = 540 ft-kips, I_x = 1330 in.^{4})

Step 3: Design the columns for strength. The gravity load on the interior column using *Manual* Table 4-1a. Restrict the columns to a W14.

For buckling out of the plane in a braced frame, K = 1.0 and L_{cy} = 12.0

Thus, with P_u = 196 kips, try

W14×43, (ϕP_n = 371 kips, I_x = 428 in.^{4}, r_x/r_y = 3.08)

which is the lightest W14 available in Table 4-1a.

Step 4: To check the column for stability in the plane, determine the effective length factor from the alignment chart with

$$G_{top} = G_{bottom} = \frac{\Sigma(I/L)_c}{\Sigma(I/L)_g} = \frac{2\left(\dfrac{428}{12.0}\right)}{\left(\dfrac{1330}{2(30.0)}\right)} = 3.21$$

Note that only one beam is capable of restraining the column and that the beam is pinned at its far end; thus, only one beam can be considered and the effective beam length is taken as twice its actual length.

Considering the stress in the column under load, the stiffness reduction factor can be determined.

$$P_u / A = 196/12.6 = 15.6 \text{ ksi}$$

Thus, since $P_u/A = P_r/A = 15.6 < 0.5F_y$, the stiffness reduction factor from Manual Table 4-13 is $\tau_b = 1.00$. The stiffness ratio then remains

$$G_{top} = G_{bottom} = 3.21$$

which yields, from the alignment chart, Figure 5.20 using Equation AISC C-A-7-2

$$K = 1.87$$

Step 5: Determine the effective length in the plane of bending.

$$\left(L_{cx}\right)_{eff} = \frac{L_{cx}}{r_x / r_y} = \frac{1.87(12.0)}{3.08} = 7.29 \text{ ft}$$

Step 6 Determine the column compressive strength from *Manual* Table 4-1a or 6-1 with $L_c = 7.29$ ft.

$$\phi P_n = 484 \text{ kips}$$

Step 7: Determine the second-order moment.

The applied wind moment is $M_u = 1.0(6.0)(12.0) = 72.0$ ft-kips and the applied force is $P_u = 196$ kips.

Considering all the moment as a translation moment and using Commentary equation C-A-8-1

$$P_{e\ story} = \frac{\pi^2 EI}{\left(K_2 L\right)^2} = \frac{\pi^2 (29{,}000)(428)}{(1.87(12.0)(12))^2} = 1690 \text{ kips}$$

Therefore, for all three columns,

$$B_2 = \frac{1}{1 - \left(\dfrac{\alpha P_{story}}{P_{e\ story}}\right)} = \frac{1}{1 - \dfrac{1.0\left(3(196)\right)}{3(1690)}} = 1.13$$

and

$$M_r = 1.13\ (72.0) = 81.4 \text{ ft-kips}$$

Step 8: Determine whether the column satisfies the interaction equation

$$\frac{P_u}{\phi P_n} = \frac{196}{484} = 0.405 > 0.2$$

Therefore, use Equation H1-1a, $\phi M_n = 222$, from *Manual* Table 3-10 or 6-1, which results in

$$\frac{P_u}{\phi P_n}+\frac{8}{9}\left(\frac{M_u}{\phi M_n}\right)\leq 1.0$$

$$0.405+\frac{8}{9}\left(\frac{81.4}{222}\right)=0.731<1.0$$

This indicates that the W14×43 is adequate for stability. The members can then be used as a starting point in a more rigorous analysis.

EXAMPLE 8.7b
Column Design with Flexible Wind Connections by ASD

Goal: Select girders and columns for a building with flexible wind connections.

Given: An intermediate story of a three-story building is given in Figure 8.24. Story height is 12 ft. The frame is braced in the direction normal to that shown. Use the ASD provisions and A992 steel.

Step 1: Determine the required forces and moments for the load combination

D + 0.75L +0.75(0.6W).

The loads shown in Figure 8.24 are the code-specified nominal loads. The required forces are calculated using tributary areas as follows.

Gravity loads on exterior columns

D = (25 k + 0.75 k/ft (15 ft)) = 36.3 kips
0.75L = 0.75 (75 k + 2.25 k/ft (15 ft)) = 81.6 kips
Total = 118 kips

Gravity loads on interior columns

D = (50 k + 0.75 k/ft (30 ft)) = 72.5 kips
0.75L = 0.75 (150 k + 2.25 k/ft (30 ft)) = 163 kips
Total = 236 kips

Gravity load on girders for the worst case, D + L

D = (0.75 k/ft (30 ft)) = 22.5 kips
L = (2.25 k/ft (30 ft)) = 67.5 kips
Total = 90.0 kips

Step 2: Design the girder for the simple beam moment assuming full lateral support using *Manual* Table 3-2 or 6-1.

$$M_a = WL/8 = 90.0(30.0)/8 = 338 \text{ ft-kips}$$

Therefore use

W21×62 (M_n/Ω = 359 ft-kips, I_x = 1330 in.^{4})

Step 3: Design the columns for Strength. The gravity load on the interior column using *Manual* Table 4-1a or 6-1. Restrict the columns to a W14.

For buckling out of the plane in a braced frame, $K = 1.0$ and $L_{cy} =$ 12.0

Thus, with $P_a = 236$ kips
try W14×43 ($P_n/\Omega = 247$ kips, $I_x = 428$ in.^{4}, $r_x/r_y = 3.08$)

which is the lightest W14 available in Table 4-1a.

Step 4: To check the column for Stability in the plane, determine the effective length factor from the alignment chart with

$$G_{top} = G_{bottom} = \frac{\Sigma(I/L)_c}{\Sigma(I/L)_g} = \frac{2\left(\dfrac{428}{12.0}\right)}{\left(\dfrac{1330}{2(30.0)}\right)} = 3.21$$

Note that only one beam is capable of restraining the column and that that beam is pinned at its far end; thus, only one beam can be considered and the effective beam length is taken as twice its actual length.

Considering the stress in the column under load, the stiffness reduction factor can be determined.

$$\frac{P_a}{A} = \frac{236}{12.6} = 18.7 \text{ ksi}$$

Thus, from the *Manual* Table 4-13, the stiffness reduction factor $\tau_b =$ 0.960. The inelastic stiffness ratio then becomes

$$G_{top} = G_{bottom} = 0.960(3.21) = 3.08$$

which yields, from the alignment chart, Figure 5.20 using Equation AISC C-A-7-2

$$K = 1.84$$

Step 5: Determine the effective length in the plane of bending.

$$\left(L_{cx}\right)_{eff} = \frac{L_{cx}}{r_x/r_y} = \frac{1.84(12.0)}{3.08} = 7.17 \text{ ft}$$

Step 6: Determine the column compressive strength from *Manual* Table 4-1a or 6-1 with $L_c = 7.17$ ft.

$$P_n/\Omega = 324 \text{ kips}$$

Step 7: Determine the second-order moment.

The applied wind moment is $M_a = 0.75(0.6)(6.0(12.0)) = 32.4$ ft-kips and the applied force is $P_a = 236$ kips.

Considering all the moment as a translation moment and using Commentary equation C-A-8-1

$$P_{e\,story} = \frac{\pi^2 EI}{(K_2L)^2} = \frac{\pi^2(29{,}000)(428)}{(1.84(12)(12))^2} = 1740 \text{ kips}$$

$$\alpha P_a = 1.6(236) = 378 \text{ kips}$$

Therefore, for all three columns,

$$B_2 = \frac{1}{1 - \dfrac{\alpha P_{story}}{P_{e\,story}}} = \frac{1}{1 - \dfrac{3(378)}{3(1740)}} = 1.28$$

and

$$M_r = 1.28(32.4) = 41.5 \text{ ft-kips}$$

Step 8: Determine whether the column satisfies the interaction equation

$$\frac{P_r}{P_n/\Omega} = \frac{236}{324} = 0.728 < 0.2$$

Therefore, use Equation H1-1a, $M_n/\Omega = 148$, from *Manual* Table 3-10 or 6-1, which results in

$$\frac{P_a}{(P_n/\Omega)} + \frac{8}{9}\left(\frac{M_a}{(M_n/\Omega)}\right) \le 1.0$$

$$0.728 + \frac{8}{9}\left(\frac{41.5}{148}\right) = 0.98 < 1.0$$

This indicates that the W14×43 is adequate for stability. These members can then be used as a starting point in a more rigorous analysis.

After an acceptable column is selected, the lateral displacement of the structure must be checked. Coverage of drift in wind moment frames is beyond the treatment intended here but is covered in the Geschwindner and Disque paper already referenced.

8.12 STABILITY BRACING DESIGN

Braces in steel structures are used to reduce the effective length of columns, reduce the unbraced length of beams, and provide overall structural stability. The discussion of columns in Chapter 5 showed how braces could be effective in reducing effective length and thereby increasing column strength. Chapter 6 demonstrated how the unbraced length of a beam influenced its strength, and earlier in this chapter the influence of sway on the stability of a structure was discussed. Every case assumed that the given bracing requirements were satisfied, but nothing was said about the strength or stiffness of the required braces. For cases when braces are not specifically included in the second-order analysis, design of braces will follow the provisions of Appendix 6 of the *Specification*. Appendix 6 treats bracing for columns and beams similarly, although the specific requirements are different. Two types of braces are defined: point bracing and panel bracing.

Point bracing controls the movement of a point on the member without interaction with any adjacent braced points. These braces would be attached to the member and then to a fixed support, such as the abutment shown in Figure 8.25b.

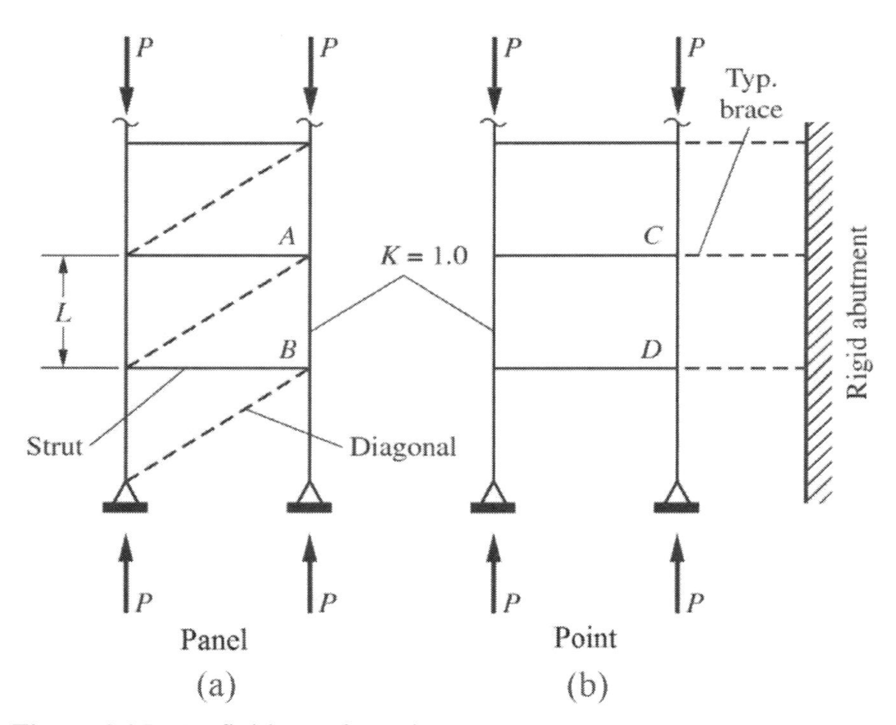

Figure 8.25 Definitions of Bracing Types. Copyright © American Institute of Steel Construction, Inc. Reprinted with Permission. All rights reserved.

Panel bracing relies on other braced points of the structure to provide support. A diagonal brace within a frame would be a panel brace, as shown in Figure 8.25a. In this case, the axial deformation of the diagonal brace is a function of the displacement at each end of the brace. Because the horizontal strut is usually a part of a very stiff floor system that has significant strength in its plane, the strength and stiffness of the diagonal element usually controls the overall behavior of this braced system.

The brace requirements of the *Specification* are intended to enable the members' being designed to reach their maximum strength based on the length between bracing points and an effective length factor, $K = 1.0$. A brace has two requirements: strength and stiffness. A brace that is inadequate in either of these respects is not sufficient to enable the member it is bracing to perform as it was designed.

8.12.1 Column Bracing

For column panel bracing, the required shear strength of the bracing system is

$$V_{br} = 0.005P_r \quad \text{(AISC A-6-1)}$$

and the required shear stiffness is

$$\beta_{br} = \frac{1}{\phi}\left(\frac{2P_r}{L_{br}}\right)\text{(LRFD)} \qquad \beta_{br} = \Omega\left(\frac{2P_r}{L_{br}}\right)\text{(ASD)} \quad \text{(AISC A-6-2)}$$

$$\phi = 0.75 \text{ (LRFD)} \qquad \Omega = 2.00 \text{ (ASD)}$$

where

L_{br} = unbraced length of the panel under consideration
P_r = required strength of the column within the panel under consideration for ASD or LRFD as appropriate for the design method being used.

For a column point brace, the required brace strength is

$$P_{br} = 0.01\ P_r \quad \text{(AISC A-6-3)}$$

and the required brace stiffness is

$$\beta_{br} = \frac{1}{\phi}\left(\frac{8P_r}{L_{br}}\right)\text{(LRFD)} \qquad \beta_{br} = \Omega\left(\frac{8P_r}{L_{br}}\right)\text{(ASD)} \quad \text{(AISC A-6-4)}$$

$$\phi = 0.75 \text{ (LRFD)} \qquad \Omega = 2.00 \text{ (ASD)}$$

where

L_{br} = laterally unbraced length adjacent to the point brace
P_r = required strength for ASD or LRFD as appropriate for the design method being used.

It should be noted that the requirements for point braces are significantly greater than those for panel braces. Thus, if a panel bracing system can be developed, it has the potential to be the more economical approach.

8.12.2 Beam Bracing

For a beam panel brace, the required shear strength of the bracing system is

$$V_{br} = 0.01\left(\frac{M_r C_d}{h_o}\right) \quad \text{(AISC A-6-5)}$$

and the required panel brace stiffness is

$$\beta_{br} = \frac{1}{\phi}\left(\frac{4M_r C_d}{L_{br} h_o}\right)\text{(LRFD)} \qquad \beta_{br} = \Omega\left(\frac{4M_r C_d}{L_{br} h_o}\right)\text{(ASD)} \qquad \text{(AISC A-6-6)}$$

$$\phi = 0.75 \text{ (LRFD)} \qquad \Omega = 2.00 \text{ (ASD)}$$

where

h_o = distance between flange centroids
C_d = 1.0 except for the brace closest to the inflection point for double curvature bending $C_d = 2.0$
L_{br} = laterally unbraced length within the panel under consideration
M_r = the largest required flexural strength of the beam within the panel under consideration

For a beam point brace, the required strength of the brace is

$$P_{br} = 0.02 M_r C_d / h_o \qquad \text{(AISC A-6-7)}$$

and the required brace stiffness is

$$\beta_{br} = \frac{1}{\phi}\left(\frac{10M_r C_d}{L_{br} h_o}\right)\text{(LRFD)} \qquad \beta_{br} = \Omega\left(\frac{10M_r C_d}{L_{br} h_o}\right)\text{(ASD)} \qquad \text{(AISC A-6-8)}$$

$$\phi = 0.75 \text{ (LRFD)} \qquad \Omega = 2.00 \text{ (ASD)}$$

where

h_o = distance between flange centroids
C_d = 1.0 except for the brace closest to the inflection point for double curvature bending $C_d = 2.0$
L_{br} = laterally unbraced length adjacent to the point brace
M_r = the largest of the required flexural strengths of the beam within the unbraced lengths adjacent to the point brace

As with column bracing, the requirements for point braces are greater than those for panel braces.

8.12.3 Frame Bracing

Frame bracing and column bracing are accomplished by the same panel and point braces and may be designed using the same stiffness and strength equations. However, the most direct approach to bracing design for frames is to include the braces in the model when a second-order analysis is carried out. When that is the case, the provisions of Appendix 6 do not need to be checked.

EXAMPLE 8.8a
Bracing Design by LRFD

Goal: Determine the required bracing for a braced frame to provide stability for the gravity load.

Given: Using the LRFD requirements, select a rod in tension to provide the point bracing shown in the center panel of the three-bay frame of Figure 8.9a to provide stability for a total gravity dead load of 113 kips and live load of 45 kips.

SOLUTION

Step 1: Determine the required brace stiffness for gravity load.

For the gravity load, the required brace stiffness is based on $1.2D + 1.6L$.

$$P_r = 1.2(113) + 1.6(45.0) = 208 \text{ kips}$$

and from Equation A-6-4

$$\beta_{br} = \frac{1}{\phi}\left(\frac{8P_r}{L_{br}}\right) = \frac{1}{0.75}\left(\frac{8(208)}{16.0}\right) = 139 \text{ kips/ft}$$

Step 2: Determine the required brace area based on required stiffness and accounting for the angle of the brace.

Based on the geometry of the brace from Figure 8.9, where θ is the angle of the brace with the horizontal and $L_r = 34.0$ ft is the length of the brace,

$$\beta_{br} = \frac{A_{br}E}{L_r}\cos^2\theta = 139 \text{ kips/ft}$$

This results in a required brace area

$$A_{br} = \frac{\beta_{br}L_r}{E\cos^2\theta} = \frac{139(34.0)}{29{,}000\left(\dfrac{30}{34}\right)^2} = 0.209 \text{ in.}^2$$

Step 3: Determine the required brace force for gravity load. The required horizontal brace force for a point brace given by Equation A-6-3 is

$$P_{br} = 0.01P_r = 0.01(208) = 2.08 \text{ kips}$$

which gives a force in the member of

$$P_{br(angle)} = 2.08(34/30) = 2.36 \text{ kips}$$

and a required area, assuming $F_y = 36$ ksi for a rod, of

$$A_{br} = \frac{P_{br(angle)}}{\phi F_y} = \frac{2.36}{0.9(36)} = 0.0728 \text{ in.}^2$$

Step 4: For the dead plus live load case,

$$A_{min} = 0.209 \text{ in.}^2$$

Step 5: Select a rod to meet the required area for the controlling case of stiffness for the dead plus live load case where $A_{min} = 0.209 \text{ in.}^2$

use a 5/8-in. rod with A = 0.307 in.^{2}

EXAMPLE 8.8b
Bracing Design by ASD

Goal: Determine the required bracing for a braced frame to provide stability for the gravity load.

Given: Using the ASD requirements, select a rod in tension to provide the point bracing shown in the center panel of the three-bay frame of Figure 8.9a to provide stability for a total gravity dead load of 113 kips and live load of 45 kips.

SOLUTION

Step 1: Determine the required brace stiffness for gravity load.

For the gravity load, the required brace stiffness is based on $D + L$.

$$P_r = 113 + 45.0 = 158 \text{ kips}$$

and from Equation A-6-4

$$\beta_{br} = \Omega\left(\frac{8P_r}{L_{br}}\right) = 2.00\left(\frac{8(158)}{16.0}\right) = 158 \text{ kips/ft}$$

Step 2: Determine the required brace area based on required stiffness and accounting for the angle of the brace.

Based on the geometry of the brace from Figure 8.9, where θ is the angle of the brace with the horizontal and L_r =34.0 ft is the length of the brace.

$$\beta_{br} = \frac{A_{br}E}{L_r}\cos^2\theta = 158 \text{ kips/ft}$$

This results in a required brace area

$$A_{br} = \frac{\beta_{br}L_r}{E\cos^2\theta} = \frac{158(34.0)}{29{,}000\left(\frac{30}{34}\right)^2} = 0.238 \text{ in.}^2$$

Step 3: Determine the required brace force for gravity load.

The required horizontal brace force for a point brace given by Equation A-6-3 is

$$P_{br} = 0.01P_r = 0.01(158) = 1.58 \text{ kips}$$

which gives a force in the member of

$$P_{br(angle)} = 1.58(34/30) = 1.79 \text{ kips}$$

and a required area, assuming $F_y = 36$ ksi for a rod, of

$$A_{br} = \frac{P_{br(angle)}}{F_y/\Omega} = \frac{1.79}{(36/1.67)} = 0.0830 \text{ in.}^2$$

Step 4: For the dead plus live load case,

$$A_{min} = 0.238 \text{ in.}^2$$

Step 5: Select a rod to meet the required area for the controlling case of stiffness for the dead plus live load case, $A_{min} = 0.238$ in.^{2}.

use a 5/8-in. rod with A = 0.307 in.^{2}

8.13 TENSION PLUS BENDING

Throughout this chapter, the case of combined compression plus bending has been treated. That is the most common case of combined loading in typical building structures. However, the *Specification* also has provisions, in Section H1.2, for combining flexure and tension. The addition of a tension force to a member already undergoing bending may be beneficial.

The interaction equations for combined tension and flexure are the same as those already discussed and given as Equations H1-1a and H1-1b. However, if the flexural strength is controlled by the limit state of lateral-torsional buckling, the addition of a tension force can increase bending strength. This is accounted for in the *Specification* by the introduction of a modification factor to be applied to C_b. Thus, for doubly symmetric members, C_b in Chapter F can be multiplied by $\sqrt{1+\alpha P_r/P_{ey}}$ for axial tension that acts concurrently with flexure, where $P_{ey} = \pi^2 EI_y/L_b^2$ and $\alpha = 1.0$ for LRFD and 1.6 for ASD, as before. The limit that M_n cannot exceed M_p still must be satisfied as it was for beam design discussed in Chapter 6.

EXAMPLE 8.9a ***Combined Tension and Bending by LRFD***

Goal: Check the given W-shape beam for combined tension and bending

Given: A W16×77 beam spans 25 ft and carries a uniform dead load of 0.92 kips/ft and a uniform live load of 2.79 kips/ft. It also carries a tension live load of 62.5 kips. The member is braced at the ends only for lateral-torsional buckling. Use A992 steel.

SOLUTION

Step 1: Determine the required moment strength

$$w_u = 1.2(0.92) + 1.6(2.79) = 5.57 \text{ kips/ft}$$

$$M_u = \frac{5.57(25)^2}{8} = 435 \text{ ft-kips}$$

Step 2: Determine the required tension strength

$$T_u = 1.6(62.5) = 100 \text{ kips}$$

Step 3: Determine the available moment strength. With L_b = 25 ft and C_b = 1.14, from Manual Table 6-1

$$\phi M_n = 1.14(383) = 437 \text{ ft-kips} < \phi M_p = 563 \text{ ft-kips}$$

Step 4: Determine the available tension strength for the limit state of yielding. Connections at the end of the member are at a location of zero moment so tension rupture will not be a factor for interaction with bending. From Table 6-1

$$\phi T_n = 1020 \text{ kips}$$

Step 5: Determine the increase to be applied to C_b when tension is applied in conjunction with moment strength determined for the lateral-torsional buckling limit state.

$$P_{ey} = \frac{\pi^2 EI}{L_b^2} = \frac{\pi^2 (29{,}000)(138)}{\left(25(12)\right)^2} = 439 \text{ kips}$$

$$\sqrt{1+\frac{\alpha P_r}{P_{ey}}} = \sqrt{1+\frac{1.0(100)}{439}} = 1.11$$

Step 6: Moment strength when considered in combination with tension

$$\phi M_n = 1.11(437) = 485 \text{ ft-kips} < \phi M_p = 563 \text{ ft-kips}$$

Step 7: Determine the interaction equation to use

$$\frac{P_r}{P_c} = \frac{100}{1020} = 0.098 < 0.2$$

Step 8: Use Equation H1-1b

$$\frac{P_r}{2P_c} + \frac{M_r}{M_c} = \frac{0.098}{2} + \frac{435}{485} = 0.049 + 0.897 = 0.95 < 1.0$$

So, the beam is adequate to carry the bending moment and tension force.

Step 9: Check the beam for bending alone, in case the tension force were not there. Use the available moment strength from Step 3.

$$\frac{M_r}{M_c} = \frac{435}{437} = 0.995 \le 1.0$$

So, the beam would just be adequate. In cases where the application of the tension force increases interaction strength and that force may not actually occur, it is important to check the member for flexure alone.

The W16×77 is adequate to carry the applied loads.

EXAMPLE 8.9b ***Combined Tension and Bending by ASD***

Goal: Check the given W-shape beam for combined tension and bending

Given: A W16×77 beam spans 25 ft and carries a uniform dead load of 0.92 kips/ft and a uniform live load of 2.79 kips/ft. It also carries a tension live load of 62.5 kips. The member is braced at the ends only for lateral-torsional buckling. Use A992 steel.

SOLUTION

Step 1: Determine the required moment strength

$$w_a = 0.92 + 2.79 = 3.71 \text{ kips/ft}$$

$$M_a = \frac{3.71(25)^2}{8} = 290 \text{ ft-kips}$$

Step 2: Determine the required tension strength

$$T_a = 62.5 \text{ kips}$$

Step 3: Determine the available moment strength. With L_b = 25 ft and C_b = 1.14, from Manual Table 6-1

$$\frac{M_n}{\Omega} = 1.14(255) = 291 \text{ ft-kips} < \frac{M_p}{\Omega} = 374 \text{ ft-kips}$$

Step 4: Determine the available tension strength for the limit state of yielding. Connections at the end of the member are at a location of zero moment so tension rupture will not be a factor for interaction with bending. From Table 6-1

$$\frac{T_n}{\Omega} = 677 \text{ kips}$$

Step 5: Determine the increase to be applied to C_b when tension is applied in conjunction with moment strength determined for the lateral-torsional buckling limit state.

$$P_{ey} = \frac{\pi^2 EI}{L_b^2} = \frac{\pi^2(29{,}000)(138)}{\left(25(12)\right)^2} = 439 \text{ kips}$$

$$\sqrt{1 + \frac{\alpha P_r}{P_{ey}}} = \sqrt{1 + \frac{1.6(62.5)}{439}} = 1.11$$

Step 6: Moment strength when considered in combination with tension

$$\frac{M_n}{\Omega} = 1.11(291) = 323 \text{ ft-kips} < \frac{M_p}{\Omega} = 374 \text{ ft-kips}$$

Step 7: Determine the interaction equation to use

$$\frac{P_r}{P_c} = \frac{62.5}{677} = 0.092 < 0.2$$

Step 8: Use Equation H1-1b

$$\frac{P_r}{2P_c} + \frac{M_r}{M_c} = \frac{0.092}{2} + \frac{290}{323} = 0.046 + 0.898 = 0.94 < 1.0$$

So, the beam is adequate to carry the bending moment and tension force.

Step 9: Check the beam for bending alone, in case the tension force were not there. Use the available moment strength from Step 3.

$$\frac{M_r}{M_c} = \frac{290}{291} = 0.997 \leq 1.0$$

So, the beam would just be adequate. In cases where the application of the tension force increases interaction strength and that force may not actually occur, it is important to check the member for flexure alone.

The W16×77 is adequate to carry the applied loads.

8.14 PROBLEMS

Unless noted otherwise, all columns should be considered pinned in a braced frame out of the plane being considered in the problem with bending about the strong axis.

1. Determine whether a W14×90, A992 column with a length of 12.5 ft is adequate in a braced frame to carry the following loads from a first-order analysis: a compressive dead load of 100 kips and live load of 300 kips, a dead load moment of 30 ft-kips and live load moment of 70 ft-kips at each end. The member is bending in reverse curvature about the strong axis. Determine by (a) LRFD and (b) ASD.

2. A W12×58, A992 is used as a 14 ft column in a braced frame to carry a compressive dead load of 60 kips and live load of 120 kips. Will this column be adequate to carry a dead load moment of 30 ft-kips and live load moment of 60 ft-kips at each end, bending the column in single curvature about the strong axis? The analysis results are from a first-order analysis. Determine by (a) LRFD and (b) ASD.

3. Determine whether a W12×190, A992 column with a length of 22 ft is adequate in a braced frame to carry the following loads from a first-order analysis: a compressive dead load of 300 kips and live load of 500 kips, a dead load moment of 50 ft-kips and live load moment of 100 ft-kips at one end, and a dead load moment of 75 ft-kips and a live load moment of 85 ft-kips at the other. The member is bending in reverse curvature about the strong axis. Determine by (a) LRFD and (b) ASD.

4. A W10×60, A992 is used as a 13 ft column in a braced frame to carry a compressive dead load of 150 kips and live load of 120 kips. Will this column be adequate to carry a dead load moment of 30 ft-kips and live load moment of 45 ft-kips at each end, bending the column in single curvature about the strong axis? The analysis results are from a first-order analysis. Determine by (a) LRFD and (b) ASD.

5. Given a W14×283, A992 42 ft column in a braced frame with a compressive dead load of 90 kips and live load of 270 kips. Maintaining a live load to dead load ratio of 3, determine the maximum live and dead load second-order moments that can be applied about the strong axis on the upper end when the lower end is pinned by (a) LRFD and (b) ASD.

6. Given a W14×132, A992 15 ft column in a braced frame with a compressive dead load of 350 kips and live load of 350 kips, and maintaining a live load to dead load ratio of 1, determine the maximum live and dead load second-order moments that can be applied about the strong axis on the upper end when the lower end is pinned by (a) LRFD and (b) ASD.

7. Reconsider the column and loadings in Problem 1 if that column were bent in single curvature by (a) LRFD and (b) ASD.

8. Reconsider the column and loadings in Problem 2 if that column were bent in reverse curvature by (a) LRFD and (b) ASD.

9. Reconsider the column and loadings in Problem 3 if that column were bent in single curvature by (a) LRFD and (b) ASD.

10. Reconsider the column and loadings in Problem 4 if that column were bent in reverse curvature by (a) LRFD and (b) ASD.

11. A 14 ft pin-ended column in a braced frame must carry a compressive dead load of 85 kips and live load of 280 kips, along with a uniformly distributed transverse dead load of 0.4 kips/ft and live load of 1.4 kips/ft. Will a W14×68, A992 member be adequate if the transverse load is applied to put bending about the strong axis? Determine by (a) LRFD and (b) ASD.

12. A pin-ended chord of a truss is treated as a member in a braced frame. Its length is 12 ft. It must carry a compressive dead load of 90 kips and live load of 170 kips, along with a uniformly distributed transverse dead load of 1.1 kips/ft and live load of 2.3 kips/ft. Will a W8×58, A992 member be adequate if the transverse load is applied to put bending about the strong axis? Determine by (a) LRFD and (b) ASD.

13. A moment frame is designed so that under a service lateral load $H = 150$ kips, the frame drifts no more than $L/400$. There are a total of 15 columns in this frame, so P_{story} is 15 times the load on this column. A 13 ft, W14×109, A992 column is to be checked. Analysis results are from a first-order analysis. The column is called upon to carry a compressive dead load of 100 kips and live load of 300 kips. This load will be taken as coming from a no-translation analysis. The top of the column is loaded with no-translation dead load moment of 25 ft-kips and a no-translation live load moment of 80 ft-kips. The translation moments applied to that column end are a dead load moment of 35 ft-kips and a live load moment of 100 ft-kips. The lower end of the column feels half of these moments. The column is bending in reverse curvature about the strong axis. Will the W14×109, A992 member be adequate to carry this loading? Analysis shows that the effective length factor in the plane of bending is 1.66. Determine by (a) LRFD and (b) ASD.

14. A W14×176, A992 member is proposed for use as a 12.5 ft column in a moment frame. The frame is designed so that under a service lateral load $H = 120$ kips, the frame drifts no more than $L/500$. The total story load, P_{story}, is 15 times the individual column load. Analysis results are from a first-order analysis. Will this member be adequate to carry a no-translation compressive dead load of 160 kips and live load of 490 kips? The top of the column is loaded with a no-translation dead load moment of 15 ft-kips and a no-translation live load moment of 30 ft-kips. The translation moments applied to that column end are a dead load moment of 80 ft-kips and a live load moment of 250 ft-kips. The column is bending about the strong axis. The lower end of the column is considered pinned and the effective length factor is taken as 1.5. Determine by (a) LRFD and (b) ASD.

15. Will a W14×61 be adequate as a 14 ft column in a moment frame with a compressive dead load of 35 kips and live load of 80 kips? One half of this compressive load is taken as a no-translation load and one half as a translation load. The top and bottom of the column are loaded with a no-translation dead load moment of 20 ft-kips and a no-translation live load moment of 55 ft-kips. The translation moments applied to the column ends are a dead load moment of 10 ft-kips and a live load moment of 50 ft-kips. Analysis results are from a first-order analysis. The frame is designed so that under a service lateral load $H = 50$ kips, the frame drifts no more than $L/300$. The total story load, P_{story}, is eight times the individual column load. The column is bent in reverse curvature about the strong axis, and $K_x = 1.3$. Determine by (a) LRFD and (b) ASD.

16. Determine whether a 10 ft braced frame W14×43, A992 column can carry a compressive dead load of 35 kips and live load of 80 kips along with a dead load moment of 20 ft-kips and live load moment of 40 ft-kips bending about the strong axis. One half of these moments are applied at the other end, bending it in single curvature. Determine by (a) LRFD and (b) ASD.

17. A two-story single bay frame is shown in Figure P8.17. The uniform live and dead loads are indicated along with the wind load. A first-order elastic analysis has yielded the results shown in the figure for the given loads and the appropriate notional loads. Assuming that the story drift is limited to height/300 under the given wind loads, determine whether the first- and second-story columns are adequate. The gravity loads will produce the no-translation results and the wind load will produce the translation results. The members are shown and are all A992 steel. Determine by (a) LRFD and (b) ASD.

18. Determine whether the columns of the two-bay unbraced frame shown in Figure P8.18 are adequate to support the given loading. Results for the first-order analysis are provided. The gravity loads will produce the no-translation results and the wind load will produce the translation results. Assume that the lateral drift under the given wind load will be limited to a maximum of 0.1 in. All members are A992 steel and the sizes are as shown. Determine by (a) LRFD and (b) ASD.

19. A nonsymmetric two-bay unbraced frame is required to support the live and dead loads given in Figure P8.19. Using the results from the first-order elastic analysis provided, assuming the axial forces provided are from the no-translation analysis and the lateral drift due to a 5 kip horizontal force is limited to 0.4 in., determine whether each column will be adequate. All members are A992 steel bending about the strong axis and the sizes are as shown. Determine by (a) LRFD and (b) ASD.

20. A 14 ft column in a moment frame must carry a compressive load of 540 kips and a moment about the strong axis of 135 ft-kips from an LRFD second-order direct analysis. Will a W14×82, A992 member be adequate if the moment is applied to put bending about the strong axis?

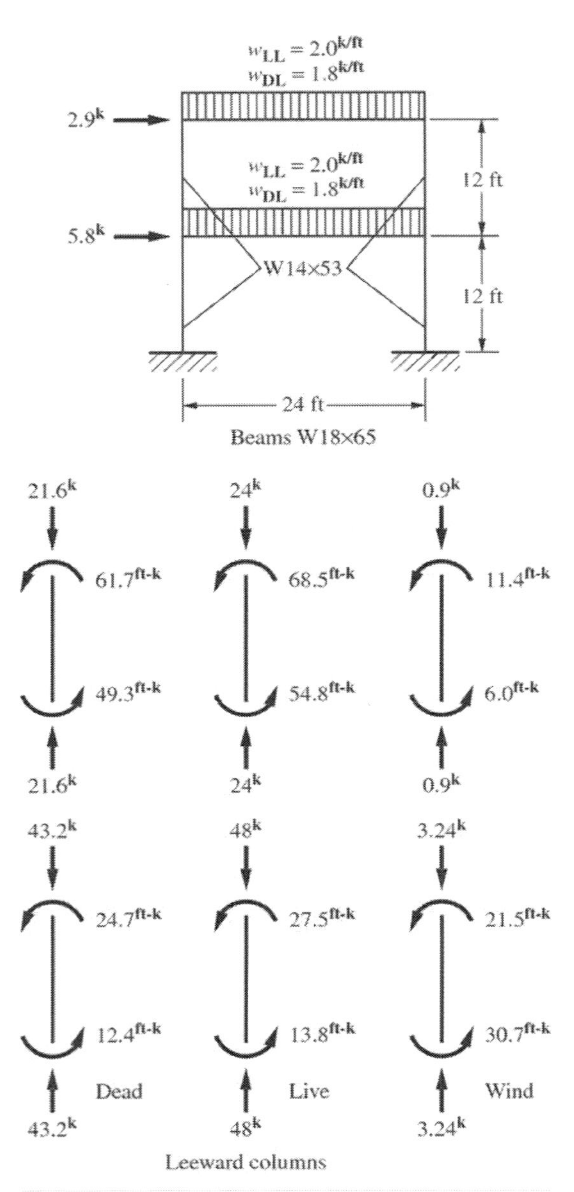

P8.17

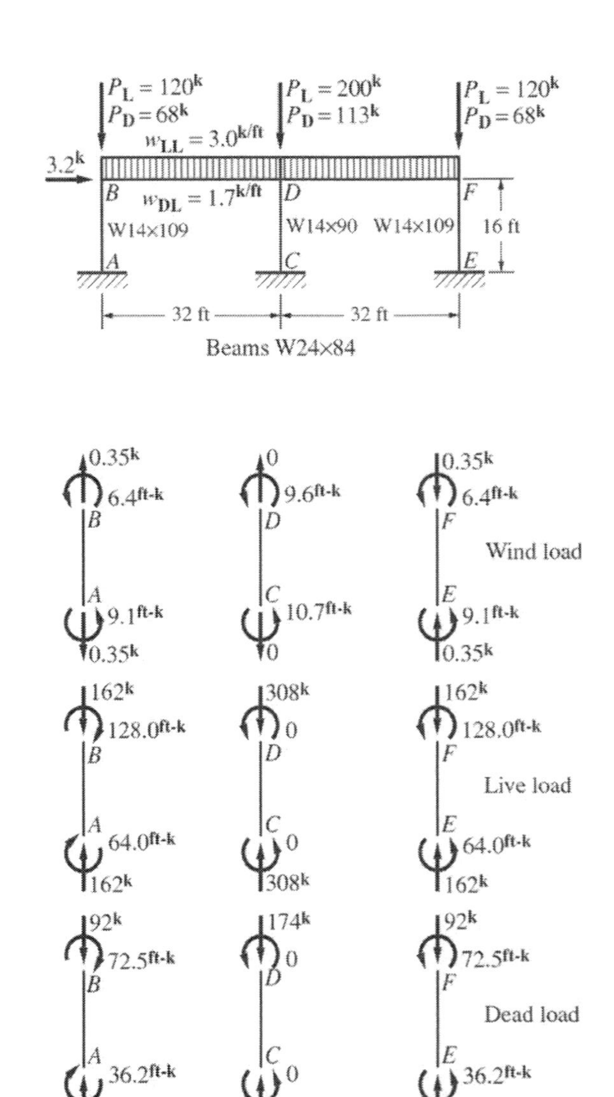

P8.18

21. A 14 ft column in a moment frame must carry a compressive load of 360 kips and a moment about the strong axis of 90 ft-kips from an ASD second-order direct analysis. Will a W14×82, A992 member be adequate if the moment is applied to put bending about the strong axis?

22. A 28 ft column in a moment frame must carry a compressive load of 110 kips and moment about the strong axis of 100 ft-kips from an LRFD second-order direct analysis. Will a W10×54, A992 member be adequate if the moment is applied to put bending about the strong axis?

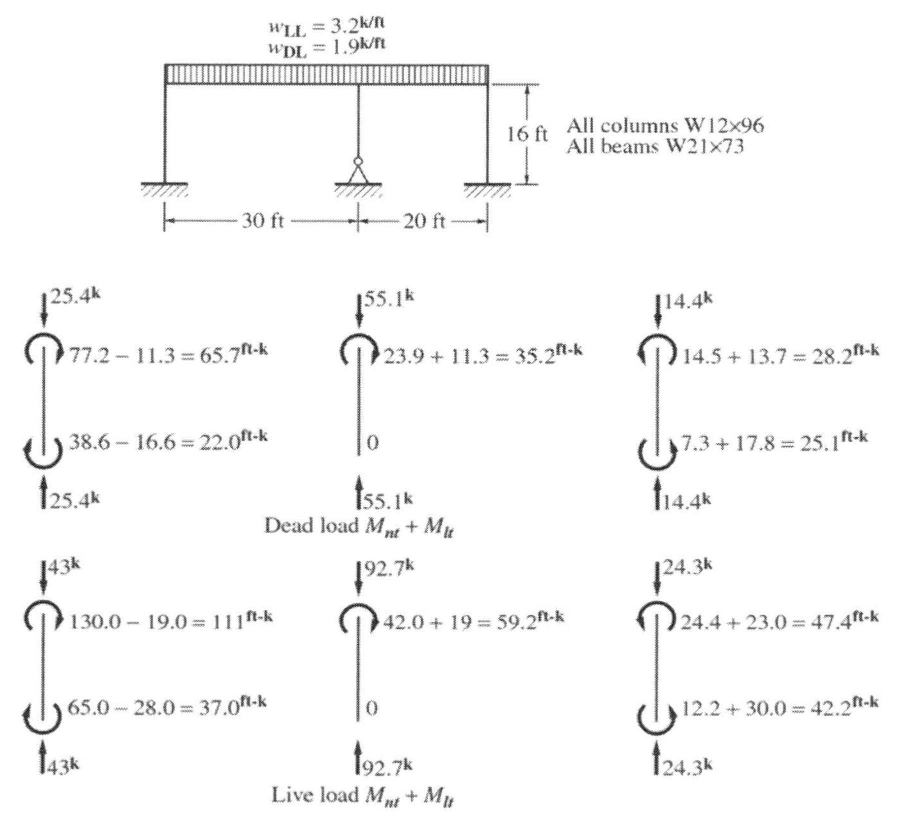

P8.19

23. A 28 ft column in a moment frame must carry a compressive load of 73 kips and moment about the strong axis of 67 ft-kips from an ASD second-order direct analysis. Will a W10×54, A992 member be adequate if the moment is applied to put bending about the strong axis?

24a. Select a W-shape for a column with a length of 16 ft. The results of a second-order direct analysis indicate that the member must carry a force of 700 kips and a moment of 350 ft-kips. Design by LRFD.

24b. Select a W-shape for a column with a length of 16 ft. The results of a second-order direct analysis indicate that the member must carry a force of 467 kips and a moment of 230 ft-kips. Design by ASD.

25a. Select a W-shape for a column with a length of 26 ft. The results of a second-order direct analysis indicate that the member must carry a force of 1100 kips and a moment of 170 ft-kips. Design by LRFD.

25b. Select a W-shape for a column with a length of 26 ft. The results of a second-order direct analysis indicate that the member must carry a force of 730 kips and a moment of 110 ft-kips. Design by ASD.

26a. Select a W-shape for a column with a length of 16 ft. The results of a second-order direct analysis indicate that the member must carry a force of 350 kips and a moment of 470 ft-kips. Design by LRFD.

26b. Select a W-shape for a column with a length of 16 ft. The results of a second-order direct analysis indicate that the member must carry a force of 230 kips and a moment of 320 ft-kips. Design by ASD.

27a. Select a W-shape for a column with a length of 18 ft. The results of a second-order direct analysis indicate that the member must carry a force of 1250 kips and a moment of 450 ft-kips. Design by LRFD.

27b. Select a W-shape for a column with a length of 18 ft. The results of a second-order direct analysis indicate that the member must carry a force of 830 kips and a moment of 300 ft-kips. Design by ASD.

28. The two-bay moment frame shown in Figure P8.28 contains a single leaning column. The results of a first-order elastic analysis for each load are given. Determine whether the exterior columns are adequate to provide stability for the frame under dead and live load. All W-shapes are given and the steel is A992. Determine by (a) LRFD and (b) ASD.

29. The two-story frame shown in Figure P8.29 relies on the left-hand columns to provide stability. Using the first-order analysis results shown, determine whether the given structure is adequate if the steel is A992.

30. The two-bay, two-story frame shown in Figure P8.30 is to be designed. Using the Live, Dead, Snow, and Wind Loads given in the figure, design the columns and beams to provide the required strength and stability by (a) LRFD and (b) ASD.

31. Select an A36 rod to provide the point bracing shown in the center panel of the three-bay frame of Figure 8.9a to provide stability for a total gravity dead load of 150 kips and live load of 60 kips. Design by (a) LRFD and (b) ASD.

32. Select an A36 rod to provide the point bracing shown in the center panel of the three-bay frame of Figure 8.9a to provide stability for a total gravity dead load of 180 kips and live load of 95 kips. Design by (a) LRFD and (b) ASD.

33. A 30 ft simply supported W24×76 beam has point bracing at third points. The beam supports a uniformly distributed dead load of 1.2 kips/ft and a uniformly distributed live load of 2.0 kips/ft. Determine the required point brace strength and the required point brace stiffness by (a) LRFD and (b) ASD.

34. A 30 ft simply supported W24×76 beam has panel bracing in three equal panels over its length. The beam supports a uniformly distributed dead load of 1.2 kips/ft and a uniformly distributed live load of 2.0 kips/ft. Determine the required shear strength of the bracing system and the required panel brace stiffness by (a) LRFD and (b) ASD.

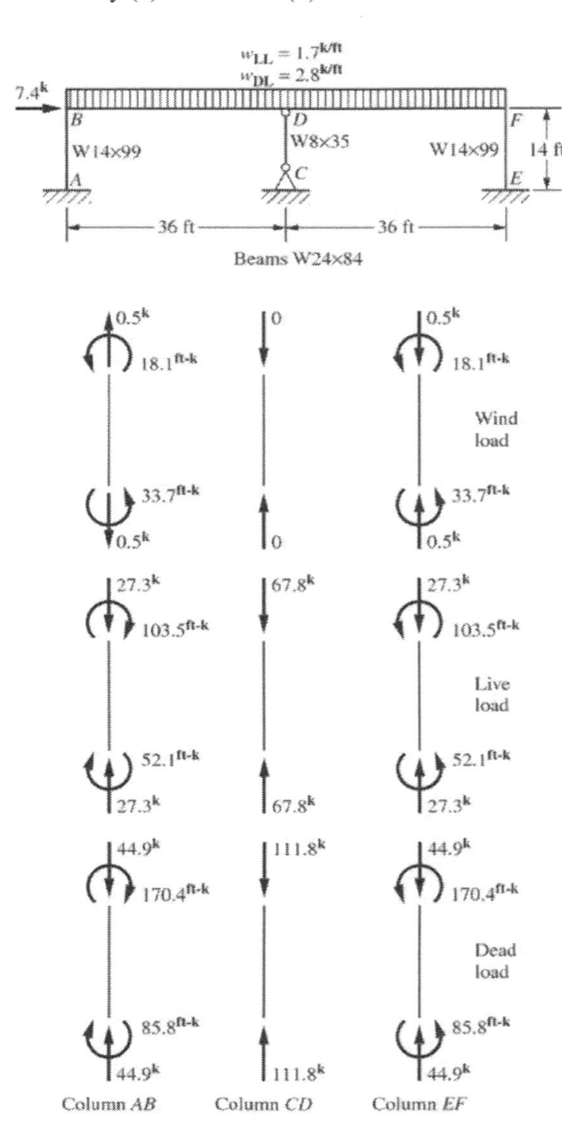

P8.28

35. A simply supported W18×86 beam spans 30 ft and carries a uniformly distributed dead load of 0.8 kips/ft and a uniformly distributed live load of 2.4 kips/ft. It also carries a tension live load of 36 kips. The member is fully braced for lateral-torsional buckling. Use A992 steel. Determine if the W18×86 is adequate for the combined tension and bending by (a) LRFD and (b) ASD.

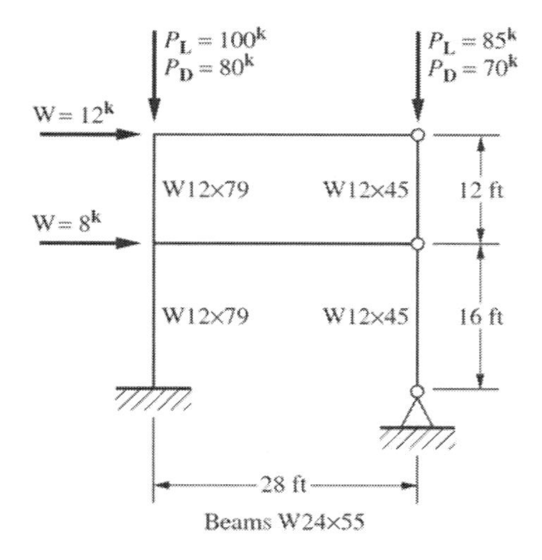

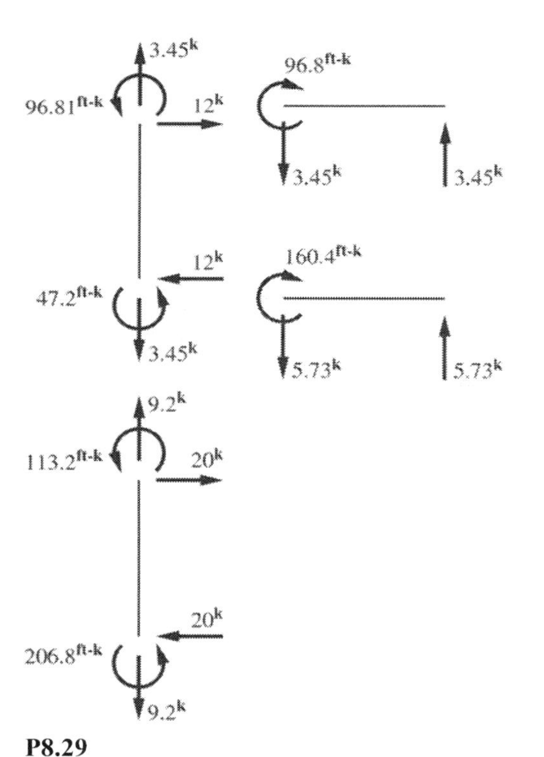

P8.29

36. A simply supported W27×84 beam spans 30 ft and carries a uniformly distributed dead load of 1.2 kips/ft and a uniformly distributed live load of 2.5 kips/ft. It also carries a tension live load of 45 kips. The member is braced at third points for lateral-torsional buckling. Use A992 steel. Determine if the W27×84 is adequate for the combined tension and bending by (a) LRFD and (b) ASD.

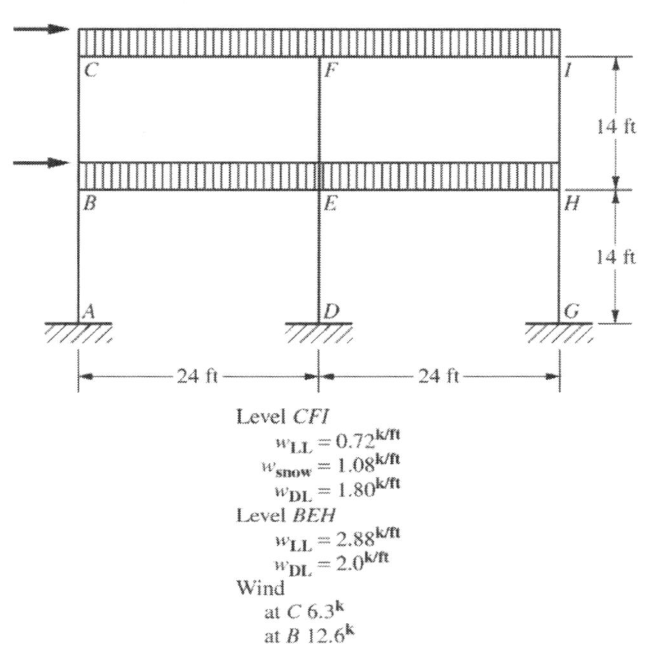

P8.30

37. Integrated Design Project – Effective Length Method

Lateral load resistance in the east-west direction is provided by two perimeter moment frames as seen in Figure 1.24. Before the forces in these members can be determined the specified wind load must be determined.

At this stage in the design, a simplified approach to wind load calculation similar to that used in Chapter 4 might yield the following loads at each level:

Roof	32.0 kips
4th Floor	59.0 kips
3rd Floor	54.0 kips
2nd Floor	50.0 kips
Total Wind Load	195.0 kips

The moment frames will share equally in carrying these loads. They will be designed using the effective length method of Appendix 7.2, and second-order effects will be incorporated using the amplified first-order analysis method of Appendix 8; thus, superposition may be used.

Before an analysis may be carried out, preliminary member sizes must be obtained. Using the gravity loads calculated in Chapter 2, select preliminary column and beam sizes without concern for the frame behavior of the structure.

With these member sizes, the analysis is to be carried out for dead load, live load, roof load, and wind load. Members are to be selected for the gravity plus wind load combination, so there should be no need to include notional loads.

Design the columns and beams for the resulting load effects and redo the analysis to check the strength of these new members and the drift of the structure. Confirm that the effective length method may be used.

38. Integrated Design Project – Direct Analysis Method
Lateral load resistance in the east-west direction is provided by two perimeter moment frames as seen in Figure 1.24. Before the forces in these members can be determined, the specified wind load must be determined. At this stage in the design, a simplified approach to wind load calculation similar to that used in Chapter 4 might yield the following loads at each level:

Roof	32.0 kips
4th Floor	59.0 kips
3rd Floor	54.0 kips
2nd Floor	50.0 kips
Total Wind Load	195.0 kips

The moment frames will share equally in carrying these loads. They will be designed using the Direct Analysis Method from Chapter C.

Before an analysis may be carried out, preliminary member sizes must be obtained. Using the gravity loads calculated in Chapter 2, select preliminary column and beam sizes without concern for the frame behavior of the structure.

With these member sizes, the analysis is to be carried out for dead load, live load, roof load, and wind load, following the general analysis requirements of Section C2. Members are to be selected for the gravity plus wind load combination, so there should be no need to include notional loads.

Design the columns and beams for the resulting load effects and redo the analysis to check the strength of these new members and the drift of the structure.

Chapter 9

Composite Construction

Salesforce Tower, Chicago IL
Photo courtesy of Magnusson Klemencic Associates

9.1 INTRODUCTION

Any structural member in which two or more materials having different stress-strain relationships are combined and called upon to work as a single member may be considered a *composite member*.

Many different types of members have been used that could be called composite. Such members, as shown in Figure 9.1, include (a) a reinforced concrete beam, (b) a precast concrete beam with a cast-in-place slab, (c) a "flitch" girder combining wood side members and a steel plate, (d) a stressed skin panel where plywood is combined with solid wood members, and (e) a steel shape combined with concrete.

The last type of member, and similar members, are what are normally thought of as composite beams in building applications. *Specification* Chapter I provides rules for design of the composite members illustrated in Figure 9.2. These members are (a) steel beams fully encased in concrete, (b) steel beams with flat soffit concrete slabs, (c) steel beams combined with concrete slabs on formed steel deck, (d) steel columns fully encased in concrete, and (e) hollow steel shapes filled with concrete.

Encased beams and filled columns, shown in Figure 9.2a and e, may not require mechanical anchorage between the steel and concrete, other than the natural bond that exists between the two materials; however, the flexural and compression members shown in Figure 9.2b, c, and d always require some form of mechanical shear connection.

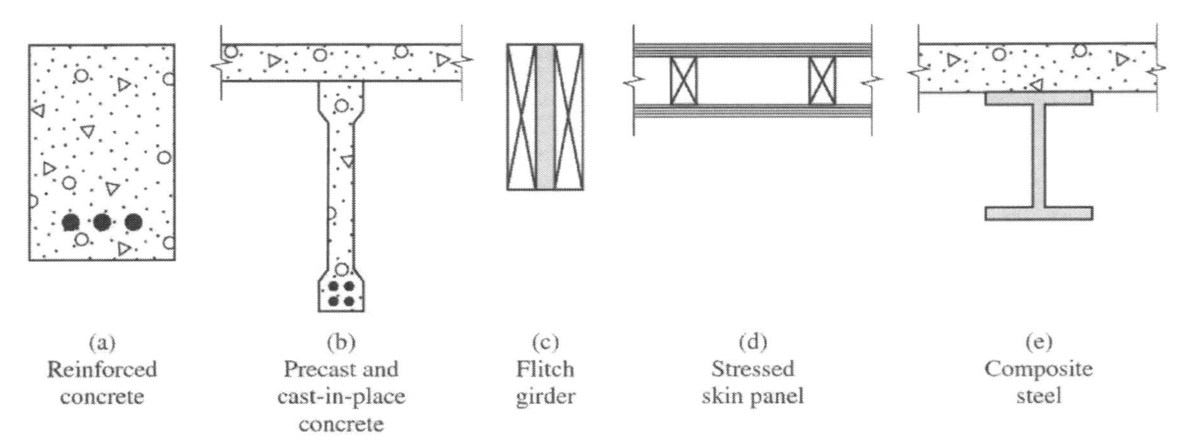

Figure 9.1 Composite Members

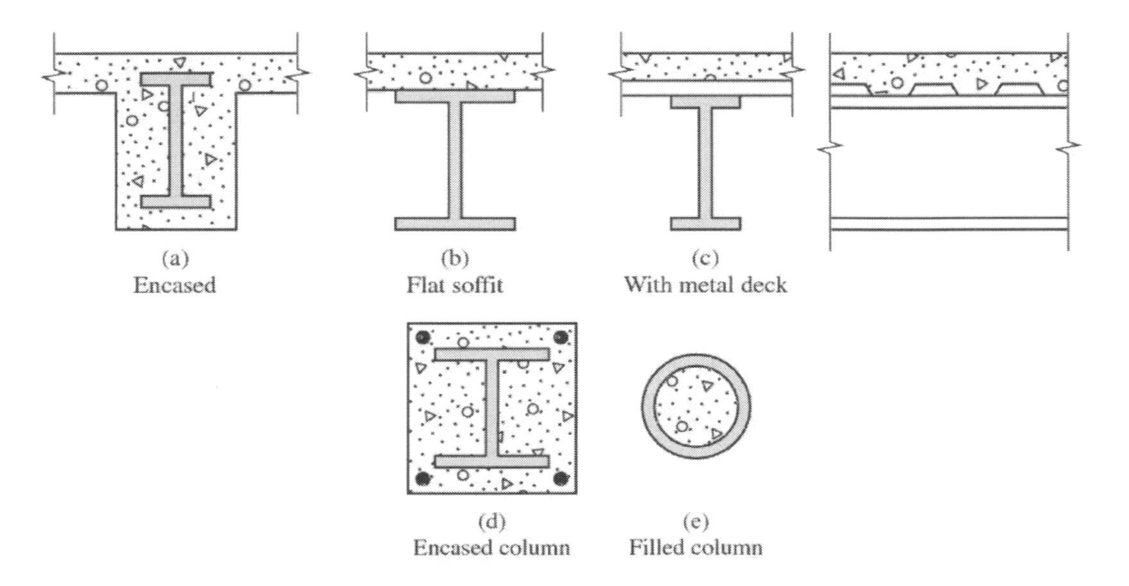

Figure 9.2 Composite Steel Beams and Columns

Regardless of the type of mechanical shear device provided, it must connect the steel and concrete to form a unit and permit the two materials to work together to resist the load. This considerably increases the strength of the bare steel shape. Composite beams were first used in bridge design in the United States around 1935. Until the invention of the shear stud, the concrete floor slab was connected to the stringer beams by means of wire spirals or channels welded to the top flange of the beam, as shown in Figure 9.3.

Figure 9.3 Composite Beam Using a Spiral Shear Connector

In the 1940s the Nelson Stud Company invented the shear stud, a headed rod welded to the steel beam by means of a special device or gun, as shown in Figure 9.4. The company did not enforce its patent but instead encouraged nonproprietary use of the system, assuming correctly that the company would get its share of business if it became popular. In a very short time, studs replaced spirals and channels, and today studs are used almost exclusively in composite beam construction.

In 1952 AISC adopted composite design rules for encased beams in its specification for building design, and in 1956 it extended them to beams with flat soffits. Although the design procedure was based on the ultimate strength of the composite section, the rules were written in terms of an allowable stress formulation, as was common for the time. Since then, that original formulation for allowable stress design for composite beams has often been criticized as being convoluted and difficult to understand.

Figure 9.4 Installation of a Shear Stud with a Stud Gun
Photo Courtesy W. Samuel Easterling

In the current *Specification*, whether for ASD or LRFD, the rules for the design of composite beams are straightforward and surprisingly simple. The ultimate flexural strength of the composite member is based on plastic stress distribution, with the ductile shear connector transferring shear between the steel section and the concrete slab.

Specification Section I1.3 gives limitations on material properties for use in composite concrete members. Concrete is limited to f_c' between 3 ksi and 10 ksi for normal-weight concrete and between 3 ksi and 6 ksi for lightweight concrete. The specified minimum yield strength of the structural steel used in calculating the strength of composite members is not to exceed 75 ksi. Similarly, for reinforcing steel the limit for calculations is 80 ksi.

This chapter discusses the design of both composite beams and composite columns. Table 9.1 lists the sections of the *Specification* and parts of the *Manual* discussed in this chapter.

Table 9.1 Sections of *Specification* and Parts of *Manual* Covered in This Chapter

	Specification
I1	General Provisions
I2	Axial Force
I3	Flexure
I5	Combined Flexure and Axial Force
I6	Load Transfer
I8	Steel Anchors
H1	Doubly and Singly Symmetric Members Subject to Flexure and Compression
	Manual
Part 3	Design of Flexural Members
Part 4	Design of Compression Members
Part 6	Design of Members Subject to Combined Forces

9.2 ADVANTAGES AND DISADVANTAGES OF COMPOSITE BEAM CONSTRUCTION

One feature of composite construction makes it particularly advantageous for use in building structures. The typical building floor system is composed of two main parts: a floor structure that carries load to supporting members, usually a concrete slab or a slab on a metal deck; and the supporting members that span the space between girders, usually steel beams or joists. The advantage of a composite floor system stems from the concrete slab doing "double duty". The concrete slab, with or without metal deck, is first designed to span between beams. Then, since the concrete is there for that purpose in one direction, it can be used to advantage in the other direction, that of the beam span. All other factors that could be identified as advantages of this type of construction can be traced back to this single feature. A composite beam takes the existing concrete slab and enables it to work with the steel beam to carry the load to the girders. Thus, the resulting system has a greater strength than would be available from the steel beam alone. The composite beam is stronger and stiffer than the noncomposite beam.

The advantage of this factor lies in the reduced weight and/or shallower member depth necessary to carry a given load compared to the bare steel beam. Because the concrete slab is in compression and the majority of the steel is in tension, the two materials are working to their best advantage. In addition, the effective beam depth has been increased from just the depth of the steel to the total distance from the top of the slab to the bottom of the steel, increasing the overall efficiency of the member.

With regard to stiffness, the composite section has an increased elastic moment of inertia compared to the bare steel beam. Although the actual calculation of the stiffness of the composite section may be approximate in many cases, the impact of the increased stiffness profoundly affects the static deflection.

The only disadvantage of composite construction is the added cost of the required shear connectors, referred to as steel headed stud anchors in the *Specification*. Because the increased strength, or reduction in required steel weight, is normally sufficient to offset the added cost of providing and installing the shear connectors, this increased cost is usually not a true disadvantage.

9.3 SHORED VERSUS UNSHORED CONSTRUCTION

Two methods of construction are available for composite beams: *shored* and *unshored construction*. Each has advantages and disadvantages, which will be discussed briefly. The difference between these two approaches to the construction of a composite beam is how the self-weight of the wet concrete is carried.

When the steel shape alone is called upon to carry the wet concrete weight during construction, the beam is considered to be unshored. In this case, the steel is stressed and it deflects. This is the simplest approach to constructing the composite beam because the formwork and/or decking is supported directly on the steel beam. Unshored construction may, however, lead to a deflection problem during the construction phase because as the wet concrete is placed, the steel beam deflects. To obtain a level slab, more concrete is placed where the beam deflection is greatest. This means that the contractor must place more concrete than required based on the specified slab thickness, and the designer needs to provide more strength than would be needed if the slab had remained of uniform thickness. The common solution is to select a beam that limits the deflection during concrete placement to an acceptable amount.

For shored construction, temporary supports called *shores* are placed under the steel beam to carry the wet concrete weight. In this case, the composite section carries the entire load after the shores are removed. No load is carried by the bare steel beam alone, and thus, no deflection occurs during concrete placement. Two factors must be considered in the selection of shored construction: (1) the additional cost, in terms of both time and money, of placing and removing the temporary shoring; and (2) the potential increase in long-term dead load deflection due to creep in the concrete caused by carrying the permanent weight of the slab, which is not considered in unshored construction but now must be considered.

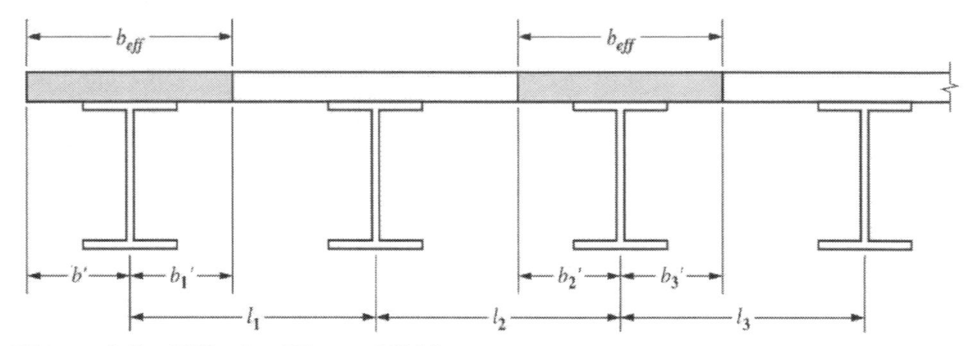

Figure 9.5 Effective Flange Width

Although elastic stress distribution and deflection under service load conditions are influenced by whether the composite beam is shored or unshored, research has shown that the ultimate strength of the composite section is independent of the shoring situation. Thus, the use of shoring is entirely a serviceability and constructability question that must be considered by both the designer and constructor. Whether ASD or LRFD provisions are used, the shored and unshored systems have the same nominal strength, as determined through application of *Specification* Section I3.

9.4 EFFECTIVE FLANGE

A cross section through a series of typical composite beams is shown in Figure 9.5. Because the concrete slab is normally part of the transverse spanning floor system, its thickness and the spacing of the steel beams are usually established prior to the design of the composite beams.

Because the ability of the slab to participate in load carrying decreases as the distance from the beam centerline increases, some limit must be established to determine the portion of the slab that can be used in the calculations to determine the strength of the composite beam. The *Specification* provides two criteria for determining the effective width of the concrete slab for an interior beam and an additional criterion for an edge beam in Section I3.1a. As shown in Figure 9.5, the effective flange width, b_{eff}, is the sum of b' values on each side of the centerline of the steel section. For an interior beam, b' is the lesser of

$$b' \leq \frac{\text{span}}{8}$$

$$b' \leq \frac{1}{2}\text{distance to the adjacent beam}$$

For an edge beam, the additional criterion is

$$b' \leq \text{distance to the edge of the slab}$$

The entire thickness of the concrete slab is available to carry a compressive force. However, the depth of the concrete used in calculations is only that required to provide sufficient area in compression to balance the force transferred by the shear connectors to the steel shape. This may be significantly less than the entire slab thickness. It should be noted that the slab thickness does not influence the effective width of the slab as it does for reinforced concrete T-beam design.

9.5 STRENGTH OF COMPOSITE BEAMS AND SLAB

Flat soffit composite beams (Figure 9.2b) are constructed using formwork that is set at the same elevation as the top of the steel section. The concrete slab is placed directly on the steel section, resulting in a flat surface at the level of the top of the steel. Composite beams with a formed steel deck (Figure 9.2c) are constructed with the steel deck resting on top of the steel beam or girder. The concrete is placed on top of the deck so that the concrete ribs and voids alternate. Provided that the portion of concrete required to balance the tension force in the steel is available above the tops of the ribs, the ultimate strength is determined similarly for the two types of composite beams. Although the steel member may be either shored or unshored, the strength of the composite member is independent of the use of shores, and the design rules are independent of the method of construction.

The flexural strength of a composite beam under positive moment where concrete is in compression is presented in Section I3.2a of the *Specification*. In this section, strength is developed for flat soffit beams. The required modifications to account for the use of metal deck are presented in the next section. For steel shapes with a web slenderness ratio of $h / t_w \leq 3.76\sqrt{E / F_y}$ (which is the case for all rolled W-, S-, and HP-shapes), the nominal moment, M_n, is determined from the plastic distribution of stress on the composite section, and

$$\phi_b = 0.90 \text{ (LRFD)} \qquad \Omega_b = 1.67 \text{ (ASD)}$$

A composite beam cross section is shown in Figure 9.6 with three possible plastic stress distributions. Regardless of the stress distribution considered, equilibrium requires that the total tension force equal the total compression force, $T = C$. In Figure 9.6a, the plastic neutral axis (PNA) is located at the top of the steel shape. The compression force developed using all of the concrete is exactly equal to the tension force developed using all of the steel. For the distribution of Figure 9.6b, the PNA is located within the steel shape. In this case, all of the concrete is taking compression, but this is not sufficient to balance the tension force that the full steel shape could provide. Thus, some of the steel shape is in compression in order to satisfy $T = C$. The plastic stress distribution shown in Figure 9.6c is what occurs when less than the full amount of concrete is needed to balance the tensile force developed in the steel shape. Here the PNA is located within the concrete. All of the concrete above the PNA is in compression and that portion of the concrete below the PNA is not used because it would be in tension and concrete is not effective in resisting tension.

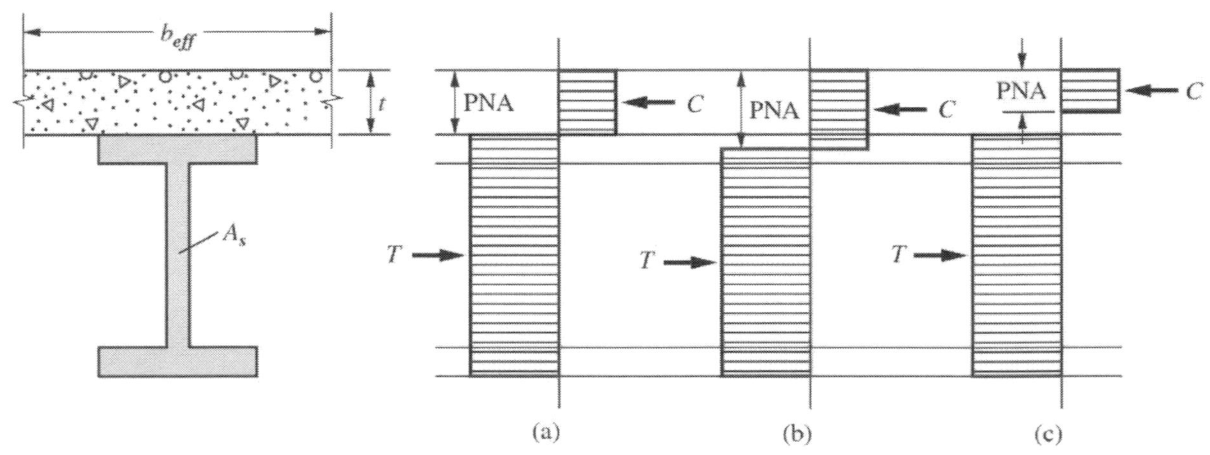

Figure 9.6 Plastic Stress Distribution

In all three cases, equilibrium of the cross section requires that the shear connectors be capable of transferring the force carried by the concrete into the steel. For the cases in Figure 9.6a and b, this is the full strength of the concrete. For the case in Figure 9.6c, this is the strength of the steel shape. Because the shear connectors are carrying the full amount of shear force required to provide equilibrium using the maximum capacity of one of the elements, this is called a *fully composite* beam. It is also possible to design a composite beam when the shear force that can be transferred by the shear connectors is less than this amount. In this case, the beam is called a *partially composite* beam. Although it has less strength than the fully composite member, it is often the most economical solution.

The *Specification* indicates that the plastic stress distribution in the concrete shall be taken as a uniform stress at a magnitude of $0.85f_c'$ for all of the concrete area above the PNA. This is the same stress magnitude as specified by ACI 318 *Building Code Requirements for Reinforced Concrete*. In addition, the distribution of stress in the steel is taken as a uniform F_y, as was the case for determining the plastic moment strength of a steel shape.

The *Specification* also provides for the use of a strain compatibility method and an elastic stress distribution method for determining the strength of a composite section. Strain compatibility should be considered when a section is of unusual geometry and elastic stress distribution should be used when the web is not compact.

9.5.1 Fully Composite Beams

Establishing which stress distribution is in effect for a particular combination of steel and concrete requires calculating the minimum compressive force as controlled by the three components of the composite beam: concrete, steel, and shear connectors.

If all of the concrete were working in compression,

$$V_c' = V' = 0.85 f_c' b_{eff} t \qquad \text{(AISC I3-1a)}$$

If all of the steel shape were working in tension,

$$V_s' = V' = F_y A_s \qquad \text{(AISC I3-1b)}$$

If the shear studs were carrying their full capacity,

$$V_q' = V' = \Sigma Q_n \qquad \text{(AISC I3-1c)}$$

where ΣQ_n is the sum of the nominal shear strengths of the shear connectors between the point on maximum positive moment and the point of zero moment. Because fully composite action is being considered first, V_q' will not control and is not considered further in this current discussion. If $V_s' \leq V_c'$, the steel is fully stressed and all or only a portion of the concrete is stressed. This is the distribution given in either Figure 9.6a or c. If $V_c' < V_s'$, the concrete is fully stressed, and the steel is called upon to carry both tension and compression to ensure equilibrium. This results in the distribution shown in Figure 9.6b. Once the proper stress distribution is known, the corresponding forces can be determined, and their point of application found. With this information, the nominal moment, M_n, can be determined by taking moments about some reference point. Because the internal forces are equivalent to a force couple, any point of reference can be used for taking moments; however, it is convenient to use a consistent reference point. These calculations use the top of the steel as the point about which moments are taken.

Determination of the PNA for the cases in Figure 9.6a and c is quite straightforward. In both cases the steel is fully stressed in tension, so it is said that the *steel controls* and it is known that the concrete must carry a compressive force equal to V_s'. Only that portion of the concrete required to resist this force will be used, so that force is defined as $C_c = 0.85 f_c' b_{eff} a$, where a defines the depth of the concrete stressed to its ultimate. Setting $V_s' = C_c$ and solving for a yields

$$a = \frac{V_s'}{0.85 f_c' b_{eff}} = \frac{F_y A_s}{0.85 f_c' b_{eff}} \qquad (9.1)$$

For the special case where V_s' is exactly equal to V_c', the value of a thus obtained is equal to the actual slab thickness, t. This is the case shown in Figure 9.6a. For all other values of a, the distribution of Figure 9.6c results. The nominal flexural strength can then be obtained by taking moments about the top of the steel so that

$$M_n = T_s\left(d/2\right) + C_c\left(t - a/2\right) \qquad (9.2)$$

where $T_s = F_y A_s$ and $C_c = T_s$ because the system must be in equilibrium.

When the concrete controls, $V_c' < V_s'$, the determination of the PNA is a bit more complex. It is best to consider this case as two separate subcases: (1) the PNA occurring within the steel flange and (2) the PNA occurring within the web. Once it is determined that V_c' controls, and thus $C_c = V_c'$, the next step is to determine the force in the steel flange and web respectively from

$$T_f = F_y b_f t_f \qquad (9.3)$$

$$T_w = T_s - 2T_f \qquad (9.4)$$

A comparison between the force in the concrete and the force in the bottom flange plus the web shows whether the PNA is in the top flange or web. If $C_c > T_w + T_f$, more tension is needed for equilibrium and the PNA must be in the top flange. If $C_c < T_w + T_f$, the PNA could be either in the web or flange. In either case, the difference between the concrete force, C_c, and the available steel force, T_s, must be divided evenly between tension and compression in order to obtain equilibrium. This allows determination of the PNA location and the nominal moment strength. Thus, with

$$A_{s-c} = \text{area of steel in compression}$$

and

$$A_s = \text{total area of steel}$$

equilibrium is given by

$$C_c + F_y A_{s-c} = T_s - F_y A_{s-c} \tag{9.5}$$

Solving for the area of steel in compression yields

$$A_{s-c} = \frac{T_s - C_c}{2F_y} \tag{9.6}$$

For the case where the PNA is in the flange, the distance from the top of the flange to the PNA is given by x, where

$$x = \frac{A_{s-c}}{b_f} \tag{9.7}$$

and for the case where the PNA is in the web,

$$x = \frac{A_{s-c} - b_f t_f}{t_w} + t_f \tag{9.8}$$

Equation 9.8 can be more easily understood if it is related to the areas being considered. The area of the web in compression is the area of steel in compression less the flange area. This web compression area is divided by the web thickness, and the result is the location of the PNA measured from the underside of the flange. Thus, x is simply the thickness of the flange plus the depth of the web in compression.

EXAMPLE 9.1
Fully Composite Beam Strength

Goal: Determine the nominal moment strength for the interior composite beam shown as Beam A in Figure 9.7. Also determine the design moment and the allowable moment.

Given: The section is a W21×44 and supports a 4.5 in. concrete slab. The dimensions are as shown. F_y = 50 ksi and f_c' = 4 ksi. Assume full composite action.

SOLUTION

Step 1: Determine the effective flange width, the minimum of

$$b_{eff} = 30.0(12 \text{ in./ft})/4 = 90.0 \text{ in.}$$

and

$$b_{eff} = (10.0 + 10.0)(12 \text{ in./ft})/2 = 120 \text{ in.}$$

Therefore use

$$b_{eff} = 90.0 \text{ in.}$$

Step 2: Determine the controlling compression force using Equations I3-1a and I3-1b.

$$V_c' = 0.85 f_c' A_c = 0.85(4.0)(90.0)(4.5) = 1380 \text{ kips}$$

$$V_s' = F_y A_s = 50.0(13.0) = 650 \text{ kips}$$

Assuming full composite action, the shear connectors must carry the smaller of V_c' and V_s'; thus,

$$V_q' = 650 \text{ kips}$$

Because V_s' is less than V_c', the PNA is in the concrete.

Step 3: Determine the PNA location using Equation 9.1.

$$a = \frac{F_y A_s}{0.85 f_c' b_{eff}} = \frac{650}{0.85(4)(90.0)} = 2.12 \text{ in.}$$

The resulting plastic stress distribution is shown in Figure 9.7c.

Step 4: Determine the nominal moment strength using Equation 9.2,

$$M_n = T_s(d/2) + C_c(t - a/2)$$

$$= 650\left(\frac{20.7}{2}\right) + 650\left(4.50 - \frac{2.12}{2}\right) = 8960 \text{ in.-kips}$$

$$M_n = \left(\frac{8960}{12}\right) = 747 \text{ ft-kips}$$

For LRFD Step 5: The design moment is

$$\phi M_n = 0.9(747) = 672 \text{ ft-kips}$$

For ASD Step 5: The allowable moment is

$$M_n/\Omega = 747/1.67 = 447 \text{ ft-kips}$$

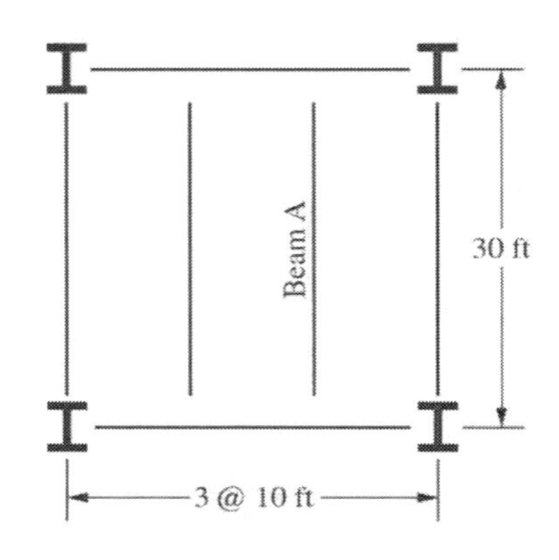

(a) Framing plan

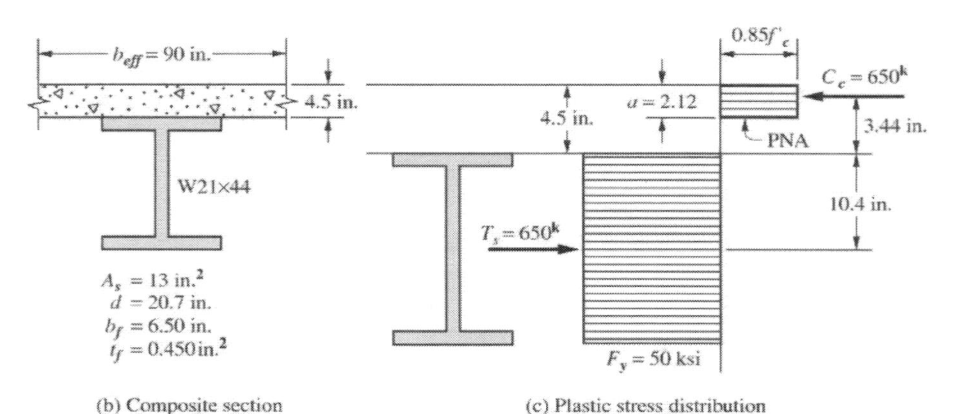

(b) Composite section (c) Plastic stress distribution

Figure 9.7 Interior Composite Beam (Examples 9.1 and 9.2)

EXAMPLE 9.2
Fully Composite Beam Strength

Goal: Determine the nominal moment strength for the interior composite beam shown as Beam A in Figure 9.7 using a larger W-shape. Also determine the design moment and the allowable moment.

Given: Use a W21×111 as shown in Figure 9.8 for the steel member and the same materials as in Example 9.1. Again, assume full composite action.

SOLUTION

Step 1: Determine the effective flange width.

The effective flange width will remain the same; thus,

$$b_{eff} = 90.0 \text{ in.}$$

Step 2: Determine the controlling compression force using Equations I3-1a and I3-1b.

$$V'_c = 0.85 f'_c A_c = 0.85(4.0)(90.0)(4.5) = 1380 \text{ kips}$$

$$V'_s = F_y A_s = 50.0(32.6) = 1630 \text{ kips}$$

Assuming full composite action,

$$V_q' = 1380 \text{ kips}$$

Because V_c' is less than V_s', the PNA is in the steel.

Step 3: Determine whether the PNA is in the steel flange or web.

$$T_f = F_y b_f t_f = 50(12.3)(0.875) = 538 \text{ kips}$$

$$T_w = F_y A_s - 2T_f = 1630 - 2(538) = 554 \text{ kips}$$

Thus,

$$C_c = V_c' = 1380 > T_f + T_w = 538 + 554 = 1090 \text{ kips}$$

Because additional tension is required to balance the compression in the concrete, the PNA is in the flange.

Step 4: Determine the area of steel in compression.

Use Equation 9.6.

$$A_{s-c} = \frac{T_s - C_c}{2F_y} = \frac{1630 - 1380}{2(50)} = 2.50 \text{ in.}^2$$

Step 5: Determine the location of the PNA in the flange.

The PNA is located down from the top of the steel by a distance x as given by Equation 9.7.

$$x = \frac{A_{s-c}}{b_f} = \frac{2.50}{12.3} = 0.203 \text{ in.}$$

The stress distribution for this PNA location is shown in Figure 9.8b.

Step 6: Determine the nominal moment strength of the composite beam.

Moments could be taken about any point to determine the nominal moment; however, a simplified mathematical model shown in Figure 9.8c makes the analysis quicker. In this case, the full area of steel is shown in tension, and the portion in compression is first removed (130 kips on the compression side) and then added in compression (another 130 kips on the compression side), shown as 2(130) = 260 kips. This results in only three forces and moment arms entering the moment equation. Thus

$$M_n = T_s\left(\frac{d}{2}\right) + C_c\left(\frac{t}{2}\right) - 2A_{s-c}F_y\left(\frac{x}{2}\right)$$

$$= 1630\left(\frac{21.5}{2}\right) + 1380\left(\frac{4.5}{2}\right) - 2(2.50)(50)\left(\frac{0.203}{2}\right) = 20{,}600 \text{ in.-kips}$$

$$M_n = \frac{20{,}600}{12} = 1720 \text{ ft-kips}$$

Step 7: For LRFD, the design moment is

$$\phi M_n = 0.9(1720) = 1550 \text{ ft-kips}$$

Step 7: For ASD, the allowable moment is

$$\frac{M_n}{\Omega} = \frac{1720}{1.67} = 1030 \text{ ft-kips}$$

9.5.2 Partially Composite Beams

The composite members considered thus far have been fully composite. This means that the shear connectors have been assumed to be capable of transferring whatever force was required for equilibrium when either the concrete or steel were fully stressed. There are many conditions where the required strength of the composite beam is less than what would result from full composite action. In particular, there are cases where the size of the steel member is dictated by factors other than the strength of the composite section. Because shear connectors compose a significant part of the cost of a composite beam, economies can result if the lower required flexural strength can be translated to a reduced number of shear connectors when the steel section and concrete geometry are already given.

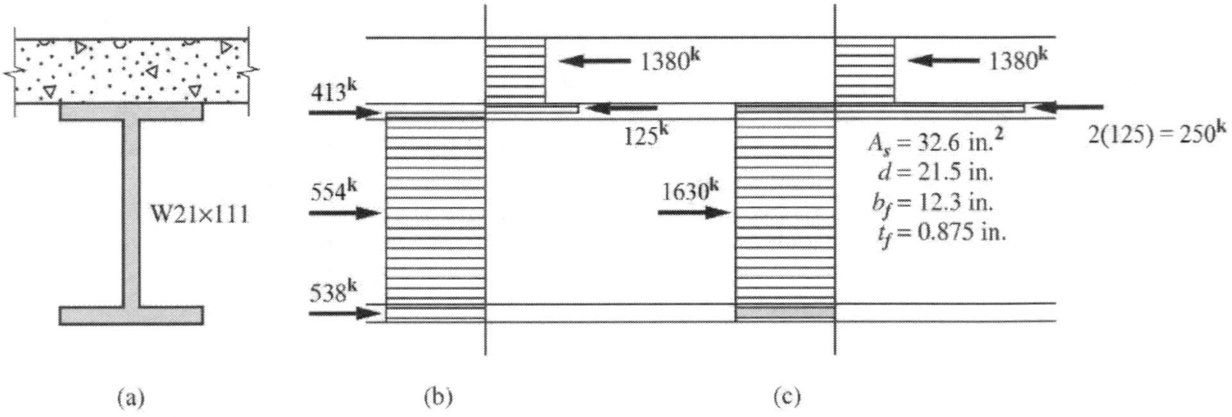

Figure 9.8 Interior Composite Beam (Example 9.2)

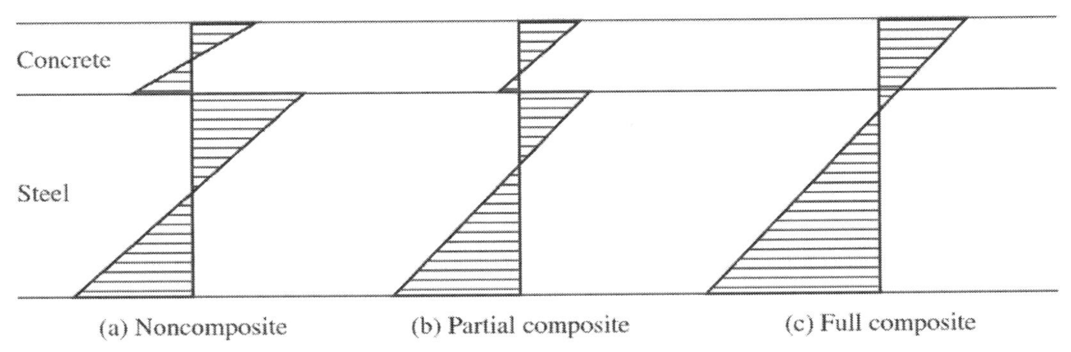

Figure 9.9 Levels of Composite Action for Elastic Behavior

If the composite section is viewed under elastic stress distributions, partial composite action can be more easily understood. Figure 9.9 shows elastic stress distributions for three cases of combined steel and concrete. The first case, Figure 9.9a, is what results when the concrete simply rests on the steel with no shear transfer between the two materials. The result is two independent members that slip past each other at the interface. If the two materials are fully connected, the elastic stress distribution is as shown in Figure 9.9c and the materials are not permitted to slip at all. If some limited amount of slip is permitted between the steel and the concrete, the resulting elastic stress distribution is similar to that shown in Figure 9.9b. This is how the partially composite beam would behave in the elastic region.

The plastic moment strength for a partially composite member is the result of a stress distribution similar to that shown in Figure 9.10. The PNA will be in the steel and the magnitude of the compression force in the concrete will be controlled by the strength of the shear connectors.

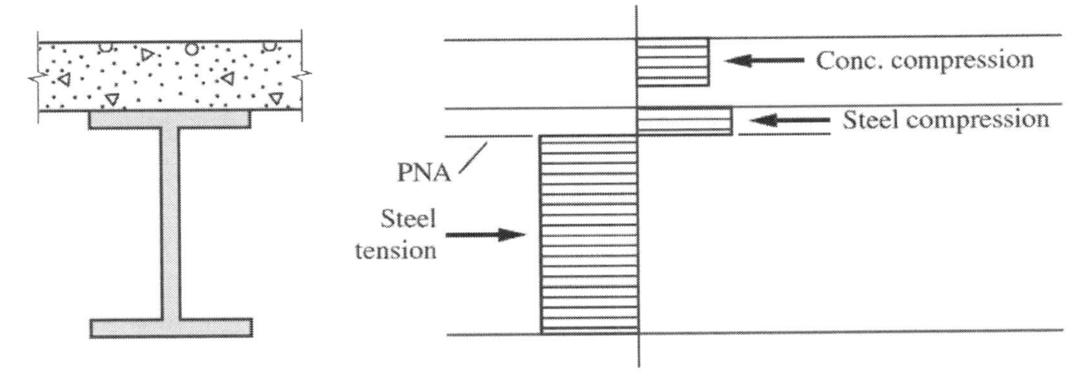

Figure 9.10 Plastic Stress Distribution for Partially Composite Action

Regardless of the final location of the PNA, the force in the concrete is limited by the strength of the shear studs. Thus, an approach combining those taken for the three cases of fully composite sections is used for the partially composite member. By the definition of partially composite members,

$$C_c = V_q' = \Sigma Q_n$$

and the depth of the concrete acting in compression is given by

$$a = \frac{\Sigma Q_n}{0.85 f_c' b_{eff}} \tag{9.9}$$

Equations 9.3 through 9.8 can then be used to determine the location of the PNA within the steel and the nominal moment can be obtained as before.

Partial composite action is generally described in terms of a percent of full composite action. Thus, a beam with shear stud strength equal to the tensile strength of the W-shape is said to be 100% composite or fully composite. A beam with shear stud strength equal to 3/4 of the tensile strength of the W-shape is said to be 75% composite or partially composite. The slip between the steel and the concrete that allows for partial composite action requires a certain amount of ductile behavior or deformation capacity of the shear studs. Although there are no set requirements in the *Specification* for determining a minimum level of ductility, the Commentary provides guidelines to ensure that the partially composite beams will behave as designed. For configurations of composite beams in normal design practice, three specific guidelines are provided that ensure sufficient ductility. Thus, if partially composite beams meet one or more of the following conditions, shear connectors are not subject to failure due to insufficient deformation capacity:

1. Beams with span not exceeding 30 ft,

2. Beams with a degree of composite action of at least 50%, or

3. Beams with an average nominal shear connector capacity of at least 16 kips per ft along their span. This corresponds to a single 3/4 in. shear stud placed at 12 in. spacing on average.

Beams that do not meet these criteria may still perform adequately but they should be checked through a detailed analysis that captures the actual deformation characteristics of the shear studs.

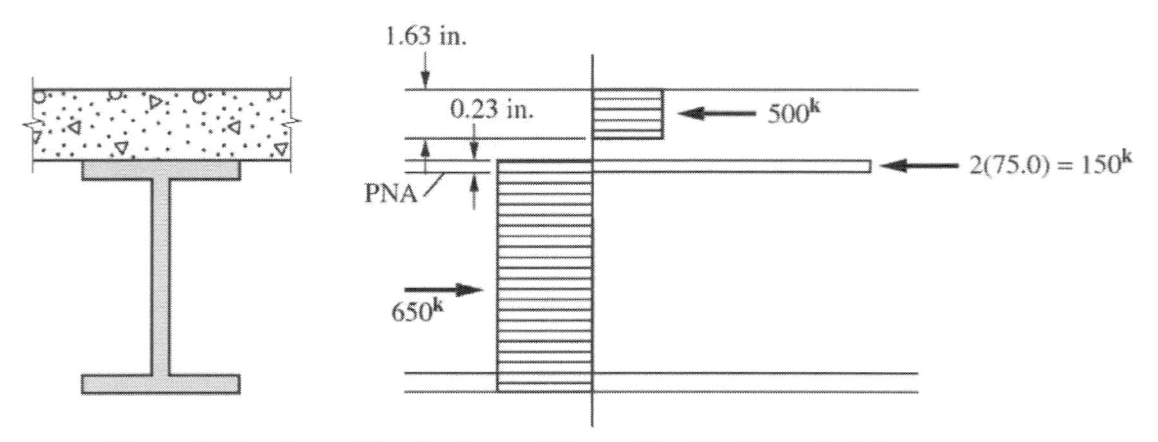

Figure 9.11 Stress Distribution and Forces Used in Example 9.3

EXAMPLE 9.3
Partially Composite Beam Strength

Goal: Determine the nominal moment strength of a partially composite beam. Also determine the design moment and the allowable moment.

Given: Consider the concrete and steel beam given in Example 9.1 and shown in Figure 9.7. In this case, however, assume that the shear connectors are capable of transferring only $V_q' = 500$ kips.

SOLUTION

Step 1: Determine the effective flange width. This is the same as determined for Example 9.1.

$$b_{eff} = 90.0 \text{ in.}$$

Step 2: Determine the controlling compression force. From Example 9.1,

$$V_c' = 1380 \text{ kips}$$
$$V_s' = 650 \text{ kips}$$

From the given data

$$V_q' = 500 \text{ kips}$$

Because the lowest value of the compressive force is given by V_q', this is a partially composite member and $C_c = V_q' = 500$ kips.

Since $V_q'/V_s' = 500/650 = 0.77 \geq 0.5$, the beam is 77% composite and also meets the requirement for sufficient connector ductility.

Step 3: Determine the depth of the concrete working in compression from Equation 9.9.

$$a = \frac{\Sigma Q_n}{0.85 f_c' b_{eff}} = \frac{500}{0.85(4)(90.0)} = 1.63 \text{ in.}$$

Step 4: Determine the area of steel in compression from Equation 9.6.

$$A_{s-c} = \frac{T_s - C_c}{2F_y} = \frac{650 - 500}{2(50)} = 1.50 \text{ in.}^2$$

Because this is less than the area of the flange, 6.50(0.450) = 2.93 in.^{2}, the PNA is in the flange.

Step 5: Determine the location of the PNA from Equation 9.7.

$$x = \frac{A_{s-c}}{b_f} = \frac{1.50}{6.50} = 0.231 \text{ in.}$$

Step 6: Determine the nominal moment strength by taking moments about the top of the steel shape using the three forces shown in Figure 9.11.

$$M_n = T_s\left(\frac{d}{2}\right) + C_c\left(t - \frac{a}{2}\right) - 2A_{s-c}F_y\left(\frac{x}{2}\right)$$

$$= 650\left(\frac{20.7}{2}\right) + 500\left(4.50 - \frac{1.63}{2}\right) - 2(1.50)(50)\left(\frac{0.231}{2}\right)$$

$$= 8550 \text{ in.-kips}$$

$$M_n = \frac{8550}{12} = 713 \text{ ft-kips}$$

Step 7: For LRFD, the design moment is

$$\phi M_n = 0.9(713) = 642 \text{ ft-kips}$$

Step 7: For ASD, the allowable moment is

$$\frac{M_n}{\Omega} = \frac{713}{1.67} = 427 \text{ ft-kips}$$

The nominal moment strength decreased from 750 ft-kips for the full composite action of Example 9.1 to 713 ft-kips for the level of partial composite action given in Example 9.3. This is approximately a 5 percent reduction in strength, corresponding to more than a 23 percent reduction in shear connector strength. In both cases, the strength of the composite beam is significantly greater than that of the bare steel beam, where the plastic moment strength of the bare steel beam is $M_p = 398$ ft-kips. It is acceptable to make comparisons at the nominal strength level because for both the bare steel beam and the composite beam, the resistance factors and safety factors are the same.

9.5.3 Composite Beam Design Tables

The force transferred between the steel and concrete governs the strength of the composite beams. The shear studs transfer that force to the concrete, so design can be linked to the total shear force, ΣQ_n. Design tables have been developed that use the shear stud strength in combination with an infinite selection of concrete areas and strengths to determine the flexural strength of the composite beam. These are given in *Manual* Table 3-18, an example of which is shown here as Figure 9.12.

The variables used in *Manual* Table 3-18 are defined in Figure 9.13. The beam is divided into seven PNA locations; five are in the flange and two are in the web. When the PNA is at the top of the flange (TFL), in position 1, the entire steel section is in tension. This is a fully composite beam. When the PNA is in the web at location 7, 25 percent of the potential steel section force is transferred to the concrete through the studs. As shown in Figure 9.13, the flange has five PNA locations and the stud strength for location 6 is one-half the difference between that at locations 5 and 7. These seven PNA locations establish corresponding stud strengths, ΣQ_n, which are also given in the tables. Designs using PNA location 7 are clearly of concern regarding the deformation capacity of the shear studs since this location represents only 25% composite action.

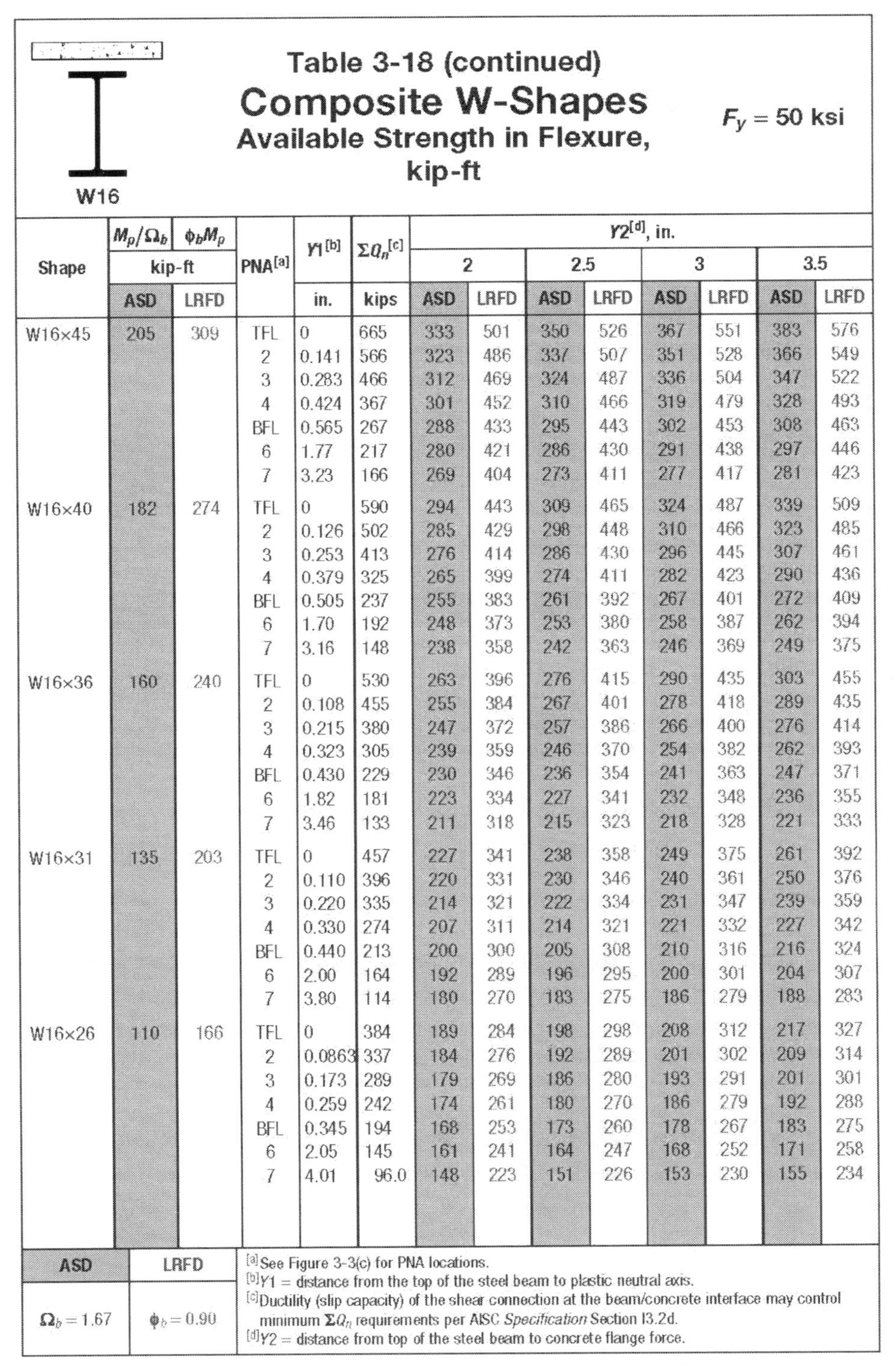

Table 3-18 (continued)
Composite W-Shapes
Available Strength in Flexure, kip-ft

W16 — $F_y = 50$ ksi

Shape	M_p/Ω_b	$\phi_b M_p$	PNA[a]	Y1[b]	ΣQ_n[c]	Y2[d], in.							
	kip-ft					2		2.5		3		3.5	
	ASD	LRFD		in.	kips	ASD	LRFD	ASD	LRFD	ASD	LRFD	ASD	LRFD
W16×45	205	309	TFL	0	665	333	501	350	526	367	551	383	576
			2	0.141	566	323	486	337	507	351	528	366	549
			3	0.283	466	312	469	324	487	336	504	347	522
			4	0.424	367	301	452	310	466	319	479	328	493
			BFL	0.565	267	288	433	295	443	302	453	308	463
			6	1.77	217	280	421	286	430	291	438	297	446
			7	3.23	166	269	404	273	411	277	417	281	423
W16×40	182	274	TFL	0	590	294	443	309	465	324	487	339	509
			2	0.126	502	285	429	298	448	310	466	323	485
			3	0.253	413	276	414	286	430	296	445	307	461
			4	0.379	325	265	399	274	411	282	423	290	436
			BFL	0.505	237	255	383	261	392	267	401	272	409
			6	1.70	192	248	373	253	380	258	387	262	394
			7	3.16	148	238	358	242	363	246	369	249	375
W16×36	160	240	TFL	0	530	263	396	276	415	290	435	303	455
			2	0.108	455	255	384	267	401	278	418	289	435
			3	0.215	380	247	372	257	386	266	400	276	414
			4	0.323	305	239	359	246	370	254	382	262	393
			BFL	0.430	229	230	346	236	354	241	363	247	371
			6	1.82	181	223	334	227	341	232	348	236	355
			7	3.46	133	211	318	215	323	218	328	221	333
W16×31	135	203	TFL	0	457	227	341	238	358	249	375	261	392
			2	0.110	396	220	331	230	346	240	361	250	376
			3	0.220	335	214	321	222	334	231	347	239	359
			4	0.330	274	207	311	214	321	221	332	227	342
			BFL	0.440	213	200	300	205	308	210	316	216	324
			6	2.00	164	192	289	196	295	200	301	204	307
			7	3.80	114	180	270	183	275	186	279	188	283
W16×26	110	166	TFL	0	384	189	284	198	298	208	312	217	327
			2	0.0863	337	184	276	192	289	201	302	209	314
			3	0.173	289	179	269	186	280	193	291	201	301
			4	0.259	242	174	261	180	270	186	279	192	288
			BFL	0.345	194	168	253	173	260	178	267	183	275
			6	2.05	145	161	241	164	247	168	252	171	258
			7	4.01	96.0	148	223	151	226	153	230	155	234

ASD	LRFD
$\Omega_b = 1.67$	$\phi_b = 0.90$

[a] See Figure 3-3(c) for PNA locations.
[b] Y1 = distance from the top of the steel beam to plastic neutral axis.
[c] Ductility (slip capacity) of the shear connection at the beam/concrete interface may control minimum ΣQ_n requirements per AISC *Specification* Section I3.2d.
[d] Y2 = distance from top of the steel beam to concrete flange force.

Figure 9.12 Composite W-Shapes: Available Strength in Flexure

$F_y = 50$ ksi

Table 3-18 (continued)
Composite W-Shapes
Available Strength in Flexure, kip-ft

W16

Shape	$Y2^{[d]}$, in.													
	4		4.5		5		5.5		6		6.5		7	
	ASD	LRFD	ASD	LRFD	ASD	LRFD	ASD	LRFD	ASD	LRFD	ASD	LRFD	ASD	LRFD
W16×45	400	601	416	626	433	651	450	676	466	701	483	726	499	751
	380	571	394	592	408	613	422	634	436	655	450	677	464	698
	359	539	370	557	382	574	394	592	405	609	417	627	429	644
	337	507	346	521	355	534	365	548	374	562	383	576	392	589
	315	473	322	483	328	493	335	503	342	513	348	523	355	533
	302	454	307	462	313	470	318	478	324	486	329	495	334	503
	286	429	290	436	294	442	298	448	302	454	306	460	310	467
W16×40	353	531	368	553	383	575	397	597	412	620	427	642	442	664
	335	504	348	523	360	542	373	561	385	579	398	598	410	617
	317	476	327	492	338	507	348	523	358	538	368	554	379	569
	298	448	306	460	314	472	322	484	330	496	338	509	347	521
	278	418	284	427	290	436	296	445	302	454	308	463	314	472
	267	401	272	409	277	416	282	423	286	430	291	438	296	445
	253	380	257	386	260	391	264	397	268	402	271	408	275	413
W16×36	316	475	329	495	342	515	356	535	369	555	382	574	395	594
	301	452	312	469	324	486	335	503	346	520	358	537	369	555
	285	429	295	443	304	457	314	471	323	486	333	500	342	514
	269	405	277	416	284	428	292	439	300	450	307	462	315	473
	253	380	259	389	264	397	270	406	276	414	281	423	287	432
	241	362	245	368	250	375	254	382	259	389	263	396	268	402
	225	338	228	343	231	348	235	353	238	358	241	363	245	367
W16×31	272	409	284	426	295	443	306	460	318	478	329	495	341	512
	260	391	270	405	280	420	290	435	299	450	309	465	319	480
	247	372	256	384	264	397	272	409	281	422	289	434	297	447
	234	352	241	362	248	373	255	383	262	393	268	404	275	414
	221	332	226	340	232	348	237	356	242	364	248	372	253	380
	208	313	212	319	216	325	221	332	225	338	229	344	233	350
	191	287	194	292	197	296	200	300	203	304	205	309	208	313
W16×26	227	341	237	356	246	370	256	384	265	399	275	413	285	428
	218	327	226	340	234	352	243	365	251	377	259	390	268	403
	208	312	215	323	222	334	229	345	237	356	244	366	251	377
	198	297	204	306	210	315	216	324	222	333	228	343	234	352
	188	282	192	289	197	296	202	304	207	311	212	318	217	326
	175	263	179	268	182	274	186	279	189	285	193	290	197	296
	158	237	160	241	163	244	165	248	167	252	170	255	172	259

ASD	LRFD	[d] $Y2$ = distance from top of the steel beam to concrete flange force.
$\Omega_b = 1.67$	$\phi_b = 0.90$	

Figure 9.12 (*Continued*) Composite W-Shapes: Available Strength in Flexure

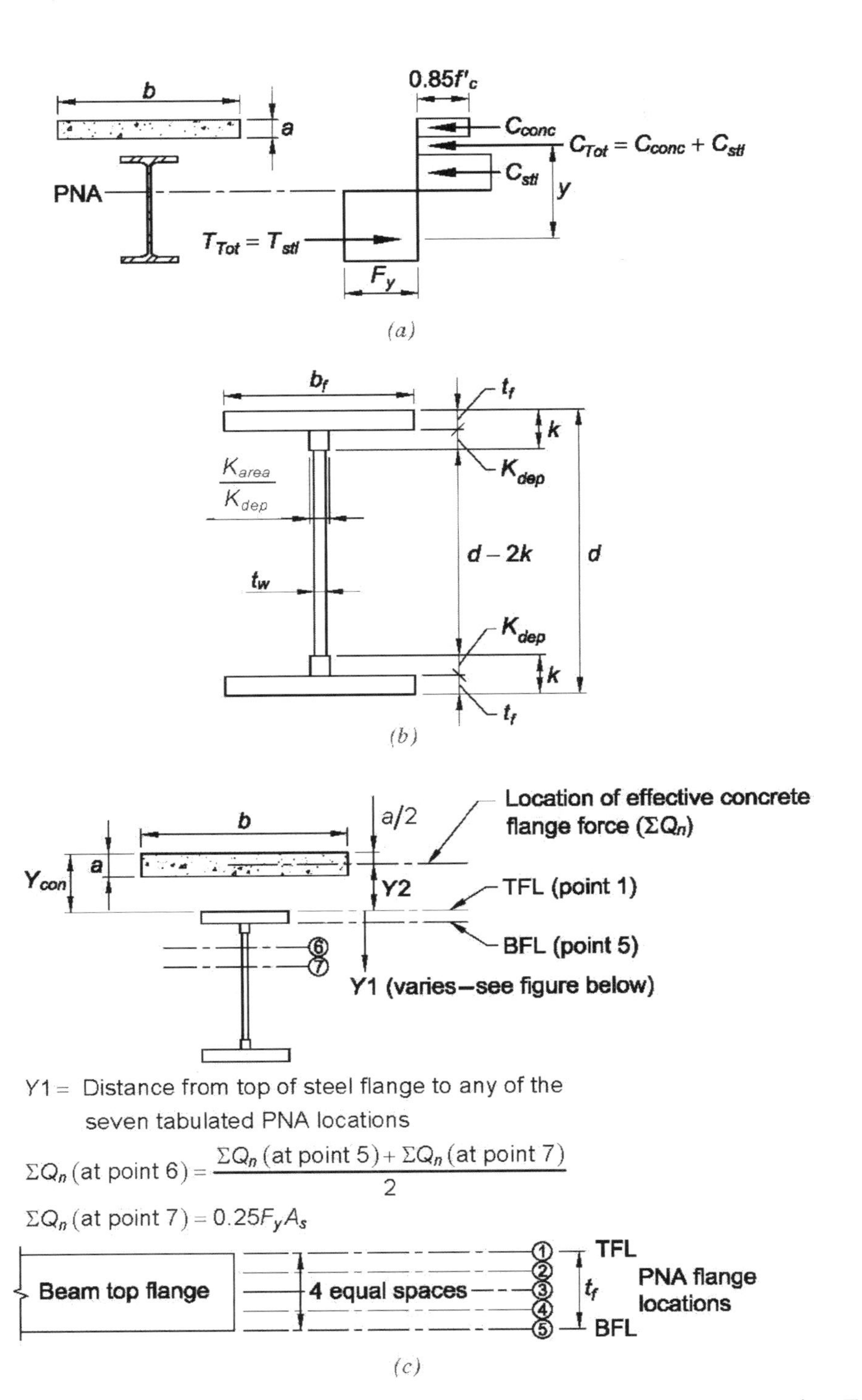

Figure 9.13 Definition of Variables for Use with Composite Beam Design Tables

The location where 50% composite action is used will vary, sometimes being in the web and sometimes in the flange and should be checked as part of the design process.

Finding the contribution of the concrete to the beam strength requires knowledge of the location of the concrete compressive force. As already discussed, the force in the concrete is equal to the force in the studs, ΣQ_n. The moment arm for that force is defined as $Y2$ in Figure 9.13. It is a function of the concrete strength and concrete geometry. These tables are quite flexible and accommodate any permitted concrete strength and effective slab width. The thickness of the slab is limited only by the maximum moment arm given in the table.

Although these tables are of most value in the design of a composite beam, they can also be used to check a particular combination. Selection of a composite beam is illustrated in Section 9.9.

EXAMPLE 9.4
Composite Beam Strength Using Tables

Goal: Determine the design flexural strength and allowable flexural strength for the fully composite W16×26.

Given The W16×26 beam is used with the metal deck and slab shown in Figure 9.14. The effective flange width is given, $b_{eff} = 60.0$ in. $f_c' = 4$ ksi. Use the portion of Table 3-18 from Figure 9.12.

SOLUTION

Step 1: Determine the controlling compression force using Equations I3-1a and I3-1b.

$$V_c' = 0.85 f_c' t b_{eff} = 0.85(4.0)(3.0)(60.0) = 612 \text{ kips}$$

$$V_s' = F_y A_s = 50.0(7.68) = 384 \text{ kips}$$

Because $V_s' < V_c'$, the steel controls and PNA is at or above the top of the flange. In Figure 9.12 this PNA location confirms that the $\Sigma Q_n = 384$ kips.

Step 2: Determine the depth of the concrete acting in compression that is needed to balance this stud force using Equation 9.9.

$$a = \frac{\Sigma Q_n}{0.85 f_c' b_{eff}} = \frac{384}{0.85(4)(60.0)} = 1.88 \text{ in.}$$

Step 3: Determine the moment arm of the compressive force from the top of the steel.

$$Y2 = t_c - \frac{a}{2} = 6.0 - \frac{1.88}{2} = 5.06 \text{ in.}$$

For LRFD

Step 4: Determine the design moment strength from Figure 9.12

Enter the table with $Y2 = 5.0$, which will be slightly conservative for an actual $Y2 = 5.06$, and $\Sigma Q_n = 384$.

For LRFD, the design moment is

$$\phi M_n = 370 \text{ ft-kips}$$

For ASD Step 4: Determine the allowable moment strength from Figure 9.12 (ASD).

Enter the table with $Y2 = 5.0$, which will be slightly conservative for an actual $Y2 = 5.06$, and $\Sigma Q_n = 384$.

For ASD the allowable moment is

$$M_n/\Omega = 246 \text{ ft-kips}$$

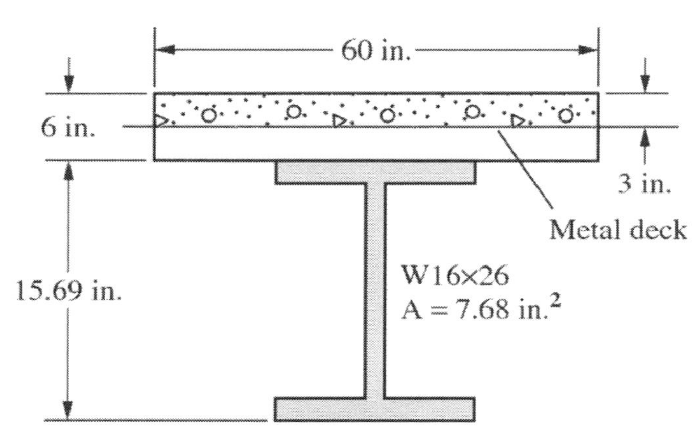

Figure 9.14 Composite Beam for Example 9.4

9.5.4 Negative Moment Strength

According to *Specification* Section I3.2b, the negative flexural strength can be taken either as that for the bare steel beam according to the provisions of Chapter F or, if the composite section satisfies the criteria, as a composite section using a plastic stress distribution. For the composite section, the concrete in tension is ignored and reinforcing steel is placed in the tension region. The resistance factor and safety factor are the same as for the positive moment case, and the nominal flexural strength is calculated assuming a plastic stress distribution similar to that for the positive moment case. The limitations of the *Specification* state that

1. The steel beam must be compact and adequately braced according to Chapter F.
2. Shear connectors (steel headed stud or steel channel anchors) are provided in the negative moment region.

3. The required longitudinal reinforcing bars are placed within the effective width of the slab and are properly developed.

9.6 SHEAR STUD STRENGTH

What are recognized in the construction industry as shear connectors are called steel anchors in the *Specification*. Two different types of shear connectors: steel studs, called steel headed stud anchors, and channels, called steel channel anchors are recognized by the *Specification*. The shear connector used almost exclusively in building construction is the steel headed stud anchor. The diameter of the studs must be 3/4 in. or less except when used in a flat soffit slab in which case 7/8 in. and 1 in. diameter studs are permitted. In addition, the stud's diameter may be no greater than 2.5 times the flange thickness of the beam to which they are welded if they are not located directly over the beam web. As a practical matter, only beams with a flange thickness that satisfies this relationship should be selected in design because there is no way to locate the beam web when installing studs in the field and some stud arrangements will require placement off the web.

The nominal strength of a single steel stud, Q_n, used in a composite beam is given in *Specification* Section I8.2a as

$$Q_n = 0.5A_{sa}\sqrt{f_c'E_c} \le R_g R_p A_{sa} F_u \qquad \text{(AISC I8-1)}$$

Table 9.2 Nominal Shear Strength, Q_n, for One Stud with $R_p = R_g = 1.0$ (kips)

	Normal-weight concrete		Lightweight concrete		
	$w_c = 145$ pcf		$w_c = 110$ pcf		
Stud diameter (in.)	$f_c' = 3$ ksi	$f_c' = 4$ ksi	$f_c' = 3$ ksi	$f_c' = 4$ ksi	Based on $F_u = 65$ ksi
3/8	5.26	6.53	4.28	5.31	7.18
1/2	9.35	11.6	7.60	9.43	12.8
5/8	14.6	18.1	11.9	14.7	19.9
3/4	21.0	26.1	17.1	21.2	28.7

where

A_{sa} = cross-sectional area of the shank of the stud, in.^{2}

f_c' = specified compressive strength of the concrete, ksi

F_u = minimum specified tensile strength of the stud, ksi

E_c = modulus of elasticity of the concrete, $w_c^{1.5}\sqrt{f_c'}$ ksi, where w_c is the unit weight of the concrete in pounds per cubic foot and f_c' is in ksi. Although this is somewhat different from the equation used by ACI 318, it provides sufficiently accurate results in this instance.

R_g, R_p = reduction factors used to bring the predicted stud strength into agreement with test results. When used in a flat soffit slab, $R_g = 1.0$ and $R_p = 0.75$.

The nominal strength of typical 3/4 in. shear studs is given, along with other results, in Table 9.2. Values are given for normal and lightweight concrete with $f_c' = 3$ ksi and 4 ksi. Values are also given for the stud strength based on the tensile strength of the stud material with $R_g = R_p = 1.0$. The reductions to be applied when these studs are incorporated into a slab on a metal deck are addressed later.

Although not normally used in practice today, channel shear connectors are also permitted by the *Specification*. The nominal strength of a channel shear connector, Q_n, is given as

$$Q_n = 0.3(t_f + 0.5t_w)l_a\sqrt{f_c' E_c} \qquad \text{(AISC I8-2)}$$

where

$t_f =$ thickness of channel flange, in.
$t_w =$ thickness of channel web, in.
$l_a =$ length of channel, in.

The strength of the channel shear connector must be developed by welding the channel to the beam flange for the force Q_n with appropriate consideration of the eccentricity of the force on the connector.

9.6.1 Number and Placement of Shear Studs

Although a shear stud serves to transfer load between the steel beam and the concrete slab, it is not necessary to place the studs in accordance with the shear diagram of the loaded beam. Tests have demonstrated that the studs have sufficient ductility to redistribute the shear load under the ultimate load condition. Therefore, in design, it is assumed that the studs share the load equally and the total shear force determined according to Section 9.5 can be transferred over the distance between maximum moment and zero moment using studs distributed along this distance. For a uniform load this results in V_q'/Q_n connectors on each side of the maximum moment at the centerline of the beam span. In the case of concentrated loads placed at the third points of the beam, the same number of studs would be used on each side of the beam between the load and the support, and a minimum number of studs would be required between the loads. This is shown in Figure 9.15.

Shear studs must be placed so that they have a minimum of 1 in. of lateral concrete cover, unless used in ribs of formed steel deck. In the direction of the shear force the minimum distance from the center of the stud to the free edge is 8 in. for normal weight concrete and 10 in. for light weight concrete. For studs not placed in steel deck ribs, the minimum center-to-center spacing is six stud diameters along the member and four stud diameters transverse to the member. When placed in metal deck ribs, the spacing is to be no less than four diameters in any direction. The maximum center-to-center stud spacing is eight times the total slab thickness or 36 in.

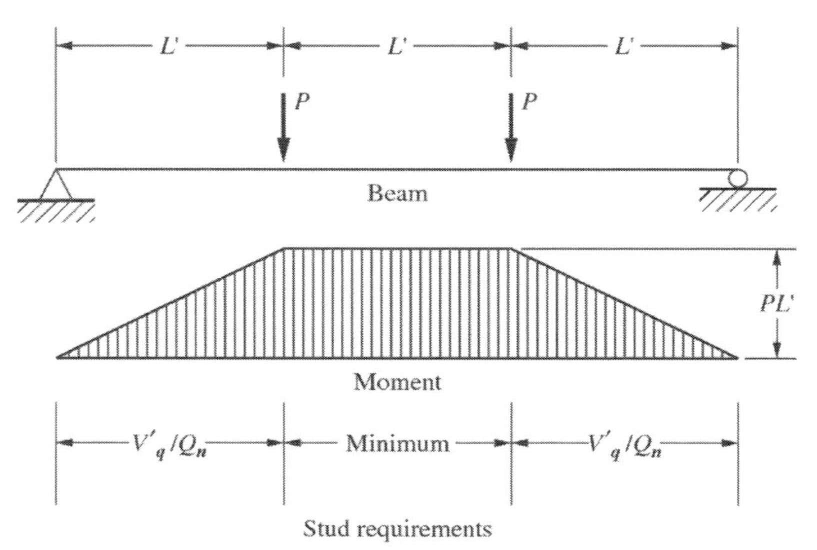

Figure 9.15 Stud Placement for Concentrated Load

EXAMPLE 9.5
Shear Stud Determination

Goal: Determine the number of 3/4 in. shear studs required over the complete beam span.

Given: Use the fully composite beam of Example 9.1. Assume normal-weight concrete and the values of Example 9.1.

SOLUTION

Step 1: Determine the strength of a single shear stud.

From Table 9.2, based on the concrete,

$$Q_n = 0.5A_{sa}\sqrt{f_c'E_c} = 26.1 \text{ kips}$$

and based on the stud without the reduction factors

$$Q_n = A_{sa}F_u = 28.7 \text{ kips}$$

For the slab in Example 9.1, the stud is welded directly to the top flange of the beam in a flat soffit slab, so

$$R_g = 1.0$$
$$R_p = 0.75$$

and

$$Q_n = R_g R_p A_{sa} F_u = 1.0(0.75)(28.7) = 21.5 \text{ kips}$$

Use the lowest Q_n, so

$$Q_n = 21.5 \text{ kips}$$

Step 2: Determine the number of studs required.

From Example 9.1

$$V_q' = \Sigma Q_n = 650 \text{ kips}$$

Thus, the number required is

$$650/21.5 = 30.2 \text{ studs}$$

Step 3: Determine the total number of studs required for the beam.

Place thirty-one 3/4 in. shear studs on each side of the beam between the maximum moment and the zero moment. Thus,

Use 62 studs for the entire beam span

Note that these calculations are independent of the use of ASD or LRFD when the shear force to be transferred is the same, as in this example, because the calculations are carried out at the nominal strength level.

9.7 COMPOSITE BEAMS WITH FORMED METAL DECK

The combination of formed steel deck and composite design is considered today to be one of the most economical methods of floor construction. The steel deck is a stay-in-place formwork for the concrete slab. Cells, which can be formed by enclosing the space below the deck and between the ribs, can be used to distribute the electrical and electronic systems of the building, contributing greatly to the overall economy of the system.

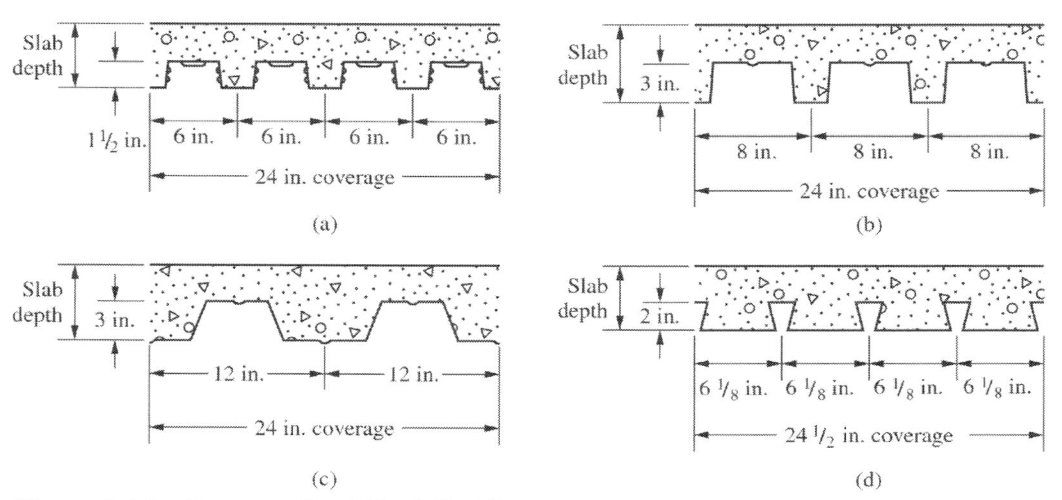

Figure 9.16 Common Steel Deck Profiles

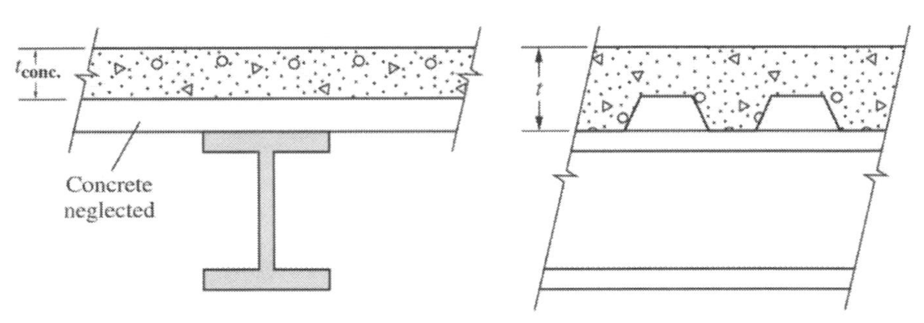

Figure 9.17 Beam with Formed Metal Deck

The *Specification* provides rules in Section I3.2c for steel decks with nominal rib heights of up to 3 in. and average rib widths of 2 in. or more. For a deck that has ribs narrower at the top than at the interface with the beam, the width of the rib used in calculations must be taken as no more than the width at the narrow portion. A deck section with this profile is shown in Figure 9.16(d) along with other common deck profiles. Studs must extend at least 1-1/2 in. above the top of the steel deck. The concrete slab thickness must be specified to provide 1/2 in. of cover over the top of the installed stud. The slab thickness above the metal deck must be at least 2.0 in., and the deck must be anchored to the supporting beam with a combination of puddle welds or other fasteners and studs at a spacing not to exceed 18 in.

9.7.1 Deck Ribs Perpendicular to Steel Beam

For beams supporting the steel deck the ribs run perpendicular to the beam, as shown in Figure 9.17. The space below the top of the rib contains concrete only in the alternating spaces, so there is no opportunity to transfer force at this level. Thus, the only concrete available for calculating the full concrete force is above the top of the deck.

The right-hand side of the inequality of Equation I8-1, without R_p or R_g, accounts for the tensile strength of the stud material. This strength must be reduced to account for the location of the stud within the concrete rib, because there is a difference in strength for a stud placed closer to the rib wall in the direction of force versus one placed closer to the rib wall away from the force. The stud strength value specified on the right-hand side of the inequality of Equation I8-1 includes two multipliers, R_g and R_p. A simplified table of the values for these adjustment factors is given in Table 9.3, and the strength of the stud controlled by F_u, with $R_p = R_g = 1.0$, is given in Table 9.2.

R_g is used to account for the number of studs in a given concrete rib. If a rib contains a single stud, R_g = 1.0; if the rib contains two studs, R_g = 0.85; and if a rib contains three or more studs, R_g= 0.7.

Table 9.3 Shear Stud Strength Adjustment Factors

Condition	R_g	R_p
No decking	1.0	0.75
Decking oriented parallel to the steel shape		
$\frac{w_r}{h_r} \geq 1.5$	1.0	0.75
$\frac{w_r}{h_r} < 1.5$	0.85**	0.75
Decking oriented perpendicular to the steel shape		
Number of studs occupying the same decking rib		
1	1.0	0.6†
2	0.85	0.6†
3 or more	0.7	0.6†

h_r = nominal rib height, in.
w_r = average width of concrete rib or haunch (as defined in Section I3.2c), in.
**For a single stud.
†This value may be increased to 0.75 when $e_{mid\text{-}ht} \geq 2$ in.

R_p is used to account for the location of the stud in the rib in either the strong or the weak position. Figure 9.18 shows the strong and weak locations of a stud in relation to the applied force. Because it is difficult to ensure that the studs are located in the strong position, it is recommended that $R_p = 0.6$ be used unless the need for the slight strength increase is critical enough to warrant the extra effort in the field to ensure that studs are placed in the strong position. When studs are placed in the strong position, which is at least 2.0 in. from the loaded side of the rib edge at mid-height of the rib, as shown in Figure 9.18, R_p can be increased to 0.75, which is the same value that was used for the flat soffit slab.

The maximum stud spacing is specified as 36 in., which is convenient because many decks have a rib spacing in multiples of 6 in.

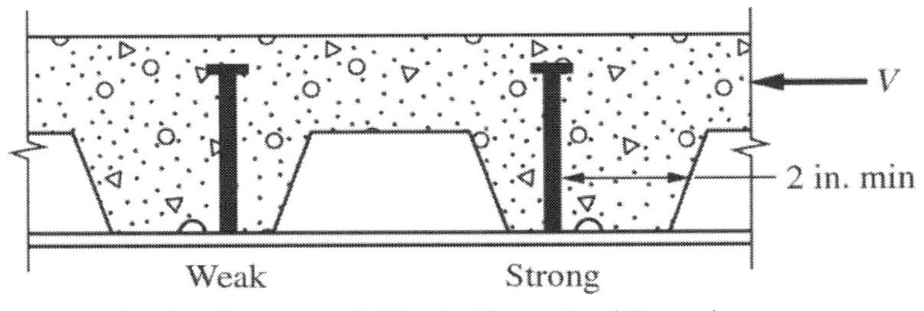

Figure 9.18 Strong and Weak Shear Stud Locations

9.7.2 Deck Ribs Parallel to Steel Beam

For girders supporting beams that carry steel deck, the ribs run parallel to the steel section, as shown in Figure 9.19. Concrete below the top of the deck can be used in calculating the composite section properties and must be used in shear stud calculations. For calculation purposes, concrete below the top of the steel deck can be neglected unless it is needed to balance the shear stud strength. The design procedure described for flat soffit beams applies here as well, provided sufficient concrete is available above the top of the steel deck as determined using Equation 9.1 or 9.9. If the concrete below the top of the steel deck must be used to balance the stud or steel beam strength, the only difference in the determination of strength is related to the changed geometry from inclusion of a portion of the ribs.

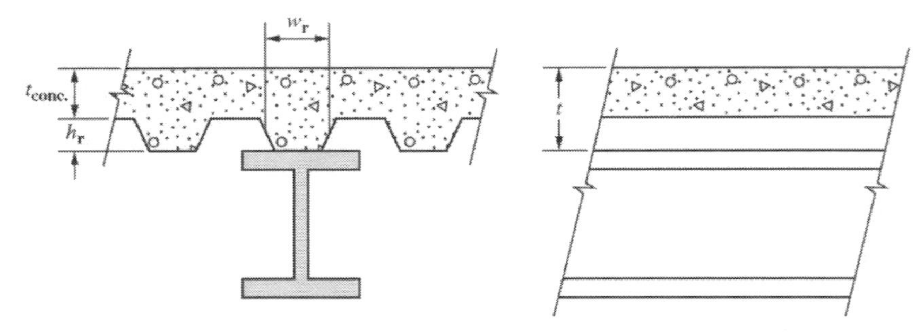

Figure 9.19 Girder with Formed Metal Deck

When the depth of the steel deck is 1-1/2 in. or greater, the average width, w_r, of the haunch or rib is not to be less than 2 in. for the first stud in the transverse row plus four stud diameters for each additional stud. If the deck rib is too narrow, the deck can be split over the beam and spaced in such a way as to allow for the necessary rib width without adversely affecting member strength.

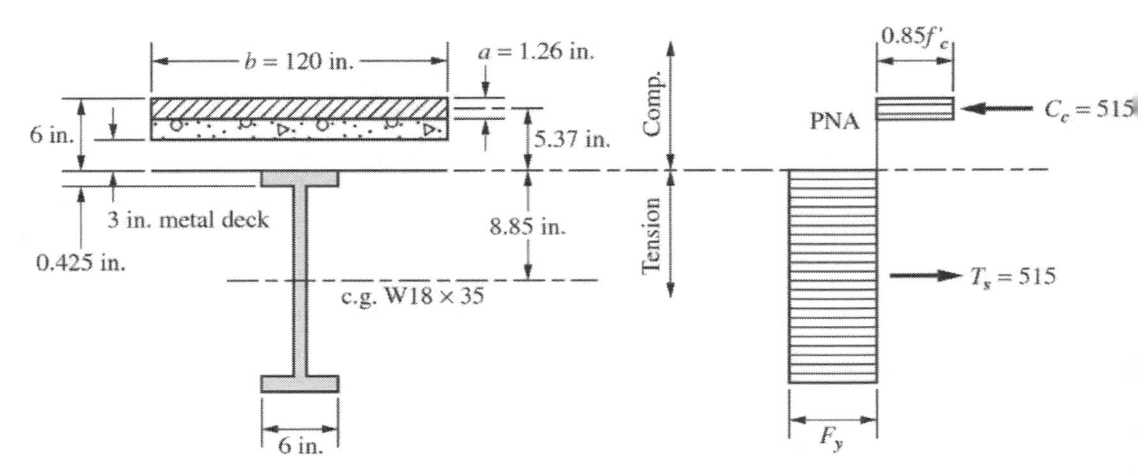

Figure 9.20 Example 9.6, case (a)

For this deck orientation, R_g is used to account for the width-to-height ratio of the deck rib. When $w_r/h_r \geq 1.5$, $R_g = 1.0$. When $w_r/h_r < 1.5$, $R_g = 0.85$.

R_p is taken as 0.75 in all cases where the deck ribs are parallel to the supporting member, just as for the flat soffit slab.

EXAMPLE 9.6
Composite Beam Design

Goal: Calculate the design moment and allowable moment and determine the stud requirements for a composite section. Carry out the calculations for the following three cases.

(**a**) Full composite action (Figure 9.20).
(**b**) Partial composite action with $\Sigma Q_n = 387$ kips, which results in the PNA at the center of the top flange of the steel beam (Figure 9.21).
(**c**) Partial composite action with $\Sigma Q_n = 260$ kips, which results in the PNA at the bottom of the top flange of the steel beam (Figure 9.22).

Given: Use a W18×35 with a 6-in. slab on a 3 in. metal deck perpendicular to the beam with the profile shown in Figure 9.16c. The beam spacing is 12 ft and the beam span is 40 ft. $f_c' = 4$ ksi and $F_y = 50$ ksi. From *Manual* Table 1-1, $A_s = 10.3$ in.^{2}, $d = 17.7$ in. and $t_f = 0.425$ in.

SOLUTION

Step 1: Determine the effective flange width.

$$b_{eff} = 40.0/4 = 10.0 \text{ ft (governs)}$$
$$b_{eff} = \text{beam spacing} = 12.0 \text{ ft}$$

Step 2: Determine the compression force using the full concrete and full steel areas using Equations I3-1a and I3-1b.

$$V_c' = 0.85 f_c' t b_{eff} = 0.85(4)(3.0)(120) = 1220 \text{ kips}$$
$$V_s' = F_y A_s = 50(10.3) = 515 \text{ kips}$$

Part a Full Composite Action (Figure 9.20)
Step 3: Determine the controlling concrete force.

For full composite action, C_c is the smaller of V_c' and V_s'; thus,

$$C_c = 515 \text{ kips}$$

Step 4: Calculate the effective depth of the concrete using Equation 9.1.

$$a = \frac{C_c}{0.85 f_c' b_{eff}} = \frac{515}{0.85(4)(120)} = 1.26 \text{ in.}$$

Because a is less than the 3 in. available in the concrete above the deck, the procedures for a flat soffit beam can be used.

Step 5: Determine the nominal moment strength.

$$M_n = T_s\left(\frac{d}{2}\right) + C_c\left(t - \frac{a}{2}\right)$$

$$= 515\left(\frac{17.7}{2}\right) + 515\left(6.0 - \frac{1.26}{2}\right) = 7320 \text{ in.-kips}$$

$$= \frac{7320}{12} = 610 \text{ ft-kips}$$

For LRFD Step 6: The design moment is

$$\phi M_n = 0.9(610) = 549 \text{ ft-kips}$$

For ASD Step 6: The allowable moment is

$$\frac{M_n}{\Omega} = \frac{610}{1.67} = 365 \text{ ft-kips}$$

Step 7: Determine the strength of a single stud.

From Table 9.2, the value of a single 3/4 in. stud with normal-weight concrete, $f_c' = 4$ ksi, is 26.1 kips based on the concrete.

However, because the studs are used in conjunction with the metal deck, a check for any required reduction must be made. For the deck profile given, the ribs are spaced at 12.0 in. on center which would permit 39 studs per beam if only one stud were placed in each rib. Thus, assuming two studs per rib with the studs placed in the weakest location, from Tables 9.2 and 9.3,

$$R_g = 0.85$$

$$R_p = 0.6$$

$$A_{sc}F_u = 28.7 \text{ kips}$$

$$R_g R_p A_{sa} F_u = 0.85(0.6)(28.7) = 14.6 \le 26.1 \text{ kips}$$

The stud strength is the lower value, based on the stud placement in the metal deck.

Step 8: Determine the number of studs required on each side of the maximum moment.

The shear that is to be transferred is 515 kips. Therefore,

$$\text{Number of studs} = 515/14.6 = 35.3$$

Step 9: Determine the total number of studs required for the beam.

Use 36 studs on each half span or 72 studs for the full beam span

With this deck profile, studs can be placed in pairs every 12 in. This will nicely accommodate the 72 studs on the 40 ft span with two studs placed in each rib.

Part b Partial Composite Action (Figure 9.21)

Step 10: Determine the controlling concrete force.

Because the value of $\Sigma Q_n = 387$ kips given is less than V_c' and V_s' as determined in Part a, ΣQ_n controls and this is a partially composite beam. With $\Sigma Q_n / V_s' = 387/515 = 0.75$ stud ductility will not be a concern.

Step 11: Calculate the effective depth of the concrete using Equation 9.9 with $C_c = \Sigma Q_n = 387$ kips.

$$a = \frac{C_c}{0.85 f_c' b_{eff}} = \frac{387}{0.85(4)(120)} = 0.949 \text{ in.}$$

Because $a < 3.0$ in., sufficient concrete is available above the metal deck as in Part a.

Step 12: Determine the area of steel in compression.

$$A_{s-c} = \frac{V_s' - C_c}{2F_y} = \frac{515 - 387}{2(50)} = 1.28 \text{ in.}^2$$

Step 13: Determine the location of the PNA in the steel.

Assume the PNA is in the flange.

$$x = \frac{A_{s-c}}{b_f} = \frac{1.28}{6.00} = 0.213 \text{ in.} < t_f = 0.425 \text{ in.}$$

Therefore, the PNA is in the flange.

Step 14: Determine the nominal moment strength.

$$M_n = T_s\left(\frac{d}{2}\right) + C_c\left(\frac{t}{2}\right) - 2A_{s-c}F_y\left(\frac{x}{2}\right)$$

$$= 515\left(\frac{17.7}{2}\right) + 387\left(6.0 - \frac{0.949}{2}\right) - 2(1.28)(50)\left(\frac{0.213}{2}\right) = 6680 \text{ in.}$$

$$= \frac{6680}{12} = 557 \text{ ft-kips}$$

For LRFD Step 15: The design moment is

$$\phi M_n = 0.9(557) = 501 \text{ ft-kips}$$

For ASD Step 15: The allowable moment is

$$\frac{M_n}{\Omega} = \left(\frac{557}{1.67}\right) = 334 \text{ ft-kips}$$

Step 16: Determine the stud requirements.

The shear to be transferred is 387 kips.

The value of stud strength previously determined in Part a for two studs per rib is

$$Q_n = 14.6 \text{ kips}$$

Step 17: Determine the required number of shear studs.

$$387/14.6 = 26.5 \text{ studs}$$

Step 18: Determine the total number of studs required for the beam.

Use 27 studs in each half span or 54 studs for the full beam span.

With 54 studs symmetrically placed in 39 available ribs, 12 ribs will remain without studs. It is possible to recalculate the stud requirements to see if the higher strength when only one stud per rib is used would work for this beam.

Since $R_g = 1.0$ for a single stud per rib, from Part a

$$R_g R_p A_{sa} F_u = 1.0(0.6)(28.7) = 17.2 \leq 26.1 \text{ kips}$$

and the number of studs required for the half span is

$$387/17.2 = 22.5$$

Thus, 46 studs are required, and this would still require two studs per rib for 8 ribs. For constructability reasons, 54 studs placed symmetrically, two studs per rib will be selected.

Part c Partial Composite Action (Figure 9.22)

Step 19: Determine the controlling concrete force.

Because the value of $\Sigma Q_n = 260$ kips given is less than V_c' and V_s', ΣQ_n controls and this is a partially composite beam. With $\Sigma Q_n / V_s' = 260/515 = 0.50$ stud ductility will not be a concern.

Step 20: Calculate the effective depth of concrete using Equation 9.9.

$$a = \frac{C_c}{0.85 f'_c b_{eff}} = \frac{260}{0.85(4)(120)} = 0.637 \text{ in.}$$

Step 21: Determine the area of steel in compression.

$$A_{s-c} = \frac{V'_s - C_c}{2F_y} = \frac{515 - 260}{2(50)} = 2.55 \text{ in.}^2$$

Step 22: Determine the location of the PNA in the steel.

Assume the PNA is in the flange.

$$x = \frac{A_{s-c}}{b_f} = \frac{2.55}{6.00} = 0.425 \text{ in.}$$

which is the flange thickness, as expected.

Step 23: Determine the nominal moment strength.

$$M_n = T_s\left(\frac{d}{2}\right) + C_c\left(\frac{t}{2}\right) - 2A_{s-c}F_y\left(\frac{x}{2}\right)$$

$$= 515\left(\frac{17.7}{2}\right) + 260\left(6.0 - \frac{0.637}{2}\right) - 2(2.55)(50)\left(\frac{0.425}{2}\right)$$

$$= 5980 \text{ in.-kips}$$

$$= \frac{5980}{12} = 498 \text{ ft-kips}$$

For LRFD Step 24: The design moment is

$$\phi M_n = 0.9(498) = 448 \text{ ft-kips}$$

For ASD Step 24: The allowable moment is

$$\frac{M_n}{\Omega} = \left(\frac{498}{1.67}\right) = 298 \text{ ft kips}$$

Step 25: Determine the stud requirements.

The shear to be transferred is 260 kips.

As before, the single stud strength is

$$Q_n = 14.6 \text{ kips}$$

Step 26: Determine the required number of shear studs.

260/14.6 = 17.8 studs.

Step 27: Determine the total number of studs required for the beam.

Use 18 studs in each half span or 36 studs for the full beam span.

However, this requires only a single stud in each rib, so the $R_g = 0.85$ does not need to be applied. Therefore, as from Part b

$$Q_n = 1.0(0.6)(28.7) = 17.2 < 26.1 \text{ kips}$$

and

$$260/17.2 = 15.1 \text{ studs}$$

Thus,

use a total of 32 studs

which will also be accommodated with one stud per rib.

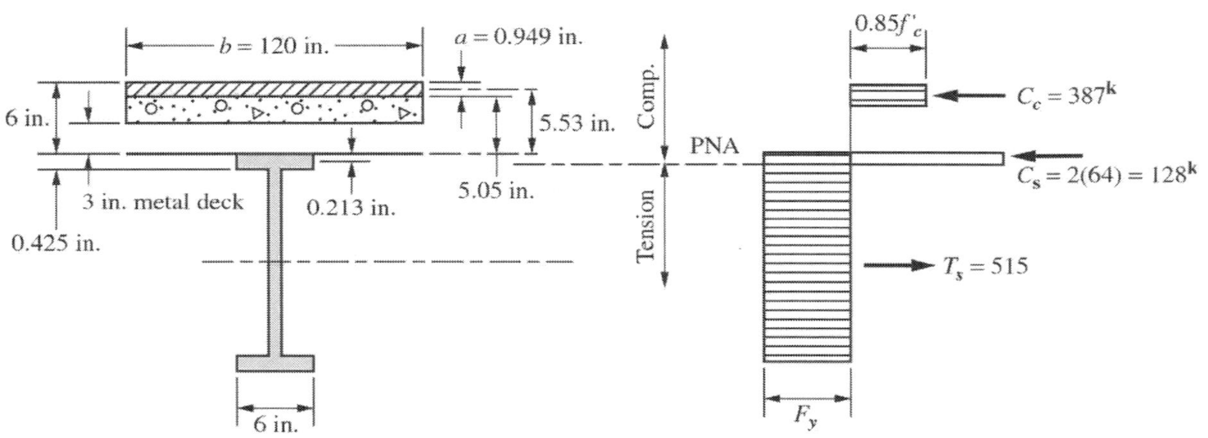

Figure 9.21 Example 9.6, case (b)

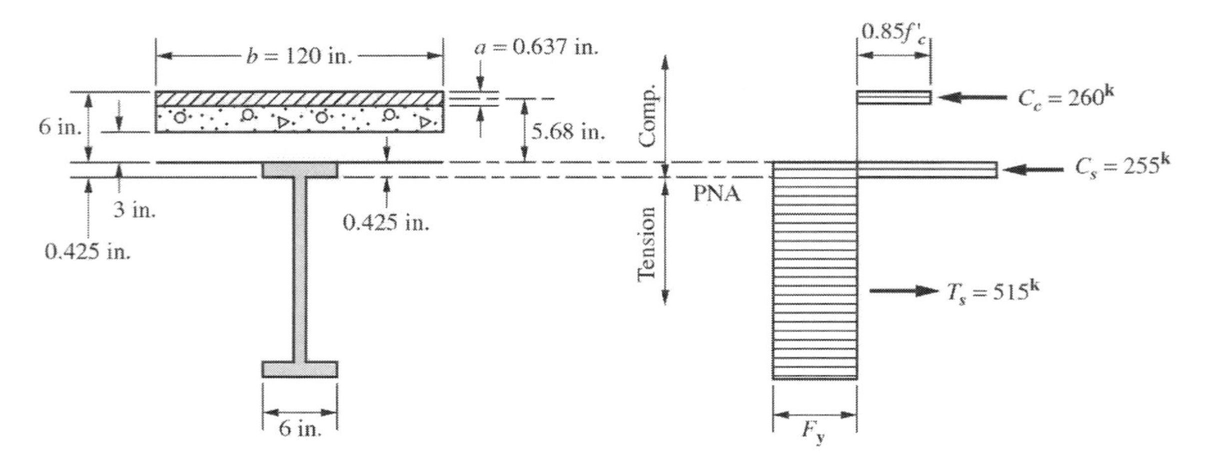

Figure 9.22 Example 9.6, case (c)

A comparison of the results from Example 9.6 shows that the magnitude of the reduction in moment strength that results from a reduction in the shear force transferred by the shear studs is less than the magnitude of shear strength reduction. For Part b of Example 9.6, the shear strength was reduced approximately 25 percent, but it produced a moment strength reduction of only approximately 9 percent. Similarly, for Part c, the 50 percent reduction in stud strength led to an 18 percent reduction in moment strength. This reinforces the idea that a partially composite beam may be more economical than a fully composite beam and suggests why they are quite common in building design.

9.8 FULLY ENCASED STEEL BEAMS

Steel beams fully encased in concrete that contributes to the strength of the final member are called *encased beams*. Such beams can be designed through one of two procedures given in *Specification* Section I3.3. The flexural strength can be calculated from the superposition of elastic stresses, considering the effects of shoring. Or, when shear connectors are provided, the flexural strength can be based on the plastic stress distribution or strain compatibility approach for the composite section. Alternatively, the strength can be calculated from the plastic stress distribution on the steel section alone. In all cases, $\phi_b = 0.9$ and $\Omega_b = 1.67$. Although encased beams are not particularly common in building construction today, these provisions will also apply when determining the flexural strength of composite beam-columns.

9.9 SELECTING A SECTION

The design of a composite beam is somewhat of a trial-and-error procedure, as are numerous other design situations. The material presented thus far in this chapter has been directed toward the determination of section strength when the cross section and concrete dimensions are known. This section addresses the preliminary selection of the steel shape to go along with a given concrete slab. This procedure is followed by a discussion of the design tables found in the *Manual*.

With an estimate of beam depth, the weight of the beam can be estimated. This is based on the assumption that the PNA is within the concrete so that the full steel section is at yield. The resulting dimensions are given in Figure 9.23.

The moment arm between the tension force in the steel and the compression force in the concrete is given by

$$\text{moment arm} = \frac{d}{2} + \left(t - \frac{a}{2}\right) \tag{9.10}$$

If the nominal moment strength is divided by the moment arm, the required tension force can be determined. If that force is divided by the steel yield stress, the required area is determined. Multiplying the required area by weight of steel, 3.4 lb/ft for each square inch, yields an estimate of the beam weight. Thus,

$$\text{beam weight} = \frac{3.4 M_n}{\left(\frac{d}{2} + t - \frac{a}{2}\right) F_y} \tag{9.11}$$

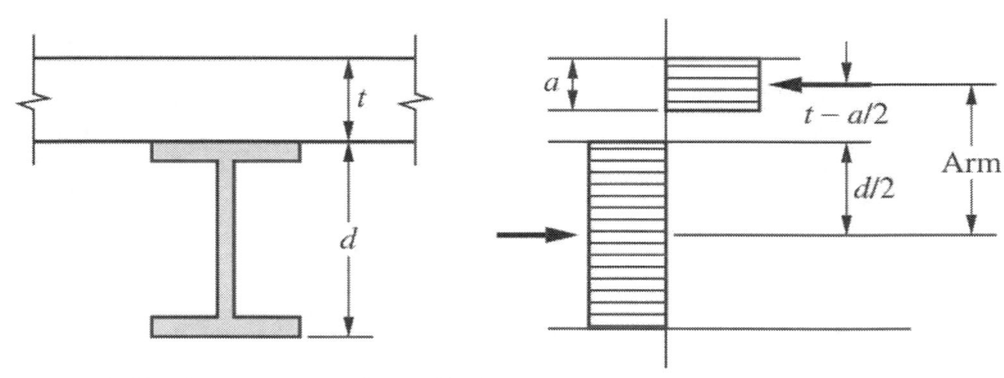

Figure 9.23 Moment Arm for Preliminary Weight Determination

To determine the beam weight by this approach, the depth of the beam must be estimated. Several approaches have been suggested for this. One simple approach is to take the span in feet and divide it by 24 to get the depth, d, in inches. Another approach to determine the depth of the total composite section ($d + t$) is to take the span in feet and divide by 16 to obtain the total depth in inches. Any reasonable approach gives a starting point. Because the thickness of the slab is determined from the design in the transverse direction, only the effective depth of the concrete is left to be determined. It is generally sufficient to assume that the effective depth of the concrete is 1 in.; therefore, $a/2 = 0.5$ in.

Although this approach to finding a starting point for composite beam design might be helpful, the design tables in the *Manual* make this much effort unnecessary. *Manual* Table 3-18 (Figure 9.12) has already been discussed in the context of determining the strength of a given combination of steel and concrete. It will now be approached from the perspective of selecting a section through the use of an example.

EXAMPLE 9.7a
Composite Beam Design by LRFD

Goal: Select an economical W-shape to be used as a composite beam.

Given: The composite beam spans 30.0 ft and is spaced at 10.0 ft from adjacent beams. It supports a 5 in. slab on a 2 in. formed steel deck with a profile similar to that shown in Figure 9.16a. The beam must carry a dead load moment of 50.0 ft-kips and a live load moment of 150 ft-kips. In addition, the bare steel beam must be checked for the dead load plus a construction live load moment of 40.0 ft-kips. $F_y = 50$ ksi and $f_c' = 3$ ksi.

SOLUTION

Step 1: Determine the required moment for the composite beam.

$$M_u = 1.2(50.0) + 1.6(150) = 300 \text{ ft-kips}$$

Step 2: Determine the starting moment arm for the concrete from the top of the steel.

Manual Table 3-18 is most effective when entered with a value for $Y2$. In order to start the design process, the moment arm for the compressive force in the concrete must be estimated. It is almost always adequate to assume, as a starting point, that $a = 1.0$ in. Thus

$$Y2 = 5 - \frac{1.0}{2} = 4.5 \text{ in.}$$

Step 3: Select potential W-shapes from Figure 9.12 and Manual Table 3-18. Since the beam span is 30 ft it will be acceptable to use partial composite action lower than 50%.

Enter the column in Figure 9.12 with $Y2 = 4.5$ and proceed down to identify the lightest weight potential section that will carry the design moment of 300 ft-kips.

W16×26 with $\phi M_n = 306$ ft kips and $\Sigma Q_n = 242$ kips, PNA location 4

Using portions of Table 3-18 from the *Manual* yields additional possibilities:

W16×31 with $\phi M_n = 319$ ft-kips and $\Sigma Q_n = 164$ kips, PNA location 6
W14×34 with $\phi M_n = 309$ ft-kips and $\Sigma Q_n = 159$ kips, PNA location 6
W14×30 with $\phi M_n = 311$ ft-kips and $\Sigma Q_n = 248$ kips, PNA location 4
W14×26 with $\phi M_n = 312$ ft-kips and $\Sigma Q_n = 332$ kips, PNA location 2

Step 4: Determine the effective flange width.

$$b_{eff} \leq \frac{30.0(12)}{4} = 90.0 \text{ in.}$$

$$b_{eff} \leq 10.0(12) = 120 \text{ in.}$$

Therefore

$$b_{eff} = 90.0 \text{ in.}$$

Step 5: Using the W16×31, determine the depth of concrete needed to balance the force in the shear studs.

For the beam span and spacing given using $C_c = \Sigma Q_n = 164$ kips and the effective flange width already determined.

$$a = \frac{C_c}{0.85 f'_c b_{eff}} = \frac{164}{0.85(3)(90.0)} = 0.715 \text{ in.}$$

Because this is less than the value $a = 1.0$ in. that was assumed to start the problem, the assumption was conservative. Design could continue with the determination of a more accurate required stud strength, or this conservative solution could be used. The required number of studs would be determined as before, accounting for the presence of any formed steel deck and its influence on the individual stud strength.

Step 6: Determine the required strength of the bare steel beam under dead load plus construction live load.

$$M_u = 1.2(50.0) + 1.6\,(40.0) = 124 \text{ ft-kips}$$

Step 7: Check to verify that the bare steel beam will provide the required strength. Since this beam has the steel deck attached to its compression flange, it will be considered as fully braced for lateral-torsional buckling under the weight of the construction loads.

From *Manual* Table 3-18, the W16×31 has a design strength of

$$\phi M_p = 203 \text{ ft-kips} > 124 \text{ ft-kips}$$

Therefore, the W16×31 is an acceptable selection for strength.

Step 8: To show what happens when the assumption for a is not quite as good, the W14×26 is considered. Again, using the ΣQ_n determined from the table and the effective flange width,

$$a = \frac{C_c}{0.85 f_c' b_{eff}} = \frac{332}{0.85(3)(90.0)} = 1.45 \text{ in.}$$

This is significantly greater than the assumed value. To consider this section further, determine a new $Y2$ such that

$$Y2 = 5 - \frac{1.45}{2} = 4.28 \text{ in.}$$

Entering *Manual* Table 3-18 with $Y2 = 4.0$ as a conservative number, $M_u = 300$ ft-kips is determined and it corresponds to the same required shear stud strength. Thus, this section also meets the strength requirements, and the design can proceed with stud selection.

EXAMPLE 9.7b
Composite Beam Design by ASD

Goal: Select an economical W-shape to be used as a composite beam.

Given: The composite beam spans 30.0 ft and is spaced at 10.0 ft from adjacent beams. It supports a 5 in. slab on a 2 in. formed steel deck with a profile similar to that shown in Figure 9.16a. The beam must carry a dead load moment of 50.0 ft-kips and a live load moment of 150 ft-kips. In addition, the bare steel beam must be checked for the dead load plus a construction live load moment of 40.0 ft-kips. $F_y = 50$ ksi and $f_c' = 3$ ksi.

SOLUTION

Step 1: Determine the required moment for the composite beam.

$$M_a = 50.0 + 150 = 200 \text{ ft-kips}$$

Step 2: Determine the starting moment arm for the concrete from the top of the steel.

Manual Table 3-18 is most effective when entered with a value for *Y*2. In order to start the design process, the moment arm for the compressive force in the concrete must be estimated. It is almost always adequate to assume, as a starting point, that $a = 1.0$ in. Thus,

$$Y2 = 5 - \frac{1.0}{2} = 4.5 \text{ in.}$$

Step 3: Select potential W-shapes from Figure 9.12 and Manual Table 3-18. Since the beam span is 30 ft it will be acceptable to use partial composite action lower than 50%.

Enter the column in Figure 9.12 with $Y2 = 4.5$ and proceed down to identify the lightest weight potential section that will carry the design moment of 200 ft-kips.

W16×26 with $M_n/\Omega = 204$ ft kips, $\Sigma Q_n = 242$ kips, PNA location 4

Using portions of Table 3-18 from the *Manual* yields additional possibilities:

W16×31 with $M_n/\Omega = 212$ ft-kips, $\Sigma Q_n = 164$ kips, PNA location 6
W14×34 with $M_n/\Omega = 205$ ft-kips, $\Sigma Q_n = 159$ kips, PNA location 6
W14×30 with $M_n/\Omega = 207$ ft-kips, $\Sigma Q_n = 248$ kips, PNA location 4
W14×26 with $M_n/\Omega = 208$ ft-kips, $\Sigma Q_n = 332$ kips, PNA location 2

Step 4: Determine the effective flange width.

$$b_{eff} \le \frac{30.0(12)}{4} = 90.0 \text{ in.}$$

$$b_{eff} \le 10.0(12) = 120 \text{ in.}$$

Therefore,

$$b_{eff} = 90.0 \text{ in.}$$

Step 5: Using the W16×31, determine the depth of concrete needed to balance the force in the shear studs.

For the beam span and spacing given using $C_c = \Sigma Q_n = 164$ kips and the effective flange width already determined.

$$a = \frac{C_c}{0.85 f'_c b_{eff}} = \frac{164}{0.85(3)(90.0)} = 0.715 \text{ in.}$$

Because this is less than the value a = 1.0 in. that was assumed to start the problem, the assumption was conservative. Design could continue with the determination of a more accurate required stud strength, or this conservative solution could be used. The required number of studs would be determined as before, accounting for the presence of any formed steel deck and its influence on the individual stud strength.

Step 6: Determine the required strength of the bare steel beam under dead load plus construction live load.

$$M_a = 50.0 + 40.0 = 90.0 \text{ ft-kips}$$

Step 7: Check to verify that the bare steel beam will support the required strength. Since this beam has the steel deck attached to its compression flange, it will be considered as fully braced for lateral-torsional buckling under the weight of the construction loads.

From *Manual* Table 3-18, the W16×31 has an allowable strength of

$$M_p / \Omega = 135 \text{ ft-kips} > 90.0 \text{ ft-kips}$$

Therefore, the W16×31 is an acceptable selection for strength.

Step 8: To show what happens when the assumption for a is not quite as good, the W14×26 is considered. Again, using the ΣQ_n determined from the table and the effective flange width,

$$a = \frac{C_c}{0.85 f'_c b_{eff}} = \frac{332}{0.85(3)(90.0)} = 1.45 \text{ in.}$$

This is significantly greater than the assumed value. To consider this section further, determine a new $Y2$ such that

$$Y2 = 5 - \frac{1.45}{2} = 4.28 \text{ in.}$$

Entering *Manual* Table 3-18 with $Y2$ = 4.0 as a conservative number, M_a = 210 ft-kips is determined and it corresponds to an increased required shear stud strength.

Thus, this section meets the strength requirements, and the design can proceed with stud selection.

Table 9.4 Shapes Selected for Example 9.7a (LRFD)

Shape	ϕM_n	ΣQ_n	Percent composite action	Weight of steel	Number of studs	Equivalent weight of studs	Equivalent total weight of beam
W16×31	319	164	35.9	930	16	160	1090
W16×26	306	242	63.0	780	24	240	1020
W14×34	309	159	31.8	1020	16	160	1180
W14×30	311	248	56.0	900	24	240	1140
W14×26	312	332	86.2	780	32	320	1100

Which of the many possible sections should be chosen as the final design depends on the overall economics of the situation. One way to compare several choices is to look at the total weight of the steel sections combined with the total quantity of studs required. To make this comparison it is often effective to assume that an installed single shear stud has the equivalent cost of 10 pounds of steel. To make this type of comparison, the five potential sections found initially in Example 9.7a are presented in Table 9.4. Here, it was assumed that $Q_n = 21.0$ kips. This means that no consideration was taken for metal deck reduction. In addition, no check was made for the assumed versus actual a dimension. This table is simply to help determine which of the potential shapes should be considered further. Based on this table, it could be said that the W16×26 with an equivalent weight of 1020 lbs should be investigated further to determine such things as actual steel headed stud anchor requirements and if deflection requirements are satisfied.

9.10 SERVICEABILITY CONSIDERATIONS

Three important serviceability considerations are associated with the design of composite floor systems: deflection during construction, vibration under service loads, and live load deflection under service loads.

9.10.1 Deflection During Construction

As discussed in Section 9.3, the *Specification* permits either shored or unshored construction. With unshored construction, the *Specification* requires that the steel section alone have adequate strength to support all loads applied prior to the concrete's attaining 75 percent of its specified strength. The bare steel beam, under the weight of these loads, deflects as an elastic member. Because of this deflection of the beam under the weight of the wet concrete, cambering of the steel beam is often specified. Cambering is the imposition of a permanent upward deflection of the beam in its unloaded state so that, under load, the downward deflection results in a beam without excessive final deflection from the original horizontal position. Predicting the necessary camber is difficult because of the varying methods and sequences of concrete placement used by different contractors as well as such factors as the end restraint provided by the beam connection. Thus, it is normal practice to camber only 70-80% of the dead load deflection. AISC Design Guide 36 Design Considerations for Camber provides additional guidance for when selecting camber. Even with camber it is often prudent for the designer to add a little extra concrete load into the design dead load and for the contractor to allow for a little extra concrete in the quantity estimate.

In the case of shored construction, deflection during construction is usually not a concern, because the shores are not removed until the concrete has achieved some strength and composite action can be counted upon. The deflection under the wet concrete for shored construction is at a minimum. On the other hand, long-term deflection due to creep of the concrete may have to be investigated because the concrete is stressed under the self-weight as a permanent load along with the sustained service loads.

9.10.2 Vibration Under Service Loads

Composite construction usually is shallower than comparable noncomposite construction and, therefore, may be more susceptible to perceived vibrations. Because vibration calculations assume that the beam behaves compositely, even when it is not a composite beam, the additional stiffness of a composite beam does not improve its vibration characteristics. If problems occur, they usually occur in applications with long spans and little damping. For instance, a large area of a department store containing only a light jewelry display and with no partition walls connecting one floor to the next might exhibit vibrations that would be perceptible to some customers. On the other hand, an office building constructed with the same floor system could contain full or partial height partitions that would provide sufficient damping to obviate any perceived vibration. Because of wide differences in human perception of vibration and many other factors, vibration problems do not lend themselves to simple solutions. AISC Design Guide 11, *Vibrations of Steel Framed Structural Systems Due to Human Activity*, provides more information and gives the designer an approach to vibration acceptance criteria, damping, and rational design techniques.

9.10.3 Live Load Deflections

Live load deflections can be a critical design consideration for many applications. Excessive deflection could cause problems with the proper fit of partitions, doors, and equipment and may also result in an unacceptable appearance, including cracking of finishes and other visible evidence of distress. Therefore, a live load deflection calculation should be carried out for most situations. As discussed in Chapter 6, this calculation is made with the service loads for which deflection is of interest, usually the nominal live loads.

Because beam deflections are a function of the stiffness of the beam, the modulus of elasticity and moment of inertia of the composite section must be determined. The true moment of inertia at the service load level for which deflections are to be calculated is not easily determined. In addition, the modulus of elasticity of the composite section must account for the interaction of steel and concrete. The normal approach is to transform the concrete into a material that behaves like steel, with the same modulus of elasticity. The moment of inertia of the new transformed section can then be determined. Transformation of the concrete into steel is accomplished by dividing the concrete area by the modular ratio, $n = E_s / E_c$, and using the same thickness. Although this seems like a fairly straightforward process, the problem is determining the thickness of the concrete that is actually participating in resisting the deflection.

One approach is to assume that the only concrete participating in deflection resistance is also that which is providing strength. Thus, whether it is a fully composite or partially composite member, a moment of inertia can be determined using the known value of *a* from the strength calculations. Because the nominal strength is calculated at the ultimate load level, the amount of concrete actually participating for service loads could be significantly more than that used in the strength calculations. Thus, a moment of inertia determined by this approach is less than what might actually be available and is called a *lower bound moment of inertia*, I_{LB}. Figure 9.24 is a sample from *Manual* Table 3-19, which gives the lower bound moment of inertia in a format that parallels the strength tables already discussed. Use of these I_{LB} values results in a conservative estimate of service load deflections.

EXAMPLE 9.8
Deflection

Goal: Determine the construction load deflection of the bare steel beam and the service load deflection for the composite beam of Example 9.7

Given: Consider the W16×26 from Example 9.7. Check the beam for a dead load of 0.45 kips/ft, a construction live load of 0.36 kips/ft, and an in-service live load of 1.35 kips/ft. Compare the construction load deflection to span/360. For the live load deflection, use the lower bound moment of inertia from Figure 9.24 and compare the calculated deflection to span/360 as a design limit.

SOLUTION

Step 1: Determine the total construction load for deflection calculations.

$$w = w_D + w_{L\text{-}const.} = 0.45 + 0.36 = 0.81 \text{ kips/ft}$$

Step 2: Determine the moment of inertia of the W16×26 from *Manual* Table 1-1 or *Manual* Table 3-19.

$$I_x = 301 \text{ in.}^4$$

Step 3: Calculate the construction load deflection and compare to span/360.

$$\Delta_{LL} = \frac{5(0.81)(30.0)^4(1728)}{384(29{,}000)(301)} = 1.69 \text{ in.} > \frac{30(12)}{360} = 1.0 \text{ in.}$$

Because the deflection exceeds our limit, cambering of the beam or shoring during construction would be required. Shoring has a significant impact on cost as well as on scheduling; therefore, it is likely that the beam would be cambered, or a larger section used. Selecting camber as the appropriate solution, if we must continue to use this W16×26, the beam should be cambered to approximately 75 percent of the dead load deflection.

Thus, 0.75(1.69) = 1.27 in., so specify a camber of 1.25 in.

Table 3-19 (continued) Lower-Bound Elastic Moment of Inertia, I_{LB}, for Plastic Composite Sections, in.^{4}

F_y = 50 ksi

I_{LB}

W16

Shape[a]	PNA[b]	Y1[c]	ΣQ_n	Y2[d], in.										
		in.	kips	2	2.5	3	3.5	4	4.5	5	5.5	6	6.5	7
W16×45 (586)	TFL	0	665	1260	1330	1400	1470	1550	1630	1720	1810	1900	1990	2090
	2	0.141	566	1200	1270	1330	1400	1470	1550	1630	1710	1790	1880	1970
	3	0.283	466	1140	1200	1260	1320	1380	1450	1520	1590	1670	1750	1830
	4	0.424	367	1060	1110	1160	1220	1270	1330	1390	1450	1520	1590	1660
	BFL	0.565	267	971	1010	1050	1090	1140	1190	1230	1290	1340	1390	1450
	6	1.77	217	917	950	986	1020	1060	1100	1140	1190	1230	1280	1330
	7	3.23	166	854	882	910	940	972	1000	1040	1070	1110	1150	1190
W16×40 (518)	TFL	0	590	1110	1170	1230	1300	1370	1440	1520	1590	1670	1760	1850
	2	0.126	502	1060	1120	1170	1240	1300	1370	1430	1510	1580	1660	1740
	3	0.253	413	1000	1050	1110	1160	1220	1280	1340	1400	1470	1540	1610
	4	0.379	325	937	980	1030	1070	1120	1170	1230	1280	1340	1400	1460
	BFL	0.505	237	856	891	927	965	1000	1050	1090	1130	1180	1230	1280
	6	1.70	192	808	837	869	901	935	971	1010	1050	1090	1130	1170
	7	3.16	148	755	779	804	831	859	888	918	949	982	1020	1050
W16×36 (448)	TFL	0	530	973	1030	1080	1140	1200	1270	1340	1410	1480	1550	1630
	2	0.108	455	933	983	1040	1090	1150	1210	1270	1330	1400	1470	1540
	3	0.215	380	886	931	979	1030	1080	1130	1190	1250	1310	1370	1440
	4	0.323	305	831	871	912	956	1000	1050	1100	1150	1200	1260	1310
	BFL	0.430	229	765	797	831	867	905	944	984	1030	1070	1120	1160
	6	1.82	181	715	743	772	802	833	866	901	936	973	1010	1050
	7	3.46	133	659	680	703	727	752	778	805	833	862	892	923
W16×31 (375)	TFL	0	457	827	874	923	974	1030	1080	1140	1200	1260	1330	1400
	2	0.110	396	795	838	884	931	981	1030	1090	1140	1200	1260	1320
	3	0.220	335	758	797	838	882	927	974	1020	1070	1130	1180	1240
	4	0.330	274	714	749	786	824	864	906	949	995	1040	1090	1140
	BFL	0.440	213	663	692	723	756	790	825	862	900	940	982	1020
	6	2.00	164	614	639	664	691	720	749	780	812	845	879	914
	7	3.80	114	556	574	594	614	636	658	681	705	730	756	783
W16×26 (301)	TFL	0	384	674	712	753	796	840	887	935	985	1040	1090	1150
	2	0.0863	337	649	686	724	763	805	849	894	941	990	1040	1090
	3	0.173	289	621	654	689	726	764	804	846	889	934	980	1030
	4	0.259	242	589	619	651	683	718	754	791	830	871	912	956
	BFL	0.345	194	551	577	604	633	663	694	727	760	795	832	869
	6	2.05	145	505	527	549	572	597	622	649	676	705	734	765
	7	4.01	96	450	466	482	499	517	535	555	575	596	617	640

[a] Value in parentheses is I_x (in.^{4}) of noncomposite steel shape.
[b] See Figure 3-3(c) for PNA locations.
[c] Y1 = distance from top of the steel beam to plastic neutral axis.
[d] Y2 = distance from top of the steel beam to concrete flange force.

Figure 9.24 Lower-Bound Elastic Moment of Inertia, I_{LB}, for Plastic Composite Sections

Step 4: Assuming that the construction load deflection issue is resolved, determine the live load deflection under the in-service live load.

From Example 9.7, the W16×26 was selected using $Y2 = 4.5$ and the resulting shear stud force was $\Sigma Q_n = 242$ kips at PNA location 4 for both LRFD and ASD.

Step 5: Determine the lower bound moment of inertia.

Using *Manual* Table 3-19 with the values given in step 4, select

$$I_{LB} = 754 \text{ in.}^4$$

Step 6: Determine the live load deflection.

$$\Delta_{LL} = \frac{5(1.35)(30.0)^4(1728)}{384(29{,}000)(754)} = 1.13 \text{ in.}$$

Step 7: Compare the calculated deflection with the given limit.

$$\Delta_{LL} = 1.13 > \frac{\text{span}}{360} = \frac{30(12)}{360} = 1.0 \text{ in.}$$

Because the calculated deflection is greater than the limiting value, the live load deflection is not acceptable based on the given criteria. This result, combined with the construction load deflection issue, would likely lead the designer to select a larger section for this situation, as was actually done in Example 9.7.

Deflection calculations are carried out under service loads and are independent of design by LRFD or ASD.

9.11 COMPOSITE COLUMNS

Composite columns in building construction have been much slower to gain acceptance than composite beams. *Specification* provisions were first provided in the 1986 LRFD *Specification* and for ASD in the 2005 *Specification*. Although the use of composite columns in buildings is still quite limited, attention to hardening of structures against blast forces will likely bring them more to the forefront.

Specification Section I2 provides for two types of composite columns: open shapes encased by concrete and hollow shapes filled with concrete. Composite columns exist at the interface between specification provisions for steel and those for concrete. For a member to qualify as a composite column under the *Specification*, it must meet the following limitations:

1. The cross-sectional area of the steel member must constitute at least 1 percent of the gross area.
2. For an encased member the concrete encasement must be reinforced with continuous longitudinal steel as well as lateral ties or spirals. The longitudinal steel area must be at least 0.004 times the gross area, and the tie area must be at least equivalent to No. 3 bars at 12 in. spacing. Tie spacing must not exceed one-half the smallest dimension of the column.
3. For filled hollow sections, strength is a function of local buckling of the walls of the hollow section. To be classified as a compact section, HSS must have a minimum wall thickness such that $b/t \leq 2.26\sqrt{E/F_y}$ for rectangular HSS and $D/t \leq 0.15(E/F_y)$ for round HSS. Noncompact and slender wall hollow sections may be used for filled composite columns provided the appropriate *Specification* provisions are met. For these shapes, the compressive strength will be limited by the limit state of local buckling.
4. The concrete strength, f_c', must be between 3 ksi and 10 ksi for normal-weight concrete and 3 ksi and 6 ksi for lightweight concrete as for beams.
5. The maximum value of F_y for structural steel shapes to be used in calculating strength is 75 ksi as for beams.
6. The maximum value of F_y for reinforcing bars to be used in calculating strength is 80 ksi as for beams.

Although these requirements are usually readily satisfied, for situations where they are not, ACI 318 should also be consulted.

To account for the effects of column slenderness on the nominal strength of a composite column, the equations found in Chapter E for steel columns are used with slight modification. Because of the combination of two dissimilar materials and the general uncertainties of composite column behavior, the resistance and safety factors are taken as

$$\phi_c = 0.75 \text{ (LRFD)} \qquad \Omega_c = 2.00 \text{ (ASD)}$$

To convert the column equations, Equations E3-2 and E3-3, for use with a composite column, multiply each occurrence of stress; the nominal stress, F_n, the yield stress, F_y, and the elastic buckling stress, F_e, by the area of the column and make the following substitutions; $F_n A = P_n$, $F_y A = P_{no}$ and $F_e A = P_e$. This is presented in *Specification* Section I2.1 for encased composite columns where

for $\dfrac{P_{no}}{P_e} \leq 2.25$

$$P_n = P_{no}\left[0.658^{\left(\frac{P_{no}}{P_e}\right)}\right] \qquad \text{(AISC I2-2)}$$

and for $\frac{P_{no}}{P_e} > 2.25$

$$P_n = 0.877 P_e \qquad \text{(AISC I2-3)}$$

Where P_{no} for the bare steel column was the yield strength of the shape, for a composite column this strength must also include the contribution of the reinforcing steel and the concrete stressed to their ultimate strength. Thus, to account for the reinforcing steel, $F_{ysr}A_{sr}$ is added and to account for the concrete, $0.85f'_cA_c$ is added, Thus,

$$P_{no} = F_y A_s + F_{ysr} A_{sr} + 0.85 f'_c A_c \qquad \text{(AISC I2-4)}$$

Where P_e for the bare steel column was the elastic buckling strength based on the stiffness, *EI*, of the steel shape, for the composite column it must be based on an effective stiffness, EI_{eff}, which incorporates the contribution of the reinforcing steel and the concrete. Thus,

$$P_e = \frac{\pi^2 EI_{eff}}{(L_c)^2} \qquad \text{(AISC I2-5)}$$

and the effective stiffness is a direct summation of the contribution of the steel shape, the reinforcing steel and an appropriate portion of the concrete, thus

$$EI_{eff} = E_s I_s + E_s I_{sr} + C_1 E_c I_c \qquad \text{(AISC I2-6)}$$

Research has shown that the contribution to stiffness from the concrete is a function of the ratio of the total steel area to the gross area of the composite member, up to 70% of the stiffness of the uncracked concrete section, given by

$$C_1 = 0.25 + 3\left(\frac{A_s + A_{sr}}{A_g}\right) \le 0.7 \qquad \text{(AISC I2-7)}$$

In these equations, the *s* subscript refers to the steel section, the *sr* subscript refers to the longitudinal reinforcing steel, and the *c* subscript refers to the concrete. With these variables defined, Equations I2-2 and I2-3 are then used to determine the nominal strength of the encased composite column.

For filled columns with compact HSS or box sections, the equivalent yield strength must account for the behavior that these columns exhibit. Thus, as for encased composite columns, the steel section is at yield and the concrete is at $0.85f'_c$ for rectangular sections and $0.95f'_c$ for round sections. This increase to $0.95f'_c$ reflects the increase in strength of the concrete due to the confinement of the circular steel section just as for spiral columns designed according to ACI 318. The reinforcing steel is transformed by the modular ratio, E_s/E_c, into a material that behaves like concrete. Thus

$$P_{no} = P_p = F_y A_s + C_2 f'_c\left(A_c + A_{sr}\left(\frac{E_s}{E_c}\right)\right) \qquad \text{(AISC I2-9a,b)}$$

where

C_2 = 0.85 for rectangular sections
= 0.95 for round sections

The effective stiffness of the filled hollow section also accounts for the confinement provided by the rectangular and round steel shape. In this case, research has shown that the contribution to the effective stiffness of the concrete is up to 90% of the stiffness of the uncracked concrete section. Thus,

$$EI_{eff} = E_s I_s + E_s I_{sr} + C_3 E_c I_c \quad \text{(AISC I2-12)}$$

and

$$C_3 = 0.45 + 3\left(\frac{A_s + A_{sr}}{A_g}\right) \leq 0.9 \quad \text{(AISC I2-13)}$$

Equations I2-2 and I2-3 are then used to determine the nominal strength for the filled compact wall column.

For what is called a noncompact wall composite column, the concrete contribution to P_{no} is reduced from $0.85f'_c$ for rectangular sections and $0.95f'_c$ for round sections to $0.7f'_c$ for both as the wall slenderness increases from λ_p to λ_r. For slender wall filled composite members, the contribution of the concrete to P_{no} remains unchanged but the contribution of the steel shape becomes a function of the critical buckling stress of the wall. These requirements are given in *Specification* Section I2.2b(b) and I2.2b(c). There are four square ASTM A500 Grade C HSS that are noncompact and one that is slender; six round HSS are noncompact and none are slender.

EXAMPLE 9.9
Composite Column Strength

Goal: Determine the nominal strength of an encased composite column. Then determine the design strength and the allowable strength.

Given: The column is composed of a W14×53 encased in 18 in.×22 in. of normal weight concrete as shown in Figure 9.25. Additional given information is as follows:

Column effective length = 15 ft.
Steel shape: F_y = 50 ksi
Reinforcing: 4- #9 bars, Gr. 60, F_y = 60 ksi. Area of 1 bar = 1.0 in.^{2}
Concrete strength: f_c' = 5 ksi
$E_c = 145^{1.5}\sqrt{5.0} = 3900$ ksi

SOLUTION

Step 1: Determine the areas of the components.

$A_s = 15.6\text{ in.}^2$ from *Manual* Table 1-1

$A_{sr} = 4(1.0) = 4.0\text{ in.}^2$

$$A_c = bh - A_s - A_{sr}$$
$$= 18.0(22.0) - 15.6 - 4.0 = 376 \text{ in.}^2$$
$$A_g = bh = 18.0(22.0) = 396 \text{ in.}^2$$

Step 2: Check the minimum steel ratios.

$$\rho_s = \frac{A_s}{A_g} = \frac{15.6}{22.0(18.0)} = 0.0394 > 0.01$$

$$\rho_{sr} = \frac{A_{sr}}{A_g} = \frac{4.0}{22.0(18.0)} = 0.0101 > 0.004$$

So, the specified minimums are satisfied.

Step 3: Determine P_{no} and P_e. From Equation I2-7

$$C_1 = 0.25 + 3\left(\frac{15.6 + 4.0}{396}\right) = 0.398 < 0.7$$

By inspection, the y-axis will be the critical buckling axis since it is the weak axis for both the steel shape and the concrete encasement.

$$I_s = I_y = 57.7 \text{ in.}^4 \text{ from } \textit{Manual} \text{ Table 1-1}$$

$$I_{sr} = \Sigma A d^2 = 4(1.0)(6.625)^2 = 176 \text{ in.}^4$$

$$I_c = \frac{hb^3}{12} - I_s - I_{sr}$$
$$= \frac{22.0(18.0)^3}{12} - 57.7 - 176 = 10{,}500 \text{ in.}^4$$

From Equation I2-4

$$P_{no} = F_y A_s + F_{ysr} A_{sr} + 0.85 f_c' A_c$$
$$= 50(15.6) + 60(4.0) + 0.85(5)(376) = 2620 \text{ kips}$$

and from Equation I2-6

$$EI_{eff} = E_s I_s + E_s I_{sr} + C_1 E_c I_c$$
$$= 29{,}000(57.7) + 29{,}000(176) + 0.398(3900)(10{,}500)$$
$$= 23.1 \times 10^6 \text{ kip-in.}^2$$

and from Equation I2-5

$$P_e = \frac{\pi^2 EI_{eff}}{(L_c)^2} = \frac{\pi^2 (23.1 \times 10^6)}{(15.0(12))^2} = 7040 \text{ kips}$$

Step 4: Determine the controlling column strength equation.

$$\frac{P_{no}}{P_e} = \frac{2620}{7040} = 0.372 < 2.25$$

Therefore, use Equation I2-2.

Step 5: Determine the nominal compressive strength.

$$P_n = P_{no}(0.658)^{P_{no}/P_e} = 2620(0.658)^{0.372} = 2240 \text{ kips}$$

For LRFD Step 6: The design compressive strength is

$$\phi P_n = 0.75(2240) = 1680 \text{ kips}$$

For ASD Step 6: The allowable compressive strength is

$$P_n/\Omega = 2240/2.00 = 1120 \text{ kips}$$

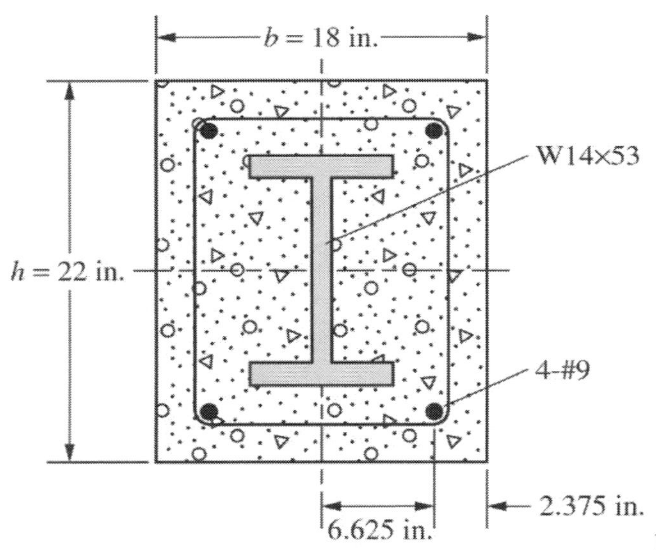

Figure 9.25 Composite Column (Example 9.9)

The Commentary indicates that the tables found in AISC Design Guide 6, *Load and Resistance Factor Design of W-Shapes Encased in Concrete*, give conservative results for encased columns by following earlier specifications. The design guide also includes suggested details for implementing encased composite columns in a project. Because filled HSS represents a limited set of possible geometries, strength tables have been developed for these shapes and are available on the AISC web page under the *Manual Companion* link at www.aisc.org/manualresources. The tables are used in exactly the same way as the column tables for the unfilled HSS column previously discussed. In cases where the filled HSS would have strength less than the unfilled HSS, the composite column tables do not provide values, and the design should revert to that for an unfilled HSS column.

EXAMPLE 9.10
HSS Composite Column Strength

Goal: Determine the available strength of a composite filled HSS column and compare it to the strength of an unfilled HSS.

Given: The column is composed of an A500 Grade C HSS7×4×1/2 filled with normal weight concrete. Additional given information is as follows:

$$L_c = 15 \text{ ft}$$
$$F_y = 50 \text{ ksi}$$
$$f_c' = 5 \text{ ksi}$$

SOLUTION

Step 1: Determine the wall slenderness of the HSS for the slenderest wall. From *Manual* Table 1-11 and the limit from Table I1.1a,

$$h/t = 12.1 < 2.26\sqrt{E/F_y} = 2.26\sqrt{29{,}000/50} = 54.4$$

So, the column is compact

Step 2: Determine the needed properties from *Manual* Table 1-11.

$$A_s = 8.81 \text{ in.}^2,\ I_y = 20.7 \text{ in.}^4,\ r_y = 1.53 \text{ in.}$$

and

$A_{sr} = 0$ since there is no reinforcing steel

$A_c = 6.0(3.0) = 18 \text{ in.}^2$ without considering the reduction in concrete area due to the rounded corners.

Confirm that there is sufficient steel to be considered a composite column

$$\rho_s = \frac{A_s}{A_g} = \frac{8.81}{7.0(4.0)} = 0.315 > 0.01$$

Step 3: Determine the equivalent yield strength of the column using Equations I2-9a and I2-9b. For the rectangular HSS, $C_2 = 0.85$.

$$P_{no} = P_p = F_y A_s + C_2 f_c'\left(A_c + A_{sr}\frac{E_s}{E_c}\right)$$
$$= 50(8.81) + 0.85(5)(18 + 0)$$
$$= 517 \text{ kips}$$

Step 4: Determine the effective stiffness. First determine C_3 from Equation I2-13.

$$C_3 = 0.45 + 3\left(\frac{A_s + A_{sr}}{A_g}\right)$$
$$= 0.45 + 3\left(\frac{8.81 + 0}{7(4)}\right)$$
$$= 1.39 > 0.9$$

Therefore $C_3 = 0.9$ and determine E_c

$$E_c = w^{1.5}\sqrt{f_c'} = 145^{1.5}\sqrt{5} = 3900 \text{ ksi}$$

Thus, from Equation I2-12

$$EI_{eff} = 29{,}000(20.7) + 0.9(3900)\left(\frac{6.0(3.0)^3}{12}\right)$$
$$= 648{,}000 \text{ kip-in.}^2$$

Step 5: Determine the elastic buckling strength from Equation I2-5

$$P_e = \frac{\pi^2 EI_{eff}}{L_c^2} = \frac{\pi^2(648{,}000)}{(12(15))^2} = 197 \text{ kips}$$

Step 6: Determine which compressive strength equation to use

$$\frac{P_{no}}{P_e} = \frac{517}{197} = 2.62 > 2.25$$

Therefore, use Equation I2-3

Step 7: Determine the nominal compressive strength

$$P_n = 0.877P_e = 0.877(197) = 173 \text{ kips}$$

For LRFD Step 8: Determine the design strength of the composite column

$$\phi P_n = 0.75(173) = 130 \text{ kips}$$

For ASD Step 8: Determine the allowable strength of the composite column

$$P_n/\Omega = 173/2.00 = 86.5 \text{ kips}$$

Step 9: To determine the strength of the unfilled HSS column, first determine the elastic buckling stress from Equation E3-4

$$F_e = \frac{\pi^2 E}{\left(\frac{L_c}{r_y}\right)^2} = \frac{\pi^2 (29{,}000)}{\left(\frac{12(15)}{1.53}\right)^2} = 20.7 \text{ ksi}$$

Step 10: Determine which nominal stress equation to use

$$\frac{F_y}{F_e} = \frac{50}{20.7} = 2.42 > 2.25$$

Therefore, use Equation E3-3

Step 11: Determine the critical stress

$$F_n = 0.877F_e = 0.877(20.7) = 18.2 \text{ ksi}$$

Step 12: Determine the nominal strength

$$P_n = 18.2(8.81) = 160 \text{ kips}$$

For LRFD Step 13: Determine the design strength

$$\phi P_n = 0.9(160) = 144 \text{ kips}$$

For ASD Step 13: Determine the allowable strength

$$P_n/\Omega = 160/1.67 = 95.8 \text{ kips}$$

Compare the available strength of the composite column and the unfilled HSS.

For LRFD Step 14: Since the unfilled HSS has a design strength of 144 kips which is greater than 130 kips for the filled HSS, use the design strength of the unfilled HSS. Thus,

$$\phi P_n = 144 \text{ kips}$$

For ASD Step 14: Since the unfilled HSS has an allowable strength of 95.8 kips which is greater than 86.5 kips for the filled HSS, Use the allowable strength of the unfilled HSS. Thus,

$$P_n/\Omega = 95.8 \text{ kips}$$

EXAMPLE 9.11
HSS Composite Column Strength

Goal: Determine the available strength of a composite filled HSS column and compare it to the strength of an unfilled HSS.

Given: The column is composed of an A500 Grade C HSS16×8×5/8 filled with normal weight concrete. Additional given information is as

follows:
Column effective length,

$L_c = 21$ ft
$F_y = 50$ ksi
$f'_c = 5$ ksi

SOLUTION

Step 1: Determine the wall slenderness of the HSS for the slenderest wall. From *Manual* Table 1-11 and the limit in Table I1.1a,

$$h/t = 24.5 < 2.26\sqrt{E/F_y} = 2.26\sqrt{29{,}000/50} = 54.4$$

So, the column is compact

Step 2: Determine the needed properties from *Manual* Table 1-11.

$$A_s = 25.7 \text{ in.}^2,\ I_y = 274 \text{ in.}^4,\ r_y = 3.27 \text{ in.}$$

and

$A_{sr} = 0$ since there is no reinforcing steel

$A_c = 14.75(6.75) = 99.6 \text{ in.}^2$ without considering the reduction in concrete area due to the rounded corners.

Confirm that there is sufficient steel to be considered a composite column.

$$\rho_s = \frac{A_s}{A_g} = \frac{25.7}{16.0(8.0)} = 0.201 > 0.01$$

Step 3: Determine the equivalent yield strength of the column using Equations I2-9a and I2-9b. For the rectangular HSS, $C_2 = 0.85$.

$$P_{no} = P_p = F_y A_s + C_2 f'_c\left(A_c + A_{sr}\frac{E_s}{E_c}\right)$$

$$= 50(25.7) + 0.85(5)(99.6 + 0)$$

$$= 1710 \text{ kips}$$

Step 4: Determine the effective stiffness. First determine C_3 using Equation I2-13.

$$C_3 = 0.45 + 3\left(\frac{A_s + A_{sr}}{A_g}\right)$$

$$= 0.45 + 3\left(\frac{25.7 + 0}{16(8)}\right)$$

$$= 1.05 > 0.9$$

Therefore $C_3 = 0.9$ and determine E_c

$$E_c = w^{1.5}\sqrt{f'_c} = 145^{1.5}\sqrt{5} = 3900 \text{ ksi}$$

Thus, the effective stiffness from Equation I2-12 is

$$EI_{eff} = 29{,}000(274) + 0.9(3900)\left(\frac{14.75(6.75)^3}{12}\right)$$

$$= 9.27 \times 10^6 \text{ kip-in.}^2$$

Step 5: Determine the elastic buckling strength from Equation I2-5

$$P_e = \frac{\pi^2 EI_{eff}}{L_c^2} = \frac{\pi^2\left(9.27\times10^6\right)}{\left(12(21)\right)^2} = 1440 \text{ kips}$$

Step 6: Determine which compressive strength equation to use

$$\frac{P_{no}}{P_e} = \frac{1710}{1440} = 1.19 < 2.25$$

Therefore, use Equation I2-2

Step 7: Determine the nominal compressive strength

$$P_n = 0.658^{1.19}(1710) = 1040 \text{ kips}$$

For LRFD Step 8: Determine the design strength of the composite column

$$\phi P_n = 0.75(1040) = 780 \text{ kips}$$

For ASD Step 8: Determine the allowable strength of the composite column

$$P_n/\Omega = 1040/2.00 = 520 \text{ kips}$$

Step 9: To determine the strength of the unfilled HSS column, first determine the elastic buckling stress using Equation E3-4

$$F_e = \frac{\pi^2 E}{\left(\frac{L_c}{r_y}\right)^2} = \frac{\pi^2(29{,}000)}{\left(\frac{12(21)}{3.27}\right)^2} = 48.2 \text{ ksi}$$

Step 10: Determine which nominal stress equation to use

$$\frac{F_y}{F_e} = \frac{50}{48.2} = 1.04 < 2.25$$

Therefore, use Equation E3-3

Step 11: Determine the critical stress

$$F_n = 0.658^{1.04}(50) = 32.4 \text{ ksi}$$

Step 12: Determine the nominal strength

$$P_n = 32.4(25.7) = 833 \text{ kips}$$

For LRFD Step 13: Determine the design strength

$$\phi P_n = 0.9(833) = 750 \text{ kips}$$

For ASD Step 13: Determine the allowable strength

$$P_n/\Omega = 833/1.67 = 499 \text{ kips}$$

Compare the available strength of the composite column and the unfilled HSS.

For LRFD Step 14: Since the unfilled HSS has a design strength of 750 kips which is less than 780 kips for the filled HSS, use the design strength of the filled HSS. Thus,

$$\phi P_n = 780 \text{ kips}$$

For ASD Step 14: Since the unfilled HSS has an allowable strength of 499 kips which is less than 520 kips for the filled HSS, Use the allowable strength of the filled HSS. Thus,

$$P_n/\Omega = 520 \text{ kips}$$

Examples 9.10 and 9.11 illustrate that filling an HSS is not a guarantee that the column strength will increase. Even in the case where the strength does increase, as in Example 9.11, the increase may not be significant enough warrant the extra cost of adding the concrete. Both of these situations result because of the increased variability of a composite column when compared to an unfilled column and the resulting reduction in resistance factor and increase in safety factor.

9.12 COMPOSITE BEAM-COLUMNS

Composite beam-columns have the same potential to occur as bare steel beam-columns. Any application where bending moment and axial force are applied simultaneously needs to be addressed according to the provisions of *Specification* Section I5.

For doubly symmetric composite beam-columns, the most common composite beam-columns found in building construction, the interaction equations of *Specification* Chapter H can be used conservatively. For a more accurate approach to determining the available strength, the interaction surface can be developed based on plastic stress distributions and the length effects modification from *Specification* Section I2 made as discussed in Section 9.11.

Figure 9.26 provides several potentially useful interaction diagrams for a composite beam-column. Curve 1 is the interaction curve based on a strain compatibility approach similar to that used for developing similar diagrams for reinforced concrete columns, without consideration of length effects. Curve 2 represents a segmented straight-line approximation based on plastic stress distributions, again without incorporating any length effects. Curve 3 is a simplification of curve 2, incorporating length effects and using only one intermediate point between pure axial strength and pure bending strength. Curve 4 results from the application of resistance or safety factors to curve 3. Curve 5 is the result of applying the interaction equations of Chapter H.

Only curves 4 and 5 in Figure 9.26 account for resistance or safety factors and the effects of length on beam-column strength. The conservatism of the Chapter H approach may not be that great compared to the curve 4 approach.

Part 6 of the *Manual* includes tables for use in determining the needed points for curve 2 of the interaction diagram in Figure 9.26. The available equations address encased W-shapes for bending about the major and minor axes, compact rectangular HSS bending about either the major or minor axis, and compact round HSS. Figure 9.27 is *Manual* Table 6-2a for an encased W-shape bending about the major axis, and Figure 9.28 is *Manual* Table 6-3 for compact rectangular HSS bending about either axis. If the curve 4 interaction of Figure 9.26 is to be constructed, only the equation identified in Figures 9.27 and 9.28 for points A, B and C need be evaluated. Then the length effects must be applied to points A and C, and the resistance or safety factors applied to all three points. This process results in the data for curve 4 which can then be used to check the strength of a composite beam-column.

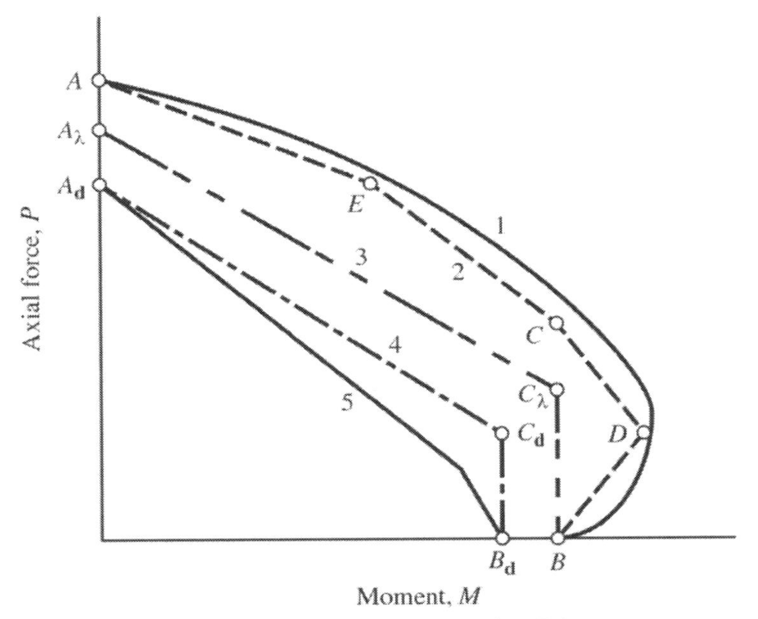

Figure 9.26 Composite Column Interaction Diagrams

EXAMPLE 9.12a
HSS Composite Column Strength under Combined Loading by LRFD

Goal: Check the interaction diagram of Chapter H for a given rectangular HSS under combined axial and flexural loading.

Given: The column is composed of an HSS8×8×1/2 filled with concrete. It must carry an axial load of P_u = 200 kips and a bending moment of M_u = 84 ft-kips. It has an effective length of 15 ft for both axes. The HSS has F_y = 50 ksi and f_c' = 5 ksi. Use the filled HSS column tables found at www.aisc.org/manualresources.

SOLUTION

Step 1: Determine the design axial strength.

From the on-line Table, with $KL_y = 15$ ft,

$$\phi P_n = 503 \text{ kips}$$

Step 2: Determine the flexural strength.

In the on-line Table, the available flexural strength can be found at the bottom of the table. Note that it is not a function of unbraced length.

$$\phi M_n = 151 \text{ ft-kips}$$

Step 3: Determine which interaction equation to use, H1-1a or H1-1b.

$$\frac{P_u}{\phi P_n} = \frac{200}{503} = 0.398 > 0.2$$

Therefore, use Equation H1-1a.

Step 4: Check the interaction equation.

$$0.398 + \frac{8}{9}\left(\frac{84.0}{151}\right) = 0.896 < 1.0$$

Since the interaction equation is less than 1.0, this column will carry the imposed load. Since the use of Equations H1-1a and H1-1b is conservative, there would be no benefit to a more complex approach.

EXAMPLE 9.12b
HSS Composite Column Strength under Combined Loading by ASD

Goal: Check the interaction diagram of Chapter H for a given rectangular HSS under combined axial and flexural loading.

Given: The column is composed of an HSS8×8×½ filled with concrete. It must carry an axial load of P_u = 133 kips and a bending moment of M_u = 56 ft-kips. It has an effective length of 15 ft for both axes. The HSS has F_y = 50 ksi and f_c' = 5 ksi. Use the filled HSS column tables found at www.aisc.org/manualresources.

SOLUTION

Step 1: Determine the design axial strength.

From the on-line Table, with $KL_y = 15$ ft,

$$P_n/\Omega = 335 \text{ kips}$$

Step 2: Determine the flexural strength.

In the on-line Table, the available flexural strength can be found at the bottom of the table. Note that it is not a function of unbraced length.

$$M_n/\Omega = 101 \text{ ft-kips}$$

Step 3: Determine which interaction equation to use, H1-1a or H1-1b.

$$P_a/(P_n/\Omega) = 133/335 = 0.397 > 0.2$$

Therefore, use Equation H1-1a.

Step 4: Check the interaction equation.

$$0.397 + \frac{8}{9}\left(\frac{56.0}{101}\right) = 0.890 < 1.0$$

Since the interaction equation is less than 1.0, this column will carry the imposed load. Since the use of Equations H1-1a and H1-1b is conservative, there would be no benefit to a more complex approach.

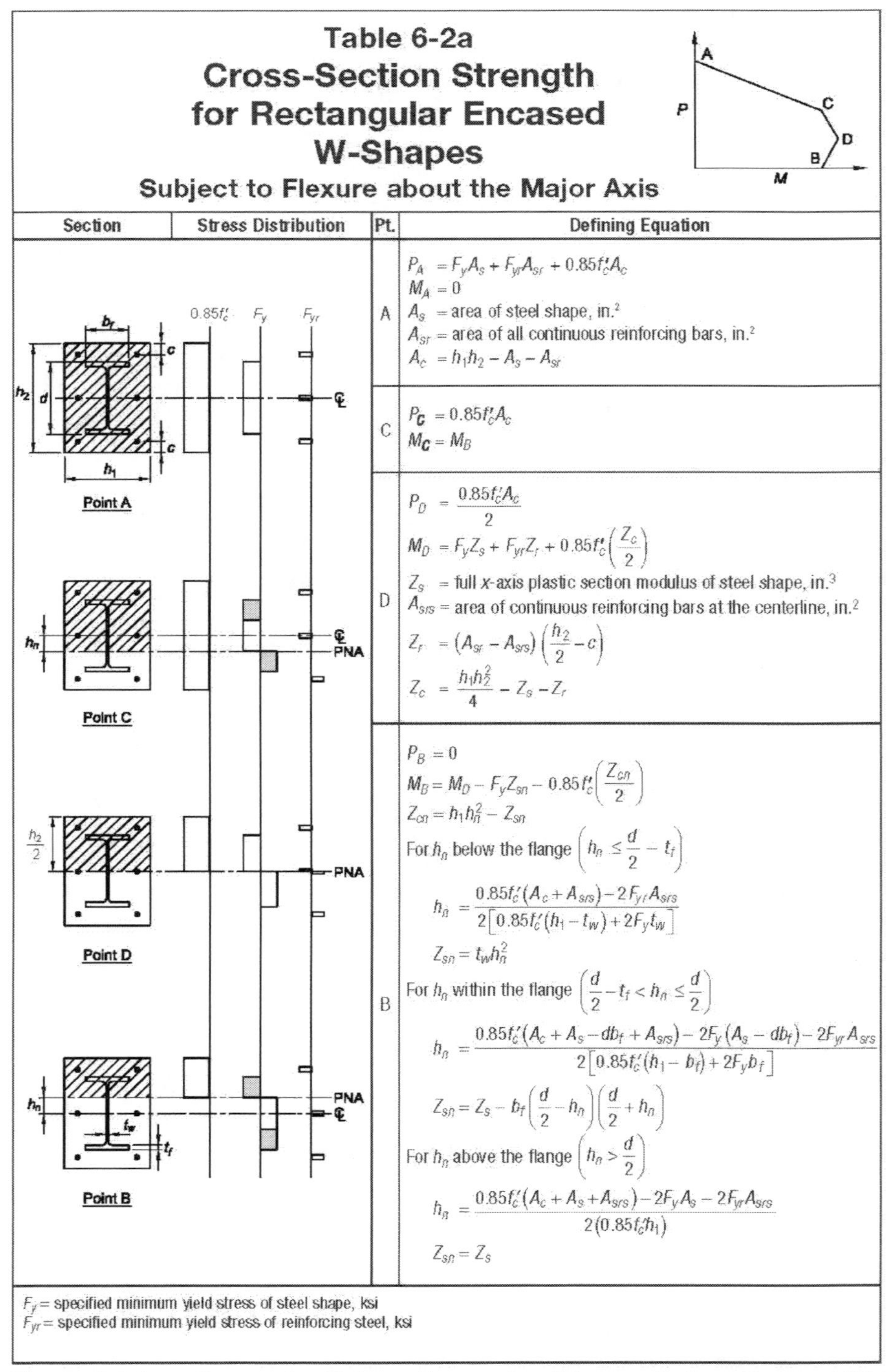

Table 6-2a
Cross-Section Strength for Rectangular Encased W-Shapes
Subject to Flexure about the Major Axis

Section	Stress Distribution	Pt.	Defining Equation
Point A	$0.85f'_c$ F_y F_{yr}	A	$P_A = F_yA_s + F_{yr}A_{sr} + 0.85f'_cA_c$ $M_A = 0$ A_s = area of steel shape, in.^{2} A_{sr} = area of all continuous reinforcing bars, in.^{2} $A_c = h_1h_2 - A_s - A_{sr}$
		C	$P_C = 0.85f'_cA_c$ $M_C = M_B$
Point C	PNA	D	$P_D = \dfrac{0.85f'_cA_c}{2}$ $M_D = F_yZ_s + F_{yr}Z_r + 0.85f'_c\left(\dfrac{Z_c}{2}\right)$ Z_s = full x-axis plastic section modulus of steel shape, in.^{3} A_{srs} = area of continuous reinforcing bars at the centerline, in.^{2} $Z_r = (A_{sr} - A_{srs})\left(\dfrac{h_2}{2} - c\right)$ $Z_c = \dfrac{h_1h_2^2}{4} - Z_s - Z_r$
Point D Point B	PNA	B	$P_B = 0$ $M_B = M_D - F_yZ_{sn} - 0.85f'_c\left(\dfrac{Z_{cn}}{2}\right)$ $Z_{cn} = h_1h_n^2 - Z_{sn}$ For h_n below the flange $\left(h_n \le \dfrac{d}{2} - t_f\right)$ $h_n = \dfrac{0.85f'_c(A_c + A_{srs}) - 2F_{yr}A_{srs}}{2\left[0.85f'_c(h_1 - t_w) + 2F_yt_w\right]}$ $Z_{sn} = t_wh_n^2$ For h_n within the flange $\left(\dfrac{d}{2} - t_f < h_n \le \dfrac{d}{2}\right)$ $h_n = \dfrac{0.85f'_c(A_c + A_s - db_f + A_{srs}) - 2F_y(A_s - db_f) - 2F_{yr}A_{srs}}{2\left[0.85f'_c(h_1 - b_f) + 2F_yb_f\right]}$ $Z_{sn} = Z_s - b_f\left(\dfrac{d}{2} - h_n\right)\left(\dfrac{d}{2} + h_n\right)$ For h_n above the flange $\left(h_n > \dfrac{d}{2}\right)$ $h_n = \dfrac{0.85f'_c(A_c + A_s + A_{srs}) - 2F_yA_s - 2F_{yr}A_{srs}}{2(0.85f'_ch_1)}$ $Z_{sn} = Z_s$

F_y = specified minimum yield stress of steel shape, ksi
F_{yr} = specified minimum yield stress of reinforcing steel, ksi

Figure 9.27 Composite Column Interaction Diagram Points: Encased W-Shape

Table 6-3
Cross-Section Strength for Composite Filled Rectangular HSS
Subject to Flexure about Either Principal Axis

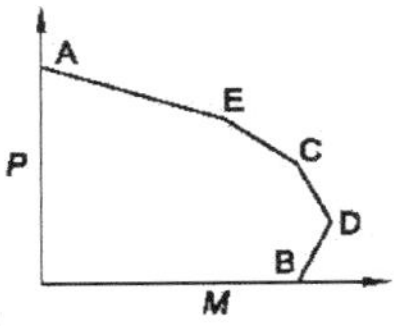

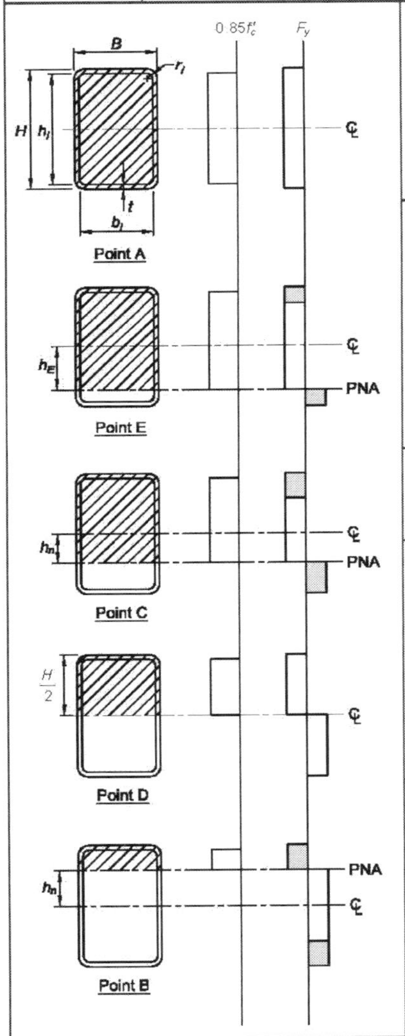

Section	Stress Distribution	Pt.	Defining Equation
Point A (B, H, h_i, b_i, t, r_i)	$0.85f'_c$, F_y, ℄	A	$P_A = F_yA_s + 0.85f'_cA_c$ $M_A = 0$ A_s = area of steel shape, in.^{2} $A_c = b_ih_i - 0.858r_i^2$ $b_i = B - 2t$ $h_i = H - 2t$ $r_i = t$
Point E (h_E)	℄, PNA	E	$P_E = \frac{0.85f'_cA_c}{2} + 0.85f'_cb_ih_E + 4F_yth_E$ $M_E = M_D - F_yZ_{sE} - 0.85f'_c\left(\frac{Z_{cE}}{2}\right)$ $Z_{cE} = b_ih_E^2$ $Z_{sE} = 2th_E^2$ $h_E = \frac{h_n}{2} + \frac{H}{4}$
Point C (h_n)	℄, PNA	C	$P_C = 0.85f'_cA_c$ $M_C = M_B$
Point D ($\frac{H}{2}$)	℄	D	$P_D = \frac{0.85f'_cA_c}{2}$ $M_D = F_yZ_s + \frac{0.85f'_cZ_c}{2}$ Z_s = full plastic section modulus of HSS, in.^{3} $Z_c = \frac{b_ih_i^2}{4} - 0.429r_i^2h_i + 0.192r_i^3$
Point B (h_n)	PNA, ℄	B	$P_B = 0$ $M_B = M_D - F_yZ_{sn} - 0.85f'_c\left(\frac{Z_{cn}}{2}\right)$ $Z_{sn} = 2th_n^2$ $Z_{cn} = b_ih_n^2$ $h_n = \frac{0.85f'_cA_c}{2(0.85f'_cb_i + 4F_yt)} \le \frac{h_i}{2}$

F_y = specified minimum yield stress of steel shape, ksi
Note: Equations in this table are applicable to single-axis bending of the shape about its x-x axis (when $H \ge B$) or about its y-y axis (when $B > H$).

Figure 9.28 Composite Column Interaction Diagram Points: Filled HSS

9.13 PROBLEMS

For all problems, assume normal weight concrete unless told otherwise.

1. Determine the location of the plastic neutral axis and the available moment strength for a flat soffit, fully composite beam composed of a W16×26 spanning 24 ft and spaced 8 ft on center, supporting a 6 in. concrete slab. Use f_c' = 4 ksi and A992 steel. Determine (a) design strength by LRFD and (b) allowable strength by ASD.

2. Determine the location of the plastic neutral axis and the available moment strength for a flat soffit, fully composite beam composed of a W16×45 spanning 20 ft and spaced 8 ft on center, supporting a 5 in. concrete slab. Use f_c' = 5 ksi and A992 steel. Determine (a) design strength by LRFD and (b) allowable strength by ASD.

3. Determine the location of the plastic neutral axis and the available moment strength for a flat soffit, fully composite beam composed of a W18×65 spanning 20 ft and spaced 6 ft on center, supporting a 5 in. concrete slab. Use f_c' = 5 ksi and A992 steel. Determine (a) design strength by LRFD and (b) allowable strength by ASD.

4. Determine the location of the plastic neutral axis and the available moment strength for a flat soffit, fully composite beam composed of a W36×330 spanning 50 ft and spaced 20 ft on center, supporting a 12 in. concrete slab. Use f_c' = 6 ksi and A992 steel. Determine (a) design strength by LRFD and (b) allowable strength by ASD.

5. Determine the location of the plastic neutral axis and the available moment strength for a flat soffit, fully composite beam composed of a W27×102 spanning 30 ft and spaced 14 ft on center, supporting a 6 in. concrete slab. Use f_c' = 6 ksi and A992 steel. Determine (a) design strength by LRFD and (b) allowable strength by ASD.

6. Determine the location of the plastic neutral axis and the available moment strength for a flat soffit, fully composite beam composed of a W10×12 spanning 12 ft and spaced 4 ft on center, supporting a 5 in. concrete slab. Use f_c' = 4 ksi and A992 steel. Determine (a) design strength by LRFD and (b) allowable strength by ASD.

7. Determine the location of the plastic neutral axis and the available moment strength for a flat soffit, fully composite beam composed of a W18×71 spanning 18 ft and spaced 5 ft on center, supporting a 4 in. concrete slab. Use f_c' = 4 ksi and A992 steel. Determine (a) design strength by LRFD and (b) allowable strength by ASD.

8. Determine the location of the plastic neutral axis and the available moment strength for a flat soffit, fully composite beam composed of a W14×48 spanning 20 ft and spaced 5 ft on center, supporting a 4 in. concrete slab. Use f_c' = 3 ksi and A992 steel. Determine (a) design strength by LRFD and (b) allowable strength by ASD.

9. Determine the location of the plastic neutral axis and the available moment strength for a flat soffit, fully composite beam composed of a W14×68 spanning 24 ft and spaced 6 ft on center, supporting a 4 in. concrete slab. Use f_c' = 3 ksi and A992 steel. Determine (a) design strength by LRFD and (b) allowable strength by ASD.

10. Determine the location of the plastic neutral axis and the available moment strength for a flat soffit, fully composite beam composed of a W18×97 spanning 20 ft and spaced 4 ft on center, supporting a 4 in. concrete slab. Use f_c' = 4 ksi and A992 steel. Determine (a) design strength by LRFD and (b) allowable strength by ASD.

11. Determine the location of the plastic neutral axis and the available moment strength for a flat soffit, fully composite beam composed of a W16×45 spanning 20 ft and spaced 5 ft on center, supporting a 4 in. concrete slab. Use f_c' = 3 ksi and A992 steel. Determine (a) design strength by LRFD and (b) allowable strength by ASD.

12. Determine the location of the plastic neutral axis and the available moment strength for a flat soffit, fully composite beam composed of a W12×30 spanning 10 ft and spaced 4 ft on center, supporting a 4 in. concrete slab. Use f_c' = 3 ksi and A992 steel. Determine (a) design strength by LRFD and (b) allowable strength by ASD.

13. Repeat Problem 1 with the shear stud capacity limited to $V_q' = 250$ kips. Determine (a) design strength by LRFD and (b) allowable strength by ASD.

14. Repeat Problem 2 with the shear stud capacity limited to $V_q' = 500$ kips. Determine (a) design strength by LRFD and (b) allowable strength by ASD.

15. Repeat Problem 3 with the shear stud capacity limited to $V_q' = 500$ kips. Determine (a) design strength by LRFD and (b) allowable strength by ASD.

16. Repeat Problem 7 with the shear stud capacity limited to $V_q' = 400$ kips. Determine (a) design strength by LRFD and (b) allowable strength by ASD.

17. Repeat Problem 8 with the shear stud capacity limited to $V_q' = 200$ kips. Determine (a) design strength by LRFD and (b) allowable strength by ASD.

18. Repeat Problem 9 with the shear stud capacity limited to $V_q' = 250$ kips. Determine (a) design strength by LRFD and (b) allowable strength by ASD.

19. Repeat Problem 4 with the shear stud capacity limited to $V_q' = 2000$ kips. Determine (a) design strength by LRFD and (b) allowable strength by ASD.

20. Repeat Problem 5 with the shear stud capacity limited to $V_q' = 900$ kips. Determine (a) design strength by LRFD and (b) allowable strength by ASD.

21. Repeat Problem 6 with the shear stud capacity limited to $V_q' = 100$ kips. Determine (a) design strength by LRFD and (b) allowable strength by ASD.

22. Repeat Problems 8 and 17 with $V_q' = 450$ kips and plot the results of all three problems as a function of V_q'. Determine (a) design strength by LRFD and (b) allowable strength by ASD.

23. Repeat Problems 9 and 18 with $V_q' = 400$ kips and plot the results of all three problems as a function of V_q'. Determine (a) design strength by LRFD and (b) allowable strength by ASD.

24. A W12 composite beam spaced every 8 ft is used to support a uniform dead load of 1.0 k/ft, including concrete and steel self-weight, and live load of 0.9 k/ft on a 20 ft span. Using a 4 in. flat soffit slab with $f_c' = 4$ ksi, 3/4 in. shear studs, and A992 steel, determine the least-weight shape and the required number of shear connectors to support the load. Design by (a) LRFD and (b) ASD.

25. A W14 composite beam is to support a uniform dead load of 1.0 k/ft, including concrete and steel self-weight, and live load of 1.4 k/ft. The beam spans 24 ft and is spaced 8 ft from adjacent beams. Using a 5 in. flat soffit slab and 3/4 in. shear studs, determine the least-weight shape to support the load if $f_c' = 4$ ksi and A992 steel is used. Design by (a) LRFD and (b) ASD.

26. A W16 composite beam spaced every 12 ft is used to support a uniform dead load of 1.2 k/ft, including concrete and steel self-weight, and live load of 1.9 k/ft on a 20 ft span. Using a 4 in. flat soffit slab with $f_c' = 5$ ksi, 3/4 in. shear studs, and A992 steel, determine the least-weight shape and the required number of shear connectors to support the load. Design by (a) LRFD and (b) ASD.

27. A W16 composite beam is to support a uniform dead load of 1.8 k/ft, including concrete and steel self-weight, and live load of 2.2 k/ft. The beam spans 26 ft and is spaced 10 ft from adjacent beams. Using a 5 in. flat soffit slab and 3/4 in. shear studs, determine the least-weight shape to support the load if $f_c' = 4$ ksi and A992 steel is used. Design by (a) LRFD and (b) ASD.

28. Compare the least-weight A992 W16 and W14 members required to support a uniform dead load of 2.4 k/ft, including concrete and steel self-weight, and live load of 3.2 k/ft. The beams span 18 ft and are spaced 12 ft on center. They support a 6 in. concrete slab with $f_c' = 4$ ksi and 3/4 in. shear studs. Design by (a) LRFD and (b) ASD.

29. Compare the least-weight A992 W18 and W16 members required to support a uniform dead load of 3.2 k/ft, including concrete and steel self-weight, and live load of 3.8 k/ft. The beams span 18 ft and are spaced 10 ft on center. They support a 5 in. concrete slab with $f_c' = 4$ ksi and 3/4 in. shear studs. Design by (a) LRFD and (b) ASD.

30. W16×31 A992 composite beams are spaced at 10 ft intervals and span 24 ft. The beams support a 2-1/2 in. metal deck perpendicular to the beam with a slab whose total thickness is 5 in. Assuming 60% composite action, determine the available moment strength and the number of 3/4 in. shear studs required. The deck has 6 in. wide ribs spaced at 12 in. Use $f_c' = 4$ ksi. Determine by (a) LRFD and (b) ASD.

31. Repeat Problem 30 using 3 in. metal deck and a 6.5 in. total slab thickness. Determine by (a) LRFD and (b) ASD.

32. Determine the available moment strength of a W18×35 A992 composite beam supporting a slab with a total thickness of 5 in. on a 3 in. metal deck perpendicular to the beam. The beam spans 28 ft and is spaced 12 ft from adjacent beams. Assume 70% composite action and determine the number of 3/4 in. shear studs required. Use $f_c' = 5$ ksi. Determine (a) design strength by LRFD and (b) allowable strength by ASD.

33. Repeat Problem 32 if the concrete used had $f_c' = 3$ ksi. Determine by (a) LRFD and (b) ASD.

34. Determine the available moment strength for a W18×46 A992 member used as a partially composite beam to support 3 in. of concrete on a 3 in. metal deck for a total slab thickness of 6 in. The metal deck is perpendicular to the beam. The beam spans 30 ft and is spaced 11 ft from adjacent beams. Shear stud strength is $V_q' = 400$ kips, $f_c' = 5$ ksi. Determine (a) design strength by LRFD and (b) allowable strength by ASD.

35. Determine the available moment strength for a W21×44 A992 member used as a partially composite beam to support 3.5 in. of concrete on a 3 in. metal deck for a total slab thickness of 6.5 in. The metal deck is perpendicular to the beam. The beam spans 30 ft and is spaced 10 ft from adjacent beams. Shear stud strength is $V_q' = 300$ kips, $f_c' = 5$ ksi. Determine (a) design strength by LRFD and (b) allowable strength by ASD.

36. A composite beam is to span 30 ft and support a 4 in. slab including a 1-1/2 in. metal deck. The deck span is 10 ft. The beam must accommodate a uniformly distributed dead load of 75 psf including the slab weight and live load of 100 psf. The deck has 2 in. ribs spaced 6 in. on center, with deck ribs perpendicular to the beam. Determine the required A992 W-shape and number of 3/4 in. shear studs. Use $f_c' = 3$ ksi. Design by (a) LRFD and (b) ASD.

37. Determine the required W-shape and number of 3/4 in. shear studs for a composite girder that spans 30 ft and supports two concentrated dead loads of 12 kips and live loads of 20 kips at the third points. The 1-1/2 in. metal deck with 2 in. ribs spaced at 6 in. on center is parallel to the girder and supports a total slab of 5 in. Use $f_c' = 4$ ksi and A992 steel. Design by (a) LRFD and (b) ASD.

38. Determine the required W-shape and number of 3/4 in. shear studs for a composite girder that spans 36 ft and supports a concentrated dead load of 22 kips and live load of 30 kips at midspan. The 3 in. metal deck with 6 in. ribs spaced at 12 in. on center is parallel to the girder and supports a total slab of 6.5 in. Use $f_c' = 5$ ksi and A992 steel. Design by (a) LRFD and (b) ASD.

39. Determine the required W-shape and number of 3/4 in. shear studs for a composite girder that spans 36 ft and supports a single concentrated dead load of 18 kips and live load of 28 kips at the left third point. The 3 in. metal deck with 6 in. ribs spaced at 12 in. on center is parallel to the girder and supports a total slab of 6.5 in. Use $f_c' = 5$ ksi and A992 steel. Design by (a) LRFD and (b) ASD.

40. Determine the live load deflection for a W24×76 A992 composite beam with a 6 in. total thickness slab on a 3 in. metal deck. The beam spans 28 ft, is spaced at 10 ft intervals, and carries a live load of 3.4 k/ft. Assume $Y2 = 5.5$ in. and $\Sigma Q_n = 394$ kips.

41. Determine the live load deflection for a W16×26 A992 composite beam supporting a 6 in. slab on a 2-1/2 in. metal deck. The beam spans 24 ft and is spaced at 8 ft on center. The live load is 2.1 k/ft. Assume $Y2 = 5.5$ in. and $\Sigma Q_n = 384$ kips.

42. Determine the live load deflection for the beam of Problem 24.

43. Determine the live load deflection for the beam of Problem 25.

44. A W12 composite beam spaced every 8 ft is used to support a uniform dead load of 1.5 k/ft, including concrete and steel self-weight, and live load of 2.5 k/ft on a 20 ft span. Using a 6 in. total thickness slab on 3 in. metal deck with 6 in. ribs spaced at 12 in., f_c' = 4 ksi, 3/4 in. shear studs, and A992 steel, determine the least-weight shape and the required number of shear connectors to support the load. Be sure to check beam shear and limit live load deflection to span/360. Design by (a) LRFD and (b) ASD.

45. A W14 composite beam is to support a uniform dead load of 1.4 k/ft, including concrete and steel self-weight, and live load of 3.2 k/ft. The beam spans 24 ft and is spaced 8 ft from adjacent beams. Using a 6.5 in. total thickness slab on 3 in. metal deck with 6 in. ribs spaced at 12 in. and 3/4 in. shear studs, determine the least-weight shape to support the load if f_c' = 4 ksi and A992 steel is used. Confirm that the beam shear strength is adequate and limit live load deflection to span/360. Design by (a) LRFD and (b) ASD.

46. A W16 composite beam spaced every 12 ft is used to support a uniform dead load of 2.0 k/ft, including concrete and steel self-weight, and live load of 3.0 k/ft on a 20 ft span. Using a 6 in. total slab thickness on 3 in. metal beck with 6 in. ribs spaced at 12 in., f_c' = 5 ksi, 3/4 in. shear studs, and A992 steel, determine the least-weight shape and the required number of shear connectors to support the load. Check that beam shear is sufficient and limit live load deflection to span/360. Design by (a) LRFD and (b) ASD.

47. Determine the available compressive strength of a 20 ft effective length 18×18 in. composite column encasing an A992 W10×68 and eight #8, Gr. 60 reinforcing bars, f_c' = 5 ksi. Each face has three bars with their centers located 2.5 in. from the face of the concrete. Determine (a) design strength by LRFD and (b) allowable strength by ASD.

48. Determine the available compressive strength of a 22×22 in. composite column with an effective length of 16 ft. The concrete encases an A992 W12×120 and eight #9, Gr. 60 bars, f_c' = 5 ksi. Each face has three bars with their centers located 2.5 in. from the face of the concrete. Determine (a) design strength by LRFD and (b) allowable strength by ASD.

49. Determine the available compressive strength of a 24×24 in. composite column with an effective length of 12 ft. The concrete encases an A992 W14×132 and eight #10, Gr. 60 bars, f_c' = 5 ksi. Each face has three bars with their centers located 2.5 in. from the face of the concrete. Determine (a) design strength by LRFD and (b) allowable strength by ASD.

50. Determine the available compressive strength of a 16×16 in. composite column with an effective length of 10 ft. The concrete encases an A992 W8×35 and 4 #10, Gr. 60 bars, f_c' = 5 ksi. Each face has two bars with their centers located 2.5 in. from the face of the concrete. Determine (a) design strength by LRFD and (b) allowable strength by ASD.

51. Determine the available compressive strength of a 22×22 in. composite column with an effective length of 14 ft. The concrete encases an A992 W14×61 and eight #10, Gr. 60 bars, f_c' = 5 ksi. Each face has three bars with their centers located 2.5 in. from the face of the concrete. Determine (a) design strength by LRFD and (b) allowable strength by ASD.

52. Determine the available compressive strength of an A500 Gr. C HSS4×4×1/2 composite column with an effective length of 8 ft. The concrete has f_c' = 4 ksi. Determine (a) design strength by LRFD and (b) allowable strength by ASD.

53. Determine the available compressive strength of an A500 Gr. C HSS8×8×3/8 composite column with an effective length of 16 ft. The concrete has f_c' = 5 ksi. Determine (a) design strength by LRFD and (b) allowable strength by ASD.

54. Determine the available compressive strength of an A500 Gr. C HSS10×5×3/8 composite column with an effective length of 12 ft. The concrete has f_c' = 5 ksi. Determine (a) design strength by LRFD and (b) allowable strength by ASD.

55. Determine the available compressive strength of an A500 Gr. C HSS12×10×1/2 composite column with an effective length of 16 ft. about the strong axis and 8 ft for the weak axis. The concrete has $f_c' = 5$ ksi. Determine (a) design strength by LRFD and (b) allowable strength by ASD.

56. Using interaction equations H1-1a and H1-1b, determine if an A500 Gr. C HSS10×6×3/16 composite column with an effective length of 20 ft will carry an axial compressive live load of 25 kips and dead load of 18 kips, and a live load moment of 10 ft-kips and a dead load moment of 8 ft-kips, both bending the member about its strong axis. Use $f'_c = 5$ ksi. Determine (a) design strength by LRFD and (b) allowable strength by ASD.

57. Using interaction equations H1-1a and H1-1b, determine if an A500 Gr. C HSS10×5×3/8 composite column with an effective length of 12 ft will carry an axial compressive live load of 80 kips and dead load of 30 kips, and a live load moment of 12 ft-kips and a dead load moment of 8 ft-kips, both bending the member about its strong axis. Use $f'_c = 5$ ksi. Determine (a) design strength by LRFD and (b) allowable strength by ASD.

58. Integrated Design Project. The framing plans shown in Figures 1.20 and 1.21 were used to design beams and girders in Chapter 6. These same gravity-only beams and girders will now be designed as composite members. Manufacturers' data on composite metal decks may be found on the Internet. For the designs carried out here, consider a 3 in. deck with a profile similar to that shown in Figure 9.16c. It will be used with 3 in. of concrete with $f_c' = 4$ ksi for a total slab thickness of 6.0 in.

Chapter 10

Connection Elements

SoFi Stadium at Hollywood Park, Inglewood, California
Photo courtesy of Walter P Moore
© SoFi Stadium at Hollywood Park

10.1 INTRODUCTION

A steel building structure is essentially a collection of individual members attached to each other to form a stable and serviceable whole, called the frame. The assumed behavior of the connection between any two members determines how the structure is analyzed to resist gravity and lateral loads. This analysis, in turn, determines the moments, shears, and axial loads for which the beams, columns, and other members are designed. It is, therefore, essential that the designer understand the basic behavior of connections.

Members are attached to each other through a variety of connecting elements, such as plates, angles, and other shapes, using mechanical fasteners or welds. The characteristics of these connecting elements and fasteners must be understood to assess the response of the complete connection. With each connection, the load transfer mechanism must be understood so that the applicable limit states of the joint can be evaluated.

Table 10.1 lists the sections of the *Specification* and parts of the *Manual* discussed in this chapter.

10.2 BASIC CONNECTIONS

The wide variety of potential member geometries and arrangements available for construction makes a listing of the corresponding connections a very demanding exercise. Every joint between members must be analyzed and designed according to the unique aspects of that joint.

Table 10.1 Sections of *Specification* and Parts of *Manual* Covered in This Chapter

	Specification
B3	Design Basis
Chapter D	Design of Members for Tension
Chapter E	Design of Members for Compression
G2	I-Shaped Members and Channels
J1	General Provisions
J2	Welds and Welded Joints
J3	Bolts, Threaded Parts, and Bolted Connections
J4	Affected Elements of Members and Connecting Elements
	Manual
Part 7	Design Considerations for Bolts
Part 8	Design Considerations for Welds
Part 9	Design of Connecting Elements

Figure 10.1 shows several examples of tension connections. The connections shown in Figure 10.1a, b, and c illustrate ways that a tension member can be spliced. In each case, the bolts are subjected to a shear force. The butt joint (Figure 10.1a) and the lap joint (Figure 10.1b) provide a connection between two members, whereas the joint shown in Figure 10.1c shows the connection of a single member to a pair of members. This type of joint can also be considered as a portion of the butt joint shown in Figure 10.1a. The joint shown in Figure 10.1d represents a hanger connected to the lower flange of a beam; in this case the connection is accomplished with a WT-shape, and the bolts are subjected to a tensile load. The connection of a tension member to a gusset plate is shown in Figure 10.1e. Here again, the bolts are subjected to a shear force. All of these examples illustrate bolted connections. Similar connections can be accomplished with welds.

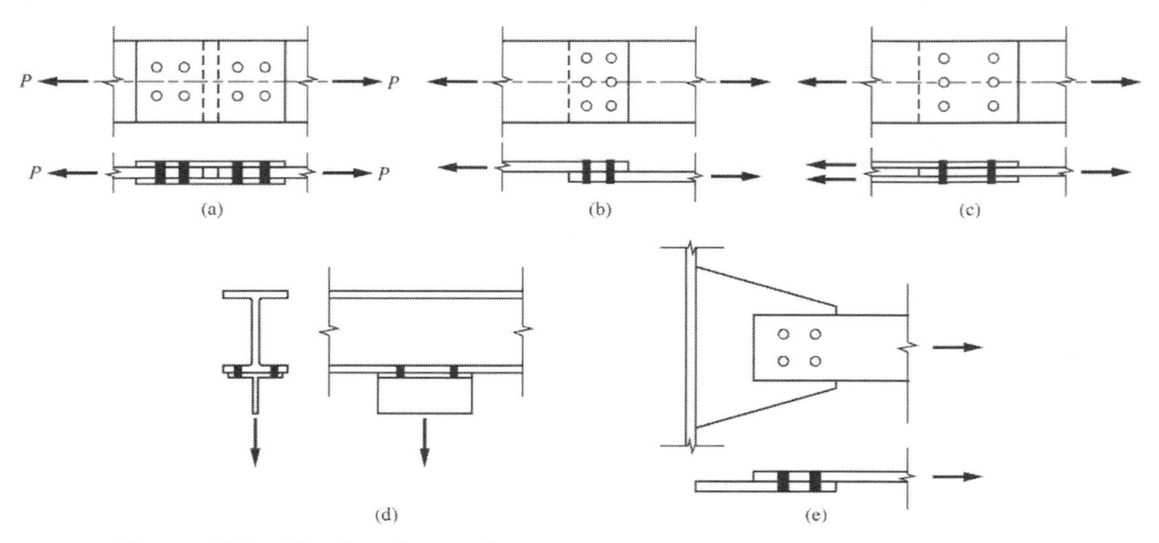

Figure 10.1 Tension Connections

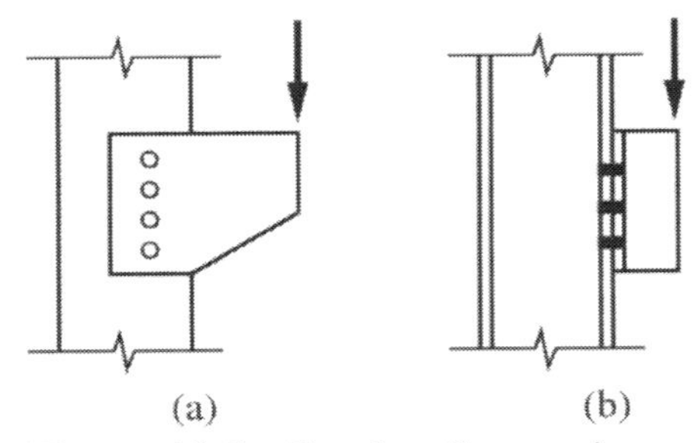

Figure 10.2 Bracket Connections

The connections illustrated in Figure 10.2 are bracket connections. The connection shown in Figure 10.2a is a bracket attached to the flange of a column. In this case, the bolts are subjected to shear due to the load and the moment in the plane of the connection when loaded as shown. The bracket shown in Figure 10.2b, when loaded as shown, subjects the bolt group to shear in the plane of the connection and a moment out of the plane that results in a tensile force in the top bolts.

10.3 BEAM-TO-COLUMN CONNECTIONS

Design of beam-to-column connections is addressed in Chapters 11 and 12. They are discussed briefly here since each connection is composed of a combination of connection elements. The connection of a beam to a column can also be accomplished in a variety of ways. Figure 10.3 illustrates several connections of W-shaped beams to W-shaped columns. The classification of these connections is a function of the forces being transferred between the members. The connections shown in Figure 10.3a through d are usually called *simple* or *shear connections* and are covered in Chapter 11, whereas those in Figure 10.3e through h are generally referred to as *fixed* or *moment connections* and covered in Chapter 12.

In normal practice beam-to-column connections are classified as simple or fixed, however; these connections actually exhibit a wide range of behaviors, as discussed in Chapter 8. This behavior can be described through a plot of the moment-rotation characteristics of a particular connection. Typical moment-rotation relationships for three beam-to-column connections are presented in Figure 10.4. When a connection is very stiff, it deforms very little, even when subjected to large moments. This type of connection is represented by curve a in Figure 10.4. At the other extreme, when a connection is quite flexible, it will rotate considerably but will not develop a significant moment, as shown by curve c in Figure 10.4. Curve b in Figure 10.4 is representative of any connection whose moment rotation behavior occurs somewhere between curves a and c. Despite some appreciable stiffness, these connections still exhibit a degree of flexibility; thus, significant rotation will occur along with significant moment resistance.

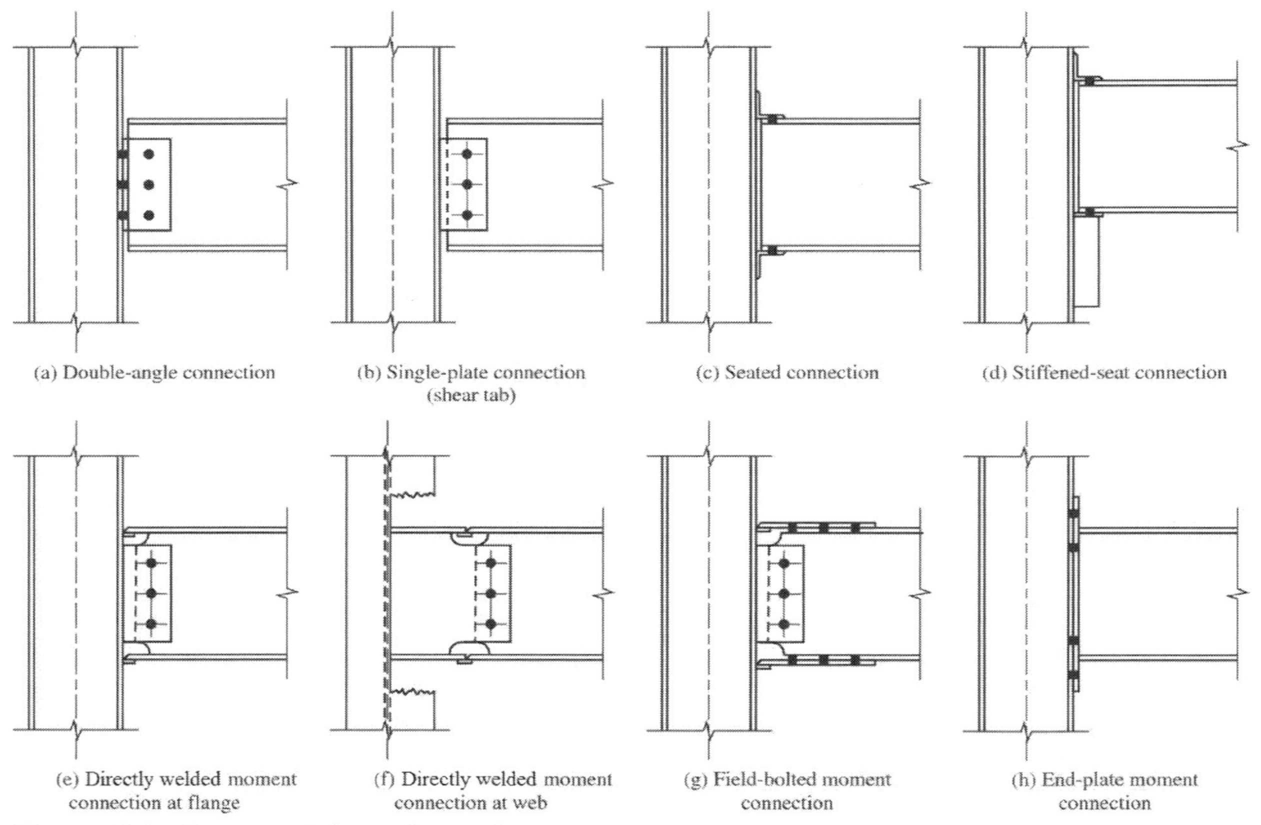

(a) Double-angle connection
(b) Single-plate connection (shear tab)
(c) Seated connection
(d) Stiffened-seat connection
(e) Directly welded moment connection at flange
(f) Directly welded moment connection at web
(g) Field-bolted moment connection
(h) End-plate moment connection

Figure 10.3 Beam-to-Column Connections

For the purposes of design, connections have usually been assumed to behave according to the simplified behaviors represented by the vertical axis of Figure 10.4 as a fixed connection and the horizontal axis of Figure 10.4 as a simple connection. Because connections do not actually behave in this way, those that follow curves a and c and exhibit a behavior close to the idealized connection are called fixed connections and simple connections, respectively. *Specification* Section B3.4 divides connections into two categories: simple connections and moment connections. The moment connection category includes fully restrained (FR) moment connections and partially restrained (PR) moment connections. FR connections transfer moment with a negligible rotation between the connected members, as shown in curve a. PR connections transfer moment between the members but the rotation is not negligible, as demonstrated by curve b.

It is the designer's responsibility to match connection behavior with the appropriate analysis model and to complete the connection design so that the actual connection behavior matches that used in the analysis. Often, this requires experience and judgment; the state of the art is such that it is usually not possible to accurately predict the M-θ for anything but the most basic of connections.

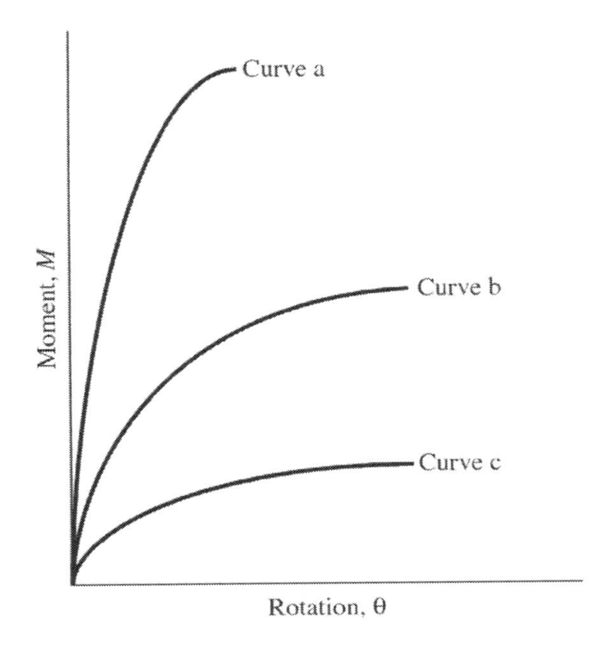

Figure 10.4 Beam-to-Column Moment-Rotation Curves

10.4 FULLY RESTRAINED CONNECTIONS

The basic assumption for frames with FR connections is that the beams and columns maintain their original geometric relationship over the entire loading history. This is normally called a *rigid* or *fixed connection*. The diagrams in Figure 10.3e through h show examples of beam-to-column connections that are usually treated as FR connections. Although they may show some small relative rotation between members (curve A in Figure 10.4), they have sufficient stiffness to justify ignoring this rotation.

Figure 10.3e shows a connection with a web plate shop-welded to the column flange and field-bolted to the beam web. The beam flanges have been beveled in the shop and are field-welded to the column. Although the beam web is not continuously connected to the column, it has been repeatedly demonstrated that this connection can adequately transfer the full plastic moment of the beam to the column. The moment strength is derived mostly from the flange connections and is equal to the flange force times beam depth. The small amount of moment in the web connection and local strain hardening in the flanges add to the connection's ability to reach the full plastic moment of the beam.

Figure 10.3f is similar to Figure 10.3e except that the beam frames into the web of the column. To ensure that this connection has adequate ductility, it is important to extend the flange connecting plates beyond the column flange and to design these plates to be a little thicker than the beam flange. Extending the connecting plate reduces the possibility of a tri-axial stress condition near the column flange tips. Thickening the plate reduces the average tension stress in the plate. It also facilitates welding to the beam flange, which may not be perfectly aligned with the plate due to mill and fabrication tolerances.

The connection illustrated in Figure 10.3g is a flange-plate connection. As with the connection shown in Figure 10.3e, the web is connected to transfer the beam shear force only. The flange forces are first transferred to the top and bottom plates and then to the column flange. This is shown as a bolted connection, but it is also possible to fabricate this as a welded connection. For fully welded connections, special care should be taken to address support during erection prior to field welding; this usually means there will be some bolts used even in a fully welded connection.

Figure 10.3h is an extended end-plate connection. For this connection, a plate is shop-welded to the end of the beam and then bolted to the column flange. Although this connection is very popular with some fabricators, others tend to avoid it. It must be fabricated with special care so that the end plates are parallel to each other. Also, it is not a very forgiving connection, and it can make erection difficult and expensive.

Fully restrained connections are covered in Chapter 12.

10.5 SIMPLE AND PARTIALLY RESTRAINED CONNECTIONS

Analysis of a frame with PR connections must account for the actual moment-rotation characteristics of the connection. These connections are now referred to as *partially restrained connections* but have historically been called semi-rigid connections. It is typically not possible to determine whether a connection should be classified as PR just by looking at it. Several connections that appear to be simple actually have the potential to resist significant moment. In the simple connection case, the analysis assumes that the connections are pinned and free to rotate. The rotation capacity of the connection must be sufficient to accommodate the simple beam rotation of the beam to which it is connected.

There are two ways in which this type of frame can be designed to resist lateral loads and to provide stability for gravity loads without a formal PR design. In one case, a positive vertical bracing system is provided, such as diagonal steel bracing or a shear wall. In the second case, lateral stability is provided by the limited restraint offered by the connections and members themselves. This type of connection is called a flexible moment connection. Flexible moment connections are designed for a limited amount of moment resistance accompanied by a significant amount of rotation. The connections are flexible enough to rotate under gravity loads so that no gravity moments are transferred to the columns. At the same time they are assumed to have sufficient strength and stiffness to resist the lateral loads and to provide frame stability. This approach to frame design is a simplification of convenience based upon experience, judgement and past usage. It was addressed in Section 8.11. Design of these flexible moment connections will follow the same approach as that for other connections to be discussed later.

The design of PR connections is more involved than the flexible moment connection approach because it requires that the frame be analyzed considering the true semi-rigid behavior of the connections. In this case, the actual M-θ curve of the connection must be known. The resulting analysis tends to be rather complex because of the nonlinear behavior of the connection. Although there are currently no commercially available computer programs for analysis of frames with PR connections, there are simplified approaches that will aid in the use of these connections.

Figure 10.3 shows examples of simple and PR connections. As mentioned earlier, it is not normally possible to tell by visual inspection whether a connection should be treated as a PR connection. Figure 10.3a shows a double-angle connection, also referred to as a *clip angle connection*. This connection has been used extensively over the years. In fact, it is usually the standard to which other simple connections are compared. Even though it is readily accepted as a simple connection, it has been shown that under certain circumstances it can be relied upon to resist some moment from lateral load.

Figure 10.3b shows a single-plate framing connection that is often referred to as a *shear tab*. Care must be taken when these connections are designed as simple connections to ensure that the elements have sufficient flexibility to accommodate the simple beam rotation. Procedures that can be followed to do this are provided in the *Manual*.

Figure 10.3c shows a seated connection and Figure 10.3d a stiffened seated connection. Either can be bolted or welded, and they are usually used to frame a beam into the web of a W-shaped column section. Although they may appear to be stiffer than the standard double-angle connection, they are designed to rotate sufficiently without transferring a moment to the column, so they can be treated as simple connections.

Simple shear connections are treated in Chapter 11.

10.6 MECHANICAL FASTENERS

The mechanical fasteners most commonly used today are bolts. The *Specification* provides for the use of common bolts and high-strength bolts. It also provides some direction for cases where bolts are to be used in conjunction with rivets in new work on historic structures. There are no provisions for rivets in new construction, however, because these connectors are no longer used in new construction of buildings.

10.6.1 Common Bolts

Common bolts are manufactured according to the ASTM A307 specification, as discussed in Section 3.6.3. When used, they are usually found in simple connections for such elements as girts, purlins, light floor beams, bracing, and other applications where the loads are relatively small. Although permitted by the *Specification*, they are not recommended for normal steel-to-steel connections and should not be used where the loads are cyclic or vibratory, or where fatigue may be a factor.

Common bolts are also called *machine*, *unfinished*, or *rough bolts*. They have square or hexagonal heads and nuts and are identified by the grade designation 307A or 307B on the heads. They are available in diameters from 1/4 in. to 4 in.

These bolts are usually installed using a spud wrench. No specified pre-tension is required. Because no clamping force is assumed, it is necessary only to tighten the nut sufficiently to prevent it from backing off of the bolt. The design shear and tensile strength are given in *Specification* Section J3, Table J3.2.

10.6.2 High-Strength Bolts

High-strength bolts are presented in *Specification* Section J3 in four groups according to their material strength. Four Group 120 high-strength bolts are currently permitted in steel structures as specified by ASTM F3125 Grades A325, A325M and F1852 and ASTM A354, Grade BC. Grade A325M is the metric equivalent of A325. These bolts all have the same minimum tensile strength, F_u = 120 ksi. One Group 144 bolt is permitted as specified by ASTM F3148 Grade 144. As the name suggests, these bolts have F_u = 144 ksi. Four Group 150 bolts are permitted as specified by ASTM F3125 Grades A490, A490M and F2280 and ASTM A354, Grade BD. Grade A490M is the metric equivalent of A490. These bolts also all have the same minimum tensile strength, F_u = 150 ksi. Two Group 200 bolts, ASTM F3043 and F3111 with minimum tensile strength of 200 ksi, are permitted. Use of Group 144 and 200 bolts is restricted by Section J3.2.

Details of the material and other properties of Groups 120 and 150 bolts are described in Section 3.6.3. Bolt strength is based on the tensile strength of the bolt material. The nominal strength of Group 150 bolts is 25 percent greater than that of Group 120 bolts. Bolts are generally referred to by their grade designation without repeating the ASTM F3125 designation. Thus, in this book that model will be followed.

All Group 120 and 150 bolts can be used for simple, FR, or PR connections and for both static and dynamic loading. Bolts have always been very popular for field installation. Their use in the shop has increased considerably with the introduction of automated equipment and tension control bolts (F1852 and F2280).

A325, A325M and F1852 bolts are available in two types. Type 1, manufactured from a medium-carbon steel, is the most commonly used. It is available in sizes ranging from 1/2 in. through 1-1/2 in. in diameter. Type 3 is a weathering steel bolt with corrosion characteristics similar to those of ASTM A242, A588, and A847 steels. Type 3 bolts are also available from 1/2 through 1-1/2 in. in diameter. They all have a minimum tensile strength F_u = 120 ksi.

A490, A490M and F2280 bolts are also available as Type 1 and Type 3 and in sizes ranging from 1/2 in. to 1-1/2 in. in diameter. All have a minimum tensile strength F_u = 150 ksi.

A325 Type 1 bolts are identified by the mark "A325" on the bolt head. Type 3 bolts have the designation "A325" underlined. A490 bolts are similar, carry the marking "A490" with the identification underlined for Type 3. Example bolt markings are shown in Figure 10.5. All bolts should also be marked with a symbol designating the manufacturer as shown in this case by SL in the figure.

A354 bolts are manufactured as quenched and tempered alloy steel in two grades, BC and BD. Any steel that meets the mechanical and chemical properties in the standard may be used. Grade BC bolts have a tensile strength of 125 or 115 ksi, based on diameter, but are included with Group 120 bolts where a tensile strength of 120 ksi is used. Grade BD bolts have a tensile strength of 150 to 173 ksi but are included with Group 150 bolts where a tensile strength of 150 ksi is used. These bolts are available in sizes ranging from 1/4 in. to 4.0 in. diameter with diameters over 2-1/2 in. having the reduced strength for Grade BC. A354 bolts are marked with the grade and may be marked with 6 radial lines, 60° apart for grade BD.

Groups 144 and 200 bolts are currently only available as proprietary products and will not be used in this book although the basic concepts of design discussed will equally apply to them.

Figure 10.6a shows the principal parts and dimensions of a high-strength bolt: head, shank, bolt length, and thread length. Figure 10.6b shows the principal parts of a tension control bolt.

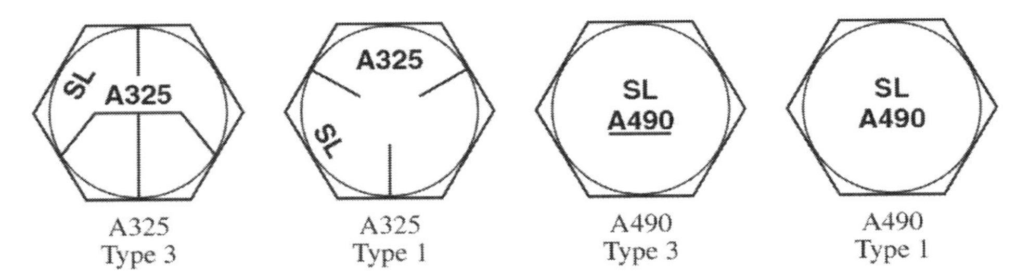

Figure 10.5 Example Bolt Identification Markings from St. Louis Screw and Bolt Company

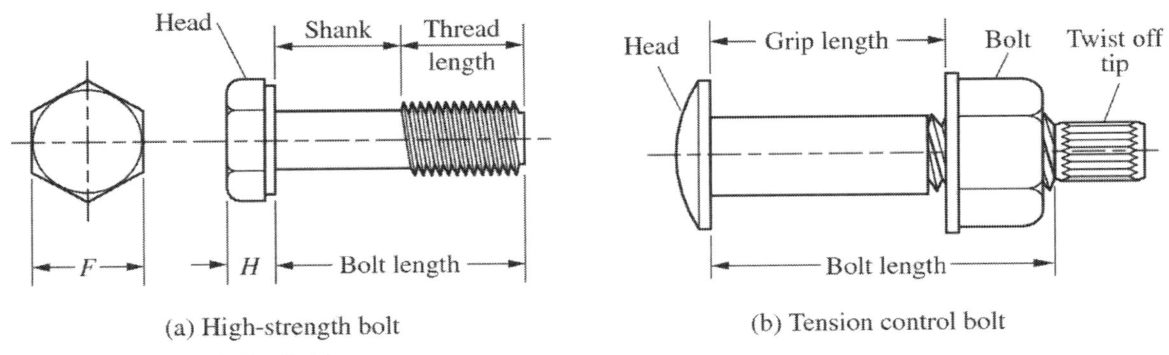

Figure 10.6 Bolt Definitions

Both A325 and A490 bolts can be installed with a spud wrench, shown in Figure 10.7a, or in cases where a clamping force is necessary, an impact wrench shown in Figure 10.7b may be used. F1852 and F2280 bolts are installed with a mechanical device that simultaneously holds the bolt shank and nut and rotates them relative to each other. The end of the bolt twists off when the prescribed tensile force is reached, ensuring the required pre-tension. Figure 10.7c shows a wrench for installation of these tension control bolts. It should be noted that placement of bolts in a connection must provide sufficient clearance for wrench access.

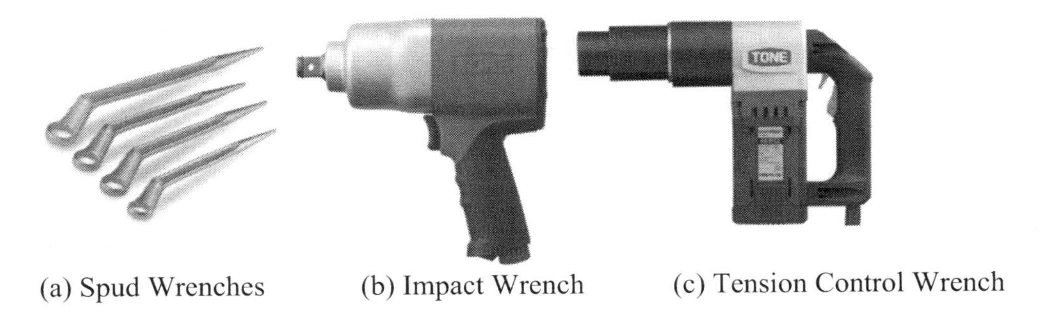

(a) Spud Wrenches (b) Impact Wrench (c) Tension Control Wrench

Figure 10.7 Wrenches Used to Install Bolts, Photos courtesy Tone Co., Ltd.

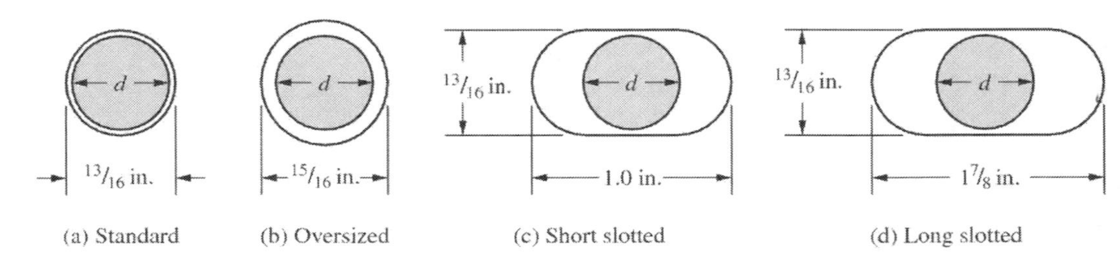

Figure 10.8 Hole Sizes for a 3/4 in. Diameter Bolt

10.6.3 Bolt Holes

Because the holes into which bolts are inserted will impact the strength of the bolts in place, it is important to address the hole requirements at this point. The *Specification* defines four types of bolt holes that are permitted in steel construction: standard, oversized, short slot, and long slot. Table 10.2 shows the nominal hole dimensions for each of these types and for bolts from 1/2 in. diameter up. Figure 10.8 shows the four hole sizes for a 3/4 in. bolt.

Standard holes or short-slotted holes transverse to the direction of load are the standard to be used unless one of the other types is permitted by the designer. This is because the other arrangements can reduce the final bolt strength and allow larger deformations from movement of bolts in holes under loading. A standard hole has a maximum diameter that is 1/16 in. greater than the bolt diameter up to 7/8 in. bolts and 1/8 in. for 1.0 in. bolts and larger to accommodate placement of the bolt. Short-slotted holes have this same dimension in one direction but are elongated in the other direction to assist in fit-up of the connection parts. Any slot longer than a short slot should be classified as a long slot, even if it is not the full length of a long slot as given in Table 10.2.

Oversized holes and long-slotted holes are specified when increased tolerance is needed to accomplish the actual connection. If a design includes other than standard holes, the requirements of *Specification* Section J3.3 for washers come into play. For simplicity in the examples in this book, only standard holes will be used.

Table 10.2 Nominal Hole Dimensions, in.

	Hole dimensions			
Bolt diameter	Standard (dia.)	Oversize (dia.)	Short slot (width × length)	Long slot (width × length)
1/2	9/16	5/8	9/16 × 11/16	9/16 × 1-1/4
5/8	11/16	13/16	11/16 × 7/8	11/16 × 1-9/16
3/4	13/16	15/16	13/16 × 1	13/16 × 1-7/8
7/8	15/16	1-1/16	15/16 × 1-1/8	15/16 × 2-3/16
1	1-1/8	1-1/4	1-1/8 × 1-5/16	1-1/8 × 2-1/2
≥1-1/8	$d + 1/8$	$d + 5/16$	$(d + 1/8) \times (d + 3/8)$	$(d + 1/8) \times (2.5 \times d)$

In addition to prescribing the sizes of bolt holes, the *Specification* gives minimum and maximum hole spacing and edge distances. Figure 10.9 shows a plate with hole spacing dimensioned with the standard variable names used in the *Specification*. The minimum hole spacing, s, for standard, oversized, or slotted holes must not be less than 2-2/3 times the bolt diameter. A spacing of 3 diameters, $3d$, is preferred. It will be shown later that even at a minimum spacing of $3d$, bolt strength may be less than what it could be if the spacing were just a little bit greater. The maximum spacing of bolts in a connection is 24 times the thickness of the connected part or 12 in. This maximum is not a strength requirement but rather one that is intended to keep the connection plies in contact and prevent any potential moisture build-up between the elements.

The minimum edge distances, l_e, specified are intended to facilitate construction and are not strength related. Table 10.3 shows the minimums from *Specification* Table J3.4. Because these dimensions will be shown to directly impact bolt strength, it is critical to provide edge distances that are compatible with the required strength of the connection. The maximum edge distance is 12 times the thickness of the connected part or 6 in. for elements not subject to corrosion, for the same reasons as for the limits on bolt spacing.

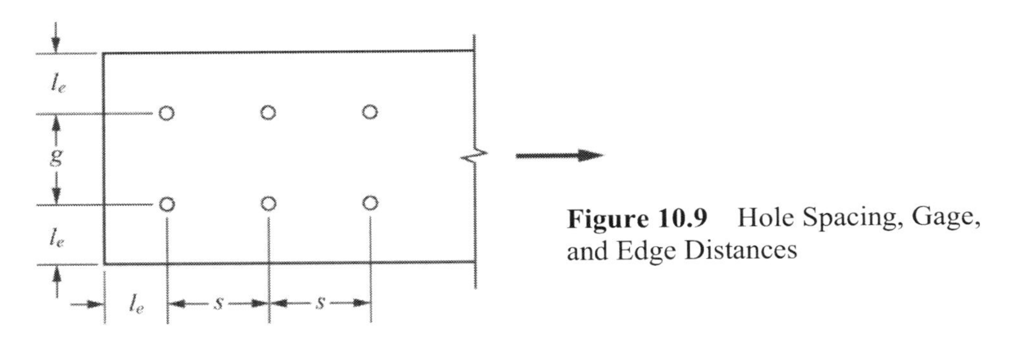

Figure 10.9 Hole Spacing, Gage, and Edge Distances

Table 10.3 Minimum Edge Distance[a] from Center of Standard Hole[b] to Edge of Connected Part

Bolt diameter (in.)	Minimum edge distance (in.)
1/2	3/4
5/8	7/8
3/4	1
7/8	1-1/8
1	1-1/4
1-1/8	1-1/2
1-1/4	1-5/8
>1-1/4	1-1/4 × d

[a] If necessary, lesser edge distances are permitted provided the appropriate provisions from Sections J3.11 and J4 are satisfied, but edge distances less than one bolt diameter are not permitted without approval from the engineer of record.
[b]For oversized or slotted holes, see *Specification* Table J3.5.

10.7 BOLT LIMIT STATES

Three basic limit states govern the response of bolts in bolted connections: shear through the shank or threads of the bolt, bearing or tearout on the elements being connected, and tension in the bolt. Cases where load reversals are expected, or fatigue is a factor, have an additional limit state to prevent slip in the connection. This limit state applies only to connections that are classified as slip-critical connections. For bolts in shear connections, the nominal connection strength is the sum of the lowest strength for the limit states of bolt shear, bearing, and tearout for each bolt in the connection.

Bolts may be installed to a snug-tight condition or pretensioned. Snug-tight installation has no required specific level of pretension. This is commonly attained after a few impacts of an impact wrench or the full effort of an ironworker with an ordinary spud wrench. The plies should be in firm contact, a condition that means the plies are solidly seated against each other, but not necessarily in continuous contact. Snug-tight bolts are permitted in connections that are classified as bearing-type connections.

Pretensioned bolts are installed to a pretension specified in Table J3.1. They must be used in connections subjected to vibratory loads where bolt loosening is a concern and in end connections of built-up members. Pretensioned bolts are required for connections designed as slip-critical connections.

10.7.1 Bolt Shear

The most common application of bolts in connections is to resist shear. Shear through the shank of the bolt is the means whereby the load, P, in Figure 10.10a is transferred from one plate to the other. In this case, the bolt is sheared along one plane. Thus, it is said to be a *bolt in single shear*. The arrangement in Figure 10.10b shows two side plates connected to a central plate. In this case, the load, P, is transferred from the center plate to the side plates and the bolt is therefore loaded in double shear. A bolt in double shear has twice the shear strength as a bolt in single shear.

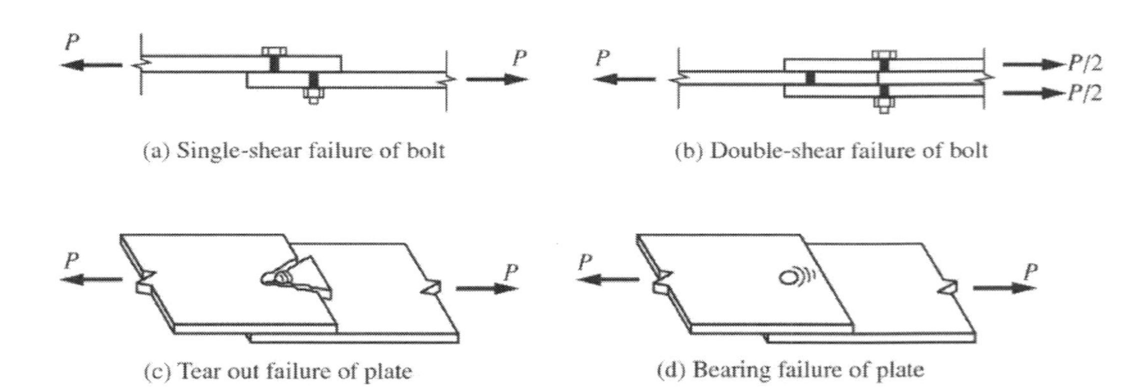

Figure 10.10 Bolt Failure Modes

For the limit state of bolt shear, the nominal strength is based on the tensile strength of the bolt and the location of the shear plane with respect to the bolt threads. Section J3.7 provides that

$$R_n = F_n A_b \qquad \text{(AISC J3-1)}$$

and

$$\phi = 0.75 \text{ (LRFD)} \qquad \Omega = 2.00 \text{ (ASD)}$$

where

F_n = shear stress, F_{nv}, from *Specification* Table J3.2
A_b = area of the bolt shank

The information in Table 10.4 is taken from *Specification* Table J3.2. Each high-strength fastener has two descriptions. The first is for cases where the threads are not excluded from the shear plane, and the second is for when the threads are excluded from the shear plane. Because, in every case, the area of the bolt shank is used to determine the nominal strength, the reduction in area when the shear plane passes through the threads is accounted for by reducing the nominal shear stress. This is done as a convenience in design so the area can be calculated using the known basic bolt diameter without having to calculate the reduction for threading. When threads are excluded from the shear plane, the bolts are called either A325-X or A490-X bolts. In these cases, $F_{nv} = 0.563F_u$. When threads are not excluded from the shear plane, the bolts are referred to as either A325-N or A490-N bolts. In these cases, $F_{nv} = 0.450F_u$. Only one value is provided for A307 bolts, and that value is based on the assumption that the threads are included in the shear plane.

Table 10.4 Nominal Stress of Fasteners and Threaded Parts, ksi

Description of fasteners	Nominal tensile stress, F_{nt}, ksi	Nominal shear stress in bearing-type connections, F_{nv}, ksi
A307 bolts	45	27
Group 120 (e.g., A325) bolts, when threads are not excluded from shear planes (N)	90	54
Group 120 (e.g., A325) bolts, when threads are excluded from shear planes (X)	90	68
Group 150 (e.g., A490) bolts, when threads are not excluded from shear planes (N)	113	68
Group 150 (e.g., A490) bolts, when threads are excluded from shear planes (X)	113	84
Threaded parts meeting the requirements of Section A3.4, when threads are not excluded from shear planes (N)	$0.75F_u$	$0.450F_u$
Threaded parts meeting the requirements of Section A3.4, when threads are excluded from shear planes (X)	$0.75F_u$	$0.563F_u$

When the designer purposefully configures the plies of the connection to exclude the bolt threads from the shear plane, the threads-excluded design values can be used. Otherwise, it is usually best to design the connection for the worst case of threads included in the shear plane.

10.7.2 Bolt Bearing and Tearout

The available strength for the limit state of bearing and tearout at bolt holes is specified in Section J3.11. Because the material strength of a bolt is greater than that of the material it is bearing on, the only bearing check is for bearing on the material of the connected parts. The *Specification* considers two limit states for bearing strength at bolt holes: the limit state based on shear in the material being connected, as shown in Figure 10.10c, and the limit state of material crushing, as shown in Figure 10.10d.

When the clear distance from the edge of the hole to the edge of the part or next hole is less than twice the bolt diameter, the limit state of shear in the plate material, also referred to as *tearout*, will control. In this case, failure occurs by a piece of material tearing out of the end of the connection as shown in Figure 10.10c or by tearing between holes in the direction of force. The nominal strength for this failure mode, R_n, is provided by shear along the two planes. From statics,

R_n = (shear strength)(2 planes)(clear distance)(material thickness)

$$R_n = 0.6F_u(2l_c)t = 1.2l_ctF_u \qquad \text{(AISC J3-6c)}$$

where

$0.6F_u$ = ultimate shear strength of the connected material, ksi
t = thickness of the material, in.
l_c = clear edge distance, measured from the edge of the hole to the edge of the material or the next hole; unlike net shear and net tension area calculations, no deduction is made at the hole edge for damage in bearing calculations

If the clear distance exceeds $2d$, crushing of the connected material, referred to as *bearing*, will be the controlling limit state, as shown in Figure 10.10d. In this case, the limit state is that of hole distortion and the calculated bolt strength will be

$$R_n = 2.4dtF_u \qquad \text{(AISC J3-6a)}$$

where

d = bolt diameter
t = connected part thickness
F_u = tensile strength of the connected part

If deformation at the bolt hole is not a design consideration at service loads, both of these limit states may be increased by one-quarter, so that for tearout

$$R_n = 1.5l_ctF_u \qquad \text{(AISC J3-6d)}$$

and for bearing

$$R_n = 3.0dtF_u \qquad \text{(AISC J3-6b)}$$

When bolts are used in a connection with long slots and the force is perpendicular to the slot, bolt strength is reduced such that for tearout

$$R_n = 1.0 l_c t F_u \quad \text{(AISC J3-6f)}$$

and for bearing

$$R_n = 2.0 d t F_u \quad \text{(AISC J3-6e)}$$

As was the case for bolt shear, the resistance and safety factors for the limit state of bolt bearing are

$$\phi = 0.75 \text{ (LRFD)} \qquad \Omega = 2.00 \text{ (ASD)}$$

10.7.3 Strength at Bolt Holes

Normal use of bolts is to hold two or more pieces of steel together. This of course requires holes in each piece to be connected. When bolts are placed through holes and a shear force is applied, their available strength will be the lowest value obtained according to the limit states of bolt shear, bearing on each connected element and tearout at the bolt hole on each connected element. The nominal strength of a connection in shear is the sum of the nominal strength of the individual bolts.

For a bolt in standard, oversized, or short slotted holes perpendicular to the direction of loading, nominal strength will be the lowest value according to Equations J3-1, J3-6a, and J3-6c. It has been shown that for bolts installed in holes with a clear distance of $2d$ or more, bearing will control over tearout. It is also possible to determine a minimum plate thickness to ensure that bolt shear will control over both bearing and tearout.

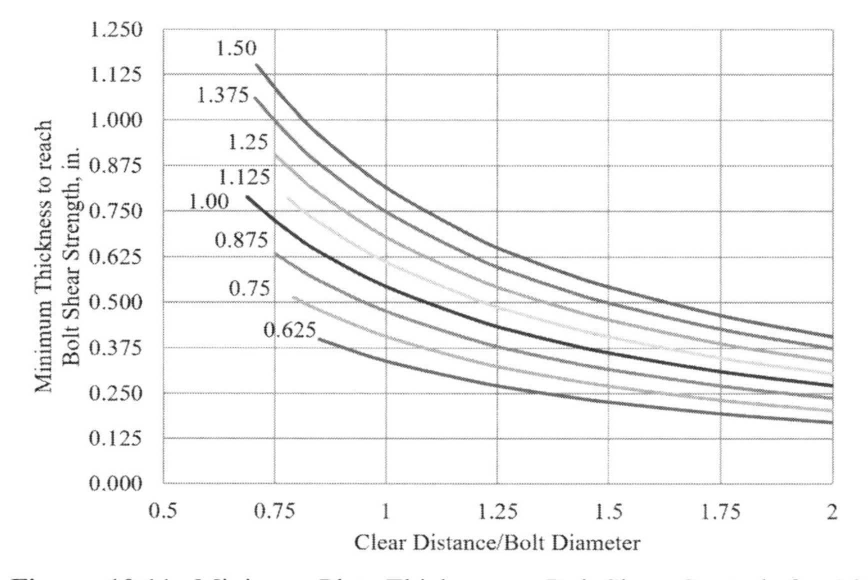

Figure 10.11 Minimum Plate Thickness so Bolt Shear Controls for A325-N bolts and $F_y = 50$ ksi

Figure 10.11 shows the minimum plate thickness for A325-N bolts in plates with $F_y = 50$ ksi as a function of clear end distance divided by bolt diameter, l_c/d for bolts with diameters from 5/8 in. to 1.5 in. diameter. The curves start at a the minimum clear distance given in Table J3.4 divided by bolt diameter and end at a clear distance of 2 bolt diameters, above which no increase in strength can be achieved.

10.7.4 Bolt Tension

For the limit state of bolt tension, strength is directly based on the tensile strength of the bolt material. Section J3.7 provides that

$$R_n = F_n A_b \qquad \text{(AISC J3-1)}$$

with

$$\phi = 0.75 \text{ (LRFD)} \qquad \Omega = 2.00 \text{ (ASD)}$$

and

F_n = tensile stress, F_{nt}, from *Specification* Table J3.2
A_b = area of the bolt shank

Table 10.4 shows the nominal tensile stress, F_{nt}, for bolts taken from *Specification* Table J3.2. Note that there is no distinction for the location of the shear plane, because the bolt is loaded axially, and the limiting stress occurs over the net tensile area. To ease calculations, the area of the bolt shank is again used, and the nominal tensile stress appropriately reduced and given as $0.75F_u$.

EXAMPLE 10.1 ***Bolt Shear Strength***

Goal: Determine the available bolt shear strength.

Given: (a) A single 3/4 in. A325-N bolt.
(b) A single 7/8 in. A490-X bolt.

SOLUTION

Part (a)

Step 1: For a single 3/4 in. A325-N bolt, determine the bolt shank area.

$$A_b = \frac{\pi d^2}{4} = \frac{\pi(0.75)^2}{4} = 0.442 \text{ in.}^2$$

Step 2: Determine the nominal shear stress.
For an A325 bolt

$$F_u = 120 \text{ ksi}$$

and for the threads included (N)

$$F_{nv} = 0.450F_u = 0.450(120) = 54 \text{ ksi}$$

For LRFD For shear, $\phi = 0.75$ and the design strength is

Step 3:

$$\phi r_n = 0.75(54)(0.442) = 17.9 \text{ kips}$$

For ASD Step 3: For shear, $\Omega = 2.00$ and the allowable strength is

$$r_n/\Omega = 54(0.442)/2.00 = 11.9 \text{ kips}$$

Part (b)

Step 4: For a single 7/8 in. A490-X bolt, determine the bolt shank area.

$$A_b = \frac{\pi d^2}{4} = \frac{\pi(0.875)^2}{4} = 0.601 \text{ in.}^2$$

Step 5: Determine the nominal shear stress.
For an A490 bolt

$$F_u = 150 \text{ ksi}$$

and for the threads excluded (X)

$$F_{nv} = 0.563F_u = 0.563(150) = 84 \text{ ksi}$$

For LRFD Step 6: For shear, $\phi = 0.75$ and the design strength is

$$\phi r_n = 0.75(84)(0.601) = 37.9 \text{ kips}$$

For ASD Step 6: For shear, $\Omega = 2.00$ and the allowable strength is

$$\frac{r_n}{\Omega} = \frac{(84)(0.601)}{2.00} = 25.2 \text{ kips}$$

Manual Table 7-1 provides single-bolt available shear strength values for a wide range of bolt sizes and strengths.

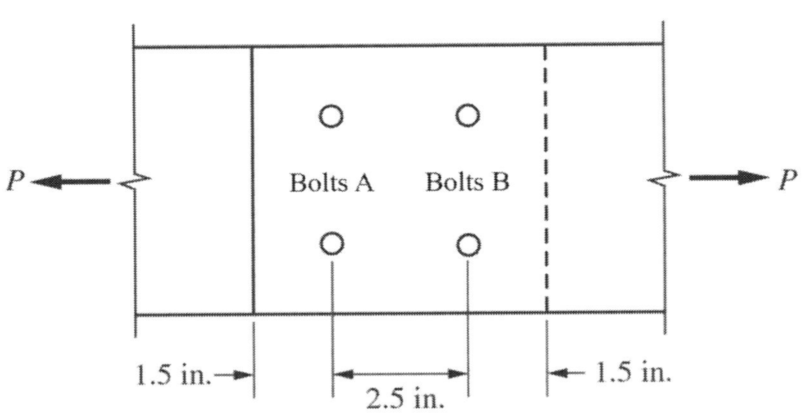

Figure 10.12 Lap Joint for Example 10.2

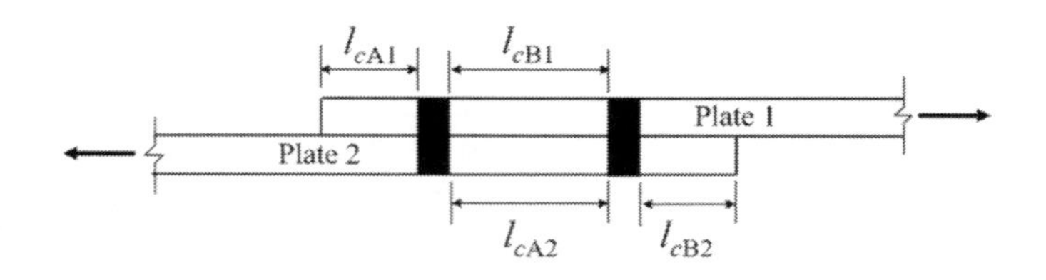

Figure 10.12 (*continued*) Lap Joint for Example 10.2

EXAMPLE 10.2 ***Lap Splice Connection Strength***

Goal: Determine the available strength for a four-bolt connection.

Given: A lap joint using 1/2 in. A36 plates is given in Figure 10.12. Use (a) 7/8 in. A325-X bolts and (b) 7/8 in. A325-N bolts. The plates are labeled plate 1 and plate 2 and the bolt pairs are labeled bolts A and bolts B. The layout of the connection is such that each bolt will have the same limiting strength. Thus, the strength of the joint is 4 times the strength of a single bolt.

SOLUTION

Part a

Step 1: Determine the nominal shear strength for 7/8 in. A325-X bolts.

$$F_{nv} = 0.563F_u = 0.563(120) = 68 \text{ ksi}$$

$$A_b = 0.601 \text{ in.}^2$$

$$r_n = F_{nv}A_b = 68(0.601) = 40.9 \text{ kips}$$

Step 2: Determine the nominal bearing strength.

Unless noted otherwise assume deformation at service load is a design consideration, thus use Equation J3-6a

$$r_n = 2.4dtF_u = 2.4(7/8)(0.50)(58) = 60.9 \text{ kips}$$

Step 3: Determine the tearout strength to the plate edge using $l_c = l_{cA1} = l_{cB2}$.

Find the clear distance from the bolt hole to the end of the member,

$$l_c = 1.5 - \left(\frac{1}{2}\right)(7/8 + 1/16) = 1.03 < 2d = 2(7/8) = 1.75 \text{ in.}$$

Since the clear distance is less than 2 bolt diameters, tearout strength will be less than bearing strength. Using Equation J3-6c

$$r_n = 1.2l_ctF_u = 1.2(1.03)(0.50)(58) = 35.8 \text{ kips}$$

Step 4: Determine the tearout strength between bolt holes using $l_c = l_{cA2} = l_{cB1}$.

Find the clear distance from one bolt hole to the other,

$$l_c = 2.5 - (7/8 + 1/16) = 1.56 < 2d = 2(7/8) = 1.75 \text{ in.}$$

Since the clear distance is less than 2 bolt diameters, tearout strength will be less than bearing strength. Using Equation J3-6c

$$r_n = 1.2 l_c t F_u = 1.2(1.56)(0.50)(58) = 54.3 \text{ kips}$$

Step 5: Determine the nominal strength of bolts A

Bolt shear	40.9 kips
Bearing	60.9 kips
Tearout on plate 1	35.8 kips
Tearout on plate 2	54.3 kips

Thus, these bolts have a nominal strength per bolt

$$r_n = 35.8 \text{ kips}$$

Determine the nominal strength of bolts B

Bolt shear	40.9 kips
Bearing	60.9 kips
Tearout on plate 1	54.3 kips
Tearout on plate 2	35.8 kips

Thus, these bolts have a nominal strength per bolt

$$r_n = 35.8 \text{ kips}$$

Therefore, all four bolts have the same strength.

Step 6: Determine the nominal strength of the 4 bolt connection.

Since all 4 bolts have the same strength, the controlling limit state is tearout to the edges of the plates, thus

$$R_n = 4r_n = 4(35.8) = 143 \text{ kips}$$

For LRFD Step 7: The design strength of the 4 bolt connection is

$$\phi R_n = 0.75(143) = 107 \text{ kips}$$

For ASD Step 7: The allowable strength for the 4 bolt connection is

$$R_n/\Omega = 143/2.00 = 71.5 \text{ kips}$$

Part b

Step 8: Determine the nominal shear strength for 7/8 in. A325-N bolts.

$$F_{nv} = 0.45F_u = 0.45(120) = 54 \text{ ksi}$$

$$A_b = 0.601 \text{ in.}^2$$

$$r_n = F_{nv}A_b = 54(0.601) = 32.5 \text{ ksi}$$

Step 9: Determine the nominal strength of the 4 bolt connection.

Since bearing and tearout are not influenced by the location of the shear plane through the threads, the nominal strengths determined in Steps 2, 3, and 4 are unchanged. Thus, the controlling limit state for each bolt in this connection is bolt shear. Thus,

$$R_n = 4r_n = 4(32.5) = 130 \text{ kips}$$

For LRFD

Step 10: The design strength of the 4 bolt connection is

$$\phi R_n = 0.75(130) = 97.5 \text{ kips}$$

For ASD

Step 10: The allowable strength for the 4 bolt connection is

$$R_n/\Omega = 130/2.00 = 65.0 \text{ kips}$$

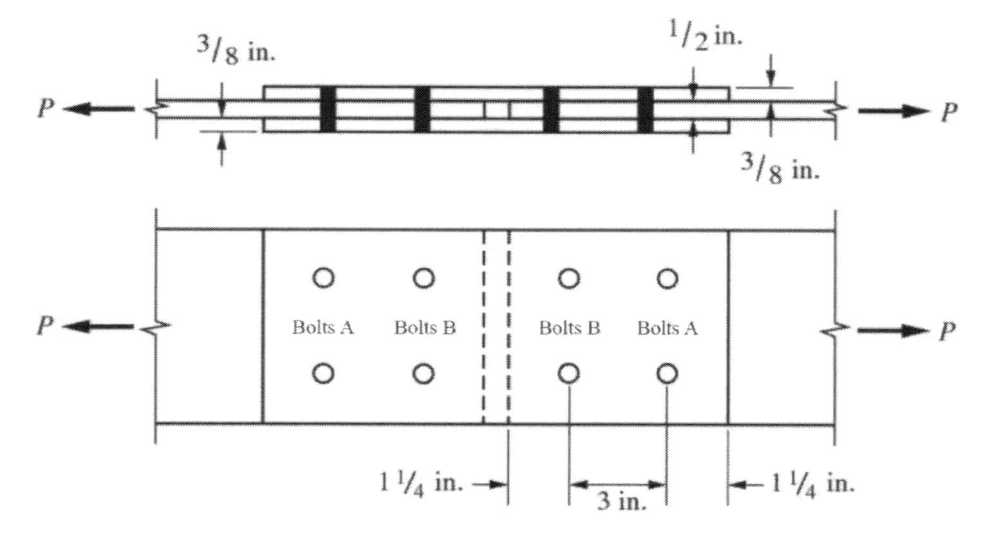

Figure 10.13 Butt Joint for Example 10.3

EXAMPLE 10.3 ***Butt Splice Connection Strength SOLUTION***

Goal: Determine the available strength of the bolts in the butt splice connection shown in Figure 10.13.

Given: Use 3/4 in. A490-N bolts and A36 plates.

Step 1: Determine bolt nominal shear strength.

$$F_{nv} = 0.450F_u = 0.450(150) = 68 \text{ ksi}$$

$$A_b = 0.442 \text{ in.}^2$$

$$r_n = F_{nv}A_b = 68(0.442) = 30.1 \text{ kips}$$

The bolts are in double shear, thus

$$r_n = 2(30.1) = 60.2 \text{ kips}$$

Step 2: Determine the nominal strength for bearing on the 1/2 in. middle plate using Equation J3-6a,

$$r_n = 2.4dtF_u = 2.4(3/4)(1/2)(58) = 52.2 \text{ kips}$$

Step 3: Determine the nominal strength for bearing on the 3/8 in. outer plate using Equation J3-6a,

$$r_n = 2.4dtF_u = 2.4(3/4)(3/8)(58) = 39.2 \text{ kips}$$

Since there are 2 outer plates, the nominal bearing strength is

$$r_n = 2(39.2) = 78.4 \text{ kips}$$

Step 4: Determine the tearout strength in the 1/2 in. middle plate

For the bolts closest to the end of the plate, bolts B,

$$l_c = 1.25 - \left(\frac{1}{2}\right)(3/4 + 1/16) = 0.844 < 2d = 2(3/4) = 1.5 \text{ in.}$$

Since the clear distance is less than 2 bolt diameters, tearout strength will be less than bearing strength. Using Equation J3-6c

$$r_n = 1.2l_c tF_u = 1.2(0.844)(0.50)(58) = 29.4 \text{ kips}$$

For the interior bolts, blots A, the clear distance between bolts is

$$l_c = 3.0 - (3/4 + 1/16) = 2.19 > 2d = 2(3/4) = 1.5 \text{ in.}$$

Since the clear distance is greater than 2 bolt diameters, tearout strength will be greater than bearing strength. To confirm this, using Equation J3-6c

$$r_n = 1.2l_c tF_u = 1.2(2.19)(0.50)(58) = 76.2 \text{ kips}$$

Step 5: Determine the tearout strength in the 3/8 in outer plate

For the bolts closest to the end of the plate, bolts A,

$$l_c = 1.25 - \left(\frac{1}{2}\right)(3/4 + 1/16) = 0.844 < 2d = 2(3/4) = 1.50 \text{ in.}$$

Since the clear distance is less than 2 bolt diameters, tearout strength will be less than bearing strength. Using Equation J3-6c

$$r_n = 1.2 l_c t F_u = 1.2(0.844)(3/8)(58) = 22.0 \text{ kips}$$

For the two plates,

$$r_n = 2(22.0) = 44.0 \text{ kips}$$

For the interior bolts, bolts B, the clear distance between bolts is

$$l_c = 3.0 - (3/4 + 1/16) = 2.19 > 2d = 2(3/4) = 1.5 \text{ in.}$$

Since the clear distance is greater than 2 bolt diameters, tearout strength will be greater than bearing strength. To confirm this, using Equation J3-6c

$$r_n = 1.2 l_c t F_u = 1.2(2.19)(3/8)(58) = 57.2 \text{ kips}$$

For the two plates,

$$r_n = 2(57.2) = 114 \text{ kips}$$

Step 6: Determine the controlling limit state and the nominal strength of each bolt.

For the bolts closest to the end of the outer plate, bolts A, the limit states and strengths are

Bolt shear	60.2 kips
Bearing on 1/2 in. plate	52.2 kips
Bearing on 2 - 3/8 in. plates	78.4 kips
Tearout on 1/2 in. plate	76.2 kips
Tearout on 2 - 3/8 in. plates	44.0 kips

Thus, these bolts have a nominal strength per bolt

$$r_n = 44.0 \text{ kips}$$

For the bolts closest to the end of the inner plate, bolts B, the limit states and strengths are

Bolt shear	60.2 kips
Bearing on 1/2 in. plate	52.2 kips
Bearing on 2 - 3/8 in. plates	78.4 kips

Tearout on 1/2 in. plate	29.4 kips
Tearout on 2 - 3/8 in. plates	114.0 kips

Thus, these bolts have a nominal strength per bolt

$$r_n = 29.4 \text{ kips}$$

Step 7: Determine the final connection nominal strength

$$R_n = 2(44.0) + 2(29.4) = 147 \text{ kips}$$

For LRFD Step 8: The design strength for the connection is

$$\phi R_n = 0.75(147) = 110 \text{ kips}$$

For ASD Step 8: The allowable strength for the connection is

$$R_n/\Omega = 147/2.00 = 73.5 \text{ kips}$$

10.7.5 Slip

The limit state of slip is associated with connections that are referred to as slip critical. *Slip-critical connections* are designed to prevent slip at the required strength. They should be used only when required, such as when the connection is subjected to fatigue, or the connection has oversized holes or slots parallel to the direction of load. Often, connections are specified to be designed as slip-critical when these conditions do not apply. Significant connection economies are lost, for no gain, when this is the case. Any slip-critical connection design must also include a check for strength as a bearing-type connection by the methods discussed in the previous sections. Bolts in slip-critical connections are designated as A325-SC or A490-SC. The nominal strength of a single bolt in a slip-critical connection is given in *Specification* Section J3.9 as

$$R_n = \mu D_u h_f T_b n_s \qquad \text{(AISC J3-4)}$$

where

μ = mean slip coefficient
= 0.30 for Class A surfaces
= 0.50 for Class B surfaces

D_u = 1.13, a multiplier that reflects the ratio of mean installed bolt pretension to specified bolt pretension.

h_f = factor for fillers
= 1.0 for one filler between connected parts
= 0.85 for more than one filler.
If multiple fillers are directly connected independently to transfer load, h_f also may be taken as 1.0

n_s = number of slip planes

T_b = minimum bolt pretension specified in Table J3.1

For connections with standard size and short-slotted holes perpendicular to the direction of load,

$$\phi = 1.00 \text{ (LRFD)} \qquad \Omega = 1.50 \text{ (ASD)}$$

For oversized and short-slotted holes parallel to the direction of the load,

$$\phi = 0.85 \text{ (LRFD)} \qquad \Omega = 1.76 \text{ (ASD)}$$

For long-slotted holes

$$\phi = 0.70 \text{ (LRFD)} \qquad \Omega = 2.14 \text{ (ASD)}$$

EXAMPLE 10.4 ***Slip-Critical Connection Strength***

Goal: Determine the available slip-critical connection strength for the connection shown in Figure 10.13. This connection was treated as a bearing-type connection in Example 10.3.

Given: Use 3/4 in. A490-SC bolts in standard holes and A36 plates with a Class A surface. There are no fillers.

SOLUTION

Step 1: Determine bolt nominal slip strength.

From *Specification* Table J3.1, the minimum bolt tension, T_b = 35 kips. From the given information, μ = 0.30, D_u = 1.13, h_f= 1.0, n_s = 2. Therefore,

$$R_n = \mu D_h h_f T_b n_s$$

$$R_n = 0.30(1.13)(1.0)(35)(2) = 23.7 \text{ kips}$$

For LRFD Step 2: Determine the design strength.
From Example 10.3, the design bearing strength of the connection was found to be

$$\phi R_n = 110 \text{ kips}$$

For the slip-critical connection,

$$\phi R_n = 1.00(4)(23.7) = 94.8 \text{ kips}$$

Thus, the design strength of this slip-critical connection is

$$\phi R_n = 94.8 \text{ kips}$$

For ASD Step 2: Determine the allowable strength.
From Example 10.3, the allowable bearing strength of the connection was found to be

$$R_n/\Omega = 73.5 \text{ kips}$$

For the slip-critical connection,

$$R_n/\Omega = 4(23.7/1.5) = 63.2 \text{ kips}$$

Thus, the allowable strength of this slip-critical connection is

$$R_n/\Omega = 63.2 \text{ kips}$$

It can be seen through Example 10.4 that a significant amount of connection strength is lost when a connection that otherwise could be designed as a bearing-type connection is forced to be designed as a slip-critical connection. A situation where the connection of Example 10.4 would need to be considered slip-critical is in a fatigue loading. By considering the resistance factor and safety factor for other hole types, it is also seen that significant strength reductions result for those slip-critical connections.

10.7.6 Combined Tension and Shear in Bearing-Type Connections

When bolts are subjected to simultaneous shear and tension, the available strength in each is diminished by the presence of the other. Experimental results indicate that the interaction between shear and tension is best represented by an elliptical relationship as shown as the solid curve in Figure 10.14. This elliptical relationship can conveniently be replaced, with little loss of accuracy, by three straight lines as shown by the dashed lines in Figure 10.14.

Specification Section J3.8 provides a nominal tensile stress modified to include the effects of shearing stress, F'_{nt}, to be used in determining the nominal bolt tensile strength such that

$$R_n = F'_{nt} A_b \qquad \text{(AISC J3-2)}$$

and

$$\phi = 0.75 \text{ (LRFD)} \qquad \Omega = 2.00 \text{ (ASD)}$$

As can be seen in Figure 10.14, when the required stress in either shear or tension is less than or equal to 30 percent of the corresponding available stress, the effects of the combined stresses can be ignored. If both required stresses exceed this 30 percent limit, the modified tensile stress is given by Equations J3-3a and J3-3b as

$$F'_{nt} = 1.3F_{nt} - \frac{F_{nt}}{F_{rv}} f_{rv} \leq F_{nt}$$

where

F_{nt} = nominal tensile stress when only tension occurs
F_{rv} = available shear stress, ϕF_{nv} for LRFD or F_{nv}/Ω for ASD
F_{nv} = nominal shear stress when only shear occurs
f_{rv} = required shear stress, either for LRFD or ASD

An example will be given in Section 11.10.

10.7.7 Combined Tension and Shear in Slip-Critical Connections

When bolts in slip-critical connections are subjected to simultaneous shear and tension, the available slip strength is diminished as the applied tensile force reduces the clamping force provided by the bolt pretension. The strength of the connection, therefore, is a linear function of the force compressing the plies. This force is the initial pre-tension, T_b, minus the applied load, T. The specified slip-critical shear value is, therefore, reduced by the factor $(1-T/T_b)$. The actual reduction factor, k_{sc}, is provided in *Specification* Equation J3-5, again with one for ASD and one for LRFD, as

$$k_{sc} = 1 - \frac{T_u}{D_u T_b n_b} \geq 0 \quad \text{(LRFD)} \qquad \text{(AISC J3-5a)}$$

$$k_{sc} = 1 - \frac{1.5T_a}{D_u T_b n_b} \geq 0 \quad \text{(ASD)} \qquad \text{(AISC J3-5b)}$$

where

T_a = required tension force using ASD load combinations
T_u = required tension force using LRFD load combinations
T_b = minimum bolt pretension
D_u = 1.13
n_b = number of bolts carrying the applied tension

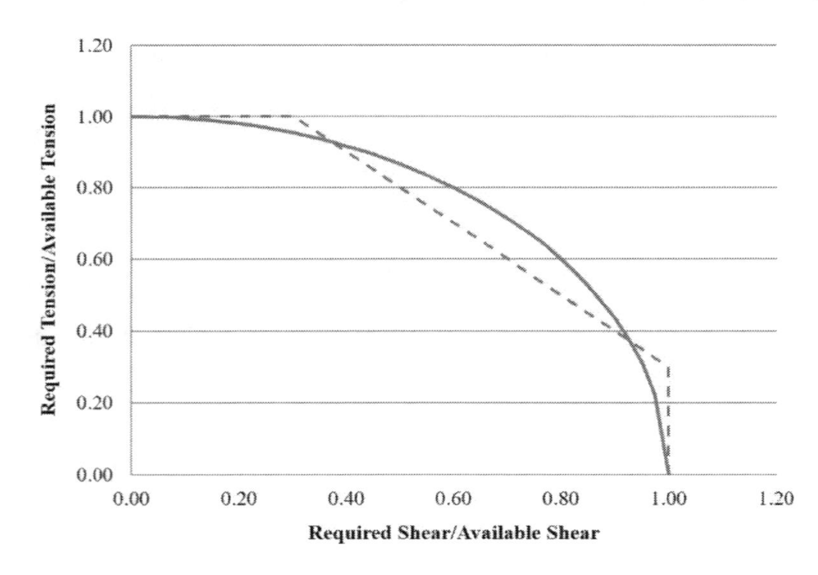

Figure 10.14 Interaction of Bolt Shear and Tension

10.8 WELDS

Welding is the process of joining steel by melting and fusing additional metal into the joint between the two pieces to be joined. The ease with which various types of steel can be joined by welding, without exhibiting cracks and other flaws, is called *weldability*. Most structural steels used today accept welding without the occurrence of unwanted defects. The American Welding Society (AWS) defines weldability as "the capacity of a metal to be welded under fabrication conditions imposed into a specific, suitably designed structure and to perform satisfactorily in the intended service."

Weldability depends primarily on the chemical composition of the steel and the thickness of the material. The impact on weldability of the various chemical elements in the composition of steel was discussed in Chapter 3.

10.8.1 Welding Processes

For structural steel, the four most popular welding processes as designated by AWS are: shielded metal arc, submerged arc, gas shielded metal arc and flux cored arc.

Shielded Metal Arc Welding

Shielded metal arc welding (*SMAW*) is one of the oldest welding processes. It is often called *manual* or *stick welding*. Figure 10.15a is a schematic representation of this welding process. A high voltage is induced between an electrode and the metal pieces that are to be joined. The electrode is the source of the metal introduced into the joint to make the weld. It is called the *consumable electrode*. When the welding operator strikes an arc between the electrode and the base metal, the resulting flow of current melts the electrode and the base metal adjacent to it. The electrode is coated with a special ceramic material called *flux*. This flux protects the molten metal from absorbing hydrogen and other impurities from the air during the welding process. When the metal cools, a permanent bond exists between the electrode material and the base material. Because the flux cools at a different rate than the metal, it separates from the weld and is easily removed from the joint.

Submerged Arc Welding

Submerged arc welding (*SAW*) is an automatic or semi-automatic process that is used primarily when long pieces of plate are to be joined. It is shown schematically in Figure 10.15b. SAW welds must be made in the near flat or horizontal position. The flux is a granular material introduced through a flexible tube on top of the electric arc. It is an economical process for applications in which the use of repetitive and automated fabrication procedures lend efficiency to the work.

Gas Shielded Metal Arc Welding

Gas shielded metal arc welding (*GMAW*) is a process in which a continuous wire is fed into the joint to be welded. The molten metal is protected from the atmosphere by gas surrounding the wire. In the field, use of this process may be limited because wind tends to blow the shielding gas away. This is the method often referred to as *MIG welding* for its use of inert gases.

Flux Cored Arc Welding

Flux cored arc welding (*FCAW*) is also a continuous-wire process, except that the wire is essentially a thin hollow tube filled with flux that protects the metal as the wire melts. It can be arranged as a semi-automatic process, and exceptionally high production rates can be attained.

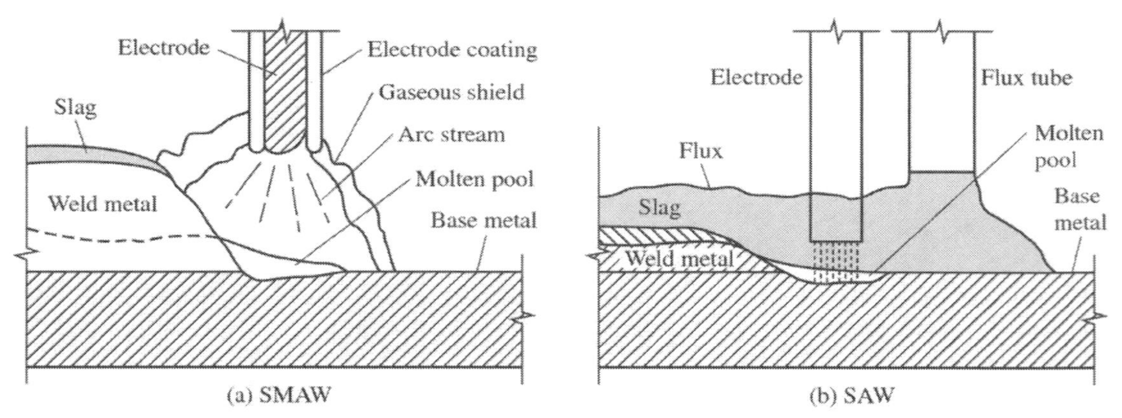

Figure 10.15 (a) Shielded Metal Arc Welding and (b) Submerged Arc Welding

10.8.2 Types of Welds

Four basic types of welds are used in steel construction: fillet welds, groove welds, plug welds, and slot welds. Fillet and groove welds are shown in Figure 10.16. Plug and slot welds fill a hole or slot with weld material to attach one piece to another.

Figure 10.16a shows a fillet weld. The leg of the weld is measured along the interface between the weld metal and the base metal. The throat of the weld is the shortest dimension and the shear plane of the weld. Because most fillet welds are symmetrical, with a 45-degree surface, the throat is usually calculated as 0.707 times the leg dimension as shown. The size of a fillet weld is given by its leg dimension, in increments of 1/16 in.

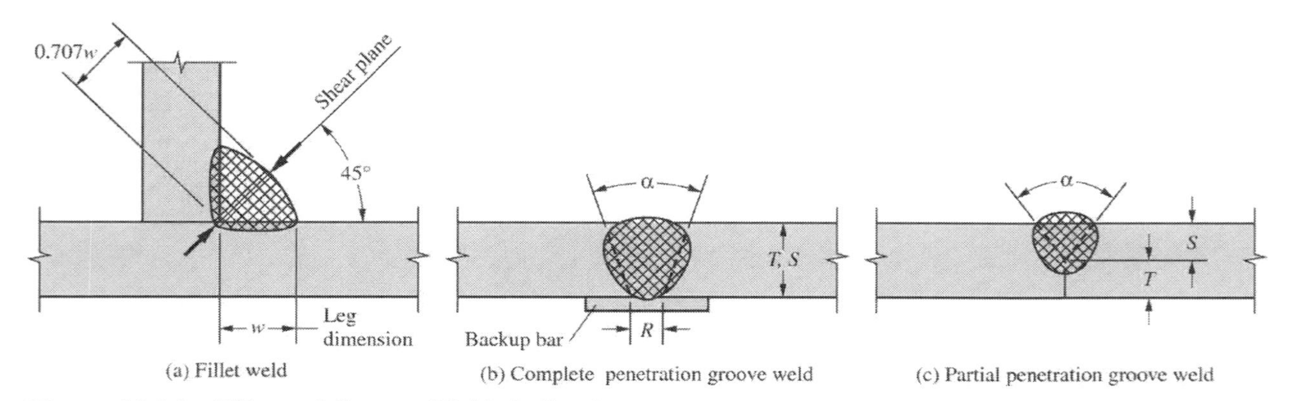

Figure 10.16 Fillet and Groove Welds in Section

A groove weld can be either a complete joint penetration (CJP) groove weld, as shown in Figure 10.16b, or a partial joint penetration (PJP) groove weld, as shown in Figure 10.16c. Both types of groove welds have joint details that have been prequalified by AWS. This prequalification means that certain weld configurations—including the root opening, R, the angle of preparation, α, and the effective thickness, S—are deemed practical to build and will carry the intended load. AWS specifies provisions for prequalifying any weld configuration if circumstances indicate that it is practical. These prequalified complete and partial joint penetration groove welds are shown in detail in *Manual* Table 8-2. The configurations shown in Figure 10.16 are schematic representations.

Both fillet welds and groove welds can be laid down in a variety of different positions depending on the orientation of the pieces to be joined. The terminology for these positions is given in Figure 10.17. Flat and horizontal welds are preferred over vertical and overhead welds.

Welds other than flat and horizontal require a slower deposition rate with lower heat and thus result in a higher cost final connection.

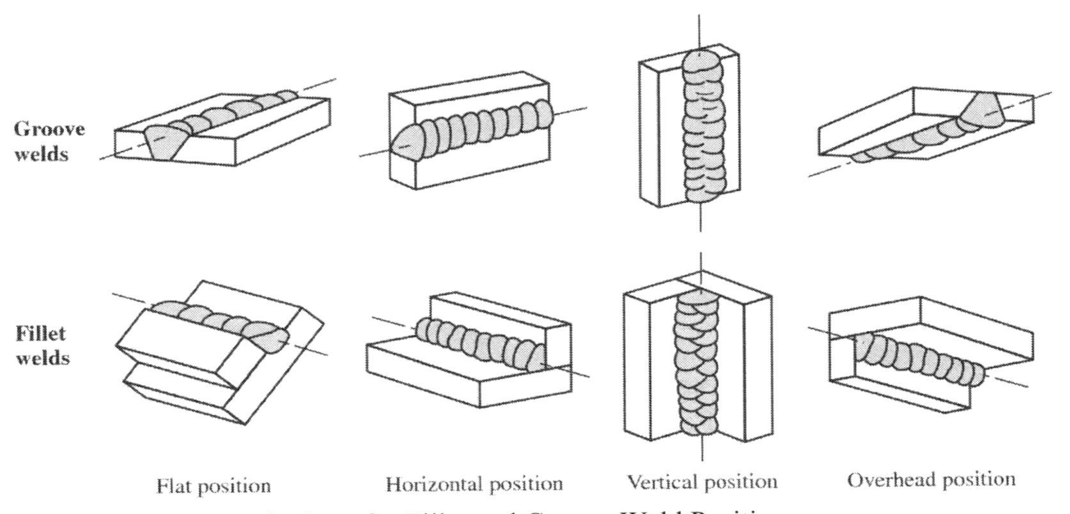

Figure 10.17 Terminology for Fillet and Groove Weld Positions

10.8.3 Weld Sizes

Specification Section J2 addresses effective areas and sizes for welds. The effective area and other limitations for groove welds are given in *Specification* Section J2.1 and Tables J2.1, J2.2 and J2.3. The effective areas of fillet welds are given in *Specification* Section J2.2a. The minimum sizes for fillet welds in joints are based on the thinner of the parts being joined as given in *Specification* Table J2.4. The maximum size of fillet welds for material less than 1/4 in. thick is the thickness of the material, whereas for material 1/4 in. thick or greater, the maximum size is the material thickness less 1/16 in.

The minimum length of fillet welds that are designed on the basis of strength is four times the weld leg size. For welds loaded longitudinally, the maximum weld length is 100 times the weld leg size for full effectiveness. Welds longer than this must use a reduced effective length as given in *Specification* Section J2.2b.

10.9 WELD LIMIT STATES

The only limit state to be considered for a weld is rupture. Yielding of the weld metal will occur, but it occurs over such a short distance that it is not a factor in connection behavior. Strain hardening occurs and rupture takes place without excessive yielding deformation.

The ultimate tensile strength of an electrode may vary from 60 to 120 ksi, depending on the specified composition. AWS classifies electrodes according to the tensile strength of the weld metal and indicates electrode strength as F_{EXX}. In this notation, the *E* represents the electrode and the *XX* represents the tensile strength. Thus, a typical electrode used to weld A992 steel would have a strength of 70 ksi and be designated as an E70 electrode.

AWS and AISC specify that for a particular grade of structural steel, as indicated by yield strength, there is a matching electrode. The user note table found in Section J2.6 summarizes the AWS provisions for matching filler metals. Both organizations further specify that the steel can be joined by welding only with the matching electrode or one that is no more than one grade higher. This is to encourage yielding in the base metal before it occurs in the weld.

10.9.1 Fillet Weld Strength

For a fillet weld as shown in Figure 10.16a, load is transferred by shear through the throat of the weld and the weld rupture strength is a function of the properties of the electrode. Shear strength provisions for welds are found in *Specification* Section J2.4 and Table J2.5, where, for fillet welds

$$R_n = F_{nw} A_{we} k_{ds} \qquad \text{(AISC J2-4)}$$

$$\phi = 0.75 \text{ (LRFD)} \qquad \Omega = 2.00 \text{ (ASD)}$$

and

F_{nw} = nominal strength of the weld metal per unit area
A_{we} = effective area of the weld
k_{ds} = directional strength increase factor = 1.0 for longitudinally loaded fillet welds.

For a longitudinally loaded fillet weld,

$$F_{nw} = 0.6 F_{EXX}$$

where

F_{EXX} = filler metal classification strength, the weld strength

Because the limit state of all fillet welds is one of shear rupture through the throat, the effective area of a symmetrical fillet weld is the width of the weld at the throat, $0.707w$, times the length of the weld, l, so that

$$A_{we} = 0.707 wl$$

The resulting nominal weld strength for a longitudinally loaded fillet weld is

$$R_n = 0.60F_{EXX}\left(0.707wl\right)$$

For the most commonly used weld electrode, $F_{EXX} = 70$ ksi, the design strength for LRFD can be determined as

$$\phi R_n = 0.75\left(0.6\left(70\right)\right)\left(0.707wl\right) = 22.27wl$$

and the allowable strength for ASD can be determined as

$$R_n/\Omega = (0.6\ (70))(\ 0.707wl)/2.00 = 14.85wl$$

It is convenient in design to use the fillet weld strength for a fillet weld with a 1/16 in. leg, which gives

Design strength for LRFD

$\phi R_n = 22.27wl = 22.27\ \left(1/16\right)\left(1.0\right) = 1.392$ kips per 1/16 in. of weld per in. of length for longitudinally loaded fillet welds and

Allowable strength for ASD

$R_n/\Omega = 14.85wl = 14.85\left(1/16\right)\left(1.0\right) = 0.928$ kips per 1/16 in. of weld per in. of length for longitudinally loaded fillet welds.

Therefore, a 1/4 in. fillet weld has a design strength of 1.392×4 (sixteenths) = 5.57 kips per inch of length and an allowable strength of 0.928×4 (sixteenths) = 3.71 kips per inch of length.

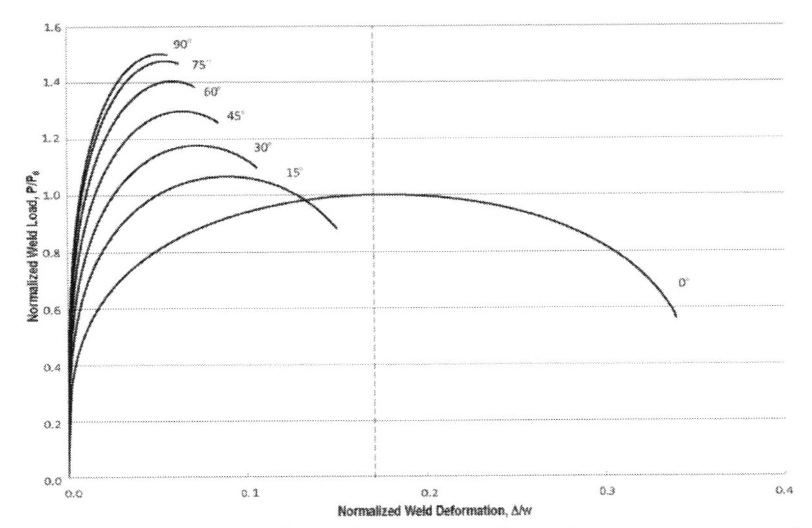

Figure 10.18 Normalized Strength vs Normalized Deformation for Welds Loaded at an Angle

Research has shown that when load is applied to a fillet weld at an angle other than along the length of the weld, more strength is available than when the load is applied longitudinally. Figure 10.18 shows that as the angle of load increases the ultimate strength increases. It also shows a corresponding reduction of deformation capacity. The vertical line in Figure 10.18 locates the maximum strength of a weld loaded longitudinally where $k_{ds} = 1.0$. Thus, welds loaded at an angle to the weld length are seen to exhibit an increase of up to 50% in strength. The *Specification* provides for this increase through the directional strength increase factor, k_{ds}, based on the angle of the load to the longitudinal axis of the weld provided strain compatibility of the various weld elements is considered. Thus,

$$k_{ds} = \left(1.0 + 0.5\sin^{1.5}\theta\right) \qquad \text{(AISC J2-5)}$$

where

θ = angle of loading measured from the weld longitudinal axis

This strength increase is intended to be used for welds or weld groups with uniform leg size in which all elements are in line or parallel and loaded through the center of gravity. When welds with different orientations are combined in the same joint, deformation of these different welds, as illustrated in Figure 10.18, must be accounted for. *Specification* Section J2.4b provides an approach for concentrically loaded fillet weld groups consisting of elements that are both longitudinal and transverse to the direction of the applied load. For this case, the nominal weld strength is taken as the larger of the simple sum of the weld strength without considering orientation, $k_{ds} = 1.0$, given by

$$R_n = R_{nwl} + R_{nwt}$$

Or the sum with $k_{ds} = 0.85$ for longitudinal welds and $k_{ds} = 1.5$ for transverse welds, Equation J2-6, given here in slightly different form as

$$R_n = 0.85R_{nwl} + 1.5R_{nwt} \qquad \text{(AISC J2-6)}$$

where

$R_{nwl} = F_{nw}A_{wel}$ = total nominal strength of the longitudinally loaded weld without considering the angle of load

$R_{nwt} = F_{nw}A_{wet}$ = total nominal strength of the transversely loaded weld without considering the angle of load

A_{wel} = effective area of longitudinally loaded weld

A_{wet} = effective area of transversely loaded weld

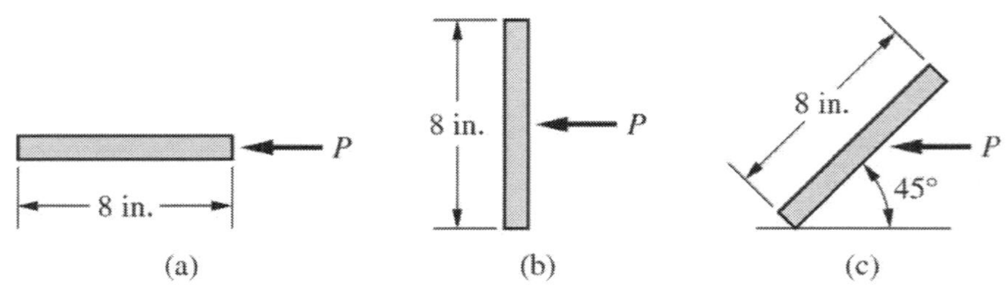

Figure 10.19 Welds for Example 10.5

The user note in Section J2.4b indicates that the strength of the weld group as a function of the sum of the lengths of the welds times the strength per unit length without consideration of the directional strength increase factor is permissible. Implementation of Equation J2-6 requires that the longitudinally loaded weld strength be discounted by 0.85 in order for the transverse welds to be increased by 1.5. When welds loaded at different angles are combined, it is critical that their deformations are compatible. From Figure 10.18 it can be seen that a weld loaded at 90° reaches its ultimate strength at a normalized deformation significantly less than the deformation at which a weld loaded at 0° reaches its ultimate strength. The weld loaded at 0° reaches only about 85% of its ultimate strength when its deformation is the same as the weld loaded at 90° is at its ultimate strength. The approach of Equation J2-6 will be most beneficial when the transverse welds are longer than the longitudinal welds.

The forgoing discussion addressed only the strength of the weld material. The material that the weld is attached to is referred to as the base material. This material must also be checked for strength according to the provisions of Sections J2.4 and J4 which will be discussed in Section 10.10.

EXAMPLE 10.5
Weld Strength and Load Angle

Goal: Determine the available strength of the three welds given in Figure 10.19.

Given: The welds are 5/16 in. welds, 8.0 in. long, and loaded (a) along the length of the weld, (b) transversely to the weld, and (c) at a 45-degree angle to the weld. Use E70 electrodes.

SOLUTION

Part a Weld Loaded Along Its Length

Step 1: Determine the number of 1/16 units for the given weld using.

The weld is five 1/16 in. units across the leg.

Step 2: Determine the strength of the weld when loaded along its length.

The strength values already discussed can be used because the weld is loaded longitudinally and $k_{ds} = 1.0$.

For LRFD Step 3: The design strength is

$$\phi R_n = 8.0(5)(1.392) = 55.7 \text{ kips}$$

For ASD Step 3: The allowable strength is

$$R_n/\Omega = 8.0(5)(0.928) = 37.1 \text{ kips}$$

Part b Weld Loaded at 90 Degrees to the Weld Length

Step 1: Determine the directional strength increase factor, k_{ds}, to account for the angle of load.

$$\begin{aligned} k_{ds} &= \left(1.0 + 0.5\sin^{1.5}\theta\right) \\ &= \left(1.0 + 0.5\sin^{1.5}(90)\right) \\ &= 1.5 \end{aligned}$$

Therefore, the strength of the weld is increased by 1.5 over what it is when loaded longitudinally.

For LRFD Step 2: The design strength is

$$\phi R_n = 1.5(55.7) = 83.6 \text{ kips}$$

For ASD Step 2: The allowable strength is

$$R_n/\Omega = 1.5(37.1) = 55.7 \text{ kips}$$

Part c Weld Loaded at 45 Degrees to the Weld Length

Step 1: Determine the directional strength increase factor, k_{ds}, to account for the angle of load.

$$\begin{aligned} k_{ds} &= \left(1.0 + 0.5\sin^{1.5}\theta\right) \\ &= \left(1.0 + 0.5\sin^{1.5}(45)\right) \\ &= 1.30 \end{aligned}$$

Therefore, the strength of the weld is increased by 1.30 over what it is when loaded longitudinally.

For LRFD Step 2: The design strength is

$$\phi R_n = 1.3(55.7) = 72.4 \text{ kips}$$

For ASD Step 2: The allowable strength is

$$R_n/\Omega = 1.3(37.1) = 48.2 \text{ kips}$$

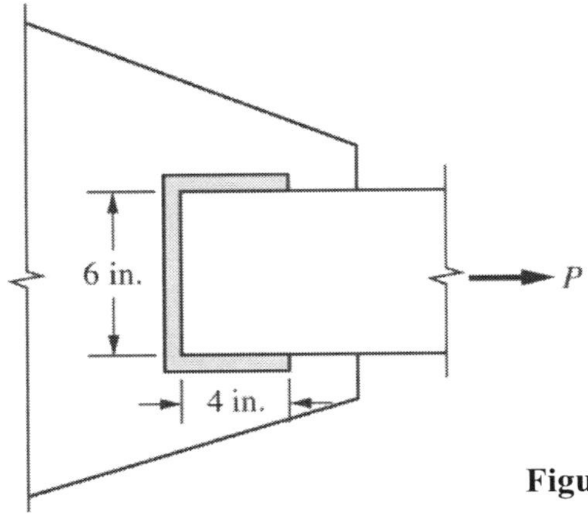

Figure 10.20 C-Shaped Weld for Example 10.6.

EXAMPLE 10.6a
Weld Strength and Load Angle by LRFD SOLUTION

Goal: Determine the design strength for C-shaped welds.

Given: A C-shaped weld group is shown in Figure 10.20 to attach a tension plate to a gusset. Use E70 electrodes and a 5/16 in. weld.

Step 1: Determine the design strength for the two 4.0 in. welds parallel to the load.

$$\phi R_{nwl} = 2(4.0)(5)(1.392) = 55.7 \text{ kips}$$

Step 2: Determine the design strength for the 6.0 in. weld transverse to the load without considering k_{sd}.

$$\phi R_{nwt} = 6(5)(1.392) = 41.8 \text{ kips}$$

Step 3: Determine the connection design strength by adding the strength based on length of the welds.

$$\phi R_n = R_{nwl} + R_{nwt} = 55.7 + 41.8 = 97.5 \text{ kips}$$

Step 4: Determine the design strength considering the added contribution of the transverse welds while reducing the contribution of the longitudinal welds so that

$$\phi R_n = 0.85 R_{nwl} + 1.5 R_{nwt} = 0.85(55.7) + 1.5(41.8) = 110 \text{ kips}$$

Step 5: Determine the weld strength by selecting the larger of the results from steps 3 and 4.

Thus, accounting for the difference between the longitudinal and transverse welds provides more strength.

$$\phi R_n = 110 \text{ kips}$$

EXAMPLE 10.6b
Weld Strength and Load Angle by ASD

Goal: Determine the design strength for C-shaped welds.

Given: A C-shaped weld group is shown in Figure 10.20 to attach a tension plate to a gusset. Use E70 electrodes and a 5/16 in. weld.

SOLUTION

Step 1: Determine the allowable strength for the two 4.0 in. welds parallel to the load.

$$R_{nwl}/\Omega = 2(4.0)(5)(0.928) = 37.1 \text{ kips}$$

Step 2: Determine the allowable strength for the 6.0-in. weld transverse to the load without considering k_{sd}.

$$R_{nwt}/\Omega = 6.0(5)(0.928) = 27.8 \text{ kips}$$

Step 3: Determine the connection allowable strength by adding the strength based on length of the welds.

$$R_n/\Omega = 37.1 + 27.8 = 64.9 \text{ kips}$$

Step 4: Determine the allowable strength considering the added contribution of the transverse welds while reducing the contribution of the longitudinal welds so that

$$R_n/\Omega = 0.85(37.1) + 1.5(27.8) = 73.2 \text{ kips}$$

Step 5: Determine the weld strength by selecting the larger of the results from steps 3 and 4. Thus, accounting for the difference between the longitudinal and transverse welds provides more strength.

$$R_n/\Omega = 73.2 \text{ kips}$$

EXAMPLE 10.7a
Weld Strength and Load Angle by LRFD

Goal: Determine the design strength for C-shaped welds.

Given: The C-shaped weld group shown in Figure 10.20 is to attach a tension plate to a gusset. However, the welds parallel to the load will be increased to 10 in. in length. Use E70 electrodes and a 5/16 in. weld.

SOLUTION

Step 1: Determine the design strength for the two 10.0 in. welds parallel to the load.

$$\phi R_{nwl} = 2(10.0)(5)(1.392) = 139 \text{ kips}$$

Step 2: Determine the design strength for the 6.0 in. weld transverse to the load without considering k_{sd}.

$$\phi R_{nwt} = 6(5)(1.392) = 41.8 \text{ kips}$$

Step 3: Determine the connection design strength by adding the strength based on length of the welds.

$$\phi R_n = R_{nwl} + R_{nwt} = 139 + 41.8 = 181 \text{ kips}$$

Step 4: Determine the design strength considering the added contribution of the transverse welds while reducing the contribution of the longitudinal welds so that

$$\phi R_n = 0.85 R_{nwl} + 1.5 R_{nwt} = 0.85(139) + 1.5(41.8) = 181 \text{ kips}$$

Step 5: Determine the weld strength by selecting the larger of the results from steps 3 and 4. Note that both approaches give the same design strength. Thus,

$$\phi R_n = 181 \text{ kips}$$

If the welds parallel to the load were any longer, considering only weld length would provide the most strength.

EXAMPLE 10.7b
Weld Strength and Load Angle by ASD

Goal: Determine the design strength for C-shaped welds.

Given: The C-shaped weld group shown in Figure 10.20 is to attach a tension plate to a gusset. However, the welds parallel to the load will be increased to 10 in. in length. Use E70 electrodes and a 5/16 in. weld.

SOLUTION

Step 1: Determine the allowable strength for the two 10.0 in. welds parallel to the load.

$$R_{nwl}/\Omega = 2(10.0)(5)(0.928) = 92.8 \text{ kips}$$

Step 2: Determine the allowable strength for the 6.0 in. weld transverse to the load without considering k_{sd}.

$$R_{nwt}/\Omega = 6.0(5)(0.928) = 27.8 \text{ kips}$$

Step 3: Determine the connection allowable strength by adding the strength based on length of the welds.

$$R_n/\Omega = R_{nwl} + R_{nwt} = 92.8 + 27.8 = 121 \text{ kips}$$

Step 4: Determine the allowable strength considering the added contribution of the transverse welds while reducing the contribution of the longitudinal welds so that

$$R_n/\Omega = 0.85 R_{nwl} + 1.5 R_{nwt} = 0.85(92.8) + 1.5(27.8) = 121 \text{ kips}$$

Step 5: Determine the weld strength by selecting the larger of the results from steps 3 and 4. Note that both approaches give the same allowable strength. Thus,

$$R_n/\Omega = 121 \text{ kips}$$

If the welds parallel to the load were any longer, considering only weld length would provide the most strength.

10.9.2 Groove Weld Strength

A groove weld can be either a complete or partial joint penetration weld as shown in Figure 10.16b and c. The complete joint penetration (CJP) groove weld is not designed in the usual sense because the weld metal is always stronger than the base metal when properly matching electrodes are used. Therefore, the strength of the base metal controls the design.

In the case of a CJP groove weld, the nominal strength of the tension joint is the product of the yield strength of the base material and the cross-sectional area of the smallest piece joined. The nominal strength of a partial joint penetration (PJP) groove weld in a tension joint is similar except that the full cross-sectional area of the joined pieces is not effective. In this case, AWS defines an effective throat dimension, S, which is a function of the configuration of the bevel as shown in Figure 10.16c and *Manual* Table 8-2.

10.10 CONNECTING ELEMENTS

The plates, angles, and other elements that go into making up a connection are called *connecting elements*. They, along with the region of the members actually involved in the connection, are treated in Section J4. There are provisions for tension, compression, flexure, shear, and block shear.

10.10.1 Connecting Elements in Tension

Although the *Specification* addresses tension in connecting elements in Section J4.1, it does not alter the basic tension provisions found in *Specification* Chapter D. This means that two limit states are to be considered, the limit state of yielding and the limit state of rupture. Again, for tension, the resistance and safety factors are different for the two limit states, so any comparison of strength must be made at the design or allowable strength level. The design strength is given by ϕR_n and the allowable strength by R_n/Ω, as is the case throughout the *Specification*. For the limit state of yielding of connecting elements

$$R_n = F_y A_g \qquad \text{(AISC J4-1)}$$

$$\phi = 0.90 \text{ (LRFD)} \qquad \Omega = 1.67 \text{ (ASD)}$$

For the limit state of rupture of connecting elements

$$R_n = F_u A_e \qquad \text{(AISC J4-2)}$$

$$\phi = 0.75 \text{ (LRFD)} \quad \Omega = 2.00 \text{ (ASD)}$$

The areas are determined as for the tension members previously considered.

10.10.2 Connecting Elements in Compression

Most connecting elements in compression are relatively short and have a fairly low slenderness ratio. In addition, determination of the appropriate effective length factor requires application of significant engineering judgment, usually amounting to making an educated guess as to an appropriate factor. With this in mind, and to simplify connection design somewhat, the *Specification* provides a simple relation in Section J4.4 for the compressive strength of connecting elements if the slenderness ratio, L_c/r, is less than or equal to 25. For this case,

$$P_n = F_y A_g \tag{AISC J4-6}$$

and the resistance and safety factors are the same as for other compression members,

$$\phi = 0.90 \text{ (LRFD)} \quad \Omega = 1.67 \text{ (ASD)}$$

If the slenderness ratio of the compression element is greater than 25, the element must be designed according to the compression member provisions of *Specification* Chapter E.

10.10.3 Connecting Elements in Flexure

Connecting elements that undergo flexure are to be designed with consideration of flexural yielding, local buckling, flexural lateral-torsional buckling, and flexural rupture. These are the same limit states to be considered for flexural members as presented in Chapter 6.

10.10.4 Connecting Elements in Shear

Member design for shear, according to Section G2, requires the consideration of the limit states of shear yielding and shear buckling. Connecting elements and the portion of members affected by the connection must be checked for the limit states of shear yielding and shear rupture according to Section J4.2. Shear yielding occurs on the gross area of the element whereas shear rupture occurs on a section containing holes. Thus, for shear yielding of the element

$$R_n = 0.6 F_y A_{gv} \tag{AISC J4-3}$$

$$\phi = 1.00 \text{ (LRFD)} \quad \Omega = 1.50 \text{ (ASD)}$$

where A_{gv} = gross area subject to shear. The resistance and safety factors for this case are the same as those for the special case of rolled I-shaped members given in *Specification* Section G2.1(a).

For the limit state of shear rupture

$$R_n = 0.6 F_u A_{nv} \tag{AISC J4-4}$$

Figure 10.21 Block Shear Failure: (a) welded tension connection; (b) bolted beam end connection, Photos courtesy of Robert Driver

$$\phi = 0.75 \text{ (LRFD)} \qquad \Omega = 2.00 \text{ (ASD)}$$

where A_{nv} = net area subjected to shear. As was the case for tension rupture, the net area is determined by removing the area of holes from the gross area.

10.10.5 Block Shear Strength

The limit state of block shear rupture can occur on the connecting elements, or the affected members as shown in Figure 10.21. Figure 10.21a shows a block shear failure in a welded connection of a tension member to the web of a W-shape. Figure 10.21b shows a block shear failure in a bolted end connection to a coped W-shape.

Block shear combines shear and tension failures into a single, complex mode of failure. Block shear was discussed in Section 4.7 as it pertained to tension members because it can be a major factor in determining tension member strength. It can also be a factor in determining the strength of a beam end reaction, depending on the connection geometry. Thus, it is repeated here. The nominal strength for the limit state of block shear rupture is

$$R_n = 0.6F_u A_{nv} + U_{bs} F_u A_{nt} \leq 0.6F_y A_{gv} + U_{bs} F_u A_{nt} \qquad \text{(AISC J4-5)}$$

where

A_{gv} = gross shear area
A_{nt} = net tension area
A_{nv} = net shear area
U_{bs} = 1.0 for uniform tension stress distribution
= 0.5 for nonuniform tension stress

The resistance and safety factors for the limit state of block shear rupture are again

$$\phi = 0.75 \text{ (LRFD)} \qquad \Omega = 2.00 \text{ (ASD)}$$

Figure 10.22 shows a single-angle tension member attached to a gusset plate and a coped beam end with the holes located in a single line. The tension area and shear area are identified for each and the area that would tear out is shaded.

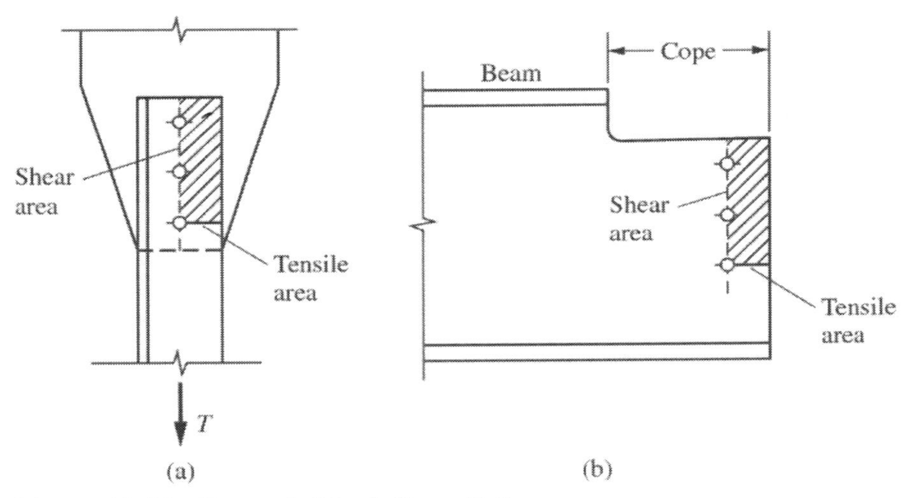

Figure 10.22 Example Block Shear Failure

A review of the block shear equation shows that the expected failure mode will always include tension rupture, whereas the shear failure mode will be the smaller of the shear rupture and the shear yield. The tension stress distribution factor, U_{bs}, is a function of the variation of the tension stress over the tension area. Figure 10.23 shows several elements and the corresponding assumed tensile stress distribution. The only case identified by the Commentary where the tensile stress distribution is not uniform is that of a coped beam with two rows of bolts, as shown in Figure 10.23g.

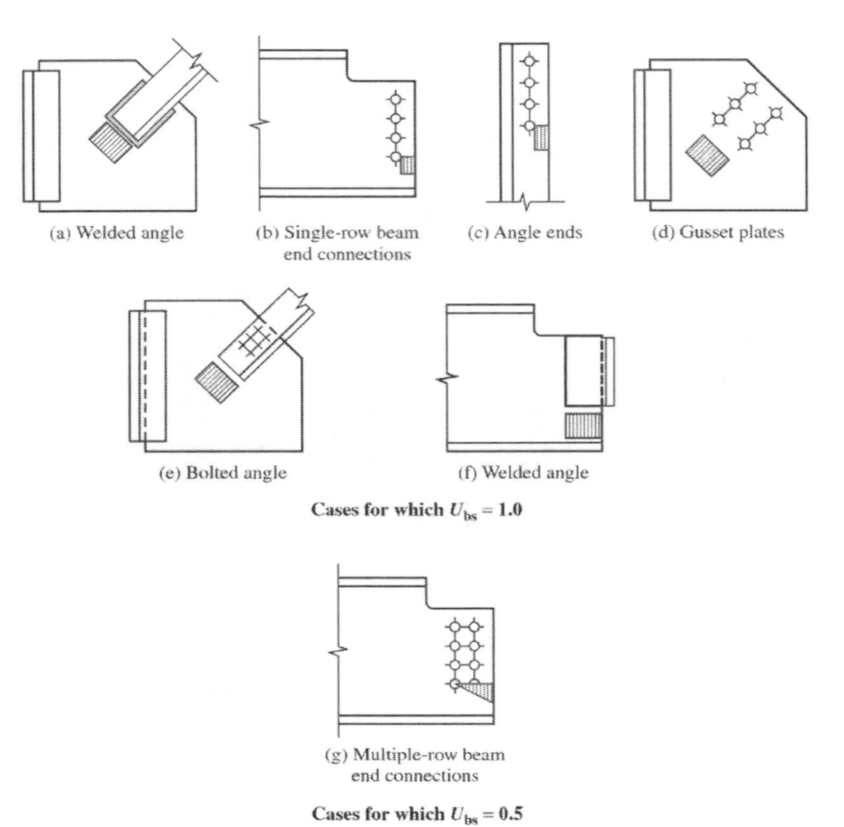

Figure 10.23 Block Shear Tensile Stress Distribution

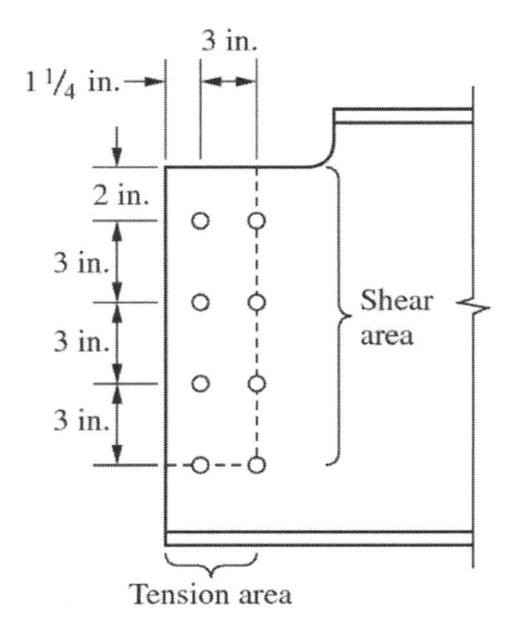

Figure 10.24 Coped Beam End for Example 10.8

EXAMPLE 10.8 ***Block Shear Strength***

Goal: Determine the block shear design strength and allowable strength for a coped beam.

Given: A coped W16×40 A992 beam end is shown in Figure 10.24. Assume that the beam has standard holes for 5/8 in. bolts.

SOLUTION

Step 1: Determine the gross and net shear areas and net tension area for the beam.

Remember from the discussion of tension members that for net area, an additional 1/16 in. must be added to the hole size to account for any hole damage from the punching operation. From *Manual* Table 1-1 $t_w = 0.305$ in.

$$A_{gv} = 11.0(0.305) = 3.36 \text{ in.}^2$$

$$A_{nv} = \left(11.0 - 3.5(5/8 + 1/8)\right)(0.305) = 2.55 \text{ in.}^2$$

$$A_{nt} = \left(4.25 - 1.5(5/8 + 1/8)\right)(0.305) = 0.953 \text{ in.}^2$$

Step 2: Determine the shear yield and rupture strength and the tension rupture strength.

For this geometry, the tensile stress distribution is nonuniform as shown in Figure 10.23g; therefore, $U_{bs} = 0.5$.

Shear yield

$$0.6F_y A_{gv} = 0.6(50)(3.36) = 101 \text{ kips}$$

Shear rupture

$$0.6F_u A_{nv} = 0.6(65)(2.55) = 99.5 \text{ kips}$$

Tensile rupture

$$U_{bs} F_u A_{nt} = 0.5(65)(0.953) = 31.0 \text{ kips}$$

Step 3: Determine the nominal block shear strength.

Because shear rupture is less than shear yield, combine the shear rupture term with the tensile rupture term. Thus,

$$R_n = 99.5 + 31.0 = 131 \text{ kips}$$

For LRFD Step 4: The design strength is

$$\phi R_n = 0.75(131) = 98.3 \text{ kips}$$

For ASD Step 4: The allowable strength is

$$R_n/\Omega = 131/2.00 = 65.5 \text{ kips}$$

10.10.6 Connecting Element Rupture Strength at Welds

The strength of connection elements at welds cannot always be determined directly. In those cases, it is often convenient to determine the minimum thickness of the connecting element required to match the rupture strength of the weld with the rupture strength of the base metal. The rupture strength of a fillet weld loaded longitudinally was given in Section 10.9.1 as

$$R_n = 0.60F_{EXX}\left(0.707wl\right)$$

If this is set equal to the shear rupture strength of the base metal, given as Equation J4.4 with the area taken as the thickness times the length of the weld,

$$R_n = 0.6F_u tl$$

and the thickness is taken as $t_{\min}$, the minimum required thickness for the base metal to match the weld, with E70 electrodes and the weld size taken as $D/16$ is

$$0.60(70)\left(0.707(D/16)l\right) = 0.6F_u t_{\min} l$$

Solving for $t_{\min}$ yields

$$t_{\min} = 3.09D/F_u \qquad (\textit{Manual } 9\text{-}6)$$

For a case where there are welds on both sides of the base metal, such as on the web of a beam with angles on both sides, the minimum web thickness will be twice that for a weld on one side. Thus,

$$t_{\min} = 6.19D/F_u \qquad (\textit{Manual } 9\text{-}7)$$

The application of these minimum thickness calculations will be illustrated in Chapters 11 and 12 as appropriate for the connections being considered.

10.11 PROBLEMS

1. Develop a table showing the nominal shear strength for A325-N bolts for the following sizes: 5/8, 3/4, 7/8, and 1 in.

2. Develop a table showing the nominal shear strength for A325-X bolts for the following sizes: 5/8, 3/4, 7/8, and 1 in.

3. Develop a table showing the nominal shear strength for A490-N bolts for the following sizes: 5/8, 3/4, 7/8, and 1 in.

4. Develop a table showing the nominal shear strength for A490-X bolts for the following sizes: 5/8, 3/4, 7/8, and 1 in.

5. Develop a table showing the design shear strength for A325-N, A325-X, A490-N, and A490-X bolts for the following sizes: 5/8, 3/4, 7/8, and 1 in.

6. Develop a table showing the allowable shear strength for A325-N, A325-X, A490-N, and A490-X bolts for the following sizes: 5/8, 3/4, 7/8, and 1 in.

7. Develop a table showing the design shear strength for F1852-N, F1852-X, F2280-N, and F2280-X bolts for the following sizes: 5/8, 3/4, 7/8, and 1 in.

8. Develop a table showing the allowable shear strength for F1852-N, F1852-X, F2280-N, and F2280-X bolts for the following sizes: 5/8, 3/4, 7/8, and 1 in.

9. Determine the available strength of the 3/4 in. A325-N bolts in the lap splice shown in Figure P10.9 with two 1/2 in. A36 plates. Determine (a) design strength by LRFD and (b) allowable strength by ASD.

10. Determine the available strength of the 3/4 in. A325-X bolts in the lap splice shown in Figure P10.9 with two 1/2 in. A36 plates. Determine (a) design strength by LRFD and (b) allowable strength by ASD.

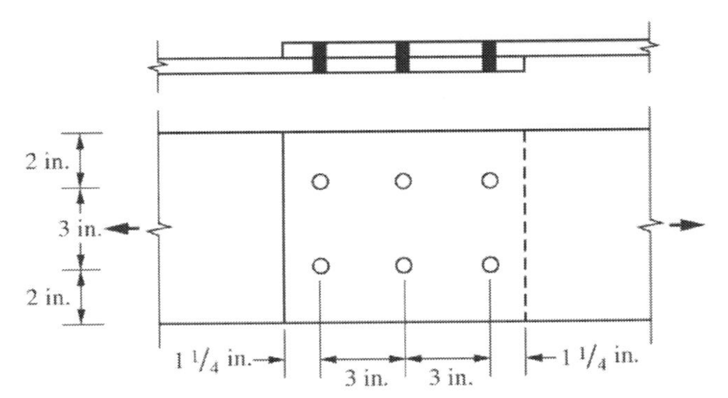

P10.9

11. Determine the available strength of the 7/8 in. A325-N bolts in the lap splice shown in Figure P10.9 with two 1/2 in. A36 plates. Determine (a) design strength by LRFD and (b) allowable strength by ASD.

12. Determine the available strength of the 3/4 in. A325-N bolts in the lap splice shown in Figure P10.9 with two 1/2 in. A572 Gr. 50 plates. Determine (a) design strength by LRFD and (b) allowable strength by ASD.

13. Determine the available strength of the 1 in. F2280-N bolts in the lap splice shown in Figure P10.9 with two 1/2 in. A36 plates. Determine (a) design strength by LRFD and (b) allowable strength by ASD.

14. Determine the available strength of the 1 in. F2280-X bolts in the lap splice shown in Figure P10.9 with two 1/2 in. A36 plates. Determine (a) design strength by LRFD and (b) allowable strength by ASD.

15. Determine the available strength of the 7/8 in. A490-N bolts in the lap splice shown in Figure P10.15 with two 1/2 in. A572 Grade 50 plates. Determine (a) design strength by LRFD and (b) allowable strength by ASD.

16. Determine the available strength of the 1-1/8 in. F2280-N bolts in the lap splice shown in Figure P10.15 with two 1/2 in. A36 plates. Determine (a) design strength by LRFD and (b) allowable strength by ASD.

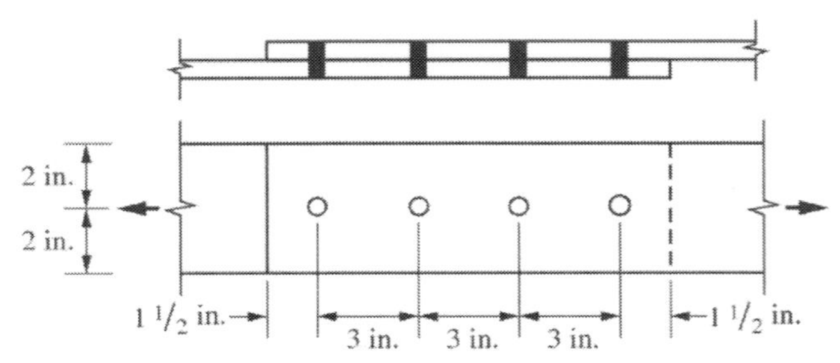

P10.15

17. Determine the available strength of the 7/8 in. A490-X bolts in the lap splice shown in Figure P10.15 with two 1/2 in. A572 Grade 50 plates. Determine (a) design strength by LRFD and (b) allowable strength by ASD.

18. Determine the available strength of the 7/8 in. A325-N bolts in the butt splice shown in Figure P10.18 with two 1/2 in. side plates and a 1 in. main plate. Use A36 plates. Determine (a) design strength by LRFD and (b) allowable strength by ASD.

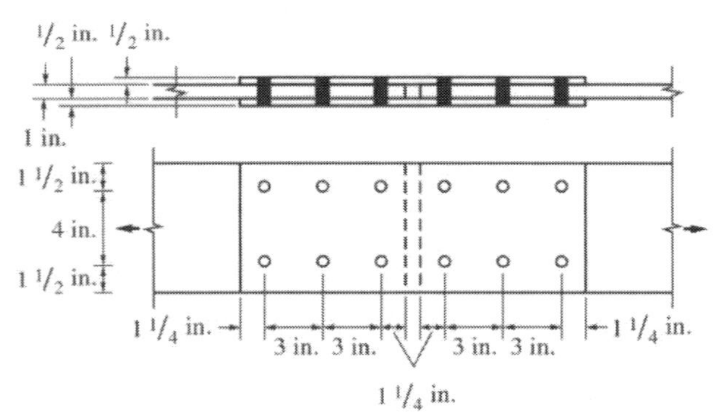

P10.18

19. Determine the available strength of the 3/4 in. A325-N bolts in the butt splice shown in Figure P10.18 with two 1/2 in. side plates and a 1 in. main plate. Use A572 Gr. 50 plates. Determine (a) design strength by LRFD and (b) allowable strength by ASD.

20. Determine the available strength of the 1-1/8 in. A490-X bolts in the butt splice shown in Figure P10.18 with two 1/2 in. side plates and a 1 in. main plate. Use A36 plates. Determine (a) design strength by LRFD and (b) allowable strength by ASD.

21. Recreate Figure 10.11 for A325-X bolts.

22. Recreate Figure 10.11 for A490-N bolts.

23. Recreate Figure 10.11 for A490-X bolts.

24. Determine the available strength of the 7/8 in. A490-SC bolts in the slip-critical lap splice shown in Figure P10.15 with two 1/2 in. A572 Gr. 50 plates, standard holes, Class B surface, and no fillers. (This is Problem 15 as a slip-critical connection.) Determine (a) design strength by LRFD and (b) allowable strength by ASD.

25. Determine the available strength of the 7/8 in. A325-SC bolts in the butt splice shown in Figure P10.18 with two 1/2 in. side plates and a 1 in. main plate. Use A36 plates, oversized holes, Class A surface, and no fillers. (This is Problem 18 as a slip-critical connection.) Determine (a) design strength by LRFD and (b) allowable strength by ASD.

26. Determine the available strength of the four 3/4 in. A325-N bolts in the single L3×3×1/2 A36 when the bolts are placed as shown in Figure P10.26. Determine (a) design strength by LRFD and (b) allowable strength by ASD.

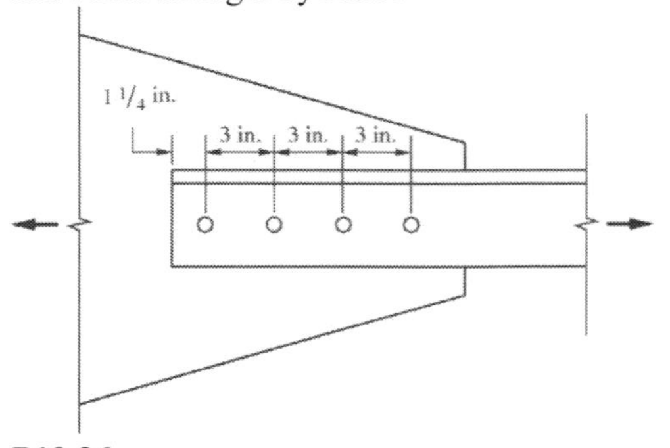

P10.26

27. Determine the available strength of the four 7/8 in. A325-N bolts in the single L3×3×1/2 A572 Gr. 50 when the bolts are placed as shown in Figure P10.26. Determine (a) design strength by LRFD and (b) allowable strength by ASD.

28. Determine the available strength of the three 3/4 in. A325-N bolts in the single L4×3×3/8 A36 when the bolts are placed as shown in Figure P10.28. Determine (a) design strength by LRFD and (b) allowable strength by ASD.

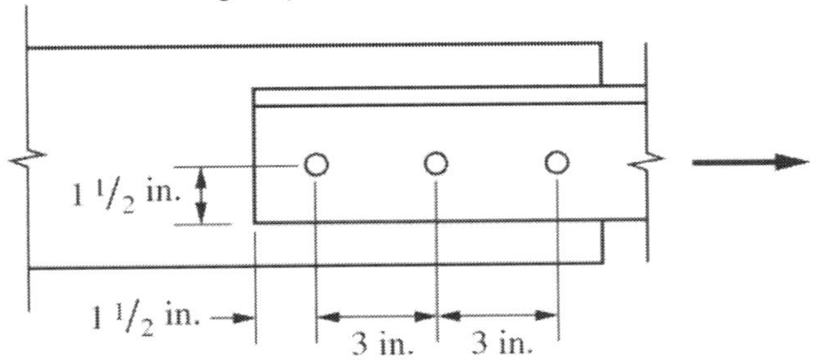

P10.28

29. Determine the available strength of the three 1 in. A325-N bolts in the single L4×3×3/8 A572 Gr. 50 when the bolts are placed as shown in Figure P10.28. Determine (a) design strength by LRFD and (b) allowable strength by ASD.

30. Determine the available strength of the six 3/4 in. A325-N bolts in a 7×1/2 in. A572 Gr. 50 plate when the bolts are placed as shown in Figure P10.30. Determine (a) design strength by LRFD and (b) allowable strength by ASD.

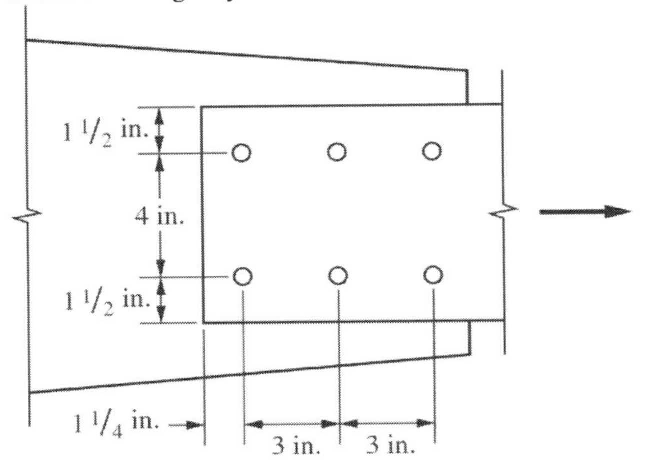

P10.30

31. Determine the available strength of the six 7/8 in. A325-N bolts in a 7×3/4 in. A36 plate when the bolts are placed as shown in Figure P10.30. Determine (a) design strength by LRFD and (b) allowable strength by ASD.

32. Determine the available strength of the six 3/4 in. A325-X bolts in a 7×3/4 in. A572 Gr. 50 plate when the bolts are placed as shown in Figure P10.30. Determine (a) design strength by LRFD and (b) allowable strength by ASD.

33. Determine the available strength of the eight 3/4 in. A490-N bolts in an A992 WT6×17.5 when the bolts are placed as shown in Figure P10.33 with a gage, g = 3.5 in.. Determine (a) design strength by LRFD and (b) allowable strength by ASD.

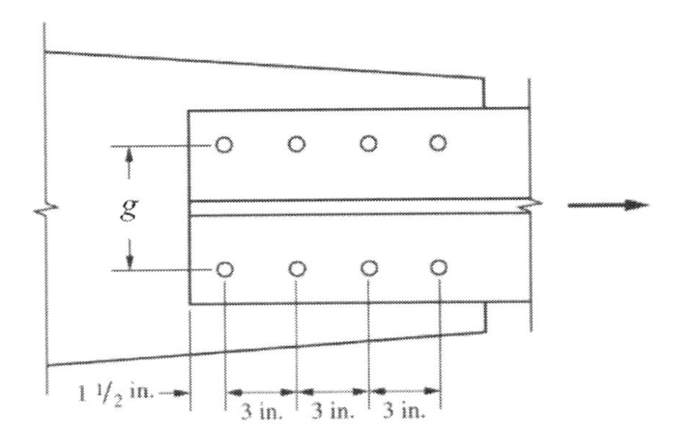

P10.33

34. Determine the available strength of the eight 7/8 in. A325-N bolts in an A992 WT5×15 when the bolts are placed as shown in Figure P10.33 with a gage, g = 2.75 in.. Determine (a) design strength by LRFD and (b) allowable strength by ASD.

35. Determine the available strength of the eight 1 in. A325-X bolts in an A992 WT12×31 when the bolts are placed as shown in Figure P10.33 with a gage, g = 3.5 in.. Determine (a) design strength by LRFD and (b) allowable strength by ASD.

36. Determine the available strength of two 3/8 in. fillet welds that are loaded parallel to their length, are 10 in. long, and are made from E70 electrodes. Determine (a) design strength by LRFD and (b) allowable strength by ASD.

37. Determine the available strength of two 1/2 in. fillet welds that are loaded parallel to their length, are 12 in. long, and are made from E70 electrodes.

Determine (a) design strength by LRFD and (b) allowable strength by ASD.

38. Determine the available strength of two 1/4 in. fillet welds that are loaded parallel to their length, are 8 in. long, and are made from E70 electrodes. Determine (a) design strength by LRFD and (b) allowable strength by ASD.

39. If the welds of Problem 36 were loaded at their centroid and at 90 degrees to the weld length, determine (a) design strength by LRFD and (b) allowable strength by ASD.

40. If the welds of Problem 36 were loaded at their centroid and at 45 degrees to the weld length, determine (a) design strength by LRFD and (b) allowable strength by ASD.

41. If the welds of Problem 38 were loaded at their centroid and at 80 degrees to the weld length, determine (a) design strength by LRFD and (b) allowable strength by ASD.

42. If the welds of Problem 38 were loaded at their centroid and at 35 degrees to the weld length, determine (a) design strength by LRFD and (b) allowable strength by ASD.

43. Plot a curve of the strength versus angle of loading for a 12 in. long 5/16 in. weld made from E70 electrodes. The plot should include angles from 0 degrees to 90 degrees.

44. Three 5/16 in. welds are grouped to form a C and are loaded at their centroid. Determine the available weld strength if the single transverse weld is 9 in. and the two longitudinal welds are each 3 in. Use E70 electrodes. Determine (a) design strength by LRFD and (b) allowable strength by ASD.

45. Repeat Problem 44 with the transverse weld at 3 in. and the two longitudinal welds at 9 in. each. Determine (a) design strength by LRFD and (b) allowable strength by ASD.

46. Three 3/8 in. welds are grouped to form a C and are loaded at their centroid. Determine the available weld strength if the single transverse weld is 8 in. and the two longitudinal welds are each 8 in. Use E70 electrodes. Determine (a) design strength by LRFD and (b) allowable strength by ASD.

47. Determine the available block shear strength for a coped W16×26 A992 beam with holes for 3/4 in. bolts as shown in Figure P10.47. Determine (a) design strength by LRFD and (b) allowable strength by ASD.

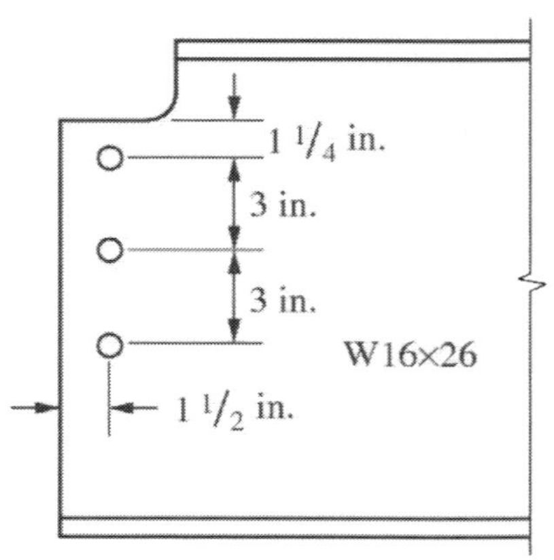

P10.47

48. Determine the available block shear strength for a coped W21×182 A992 beam with holes for 3/4 in. bolts as shown in Figure P10.48. Determine (a) design strength by LRFD and (b) allowable strength by ASD.

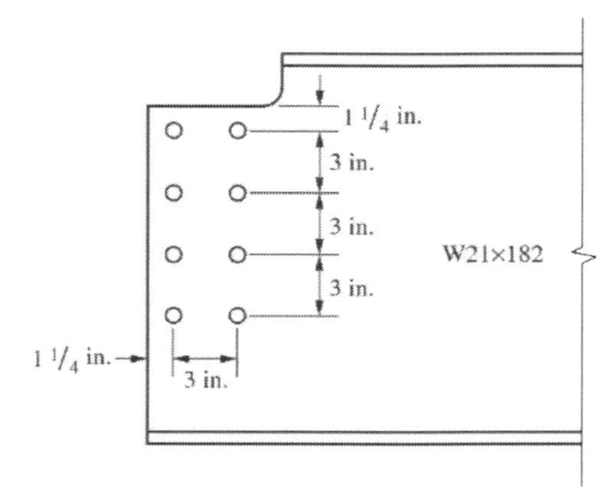

P10.48

49. Determine the available block shear strength for a coped W24×146 A992 beam with holes for 7/8 in. bolts as shown in Figure P10.49. Determine (a) design strength by LRFD and (b) allowable strength by ASD.

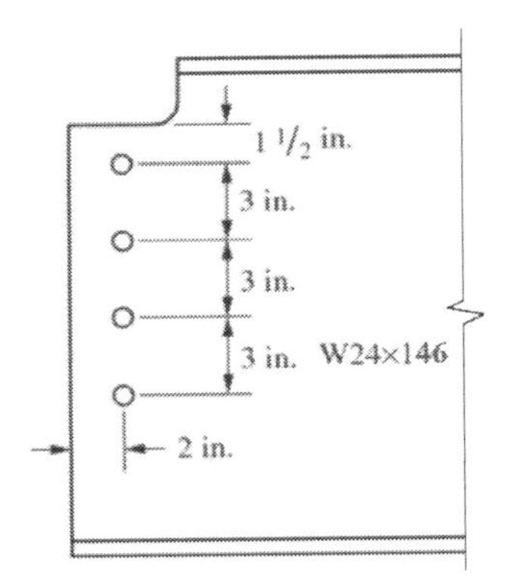

P10.49

50. Determine the available block shear strength for a coped W27×194 A992 beam with holes for 7/8 in. bolts as shown in Figure P10.50. Determine (a) design strength by LRFD and (b) allowable strength by ASD.

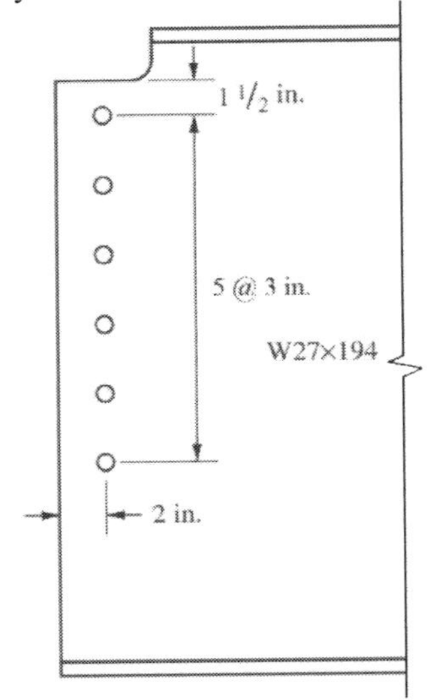

P10.50

51. Determine the available block shear strength for a coped W30×261 A992 beam with holes for 7/8 in. bolts as shown in Figure P10.51. Determine (a) design strength by LRFD and (b) allowable strength by ASD.

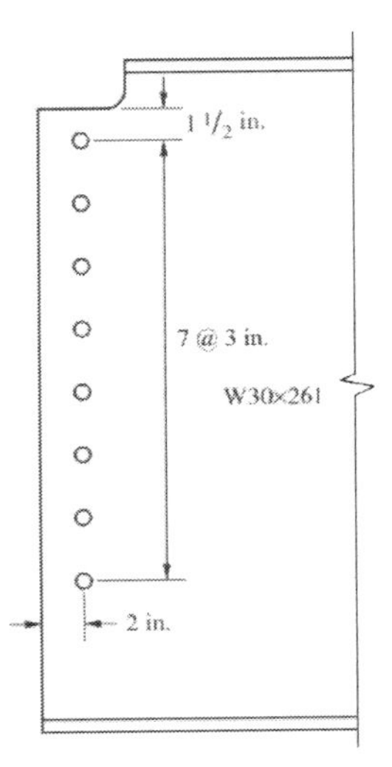

P10.51

52. Determine the available block shear strength for a coped W30×261 A992 beam with holes for 3/4 in. bolts as shown in Figure P10.52. Determine (a) design strength by LRFD and (b) allowable strength by ASD.

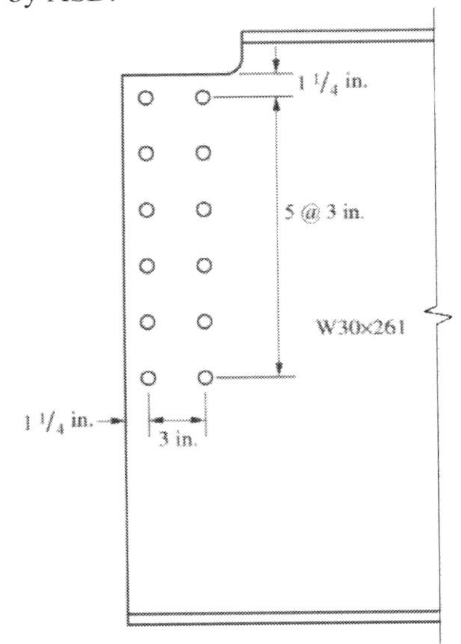

P10.52

53. Determine the available block shear strength for a coped W44×230 A992 beam with holes for 12 - 1 in. bolts spaced at 3.0 in., with horizontal edge distance of 2.0 in. and located 1-1/2 in. down from the cope. Determine (a) design strength by LRFD and (b) allowable strength by ASD.

Chapter 11

Simple Connections

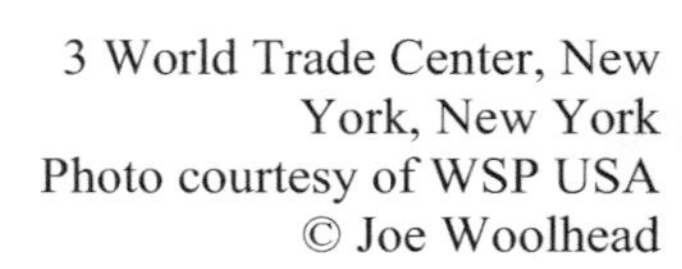

3 World Trade Center, New York, New York
Photo courtesy of WSP USA
© Joe Woolhead

11.1 TYPES OF SIMPLE CONNECTIONS

This chapter addresses two types of simple connections, beam shear connections and bracing connections. Both are commonly designed as pinned joints – simple, in the terminology of the AISC Specification. The connecting elements and the connectors required for these connections have already been discussed in Chapter 10. The limit states that control the connection have also been discussed individually, although their link to connection design may not yet be completely clear. Connection design is a combination of element and connector selection with a checking of all appropriate limit states. The goal is to obtain a connection with sufficient strength and the appropriate stiffness to carry the load in a manner consistent with the model used in the structural analysis. In addition to these simple shear connections, beam bearing plates and column base plates will be discussed.

The limit states to be considered for a particular connection depend on the connection elements, the connection geometry, and the load path. They will be identified in the following sections as each connection type is considered. A summary of the potential limit states at this time, however, may prove useful. For bolts, the limit states of tensile rupture, shear rupture, bearing and tearout, as well as slip, will be considered. For welds, the only limit state to be considered is shear rupture, although weld group geometry will add some complexity to that consideration. For connecting elements, the limit states are tension yielding and rupture, compression yielding and buckling, shear yielding and rupture, block shear rupture, and the full range of flexural limit states.

Table 11.1 lists the sections of the *Specification* and parts of the *Manual* discussed in this chapter.

11.2 SIMPLE SHEAR CONNECTIONS

A significant number of potential connection geometries are associated with the various types of members to be connected. Five of the most commonly used simple shear connections are described in the following sections with design examples following. These connections are shown in Figure 11.1a through e as double-angle, single-angle, single-plate (commonly called a shear tab), unstiffened seated, and stiffened seated connections. Part 10 of the *Manual* includes many tables that can simplify connection design; however, the examples presented here show the required calculations to improve understanding of the tabular values and to illustrate the calculations for cases that are not tabulated. Once a calculation has been sufficiently demonstrated, the *Manual* tables can be used. Additional examples of connection design may be found in the Steel Construction Manual Companion which includes Design Examples and Design Tables not included in the printed *Manual*.

Table 11.1 Sections of *Specification* and Parts of *Manual* Covered in This Chapter

	Specification
B4.3	Gross and Net Area Determination
J2	Welds and Welded Joints
J3	Bolts, Threaded Parts, and Bolted Connections
J4	Affected Elements of Members and Connecting Elements
J10	Flanges and Webs with Concentrated Forces
	Manual
Part 3	Design of Flexural Members
Part 7	Design Considerations for Bolts
Part 8	Design Considerations for Welds
Part 9	Design of Connecting Elements
Part 10	Design of Simple Shear Connections
Part 14	Design of Beam Bearing Plates, Column Base Plates, Anchor Rods, and Column Splices
Part 15	Design of Hanger Connections, Bracket Plates, and Crane-Rail Connections

Several design considerations apply to all of the shear connections to be discussed, and in some cases to other types of connections. It is helpful to address these before dealing with the specific connection. The first issue to consider is the location of the hinge within the connecting elements. It is critical that this hinge can actually occur in the real connection, because the analytical model of the connection assumes it behaves as a hinge or pin. The location of the hinge determines what forces and moments, if any, the individual elements must be designed for. Although these are simple shear connections, some elements may be required to resist moments, depending on an assumed eccentricity of forces in these elements. In all cases, the hinge is located at the most flexible point within the connection. This may be at the face of the supporting member or at some other point within the connection. Where eccentricities must be considered, they will be discussed in the examples. Several general design guidelines help to ensure that the connection behaves as desired. In most cases, this means that the hinge is located at the face of the supporting member.

For double-angle connections, angle thickness should be limited to a maximum of 5/8 in. The bolts in the outstanding legs, those connecting to the supporting member, are spaced at a gage that is appropriate for the angle size and to allow for the necessary flexibility; for welded outstanding legs, the vertical welds are spaced by the angle size, which is chosen to allow for the necessary flexibility. These characteristics will ensure that the connection behaves as a simple connection through bending of the outstanding legs.

For simple beam connections, the permitted tolerance for beam length must be considered. Although this tolerance is not normally a consideration for member design, it becomes important when the details of connecting members are considered. Beam length tolerance is $\pm 1/4$ in. To accommodate fit-up, beams are held back 1/2 in. from the face of the supporting member. Then, when considering the edge distance from a bolt hole to the end of the member, the distance used in calculations should be taken as 1/4 in. less than that actually detailed in cases where an under run in beam length might yield a lower strength.

It is also helpful to remember the considerations for hole sizes. First, standard holes are sized 1/16 in. larger than the bolt to be inserted for bolts up to 7/8 in. and 1/8 in. larger for 1 in. and larger bolts. Then, when considering net sections for the limit states of tension rupture or shear rupture, *Specification* Section B4.3 requires that an additional 1/16 in. be deducted to account for any material damage resulting from the process of making the hole. When a clear distance is calculated for the limit state of tearout, the actual hole size is used.

These design considerations are included in the examples to follow.

11.3 DOUBLE-ANGLE CONNECTIONS: BOLTED-BOLTED

A double-angle shear connection, as shown in Figure 11.1a, is perhaps the most common simple shear connection used in steel construction. It is a fairly simple connection to fabricate and also provides for fairly easy erection. When double-angle connections are to be installed back-to-back, there may be some problems, particularly when the supporting member is a column web. In the case of attachment to a column web, the safety requirements of OSHA call for special attention because the connection itself doesn't provide a way to support the first beam erected while the crane is unhooked from it to lift the second beam into place. One solution is to stagger the double angles vertically to allow two holes that are not shared by both connections. This connection can easily accommodate variations in beam length within acceptable tolerances.

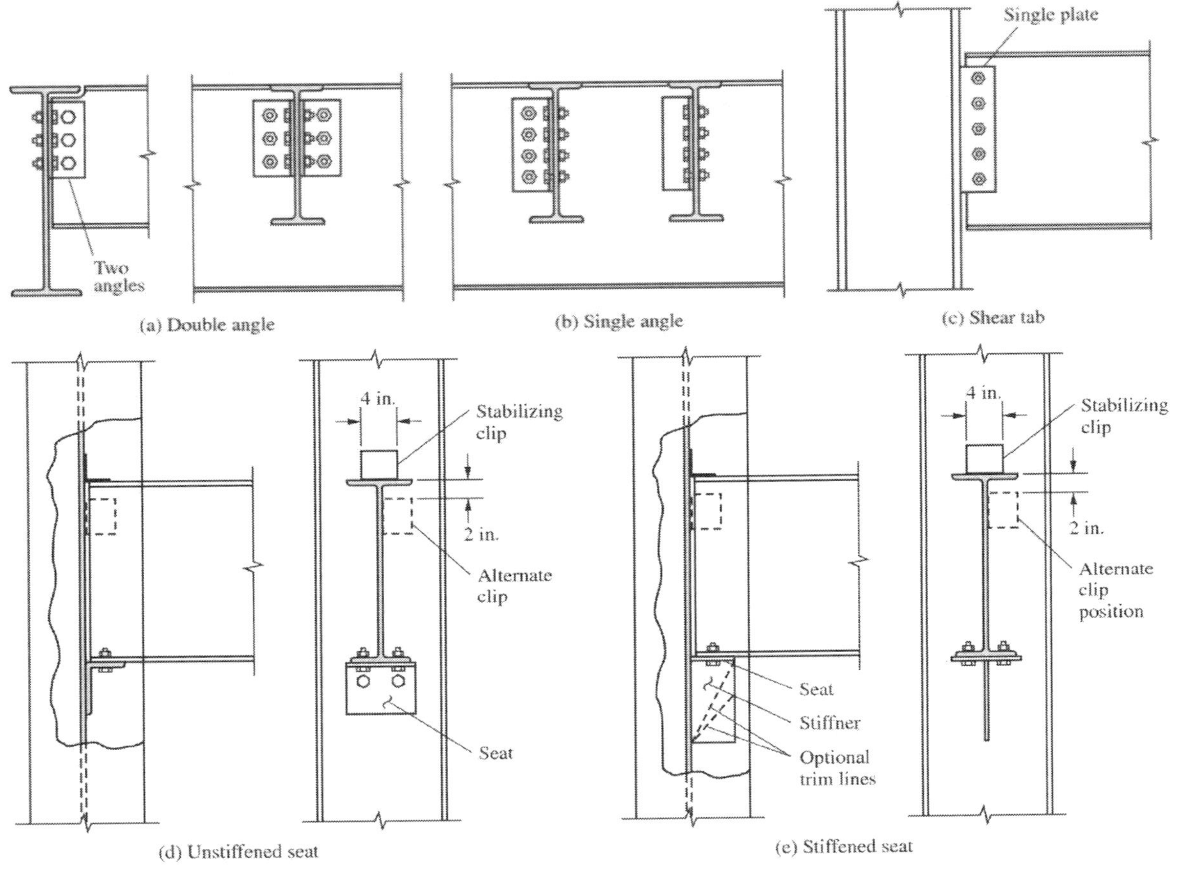

Figure 11.1 Simple Shear Connections

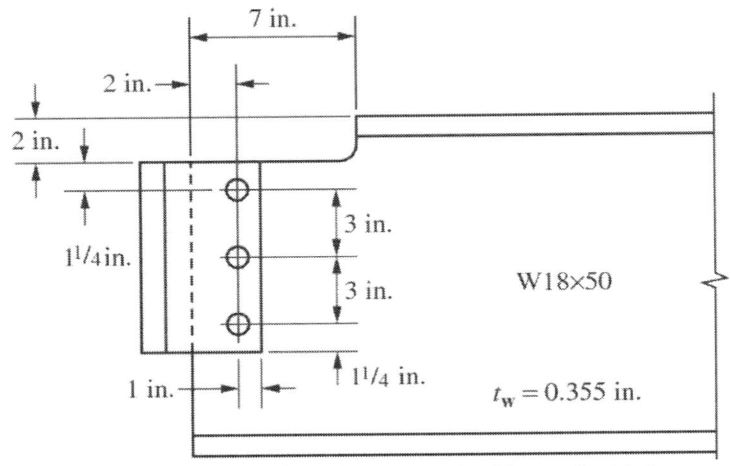

Figure 11.2 Connection Geometry for Example 11.1

The double-angle shear connection must be checked for the following limit states, grouped according to the elements that make up the connection: bolts, beam web, angles, and supporting member:

1. Bolts
   a. Shear rupture
2. Beam
   a. Bolt bearing and tearout on beam web
   b. Shear yielding of the web
   c. Block shear on coped beam web
   d. Coped beam flexural strength
3. Angles
   a. Bolt bearing and tearout on angles
   b. Shear rupture
   c. Shear yielding
   d. Block shear
4. Supporting member
   a. Bolt bearing

Each of these limit states has been addressed in this book. In the examples that follow, these limit state checks are combined into a complete connection design.

EXAMPLE 11.1a
Bolted-Bolted Double-Angle Shear Connection by LRFD

Goal: Design a bolted-bolted double-angle shear connection for a W18×50 beam.

Given: The W18×50 beam is connected to the web of a W24×207 and the connection must provide a required strength R_u = 83.0 kips. The beam and column are A992 and the 3-1/2×3-1/2×3/8 angles are A572 Gr. 50. The beam flange is coped 2 in. Use 7/8 in. A325-N bolts in standard holes in the legs on the beam web and short slots on the outstanding legs. The basic starting geometry is given in Figure 11.2.

SOLUTION

Bolt Shear

Step 1: Determine the number of bolts required based on the shear rupture strength of the bolts.

From *Manual* Table 7-1, the design shear strength per bolt is

$$\phi r_n = 24.3 \text{ kips}$$

Because the bolts are in double shear, n_s = 2 and the design shear strength of a single bolt is

$$n_s \phi r_n = 2(24.3) = 48.6 \text{ kips}$$

Values for double shear are also found in *Manual* Table 7-1. Because of rounding, the number in the table is slightly greater at 48.7 kips. The total number of bolts required is

$$N = \frac{R_u}{n_s(\phi r_n)} = \frac{83.0}{48.6} = 1.71$$

Therefore, based only on the limit state of bolt shear, try two bolts.

Bolt Strength at Holes in Beam Web

Step 2: Determine bolt bearing and tearout strength on the beam web.

For the W18×50 beam, $t_w = 0.355$ in.

For the two-bolt connection, the top bolt is 1.25 in. from the edge of the beam cope and the second bolt is spaced 3.0 in. from the first.

For the top bolt, the clear distance is

$$l_c = l_{cv} - \frac{1}{2}d_h = 1.25 - \frac{1}{2}(7/8 + 1/16) = 0.781 < 2(7/8) = 1.75$$

Since the clear distance is less than $2d_b$, tearout controls over bearing. The nominal bolt strength for tearout is

$$R_n = 1.2l_c tF_u = 1.2(0.781)(0.355)(65) = 21.6 \text{ kips}$$

and the design strength is

$$\phi R_n = 0.75(21.6) = 16.2 \text{ kips}$$

Since this is less than the design shear strength of the bolt in double shear it will control the strength of this bolt.

For the second bolt

$$l_c = s - d_h = 3.0 - (7/8 + 1/16) = 2.06 > 2(7/8) = 1.75 \text{ in.}$$

Since the clear distance is greater than $2d_b$, bearing controls over tearout. Thus, the nominal bolt bearing strength is

$$R_n = 2.4dtF_u = 2.4(7/8)(0.355)(65) = 48.5 \text{ kips}$$

and the design strength for bearing on the beam web is

$$\phi R_n = 0.75(48.5) = 36.4 \text{ kips}$$

Since this is less than the shear strength of the bolt in double shear, it will control the strength of this bolt.

Therefore, the connection design strength, if only considering the beam web, is based on one bolt limited by tearout and one bolt limited by bearing. Thus,

$$\phi R_n = (16.2 + 36.4) = 52.6 < 83.0 \text{ kips}$$

and the two-bolt connection cannot support the load.

Step 3: Determine the number of bolts required considering bearing and tearout on the web.

Adding a third bolt spaced at 3.0 in., which means that bearing controls over tearout, gives a connection design strength for bolt bearing and tearout of

$$\phi R_n = (16.2 + 2(36.4)) = 89.0 > 83.0 \text{ kips}$$

Therefore a three bolt connection will be considered further.

Bolt Strength at Holes in Angle Leg

Step 4: Determine bolt bearing and tearout strength on the A572 Gr. 50 angle. The angle legs on the beam and on the supporting member are identical and each angle takes one half of the load.

For the bottom edge bolt, the clear distance is

$$l_c = l_{ev} - \frac{1}{2}d_h = 1.25 - \frac{1}{2}(7/8 + 1/16) = 0.781 < 2(7/8) = 1.75 \text{ in.}$$

Since the clear distance is less than $2d_b$, tearout controls over bearing. The nominal bolt strength for tearout is

$$R_n = 1.2 l_c t F_u = 1.2(0.781)(3/8)(65) = 22.8 \text{ kips}$$

and the design strength is

$$\phi R_n = 0.75(22.8) = 17.1 \text{ kips}$$

Since this is less than the design shear strength of the bolt in single shear, it will control the strength of this bolt

For the second and third bolts, the clear distance is

$$l_c = s - d_h = 3.0 - (7/8 + 1/16) = 2.06 > 2(7/8) = 1.75 \text{ in.}$$

Since the clear distance is greater than $2d_b$, bearing controls over tearout. The nominal bolt strength for bearing is

$$R_n = 2.4 d t F_u = 2.4(7/8)(3/8)(65) = 51.2 \text{ kips}$$

and the design strength for bearing is

$$\phi R_n = 0.75(51.2) = 38.4 \text{ kips}$$

Since this is greater than the design shear strength of the bolt in single shear, bolt shear will control the strength of these two bolts.

Connection Strength at Beam

Step 5: Thus far we have looked at the various limit states in a particular element. However, the controlling strength for each bolt on the beam must be considered

Top bolt	
Bolt double shear	48.6 kips
Web tearout	16.2 kips
Angle bearing ×2	76.8 kips
Middle bolt	
Bolt double shear	48.6 kips
Web bearing	36.4 kips
Angle bearing ×2	76.8 kips
Bottom bolt	
Bolt double shear	48.6 kips
Web bearing	36.4 kips
Angle tearout ×2	34.2 kips

Therefore, the strength of the connection at the supported beam is the sum of web tearout for the top bolt plus web bearing for the middle bolt plus angle tearout for the bottom bolt. Thus,

$$\phi R_n = 16.2 + 36.4 + 34.2 = 86.8 > 83.0 \text{ kips}$$

Bolt Strength at Holes in Supporting Member

Step 6: Consider the outstanding legs of the angles attached to the supporting member.

In this case, the bolts are in single shear but there are twice as many bolts, so the load per bolt is half of the load for each of the bolts in the beam web. If the supporting member thickness is at least one-half of the beam web thickness and the strengths are the same, the bolts in the supporting member will be satisfactory. For the W24×207, clear distances result in all three bolts being controlled by bearing over tearout. Thus, with $t_w = 0.870$ in.,

$$R_n = 2.4dtF_u = 2.4(7/8)(0.870)(65) = 119 \text{ kips}$$

and the design strength is

$$\phi R_n = 0.75(119) = 89.3 \text{ kips}$$

Connection Strength at Supporting Member

Step 7: Determine the controlling strength for each bolt at the supporting member

Top bolt		
	Bolt single shear	24.3 kips
	Web bearing	89.3 kips
	Angle tearout	17.1 kips
Middle bolt		
	Bolt single shear	24.3 kips
	Web bearing	89.3 kips
	Angle bearing	38.4 kips
Bottom bolt		
	Bolt single shear	24.3 kips
	Web bearing	89.3 kips
	Angle bearing	38.4 kips

Therefore, the strength of the connection at the supporting member is the sum of angle tearout for the top bolt plus bolt single shear for the other two bolts. Thus,

$$\phi R_n = 17.1 + 24.3 + 24.3 = 65.7 > 83.0/2 = 41.5 \text{ kips}$$

Step 8: Evaluate the minimum depth of the connection.

The beam web connection should be at least half the depth of the beam web, measured as the distance between the fillets, T, given in *Manual* Table 1-1. This requirement is to prevent twisting of the simple supports. For W18×50, T = 15-1/2 in., so the minimum angle depth should be 7-3/4 in. Thus, the 8-1/2 in. long angle will provide an acceptable connection depth and will also fit within the 15.5 in. available flat length.

Remaining Beam Limit States

Step 9: Check shear yield of the beam web.

Note that this is a check that should be carried out during the beam design process. At the point of connection design it is too late to find out that the beam will not be adequate. Nonetheless, it is good practice to verify it in the connection design process. From *Manual* Table 3-2,

$$\phi V_n = 192 \text{ kips} > 83.0 \text{ kips}$$

Step 10: Check shear yield of the beam web at the cope

This check is carried out according to Section J4.2 where the full area of the remaining web is stressed to F_y and ϕ=1.0. Thus

$$\begin{aligned}\phi V_n &= \phi 0.6 F_y h t_w \\ &= 1.0(0.6(50))(18.0-2.0)(0.355) \\ &= 170 \text{ kips } > 83.0 \text{ kips}\end{aligned}$$

Therefore, shear yield strength at the cope is adequate

Step 11: Check shear rupture of the beam web at the cope

This check is carried out according to Section J4.2 where the effective area, A_e, the full area of the remaining web less the bolt holes, is stressed to F_u and ϕ=0.75. Thus

$$\begin{aligned}\phi V_n &= \phi 0.6 F_u A_e \\ &= 0.75(0.6(65))\left(18.0-2.0-3\left(\frac{7}{8}+\frac{1}{8}\right)\right)(0.355) \\ &= 135 \text{ kips } > 83.0 \text{ kips}\end{aligned}$$

Therefore, shear rupture at the cope is adequate

Step 12: Check the beam web for block shear.

The equations for block shear rupture are found in *Specification* Section J4.3 and were presented in Section 10.10.5.

First calculate the areas, remembering to account for the 1/4 in. beam length under run tolerance in the tension area calculation, so that l_{eh} = 2.0 – 0.25 = 1.75 in.:

$$\begin{aligned}A_{nt} &= \left(l_{eh} - \frac{1}{2}(d_h + 1/16)\right)t_w \\ &= \left(1.75 - \frac{1}{2}(7/8 + 1/8)\right)(0.355) = 0.444 \text{ in.}^2\end{aligned}$$

$$\begin{aligned}A_{gv} &= l t_w \\ &= 7.25(0.355) = 2.57 \text{ in.}^2\end{aligned}$$

$$\begin{aligned}A_{nv} &= \left(l - (n-0.5)(d_h + 1/16)\right)t_w \\ &= \left(7.25 - 2.5(7/8 + 1/8)\right)(0.355) = 1.69 \text{ in.}^2\end{aligned}$$

Determine the tension rupture strength

$$F_u A_{nt} = 65(0.444) = 28.9 \text{ kips}$$

Consider shear yield and shear rupture and select the least nominal strength; thus,

$$0.6F_yA_{gv} = 0.6(50)(2.57) = 77.1 \text{ kips}$$

$$0.6F_uA_{nv} = 0.6(65)(1.69) = 65.9 \text{ kips}$$

Selecting the shear rupture term because it is smaller and combining it with the tension rupture term gives the connection block shear design strength, and recalling that U_{bs} = 1.0 for the case of uniform tensile stress distribution, we have

$$\phi R_n = 0.75(65.9 + 1.0(28.9)) = 0.75(94.8) = 71.1 < 83.0 \text{ kips}$$

Thus, the given three-bolt connection is not adequate, with block shear being the critical limit state to this point in our calculations.

Step 13: Revise the connection to meet the block shear strength requirements.

Consideration could be given to increasing the number of bolts and thereby increasing the length of the connection. However, because bolt shear required only two bolts, this might not be a particularly economical solution. If the connection were to be lowered on the beam end so that the distance from the center of the top bolt to the edge of the cope were 2.5 in., the connection would have more block shear strength in the beam web. It should also be noted that the clear end distance for the top bolt in the beam web would increase and thus increase the tearout strength of that top bolt. There is no need to consider that increase since at best it could contribute to reducing the number of bolts and this would have a negative impact on the block shear calculations being addressed in this step.

Thus, the new shear areas, with the distance from the bottom bolt to the beam cope of 8.5 in., become

$$A_{gv} = lt_w$$

$$= 8.50(0.355) = 3.02 \text{ in.}^2$$

$$A_{nv} = (l - (n - 0.5)(d_h + 1/16))t_w$$

$$= (8.50 - 2.5(7/8 + 1/8))(0.355) = 2.13 \text{ in.}^2$$

and the nominal shear yield and rupture strengths become

$$0.6F_yA_{gv} = 0.6(50)(3.02) = 90.6 \text{ kips}$$

$$0.6F_uA_{nv} = 0.6(65)(2.13) = 83.1 \text{ kips}$$

Selecting the shear rupture term because it is smaller and combining it with the tension rupture term gives the resulting block shear design strength as

$$\phi R_n = 0.75\left(83.1 + 1.0\left(28.9\right)\right) = 0.75\left(112\right) = 84.0 > 83.0 \text{ kips}$$

Step 14: Check the flexural strength of the coped beam.

It is a good idea to check this limit state during the initial design of the beam. It should be anticipated during the beam design stage that a coped connection will be required, and it is at that stage that a change in beam section can most readily be accommodated.

Flexural strength of the coped beam is not addressed in the *Specification* directly but is covered in Part 9 of the *Manual*. The moment in the coped beam is taken as the shear force times the eccentricity from the face of the support to the edge of the cope, taken as cope + setback = 7.0 + 0.50 = 7.5 in. in this example. Thus,

$$M_u = R_u e = 83.0\left(7.5\right) = 623 \text{ in.-kips} \qquad (\textit{Manual } 9\text{-}9)$$

To determine the flexural strength of the coped beam, the web slenderness must be checked. This requires determination of the plate buckling coefficient, k, and the buckling adjustment factor, f. The web slenderness is a function of the beam depth less the cope depth, h_c = 18.0-2.0=16.0 in. The web slenderness of this coped beam is

$$\lambda = h_c / t_w = 16.0/0.355 = 45.1$$

The limiting web slenderness is

$$\lambda_p = 0.475\sqrt{\frac{k_1 E}{F_y}} \qquad (\textit{Manual } 9\text{-}18)$$

where

$$k_1 = fk \qquad (\textit{Manual } 9\text{-}14)$$

The buckling adjustment factor, f, is a function of the ratio of the cope length, c, to the beam depth, d, and the plate buckling coefficient, k, is a function of the ratio of cope length, c, to the reduced beam depth, h_c. For this example, with $c/d = 7.0/18.0 = 0.389 < 1.0$

$$f = \frac{2c}{d} = \frac{2(7.0)}{18.0} = 0.778 \qquad (\textit{Manual } 9\text{-}20\text{a})$$

and with $c/h_c = 7.0/16.0 = 0.438 \leq 1.0$

$$k = 2.2\left(\frac{h_c}{c}\right)^{1.65} = 2.2\left(\frac{16.0}{7.0}\right)^{1.65} = 8.61 \quad (\textit{Manual } 9\text{-}19\text{a})$$

Thus,

$$k_1 = fk = 0.778(8.61) = 6.70 \geq 1.61 \qquad (\textit{Manual } 9\text{-}14)$$

and

$$\lambda_p = 0.475\sqrt{\frac{k_1 E}{F_y}} = 0.475\sqrt{\frac{6.70(29{,}000)}{50}} = 29.6$$

Since λ = 45.1 is between λ_p and $2\lambda_p$ the net elastic and net plastic section moduli will be needed. The net elastic section modulus is taken from *Manual* Table 9-2a. With the depth of the cope, d_c = 2.0 in.,

$$S_c = 23.4 \text{ in.}^3$$

The net plastic section modulus is provided in the *Manual* Table 9.2b. With the depth of the cope, d_c = 2.0 in.,

$$Z_c = 42.5 \text{ in.}^3$$

and from

$$M_n = M_p - (M_p - M_y)\left(\frac{\lambda}{\lambda_p} - 1\right) \qquad (\textit{Manual } 9\text{-}11)$$

with

$$M_p = F_y Z_{net} = 50(42.5) = 2130 \text{ in.-kips}$$

$$M_y = F_y S_{net} = 50(23.4) = 1170 \text{ in-kips}$$

The nominal moment strength at the cope is

$$M_n = 2130 - (2130 - 1170)\left(\frac{45.1}{29.6} - 1\right) = 1630 \text{ in.-kips}$$

and the design strength is

$$\phi M_n = 0.9(1630) = 1470 \text{ in.-kips} > 623 \text{ in.-kips}$$

So the coped beam has sufficient flexural strength.

Angles

Step 15: Check the angles for shear rupture using Equation J4-4.

The net area of the angle on the vertical shear plane is

$$A_{nv} = \left(l - n\left(d_h + 1/16\right)\right)t_a = \left(8.5 - 3\left(7/8 + 1/8\right)\right)\left(3/8\right) = 2.06 \text{ in.}^2$$

and the design shear rupture strength for two angles is

$$\phi V_n = 2\left(\phi 0.6 F_u A_{nv}\right) = 2\left(0.75\left(0.6\left(65\right)\left(2.06\right)\right)\right) = 121 \text{ kips} > 83.0 \text{ kips}$$

So the angles are adequate for shear rupture.

Step 16: Check the angles for shear yield using Equation J4-3.

The gross area of the angle on the vertical shear plane is

$$A_{gv} = 8.5\left(3/8\right) = 3.19 \text{ in.}^2$$

and the design shear yield strength for two angles is

$$\phi V_n = 2\left(\phi 0.6 F_y A_{gv}\right) = 2\left(1.0\left(0.6\left(50\right)\left(3.19\right)\right)\right) = 191 \text{ kips} > 83.0 \text{ kips}$$

So the angles are also adequate for shear yield.

Step 17 Check the angles for block shear.

The equations for block shear in the angle are the same as those for the web and as presented in Section 10.10.5.

First calculate the areas,

$$\begin{aligned}
A_{nt} &= \left(l_{eh} - \frac{1}{2}\left(d_h + 1/16\right)\right)t_w \\
&= \left(1.0 - \frac{1}{2}\left(7/8 + 1/8\right)\right)\left(3/8\right) = 0.188 \text{ in.}^2 \\
A_{gv} &= l t_w \\
&= 7.25\left(3/8\right) = 2.72 \text{ in.}^2 \\
A_{nv} &= \left(l - \left(n - 0.5\right)\left(d_h + 1/16\right)\right)t_w \\
&= \left(7.25 - 2.5\left(7/8 + 1/8\right)\right)\left(3/8\right) = 1.78 \text{ in.}^2
\end{aligned}$$

Determine the tension rupture strength

$$F_u A_{nt} = 65\left(0.188\right) = 12.2 \text{ kips}$$

Consider shear yield and shear rupture, and select the least nominal strength; thus,

$$0.6F_yA_{gv} = 0.6(50)(2.72) = 81.6 \text{ kips}$$

$$0.6F_uA_{nv} = 0.6(65)(1.78) = 69.4 \text{ kips}$$

Selecting the shear rupture term because it is smaller and combining it with the tension rupture term gives a connection block shear design strength of the double angle—again, U_{bs} = 1.0 for this case of uniform tensile stress distribution—of

$$\phi R_n = 2(0.75(69.4 + 12.2)) = 122 \text{ kips} > 83.0 \text{ kips}$$

Step 18: Present the final connection design.

The three-bolt connection, revised as shown in Figure 11.3, is adequate to carry the imposed load of 83.0 kips.

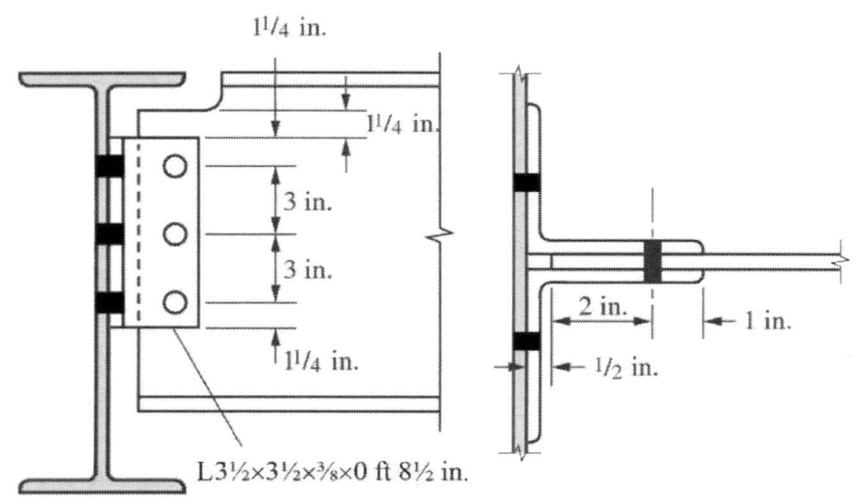

Figure 11.3 Final Connection Design for Example 11.1

EXAMPLE 11.1b
Bolted-Bolted Double-Angle Shear Connection by ASD

Goal: Design a bolted-bolted double-angle shear connection for a W18×50 beam.

Given: The W18×50 beam is connected to the web of a W24×207 and the connection must provide a required strength R_a = 55.0 kips. The beam and column are A992 and the 3-1/2×3-1/2×3/8 angles are A572 Gr. 50. The beam flange is coped 2 in. Use 7/8 in. A325-N bolts in standard holes in the legs on the beam web and short slots on the outstanding legs. The basic starting geometry is given in Figure 11.2.

SOLUTION

Bolt Shear

Step 1: Determine the number of bolts required based on the shear rupture strength of the bolts.

From *Manual* Table 7-1, the allowable shear strength per bolt is

$$r_n/\Omega = 16.2 \text{ kips}$$

Because the bolts are in double shear, $n_s = 2$ and the allowable shear strength of a single bolt is

$$n_s\, r_n/\Omega = 2(16.2) = 32.4 \text{ kips}$$

Values for double shear are also found in *Manual* Table 7-1. Because of rounding, the number in the table is slightly greater at 32.5 kips. The total number of bolts required is

$$N = \frac{R_a}{n_s(r_n/\Omega)} = \frac{55.0}{32.4} = 1.70$$

Therefore, based only on the limit state of bolt shear, try two bolts.

Bolt Strength at Holes in Beam Web

Step 2: Determine bolt bearing and tearout strength on the beam web.

For the W18×50 beam, $t_w = 0.355$ in.

For the two-bolt connection, the top bolt is 1.25 in. from the edge of the beam cope and the second bolt is spaced 3.0 in. from the first.

For the top bolt, the clear distance is

$$l_c = l_{cv} - \frac{1}{2}d_h = 1.25 - \frac{1}{2}(7/8 + 1/16) = 0.781 < 2(7/8) = 1.75$$

Since the clear distance is less than $2d_b$, tearout controls over bearing. The nominal bolt strength for tearout is

$$R_n = 1.2 l_c t F_u = 1.2(0.781)(0.355)(65) = 21.6 \text{ kips}$$

and the allowable strength is

$$R_n/\Omega = 21.6/2.00 = 10.8 \text{ kips}$$

Since this is less than the allowable shear strength of the bolt in double shear it will control the strength of this bolt.

For the second bolt

$$l_c = s - d_h = 3.0 - (7/8 + 1/16) = 2.06 > 2(7/8) = 1.75 \text{ in.}$$

Since the clear distance is greater than $2d_b$, bearing controls over tearout. Thus, the nominal bolt bearing strength is

$$R_n = 2.4dtF_u = 2.4(7/8)(0.355)(65) = 48.5 \text{ kips}$$

and the allowable strength for bearing on the beam web is

$$R_n/\Omega = 48.5/2.00 = 24.3 \text{ kips}$$

Since this is less than the shear strength of the bolt in double shear, it will control the strength of this bolt.

Therefore, the connection allowable strength, if only considering the beam web, is based on one bolt limited by tearout and one bolt limited by bearing. Thus,

$$R_n/\Omega = (10.8 + 24.3) = 35.1 < 55.0 \text{ kips}$$

and the two-bolt connection cannot support the load.

Step 3: Determine the number of bolts required considering bearing and tearout on the web.

Adding a third bolt spaced at 3.0 in., which means that bearing controls over tearout, gives a connection allowable strength for bolt bearing and tearout of

$$R_n/\Omega = (10.8 + 2(24.3)) = 59.4 > 55.0 \text{ kips}$$

Therefore, a three bolt connection will be considered further.

Bolt Strength at Holes in Angle Leg

Step 4: Determine bolt bearing and tearout strength on the A572 Gr. 50 angle. The angle legs on the beam and on the supporting member are identical and each angle takes one half of the load.

For the bottom edge bolt, the clear distance is

$$l_c = l_{ev} - \frac{1}{2}d_h = 1.25 - \frac{1}{2}(7/8 + 1/16) = 0.781 < 2(7/8) = 1.75 \text{ in.}$$

Since the clear distance is less than $2d_b$, tearout controls over bearing. The nominal bolt strength for tearout is

$$R_n = 1.2l_ctF_u = 1.2(0.781)(3/8)(65) = 22.8 \text{ kips}$$

and the allowable strength is

$$R_n/\Omega = 22.8/2.00 = 11.4 \text{ kips}$$

Since this is less than the allowable shear strength of the bolt in single shear, it will control the strength of this bolt

For the second and third bolts, the clear distance is

$$l_c = s - d_h = 3.0 - (7/8 + 1/16) = 2.06 > 2(7/8) = 1.75 \text{ in.}$$

Since the clear distance is greater than $2d_b$, bearing controls over tearout. The nominal bolt strength for bearing is

$$R_n = 2.4dtF_u = 2.4(7/8)(3/8)(65) = 51.2 \text{ kips}$$

and the allowable strength for bearing is

$$R_n/\Omega = 51.2/2.00 = 25.6 \text{ kips}$$

Since this is greater than the allowable shear strength of the bolt in single shear, bolt shear will control the strength of these two bolts

Connection Strength at Beam

Step 5: Thus far we have looked at the various limit states in a particular element. However, the controlling strength for each bolt on the beam must be considered.

Top bolt		
	Bolt double shear	32.4 kips
	Web tearout	10.8 kips
	Angle bearing ×2	51.2 kips
Middle bolt		
	Bolt double shear	32.4 kips
	Web bearing	24.3 kips
	Angle bearing ×2	51.2 kips
Bottom bolt		
	Bolt double shear	32.4 kips
	Web bearing	24.3 kips
	Angle tearout ×2	22.8 kips

Therefore, the allowable strength of the connection at the supported beam is the sum of web tearout for the top bolt plus web bearing for the middle bolt plus angle tearout for the bottom bolt. Thus,

$$R_n/\Omega = 10.8 + 24.3 + 22.8 = 57.9 > 55.0 \text{ kips}$$

Bolt Strength at Holes in Supporting Member

Step 6: Consider the outstanding legs of the angles attached to the supporting member.

In this case, the bolts are in single shear but there are twice as many bolts, so the load per bolt is half of the load for each of the bolts in the beam web. If the supporting member thickness is at least one-half of the beam web thickness and the strengths are the same, the bolts in the supporting member will be satisfactory. For the W24×207, clear distances result in all three bolts being controlled by bearing over tearout. Thus, with $t_w = 0.870$ in.,

$$R_n = 2.4dtF_u = 2.4(7/8)(0.870)(65) = 119 \text{ kips}$$

and the allowable strength is

$$R_n/\Omega = 119/2.00 = 59.5 \text{ kips}$$

Connection Strength at Supporting Member

Step 7: Determine the controlling strength for each bolt at the supporting member

Top bolt

Bolt single shear	16.2 kips
Web bearing	59.5 kips
Angle tearout	11.4 kips

Middle bolt

Bolt single shear	16.2 kips
Web bearing	59.5 kips
Angle bearing	25.6 kips

Bottom bolt

Bolt single shear	16.2 kips
Web bearing	59.5 kips
Angle bearing	25.6 kips

Therefore, the allowable strength of the connection at the beam web is

$$R_n/\Omega = 11.4 + 16.2 + 16.2 = 43.8 > 55.0/2 = 27.5 \text{ kips}$$

Step 8: Evaluate the minimum depth of the connection.

The beam web connection should be at least half the depth of the beam web, measured as the distance between the fillets, T, given in *Manual* Table 1-1. This requirement is to prevent twisting of the simple supports. For W18×50, T = 15-1/2 in., so the minimum angle depth should be 7-3/4 in. Thus, the 8-1/2 in. long angle will provide an acceptable connection depth and will also fit within the 15.5 in. available flat length.

Remaining Beam Limit States

Step 9: Check shear yield of the beam web.

Note that this is a check that should be carried out during the beam design process. At the point of connection design, it is too late to find out that the beam will not be adequate. Nonetheless, it is good practice to verify it in the connection design process. From *Manual* Table 3-2,

$$V_n/\Omega = 128 \text{ kips} > 55.0 \text{ kips}$$

Step 10: Check shear yield of the beam web at the cope

This check is carried out according to Section J4.2 where the full area of the remaining web is stressed to F_y and $\Omega = 1.5$. Thus

$$\begin{aligned} V_n/\Omega &= 0.6F_y h t_w/\Omega \\ &= (0.6(50))(18.0-2.0)(0.355)/1.50 \\ &= 114 \text{ kips} > 55.0 \text{ kips} \end{aligned}$$

Therefore, the allowable shear yield strength at the cope is adequate

Step 11: Check shear rupture of the beam web at the cope

This check is carried out according to Section J4.2 where the effective area, A_e, the full area of the remaining web less the bolt holes, is stressed to F_u and $\Omega = 2.0$. Thus

$$\begin{aligned} V_n/\Omega &= 0.6F_u A_e/\Omega \\ &= (0.6(65))\left(18.0-2.0-3\left(\frac{7}{8}+\frac{1}{8}\right)\right)(0.355)/2.00 \\ &= 90.0 \text{ kips} > 55.0 \text{ kips} \end{aligned}$$

Therefore, the allowable shear rupture strength at the cope is adequate

Step 12: Check the beam web for block shear.

The equations for block shear rupture are found in *Specification* Section J4.3 and were presented in Section 10.10.5.

First calculate the areas, remembering to account for the 1/4 in. beam length under run tolerance in the tension area calculation, so that $l_{eh} =$ $2.0 - 0.25 = 1.75$ in.:

$$\begin{aligned} A_{nt} &= \left(l_{eh} - \frac{1}{2}(d_h + 1/16)\right)t_w \\ &= \left(1.75 - \frac{1}{2}(7/8 + 1/8)\right)(0.355) = 0.444 \text{ in.}^2 \end{aligned}$$

$$A_{gv} = lt_w$$
$$= 7.25(0.355) = 2.57 \text{ in.}^2$$
$$A_{nv} = \left(l - (n - 0.5)(d_h + 1/16)\right)t_w$$
$$= \left(7.25 - 2.5(7/8 + 1/8)\right)(0.355) = 1.69 \text{ in.}^2$$

Determine the tension rupture strength.

$$F_u A_{nt} = 65(0.444) = 28.9 \text{ kips}$$

Consider shear yield and shear rupture and select the least nominal strength; thus,

$$0.6F_y A_{gv} = 0.6(50)(2.57) = 77.1 \text{ kips}$$
$$0.6F_u A_{nv} = 0.6(65)(1.69) = 65.9 \text{ kips}$$

Selecting the shear rupture term because it is smaller and combining it with the tension rupture term gives the connection block shear allowable strength, and recalling that U_{bs} = 1.0 for the case of uniform tensile stress distribution, we have

$$R_n/\Omega = \left(65.9 + 1.0(28.9)\right)/2.00 = 47.4 < 55.0 \text{ kips}$$

Thus, the given three-bolt connection is not adequate, with block shear being the critical limit state to this point in our calculations.

Step 13: Revise the connection to meet the block shear strength requirements.

Consideration could be given to increasing the number of bolts and thereby increasing the length of the connection. However, because bolt shear required only two bolts, this might not be a particularly economical solution. If the connection were to be lowered on the beam end so that the distance from the center of the top bolt to the edge of the cope were 2.5 in., the connection would have more block shear strength in the beam web. It should also be noted that the clear end distance for the top bolt in the beam web would increase and thus increase the tearout strength of that top bolt. There is no need to consider that increase since at best it could contribute to reducing the number of bolts and this would have a negative impact on the block shear calculations being addressed in this step.

Thus, the new shear areas, with the distance from the bottom bolt to the beam cope of 8.5 in., become

$$A_{gv} = lt_w$$
$$= 8.50(0.355) = 3.02 \text{ in.}^2$$

$$A_{nv} = \left(l - (n - 0.5)(d_h + 1/16)\right)t_w$$
$$= \left(8.50 - 2.5(7/8 + 1/8)\right)(0.355) = 2.13 \text{ in.}^2$$

and the nominal shear yield and rupture strengths become

$$0.6F_y A_{gv} = 0.6(50)(3.02) = 90.6 \text{ kips}$$
$$0.6F_u A_{nv} = 0.6(65)(2.13) = 83.1 \text{ kips}$$

Selecting the shear rupture term because it is smaller and combining it with the tension rupture term gives the resulting block shear allowable strength as

$$R_n/\Omega = \left(83.1 + 1.0(28.9)\right)/2.00 = 112/2.00 = 56.0 > 55.0 \text{ kips}$$

Step 14: Check the flexural strength of the coped beam.

It is a good idea to check this limit state during the initial design of the beam. It should be anticipated during the beam design stage that a coped connection will be required, and it is at that stage that a change in beam section can most readily be accommodated.

Flexural strength of the coped beam is not addressed in the *Specification* directly but is covered in Part 9 of the *Manual*. The moment in the coped beam is taken as the shear force times the eccentricity from the face of the support to the edge of the cope, taken as cope + setback = 7.0 + 0.50 = 7.5 in. in this example. Thus,

$$M_a = R_a e = 55.0(7.5) = 413 \text{ in.-kips} \qquad (\textit{Manual } 9\text{-}9)$$

To determine the flexural strength of the coped beam, the web slenderness must be checked. This requires determination of the plate buckling coefficient, k, and the buckling adjustment factor, f. The web slenderness is a function of the beam depth less the cope depth, h_c = 18.0 – 2.0 = 16.0 in. The web slenderness of this coped beam is

$$\lambda = h_c/t_w = 16.0/0.355 = 45.1$$

The limiting web slenderness is

$$\lambda_p = 0.475\sqrt{\frac{k_1 E}{F_y}} \qquad (\textit{Manual } 9\text{-}18)$$

where

$$k_1 = fk \qquad (\textit{Manual } 9\text{-}14)$$

The buckling adjustment factor, f, is a function of the ratio of the cope length, c, to the beam depth, d, and the plate buckling coefficient, k, is a function of the ratio of cope length, c, to the reduced beam depth, h_o. For this example, with $c/d = 7.0/18.0 = 0.389 \leq 1.0$

$$f = \frac{2c}{d} = \frac{2(7.0)}{18.0} = 0.778 \qquad \text{(\textit{Manual} 9-20a)}$$

and with $c/h_c = 7.0/16.0 = 0.438 \leq 1.0$

$$k = 2.2\left(\frac{h_c}{c}\right)^{1.65} = 2.2\left(\frac{16.0}{7.0}\right)^{1.65} = 8.61 \qquad \text{(\textit{Manual} 9-19a)}$$

Thus,

$$k_1 = fk = 0.778(8.61) = 6.70 \geq 1.61 \qquad \text{(\textit{Manual} 9-14)}$$

and

$$\lambda_p = 0.475\sqrt{\frac{k_1 E}{F_y}} = 0.475\sqrt{\frac{6.70(29{,}000)}{50}} = 29.6$$

Since λ = 45.1 is between λ_p and $2\lambda_p$ the net elastic and net plastic section modulus will be needed. The net elastic section modulus is taken from *Manual* Table 9-2a. With the depth of the cope, d_c = 2.0 in.,

$$S_c = 23.4 \text{ in.}^3$$

The net plastic section modulus is provided in the *Manual* Table 9.2b. With the depth of the cope, d_c = 2.0 in.,

$$Z_c = 42.5 \text{ in.}^3$$

and from

$$M_n = M_p - (M_p - M_y)\left(\frac{\lambda}{\lambda_p} - 1\right) \qquad \text{(\textit{Manual} 9-11)}$$

with

$$M_p = F_y Z_{net} = 50(42.5) = 2130 \text{ in.-kips}$$

$$M_y = F_y S_{net} = 50(23.4) = 1170 \text{ in-kips}$$

The nominal moment strength at the cope is

$$M_n = 2130 - (2130 - 1170)\left(\frac{45.1}{29.6} - 1\right) = 1630 \text{ in.-kips}$$

and the allowable strength is

$$M_n/\Omega = 1630/1.67 = 976 \text{ in.-kips} > 413 \text{ in.-kips}$$

So the coped beam has sufficient flexural strength.

Angles

Step 15: Check the angles for shear rupture using Equation J4-4.

The net area of the angle on the vertical shear plane is

$$A_{nv} = \left(l - n\left(d_h + 1/16\right)\right)t_a = \left(8.5 - 3\left(7/8 + 1/8\right)\right)\left(3/8\right) = 2.06 \text{ in.}^2$$

and the allowable shear rupture strength for two angles is

$$V_n/\Omega = 2\left(0.6F_u A_{nv}/\Omega\right) = 2\left(\left(0.6(65)(2.06)\right)/2.00\right)$$
$$= 80.3 \text{ kips} > 55.0 \text{ kips}$$

So the angles are adequate for shear rupture.

Step 16: Check the angles for shear yield using Equation J4-3.

The gross area of the angle on the vertical shear plane is

$$A_{gv} = 8.5\left(3/8\right) = 3.19 \text{ in.}^2$$

and the allowable shear yield strength for two angles is

$$V_n/\Omega = 2\left(0.6F_y A_{gv}/\Omega\right) = 2\left(\left(0.6(50)(3.19)\right)/1.5\right)$$
$$= 128 \text{ kips} > 55.0 \text{ kips}$$

So the angles are also adequate for shear yield.

Step 17 Check the angles for block shear.

The equations for block shear in the angle are the same as those for the web and as presented in Section 10.10.5.

First calculate the areas,

$$A_{nt} = \left(l_{eh} - \frac{1}{2}\left(d_h + 1/16\right)\right)t_w$$
$$= \left(1.0 - \frac{1}{2}\left(7/8 + 1/8\right)\right)\left(3/8\right) = 0.188 \text{ in.}^2$$
$$A_{gv} = lt_w$$
$$= 7.25\left(3/8\right) = 2.72 \text{ in.}^2$$
$$A_{nv} = \left(l - (n - 0.5)\left(d_h + 1/16\right)\right)t_w$$
$$= \left(7.25 - 2.5\left(7/8 + 1/8\right)\right)\left(3/8\right) = 1.78 \text{ in.}^2$$

Determine the tension rupture strength

$$F_u A_{nt} = 65\left(0.188\right) = 12.2 \text{ kips}$$

Consider shear yield and shear rupture, and select the least nominal strength; thus,

$$0.6F_yA_{gv} = 0.6(50)(2.72) = 81.6 \text{ kips}$$

$$0.6F_uA_{nv} = 0.6(65)(1.78) = 69.4 \text{ kips}$$

Selecting the shear rupture term because it is smaller and combining it with the tension rupture term gives a connection block shear allowable strength of the double angle—again, $U_{bs} = 1.0$ for this case of uniform tensile stress distribution—of

$$R_n/\Omega = 2((69.4 + 12.2)/2.00) = 81.6 \text{ kips} > 55.0 \text{ kips}$$

Step 18: Present the final connection design.

The three-bolt connection, revised as shown in Figure 11.3, is adequate to carry the imposed load of 55.0 kips.

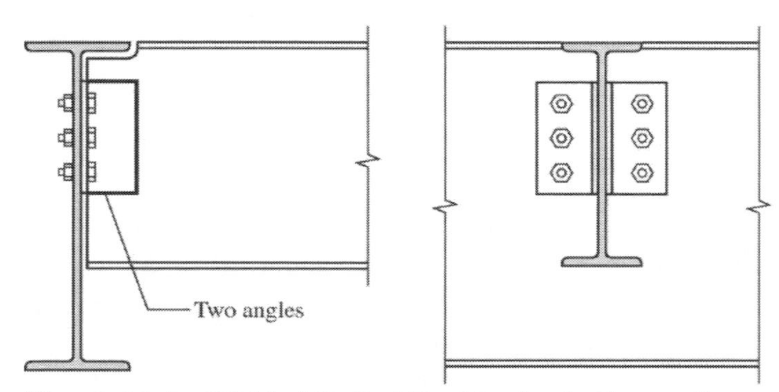

Figure 11.4 Welded-Bolted Double-Angle Connection

11.4 DOUBLE-ANGLE CONNECTIONS: WELDED-BOLTED

The double-angle shear connection can also be constructed by combining welding and bolting. In this case the angles are welded to the beam web, as shown in Figure 11.4.

The limit states to be considered are

1. Bolts
   - a. Shear rupture
2. Weld
   - a. Rupture
3. Beam
   - a. Shear yielding of the web
   - b. Block shear on coped beam web
   - c. Coped beam flexural strength

- **d.** Web strength at the weld
4. Angles
   - **a.** Bolt bearing and tearout on angles
   - **b.** Shear rupture
   - **c.** Shear yield
   - **d.** Block shear
5. Supporting member
   - **a.** Bolt bearing and tearout

The limit states that were not considered for the bolted-bolted connection from Section 11.3 are those associated with the weld. These include block shear of the beam web as a result of the welded connection; weld rupture, which is influenced by the eccentricity of the force on the weld group; and the strength of the beam web at the weld.

Block shear for a welded connection differs only slightly from block shear for a bolted connection. The difference is in the lack of holes to be deducted in determining the net area. Thus, the net shear area and gross shear area are the same. As a result, yielding is the controlling shear term in the block shear equation for this type of welded connection.

Weld rupture is a much more complex limit state to incorporate in this type of connection design. Chapter 10 discussed the strength of a weld loaded at its centroid and at any angle. The welds in the double-angle connection are loaded parallel to the length on one side of the angle and perpendicular to their lengths on the other two sides. Unfortunately, these welds are not loaded through their centroid, so an eccentricity exists and the simplified approach to combining them, previously shown in Chapter 10, cannot be used. The *Manual* uses the instantaneous center of rotation method to determine weld strength in cases like this. This approach accounts for the loading at an angle to the weld as well as the eccentricity of the load to the weld group. It requires an iterative solution to a nonlinear problem as discussed in *Manual* Part 8. Figure 11.5 shows a C-shaped weld with the geometric variables labeled. In the typical connection design, the geometry can be set and Manual Table 8-8 can be used to determine the weld group strength. The application of this table is shown in Example 11.2.

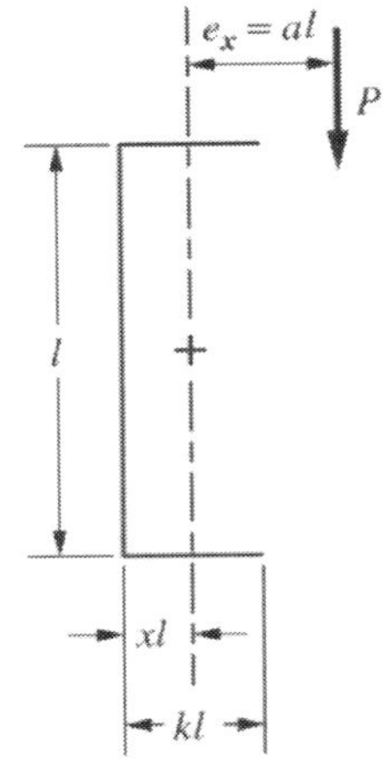

Figure 11.5 C-Shaped Weld Group

The beam web strength at the weld is also a bit difficult to calculate. The usual approach is to determine the total strength of the weld and then proportion that force to the web based on a one-inch length of web and one-inch length of weld. This, too, is illustrated in Example 11.2.

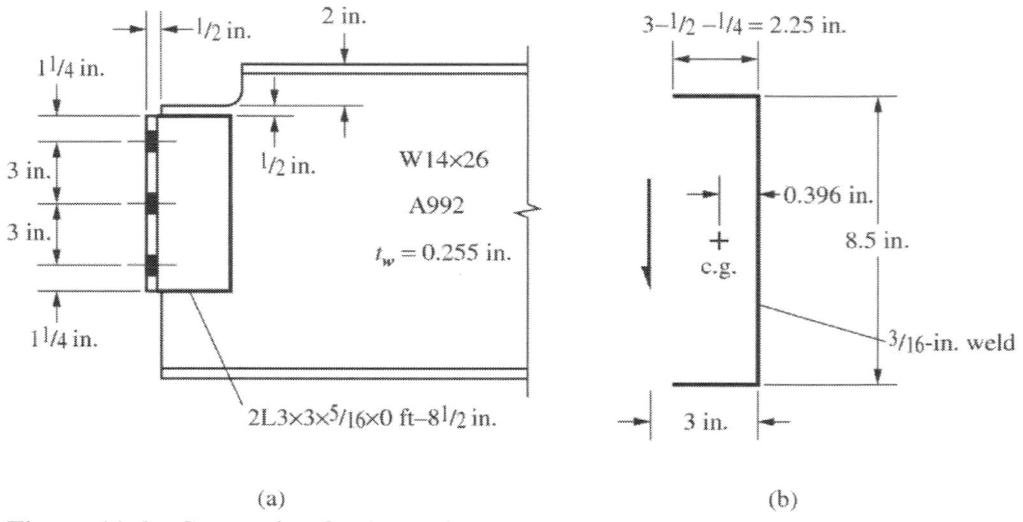

Figure 11.6 Connection for Example 11.2

EXAMPLE 11.2 ***Welded-Bolted Double-Angle Shear Connection***

Goal: Determine the available strength of the welded-bolted connection shown in Figure 11.6a.

Given: Determine the design strength and allowable strength of the connection shown in Figure 11.6a for the three new limit states discussed for the welded-bolted double-angle connection. Although there are other limit states that still must be checked for a complete design, this example will be limited to these three newly introduced limit states. For the given 3/16 in. weld use E70 electrodes.

SOLUTION

Step 1: Determine the nominal strength for the limit state of block shear in the beam web.

For the tension area, the length is found by taking the 3.0 in. angle leg and subtracting the 1/2 in. setback and the 1/4 in. potential beam under run tolerance.
Thus,

$$A_{nt} = (l_a - 0.50 - 0.25)t_w = (3.0 - 0.75)(0.255) = 0.574 \text{ in.}^2$$

For the gross shear area, the angle is 8.5 in. long and set down from the cope 1/2 in. Thus, the length for block shear calculation is

$l = 8.5 + 0.5 = 9.0$ in.

$$A_{gv} = lt_w = 9.0(0.255) = 2.30 \text{ in.}^2$$

Therefore, with $U_{bs} = 1.0$, the nominal block shear strength is

$$R_n = 0.6F_y A_{gv} + U_{bs} F_u A_{nt} = 0.6(50)(2.30) + 1.0(65)(0.574) = 106 \text{ kips}$$

For LRFD Step 2: The design block shear strength is

$$\phi R_n = 0.75(106) = 79.5 \text{ kips}$$

For ASD Step 2: The allowable block shear strength is

$$R_n/\Omega = 106/2.00 = 53.0 \text{ kips}$$

Step 3: Determine the nominal strength for the limit state of weld rupture.

The geometry of the weld is given in Figure 11.6b. The minimum angle thickness is the weld plus 1/16 in. *Manual* Equation 9-6, discussed in Chapter 10, shows that when welding with properly matched electrodes, this minimum angle thickness will always have sufficient strength to match the weld rupture strength.

Thus, the 5/16 in. angle has sufficient strength to support the 3/16 in. weld. Using terms as defined for *Manual* Table 8-8, the angle is 8.5 in. long, so the weld length is $l = 8.5$ in. The leg of the angle is 3.0 in. and the weld length is $kl = 3.0 - 1/2$ in. setback $- 1/4$ in. under run $= 2.25$ in. Thus,

$$k = \frac{2.25}{8.5} = 0.265$$

From *Manual* Table 8-8, for the load parallel to the weld along the angle toe, loading angle = 0°, the location of the weld centroid can be determined. Enter the table with $k = 0.265$ and interpolate for x from the values at the bottom of the table, which yields $x = 0.0466$. With this, the weld centroid is determined as

$$xl = 0.0466(8.5) = 0.396 \text{ in.}$$

The eccentricity of the force is then determined as

$$a = \frac{e_x}{l} = \frac{(3.0 - 0.396)}{8.5} = 0.306$$

Using this value for a and the previously determined value for k, the value of C can be determined from the table as C = 2.62. As indicated in the table, the nominal strength of the weld group is then

$$R_n = CC_1Dl$$

C has been determined above. C_1 represents the electrode strength and is 1.0 for the E70XX electrodes used here. D is the number of sixteenths of an inch in the fillet weld size, and l is the defined length of the weld group. Thus, for each angle with this weld

$$R_n = CC_1Dl = 2.62(1.0)(3)(8.5) = 66.8 \text{ kips}$$

For LRFD Step 4: The design strength of the welds on the double angle connection is

$$\phi R_n = 2(\phi R_n) = 2(0.75(66.8)) = 100 \text{ kips}$$

Step 5: Determine the design shear rupture strength for the beam web at the weld. Using *Manual* Equation 9-7 determine the minimum web thickness to balance the 3/16 in. weld on both sides of the web.

$$t_{\min} = \frac{6.19D}{F_u} = \frac{6.19(3)}{65} = 0.286 \text{ in.}$$

Since the beam web thickness $t_w < t_{\min}$, an approach that determines an effective weld length is used.

The design rupture strength of the 3/16 in. weld of unit length on both sides of the web, using the weld design strength determined in Chapter 10, is twice the weld strength of a single weld, thus,

$$\text{design weld strength per unit length} = 2(3)(1.392) = 8.35 \text{ kips/in.}$$

Using this strength, the effective length of the weld is weld strength divided by strength per inch. Thus,

$$\text{effective weld length} = \frac{100}{8.35} = 12.0 \text{ in.}$$

The design shear rupture strength of a unit length of the beam web, using Equation J4-4, is

$$\phi 0.6F_u t_w = 0.75(0.6(65)(0.255)) = 7.46 \text{ kips/in.}$$

Therefore, the beam web design rupture strength at the weld, using the weld effective length, is

$$\phi R_n = (12.0)(7.46) = 89.5 \text{ kips}$$

Step 6: Determine the controlling limit state for the three limit states considered in this example.

For the three limit states considered in this example, the connection is controlled by the limit state of block shear. Therefore,

$$\phi R_n = 79.5 \text{ kips}$$

For ASD Step 4: The allowable strength of the welds on the double angle connection is

$$R_n/\Omega = 2(R_n/\Omega) = 2(66.8/2.00) = 66.8 \text{ kips}$$

Step 5: Determine the allowable shear rupture strength for the beam web at the weld. Using *Manual* Equation 9-7 determine the minimum web thickness to balance the 3/16 in. weld on both sides of the web.

$$t_{min} = \frac{6.19D}{F_u} = \frac{6.19(3)}{65} = 0.286 \text{ in.}$$

Since the beam web thickness $t_w < t_{min}$, an approach that determines an effective weld length is used.

The allowable rupture strength of the 3/16 in. weld of unit length on both sides of the web, using the weld allowable strength determined in Chapter 10, is twice the weld strength of a single weld, thus

$$\text{allowable weld strength per unit length} = 2(3)(0.928) = 5.57 \text{ kips/in.}$$

Using this strength, the effective length of the weld is weld strength divided by strength per inch. Thus,

$$\text{effective weld length} = \frac{66.8}{5.57} = 12.0 \text{ in.}$$

The allowable shear rupture strength of a unit length of the beam web, using Equation J4-4, is

$$0.6F_u t_w/\Omega = 0.6(65)(0.255)/2.00 = 4.97 \text{ kips/in.}$$

Therefore, the beam web allowable rupture strength at the weld, using the weld effective length, is

$$R_n/\Omega = (12.0)(4.97) = 59.6 \text{ kips}$$

Step 6: Determine the controlling limit state for the three limit states considered in this example.

For the three limit states considered in this example, the connection is controlled by the limit state of block shear. Therefore,

$$R_n/\Omega = 53.0 \text{ kips}$$

11.5 DOUBLE-ANGLE CONNECTIONS: BOLTED-WELDED

This connection is shown in Figure 11.7, where the angles are bolted to the beam web and welded to the supporting member. The beam has a cope on the tension flange to permit the beam to be inserted in the space between the double angles, which are shop-welded to the supporting member, like a knife being inserted into its sheath. This connection might be used on a beam-to-column flange connection, but it would not be used as a beam-to-girder connection because of the interference of the girder flange.

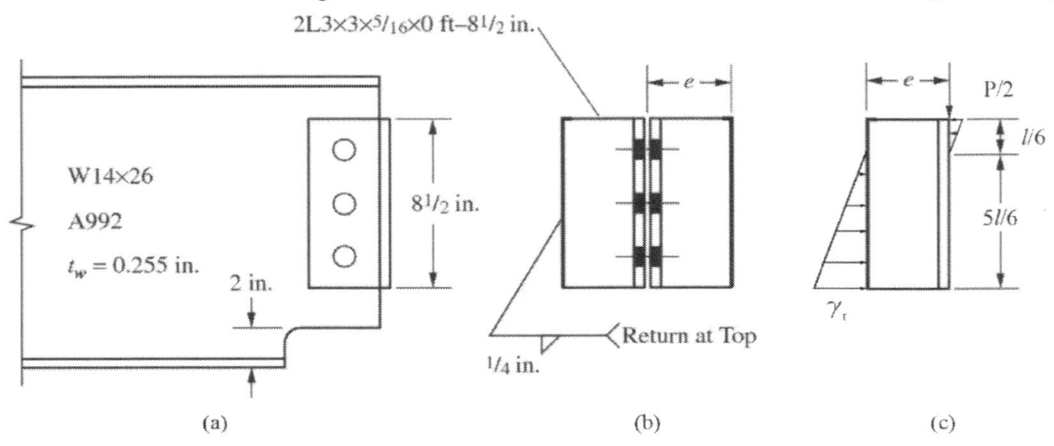

Figure 11.7 Double-Angle Bolted-Welded Connection for Example 11.3

For this connection, the new limit state of weld strength on the outstanding legs of the angles must be considered. In addition, consideration must be given to the flexural strength of the coped beam with the tension flange removed.

The strength of the weld on the outstanding leg of the angle may be determined by the instantaneous center of rotation method or an elastic method that assumes a uniform shear stress on the weld and a linearly varying tension stress over the lower five-sixths of the angle, with the top one-sixth of the angles in compression against each other through the beam web.

Using the geometry shown in Figure 11.7c, where e is taken as the width of the outstanding leg of the angle, moment equilibrium in the plane of the load, P, gives, for a load of $P/2$ on each angle

$$\frac{r_{rm}(5l/6)}{2}\left(\frac{2l}{3}\right) = \frac{Pe}{2}$$

where r_{rm} is the maximum horizontal force per unit length at the bottom of the weld due to moment. Solving for r_{rm} results in

$$r_{rm} = \frac{9Pe}{5l^2}$$

The direct vertical shear component is determined by dividing the applied force by the length of weld as

$$r_{rv} = \frac{P}{2l}$$

These two orthogonal components at the bottom of the weld can then be combined by finding the resultant as

$$r_r = \sqrt{r_{rm}^2 + r_{rv}^2} = \sqrt{\left(\frac{9Pe}{5l^2}\right)^2 + \left(\frac{P}{2l}\right)^2} = \frac{P\sqrt{l^2 + 12.96e^2}}{2l^2} \qquad (\textit{Manual}\ 8\text{-}18)$$

If this strength per unit length due to an applied load is set equal to the nominal weld strength per unit length, as shown in Chapter 10, P then becomes the nominal strength, R_n. Thus,

$$\frac{R_n\sqrt{l^2 + 12.96e^2}}{2l^2} = 0.6F_{EXX}\left(0.707w\right) = 1.856D$$

Solving for R_n and multiplying by ϕ or dividing by Ω results in the available strength equations given in Part 8 of the *Manual* as

$$\phi R_n = 2\left[\frac{1.392Dl}{\sqrt{1 + \frac{12.96e^2}{l^2}}}\right] \text{(LRFD)} \quad \text{and} \quad R_n/\Omega = 2\left[\frac{0.928Dl}{\sqrt{1 + \frac{12.96e^2}{l^2}}}\right] \text{(ASD)} \quad (\textit{Manual}\ 8\text{-}19)$$

where l is the length of the angle, e is the width of the outstanding leg, and D is the number of sixteenths-of-an-inch in weld size. The elastic method illustrated here is a bit more conservative than the instantaneous center of rotation method, which is easy to use when the additional strength it provides is important to consider.

EXAMPLE 11.3 ***Bolted-Welded Double-Angle Shear Connection***

Goal: Determine the available strength of the welds for a bolted-welded double-angle connection.

Given: The bolted-welded double-angle connection is shown in Figure 11.7. Assume 1/4 in. welds with a 1/2 in. return on top as shown.

SOLUTION

For LRFD Step 1: Determine the design strength.

With the given information, $D = 4$, $l = 8.5$ in., and $e = 3.0$ in.; thus, from Equation 8-19a

$$\phi R_n = 2\left[\frac{1.392Dl}{\sqrt{1+\frac{12.96e^2}{l^2}}}\right] = 2\left[\frac{1.392(4)(8.5)}{\sqrt{1+\frac{12.96(3)^2}{(8.5)^2}}}\right]$$

$$= 58.5 \text{ kips}$$

For ASD Step 1: Determine the allowable strength.

With the given information, $D = 4$, $l = 8.5$ in., and $e = 3.0$ in.; thus, from Equation 8-19b

$$R_n/\Omega = 2\left[\frac{0.928Dl}{\sqrt{1+\frac{12.96e^2}{l^2}}}\right] = 2\left[\frac{0.928(4)(8.5)}{\sqrt{1+\frac{12.96(3)^2}{(8.5)^2}}}\right]$$

$$= 39.0 \text{ kips}$$

11.6 DOUBLE ANGLE CONNECTIONS: WELDED-WELDED

The double-angle welded-welded connection is not a particularly common connection because it requires temporary support for field welding, which usually means bolts are used. This tends to favor a bolted field joint over a welded field joint. Nonetheless, if it is desirable for a particular situation, the limit states are those that have already been discussed. The procedures are the same and all potential limit states must be checked.

11.7 SINGLE-ANGLE CONNECTIONS

The single-angle connections shown in Figure 11.1b represent a bolted-bolted connection and a bolted-welded connection. In both cases, the connection is shown as a beam-to-girder connection. This is a particularly efficient connection because it eliminates erection problems when transverse beams frame into a girder at the same point on opposite sides of the girder web by shop-attaching the angles to the girder. It is also efficient because it has fewer parts than the double-angle connection. This connection is growing in popularity with both fabricators and erectors. It aids in erection efficiency because the beam can be installed from one side with the angle pre-attached to the girder. The disadvantages of this connection are that the components, such as angles, bolts, and welds, are larger than for the double-angle connection. For this connection, the bolts in the beam web are in single shear and, if the controlling limit state were bolt shear, it would require twice as many bolts than if it were a double-angle connection. Because all of the beam force must pass through only one angle, the angle likely needs to be larger. Greater weld size and weld length are also required. However, the single angle often is still the best choice in many situations, particularly when limit states such as block shear might otherwise control the strength of the connection.

The single-angle connection easily behaves as a simple shear connection, as it is modeled in analysis, because it is more flexible than the previously considered double-angle connections. Because of this increased flexibility, however, this connection is not recommended for laterally unsupported beams that rely on their end connections for lateral stability.

The limit states to be checked for the single-angle connection are the same as those for the double-angle connection, with some modifications and additions. The major modifications involve the eccentricities induced in the connecting elements. For the supported beam, as long as there is only one row of bolts, no eccentricities are considered, and this portion of the connection is treated the same as for the double-angle connections. For the outstanding leg, the bolts or welds must be designed to account for the connection eccentricity. This eccentricity also adds the limit states of flexural yielding and flexural rupture for the outstanding leg of the angle.

Figure 11.8a and b shows how the eccentricity for the outstanding leg, e_a, is measured for a bolted single angle, e_b, and a welded single angle, e_w Note that the eccentricity is measured, in both cases, from the centerline of the supported beam web. Figures 11.8c and d illustrate, with a bold line, the location and cross section for the moment in the angle. For the limit state of flexural yielding, the full length of the angle is included in the calculation of the plastic section modulus, *Z*, (Figure 11.8c). For the limit state of flexural rupture the full length of the angle less the length removed by the holes is used in the calculation of the net plastic section modulus, Z_{net}, (Figure 11.8d).

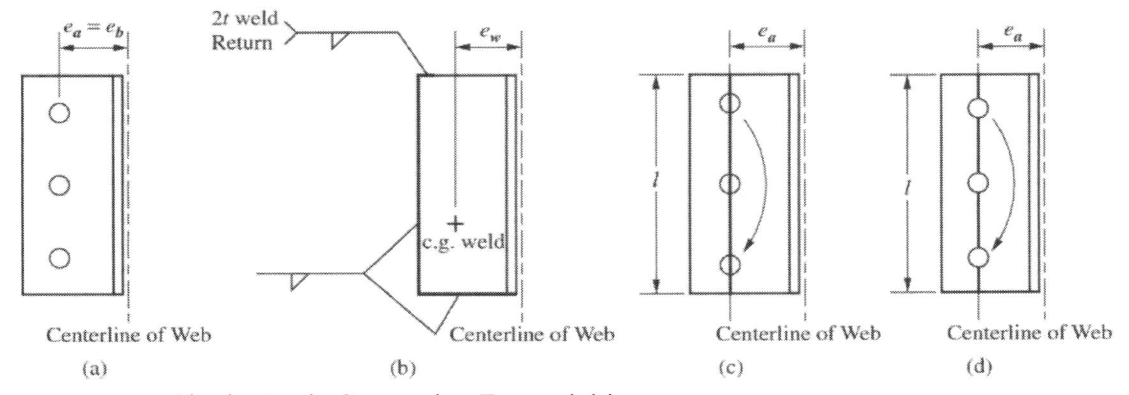

Figure 11.8 Single-Angle Connection Eccentricities

The limit state of flexural yielding is calculated based on the plastic moment of the element. Thus, the plastic section modulus for the angle leg without holes, $Z = t_p l^2/4$, is used and the nominal moment strength is given as

$$M_n = R_n e_a = F_y Z$$

For the flexural yielding limit state, $\phi = 0.9$ and $\Omega = 1.67$.

For the limit state of flexural rupture, the plastic section modulus of the net section is needed. This can be either determined by calculation or obtained from *Manual* Table 15-2. The resulting nominal moment strength is

$$M_n = R_n e_a = F_u Z_{net}$$

For the flexural rupture limit state, $\phi = 0.75$ and $\Omega = 2.00$.

To account for the effects of load eccentricity on the bolt group, an equivalent number of bolts must be determined. This can be accomplished using *Manual* Table 7-6, and the procedure is illustrated in Example 11.4. To account for the eccentricity effects on the weld group, *Manual* Table 8-11 can be used. This table is for an L-shaped weld group with the welds on only two sides. The top of the leg is kept free to ensure sufficient rotation capacity. Application of this table is also demonstrated in Example 11.4.

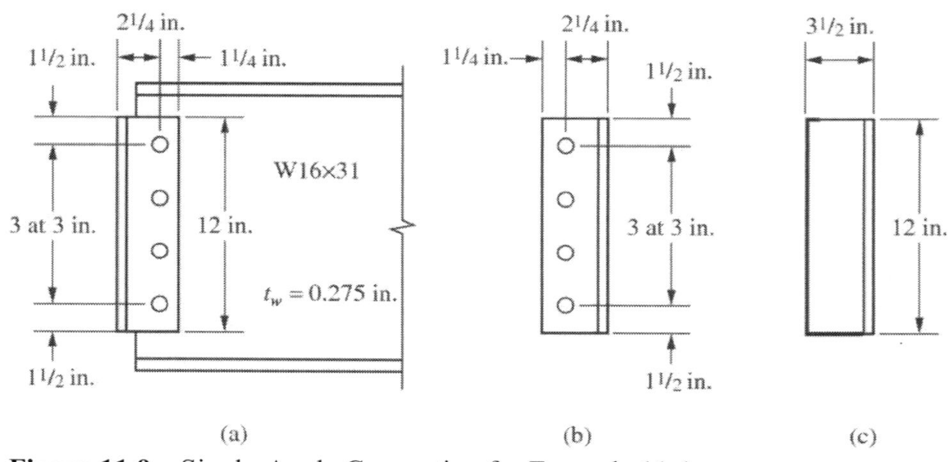

Figure 11.9 Single-Angle Connection for Example 11.4

EXAMPLE 11.4a
Single-Angle Shear Connection by LRFD

Goal: Determine the design strength of a bolted-bolted and a bolted-welded single-angle connection.

Given: A single-angle connection is shown in Figure 11.9 for the bolted outstanding leg case (Figure 11.9b) and the welded outstanding leg case (Figure 11.9c). The angle is A572 Gr. 50 steel, L3-1/2×3-1/2×3/8×1 ft – 0 in. The bolts are 3/4 in. A325-N, the weld is 3/16 in. with E70 electrodes and the beam is a W16×31, A992 steel.

SOLUTION

Part (a): Supported Beam

Step 1: Determine the bolt shear rupture strength. From *Manual* Table 7-1,

$$\phi r_n = 17.9 \text{ kips}$$

Step 2: Determine the bolt bearing and tearout strength on the angle.

The bottom bolt is 1.5 in. from the bottom of the angle, with the remaining bolts spaced at 3.0 in. Determine the clear distances for each of the bolts.

For the bottom bolt

$$l_c = l_{ev} - \frac{1}{2}d_h = 1.5 - \frac{1}{2}(3/4 + 1/16) = 1.09 < 2(3/4) = 1.5 \text{ in.}$$

Thus, tearout controls over bearing, and the nominal bolt strength is

$$R_n = 1.2 l_c t F_u = 1.2(1.09)(0.375)(65) = 31.9 \text{ kips}$$

The design bolt strength for tearout is

$$\phi R_n = 0.75(31.9) = 23.9 \text{ kips}$$

For the other bolts

$$l_c = s - d_h = 3.0 - (3/4 + 1/16) = 2.19 > 2(3/4) = 1.5 \text{ in.}$$

Therefore, bearing controls over tearout, and the nominal bolt strength is

$$R_n = 2.4 d t F_u = 2.4(3/4)(0.375)(65) = 43.9 \text{ kips}$$

The design bolt strength for bearing is

$$\phi R_n = 0.75(43.9) = 32.9 \text{ kips}$$

Step 3: Determine the bolt bearing and tearout strength on the supported beam web with $t_w = 0.275$ in.

Because the clear distance is greater than two times the bolt diameter for each of the four bolts, the nominal bearing strength is

$$R_n = 2.4 d t F_u = 2.4(3/4)(0.275)(65) = 32.2 \text{ kips}$$

and the design bearing strength

$$\phi R_n = 0.75(32.2) = 24.2 \text{ kips}$$

Step 4: Determine the connection design shear strength on the supported beam based on the bolts.

Bottom bolt

Bolt single shear	17.9 kips
Angle tearout	23.9 kips
Web bearing	24.2 kips

Other 3 bolts

Bolt single shear	17.9 kips
Angle bearing	32.9 kips
Web bearing	24.2 kips

Since all bolts are limited by bolt shear rupture

$$\phi R_n = 4(17.9) = 71.6 \text{ kips}$$

Step 5: Determine the shear yield strength of the angle. The gross shear area of the angle is

$$A_{gv} = lt_a = 12.0(0.375) = 4.50 \text{ in.}^2$$

and the design shear strength is

$$\phi V_n = \phi 0.6 F_y A_{gv} = 1.0(0.6(50)(4.50)) = 135 \text{ kips}$$

Step 6: Determine the shear rupture strength of the angle. The net shear area of the angle is

$$A_{nv} = (l - n(d_h + 1/16))t_a = (12.0 - 4(3/4 + 1/8))(0.375) = 3.19 \text{ in.}^2$$

and the design shear strength is

$$\phi V_n = \phi 0.6 F_u A_{nv} = 0.75(0.6(65)(3.19)) = 93.3 \text{ kips}$$

Step 7: Determine the block shear strength of the angle.

First calculate the areas.

$$A_{nt} = \left(l_{eh} - \frac{1}{2}(d_h + 1/16)\right)t_w$$

$$= \left(1.25 - \frac{1}{2}(3/4 + 1/8)\right)(0.375) = 0.305 \text{ in.}^2$$

$$A_{gv} = lt_w$$

$$= 10.5(0.375) = 3.94 \text{ in.}^2$$

$$A_{nv} = (l - (n - 0.5)(d_h + 1/16))t_w$$

$$= (10.5 - 3.5(3/8 + 1/8))(0.375) = 2.79 \text{ in.}^2$$

Determine the tension rupture strength

$$F_u A_{nt} = 65(0.305) = 19.8 \text{ kips}$$

Consider shear yield and shear rupture and select the one with the least strength; thus,

$$0.6F_y A_{gv} = 0.6(50)(3.94) = 118 \text{ kips}$$

$$0.6F_u A_{nv} = 0.6(65)(2.79) = 109 \text{ kips}$$

Selecting the shear rupture term because it is smaller and combining it with the tension rupture term, with U_{bs} = 1.0 for this case of uniform tensile stress distribution, gives

$$\phi R_n = 0.75(109 + 1.0(19.8)) = 96.6 \text{ kips}$$

Step 8: Determine the design strength of the leg attached to the supported member.

The design strength is controlled by bolt shear, where

$$\phi R_n = 71.6 \text{ kips}$$

Part (b): Bolted Outstanding Leg

Step 9: Check the eccentric shear using the instantaneous center of rotation method tabulated in *Manual* Table 7-6 to account for the effects of eccentricity on the bolt group.

The eccentricity of the load on the line of bolts is calculated as the distance of the bolts from the angle heel plus one-half of the beam web as shown in Figure 11-9b; thus,

$$e_x = 2.25 + \frac{0.275}{2} = 2.39 \text{ in.}$$

For the four-bolt connection with bolt spacing of 3.0 in., *Manual* Table 7-6 with the load at an angle of zero gives, by interpolation, the effective number of bolts as C = 3.12.

For simplicity, the supporting member is not provided and can be assumed to not control for this problem. Therefore, determine the bolt strength based only on the angle. For bearing and tearout on the angle, the available strength is the same as for the leg attached to the supported beam. Thus,

Top bolt

Bolt single shear	17.9 kips
Angle tearout	23.9 kips

Other 3 bolts

Bolt single shear	17.9 kips
Angle bearing	32.9 kips

Therefore, bolt shear controls all four bolts and the strength at the supported member is

$$\phi R_n = C\phi r_n = 3.12(17.9) = 55.8 \text{ kips}$$

As expected, the strength of the bolts in the outstanding leg is less than that in the leg on the supported beam because the outstanding leg must accommodate an eccentricity that is not present in the leg on the beam.

Step 10: Determine the flexural yield strength of the outstanding leg.

The plastic section modulus is determined for the rectangle formed by the length and thickness of the angle, and the nominal moment strength is determined by multiplying the plastic section modulus by the yield stress; thus,

$$Z = \frac{(0.375)(12^2)}{4} = 13.5 \text{ in.}^3$$

and

$$M_n = F_y Z = 50(13.5) = 675 \text{ in.-kips}$$

Because the moment is the shear force times the eccentricity, using the same eccentricity calculated for the bolt group,

$$\phi R_n = \frac{\phi M_n}{e} = \frac{0.9(675)}{2.39} = 254 \text{ kips}$$

Step 11: Determine the flexural rupture strength of the outstanding leg.

The net plastic section modulus is determined for the rectangle less the holes. Although this can readily be calculated, it can also be obtained from *Manual* Table 15-2, where $Z_{net} = 9.56 \text{ in.}^3$ Thus,

$$M_n = F_u Z_{net} = 65(9.56) = 621 \text{ in.-kips}$$

and

$$\phi R_n = \frac{\phi M_n}{e} = \frac{0.75(621)}{2.39} = 195 \text{ kips}$$

Step 12: Determine the controlling limit state strength for the bolted outstanding leg.

For the bolted outstanding leg, the strength is controlled by shear strength of the bolts with eccentricity, where

$$\phi R_n = 55.8 \text{ kips}$$

Because this is less than the value for the leg attached to the beam, this is the design strength of the bolted-bolted single-angle connection as long as the supporting member has sufficient bearing and tearout strength.

Part (c): Welded Outstanding Leg

Step 13: Determine the eccentric weld rupture strength.

Manual Table 8-10, which uses the instantaneous center of rotation method, will be used to determine the eccentric weld rupture strength. The weld for the single-angle connection is applied to the vertical toe and the bottom edge of the angle, not the top. This ensures that the angle is sufficiently flexible to behave as a simple connection as modeled in the analysis of a simple beam. *Manual* Table 8-10 for load at an angle of zero degrees shows this weld on the top or the bottom. Thus, the table can be used for either arrangement because the geometry is the same whether the horizontal weld is at the top or bottom of the connection.

Based on the dimensions given in Figure 11.9c, $l = 12.0$, $kl = 3.5$, and thus $k = 0.292$. The weld is given as a 3/16 in. weld with E70 electrodes. From *Manual* Table 8-10, interpolating between $k = 0.2$ and $k = 0.3$ yields $x = 0.0336$.

Therefore, the eccentricity is

$$e_x = kl + \frac{t_w}{2} - xl = 3.5 + \frac{0.275}{2} - 0.0336(12.0) = 3.23 \text{ in.}$$

and

$$a = \frac{e_x}{l} = \frac{3.23}{12.0} = 0.269$$

With a double interpolation between $k = 0.2$ and 0.3 and $a = 0.25$ and 0.30, the coefficient C is determined as

$$C = 2.21$$

Therefore, with $C_1 = 1.0$ for an E70*XX* electrode, the nominal weld strength is

$$R_n = CC_1Dl = 2.21(1.0)(3)(12.0) = 79.6 \text{ kips}$$

and the design strength is

$$\phi R_n = 0.75(79.6) = 59.7 \text{ kips}$$

Step 14: Determine the design strength for the limit state of flexural yielding.

For the limit state of flexural yielding of the angle, the strength is determined as was shown for the bolted outstanding leg in step 10;

thus,

$$\phi R_n = 254 \text{ kips}$$

Step 15: Determine the controlling limit state strength for the welded outstanding legs.

For the welded outstanding leg, the design strength is controlled by eccentric shear on the weld, where

$$\phi R_n = 59.7 \text{ kips}$$

Because this is less than the value for the leg attached to the beam, this is the design strength of the bolted-welded single-angle connection as long as the supporting member has sufficient rupture strength. For both bolted and welded outstanding leg arrangements the outstanding leg controlled the strength of the connection.

EXAMPLE 11.4b
Single-Angle Shear Connection by ASD

Goal: Determine the allowable strength of a bolted-bolted and a bolted-welded single-angle connection.

Given: A single-angle connection is shown in Figure 11.9 for the bolted outstanding leg case (Figure 11.9b) and the welded outstanding leg case (Figure 11.9c). The angle is A572 Gr. 50 steel, L3-1/2×3-1/2×3/8×1 ft – 0 in. The bolts are 3/4 in. A325-N, the weld is 3/16 in. with E70 electrodes and the beam is a W16×31, A992 steel.

SOLUTION

Part (a): Supported Beam

Step 1: Determine the bolt shear rupture strength. From *Manual* Table 7-1,

$$r_n/\Omega = 11.9 \text{ kips}$$

Step 2: Determine the bolt bearing and tearout strength on the angle.

The bottom bolt is 1.5 in. from the bottom of the angle, with the remaining bolts spaced at 3.0 in. Determine the clear distances for each of the bolts.

For the bottom bolt

$$l_c = l_{ev} - \frac{1}{2}d_h = 1.5 - \frac{1}{2}(3/4 + 1/16) = 1.09 < 2(3/4) = 1.5 \text{ in.}$$

Thus, tearout controls over bearing, and the nominal bolt strength is

$$R_n = 1.2 l_c t F_u = 1.2(1.09)(0.375)(65) = 31.9 \text{ kips}$$

The allowable bolt strength for tearout is

$$R_n/\Omega = 31.9/2.00 = 16.0 \text{ kips}$$

For the other bolts

$$l_c = s - d_h = 3.0 - (3/4 + 1/16) = 2.19 > 2(3/4) = 1.5 \text{ in.}$$

Therefore, bearing controls over tearout, and the nominal bolt strength is

$$R_n = 2.4dtF_u = 2.4(3/4)(0.375)(65) = 43.9 \text{ kips}$$

The allowable bolt strength for bearing is

$$R_n/\Omega = 43.9/2.00 = 22.0 \text{ kips}$$

Step 3: Determine the bolt bearing and tearout strength on the supported beam web.

Because the clear distance is greater than two times the bolt diameter for each of the four bolts, the nominal bearing strength is

$$R_n = 2.4dtF_u = 2.4(3/4)(0.275)(65) = 32.2 \text{ kips}$$

and the allowable bearing strength

$$R_n/\Omega = 32.2/200 = 16.1 \text{ kips}$$

Step 4: Determine the connection allowable shear strength on the supported beam based on the bolts.

Bottom bolt

Bolt single shear	11.9 kips
Angle tearout	16.0 kips
Web bearing	16.1 kips

Other 3 bolts

Bolt single shear	11.9 kips
Angle bearing	22.0 kips
Web bearing	16.1 kips

Since all bolts are limited by bolt shear rupture

$$R_n/\Omega = 4(11.9) = 47.6 \text{ kips}$$

Step 5: Determine the shear yield strength of the angle. The gross shear area of the angle is

$$A_{gv} = lt_a = 12.0(0.375) = 4.50 \text{ in.}^2$$

and the allowable shear strength is

$$V_n/\Omega = 0.6F_yA_{gv}/\Omega = \left(0.6(50)(4.50)\right)/1.50 = 90.0 \text{ kips}$$

Step 6: Determine the shear rupture strength of the angle. The net shear area of the angle is

$$A_{nv} = \left(l - n\left(d_h + 1/16\right)\right)t_a = \left(12.0 - 4\left(3/4 + 1/8\right)\right)(0.375) = 3.19 \text{ in.}^2$$

and the allowable shear strength is

$$V_n/\Omega = 0.6F_uA_{nv}/\Omega = \left(0.6(65)(3.19)\right)/2.00 = 62.2 \text{ kips}$$

Step 7: Determine the block shear strength of the angle.

First calculate the areas.

$$\begin{aligned} A_{nt} &= \left(l_{eh} - \frac{1}{2}\left(d_h + 1/16\right)\right)t_w \\ &= \left(1.25 - \frac{1}{2}\left(3/4 + 1/8\right)\right)(0.375) = 0.305 \text{ in.}^2 \\ A_{gv} &= lt_w \\ &= 10.5(0.375) = 3.94 \text{ in.}^2 \\ A_{nv} &= \left(l - (n - 0.5)\left(d_h + 1/16\right)\right)t_w \\ &= \left(10.5 - 3.5\left(3/4 + 1/8\right)\right)(0.375) = 2.79 \text{ in.}^2 \end{aligned}$$

Determine the tension rupture strength

$$F_uA_{nt} = 65(0.305) = 19.8 \text{ kips}$$

Consider shear yield and shear rupture and select the one with the least strength; thus,

$$\begin{aligned} 0.6F_yA_{gv} &= 0.6(50)(3.94) = 118 \text{ kips} \\ 0.6F_uA_{nv} &= 0.6(65)(2.79) = 109 \text{ kips} \end{aligned}$$

Selecting the shear rupture term and combining it with the tension rupture term, with $U_{bs} = 1.0$ for this case of uniform tensile stress distribution, gives

$$R_n/\Omega = \left(109 + 1.0(19.8)\right)/2.00 = 64.4 \text{ kips}$$

Step 8: Determine the allowable strength of the leg attached to the supported member.

The allowable strength is controlled by bolt shear, where

$$R_n/\Omega = 47.6 \text{ kips}$$

Part (b): Bolted Outstanding Leg

Step 9: Check the eccentric shear using the instantaneous center of rotation method tabulated in *Manual* Table 7-6 to account for the effects of eccentricity on the bolt group.

The eccentricity of the load on the line of bolts is calculated as the distance of the bolts from the angle heel plus one-half of the beam web as shown in Figure 11-9b; thus,

$$e_x = 2.25 + \frac{0.275}{2} = 2.39 \text{ in.}$$

For the four-bolt connection with bolt spacing of 3.0 in., *Manual* Table 7-6 with the load at an angle of zero gives, by interpolation, the effective number of bolts as $C = 3.12$.

The supporting member is not provided. Therefore, determine the bolt strength based only on the angle. For bearing and tearout on the angle, the available strength is the same as for the leg attached to the supported beam. Thus,

Top bolt

Bolt single shear	11.9 kips
Angle tearout	16.0 kips

Other 3 bolts

Bolt single shear	11.9 kips
Angle bearing	22.0 kips

Therefore, bolt shear controls all four bolts and the strength at the supported member is

$$R_n/\Omega = Cr_n/\Omega = 3.12(11.9) = 37.1 \text{ kips}$$

As expected, the strength of the bolts in the outstanding leg is less than that in the leg on the supported beam because the outstanding leg must accommodate an eccentricity that is not present in the leg on the beam.

Step 10: Determine the flexural yield strength of the outstanding leg.

The plastic section modulus is determined for the rectangle formed by the length and thickness of the angle, and the nominal moment strength is determined by multiplying the plastic section modulus by the yield stress; thus,

$$Z = \frac{(0.375)(12^2)}{4} = 13.5 \text{ in.}^3$$

and

$$M_n = F_y Z = 50(13.5) = 675 \text{ in.-kips}$$

Because the moment is the shear force times the eccentricity, using the same eccentricity calculated for the bolt group,

$$R_n/\Omega = \frac{M_n/\Omega}{e} = \frac{(675/1.67)}{2.39} = 169 \text{ kips}$$

Step 11: Determine the flexural rupture strength of the outstanding leg.

The net plastic section modulus is determined for the rectangle less the holes. Although this can readily be calculated, it can also be obtained from *Manual* Table 15-2, where $Z_{net} = 9.56 \text{ in.}^3$. Thus,

$$M_n = F_u Z_{net} = 65(9.56) = 621 \text{ in.-kips}$$

and

$$R_n/\Omega = \frac{M_n/\Omega}{e} = \frac{(621/2.00)}{2.39} = 130 \text{ kips}$$

Step 12: Determine the controlling limit state strength for the bolted outstanding legs.

For the bolted outstanding leg, the strength is controlled by shear strength of the bolts with eccentricity, where

$$R_n/\Omega = 37.1 \text{ kips}$$

Because this is less than the value for the leg attached to the beam, this is the allowable strength of the bolted-bolted single-angle connection as long as the supporting member has sufficient bearing and tearout strength.

Part (c): Welded Outstanding Leg

Step 13: Determine the eccentric weld rupture strength.

Manual Table 8-10, which uses the instantaneous center of rotation method, will be used to determine the eccentric weld rupture strength. The weld for the single-angle connection is applied to the vertical toe and the bottom edge of the angle, not the top. This ensures that the angle is sufficiently flexible to behave as a simple connection as modeled in the analysis of a simple beam. *Manual* Table 8-10 for load at an angle of zero degrees shows this weld on the top or the bottom. Thus, the table can be used for either arrangement because the geometry is the same whether the horizontal weld is at the top or bottom of the connection.

Based on the dimensions given in Figure 11.9c, l = 12.0, kl = 3.5, and thus k = 0.292. The weld is given as a 3/16 in. weld with E70 electrodes. From *Manual* Table 8-10, interpolating between $k = 0.2$ and $k = 0.3$ yields $x = 0.0336$.

Therefore, the eccentricity is

$$e_x = kl + \frac{t_w}{2} - xl = 3.5 + \frac{0.275}{2} - 0.0336(12.0) = 3.23 \text{ in.}$$

and

$$a = \frac{e_x}{l} = \frac{3.23}{12.0} = 0.269$$

With a double interpolation between $k = 0.2$ and 0.3 and $a = 0.25$ and 0.30, the coefficient C is determined as

$$C = 2.21$$

Therefore, with C_1 = 1.0 for an E70*XX* electrode, the nominal weld strength is

$$R_n = CC_1Dl = 2.21(1.0)(3)(12.0) = 79.6 \text{ kips}$$

and the allowable strength is

$$R_n/\Omega = 79.6/2.00 = 39.8 \text{ kips}$$

Step 14: Determine the allowable strength for the limit state of flexural yielding.

For the limit state of flexural yielding of the angle, the allowable strength is determined as was shown for the bolted outstanding leg in step 10; thus,

$$R_n/\Omega = 169 \text{ kips}$$

Step 15: Determine the controlling limit state strength for the welded outstanding legs.

For the welded outstanding leg, the allowable strength is controlled by eccentric shear on the weld, where

$$R_n/\Omega = 39.8 \text{ kips}$$

Because this is less than the value for the leg attached to the beam, this is the allowable strength of the bolted-welded single-angle connection as long as the supporting member has sufficient rupture strength. For both bolted and welded outstanding leg arrangements the outstanding leg controlled the strength of the connection.

11.8 SINGLE-PLATE SHEAR CONNECTIONS

The single-plate shear connection, also called a *shear tab connection*, is shown in Figure 11.1c. It consists of a plate that is shop-welded to the support and field-bolted to the beam and is similar to the single-angle connection when it comes to erection. The shear tab consists of only a single plate, which is about as simple as can be expected. It is welded to the supporting member and must be bolted to the supported beam in order to accommodate the required rotation. Even when bolted to the beam, this connection is stiffer than the single- or double-angle connection and requires careful detailing to ensure sufficient flexibility.

The behavior of this connection is similar to that of a double-angle connection except that it achieves its rotation capacity through the bending of the tab and deformation of the plate or beam web in bearing at the bolt holes as rotation occurs. Because of the complexity of assessing some of the limit states for this connection, AISC has developed two design approaches—including a somewhat prescriptive approach for what is called the *conventional configuration*, and a detailed limit states checking procedure for all others, referred to as the *extended configuration*.

The limit states that must be checked are the same for either configuration; the difference is that in the conventional configuration, physical limitations have been set so that most of those limit states do not govern. The potential limit states are

1. Bolts
   - **a.** Shear rupture
2. Beam
   - **a.** Bearing and tearout on the web
   - **b.** Shear yielding of web
3. Plate
   - **a.** Bearing and tearout on the plate
   - **b.** Elastic yield moment
   - **c.** Flexural rupture
   - **d.** Shear yielding
   - **e.** Shear rupture

  - **f.** Block shear rupture
  - **g.** Lateral-torsional buckling
  - **h.** Plastic flexural yielding with shear interaction

4. Weld
  - **a.** Weld rupture with eccentricity

Of these 11 limit states, those associated with flexure and buckling of the plate are new to the discussion of simple connection design, and the weld rupture limit state is treated a little bit differently than those weld limit states already discussed.

The conventional configuration of the shear tab results in a connection that is very simple to design. This is the type of connection that is treated here. For other configurations, detailed procedures are given in Part 10 of the *Manual*. The dimensional limitations of the conventional shear tab lead to the following requirements:

1. There can be only a single vertical row containing a total of 2 to 12 bolts.
2. The distance from the bolt line to the weld line cannot exceed 3-1/2 in.
3. Only standard or short slotted holes transverse to the direction of the load can be used.
4. The vertical edge distance, l_{ev}, must satisfy the *Specification* minimums from Table J3.4 and l_{eh} should be greater than or equal to 2 times the diameter of the bolt for both the plate and beam web.
5. For standard holes, either the plate or beam web must have $t \leq (d_b/2 + 1/16)$ for 2 to 5 bolts, or $t \leq (d_b/2 - 1/16)$ for 6 to 12 bolts to accommodate simple beam rotation. These are given in *Manual* Table 10-9 along with the thickness requirements for short-slotted holes and the required minimum eccentricity to be included.

Once these limitations are satisfied, the connection need be checked only for the following limit states using the eccentricity specified in *Manual* Table 10-9

a. Eccentric bolt shear rupture
b. Bolt bearing and tearout assuming a concentrically applied load
c. Block shear rupture of plate
d. Plate shear yielding
e. Plate shear rupture

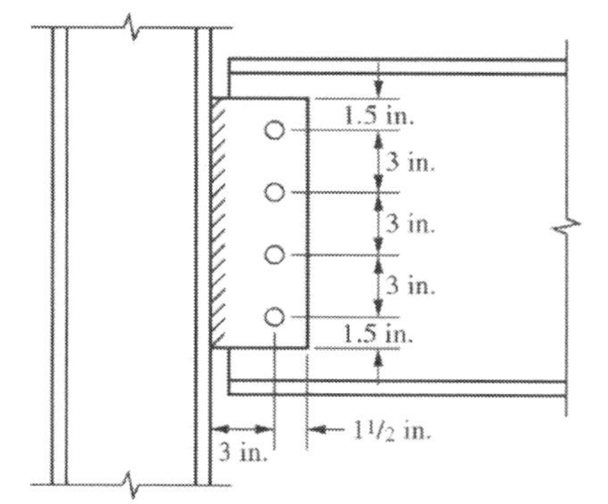

Figure 11.10 Shear Tab Connection for Example 11.5

The *Manual* treatment of bolt strength at holes in the conventional configuration single plate shear connection is different from that in other bolted connections. Here, eccentricity is applied to the bolt group shear strength but is neglected for bearing and tearout on the plate and beam web. The justification for this appears to be an attempt at simplicity and an approach that maintains the goal to have a plate that is thick enough for strength but also thin enough to allow for the deformations that produce the ductility of the connection. However, eccentricity is not neglected by all connection designers. A more conservative yet still simple approach is to select the least strength bolt based on bolt shear, bearing and tearout on the plate, and bearing and tearout on the beam web and then apply the eccentric bolt group multiplier to that strength.[1] Example 11.5 will illustrate the *Manual* approach to design of a conventional configuration single plate shear connection with this more conservative approach given in the last step for comparison.

EXAMPLE 11.5a
Shear Tab Conventional Configuration by LRFD

Goal: Determine the design strength of a conventional configuration shear tab connection.

Given The shear tab connection is given in Figure 11.10. The beam is a W16×50 A992 framing into the flange of a W14×90 A992 column with an A572 Gr. 50 PL1/4×4-1/2×1 ft - 0 in. plate. Use four 3/4 in. A325-N bolts in standard holes.

SOLUTION

Step 1: Determine whether the given shear tab meets the limitations for the conventional configuration.

Limitations for the conventional configuration:

1. 4 bolts—is between 2 and 12

2. $a = 3.0$ in.—does not exceed 3½ in.

3. Standard holes—standard or short-slotted holes are permitted

4. $l_{ev} = 1.5$ in. > 1.0 in. from Table J3.4
$l_{eh} = 1.5 \text{ in.} \geq 2d_b = 2(3/4) = 1.5 \text{ in.}$

5. $t_{plate} = 1/4 \text{ in.} < \left(\dfrac{d_b}{2} + 1/16\right) = \left(\dfrac{3/4}{2} + 1/16\right) = 7/16 \text{ in.}$

Step 2: Determine the bolt shear strength considering eccentricity.

From *Manual* Table 7-1,

$$\phi R_n = 17.9 \text{ kips/bolt}$$

From *Manual* Table 10-9 the eccentricity to be included is

$$e = a/2 = 3.0/2 = 1.5 \text{ in.}$$

[1] Muir, L. and Thornton, W. A., "The Development of a new Design Procedure for Conventional Single-Plate Shear Connections," *Engineering Journal*, AISC, Vol. 48, No. 2, pp. 141-152.

From *Manual* Table 7-6 the equivalent number of bolts to account for eccentricity in the connection can be determined by interpolating between C = 3.75 for $e_x = 1.0$ in. and C = 3.32 for $e_x = 2.0$ in. Thus, $C = 3.54$ and

$$\phi R_n = C\phi r_n = 3.54(17.9) = 63.4 \text{ kips}$$

Step 3: Determine the bolt bearing and tearout strength on the plate assuming it is concentrically loaded.

For the bottom bolt

$$l_c = l_{ev} - \frac{1}{2}d_h = 1.5 - \frac{1}{2}(3/4 + 1/16) = 1.09 < 2(3/4) = 1.5$$

Thus, tearout controls over bearing, and the nominal bolt strength is

$$R_n = 1.2 l_c t_p F_u = 1.2(1.09)(0.250)(65) = 21.3 \text{ kips}$$

The design bolt tearout strength is

$$\phi R_n = 0.75(21.3) = 16.0 \text{ kips}$$

For the other bolts

$$l_c = s - d_h = 3.0 - (3/4 + 1/16) = 2.19 > 2(3/4) = 1.5 \text{ in.}$$

Therefore, bearing controls over tearout, and the nominal bolt strength is

$$R_n = 2.4 d t_p F_u = 2.4(3/4)(0.250)(65) = 29.3 \text{ kips}$$

The design bolt bearing strength is

$$\phi R_n = 0.75(29.3) = 22.0 \text{ kips}$$

Step 4: Determine the bolt bearing and tearout strength on the web.

For the beam web, the material is A992, the web thickness is 0.380 in. and the member is not coped. Because the bolts will not tearout and the thickness of the web is greater than the thickness of the plate, the beam web will not control. However, strength will still be determined for consistency.

For each bolt in bearing

$$R_n = 2.4 d t_w F_u = 2.4(3/4)(0.380)(65) = 44.5 \text{ kips}$$

and the design strength in the web is

$$\phi R_n = 0.75(44.5) = 33.4 \text{ kips}$$

Step 5: Determine the connection strength based on the bolts in holes considering only the plate and web without considering eccentricity,

Bottom bolt

Plate tearout	16.0 kips
Web bearing	33.4 kips

Other 3 bolts

Plate bearing	22.0 kips
Web bearing	33.4 kips

Thus, the connection strength based on the bolts in holes without considering eccentricity or bolt shear is the sum of plate tearout for the bottom bolt plus 3 times the plate bearing for the other bolts. Thus,

$$\phi R_n = 16.0 + 3(22.0) = 82.0 \text{ kips}$$

This is greater than the strength determined for bolt shear, including eccentricity, in Step 2. Thus, connection strength based on the bolts is as given in Step 2,

$$\phi R_n = 63.4 \text{ kips}$$

Step 6: Determine the design block shear strength of the plate.

Calculating the areas,

$$A_{nt} = \left(l_{eh} - \frac{1}{2}(d_h + 1/16)\right)t_w$$

$$= \left(1.5 - \frac{1}{2}(3/4 + 1/8)\right)(0.250) = 0.266 \text{ in.}^2$$

$$A_{gv} = lt_w$$

$$= 10.5(0.250) = 2.63 \text{ in.}^2$$

$$A_{nv} = \left(l - (n - 0.5)(d_h + 1/16)\right)t_w$$

$$= \left(10.5 - 3.5(3/4 + 1/8)\right)(0.250) = 1.86 \text{ in.}^2$$

Determine the tension rupture strength

$$F_u A_{nt} = 65(0.266) = 17.3 \text{ kips}$$

Consider shear yield and shear rupture and select the one with least strength; thus,

$$0.6F_y A_{gv} = 0.6(50)(2.63) = 78.9 \text{ kips}$$

$$0.6F_u A_{nv} = 0.6(65)(1.86) = 72.5 \text{ kips}$$

Selecting the shear rupture term and combining it with the tension rupture term gives a connection design block shear strength, with U_{bs} = 1.0, of

$$\phi R_n = 0.75(72.5 + 1.0(17.3)) = 67.4 \text{ kips}$$

Step 7: Determine the design shear yield strength of the plate. The gross shear area is

$$A_{gv} = lt_p = 12.0(0.250) = 3.00 \text{ in.}^2$$

and

$$\phi V_n = \phi 0.6 F_y A_{gv} = 1.0(0.6(50)(3.00)) = 90.0 \text{ kips}$$

Step 8: Determine the design shear rupture strength of the plate. The net shear area is

$$A_{nv} = (l - n(d_h + 1/16))t_p = (12.0 - 4(3/4 + 1/8))(0.250) = 2.13 \text{ in.}^2$$

and

$$\phi V_n = \phi 0.6 F_u A_{nv} = 0.75(0.6(65)(2.13)) = 62.3 \text{ kips}$$

Step 9 Determine the connection design strength based on the plate

Block shear	67.4 kips
Shear yield	90.0 kips
Shear rupture	62.3 kips

Thus, the connection design strength for the plate is based on shear rupture of the plate.

$$\phi R_n = 62.3 \text{ kips}$$

Step 10: Determine the required weld size and minimum thickness of the supporting column flange and confirm that these are met.

The conventional configuration requires that the plate be welded to the supporting member through a pair of fillet welds on each side of the plate, with the weld leg width $w = (5/8)t_p$. Therefore, use a 3/16 in. fillet weld. This develops the strength of either a 36 ksi or 50 ksi plate and therefore does not require any further limit states checking. However, this does not check for the strength of the supporting member.

Step 11: Determine the minimum thickness of the supporting member to match the weld rupture strength.

Using the minimum thickness given by *Manual* Equation 9-6

$$t_{min} = \frac{3.09D}{F_u} = \frac{3.09(3)}{65} = 0.143 \text{ in.}$$

This minimum is significantly less than the W14×90 flange thickness, $t_f = 0.710$ in.

Even if the minimum thickness had not been exceeded by the column flange thickness, the fact that the plate is welded to the column flange at the column flange-web junction would provide sufficient base metal strength.

Step 12: Determine the controlling limit state and design strength of the connection.

The design strength is controlled by the shear rupture of the plate as found in Step 9, where

$$\phi R_n = 62.3 \text{ kips}$$

Step 13: Using the more conservative approach to determine bolt strength discussed above, select the single bolt with the least strength from Step 5 and bolt shear. For the bottom bolt in the plate, tearout controls. Thus,

$$\phi r_n = 16.0 \text{ kips}$$

Applying the factor from Step 2 for the equivalent number of bolts, the strength for bolts in holes, accounting for eccentricity is

$$\phi R_n = C\phi r_n = 3.54(16.0) = 56.6 \text{ kips}$$

Comparing this to the strength for bolts in holes determined in Step 5 shows this to be a bit more conservative approach.

Then, comparing this to the controlling limit state determined in Step 12 shows that, with this approach, the controlling limit state is now bolt strength in holes. Thus,

$$\phi R_n = 56.6 \text{ kips}$$

EXAMPLE 11.5b
Shear Tab Conventional Configuration by ASD

Goal: Determine the allowable strength of a conventional configuration shear tab connection.

Given The shear tab connection is given in Figure 11.10. The beam is a W16×50 A992 framing into the flange of a W14×90 A992 column with an A572 Gr. 50 PL1/4×4-1/2×1 ft - 0 in. plate. Use four 3/4 in. A325-N bolts in standard holes.

SOLUTION

Step 1: Determine whether the given shear tab meets the limitations for the conventional configuration.

Limitations for the conventional configuration:

1. 4 bolts—is between 2 and 12

2. $a = 3.0$ in.—does not exceed 3-1/2 in.

3. Standard holes—standard or short-slotted holes are permitted

4. $l_{ev} = 1.5$ in. > 1.0 in. from Table J3.4

$l_{eh} = 1.5 \text{ in.} \geq 2d_b = 2(3/4) = 1.5 \text{ in.}$

5. $t_{plate} = 1/4 \text{ in.} < \left(\frac{d_b}{2} + 1/16\right) = \left(\frac{3/4}{2} + 1/16\right) = 7/16 \text{ in.}$

Step 2: Determine the bolt shear strength considering eccentricity.

From *Manual* Table 7-1,

$$r_n/\Omega = 11.9 \text{ kips/bolt}$$

From *Manual* Table 10-9 the eccentricity to be included is

$$e = a/2 = 3.0/2 = 1.5 \text{ in.}$$

From *Manual* Table 7-6 the equivalent number of bolts to account for eccentricity in the connection can be determined by interpolating between $C = 3.75$ for $e_x = 1.0$ in. and $C = 3.32$ for $e_x = 2.0$ in. Thus, $C = 3.54$ and

$$R_n/\Omega = C r_n/\Omega = 3.54(11.9) = 42.1 \text{ kips}$$

Step 3: Determine the bolt bearing and tearout strength on the plate assuming it is concentrically loaded.

For the bottom bolt

$$l_c = l_{ev} - \frac{1}{2}d_h = 1.5 - \frac{1}{2}(3/4 + 1/16) = 1.09 < 2(3/4) = 1.5$$

Thus, tearout controls over bearing, and the nominal bolt strength is

$$R_n = 1.2 l_c t_p F_u = 1.2(1.09)(0.250)(65) = 21.3 \text{ kips}$$

The allowable bolt tearout strength is

$$R_n/\Omega = 21.3/2.00 = 10.7 \text{ kips}$$

Since this is less than the bolt shear strength, it will control the strength of this bolt.

For the other bolts

$$l_c = s - d_h = 3.0 - (3/4 + 1/16) = 2.19 > 2(3/4) = 1.5 \text{ in.}$$

Therefore, bearing controls over tearout, and the nominal bolt strength is

$$R_n = 2.4dt_pF_u = 2.4(3/4)(0.250)(65) = 29.3 \text{ kips}$$

The allowable bolt bearing strength is

$$R_n/\Omega = 29.3/2.00 = 14.7 \text{ kips}$$

Step 4: Determine the bolt bearing and tearout strength on the web.

For the beam web, the material is A992, the web thickness is 0.380 in. and the member is not coped. Because the bolts will not tearout and the thickness of the web is greater than the thickness of the plate, the beam web will not control. However, strength will still be determined for consistency.

For each bolt in bearing

$$R_n = 2.4dt_wF_u = 2.4(3/4)(0.380)(65) = 44.5 \text{ kips}$$

and the allowable strength in the web is

$$R_n/\Omega = 44.5/2.00 = 22.3 \text{ kips}$$

Step 5: Determine the allowable connection strength based on the bolts in holes considering only the plate and web without considering eccentricity,

Bottom bolt

Plate tearout	10.7 kips
Web bearing	22.3 kips

Other 3 bolts

Plate bearing	14.7 kips
Web bearing	22.3 kips

Thus, the connection strength based on the bolts in holes without considering eccentricity or bolt shear is the sum of plate tearout for the bottom bolt plus 3 times the plate bearing for the other bolts. Thus,

$$R_n/\Omega = 10.7 + 3(14.7) = 54.8 \text{ kips}$$

This is greater than the strength determined for bolt shear, including eccentricity, in Step 2. Thus, connection strength based on the bolts is as given in Step 2,

$$R_n/\Omega = 42.1 \text{ kips}$$

Step 6: Determine the block shear strength of the plate.

Calculating the areas,

$$A_{nt} = \left(l_{eh} - \frac{1}{2}(d_h + 1/16)\right)t_w$$
$$= \left(1.5 - \frac{1}{2}(3/4 + 1/8)\right)(0.250) = 0.266 \text{ in.}^2$$
$$A_{gv} = lt_w$$
$$= 10.5(0.250) = 2.63 \text{ in.}^2$$
$$A_{nv} = (l - (n - 0.5)(d_h + 1/16))t_w$$
$$= (10.5 - 3.5(3/4 + 1/8))(0.250) = 1.86 \text{ in.}^2$$

Determine the tension rupture strength

$$F_u A_{nt} = 65(0.266) = 17.3 \text{ kips}$$

Consider shear yield and shear rupture and select the one with least strength; thus,

$$0.6F_y A_{gv} = 0.6(50)(2.63) = 78.9 \text{ kips}$$
$$0.6F_u A_{nv} = 0.6(65)(1.86) = 72.5 \text{ kips}$$

Selecting the shear rupture term and combining it with the tension rupture term gives a connection allowable block shear strength, with $U_{bs} = 1.0$, of

$$R_n/\Omega = (72.5 + 1.0(17.3))/2.00 = 44.9 \text{ kips}$$

Step 7: Determine the allowable shear yield strength of the plate. The gross shear area is

$$A_{gv} = lt_p = 12.0(0.250) = 3.00 \text{ in.}^2$$

and

$$V_n/\Omega = 0.6F_y A_{gv}/\Omega = (0.6(50)(3.00))/1.50 = 60.0 \text{ kips}$$

Step 8: Determine the allowable shear rupture strength of the plate. The net shear area is

$$A_{nv} = (l - n(d_h + 1/16))t_p = (12.0 - 4(3/4 + 1/8))(0.250) = 2.13 \text{ in.}^2$$

and

$$V_n/\Omega = 0.6F_u A_{nv}/\Omega = (0.6(65)(2.13))/2.00 = 41.5 \text{ kips}$$

Step 9: Determine the connection allowable strength based on the plate.

Block shear	44.9 kips
Shear yield	60.0 kips
Shear rupture	41.5 kips

Thus, the connection design strength for the plate is based on shear rupture of the plate.

$$R_n/\Omega = 41.5 \text{ kips}$$

Step 10: Determine the required weld size and minimum thickness of the supporting column flange and confirm that these are met.

The conventional configuration requires that the plate be welded to the supporting member through a pair of fillet welds on each side of the plate, with the weld leg width $w = (5/8)t_p$. Therefore, use a 3/16 in. fillet weld. This develops the strength of either a 36 ksi or 50 ksi plate and therefore does not require any further limit states checking. However, this does not check for the strength of the supporting member.

Step 11: Determine the minimum thickness of the supporting member to match the weld rupture strength.

Using the minimum thickness given by *Manual* Equation 9-6

$$t_{\min} = \frac{3.09D}{F_u} = \frac{3.09(3)}{65} = 0.143 \text{ in.}$$

This minimum is significantly less than the W14×90 flange thickness, $t_f = 0.710$ in.

Even if the minimum thickness had not been exceeded by the column flange thickness, the fact that the plate is welded to the column flange at the column flange-web junction would provide sufficient base metal strength.

Step 12: Determine the controlling limit state and allowable strength of the connection.

The allowable strength is controlled by shear rupture of the plate, where

$$R_n/\Omega = 41.5 \text{ kips}$$

Step 13: Using the more conservative approach to determine bolt strength discussed above, select the bolt with the least strength from Step 5 and bolt shear. For the bottom bolt in the plate, tearout controls.

$$r_n/\Omega = 10.7 \text{ kips}$$

Applying the factor from Step 2 for the equivalent number of bolts, the strength for bolts in holes, accounting for eccentricity is

$$\phi R_n = C\phi r_n = 3.54(10.7) = 37.9 \text{ kips}$$

Comparing this to the strength for bolts in holes determined in Step 5 shows this to be a bit more conservative approach.

Then, comparing this to the controlling limit state determined in Step 12 shows that, with this approach, the controlling limit state is now bolt strength in holes. Thus

$$\phi R_n = 37.9 \text{ kips}$$

11.9 SEATED CONNECTIONS

An unstiffened seated connection is shown in Figure 11.1d, and a stiffened seated connection is shown in Figure 11.1e. These connections are typically used to attach a beam to the web of a column. They can also be used to add capacity to other types of existing connections in a retrofit situation. Because of their simplicity, they are fairly easy connections to erect. They have very few parts: a seat angle, a connection to the supporting member through welds or bolts, a limited connection to the supported member, and a top connection to ensure stability of the supported beam. All of the force is transferred through bearing of the beam on the seat and then through the connection of the seat to the supporting member. When the seat lacks sufficient strength in bending of the top leg, it can be stiffened to produce the stiffened seated connection.

The seat can be welded or, as is usually the case, bolted to the supporting member. The connection to the supported member is not designed for a specific strength when only a vertical force is being transferred. The seated connection performs excellently as a simple connection. It can rotate sufficiently about the bottom of the beam without imposing any significant moment to the supporting member.

The simplicity of this connection results in relatively few limit states to be checked. Because the transfer of force between the beam and seat is through the bearing of the beam on the seat, the limit states of beam web yielding and beam web local crippling must be checked. These limit states were introduced in Section 6.15. The outstanding leg of the seat angle must be checked for the limit states of flexural yielding and shear yielding, and the connection to the supporting member, bolts, or welds must be checked for their appropriate limit states. In summary, the potential limit states are

1. Beam
- **a.** Web local yielding
- **b.** Web local crippling

2. Seat angle
- **a.** Flexural yielding
- **b.** Shear yielding

3. Connector

a. Bolt or weld shear with eccentricity

The nominal strength for the limit state of web local yielding was discussed in Section 7.4.2. If the strength equation for a force applied at the end of a member, Equation J10-3 is rewritten to solve for the minimum required bearing length, it can be used for design. Thus, from *Specification* Section J10.2, for the limit state of web local yielding

$$l_{b\ \min} = \frac{R_n}{F_y t_w} - 2.5k \tag{11.1}$$

For the limit state of web local yielding, $\phi = 1.00$ and $\Omega = 1.50$.

Similarly, from *Specification* Section J10.3, for the limit state of web local crippling, an equation for the minimum required bearing length can be determined. However, there are three different strength equations, with the choice depending on the relationship l_b/d. Because it is most likely that l_b will be less than $d/2$ and very likely it will be less than $0.2d$, only *Specification* Equation J10-5a is treated here. The other situations can both be handled in the same way if need be. Thus, rearranging Equation J10-5a to solve for the minimum required bearing length for the limit state of web local crippling yields, for $l_b/d \le 0.2$,

$$l_{b\ \min} = \frac{d}{3}\left(\frac{R_n}{0.40t_w^2}\sqrt{\frac{t_w}{EF_y t_f}} - 1\right)\left(\frac{t_f}{t_w}\right)^{1.5} \tag{11.2}$$

For the limit state of web local crippling, $\phi = 0.75$ and $\Omega = 2.00$.

Because the unstiffened angle is a very flexible connection, the load levels usually considered are quite low. This tends to result in very small minimum required bearing lengths and, in some calculations, a negative minimum required bearing length. To offset this potential problem, the minimum bearing length for seated connections is taken as k_{det}. A review of *Manual* Table 1-1 shows two values for k: k_{des} is a dimension used in design calculations and is the smaller of the two values, and k_{det} is a dimension normally used in detailing and is used here because it is the larger of these two values. These two k's are the result of differences in production by different mills. They represent the extremes of the values actually found and the selection is made within a calculation to give a conservative answer.

The outstanding leg of the angle must be capable of supporting the beam reaction applied at an eccentricity from the critical section of the angle. This requires checking the limit state of flexural yielding of the leg. The angle shown in Figure 11.11a is an unstiffened seat angle. The critical section for both flexure and shear is taken as 3/8 in. out from the face of the vertical leg, the radius of the fillet. The eccentricity is measured from this line to the midpoint of the minimum required bearing length of the beam on the angle, l_b. Thus, the eccentricity is

$$e = l_b/2 + 1/2 \text{ in. setback } + 1/4 \text{ in. under run } - (t_a + 3/8) = l_b/2 + 3/8 - t_a \tag{11.3}$$

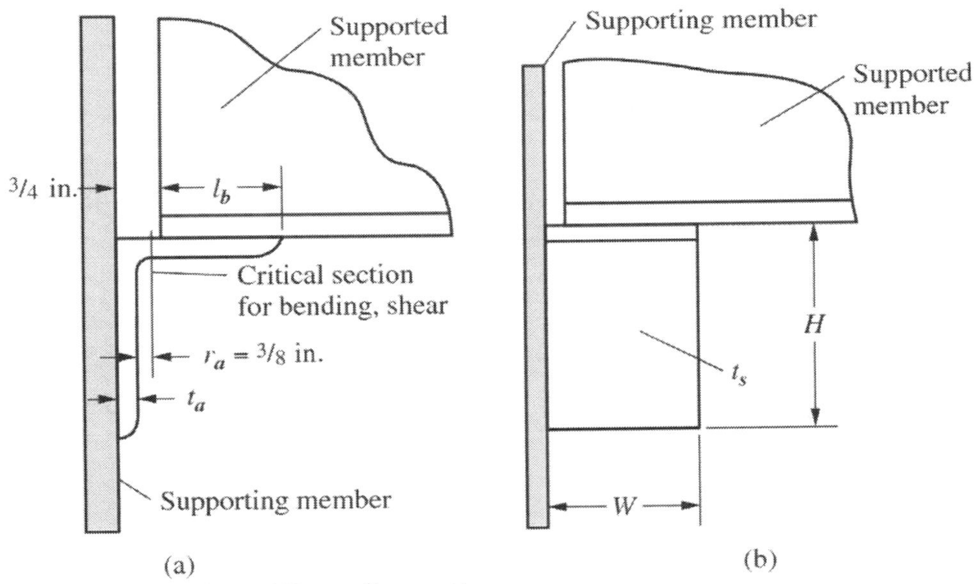

Figure 11.11 Seated Beam Connection

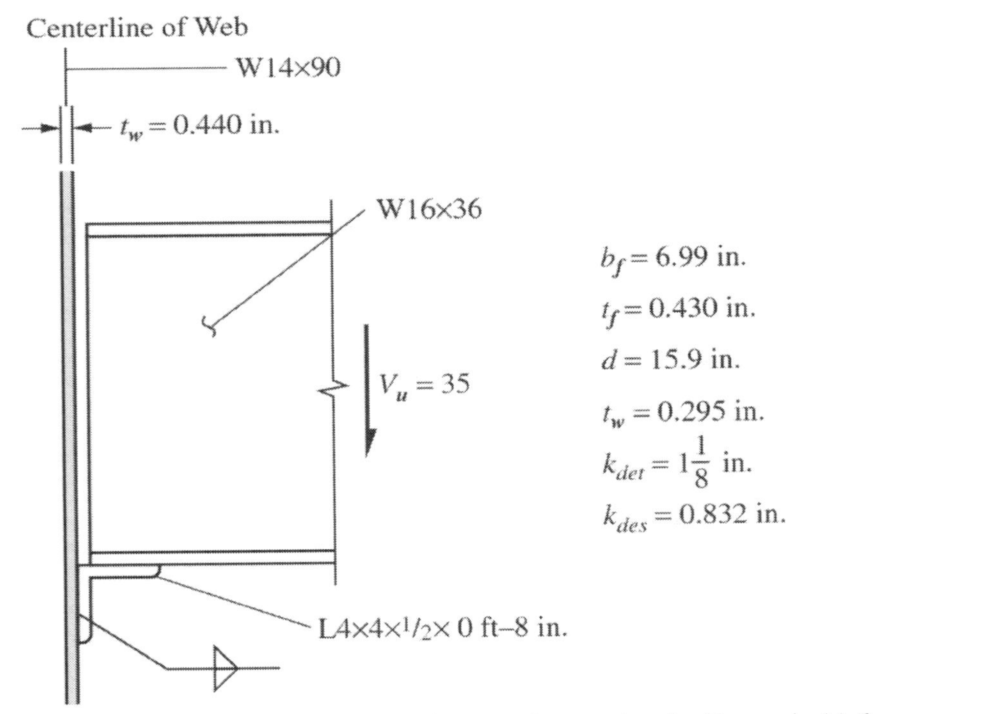

Figure 11.12 Unstiffened Seated Beam Connection for Example 11.6

The nominal beam reaction strength, for the limit state of flexural yielding, is based on the plastic section modulus of the seat leg of the angle. Thus,

$$Z = \frac{l_a t_a^2}{4}$$

and

$$R_n = \frac{M_n}{e} = \frac{F_y Z}{e} = \frac{F_y\left(\frac{l_a t_a^2}{4}\right)}{e} \tag{11.4}$$

Because this is based on flexural yielding, $\phi = 0.9$ and $\Omega = 1.67$. Equation 11.4 can then be solved for the thickness of the angle as a function of the required nominal reaction as

$$t_a = \sqrt{\frac{4R_n e}{F_y l_a}} \tag{11.5}$$

For the limit state of shear yielding, the gross area of the angle leg at this same location is used. Because this is beyond the angle fillet, the angle thickness is used to determine the gross area and

$$R_n = 0.6F_y l_a t_a \tag{11.6}$$

For the limit state of shear yielding, $\phi = 1.0$ and $\Omega = 1.5$.

EXAMPLE 11.6a
Seated Connection Design by LRFD

Goal: Design an unstiffened welded seated connection.

Given: An unstiffened welded seated connection is shown in Figure 11.12. A W16×36 beam is framing into a W14×90 column. The beam has an LRFD required strength of $R_u = 35$ kips. The beam and column are A992 and the angle is A572 Gr. 50. From *Manual* Table 1-1, for the W16×36, k_{det} = 1-1/8 in. and k_{des} = 0.832 in.

SOLUTION

Step 1: Determine the minimum required bearing length for web yielding using Equation 11.1.

$$l_{b\ \min} = \frac{R_n}{F_y t_w} - 2.5k = \frac{35/(1.0)}{(50)(0.295)} - 2.5(0.832)$$
$$= 0.293\text{ in.} < k_{det} = 1\tfrac{1}{8}\text{ in.}$$

Step 2: Determine the minimum required bearing length for web local crippling using Equation 11.2, assuming that $l_b/d \le 0.2$.

$$l_{b\ \min} = \frac{d}{3}\left(\frac{R_n}{0.40t_w^2}\sqrt{\frac{t_w}{EF_y t_f}} - 1\right)\left(\frac{t_f}{t_w}\right)^{1.5}$$

$$= \frac{15.9}{3}\left(\frac{35/0.75}{(0.40)(0.295)^2}\sqrt{\frac{0.295}{29{,}000(50)(0.430)}} - 1\right)\left(\frac{0.430}{0.295}\right)^{1.5}$$

$$= -0.726 < k_{\text{det}}$$

Thus, $l_b = k_{det} = 1.125$ in. and $l_b/d = 1.125/15.9 = 0.0708 \leq 0.2$, so the correct equation has been used and the bearing length is taken as the minimum required length of 1.125 in.

Step 3: Determine the eccentricity to be used in calculating angle thickness.

Assume an angle thickness of 1/2 in. From Equation 11.3

$$e = l_b/2 + 3/8 - t_a = 1.125/2 + 3/8 - 1/2 = 0.438 \text{ in.}$$

Step 4: Determine the minimum required angle thickness based on the limit state of flexural yielding.

Assume an 8.0 in. long angle so that it extends beyond the beam flange on both sides and use Equation 11.5. Thus

$$t_{req} = \sqrt{\frac{4R_n e}{F_y l}} = \sqrt{\frac{4(35/0.9)(0.438)}{(50)(8.0)}} = 0.413 \text{ in.}$$

Step 5: Determine the minimum angle thickness for the limit state of shear yielding using Equation 11.6. Thus,

$$t_{\min} = \frac{R_n}{0.6F_y L} = \frac{(35/1.0)}{0.6(50)(8.0)} = 0.146 \text{ in.}$$

Step 6: Check the selected angle thickness.

The 1/2 in. angle provides a thickness greater than each of the minimums determined in steps 4 and 5. Thus, the 1/2 in. angle is adequate.

Step 7: Determine the required weld size.

For parallel welds along the sides of the seat angle, use *Manual* Table 8-4 with the angle of the load equal to 0 degrees to account for the eccentricity.

The eccentricity for the weld is taken from the center of bearing to the face of the supporting column, which yields

$$e = \frac{3}{4} + \frac{1.125}{2} = 1.31 \text{ in.}$$

Assuming an angle with a 4.0 in. vertical leg,

$$a = \frac{e}{l} = \frac{1.31}{4.0} = 0.328$$

From *Manual* Table 8-4, the coefficient is determined for $k = 0$ through interpolation as $C = 2.97$, so the minimum weld is

$$D_{\min} = \frac{R_u}{\phi C C_1 l} = \frac{35}{0.75(1.0(2.97)(4.0))} = 3.93$$

Therefore, the calculated minimum weld is 1/4 in. and the angle is an L4×4×1/2, as shown in Figure 11.12.

Note: In addition to the seat angle, a top clip angle is needed to provide lateral stability. This angle is not normally designed to support any load and is usually a 1/4 in. angle attached with two bolts to the beam and two bolts to the supporting member.

EXAMPLE 11.6b
Seated Connection Design by ASD

Goal: Design an unstiffened welded seated connection.

Given: An unstiffened welded seated connection is shown in Figure 11.12. A W16×36 beam is framing into a W14×90 column. The beam has an ASD required strength of $R_a = 23\text{ kips}$. The beam and column are A992 and the angle is A572 Gr. 50. From *Manual* Table 1-1, for the W16×36, k_{det} = 1-1/8 in. and k_{des} = 0.832 in.

SOLUTION

Step 1: Determine the minimum required bearing length for web yielding using Equation 11.1.

$$l_{b\ \min} = \frac{R_n}{F_y t_w} - 2.5k = \frac{1.5(23)}{(50)(0.295)} - 2.5(0.832)$$

$$= 0.259\text{ in.} < k_{det} = 1 - 1/8 \text{ in.}$$

Step 2: Determine the minimum required bearing length for web local crippling using Equation 11.2, assuming that $l_b/d \le 0.2$.

$$l_{b\ \min} = \frac{d}{3}\left(\frac{R_n}{0.40t_w^2}\sqrt{\frac{t_w}{EF_y t_f}} - 1\right)\left(\frac{t_f}{t_w}\right)^{1.5}$$

$$= \frac{15.9}{3}\left(\frac{2.00(23)}{(0.40)(0.295)^2}\sqrt{\frac{0.295}{29{,}000(50)(0.430)}} - 1\right)\left(\frac{0.430}{0.295}\right)^{1.5}$$

$$= -0.849 < k_{\text{det}}$$

Thus, $l_b = k_{det} = 1.125$ in. and $l_b/d = 1.125/15.9 = 0.0708 \leq 0.2$, so the correct equation has been used and the bearing length is taken as the minimum required length of 1.125 in.

Step 3: Determine the eccentricity to be used in calculating angle thickness.

Assume an angle thickness of 1/2 in. From Equation 11.3

$$e = l_b/2 + \ 3/8 - t_a = 1.125/2 + 3/8 - 1/2 = 0.438 \text{ in.}$$

Step 4: Determine the minimum required angle thickness based on the limit state of flexural yielding.

Assume an 8.0 in. long angle so that it extends beyond the beam flange on both sides and use Equation 11.5. Thus

$$t_{req} = \sqrt{\frac{4R_n e}{F_y l}} = \sqrt{\frac{4(1.67(23))(0.438)}{50(8.0)}} = 0.410 \text{ in.}$$

Step 5: Determine the minimum angle thickness for the limit state of shear yielding using Equation 11.6. Thus,

$$t_{\min} = \frac{R_n}{0.6F_y L} = \frac{1.50(23)}{0.6(50)(8.0)} = 0.144 \text{ in.}$$

Step 6: Check the selected angle thickness.

The 1/2 in. angle provides a thickness greater than each of the minimums determined in steps 4 and 5. Thus, the 1/2 in. angle is adequate.

Step 7: Determine the required weld size.

For parallel welds along the sides of the angle, use *Manual* Table 8-4 with the angle of the load equal to 0 degrees to account for the eccentricity.

The eccentricity for the weld is taken from the center of bearing to the face of the supporting column, which yields

$$e = \frac{3}{4} + \frac{1.125}{2} = 1.31 \text{ in.}$$

Assuming an angle with a 4.0 in. vertical leg,

$$a = \frac{e}{l} = \frac{1.31}{4.0} = 0.328$$

From *Manual* Table 8-4, the coefficient is determined for $k = 0$ through interpolation as $C = 2.97$, so the minimum weld is

$$D_{\min} = \frac{\Omega R_u}{CC_1 l} = \frac{2.00(23)}{1.0(2.97)(4.0)} = 3.87$$

Therefore, the calculated minimum weld is 1/4 in. and the angle is an L4×4×1/2, as shown in Figure 11.12.

Note: In addition to the seat angle, a top clip angle is needed to provide lateral stability. This angle is not normally designed to support any load and is usually a 1/4 in. angle attached with two bolts to the beam and two bolts to the supporting member.

A stiffened seated connection is shown in Figure 11.1e, and a detail of the stiffened seated connection is shown in Figure 11.11b. This type of connection is used when the loads are too large to be supported by an unstiffened seat. The stiffener can be a single plate, the stem of a tee, or the back-to-back legs of a pair of angles. A plate on top of the stiffener provides the bearing surface and the location of the bolts required to attach the beam flange to the seat. The limit states for this connection are the same as for the unstiffened connection already discussed but also include the limit state of punching shear in the supporting member because of the potential for a large force from the stiffener being carried to the support.

Punching shear on a column web will not be critical if the following parameters set in Part 10 of the *Manual* are met:

1. The simplified approach is applicable to columns in the following depths with weights per foot no less than: W14×43, W12×40, W10×33, W8×24, W6×20, and W5×16.

2. The width of the stiffener W is no greater than 7.0 in.

3. The beam is bolted, not welded, to the bearing plate at a point no greater than $W/2$ or 2-5/8 in. from the column face.

4. The top angle has a minimum thickness of 1/4 in.

The eccentricity of the beam reaction is taken as $0.8W$ in determining the strength of the weld or bolt group connecting the seat to the supporting member. Part 10 of the *Manual* provides tables for the design of stiffened seated connections.

11.10 LIGHT BRACING CONNECTIONS

Bracing connections have as many potential variations as do the simple shear connections discussed above. Figure 11.13a shows a bolted-welded tension brace connection, and Figure 11.13b shows a welded-bolted tension brace connection. It is also permissible to have a welded-welded connection or a bolted-bolted connection, although these are not illustrated here.

The limit states for these connections have already been addressed. For the bolted-welded connection, shown in Figure 11.13a, they are

1. Angles
- **a.** Tension yielding
- **b.** Tension rupture
- **c.** Bolt bearing and tearout
- **d.** Block shear rupture

2. Bolts
- **a.** Shear rupture

3. Gusset plate
- **a.** Tension yielding
- **b.** Tension rupture
- **c.** Bolt bearing and tearout

4. Welds
- **a.** Weld rupture for combined tension and shear

For the welded-bolted connection where the gusset is replaced with a tee, Figure 11.13b, the limit states are

1. Angles
- **a.** Tension yielding
- **b.** Tension rupture

2. Welds
- **a.** Weld rupture

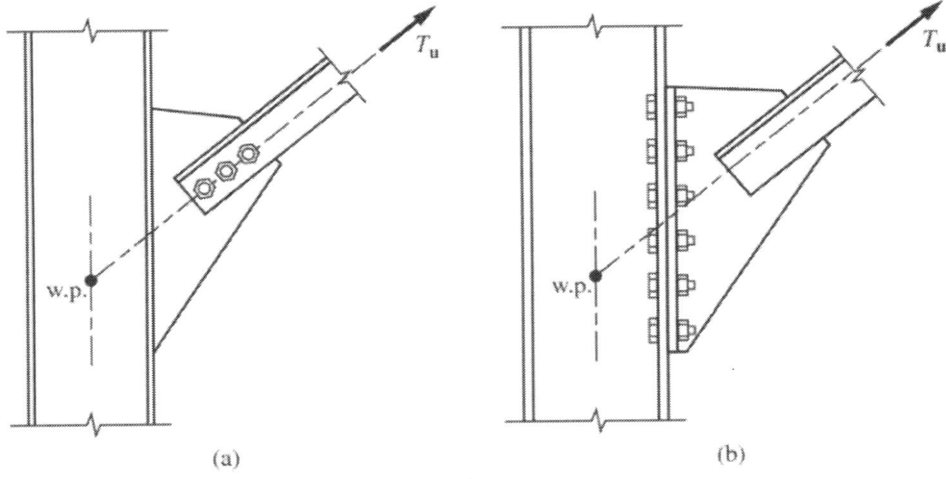

Figure 11.13 Light Bracing Connections

3. Tee stem
- **a.** Tension yielding
- **b.** Tension rupture
- **c.** Block shear rupture
- **d.** Shear yielding

4. Tee flange
- **a.** Flange bending
- **b.** Shear yielding
- **c.** Shear rupture
- **d.** Bolt bearing and tearout
- **e.** Block shear rupture

5. Bolts
- **a.** Combined shear and tension

6. Column flange
- **a.** Flange bending
- **b.** Bolt bearing and tearout

7. Column web
- **a.** Web local yielding

Although they appear to be simple connections, light bracing connections require checking for quite a number of different limit states. Three of these limit states have not previously been addressed: (1) tension rupture and tension yield within the gusset, (2) limit state of bolt rupture due to combined shear and tension, and (3) high-strength bolts in tension with prying action.

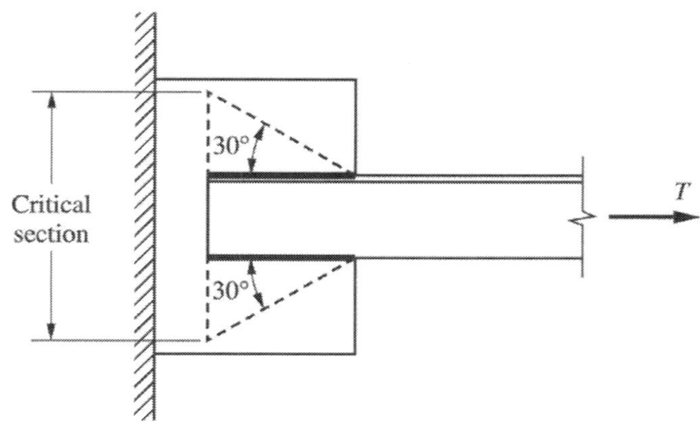

Figure 11.14 Whitmore Section in a Gusset Plate

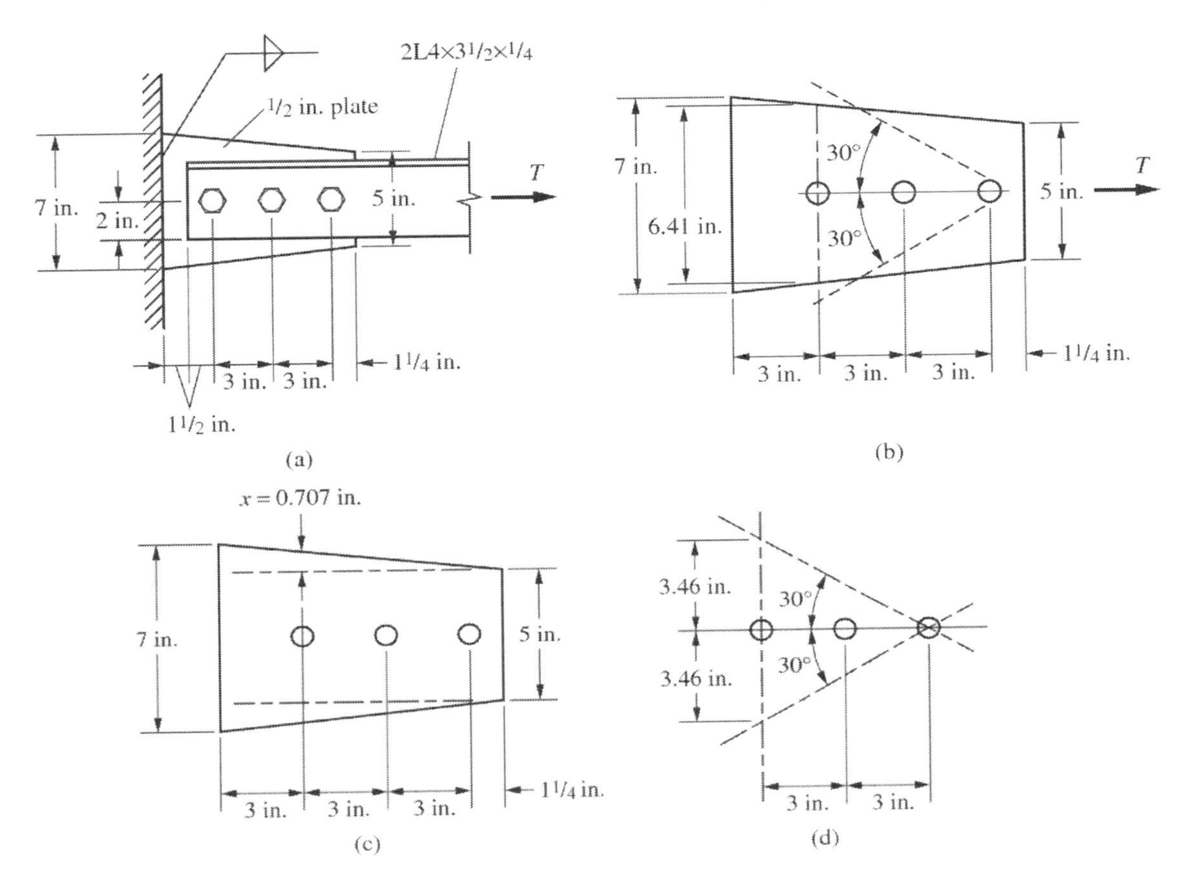

Figure 11.15 Tapered Gusset Plate for Example 11.7

Figure 11.14 shows a single-angle brace attached to a gusset plate with equal length welds along the sides of the angle. Research has shown that the distribution of stresses from the brace through the welds into the gusset is such that the entire width of the gusset is not effective if it exceeds the width defined by a 30-degree angle spreading to both sides from the beginning of the connection to the end of the connection. This width is defined as the Whitmore section and identified in Figure 11.14 as the critical section. When the gusset is wider than the Whitmore section, only the Whitmore section can be considered to resist the force. When the Whitmore section is wider than the available plate dimension, only the width of the plate at the connection end can be considered to resist the force. If the connection is bolted rather than welded as shown, the Whitmore section distribution starts at the first bolt and proceeds to the last bolt in the connection. This is illustrated in Example 11.7.

EXAMPLE 11.7
Tapered Gusset Plate Strength

Goal: Determine the available strength of a gusset plate connected to a double-angle brace. Consider only tension yielding and tension rupture of the plate.

Given: The gusset plate portion of the connection is shown in Figure 11.15. The brace is a double angle bolted with 3/4 in. bolts to a tapered gusset with the dimensions as shown. The plate is A572 Gr. 50 steel.

SOLUTION

Step 1: Determine the width of the plate at the location of the last bolt.

This is also the location of the Whitmore section. By proportions, as seen in Figure 11.15c,

$$\frac{x}{1.0} = \frac{7.25}{10.25}$$

Therefore,

$$x = 0.707 \text{ in.}$$

and the width of the plate at this location is

$$l = 5.0 + 2(0.707) = 6.41 \text{ in.}$$

Step 2: Determine the width of the Whitmore section using the geometry shown in Figure 11.15d.

$$l_W = 2(6\tan(30)) = 6.93 \text{ in.}$$

Step 3: Determine the width to be used to determine the plate strength.

Because the actual plate width at this location is less than the Whitmore section, the actual plate width is used to determine the strength of the plate.

Step 4: Determine the nominal strength for the limit state of plate yielding.

The gross area at the critical location is

$$A_g = 6.41(1/2) = 3.21 \text{ in.}^2$$

and the nominal tensile strength is

$$T_n = F_y A_g = 50(3.21) = 161 \text{ kips}$$

For LRFD Step 5: The design tensile strength for the limit state of yielding is

$$\phi T_n = 0.9(161) = 145 \text{ kips}$$

For ASD Step 5: The allowable tensile strength for the limit state of yielding is

$$T_n/\Omega = 161/1.67 = 96.4 \text{ kips}$$

Step 6: Determine the nominal strength for the limit state of plate rupture.

The effective net area at the same location is

$$A_e = A_n = \left(l - \left(d_h + 1/16\right)\right)t_p = \left(6.41 - \left(3/4 + 1/8\right)\right)\left(1/2\right) = 2.77 \text{ in.}^2$$

Therefore, the nominal tensile strength is

$$T_n = F_u A_e = 65\left(2.77\right) = 180 \text{ kips}$$

For LRFD Step 7: The design tensile strength for the limit state of rupture is

$$\phi T_n = 0.75\left(180\right) = 135 \text{ kips}$$

For ASD Step 7: The allowable tensile strength for the limit state of rupture is

$$T_n/\Omega = 180/2.00 = 90.0 \text{ kips}$$

For LRFD Step 8: Determine the design tensile strength of the gusset plate based on the controlling limit state.

The design tensile strength of the gusset is

$$\phi T_n = 135 \text{ kips}$$

based on the limit state of rupture.

For ASD Step 8: Determine the allowable tensile strength of the gusset plate based on the controlling limit state.

The allowable tensile strength of the gusset is

$$T_n/\Omega = 90.0 \text{ kips}$$

based on the limit state of rupture.

EXAMPLE 11.8
Uniform-Width Gusset Plate Strength

Goal: Determine the available strength of a gusset plate of uniform width. Consider only tension yielding and tension rupture of the plate.

Given: Determine the available strength of a gusset plate for the same situation as in Example 11.7, except that the plate has a uniform width of 8.0 in. The plate is again A572 Gr. 50 steel and the bolts are 3/4 in.

SOLUTION

Step 1: Determine the controlling width at the critical section.

The width of the plate at the location of the last bolt is given as 8.0 in.

The width of the Whitmore section is determined using the same geometry as shown in Figure 11.15d, which again yields

$$l_W = 2\left(6\tan\left(30\right)\right) = 6.93 \text{ in.}$$

In this case, the actual plate width is greater than the Whitmore section, so the Whitmore section width is used to determine the strength of the plate.

Step 2: Determine the nominal tensile strength for the limit state of plate yielding.

The gross area at the critical location is

$$A_g = 6.93\left(1/2\right) = 3.47 \text{ in.}^2$$

The nominal tensile strength is

$$T_n = F_y A_g = 50\left(3.47\right) = 174 \text{ kips}$$

For LRFD Step 3: The design tensile strength for yielding is

$$\phi T_n = 0.9\left(174\right) = 157 \text{ kips}$$

For ASD Step 3: The allowable tensile strength for yielding is

$$T_n/\Omega = 174/1.67 = 104 \text{ kips}$$

Step 4: Determine the nominal strength for the limit state of plate rupture.

The effective net area at the same location is

$$A_e = A_n = \left(l_W - \left(d_h + 1/16\right)\right)t_p = \left(6.93 - \left(3/4 + 1/8\right)\right)\left(1/2\right) = 3.03 \text{ in.}^2$$

Therefore,

$$T_n = F_u A_e = 65(3.03) = 197 \text{ kips}$$

For LRFD Step 5: The design tensile strength for tension rupture is

$$\phi T_n = 0.75(197) = 148 \text{ kips}$$

For ASD Step 5: The allowable tensile strength for tension rupture is

$$T_n/\Omega = 197/2.00 = 98.5 \text{ kips}$$

For LRFD Step 6: Determine the design tensile strength of the gusset plate based on the controlling limit state.

The design tensile strength of the uniform-width gusset is

$$\phi T_n = 148 \text{ kips}$$

based on the limit state of rupture.

For ASD Step 6: Determine the allowable tensile strength of the gusset plate based on the controlling limit state.

The allowable tensile strength of the uniform-width gusset is

$$T_n/\Omega = 98.5 \text{ kips}$$

based on the limit state of rupture.

Note: There is no advantage to using a plate wider than the Whitmore section, 6.93 in.

The next limit state to address concerns high-strength bolts in combined shear and tension. This was briefly addressed in Sections 10.7.6 and 10.7.7. A bolt loaded in combined shear and tension has a reduced ability to resist shear in a bearing-type connection due to the presence of tension. In a slip-critical connection, the tension reduces the contact force and, thus, lowers the shear required to cause the connection to slip. These reductions must be accounted for in the design of connections where these combined limit states occur.

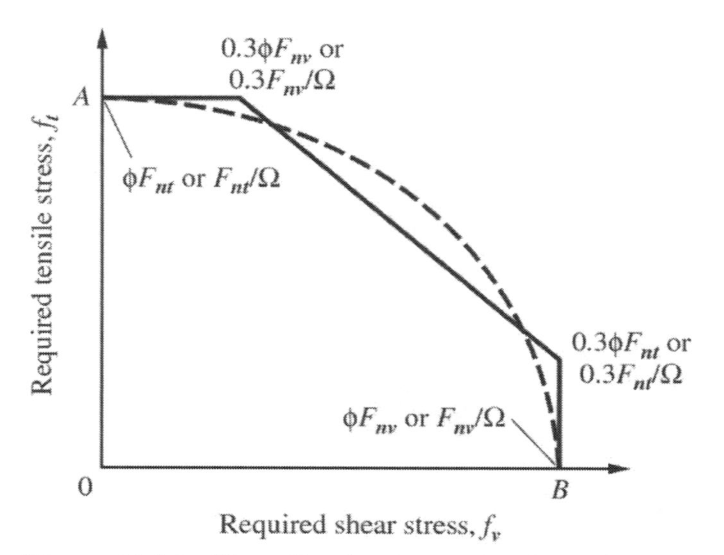

Figure 11.16 Shear-Tension Interaction for Bolts

Tests have shown that the interaction of shear and tension in a bearing-type connection can be fairly well predicted through an elliptical interaction curve. However, for simplicity, the *Specification* has adopted three straight lines to approximate the ellipse. Both the ellipse and straight-line interaction are shown in Figure 11.16. Section J3.8 gives two equations for this interaction, one for ASD and one for LRFD, where the nominal tensile stress including the effects of shear-tension interaction is

$$F'_{nt} = 1.3F_{nt} - \frac{F_{nt}}{\phi F_{nv}} f_{rv} \leq F_{nt} \quad \text{(LRFD)} \qquad \text{(AISC J3-3a)}$$

$$F'_{nt} = 1.3F_{nt} - \frac{\Omega F_{nt}}{F_{nv}} f_{rv} \leq F_{nt} \quad \text{(ASD)} \qquad \text{(AISC J3-3b)}$$

where

F_{nt} = nominal tensile stress for tension alone from *Specification* Table J3.2
F_{nv} = nominal shear stress for shear alone from *Specification* Table J3.2
F_{rv} = required shear stress

These two equations can be combined and written in terms of nominal strength if the required shear stress, f_v, is combined with ϕ and Ω to give the nominal shear stress including the effects of shear-tension interaction. Thus

$$F'_{nv} = \frac{f_{rv}}{\phi} \text{ (LRFD) or } F'_{nv} = \Omega f_{rv} \text{ (ASD)}$$

so that

$$F'_{nt} = 1.3F_{nt} - \frac{F_{nt}}{F_{nv}} F'_{nv} \leq F_{nt} \qquad (11.7)$$

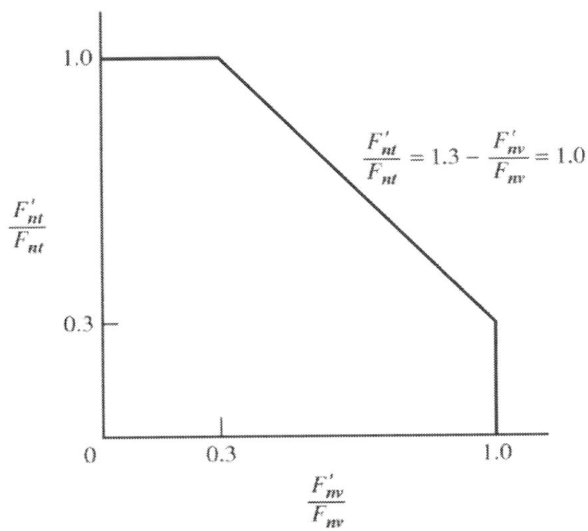

Figure 11.17 Modified Shear-Tension Interaction for Bolts

If Equation 11.7 is divided by F_{nt}, another form of the interaction equation results as

$$\frac{F'_{nt}}{F_{nt}} = 1.3 - \frac{F'_{nv}}{F_{nv}} \leq 1.0 \tag{11.8}$$

This relationship is shown in Figure 11.17.

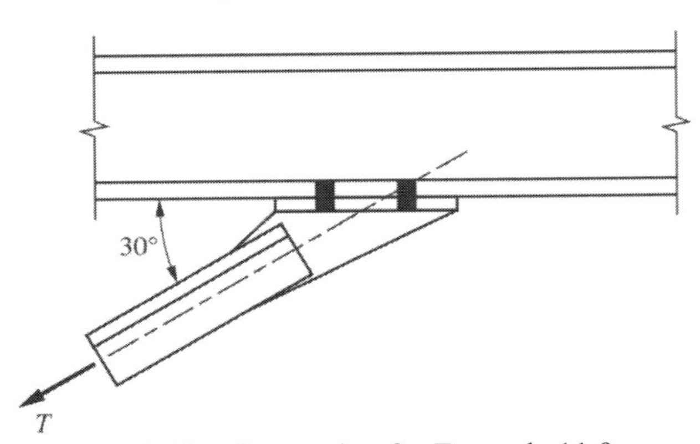

Figure 11.18 Connection for Example 11.9

EXAMPLE 11.9a ***Bolts in Combined Shear and Tension in Bearing Type Connections by LRFD***

Goal: Determine the strength of a connection using bolts in combined shear and tension and compare it to the applied load.

Given: An inclined hanger that supports a dead load of 10 kips and a live load of 50 kips is shown in Figure 11.18. The connection uses four 1.0 in. A325-N bolts. The area of a 1.0 in. bolt is 0.785 in.^{2}.

SOLUTION

Step 1: Determine the required strength for the appropriate load combination.

$$R_u = 1.2(10.0) + 1.6(50.0) = 92.0 \text{ kips}$$

Step 2: Determine the force assigned to each bolt in tension and shear.

$$\text{Bolt tension } = \sin(30)\left(\frac{92.0}{4}\right) = 11.5 \text{ kips}$$

$$\text{Bolt shear } = \cos(30)\left(\frac{92.0}{4}\right) = 19.9 \text{ kips}$$

$$\text{Bolt tension stress } = f_{rt} = \frac{11.5}{0.785} = 14.6 \text{ ksi}$$

$$\text{Bolt shear stress } = f_{rv} = \frac{19.9}{0.785} = 25.4 \text{ ksi}$$

Step 3: Determine the reduced nominal tensile stress.

The nominal shear and tensile stress values are, from *Specification* Table J3.2,

$$F_{nv} = 54 \text{ ksi}$$

$$F_{nt} = 90 \text{ ksi}$$

and the nominal shear stress including the effects of tension-shear interaction is

$$F'_{nv} = \frac{f_{rv}}{\phi} = \frac{25.4}{0.75} = 33.9 \text{ ksi}$$

Thus, from Equation 11.7

$$F'_{nt} = 1.3F_{nt} - \frac{F_{nt}}{F_{nv}}F'_{nv} = 1.3(90) - \left(\frac{90}{54}\right)(33.9) = 60.5 \le 90$$

Step 4: Check the design tensile stress versus the required tensile stress.

$$\phi F'_{nt} = 0.75(60.5) = 45.4 > 14.6 \text{ ksi}$$

Thus, by LRFD, the bolts are adequate.

EXAMPLE 11.9b
Bolts in Combined Shear and Tension in Bearing Type Connections by ASD SOLUTION

Goal: Determine the strength of a connection using bolts in combined shear and tension and compare it to the applied load.

Given: An inclined hanger that supports a dead load of 10 kips and a live load of 50 kips is shown in Figure 11.18. The connection uses four 1.0 in. A325-N bolts. The area of a 1.0 in. bolt is 0.785 in.^{2}

Step 1: Determine the required strength for the appropriate load combination.

$$R_a = 10.0 + 50.0 = 60.0 \text{ kips}$$

Step 2: Determine the force assigned to each bolt in tension and shear.

$$\text{Bolt tension } = \sin(30)\left(\frac{60.0}{4}\right) = 7.50 \text{ kips}$$

$$\text{Bolt shear } = \cos(30)\left(\frac{60.0}{4}\right) = 13.0 \text{ kips}$$

$$\text{Bolt tension stress } = f_{rt} = \frac{7.50}{0.785} = 9.55 \text{ ksi}$$

$$\text{Bolt shear stress } = f_{rv} = \frac{13.0}{0.785} = 16.6 \text{ ksi}$$

Step 3: Determine the reduced nominal tensile stress.

The nominal shear and tensile stress values are, from *Specification* Table J3.2,

$$F_{nv} = 54 \text{ ksi}$$
$$F_{nt} = 90 \text{ ksi}$$

and the nominal shear stress including the effects of tension-shear interaction is

$$F'_{nv} = \Omega f_{rv} = 2.00(16.6) = 33.2 \text{ ksi}$$

Thus, from Equation 11.7

$$F'_{nt} = 1.3F_{nt} - \frac{F_{nt}}{F_{nv}} F'_{nv} = 1.3(90) - \left(\frac{90}{54}\right)(33.2) = 61.7 \le 90$$

Step 4: Check the allowable tensile stress versus the required tensile stress.

$$F'_{nt}/\Omega = 61.7/2.00 = 30.9 > 9.55 \text{ ksi}$$

Thus, by ASD, the bolts are adequate.

When high-strength bolts are installed with an initial pre-tension, they act as a clamp, holding the two connected elements together. Figure 11.19 shows a typical tension hanger where the bolts are expected to carry the applied tension load. Any pre-tension from the bolt actually causes a compressive force to develop between the connected parts. Application of the applied load reduces the contact force but has little effect on the bolt tension, as long as contact is maintained between the plates. Once the plates are separated, the initial conditions have no influence and the bolt force, *2B*, must equal the applied load, *2T*.

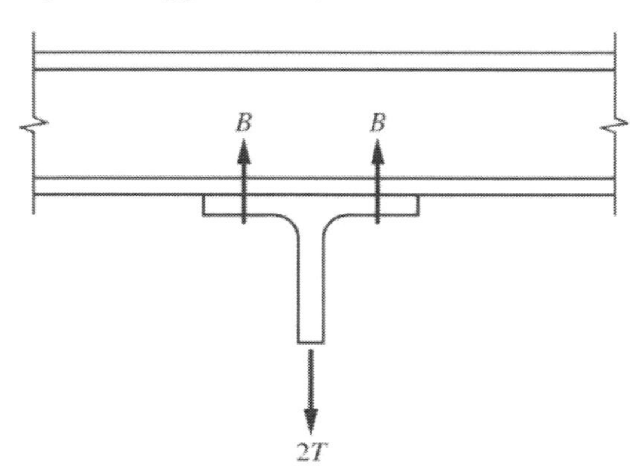

Figure 11.19 Hanger Connection with Bolts in Tension

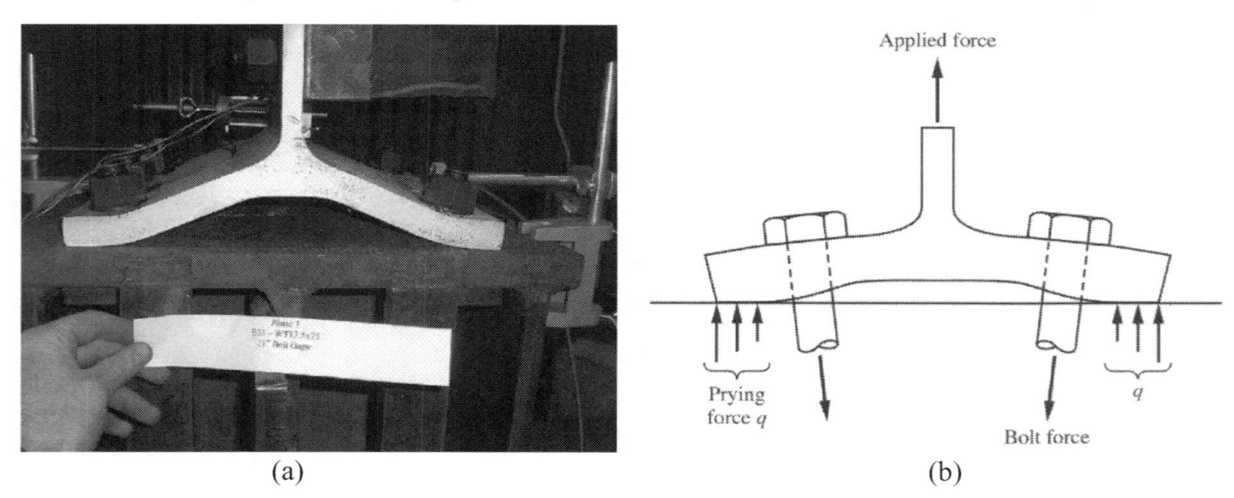

Figure 11.20 Tee Deformation with Prying Action
Part a courtesy Patrick McManus

If the attached element, in this case the flange of the tee, is permitted to deform, as shown in Figure 11.20a, additional forces develop at the tips of the flange. These additional forces, q, are shown in Figure 11.20b and are the result of prying action. They are called the prying forces. There is a relationship between the thickness of the flange and the prying force. When the thickness is large, the plate does not bend and no prying action takes place. When the thickness is small, bending of the plate may be extensive and the prying force may be large. Prying action may be completely eliminated in a design by selecting a sufficiently thick plate, although this may not be a practical solution.

The details for design of this type of connection including prying action are given in Part 9 of the *Manual*. It suggests that the minimum plate thickness needed to eliminate prying action be determined. If this is a reasonable thickness, no further action is required. If this thickness is not reasonable for the details of the design, a design that takes into account prying action should be undertaken with the goal of having a reasonable combination of strength and stiffness that results in an economical connection.

Figure 11.21 shows a WT section used as a hanger attached to a supporting member with bolts. The dimensions given are used to determine the relationship between the flexural strength of the flange and the applied load. The applied load is $2T$, so the load per bolt is T. It is not a simple matter to determine the actual moment in the flange. The design approach recognizes that the bolt head provides some stiffening to the flange so that the moment arm, b', taken as the distance from the stem to the edge of the bolt, will provide a good representation of the required moment. If there is no prying action, $q = 0$ and the required moment at the face of the stem is Tb'. It has also been found through tests that the flexural strength is best predicted by use of F_u rather than F_y. So, using a tributary width of plate associated with each bolt, p, the nominal moment strength is

$$M_n = F_u Z = F_u \frac{pt^2}{4} \tag{11.9}$$

$$\phi = 0.9 \text{ (LRFD)} \qquad \Omega = 1.67 \text{ (ASD)}$$

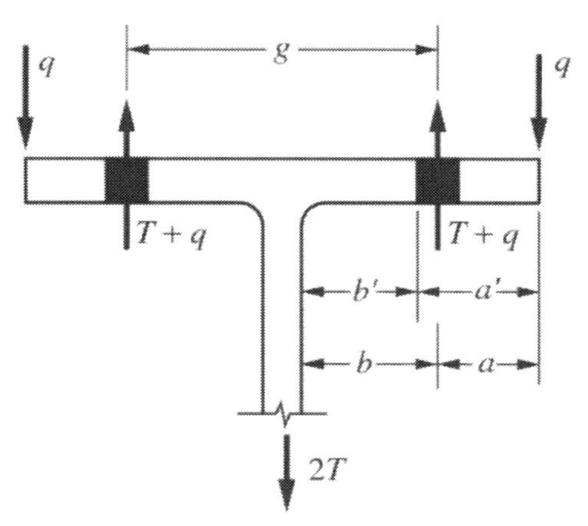

Figure 11.21 Force Equilibrium Considering Prying Action

Setting the required strength equal to the available strength, the minimum thickness required to eliminate prying action, $q = 0$, can be determined. Using the available tensile strength per bolt, T_c, for LRFD or ASD as appropriate, the *Manual* provides the equations for minimum thickness, t_c, so that prying action does not occur as

$$t_c = \sqrt{\frac{4T_c b'}{\phi p F_u}} \text{ (LRFD)} \quad t_c = \sqrt{\frac{4\Omega T_c b'}{p F_u}} \text{ (ASD)} \qquad (\textit{Manual } 9\text{-}30\text{a, b})$$

Since it is unlikely that the required bolt tensile strength will exactly equal the available bolt tensile strength, T_c in *Manual* Equation 9-30 will be taken as T_r in the following examples.

EXAMPLE 11.10a
Hanger Connection by LRFD

Goal: Determine whether the WT hanger connection is adequate to be designed without considering prying action.

Given: A WT9×48.5 section, A992 steel, is used as shown in Figure 11.21 to carry a dead load of 15 kips and a live load of 45 kips. Four 7/8 in. diameter A325 bolts are used in a 9 in. long fitting.

SOLUTION

Step 1: Determine the moment arm, b', from the web to the edge of the bolt. Based on the properties of the section.

$t_f = 0.870$ in., $t_w = 0.535$ in., $b_f = 11.1$ in., gage = 5-1/2 in., $p = 9.0/2 = 4.5$ in.

$$b = \frac{\text{gage} - t_w}{2} = \frac{(5.5 - 0.535)}{2} = 2.48 \text{ in.}$$

$$b' = b - \frac{d_b}{2} = 2.48 - \frac{7/8}{2} = 2.04 \text{ in.}$$

Step 2: Determine the force per bolt.

$$T_r = T_u = 1.2(15.0) + 1.6(45.0) = 90.0 \text{ kips}$$

$$T_r = T_u = \frac{90.0}{4} = 22.5 \text{ kips/bolt}$$

Step 3: Determine the minimum flange thickness to ignore prying action using *Manual* Equation 9-30a.

$$t_c = \sqrt{\frac{4T_r b'}{\phi p F_u}} = \sqrt{\frac{4(22.5)(2.04)}{0.9(4.5)(65)}} = 0.835 \text{ in.}$$

Step 4: Compare the available thickness with the required thickness.

$$t_c = 0.835 < t_f = 0.870 \text{ in.}$$

Because the actual flange thickness is greater than the minimum, the WT9×48.5 does not require consideration of prying action.

EXAMPLE 11.10b
Hanger Connection by ASD

Goal: Determine whether the WT hanger connection is adequate to be designed without considering prying action.

Given: A WT9×48.5 section, A992 steel, is used as shown in Figure 11.21 to carry a dead load of 15 kips and a live load of 45 kips. Four 7/8 in. diameter A325 bolts are used in a 9 in. long fitting.

SOLUTION

Step 1: Determine the moment arm, b', from the web to the edge of the bolt. Based on the properties of the section.

$$t_f = 0.870 \text{ in.}, t_w = 0.535 \text{ in.}, b_f = 11.1 \text{ in.},$$
$$\text{gage} = 5\text{-}1/2 \text{ in.}, p = 9.0/2 = 4.5 \text{ in.}$$

$$b = \frac{\text{gage} - t_w}{2} = \frac{(5.5 - 0.535)}{2} = 2.48 \text{ in.}$$

$$b' = b - \frac{d_b}{2} = 2.48 - \frac{7/8}{2} = 2.04 \text{ in.}$$

Step 2: Determine the force per bolt.

$$T_r = T_a = 15.0 + 45.0 = 60.0 \text{ kips}$$

$$T_r = T_a = \frac{60.0}{4} = 15.0 \text{ kips / bolt}$$

Step 3: Determine the minimum flange thickness to ignore prying action using *Manual* Equation 9-30b.

$$t_c = \sqrt{\frac{4\Omega T_a b'}{pF_u}} = \sqrt{\frac{4(1.67)(15.0)(2.04)}{4.5(65)}} = 0.836 \text{ in.}$$

Step 4: Compare the available thickness with the required thickness.

$$t_c = 0.836 < t_f = 0.870 \text{ in.}$$

Because the actual flange thickness is greater than the minimum, the WT9×48.5 does not require consideration of prying action.

11.11 BEAM BEARING PLATES AND COLUMN BASE PLATES

The connections discussed throughout this chapter transfer force through a series of connecting elements to a supporting member. Two other types of simple connections deserve mention here, the beam bearing plate and the column base plate, both addressed in Part 14 of the *Manual*. These plates transfer a force through direct bearing from one member to another member or directly to a support. Although these plates are used in two very different applications, their actual behavior is quite similar. Some aspects of beam bearing were discussed in Section 6.15, and column base plates were briefly discussed in Section 5.11.

For design of the plate, three properties must be determined: the width and breadth, which results in an appropriate area, and the thickness. The area of the plate is determined by assessing the limit states of the supporting member or material and those of the member applying the force to the plate. The thickness of the plate is determined through the limit state of flexural yielding of the plate.

To determine the required plate thickness for either type of plate, two primary assumptions are made: (1) the plate exerts a uniform pressure on the supporting material, and (2) the plate is treated as a cantilevered strip that is 1.0 in. wide. For a bending cross section 1.0 in. wide with a thickness, t_p, the nominal flexural strength for the limit state of yielding is

$$M_n = F_y Z = F_y\left(\frac{1.0t_p^2}{4}\right) \tag{11.10}$$

For a uniform contact pressure between the plate and the supporting material, f_p, and a cantilever length, l, the required moment strength for the cantilever is

$$M_r = \frac{f_p l^2}{2} \tag{11.11}$$

For LRFD, the required plate thickness can be obtained by setting the design moment equal to the required moment, where the required moment is obtained using $f_p = f_u$ and $\phi = 0.9$; thus,

$$\phi M_n = M_r$$

$$\frac{\phi F_y t_p^2}{4} = \frac{f_u l^2}{2}$$

which yields

$$t_p = 1.49l\sqrt{\frac{f_u}{F_y}} \quad (\text{LRFD}) \tag{11.12a}$$

Similarly, for ASD, using $f_p = f_a$ and $\Omega = 1.67$,

$$M_n/\Omega = M_r$$

$$\frac{F_y t_p^2}{4\Omega} = \frac{f_a l^2}{2}$$

which yields

$$t_p = 1.83l\sqrt{\frac{f_a}{F_y}} \quad (\text{ASD}) \tag{11.12b}$$

The determination of the cantilever distance, l, to be used in the case of a beam bearing plate or a column base plate is addressed in *Manual* Part 14. For a beam bearing plate, shown in Figure 11.22, the cantilever distance is measured from the point identified as the k distance from the web centerline to the edge of the plate, noted in the figure as n.

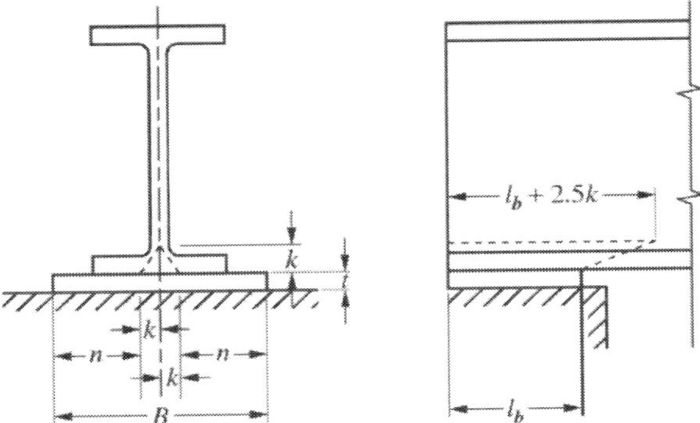

Figure 11.22 Beam Bearing Plate

Manual Table 1-1 provides two values for the section property k, k_{des} and k_{det}. The smaller value, k_{des}, called k-design, should be chosen so that, conservatively, a longer cantilever distance is used in the design. For column base plates, the cantilever dimensions are based on an assumption for the point under the column section where cantilever action begins. This is shown in Figure 11.23, where the cantilever dimension, n, is measured from a line $0.1b_f$ in from the flange tips and the cantilever dimension, m, is measured from a line $0.025d$ in from the flange face.

A third cantilever dimension is also given to account for those cases where the cantilever from the web, in the area between the flanges, is critical. The *Manual* gives this as $\lambda n'$, where $n' = \sqrt{b_f d}/4$ and λ can be conservatively taken as 1.0.

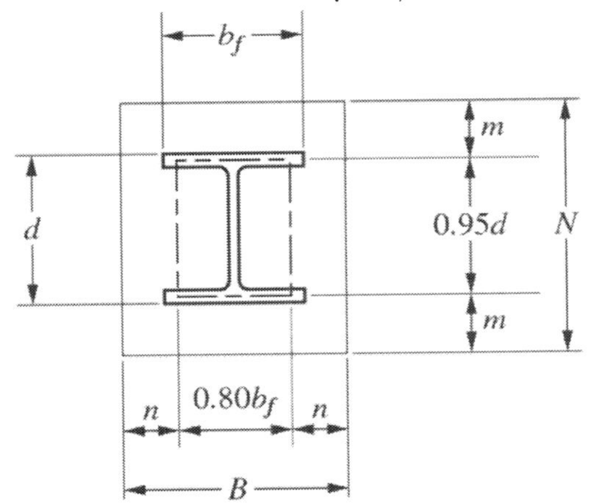

Figure 11.23 Column Base Plate

The design bearing strength, $\phi_c P_p$, and the allowable bearing strength, P_p/Ω_c, under a beam bearing plate or a column base plate are each a function of the material that supports the plate. For a concrete support, *Specification* Section J8 provides that when the plate is the same size as the concrete support,

$$P_p = 0.85 f'_c A_1 \qquad \text{(AISC J8-1)}$$

and when the plate is less than the full area of the support,

$$P_p = 0.85 f'_c A_1 \sqrt{\frac{A_2}{A_1}} \leq 1.7 f'_c A_1 \qquad \text{(AISC J8-2)}$$

where A_1 = area of the plate; A_2 = maximum area of the support that is geometrically the same as the plate; f'_c = compressive strength of the concrete; $\phi_c = 0.65$ for LRFD; and $\Omega_c = 2.31$ for ASD. The limit on Equation J8-2 means that $\sqrt{A_2/A_1} \leq 2.0$. Thus, A_2 may be up to 4 times A_1. Concrete bearing strength was also discussed in Section 5.11.

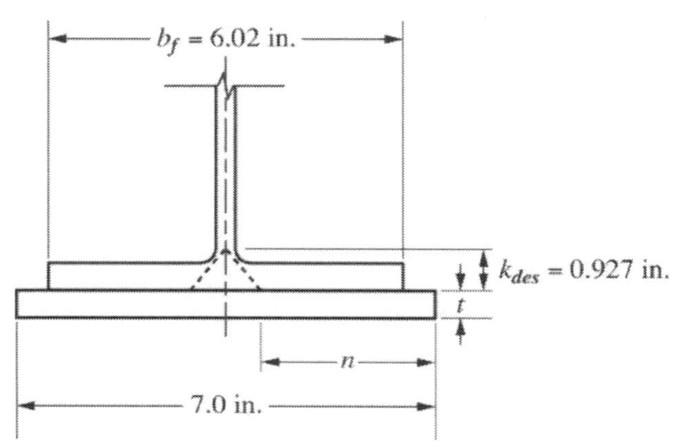

Figure 11.24 Beam Bearing Plate for Example 11.11

EXAMPLE 11.11a
Beam Bearing Plate by LRFD

Goal: Determine the required thickness of a beam bearing plate.

Given: A W18×40 A992 beam is supported by a 7×4 in. A572 Gr. 50 bearing plate as shown in Figure 11.24. The beam end reaction is $R_u = 50.0$ kips. From Manual Table 1-1, $k_{des} = 0.927$ in.

SOLUTION

Step 1: Determine the pressure under the plate.

$$f_u = \frac{50.0}{7.0(4.0)} = 1.79 \text{ ksi}$$

Step 2: Determine the cantilever dimension.

$$l = n = 3.5 - 0.927 = 2.57 \text{ in.}$$

Step 3: Determine the minimum plate thickness using Equation 11.12a

$$t_p = 1.49 l \sqrt{\frac{f_u}{F_y}} = 1.49(2.57)\sqrt{\frac{1.79}{50}} = 0.725 \text{ in.}$$

Step 4: Select the required plate thickness.

Use a 3/4 in. plate

EXAMPLE 11.11b
Beam Bearing Plate by ASD

Goal: Determine the required thickness of a beam bearing plate.

Given: A W18×40 A992 beam is supported by a 7×4 in. A572 Gr. 50 bearing plate as shown in Figure 11.24. The beam end reaction is R_a = 34.0 kips. From Manual Table 1-1, k_{des} = 0.927 in.

SOLUTION

Step 1: Determine the pressure under the plate.

$$f_a = \frac{34.0}{7.0(4.0)} = 1.21 \text{ ksi}$$

Step 2: Determine the cantilever dimension.

$$l = n = 3.5 - 0.927 = 2.57 \text{ in.}$$

Step 3: Determine the minimum plate thickness using Equation 11.12b.

$$t_p = 1.83l\sqrt{\frac{f_a}{F_y}} = 1.83(2.57)\sqrt{\frac{1.21}{50}} = 0.732 \text{ in.}$$

Step 4: Select the required plate thickness.

Use a 3/4 in. plate

Bearing on the beam web must also be checked for the limit states of web local yielding and web local crippling. These limit states were discussed in Section 6.15 and illustrated in Example 6.21.

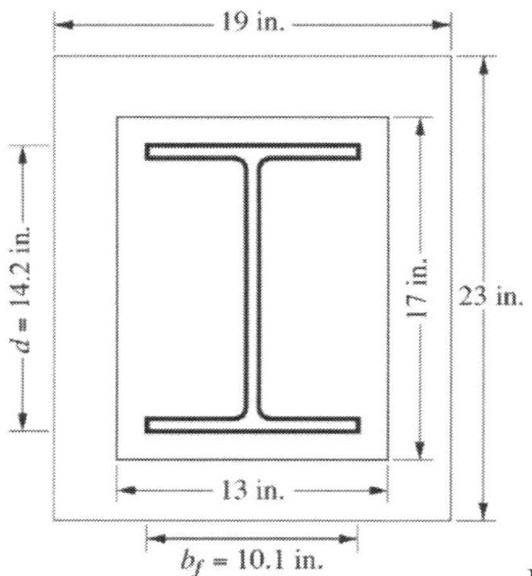

Figure 11.25 Column Base Plate for Example 11.12

EXAMPLE 11.12a
Column Base Plate by LRFD

Goal: Determine the required thickness of a column base plate.

Given: A W14×74 A992 column is supported by a 13×17 in. A572 Gr. 50 base plate, which itself is supported on a 19×23 in. concrete pier as shown in Figure 11.25. The column carries a design load P_u = 620 kips and the concrete has a compressive strength of $f_c' = 5$ ksi .

SOLUTION

Step 1: Determine the design strength of the concrete under the plate.

Area of the base plate

$$A_1 = 13(17) = 221 \text{ in.}^2$$

Area of the concrete support

$$A_2 = 19(23) = 437 \text{ in.}^2$$

Thus, since $\sqrt{437/221} = 1.41 < 2.0$, the nominal bearing strength using Equation J8-2 is

$$P_n = P_p = 0.85 f_c' A_1 \sqrt{\frac{A_2}{A_1}} = 0.85(5)(221)\sqrt{\frac{437}{221}} = 939(1.41) = 1320 \text{ kips}$$

Step 2: Determine if the design bearing strength is sufficient to support the load.

$$\phi P_p = 0.65(1320) = 858 \text{ kips} > 620 \text{ kips}$$

Thus, the concrete strength is adequate.

Step 3: Determine the pressure under the plate.

$$f_u = \frac{620}{13(17)} = 2.81 \text{ ksi}$$

Step 4: Determine the cantilever dimension.

$$n = \frac{B - 0.8b_f}{2} = \frac{13.0 - 0.8(10.1)}{2} = 2.46 \text{ in.}$$

$$m = \frac{N - 0.95d}{2} = \frac{17.0 - 0.95(14.2)}{2} = 1.76 \text{ in.}$$

and

$$n' = \frac{\sqrt{b_f d}}{4} = \frac{\sqrt{10.1(14.2)}}{4} = 2.99$$

Thus, the maximum cantilever dimension is

$$l = 2.99 \text{ in.}$$

Step 5: Determine the minimum plate thickness using Equation 11.12a.

$$t_p = 1.49l\sqrt{\frac{f_u}{F_y}} = 1.49(2.99)\sqrt{\frac{2.81}{50}} = 1.06\text{in.}$$

Step 6: Select the required plate thickness. Convention is to specify thickness in increments of 1/4 in. for plates over 1.0 in. Thus,

Use a 1-1/4 in. plate

Note: A more efficient plate dimension could be selected if the B and N dimensions were adjusted so that m and n were closer to being equal and the conservative assumption of λ = 1.0 was not made. *Manual* Part 14 includes a complete discussion of the use of λ in determining base plate required thickness.

EXAMPLE 11.12b
Column Base Plate by ASD

Goal: Determine the required thickness of a column base plate.

Given: A W14×74 A992 column is supported by a 13×17 in. A572 Gr. 50 bearing plate as shown in Figure 11.25. The column carries an allowble load P_a = 410 kips.

SOLUTION

Step 1: Determine the allowable strength of the concrete under the plate.

Area of the base plate

$$A_1 = 13(17) = 221 \text{ in.}^2$$

Area of the concrete support

$$A_2 = 19(23) = 437 \text{ in.}^2$$

Thus, since $\sqrt{437/221} = 1.41 < 2.0$, the nominal bearing strength using Equation J8-2 is

$$P_n = P_p = 0.85 f_c' A_1\sqrt{\frac{A_2}{A_1}} = 0.85(5)(221)\sqrt{\frac{437}{221}} = 939(1.41) = 1320 \text{ kips}$$

Step 2: Determine if the allowable bearing strength is sufficient to support the load.

$$P_p/\Omega = 1320/2.31 = 571 \text{ kips} > 410 \text{ kips}$$

Thus, the concrete strength is adequate.

Step 3: Determine the pressure under the plate.

$$f_a = \frac{410}{13(17)} = 1.86 \text{ ksi}$$

Step 4: Determine the cantilever dimension.

$$n = \frac{B - 0.8b_f}{2} = \frac{13.0 - 0.8(10.1)}{2} = 2.46 \text{ in.}$$

$$m = \frac{N - 0.95d}{2} = \frac{17.0 - 0.95(14.2)}{2} = 1.76 \text{ in.}$$

and

$$\lambda n' = \frac{\sqrt{b_f d}}{4} = (1.0)\frac{\sqrt{10.1(14.2)}}{4} = 2.99$$

Thus, the maximum cantilever dimension is

$$l = 2.99 \text{ in.}$$

Step 5: Determine the minimum plate thickness using Equation 11.12b.

$$t_p = 1.83l\sqrt{\frac{f_a}{F_y}} = 1.83(2.99)\sqrt{\frac{1.86}{50}} = 1.06 \text{ in.}$$

Step 6: Select the required plate thickness. Convention is to specify thickness in increments of 1/4 in. for plates over 1.0 in. Thus,

Use a 1-1/4 in. plate

Note: A more efficient plate dimension could be selected if the B and N dimensions were adjusted so that m and n were closer to being equal and the conservative assumption of λ=1.0 was not made. Manual Part 14 includes a complete discussion of the use of λ in determining base plate required thickness.

11.12 PROBLEMS

1. Design an all-bolted double-angle connection for a W18×65 A992 beam to carry a dead load reaction of 15 kips and a live load reaction of 45 kips. Use 5/16 in. A572 Gr. 50 angles and 3/4 in. A325-N bolts in standard holes. The uncoped beam is connected to the flange of a W14×109 A992 column. Design by (a) LRFD and (b) ASD.

2. Design an all-bolted double-angle connection for a W27×114 A992 beam to carry a dead load reaction of 30 kips and a live load reaction of 90 kips. Use 5/16 in. A572 Gr. 50 angles and 3/4 in. A325-N bolts in standard holes. The uncoped beam is connected to the web of a W36×135 A992 girder. Design by (a) LRFD and (b) ASD.

3. Design an all-bolted double-angle connection for a W24×146 A992 beam to carry a dead load reaction of 25 kips and a live load reaction of 75 kips. Use 5/16 in. A572 Gr. 50 angles and 3/4 in. A325-N bolts in standard holes. The uncoped beam is connected to the flange of a W14×132 A992 column. Design by (a) LRFD and (b) ASD.

4. Design an all-bolted double-angle connection for a W16×77 A992 beam to carry a dead load reaction of 20 kips and a live load reaction of 60 kips. Use 5/16 in. A572 Gr. 50 angles and 3/4 in. A325-N bolts in standard holes. The beam is uncoped and the supporting member is not critical. Design by (a) LRFD and (b) ASD.

5. Design an all-bolted double-angle connection for a W18×143 A992 beam to carry a dead load reaction of 25 kips and a live load reaction of 75 kips. Use 5/16 in. A572 Gr. 50 angles and 3/4 in. A325-N bolts in standard holes. The beam is uncoped and the supporting member is not critical. Design by (a) LRFD and (b) ASD.

6. Design an all-bolted double-angle connection for a W8×40 A992 beam to carry a dead load reaction of 8 kips and a live load reaction of 24 kips. Use 5/16 in. A572 Gr. 50 angles and 3/4 in. A325-N bolts in standard holes. The beam is uncoped and the supporting member is not critical. Design by (a) LRFD and (b) ASD.

7. Design an all-bolted double-angle connection for a W30×191 A992 beam spanning 40 ft and carrying a total uniformly distributed dead load of 60 kips and live load of 180 kips. Use 5/16 in. A572 Gr. 50 angles and 3/4 in. A325-N bolts in standard holes. The beam is uncoped and the supporting member is not critical. Design by (a) LRFD and (b) ASD.

8. Design an all-bolted double-angle connection for a W18×76 A992 beam to support a dead load reaction of 16 kips and a live load reaction of 48 kips. Use 5/16 in. A572 Gr. 50 angles and 3/4 in. A325-N bolts in standard holes. The beam is uncoped and the supporting member is not critical. Design by (a) LRFD and (b) ASD.

9. Design an all-bolted double-angle connection for a W21×73 A992 beam spanning 20 ft and carrying a total uniformly distributed dead load of 28 kips and live load of 84 kips. Use 5/16 in. A572 Gr. 50 angles and 3/4 in. A325-N bolts in standard holes. The beam is uncoped and the supporting member is not critical. Design by (a) LRFD and (b) ASD.

10. Design an all-bolted double-angle connection for a W30×191 A992 beam spanning 40 ft and carrying a total uniformly distributed dead load of 60 kips and live load of 180 kips. Use 5/16 in. A572 Gr. 50 angles and 3/4 in. A325-N bolts in standard holes. The beam is coped so that the edge distance is 1-1/4 in. The supporting member is not critical. Design by (a) LRFD and (b) ASD.

11. Design an all-bolted double-angle connection for a W18×76 A992 beam to support a dead load reaction of 16 kips and a live load reaction of 48 kips. Use 5/16 in. A572 Gr. 50 angles and 3/4 in. A325-N bolts in standard holes. The beam is coped so that the edge distance is 1-1/4 in. The supporting member is not critical. Design by (a) LRFD and (b) ASD.

12. Design an all-bolted double-angle connection for a W21×73, A992 beam spanning 20 ft and carrying a total uniformly distributed dead load of 28 kips and live load of 84 kips. Use 5/16 in. A572 Gr. 50 angles and 3/4 in. A325-N bolts in standard holes. The beam is coped so that the edge distance is 1¼ in. The supporting member is not critical. Design by (a) LRFD and (b) ASD.

13. Design an all-bolted double-angle connection for a W24×84, A992 beam to support a dead load reaction of 25 kips and a live load reaction of 75 kips. Use 5/16 in. A572 Gr. 50 angles and 3/4 in. A325-N bolts in standard holes. The beam is coped so that the edge distance is 1-1/4 in. The supporting member is not critical. Design by (a) LRFD and (b) ASD.

14. Design an all-bolted double-angle connection for a W12×87 A992 beam to support a dead load reaction of 14 kips and a live load reaction of 42 kips. Use 5/16 in. A572 Gr. 50 angles and 3/4 in. A325-N bolts in standard holes. The beam is coped so that the edge distance is 1-1/4 in. The supporting member is not critical. Design by (a) LRFD and (b) ASD.

15. Design an all-bolted double-angle connection for a W16×77 A992 beam spanning 20 ft and carrying a total uniformly distributed dead load of 23 kips and live load of 69 kips. Use 5/16 in. A572 Gr. 50 angles and 3/4 in. A325-N bolts in standard holes. The beam is coped so that the edge distance is 1-1/4 in. The supporting member is not critical. Design by (a) LRFD and (b) ASD.

16. Design an all-bolted double-angle connection for a W16×77 A992 beam to carry a dead load reaction of 20 kips and a live load reaction of 60 kips. Use 5/16 in. A572 Gr. 50 angles and 3/4 in. A325-N bolts in standard holes. The beam is coped so that the edge distance is 1-1/4 in. and the supporting member is not critical. Design by (a) LRFD and (b) ASD.

17. Design an all-bolted double-angle connection for a W18×143 A992 beam to carry a dead load reaction of 25 kips and a live load reaction of 75 kips. Use 5/16 in. A572 Gr. 50 angles and 3/4 in. A325-N bolts in standard holes. The beam is coped so that the edge distance is 1-1/4 in. and the supporting member is not critical. Design by (a) LRFD and (b) ASD.

18. Design an all-bolted double-angle connection for a W8×40 A992 beam to carry a dead load reaction of 7.5 kips and a live load reaction of 22.5 kips. Use 5/16 in. A572 Gr. 50 angles and 3/4 in. A325-N bolts in standard holes. The beam is coped so that the edge distance is 1-1/4 in. and the supporting member is not critical. Design by (a) LRFD and (b) ASD.

19. Design a welded-bolted double-angle connection for an uncoped W18×65 A992 beam to carry a dead load reaction of 15 kips and a live load reaction of 45 kips. Use 5/16 in. A572 Gr. 50 angles and 70 ksi welding electrodes to connect to the web of the supported beam and 3/4 in. A325-N bolts in standard holes to connect to the supporting member. The beam is connected to the flange of a W14×109 A992 column. Design by (a) LRFD and (b) ASD.

20. Design a welded-bolted double-angle connection for an uncoped W27×114 A992 beam to carry a dead load reaction of 30 kips and a live load reaction of 90 kips. Use 5/16 in. A572 Gr. 50 angles and 70 ksi welding electrodes to connect to the web of the supported beam and 3/4 in. A325-N bolts in standard holes to connect to the supporting member. The beam is connected to the web of a W36×135 A992 girder. Design by (a) LRFD and (b) ASD.

21. Design a welded-bolted double-angle connection for an uncoped W24×146 A992 beam to carry a dead load reaction of 25 kips and a live load reaction of 75 kips. Use 5/16 in. A572 Gr. 50 angles and 70 ksi welding electrodes to connect to the web of the supported beam and 3/4 in. A325-N bolts in standard holes to connect to the supporting member. The beam is connected to the flange of a W14×132 A992 column. Design by (a) LRFD and (b) ASD.

22. Design a welded-bolted double-angle connection for an uncoped W33×241 A992 beam to carry a dead load reaction of 50 kips and a live load reaction of 150 kips. Use 5/16 in. A572 Gr. 50 angles and 70 ksi welding electrodes to connect to the web of the supported beam and 7/8 in. A490-N bolts in standard holes to connect to the supporting member. The beam is connected to the flange of a W14×370 A992 column. Design by (a) LRFD and (b) ASD.

23. Design a welded-bolted double-angle connection for a coped W18×76 A992 beam to support a dead load reaction of 16 kips and a live load reaction of 48 kips. Assume the beam is coped so that the edge distance is 1-1/4 in. Use 5/16 in. A572 Gr. 50 angles and 70 ksi welding electrodes to connect to the web of the supported beam and 3/4 in. A325-N bolts in standard holes to connect to the supporting member. The supporting member is not critical. Design by (a) LRFD and (b) ASD.

24. Design a welded-bolted double-angle connection for a coped W12×87 A992 beam to support a dead load reaction of 14 kips and a live load reaction of 42 kips. Assume the beam is coped so that the edge distance is 1-1/4 in. Use 5/16 in. A572 Gr. 50 angles and 70 ksi welding electrodes to connect to the web of the supported beam and 3/4 in. A325-N bolts in standard holes to connect to the supporting member. The supporting member is not critical. Design by (a) LRFD and (b) ASD.

25. Design a welded-bolted double-angle connection for a coped W16×77 A992 beam spanning 20 ft and carrying a total uniformly distributed dead load of 23 kips and live load of 69 kips. Assume the beam is coped so that the edge distance is 1-1/4 in. Use 5/16 in. A572 Gr. 50 angles and 70 ksi welding electrodes to connect to the web of the supported beam and 3/4 in. A325-N bolts in standard holes to connect to the supporting member. The supporting member is not critical. Design by (a) LRFD and (b) ASD.

26. Design a welded-bolted double-angle connection for a coped W21×73 A992 beam spanning 25 ft and carrying a total uniformly distributed dead load of 30 kips and live load of 90 kips. Assume the beam is coped so that the edge distance is 1-1/4 in. Use 5/16 in. A572 Gr. 50 angles and 70 ksi welding electrodes to connect to the web of the supported beam and 3/4 in. A325-N bolts in standard holes to connect to the supporting member. The supporting member is not critical. Design by (a) LRFD and (b) ASD.

27. Design a bolted-bolted single-angle connection for an uncoped W18×65 A992 beam to carry a dead load reaction of 8 kips and a live load reaction of 24 kips using 3/8 in. A572 Gr. 50 angles and 3/4 in. A325-N bolts in standard holes. The beam is connected to the web of a W36×150 A992 girder. Design by (a) LRFD and (b) ASD.

28. Design a bolted-bolted single-angle connection for an uncoped W21×62 A992 beam to carry a dead load reaction of 10 kips and a live load reaction of 20 kips using 3/8 in. A572 Gr. 50 angles and 3/4 in. A325-N bolts in standard holes. The beam is connected to the web of a W33×201 A992 girder. Design by (a) LRFD and (b) ASD.

29. Design a bolted-welded single-angle connection for a coped W12×87 A992 beam to support a dead load reaction of 7 kips and a live load reaction of 21 kips. Assume the beam is coped so that the edge distance is 1-1/4 in. Use 3/8 in. A572 Gr. 50 angles, E70 electrodes, and 3/4 in. A325-N bolts in standard holes. Determine the minimum thickness for the web of the supporting A992 girder. Design by (a) LRFD and (b) ASD.

30. Design a bolted-welded single-angle connection for a coped W14×82 A992 beam to support a dead load reaction of 6 kips and a live load reaction of 20 kips. Assume the beam is coped so that the edge distance is 1-1/4 in. Use 3/8 in. A572 Gr. 50 angles, E70 electrodes, and 3/4 in. A325-N bolts in standard holes. Determine the minimum thickness for the web of the supporting A992 girder. Design by (a) LRFD and (b) ASD.

31. Design a welded-welded single-angle connection for a coped W16×77 A992 beam spanning 20 ft and carrying a total uniformly distributed dead load of 12 kips and live load of 36 kips. Assume the beam is coped so that the edge distance is 1-1/4 in. Use 3/8 in. A572 Gr. 50 angles and E70 electrodes. Design by (a) LRFD and (b) ASD.

32. Design a welded-welded single-angle connection for a coped W16×57 A992 beam spanning 20 ft and carrying a total uniformly distributed dead load of 16 kips and live load of 48 kips. Assume the beam is coped so that the edge distance is 1-1/4 in. Use 3/8 in. A572 Gr. 50 angles and E70 electrodes. Design by (a) LRFD and (b) ASD.

33. Design a shear tab connection for an uncoped W18×65 A992 beam to carry a dead load reaction of 10 kips and live load reaction of 30 kips. Use a 3/8 in. thick A572 Gr. 50 shear tab and 3/4 in. A325-N bolts. Assume that the supporting member is not critical. Design by (a) LRFD and (b) ASD.

34. Design a shear tab connection for an uncoped W27×114, A992 beam to carry a dead load reaction of 15 kips and a live load reaction of 45 kips. Use a 3/8 in. thick A572 Gr. 50 shear tab and 3/4 in. A325-N bolts. Assume that the supporting member is not critical. Design by (a) LRFD and (b) ASD.

35. Design a shear tab connection for a coped W21×73 A992 beam spanning 20 ft and carrying a total uniformly distributed dead load of 23 kips and live load of 70 kips. Assume that the edge distance at the cope is 1-1/4 in. Use a 3/8 in. thick A572 Gr. 50 shear tab and 3/4 in. A325-N bolts. Assume that the supporting member is not critical. Design by (a) LRFD and (b) ASD.

36. Design a shear tab connection for a coped W18×76 A992 beam to carry a dead load reaction of 10 kips and a live load reaction of 30 kips. Assume an edge distance of 1-1/4 in. Use a 3/8 in. thick A572 Gr. 50 shear tab and 3/4 in. A325-N bolts. Assume that the supporting member is not critical. Design by (a) LRFD and (b) ASD.

37. Repeat Problem 33 using an A36 shear tab and compare the required plate with the one determined in the solution to Problem 33. Design by (a) LRFD and (b) ASD.

38. Repeat Problem 35 using an A36 shear tab and compare the required plate with the one determined in the solution to Problem 35. Design by (a) LRFD and (b) ASD.

39. Design a welded seated connection for a W16×26 A992 beam framing into the web of a W14×99 A992 column. The seat must carry a dead load reaction of 6 kips and a live load reaction of 18 kips. Use an equal-leg A572 Gr. 50 angle and E70 electrode. Design by (a) LRFD and (b) ASD.

40. Design a welded seated connection for a W18×40 A992 beam framing into the web of a W14×109 A992 column. The seat must carry a dead load reaction of 8 kips and a live load reaction of 24 kips. Use an equal-leg A572 Gr. 50 angle and E70 electrode. Design by (a) LRFD and (b) ASD.

41. Design a welded seated connection for a W16×50 A992 beam framing into the web of a W14×90 A992 column. The seat must carry a dead load reaction of 9 kips and a live load reaction of 27 kips. Use an equal-leg A572 Gr. 50 angle and E70 electrode. Design by (a) LRFD and (b) ASD.

42. Design a bolted seated connection for a W16×50 A992 beam framing into the web of a W14×90 A992 column. The seat must carry a dead load reaction of 9 kips and a live load reaction of 27 kips. Use an unequal-leg A572 Gr. 50 angle and 3/4 in. A325-N bolts. Design by (a) LRFD and (b) ASD.

43. Determine the available strength of a 5×12×1 in. A572 Gr. 50 gusset plate with a single line of four 3/4 in. A325-N bolts connected to a 2L5×5×3/8 tension member. The gusset plate is welded perpendicular to the axis of the member with welds from E70 electrodes. Design by (a) LRFD and (b) ASD.

44. Determine the strength of the gusset plate connection given in Problem 43 if the gusset plate tapers from 5 in. to 8 in.

45. Design the connection for an A572 Gr. 50 double-angle tension member connected to a uniform-width A572 Gr. 50 gusset plate. The angles are L4×4×1/2 and carry a dead load of 10 kips and a live load of 30 kips. The angles are connected to the gusset plate by a single line of 3/4 in. A325-N bolts. The gusset plate is welded perpendicular to the axis of the member with welds from E70 electrodes. Design by (a) LRFD and (b) ASD.

46. Design the connection for an A572 Gr. 50 double-angle tension member connected to a uniform-width A572 Gr. 50 gusset plate. The angles are L6×4×1/2 with the long legs back-to-back and attached to the gusset plate. They carry a dead load of 20 kips and a live load of 60 kips. The angles are connected to the gusset plate by a single line of 3/4 in. A325-N bolts. The gusset plate is welded perpendicular to the axis of the member with welds from E70 electrodes. Design by (a) LRFD and (b) ASD.

47. An inclined WT hanger is used to support a tension member carrying a dead load of 8 kips and a live load of 24 kips. The force is applied at an angle of 45 degrees from the horizontal and is transferred by four 3/4 in. A325-N bolts. Determine whether the bolts have sufficient strength to carry the applied load by (a) LRFD and (b) ASD.

47. An inclined WT hanger is used to support a tension member carrying a dead load of 10 kips and a live load of 30 kips. The force is applied at an angle of 30 degrees from the horizontal and is transferred by four 5/8 in. A325-N bolts. Determine whether the bolts have sufficient strength to carry the applied load by (a) LRFD and (b) ASD.

48. If the load on the hanger of Problem 46 were applied at 90 degrees from the horizontal, determine whether the bolts have sufficient strength to carry the applied load, if there were no prying action, by (a) LRFD and (b) ASD.

49. A 9 in. long WT7×24, A992 steel, is used as a tension hanger with four 3/4 in. A325-N bolts in the flanges similar to the setup shown in Figure 11.19. The hanger must resist a dead load of 11 kips and a live load of 33 kips. Determine whether prying action must be included to detèrmine the connection strength by (a) LRFD and (b) ASD.

50. A 6 in. long WT7×88, A992 steel, is used as a tension hanger with four 3/4 in. A325-N bolts in the flanges similar to the setup shown in Figure 11.19. The hanger must resist a dead load of 10 kips and a live load of 30 kips. Determine whether prying action must be included to determine the connection strength by (a) LRFD and (b) ASD.

51. A 6 in. long WT6×36, A992 steel, is used as a tension hanger with four 3/4 in. A325-N bolts in the flanges similar to the setup shown in Figure 11.19. Determine the available strength if prying action is not to be included in the design. Determine by (a) LRFD and (b) ASD.

52. Determine the required thickness of an A572 Gr. 50 beam bearing plate to support an A992 W16×77 beam carrying a dead load reaction of 10 kips and a live load reaction of 30 kips. The plate is 12.0 in. wide and 6.0 in. along the beam web. Design by (a) LRFD and (b) ASD.

53. Determine the required thickness of an A572 Gr. 50 beam bearing plate to support an A992 W18×65 beam carrying a dead load reaction of 15 kips and a live load reaction of 45 kips. The plate is 7.5 in. wide and 10.0 in. along the beam web. Design by (a) LRFD and (b) ASD.

54. Determine the required thickness of a 24 in.×24 in. A572 Gr. 50 column base plate that supports an A992 W14×193 column and carries a dead load of 100 kips and a live load of 300 kips. The area of concrete providing support is the same as the area of the steel base plate. For the concrete, f'_c = 5 ksi. Design by (a) LRFD and (b) ASD.

55. Determine the required thickness of a 17 in.×15 in. A572 Gr. 50 column base plate that supports an A992 W12×79 column and carries a dead load of 130 kips and a live load of 390 kips. The concentric area of concrete providing support is the 4 times the area of the steel base plate. For the concrete, f'_c = 5 ksi. Design by (a) LRFD and (b) ASD.

56. Integrated Design Project. Design double-angle shear connections for the gravity beams and girders designed in Chapter 6. Some of these connections will be to girder webs, some to column webs, and some to column flanges.

Reconsider the types of shear connections, and design using single-plate connections.

Chapter 12

Moment Connections

United States Olympic & Paralympic Museum, Colorado Springs, Colorado
Photo courtesy of KL&A Engineers & Builders

12.1 TYPES OF MOMENT CONNECTIONS

Although they are called moment connections, these connections are expected to transfer both shear and moment between the connected members. Moment connections are defined in *Specification* Section B3.4b as fully restrained, Type FR, or partially restrained, Type PR. The impact of these connection types on the behavior of a steel frame was discussed in Chapter 8. For fully restrained connections, the moment is transferred while the relative rotation of the members remains at zero. For partially restrained connections, the moment is transferred while some predictable relative rotation is permitted. For the simple shear connection as discussed in Chapter 11, no moment is expected to be transferred and the connection is assumed to rotate freely. Figure 8.21, redrawn here as Figure 12.1, shows three moment-rotation curves for connections with distinctly different behavior. The rigid and simple connection behavior shown in Figure 12.1 illustrates that real connection behavior does not exactly follow the ideal behavior, as demonstrated by the vertical axis for a rigid connection and the horizontal axis for the simple connection.

Five moment connections that are commonly used for connecting beams framing into the strong axis of columns are illustrated in Figure 12.2. For the first four examples, which include the direct-welded flange (Figure 12.2a), the welded flange plate (Figure 12.2b), the bolted flange plate (Figure 12.2c), and the bolted tee (Figure 12.2d), shear is transferred through a web connection similar to those discussed in Chapter 11, whereas the moment is transferred through the various flange connections. In the extended end plate connection, Figure 12.2e, shear and moment are combined and transferred through the connecting plate and bolts.

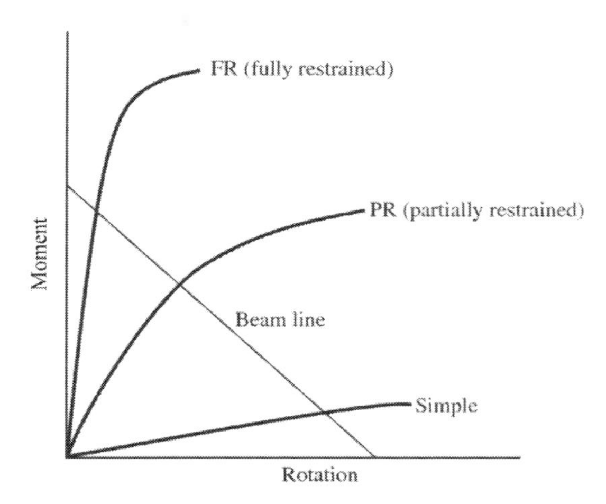

Figure 12.1 Connection Behavior

Because shear is resisted primarily by the web of a wide-flange beam, it is logical that the shear force is transferred through the web connection to the supporting member. Similarly, because the moment is resisted primarily through the flange of a wide-flange beam for the typical beam depths of rolled shapes, the flange connections transfer the moment to the supporting member. Because of the moment resistance provided by the flanges, there is no need to consider eccentricity in the design of the web shear connection. Thus, the web plate or angles are sized to resist only shear, simplifying the connection design. The flange connection is designed to resist the full moment, even though the flanges do not actually carry this full moment. Through strain hardening, and in combination with some uncalculated moment strength of the web connection, the flange connections are capable of developing the full moment.

The welded flange connection, Figure 12.2a, is the most direct moment connection and requires the fewest parts. The flanges are field-welded to the supporting member with complete joint penetration groove welds. The web connection is usually a single plate welded to the column and bolted to the beam. In this arrangement, the flange force, P_f, is determined by dividing the moment by the distance between flange midpoints. Thus,

$$P_f = \frac{M_r}{(d - t_f)} \tag{12.1}$$

The flange plated connections, Figure 12.2b and c, transfer the flange forces to the corresponding plates through either bolt shear or weld shear. The plate force is then transferred to the supporting member through welds. The flange plate connectors, bolts or welds, are sized to resist the force developed at the plate-flange interface. Thus,

$$P_f = \frac{M_r}{d} \tag{12.2}$$

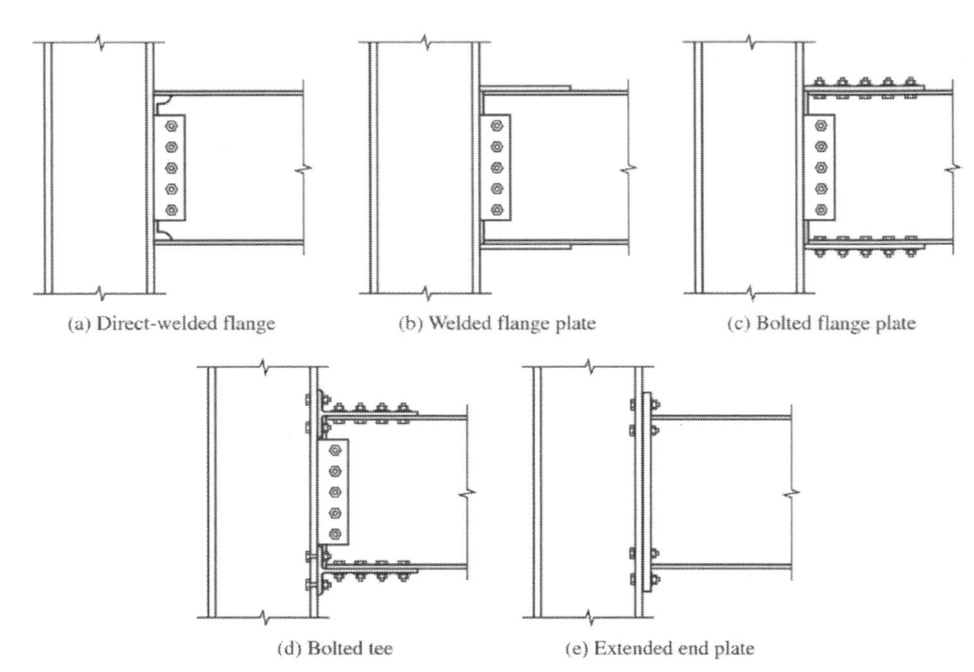

Figure 12.2 Moment Connections

Table 12.1 Sections of *Specification* and Parts of *Manual* Covered in This Chapter

	Specification
B4.1	Classification of Sections for Local Buckling
B4.3	Gross and Net Area Determination
D3	Effective Net Area
F13	Proportions of Beams and Girders
J2	Welds and Welded Joints
J3	Bolts, Threaded Parts, and Bolted Connections
J4	Affected Elements of Members and Connecting Elements
J10	Flanges and Webs with Concentrated Forces
	Manual
Part 7	Design Considerations for Bolts
Part 8	Design Considerations for Welds
Part 9	Design of Connecting Elements
Part 10	Design of Simple Shear Connections
Part 15	Design of Hanger Connections, Bracket Plates, and Crane-Rail Connections

The bolted tee connection, Figure 12.2d, is needed when the connection to the supporting member must be bolted. Although this connection is not as clean and simple as the flange plate connections, it provides a solution when there is a compelling reason to require an all-bolted connection. The connection to the beam flange is treated as with the flange plate connections, and the connection to the support is treated similarly as the tension connection discussed in Chapter 11.

The extended end plate connection, shown in Figure 12.2e, may take a variety of forms. The end plate is fully welded to the end of the beam and then bolted to the support. The end plate must extend beyond the beam flange on the tension side so that a minimum of four bolts can be symmetrically spaced, with the flange located at the bolt centroid. If an extended end plate connection is called upon to resist a moment that is always acting in the same direction, it may be extended on only one side. However, if the moment is expected to reverse, the plate must be extended beyond both the top and bottom flanges.

Table 12.1 lists the sections of the *Specification* and parts of the *Manual* discussed in this chapter.

12.2 LIMIT STATES

The limit states that control the strength of these connections are the same as those already considered for the shear connections. Their specific application depends on the complete connection geometry and the forces that the elements are expected to carry. These limit states include

1. Bolts
- **a.** Shear rupture
- **b.** Tension
- **c.** Shear-tension interaction
- **d.** Bearing and tearout

2. Welds
- **a.** Tension rupture
- **b.** Shear rupture

3. Plates
- **a.** Compression buckling
- **b.** Tension yielding
- **c.** Tension rupture
- **d.** Shear yielding
- **e.** Shear rupture
- **f.** Block shear

4. Beam
- **a.** Flexure of reduced section
- **b.** Shear yield
- **c.** Shear rupture

In addition to these limit states, which are all associated with the beam side of the connection, the designer must consider the impact of the connection on the column to which it is attached. Here limit states include

5. Column
- **a.** Flange local bending
- **b.** Web local yielding
- **c.** Web local crippling
- **d.** Web compression buckling
- **e.** Web panel zone shear

The moment that the connection transfers into the column should have already been accounted for in the column design.

12.3 MOMENT CONNECTION DESIGN

Design of moment connections is presented in two parts. First, examples are given for a direct-welded beam-to-column connection, a welded flange plate connection, and a bolted flange plate connection. These examples treat the beam side of the connection without considering the column to which the connection is attached. This is followed by a discussion of the limit states associated with the column, and examples are given to illustrate that design process.

12.3.1 Direct-Welded Flange Connection

The direct-welded flange moment connection provides an FR connection with very few connecting elements. As the name implies, the flanges are directly welded to the supporting member, usually the flange of a column. These welds are either complete joint penetration groove welds or a pair of fillet welds on each side of the beam flanges. The groove weld provides a weld that can be made in the downward position for both flanges, whereas the fillet welds require overhead welding on the bottom of each flange. The only limit state to consider for the flange connection is tension or shear rupture of the weld.

The web connection is usually made with a single plate, the same as for the shear tab simple connection discussed in Chapter 11. However, unlike the shear tab connection, the web plate in this FR connection does not need to take into account any eccentricity because the flanges are designed to carry all of the moment. The limit states for the web connection are those previously discussed for the shear tab.

Although the *Specification* specifically identifies fully restrained moment connections as Type FR moment connections, it is normal engineering practice to simply refer to them as moment connections. Thus, unless specified as Type PR connections, all moment connections are assumed to be Type FR. Since that is normal practice, that is the approach taken in this book.

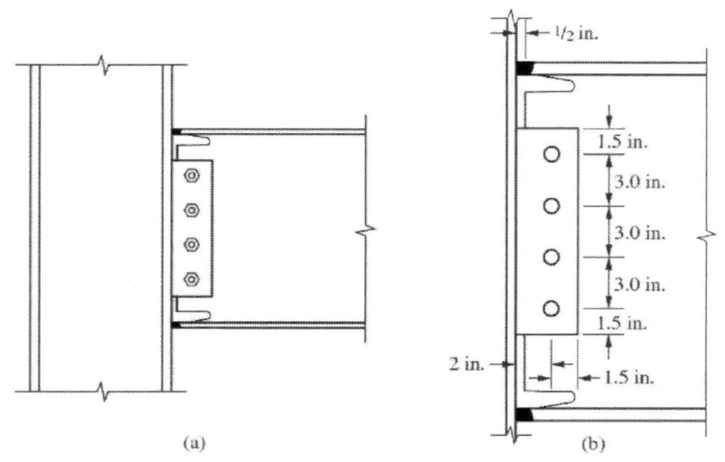

Figure 12.3 Connection for Example 12.1

EXAMPLE 12.1a
Direct-Welded Moment Connection by LRFD

Goal: Design a direct-welded beam-to-column moment connection.

Given: A direct-welded beam-to-column moment connection is shown in Figure 12.2a. The beam is a W24×76 and the column is a W14×109. Bolts are 3/4 in. A325-N and the electrodes are E70. The shapes are A992 steel, and the plate is A572 Gr. 50. The required strength is M_u = 500 ft-kips and V_u = 60.0 kips.

SOLUTION

Step 1: Obtain the beam and column properties from *Manual* Table 1-1.

Beam: W24×76
d = 23.9 in., b_f = 8.99 in., t_w = 0.440 in., t_f = 0.680 in., Z = 200 in.^{3}

Column: W14×109
d = 14.3 in., b_f = 14.6 in., t_w = 0.525 in., t_f = 0.860 in.

Step 2: Check the flexural strength of the beam.

This check should have been made during design of the beam. The beam section is not reduced because of bolt holes in the flange, so M_p can be determined using the gross section plastic section modulus and the design strength is

$$\phi M_n = \phi M_p = \phi F_y Z = \frac{0.9(50)(200)}{12} = 750 \text{ ft-kips} > 500 \text{ ft-kips}$$

Thus, the flexural strength is adequate.

Step 3: Design the flange-to-column weld.

The flange-to-column weld can be either a complete joint penetration (CJP) groove weld or fillet welds. CJP welds are used in this example. E70 electrodes match the strength of the plates, as indicated by the User Note in *Specification* Section J2.6. Therefore, the CJP welds will develop the full strength of the beam flanges without tension rupture of the weld and no further calculations are needed.

Step 4: Design the web plate.

First consider the shear rupture strength of the bolts to determine the minimum number of bolts required. For a 3/4 in. A325-N bolt, ϕr_n = 17.9 kips; therefore, the minimum number of bolts based on bolt shear is

$$n = \frac{60}{17.9} = 3.35$$

Thus, try a four-bolt connection with bolt spacing of 3.0 in. and end distances of 1.5 in. Thus, l = 12.0 in., which is greater than $T/2$ = 10.0 in. Try a plate with t = 3/8 in. See Figure 12.3.

Step 5: Determine the bolt bearing strength.

Since the beam web is thicker than the plate, bearing and tearout of the plate will be more critical than the beam web.

For the last bolt, determine the clear distance.

$$l_c = l_{ev} - \frac{1}{2}d_h = 1.5 - \frac{1}{2}(3/4 + 1/16) = 1.09 < 2(3/4) = 1.5 \text{ in.}$$

Thus, tearout controls over bearing and the bolt nominal strength is

$$R_n = 1.2 l_c t F_u = 1.2(1.09)(0.375)(65) = 31.9 \text{ kips}$$

The design strength for tearout is

$$\phi R_n = 0.75(31.9) = 23.9 \text{ kips}$$

Since this is greater than the design shear rupture strength, shear rupture will control for this bolt.

For the other bolts

$$l_c = s - d_h = 3.0 - (3/4 + 1/16) = 2.19 > 2(3/4) = 1.5 \text{in.}$$

Therefore, bearing will control over tearout, and the bolt nominal strength is

$$R_n = 2.4 d t F_u = 2.4(3/4)(0.375)(65) = 43.9 \text{ kips}$$

The design strength for bearing is

$$\phi R_n = 0.75(43.9) = 32.9 \text{ kips}$$

Since this is greater than the design shear rupture strength, shear rupture will control for these bolts also.

Thus, for the four-bolt connection, the design strength is

$$\phi R_n = 4(17.9) = 71.6 > 60.0 \text{ kips}$$

as expected from step 4.

Step 6: Check the plate for shear yield.

The gross shear area is

$$A_{gv} = tl = 0.375(12.0) = 4.50 \text{ in.}^2$$

and the design shear yield strength is

$$\phi V_n = \phi 0.6 F_y A_{gv} = 1.0(0.6(50)(4.50)) = 135 > 60.0 \text{ kips}$$

Step 7: Check the plate for shear rupture.

The net shear area is

$$A_{nv} = (l - n(d_h + 1/16))t = (12.0 - 4(3/4 + 1/8))(0.375) = 3.19 \text{ in.}^2$$

And the design shear rupture strength is

$$\phi V_n = \phi 0.6 F_u A_{nv} = 0.75(0.6(65)(3.19)) = 93.3 > 60.0 \text{ kips}$$

Step 8: Check the block shear of the plate. First calculate the areas.

$$A_{nt} = \left(l_{eh} - \frac{1}{2}(d_h + 1/16) \right) t_w$$

$$= \left(1.5 - \frac{1}{2}(3/4 + 1/8) \right)(0.375) = 0.398 \text{ in.}^2$$

$$A_{gv} = l t_w$$

$$= 10.5(0.375) = 3.94 \text{ in.}^2$$

$$A_{nv} = (l - (n - 0.5)(d_h + 1/16)) t_w$$

$$= (10.5 - 3.5(3/4 + 1/8))(0.375) = 2.79 \text{ in.}^2$$

Determine the tension rupture strength

$$F_u A_{nt} = 65(0.398) = 25.9 \text{ kips}$$

Consider shear yield and shear rupture and select the one with least strength; thus,

$$0.6 F_y A_{gv} = 0.6(50)(3.94) = 118 \text{ kips}$$

$$0.6 F_u A_{nv} = 0.6(65)(2.79) = 109 \text{ kips}$$

Selecting the shear rupture term and combining it with the tension rupture term gives a connection block shear strength, with $U_{bs} = 1.0$, of

$$\phi R_n = 0.75(109 + 25.9) = 101 \text{ kips} > 60.0 \text{ kips}$$

Step 9: Select the plate-to-column weld.

Based on a fillet weld on each side of the plate, with a weld design strength of 1.392 kips per 1/16 in. of weld per in. of length

$$D = \frac{60.0}{(2(1.392)(12.0))} = 1.80 \text{ sixteenths of in.}$$

Therefore, use a 3/16 in. weld, the minimum weld for the 3/8 in. plate, as given in *Specification* Table J2.4.

Step 10: Develop the final design.

Figure 12.3 shows the final design using
Four 3/4 in. A325-N bolts in a 3/8×3-1/2 ×1.0 ft-0 in. plate

The impact of the connection on the column flange and web must be checked. This will be addressed in Section 12.4.

EXAMPLE 12.1b
Direct-Welded Moment Connection by ASD

Goal: Design a direct-welded beam-to-column moment connection.

Given: A direct-welded beam-to-column moment connection is shown in Figure 12.2a. The beam is a W24×76 and the column is a W14×109. Bolts are 3/4 in. A325-N and the electrodes are E70. The shapes are A992 steel, and the plate is A572 Gr. 50. The required strength is M_a = 333 ft-kips and V_a = 40.0 kips.

SOLUTION

Step 1: Obtain the beam and column properties from *Manual* Table 1-1.
Beam: W24×76
$d = 23.9$ in., $b_f = 8.99$ in., $t_w = 0.440$ in., $t_f = 0.680$ in., $Z = 200$ in.^{3}

Column: W14×109
$d = 14.3$ in., $b_f = 14.6$ in., $t_w = 0.525$ in., $t_f = 0.860$ in.

Step 2: Check the flexural strength of the beam.

This check should have been made during design of the beam. The beam section is not reduced because of bolt holes in the flange, so M_p can be determined using the gross section plastic section modulus and the allowable strength is

$$M_n/\Omega = M_p/\Omega = F_y Z/\Omega = \frac{(50)(200)/1.67}{12} = 499 \text{ ft-kips} > 333 \text{ ft-kips}$$

Thus, the flexural strength is adequate.

Step 3: Design the flange-to-column weld.

The flange-to-column weld can be either a complete joint penetration (CJP) groove weld or fillet welds. CJP welds are used in this example. E70 electrodes match the strength of the plates, as indicated by the User Note in *Specification* Section J2.6. Therefore, CJP welds will develop the full strength of the beam flanges and no further calculations are needed.

Step 4: Design the web plate.

First consider the shear rupture strength of the bolts to determine the minimum number of bolts required.

For a 3/4 in. A325-N bolt, $r_n/\Omega = 11.9$ kips; therefore, the minimum number of bolts based on bolt shear is

$$n = \frac{40}{11.9} = 3.36$$

Thus, try a four-bolt connection with bolt spacing of 3.0 in. and end distances of 1.5 in. Thus, l = 12.0 in., which is greater than $T/2$ = 10.0 in. Try a plate with t = 3/8 in. See Figure 12.3.

Step 5: Determine the bolt bearing strength.

Since the beam web is thicker and has a higher yield strength than the plate, bearing and tearout of the plate will be more critical than the beam web.

For the last bolt, determine the clear distance.

$$l_c = l_{ev} - \frac{1}{2}d_h = 1.5 - \frac{1}{2}(3/4 + 1/16) = 1.09 < 2(3/4) = 1.5 \text{ in.}$$

Thus, tearout controls over bearing and the bolt nominal strength is

$$R_n = 1.2 l_c t F_u = 1.2(1.09)(0.375)(65) = 31.9 \text{ kips}$$

The allowable strength for tearout is

$$R_n/\Omega = 31.9/2.00 = 16.0 \text{ kips}$$

Since this is greater than the allowable shear rupture strength, shear rupture will control for this bolt.

For the other bolts

$$l_c = s - d_h = 3.0 - (3/4 + 1/16) = 2.19 > 2(3/4) = 1.5 \text{in.}$$

Therefore, bearing will control over tearout, and the bolt nominal strength is

$$R_n = 2.4dtF_u = 2.4(3/4)(0.375)(65) = 43.9 \text{ kips}$$

The allowable strength for bearing is

$$R_n/\Omega = 43.9/2.00 = 22.0 \text{ kips}$$

Since this is greater than the allowable shear rupture strength, shear rupture will control for these bolts also.

Thus, for the four-bolt connection, the allowable strength is

$$R_n/\Omega = 4(11.9) = 47.6 > 40.0 \text{ kips}$$

as expected from step 4.

Step 6: Check the plate for shear yield.

The gross shear area is

$$A_{gv} = tl = 0.375(12.0) = 4.50 \text{ in.}^2$$

and the allowable shear yield strength is

$$V_n/\Omega = 0.6F_y A_{gv}/\Omega = (0.6(50)(4.50))/1.50 = 90.0 > 40.0 \text{ kips}$$

Step 7: Check the plate for shear rupture.

The net shear are is

$$A_{nv} = (l - n(d_h + 1/16))t = (12.0 - 4(3/4 + 1/8))(0.375) = 3.19 \text{ in.}^2$$

and the allowable shear rupture strength is

$$V_n/\Omega = 0.6F_u A_{nv}/\Omega = (0.6(65)(3.19))/2.00 = 62.2 > 40.0 \text{ kips}$$

Step 8: Check the block shear of the plate. First calculate the required areas.

$$A_{nt} = \left(l_{eh} - \frac{1}{2}(d_h + 1/16)\right)t_w$$

$$= \left(1.5 - \frac{1}{2}(3/4 + 1/8)\right)(0.375) = 0.398 \text{ in.}^2$$

$$A_{gv} = lt_w$$

$$= 10.5(0.375) = 3.94 \text{ in.}^2$$

$$A_{nv} = \left(l - (n - 0.5)(d_h + 1/16)\right)t_w$$

$$= \left(10.5 - 3.5(3/4 + 1/8)\right)(0.375) = 2.79 \text{ in.}^2$$

Determine the tension rupture strength

$$F_u A_{nt} = 65(0.398) = 25.9 \text{ kips}$$

Consider shear yield and shear rupture and select the one with least strength; thus,

$$0.6F_y A_{gv} = 0.6(50)(3.94) = 118 \text{ kips}$$

$$0.6F_u A_{nv} = 0.6(65)(2.79) = 109 \text{ kips}$$

Selecting the shear rupture term and combining it with the tension rupture term gives a connection block shear strength, with U_{bs} = 1.0, of

$$R_n/\Omega = (109 + 25.9)/2.00 = 67.5 \text{ kips} > 40.0 \text{ kips}$$

Step 9: Select the plate-to-column weld.

Based on a fillet weld on each side of the plate, with a weld design strength of 0.928 kips per 1/16 in. of weld per in. of length

$$D = \frac{40.0}{(2(0.928)(12.0))} = 1.80 \text{ sixteenths of in.}$$

Therefore, use a 3/16 in. weld, the minimum weld for the 3/8 in. plate, as given in *Specification* Table J2.4.

Step 10: Develop the final design.

Figure 12.3 shows the final design using

Four 3/4 in. A325-N bolts in a 3/8×3-1/2×1.0 ft-0 in. plate

The impact of the connection on the column flange and web must be checked. This will be addressed in Section 12.4.

12.3.2 Welded Flange Plate Connection

The welded flange plate connection replaces the direct-welded flanges with plates that are welded to the supported beam flange in the field and to the supporting column in the shop. The web connection is usually the typical single-plate shear connection. To accommodate downhand welding positions for the plate-to-beam flange welds in the field, the top flange plate must be kept to a width at least 1.0 in. less than the beam flange width and the bottom flange plate must be at least 1.0 in. greater in width than the beam flange. The actual width difference varies with weld size and applicable tolerances; see *Manual* Figure 8-13, which provides minimum shelf dimensions for specific fillet weld sizes.

The limit states associated with the tension flange plate are yielding, rupture, and block shear, whereas those associated with the compression flange plate are yielding, local plate buckling, and compression buckling. These limit states are evaluated in Example 12.2.

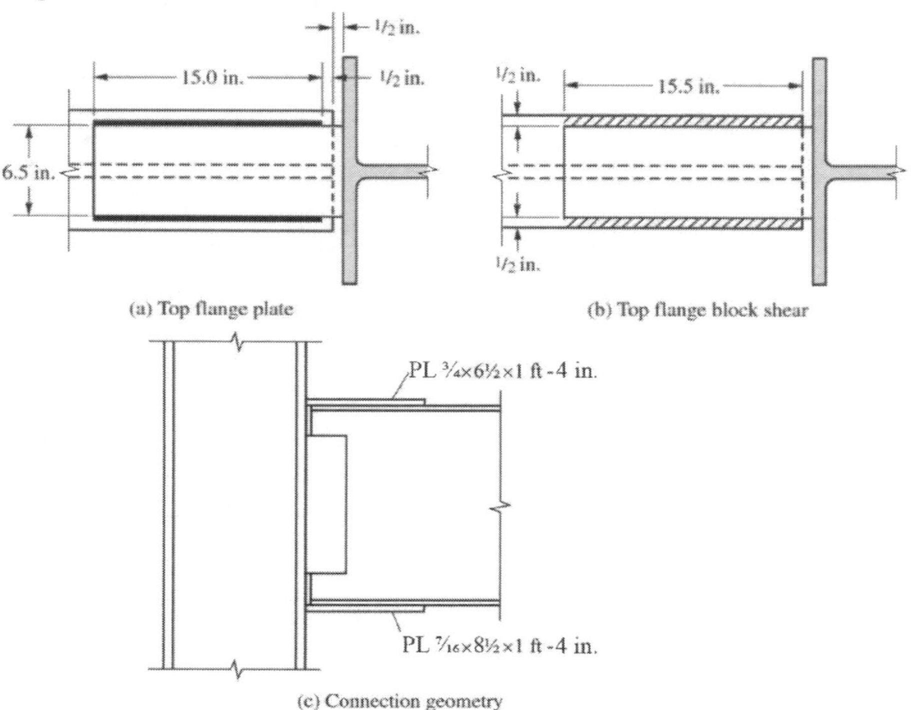

Figure 12.4 Connection for Example 12.2

EXAMPLE 12.2a ***Welded Flange Plate Moment Connection by LRFD***	**Goal:**	Design a welded flange plate beam-to-column moment connection.
	Given:	A welded flange plate beam-to-column moment connection is shown in Figure 12.2b. The beam is a W18×50 and the column is a W14×90. Bolts are 7/8 in. A325-N and the electrodes are E70. The shapes are A992 steel, and the plates are A572 Gr. 50. The LRFD required strength is M_u = 250 ft-kips and V_u = 45 kips. Assume the

moment will cause the top flange to be in tension and the bottom flange to be in compression.

SOLUTION

Step 1: Obtain the beam and column properties from *Manual* Table 1-1.

Beam: W18×50
$d = 18.0$ in., $b_f = 7.50$ in., $t_w = 0.355$ in., $t_f = 0.570$ in.

Column: W14×90
$d = 14.0$ in., $b_f = 14.5$ in., $t_w = 0.440$ in., $t_f = 0.710$ in.

Step 2: Determine the force to be carried in each flange plate.

Since the plate thickness is not yet known, conservatively assume the moment arm is the depth of the beam. Thus, using Equation 12.2 gives

$$P_u = \frac{M_u}{d} = \frac{250(12)}{18.0} = 167 \text{ kips}$$

Step 3: Determine the minimum plate area based on the limit state of yielding for the tension flange plate.

$$A_p = \frac{P_u}{\phi F_y} = \frac{167}{0.9(50)} = 3.71 \text{ in.}^2$$

The top flange plate should be narrower than the beam flange by at least 1.0 in. to facilitate welding in the down position. Therefore, try a 3/4 × 6.5 in. plate. $A_p = 4.88$ in.^{2}. Although this is not the thinnest possible selection, use of this plate allows for consideration of the reduction in rupture strength.

Step 4: Check the plate for tension rupture.

Figure 12.4a illustrates the expected weld pattern and the top flange plate. The shear lag factor, U, for a welded joint is given in *Specification* Table D3.1 case 4 as

$$U = \frac{3l^2}{3l^2 + w^2}\left(1 - \frac{\bar{x}}{l}\right)$$

In order to establish a starting point for checking plate rupture, assume that the weld length will be at least equal to the twice the plate width. That will give

$$U = \frac{3l^2}{3l^2 + w^2}\left(1 - \frac{\bar{x}}{l}\right) = \left(\frac{3(2w)^2}{3(2w)^2 + w^w}\right)\left(1 - \frac{3/8}{13.0}\right)$$
$$= 0.923(0.971) = 0.896$$

This will likely be a lower bound on the actual U for this connection. Thus, with $A_n = A_p$ for a welded plate using Equation D3-1

$$A_e = A_n U = 4.88(0.896) = 4.37 \text{ in.}^2$$

and using Equation J4-2

$$P_n = F_u A_e = 65(4.37) = 284 \text{ kips}$$

Thus,

$$\phi P_n = 0.75(284) = 213 > 167 \text{ kips}$$

The tension rupture limit state will not control as long as the final weld length is at least equal to twice the plate width.

Step 5: Select the fillet weld size based on weld rupture.

The minimum size fillet weld for a 3/4 in. plate attached to a 0.570 in. beam flange, based on *Specification* Table J2.4, is 1/4 in. Therefore, determine the required length of a pair of 1/4 in. fillet welds on the sides of the flange plate.

$$l = \frac{167}{2(1.392(4))} = 15.0 \text{ in.}$$

This length of weld appears to be reasonable for this connection and it exceeds twice the 6.5 in. plate width. The length of the top plate is determined by taking into account a 1/4 in. underrun tolerance in the length of the beam, the specified 1/2 in. setback of the beam from the column face, and a weld termination set back equal to the weld size. Thus, the plate will be

$$l = 15.0 + 1/4 + 1/2 + 1/4 = 16.0 \text{ in.}$$

Thus, the top plate is 3/4×6.5×1.0 ft – 4.0 in, as shown in Figure 12.4c.

Step 6: Check the rupture strength of the base material at the welds for the tension flange.

For the plate

$$t_{\min} = \frac{3.09D}{F_u} = \frac{3.09(4)}{65} = 0.190 \text{ in.} < t_p = 3/4 \text{ in.}$$

For the beam flange

$$t_{\min} = \frac{3.09D}{F_u} = \frac{3.09(4)}{65} = 0.190 \text{ in.} < t_f = 0.570 \text{ in.}$$

Step 7: Consider the block shear rupture of the top flange of the beam.

Because the plate is welded to the flange, the critical shear limit state is shear yielding. For the two blocks on each side of the web, as shown in Figure 12.4b, the areas are

$$A_{gv} = 15.5(0.570) = 8.84 \text{ in.}^2$$

$$A_{nt} = 0.5(0.570) = 0.285 \text{ in.}^2$$

The nominal strength for tension rupture is

$$F_u A_{nt} = 65(0.285) = 18.5 \text{ kips}$$

and the nominal strength for shear yield is

$$0.6F_y A_{gv} = 0.6(50)8.84 = 265 \text{ kips}$$

Thus, the design strength for block shear of the beam flange is

$$\phi R_n = 2\big(0.75(265 + 18.5)\big) = 425 \text{ kips} > 167 \text{ kips}$$

Step 8: Determine the required compression flange plate.

This plate must be checked for local buckling and overall buckling. As a starting point, assume that yielding will be the controlling limit state. Thus, the same area will be required as for the tension plate. However, this plate should be 1.0 in. wider than the beam flange so that the welds can again be placed in the downward position. Assume a plate width of 8.5 in. Thus,

$$t_p = \frac{3.71}{8.5} = 0.436 \text{ in.}$$

Select a 7/16 in. plate for further consideration.

Step 9: Check the compression plate for local buckling.

Local buckling of the compression plate is checked against the width-to-thickness limits from *Specification* Table B4.1a. The width of plate between welds is treated as a stiffened plate and the width that projects beyond the weld to the flange tip is treated as an unstiffened plate.

For the stiffened plate, case 7,

$$\frac{b}{t}=\frac{7.5}{0.438}=17.1<1.40\sqrt{E/F_y}=1.40\sqrt{29{,}000/50}=33.7$$

For the unstiffened plate, case 1,

$$\frac{b}{t}=\frac{0.5}{0.438}=1.14<0.56\sqrt{E/F_y}=0.56\sqrt{29{,}000/50}=13.5$$

So the plate strength is not limited by local buckling.

Step 10: Determine the compressive strength of the plate. Since the force to be transferred between the beam and the compression plate is the same as that transferred for the tension plate, use the same ¼ in. welds 15.0 in. long on each side.
The plate has a length for compression buckling of 1.0 in. from the column flange to the end of the weld, as shown in Figure 12.4a. The effective length factor is taken as 0.65, the value recommended in the Commentary for a fixed-fixed column. Determine the slenderness ratio for this plate.

$$r=\sqrt{\frac{I}{A}}=\sqrt{\frac{h^2}{12}}=\sqrt{\frac{(0.438)^2}{12}}=0.126$$

$$\frac{l_c}{r}=\frac{KL}{r}=\frac{0.65(1.0)}{0.126}=5.16$$

For compression elements that are part of connections, *Specification* Section J4.4 indicates that, when the slenderness ratio is less than 25, $F_n = F_y$. Thus, the selection of this plate for yielding, as was originally done, is correct and the 8.5×7/16 plate is acceptable for the compression limit states.

Step 11: Check the rupture strength of the base material at the welds for the compression flange.

For the plate

$$t_{\min}=\frac{3.09D}{F_u}=\frac{3.09(4)}{65}=0.190\text{ in.}<t_p=7/16\text{ in.}$$

For the beam flange the calculation is the same as it was at the tension flange

$$t_{\min} = \frac{3.09D}{F_u} = \frac{3.09(4)}{65} = 0.190 \text{ in.} < t_f = 0.570 \text{ in.}$$

Step 12: Determine the welds required to connect the flange plates to the column flange.

The force to be transferred is the same for both plates. It can be determined by dividing the design moment by the beam depth plus the average of the tension and compression plate thicknesses. Thus,

$$P_u = \frac{M_u}{(d + t_{pt}/2 + t_{pc}/2)} = \frac{250(12)}{(18.0 + 0.750/2 + 0.438/2)} = 161 \text{ kips}$$

A comparison of the plate width with the column flange width shows that both plates are compatible because each plate is narrower than the column flange width, $b_f = 14.5$ in. The force is perpendicular to the weld, so by Equation J2-5 the weld strength can be increased by 1.5. Thus, for fillet welds on both the top and bottom of the plate

$$D = \frac{161}{1.5(1.392)(2b_p)} = \frac{38.6}{b_p}$$

For the top flange plate

$$D = \frac{38.6}{6.5} = 5.94 \text{ sixteenths of in.}$$

Therefore, use a pair of 3/8 in. welds.

For the bottom flange plate

$$D = \frac{38.6}{8.5} = 4.54 \text{ sixteenths of in.}$$

Therefore, use a pair of 5/16 in. welds.

The minimum thickness of the column flange to match the rupture strength of the larger weld is given by *Manual* Equation 9-6 as

$$t_{\min} = \frac{3.09D}{F_u} = \frac{3.09(6)}{65} = 0.285 < t_f = 0.710 \text{ in.}$$

So the column flange has sufficient thickness to transfer the force.

Step 13: Develop the final design.

The web connection design that was demonstrated in Example 12.1a must also be carried out here. In addition, the impact of the connection on the column flange and web must be checked. This will be addressed in Section 12.4.

The final geometry for the welded flange plate connection is shown in Figure 12.4c.

EXAMPLE 12.2b
Welded Flange Plate Moment Connection by ASD

Goal: Design a welded flange plate beam-to-column moment connection.

Given: A welded flange plate beam-to-column moment connection is shown in Figure 12.2b. The beam is a W18×50 and the column is a W14×90. Bolts are 7/8 in. A325-N and the electrodes are E70. The shapes are A992 steel, and the plates are A572 Gr. 50. The ASD required strength is M_a = 167 ft-kips and V_a = 30 kips. Assume the moment will cause the top flange to be in tension and the bottom flange to be in compression.

SOLUTION

Step 1: Obtain the beam and column properties from *Manual* Table 1-1.

Beam: W18×50
d = 18.0 in., b_f = 7.50 in., t_w = 0.355 in., t_f = 0.570 in.

Column: W14×90
d = 14.0 in., b_f = 14.5 in., t_w = 0.440 in., t_f = 0.710 in.

Step 2: Determine the force to be carried in each flange plate.

Since the plate thickness is not yet known, conservatively assume the moment arm is the depth of the beam. Thus, using Equation 12.2 gives

$$P_a = \frac{M_a}{d} = \frac{167(12)}{18.0} = 111 \text{ kips}$$

Step 3: Determine the minimum plate area based on the limit state of yielding for the tension flange plate.

$$A_p = \frac{P_a}{F_y/\Omega} = \frac{111}{50/1.67} = 3.71 \text{ in.}^2$$

The top flange plate should be narrower than the beam flange by at least 1.0 in. to facilitate welding in the down position. Therefore, try a 3/4×6.5 in. plate. A_p = 4.88 in.^{2}. Although this is not the thinnest possible selection, use of this plate allows for consideration of the reduction in rupture strength.

Step 4: Check the plate for tension rupture.

Figure 12.4a illustrates the expected weld pattern and the top flange plate. The shear lag factor, U, for a welded joint is given in *Specification* Table D3.1 case 4 as

$$U = \frac{3l^2}{3l^2 + w^2}\left(1 - \frac{\overline{x}}{l}\right)$$

In order to establish a starting point for checking plate rupture, assume that the weld length will be at least equal to twice the plate width. That will give

$$U = \frac{3l^2}{3l^2 + w^2}\left(1 - \frac{\overline{x}}{l}\right) = \left(\frac{3(2w)^2}{3(2w)^2 + w^2}\right)\left(1 - \frac{3/8}{13.0}\right)$$
$$= 0.923(0.971) = 0.896$$

This will likely be a lower bound on the actual U for this connection. Thus, with $A_n = A_p$ for a welded plate using Equation D3-1

$$A_e = A_n U = 4.88(0.896) = 4.37 \text{ in.}^2$$

and using Equation J4-2

$$P_n = F_u A_e = 65(4.37) = 284 \text{ kips}$$

Thus,

$$P_n/\Omega = 284/2.00 = 142 > 111 \text{ kips}$$

The tension rupture limit state will not control as long as the final weld length is at least twice the plate width.

Step 5: Select the fillet weld size based on weld rupture.

The minimum size fillet weld for a 3/4 in. plate attached to a 0.570 in. beam flange, based on *Specification* Table J2.4, is 1/4 in. Therefore, determine the required length of a pair of 1/4 in. fillet welds on the sides of the flange plate.

$$l = \frac{111}{2(0.928(4))} = 15.0 \text{ in.}$$

This length of weld appears to be reasonable for this connection and it exceeds twice the 6.5 in. plate width. The length of the top plate is determined by taking into account a 1/4 in. underrun tolerance in the length of the beam, the specified 1/2 in. setback of the beam from the column face, and a weld termination set back equal to the weld size. Thus, the plate will be

$$l = 15.0 + 1/4 + 1/2 + 1/4 = 16.0 \text{ in.}$$

Thus, the top plate is 3/4×6.5×1.0 ft – 4.0 in, as shown in Figure 12.4c.

Step 6: Check the rupture strength of the base material at the welds for the tension flange.

For the plate

$$t_{\min} = \frac{3.09D}{F_u} = \frac{3.09(4)}{65} = 0.190 \text{ in.} < t_p = 3/4 \text{ in.}$$

For the beam flange

$$t_{\min} = \frac{3.09D}{F_u} = \frac{3.09(4)}{65} = 0.190 \text{ in.} < t_f = 0.570 \text{ in.}$$

Step 7: Consider the block shear rupture of the top flange of the beam.

Because the plate is welded to the flange, the critical shear limit state is shear yielding. For the two blocks on each side of the web, as shown in Figure 12.4b, the areas are

$$A_{gv} = 15.5(0.570) = 8.84 \text{ in.}^2$$

$$A_{nt} = 0.5(0.570) = 0.285 \text{ in.}^2$$

The nominal strength for tension rupture is

$$F_u A_{nt} = 65(0.285) = 18.5 \text{ kips}$$

and the nominal strength for shear yield is

$$0.6F_y A_{gv} = 0.6(50)(8.84) = 265 \text{ kips}$$

Thus, the allowable strength for block shear of the beam flange is

$$R_n/\Omega = 2\left((265 + 18.5)/2.00\right) = 284 \text{ kips} > 111 \text{ kips}$$

Step 8: Determine the required compression flange plate.

This plate must be checked for local buckling and overall buckling.

As a starting point, assume that yielding will be the controlling limit state. Thus, the same area will be required as for the tension plate. However, this plate should be 1.0 in. wider than the beam flange so that the welds can again be placed in the downward position. Assume a plate width of 8.5 in. Thus,

$$t_p = \frac{3.71}{8.5} = 0.436 \text{ in.}$$

Select a 7/16 in. plate for further consideration.

Step 9: Check the compression plate for local buckling.

Local buckling of the compression plate is checked against the width-to-thickness limits from *Specification* Table B4.1a. The width of plate between welds is treated as a stiffened plate and the width that projects beyond the weld to the flange tip is treated as an unstiffened plate. For the stiffened plate, case 7,

$$\frac{b}{t} = \frac{7.5}{0.438} = 17.1 < 1.40\sqrt{E/F_y} = 1.40\sqrt{29{,}000/50} = 33.7$$

For the unstiffened plate, case 1,

$$\frac{b}{t} = \frac{0.5}{0.438} = 1.14 < 0.56\sqrt{E/F_y} = 0.56\sqrt{29{,}000/50} = 13.5$$

So, the plate strength is not limited by local buckling.

Step 10: Determine the compressive strength of the plate. Since the force to be transferred between the beam and the compression plate is the same as that transferred for the tension plate, use the same 1/4 in. welds 15.0 in. long on each side.

The plate has a length for compression buckling of 1.0 in. from the column flange to the end of the weld, as shown in Figure 12.4a. The effective length factor is taken as 0.65, the value recommended in the Commentary for a fixed-fixed column. Determine the slenderness ratio for this plate.

$$r = \sqrt{\frac{I}{A}} = \sqrt{\frac{h^2}{12}} = \sqrt{\frac{(0.438)^2}{12}} = 0.126$$

$$\frac{l_c}{r} = \frac{KL}{r} = \frac{0.65(1.0)}{0.126} = 5.16$$

For compression elements that are part of connections, *Specification* Section J4.4 indicates that, when the slenderness ratio is less than 25, $F_n = F_y$. Thus, the selection of this plate for yielding, as was originally done, is correct and the 8.5×7/16 plate is acceptable for the compression limit states.

Step 11: Check the rupture strength of the base material at the welds for the compression flange.

For the plate

$$t_{\min} = \frac{3.09D}{F_u} = \frac{3.09(4)}{65} = 0.190 \text{ in.} < t_p = 7/16 \text{ in.}$$

For the beam flange the calculation is the same as it was at the tension flange

$$t_{\min} = \frac{3.09D}{F_u} = \frac{3.09(4)}{65} = 0.190 \text{ in.} < t_f = 0.570 \text{ in.}$$

Step 12: Determine the welds required to connect the flange plates to the column flange.

The force to be transferred is the same for both plates. It can be determined by dividing the allowable moment by the beam depth plus the average of the tension and compression plate thicknesses.

Thus,

$$P_a = \frac{M_a}{\left(d + t_{pt}/2 + t_{pc}/2\right)} = \frac{167(12)}{\left(18.0 + 0.750/2 + 0.438/2\right)} = 108 \text{ kips}$$

A comparison of the plate width with the column flange width shows that both plates are compatible because each plate is narrower than the column flange width, $b_f = 14.5$ in. The force is perpendicular to the weld, so by Equation J2-5 the weld strength can be increased by 1.5. Thus, for fillet welds on both the top and bottom of the plate

$$D = \frac{108}{1.5(0.928)(2b_p)} = \frac{38.8}{b_p}$$

For the top flange plate

$$D = \frac{38.8}{6.5} = 5.97 \text{ sixteenths of in.}$$

Therefore, use a pair of 3/8 in. welds.

For the bottom flange plate

$$D = \frac{38.8}{8.5} = 4.56 \text{ sixteenths of in.}$$

Therefore, use a pair of 5/16 in. welds.

The minimum thickness of the column flange to match the rupture strength of the larger weld is given by *Manual* Equation 9-6 as

$$t_{\min} = \frac{3.09D}{F_u} = \frac{3.09(6)}{65} = 0.285 < t_f = 0.710 \text{ in.}$$

So the column flange has sufficient thickness to transfer the force.

Step 13: Develop the final design.

The web connection design that was demonstrated in Example 12.1a must also be carried out here. In addition, the impact of the connection on the column flange and web must be checked. This will be addressed in Section 12.4.

The final geometry for the welded flange plate connection is shown in Figure 12.4c.

12.3.3 Bolted Flange Plate Connection

The bolted flange plate connection is similar to the welded flange plate connection except that the attachment of the plate to the beam flange is through bolts. The addition of bolts to the beam tension flange means that a new limit state, the flexural strength of the beam based on rupture of the tension flange, must be assessed. The other limit states that result from the use of bolts have been described several times and are applicable again here.

The bolted flange plate connection is an effective connection from the erection standpoint. The plates can be shop-welded to the column flange, with the beam inserted between the plates and bolted in the field. To accommodate this field erection process, the top plate is usually set a bit high and a filler plate used once the beam is in place.

The following example demonstrates the limit state checks associated with the transfer of the flange force, as was done for Example 12.2. The web connection will not be designed because no new limit states are to be considered.

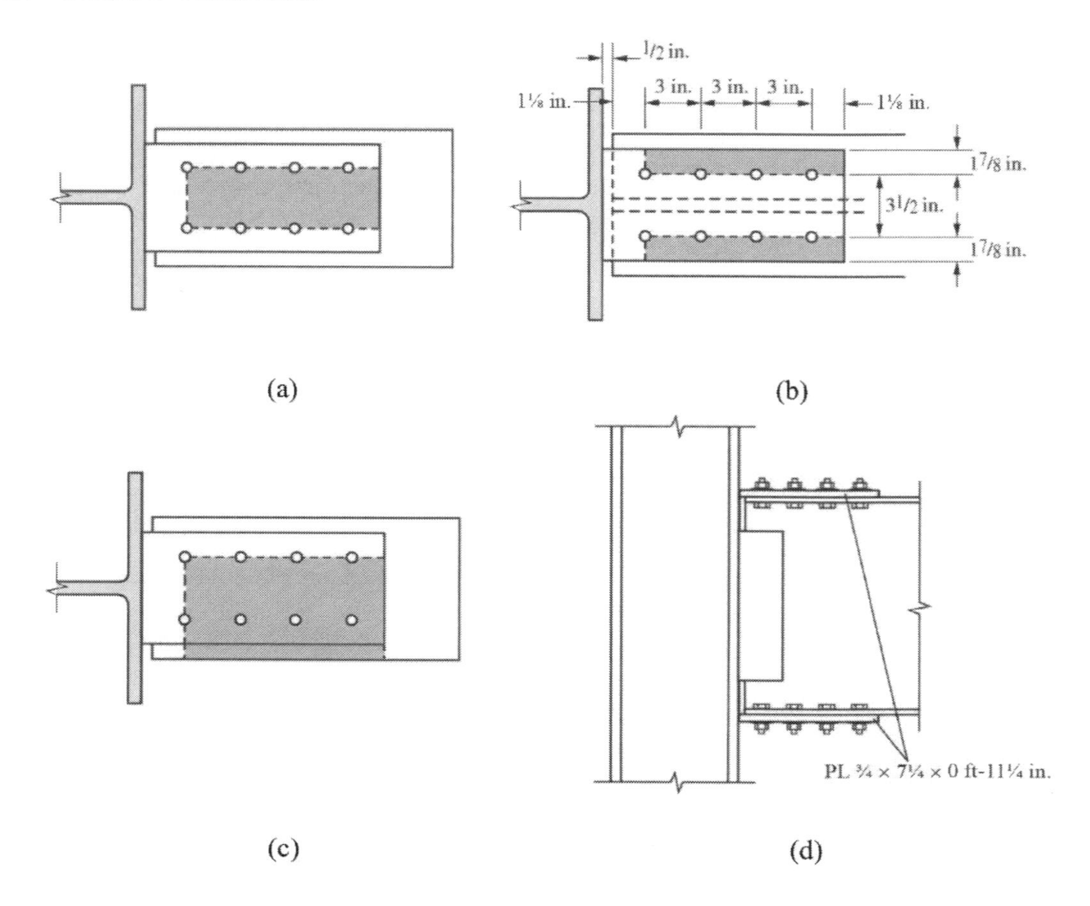

Figure 12.5 Connection for Example 12.3

EXAMPLE 12.3a
Bolted Flange Plate Moment Connection by LRFD

Goal: Design a bolted flange plate beam-to-column moment connection.

Given: A bolted flange plate beam-to-column moment connection is shown in Figure 12.2c. This connection is to be designed for the same conditions as those in Example 12.2a. The beam is a W18×50 and the column is a W14×90. Bolts are 7/8 in. A325-N and the electrodes are E70. The shapes are A992 steel and the plate is A572 Gr. 50. The required strength is $M_u = 250$ ft-kips and $V_u = 45.0$ kips. Assume the moment will cause the top flange to be in tension and the bottom flange to be in compression.

SOLUTION

Step 1: Determine the beam and column properties.

The member dimensions are the same as those given for Example 12.2a. In addition, for the W18×50, from *Manual* Table 1-1, $S_x = 88.9 \text{ in.}^3$.

Step 2: Check the reduced beam section for flexure.

Although the connection has not yet been designed, it is known that at a section through the connection there will be two bolt holes in the tension flange. This may reduce the strength of the beam below the required strength. If that is the case, there will be no reason to continue with this connection design. Thus, the provisions of *Specification* Section F13.1 must be applied for the limit state of rupture of the tension flange.

Determine the gross and net areas of the tension flange.

$$A_{fg} = b_f t_f = 7.5(0.570) = 4.28 \text{ in.}^2$$

$$A_{fn} = \left(b_f - n\left(d_h + 1/16\right)\right)t_f = \left(7.5 - 2\left(7/8 + 1/8\right)\right)(0.570) = 3.14 \text{ in.}^2$$

Check the yield stress to tensile strength ratio to determine a value for Y_t.

$$\frac{F_y}{F_u} = \frac{50}{65} = 0.76 < 0.8$$

Therefore, for all A992 shapes $Y_t = 1.0$, and for this W18×50 beam with a pair of holes for 7/8 in. bolts,

$$Y_t F_y A_{fg} = 1.0(50)(4.28) = 214 \text{ kips}$$

$$F_u A_{fn} = 65(3.14) = 204 \text{ kips}$$

Because $F_u A_{fn} < Y_t F_y A_{fg}$, the nominal moment strength is limited by *Specification* Equation F13-1 to

$$M_n = \frac{F_u A_{fn}}{A_{fg}} S_x = \frac{204}{4.28}(88.9)\left(\frac{1}{12}\right) = 353 \text{ ft-kips}$$

and

$$\phi M_n = 0.9(353) = 318 > 250 \text{ ft-kips}$$

So the beam flexural strength is adequate.

Step 3: Check the flange plate for tension yield. Use the force to be carried in each plate determined in step 2 of Example 12.2a, $P_u = 167$ kips

The flange plate will likely be similar to the one used in Example 12.2a. Try a 7-1/4×3/4 plate. The gross area is

$$A_g = 7.25(0.75) = 5.44 \text{ in.}^2$$

and the design yield strength is

$$\phi R_n = \phi F_y A_g = 0.9(50)(5.44) = 245 > 167 \text{ kips}$$

Step 4: Check the plate for tension rupture.

The net area is

$$A_n = (7.25 - 2(7/8 + 1/8))(0.750) = 3.94 \text{ in.}^2$$

and the design rupture strength is

$$\phi R_n = \phi F_u A_n = 0.75(65)(3.94) = 192 > 167 \text{ kips}$$

So the plate size is adequate based on tension.

Step 5: Determine the number of bolts required based on the bolt shear rupture strength.

First consider the shear rupture strength of the bolts to determine the minimum number of bolts required.

For a 7/8 in. bolt, from *Manual* Table 7-1, $\phi r_n = 24.3$ kips; therefore, the minimum number of bolts is

$$n = \frac{167}{24.3} = 6.87$$

Thus, try an eight-bolt connection with bolt spacing of 3.0 in. and end distances equal to the minimum permitted by Table J3.4 of 1-1/8 in. for the end of the plate and the beam end as illustrated in Figure 12.5b.

Step 6: Determine the bolt bearing and tearout strength on the plate.

For the bolt closest to the end of the plate, the clear distance is

$$l_c = 1.125 - \frac{1}{2}(7/8 + 1/16) = 0.656 \text{ in.} < 2(7/8) = 1.75 \text{ in.}$$

Since the clear distance is less than $2d_b$, tearout controls over bearing. The nominal bolt strength for tearout is

$$R_n = 1.2 l_c t F_u = 1.2(0.656)(0.750)(65) = 38.4 \text{ kips}$$

and the design tearout strength is

$$\phi R_n = 0.75(38.4) = 28.8 \text{ kips}$$

For the other bolts the spacing is 3.0 in. and the clear distance is

$$l_c = s - d_h = 3.0 - (7/8 + 1/16) = 2.06 > 2(7/8) = 1.75 \text{ in.}$$

Since the clear distance is greater than $2d_b$, bearing controls over tearout. The nominal bolt strength for bearing is

$$R_n = 2.4dtF_u = 2.4(7/8)(0.750)(65) = 102 \text{ kips}$$

and the design strength per bolt for bearing on the flange plate is

$$\phi R_n = 0.75(102) = 76.5 \text{ kips}$$

Step 7: Determine the bolt bearing and tearout strength on the beam flange

For the bolt closest to the end of the flange, the clear distance is

$$l_c = 1.125 - \frac{1}{2}(7/8 + 1/16) = 0.656 \text{ in.} < 2(7/8) = 1.75 \text{ in.}$$

Since the clear distance is less than $2d_b$, tearout controls over bearing. The nominal bolt strength for tearout is

$$R_n = 1.2l_c tF_u = 1.2(0.656)(0.570)(65) = 29.2 \text{ kips}$$

and the design tearout strength is

$$\phi R_n = 0.75(29.2) = 21.9 \text{ kips}$$

For the other bolts the spacing is 3.0 in. and the clear distance is

$$l_c = s - d_h = 3.0 - (7/8 + 1/16) = 2.06 > 2(7/8) = 1.75 \text{ in.}$$

Since the clear distance is greater than $2d_b$, bearing controls over tearout. The nominal bolt strength for bearing is

$$R_n = 2.4dtF_u = 2.4(7/8)(0.570)(65) = 77.8 \text{ kips}$$

and the design strength per bolt for bearing on the flange plate is

$$\phi R_n = 0.75(77.8) = 58.4 \text{ kips}$$

Step 8: The controlling strength for each bolt in the connection must be determined.

2 bolts closest to end of plate	
Bolt shear	24.3 kips
Plate tearout	28.8 kips
Flange bearing	58.4 kips
4 interior bolts	
Bolt shear	24.3 kips
Plate bearing	76.5 kips
Flange bearing	58.4 kips
2 bolts closest to end of flange	
Bolt shear	24.3 kips
Plate bearing	76.5 kips
Flange tearout	21.9 kips

Therefore, the strength of the connection at the tension flange is the sum of bolt shear for 6 bolts and flange tearout for the 2 bolts closest to the end of the beam flange. Thus,

$$\phi R_n = 6(24.3) + 2(21.9) = 190 > 167 \text{ kips}$$

Step 9: Check the plate for block shear rupture.

Check the plate for block shear using the geometry shown in Figure 12.5. There are three possible block shear failure patterns shown, one with the center portion failing in tension, Figure 12.5a, one with the two outside portions failing in tension, Figure 12.5b, and one with only one shear failure plane and the tension failure over the middle portion with one outside portion, Figure 12.5c. The worst case must be identified. For the first two possibilities, the critical tension area for block shear will be the one associated with the least tension width since they both have the same shear areas. In this case it will be for the middle 3½ in. section as shown in Figure 12.5a, and the critical net tension area is

$$\begin{aligned} A_{nt} &= \left(b - \left(d_h + 1/16\right)\right)t_p \\ &= \left(3.5 - \left(7/8 + 1/8\right)\right)(0.750) = 1.88 \text{ in.}^2 \end{aligned}$$

and the shear areas for the two shear planes are

$$\begin{aligned} A_{gv} &= 2lt_w \\ &= 2\left(10.125(0.750)\right) = 15.2 \text{ in.}^2 \\ A_{nv} &= 2\left(l - (n - 0.5)\left(d_h + 1/16\right)\right)t_w \\ &= 2\left(10.125 - 3.5\left(7/8 + 1/8\right)\right)(0.750) = 9.94 \text{ in.}^2 \end{aligned}$$

Determine the tension rupture strength

$$F_u A_{nt} = 65(1.88) = 122 \text{ kips}$$

Consider the shear yield and shear rupture and select the one with least strength; thus,

$$\begin{aligned} 0.6F_y A_{gv} &= 0.6(50)(15.2) = 456 \text{ kips} \\ 0.6F_u A_{nv} &= 0.6(65)(9.94) = 388 \text{ kips} \end{aligned}$$

Selecting the shear rupture term and combining it with the tension rupture term gives a connection design block shear strength, with $U_{bs} = 1.0$, of

$$\phi R_n = 0.75\left(388 + 1.0(122)\right) = 0.75(510) = 383 > 167 \text{ kips}$$

Since the shear contribution is so much greater than the tension contribution, the pattern shown in Figure 12.5c should be checked. For a tension width equal to 5.38 in., the net tension area is

$$A_{nt} = \left(b - 1.5\left(d_h + 1/16\right)\right)t_p$$
$$= \left(5.38 - 1.5\left(7/8 + 1/8\right)\right)(0.750) = 2.91 \text{ in.}^2$$

and the shear areas for a single shear plane are

$$A_{gv} = lt_w$$
$$= 10.125(0.750) = 7.59 \text{ in.}^2$$
$$A_{nv} = \left(l - (n - 0.5)\left(d_h + 1/16\right)\right)t_w$$
$$= \left(10.125 - 3.5\left(7/8 + 1/8\right)\right)(0.750) = 4.97 \text{ in.}^2$$

Determine the tension rupture strength

$$F_u A_{nt} = 65(2.91) = 189 \text{ kips}$$

Consider shear yield and shear rupture and select the one with the least strength; thus,

$$0.6F_y A_{gv} = 0.6(50)(7.59) = 228 \text{ kips}$$
$$0.6F_u A_{nv} = 0.6(65)(4.97) = 194 \text{ kips}$$

Selecting the shear rupture term and combining it with the tension rupture term gives a connection design block shear strength, with $U_{bs} = 1.0$, of

$$\phi R_n = 0.75\left(194 + 1.0(189)\right) = 0.75(383) = 287 > 167 \text{ kips}$$

Of the three block shear patterns illustrated, this is clearly the most critical. However, it should be noted from Step 4 that tension rupture of the plate is more critical than any of the block shear failure patterns. When the shear strength of the block shear patterns is larger than all the potential tension pattern strengths, tension rupture of the plate will likely control over any of the potential block shear modes. Thus, for typical flange plate widths, any time that the pattern illustrated in Figure 12.5c would be the controlling block shear pattern, tension rupture of the plate will be more critical.

Step 10: Check the beam flange for block shear.

In this case, the beam web adds enough additional shear area beyond that in the flange that it prevents a block shear failure in a pattern similar to Figures 12.5a and 12.5c. Thus, check a pattern similar to that shown in Figure 12.5b where the flange width of 7.5 in. yields two tension widths of 2.0 in. The net tension area is

$$A_{nt} = 2\left(b - \frac{1}{2}(d_h + 1/16)\right)t_p$$

$$= 2\left(2.00 - \frac{1}{2}(7/8 + 1/8)\right)(0.570) = 1.71\text{ in.}^2$$

and the shear areas for the two shear planes are

$$A_{gv} = 2lt_w$$

$$= 2(10.125(0.570)) = 11.5\text{ in.}^2$$

$$A_{nv} = 2(l - (n - 0.5)(d_h + 1/16))t_w$$

$$= 2(10.125 - 3.5(7/8 + 1/8))(0.570) = 7.55\text{ in.}^2$$

Determine the tension rupture strength

$$F_u A_{nt} = 65(1.71) = 111\text{ kips}$$

Consider the shear yield and shear rupture and select the one with least strength; thus,

$$0.6F_y A_{gv} = 0.6(50)(11.5) = 345\text{ kips}$$

$$0.6F_u A_{nv} = 0.6(65)(7.55) = 294\text{ kips}$$

Selecting the shear rupture term and combining it with the tension rupture term gives a connection design block shear strength, with $U_{bs} = 1.0$, of

$$\phi R_n = 0.75(294 + 1.0(111)) = 0.75(405) = 304 > 167\text{ kips}$$

Step 11: Check the compression plate for local buckling.

Try the same plate as was used for the tension plate using the geometry given in Figure 12.5a. Check the plate for local buckling in a similar fashion as for the welded plate. In this case, the stiffened plate width is the distance between the bolt lines, and the unstiffened width is from the bolt line to the free edge. Thus, for the stiffened plate, case 7 in Table B4.1a,

$$\frac{b}{t} = \frac{3.5}{0.750} = 4.67 < 1.40\sqrt{E/F_y} = 1.40\sqrt{29{,}000/50} = 33.7$$

For the unstiffened plate, case 1 in Table B4.1a,

$$\frac{b}{t} = \frac{1.875}{0.750} = 2.50 < 0.56\sqrt{E/F_y} = 0.56\sqrt{29{,}000/50} = 13.5$$

So the plate strength is not limited by local buckling.

Step 12: Check the compression plate for buckling over its length.

The distance from the column flange to the first bolt is taken as the buckling length of the plate; thus, $L = 1.125 + 0.50 = 1.625$ in. Assuming the effective length factor of a fixed-fixed column, $k = 0.65$, is appropriate,

$$r = \sqrt{\frac{h^2}{12}} = \sqrt{\frac{(0.750)^2}{12}} = 0.217 \text{ in.}$$

and

$$\frac{L_c}{r} = \frac{KL}{r} = \frac{0.65(1.625)}{0.217} = 4.87 < 25$$

Thus, according to Section J4.4, $F_n = F_y$ and the strength is the same as for the tension yield limit state.

Step 13: Determine the welds required to connect the flange plates to the column flange.

The force to be transferred is the same for both plates. A comparison of the plate width with the column flange width shows that they are compatible because the plates are narrower than the column flange width, $b_f = 14.5$ in. In addition, the force is perpendicular to the weld, so the weld strength can be increased by 1.5 according to Equation J2-5. Thus, for fillet welds on both the top and bottom of the plate,

$$D = \frac{167}{1.5(1.392)(2b_p)} = \frac{40.0}{b_p}$$

For both flange plates

$$D = \frac{40.0}{7.25} = 5.52 \text{ sixteenths of in.}$$

Therefore, use a pair of 3/8 in. welds, which exceeds the minimum for this plate thickness.

Step 14: Consider the web connection.

The same web shear connection as was used in the welded flange plate connection could be used in this connection.

Step 15: Develop the final design.

The flange plates of this connection are PL3/4×7-1/4×0 ft- 11.75 in. with eight 7/8 in. A325-N bolts and a pair of 3/8 fillet welds, as shown in Figure 12.5d. As was the case for the previous connection, the impact of the connection on the column flange and web must be checked. This will be addressed in Section 12.4.

EXAMPLE 12.3b
Bolted Flange Plate Moment Connection by ASD

Goal: Design a bolted flange plate beam-to-column moment connection.

Given: A bolted flange plate beam-to-column moment connection is shown in Figure 12.2c. This connection is to be designed for the same conditions as those in Example 12.2b. The beam is a W18×50 and the column is a W14×90. Bolts are 7/8 in. A325-N and the electrodes are E70. The shapes are A992 steel, and the plate is A572 Gr. 50. The required strength is $M_a = 167$ ft-kips and $V_a = 30.0$ kips. Assume the moment will cause the top flange to be in tension and the bottom flange to be in compression.

SOLUTION

Step 1: Determine the beam and column properties.

The member dimensions are the same as those given for Example 12.2b. In addition, for the W18×50, from *Manual* Table 1-1, $S_x = 88.9\text{ in.}^3$.

Step 2: Check the reduced beam section for flexure.

Although the connection has not yet been designed, it is known that at a section through the connection there will be two bolt holes in the tension flange. This may reduce the strength of the beam below the required strength. If that is the case, there will be no reason to continue with this connection design. Thus, the provisions of *Specification* Section F13.1 must be applied for the limit state of rupture of the tension flange.

Determine the gross and net areas of the tension flange.

$$A_{fg} = b_f t_f = 7.5(0.570) = 4.28\text{ in.}^2$$

$$A_{fn} = \left(b_f - n\left(d_h + 1/16\right)\right)t_f = \left(7.5 - 2\left(7/8 + 1/8\right)\right)(0.570) = 3.14\text{ in.}^2$$

Check the yield stress to tensile strength ratio to determine a value for Y_t.

$$\frac{F_y}{F_u} = \frac{50}{65} = 0.76 < 0.8$$

Therefore, for all A992 shapes $Y_t = 1.0$, and for this W18×50 beam with a pair of holes for 7/8 in. bolts,

$$Y_t F_y A_{fg} = 1.0(50)(4.28) = 214 \text{ kips}$$

$$F_u A_{fn} = 65(3.14) = 204 \text{ kips}$$

Because $F_u A_{fn} < Y_t F_y A_{fg}$, the nominal moment strength is limited by *Specification* Equation F13-1 to

$$M_n = \frac{F_u A_{fn}}{A_{fg}} S_x = \frac{204}{4.28}(88.9)\left(\frac{1}{12}\right) = 353 \text{ ft-kips}$$

and

$$M_n/\Omega = 353/1.67 = 211 > 167 \text{ ft-kips}$$

So, the flexural strength is adequate.

Step 3: Check the flange plate for tension yield. Use the force to be carried in each plate determined in step 2 of Example 12.2b, $P_a = 111$ kips

The flange plate will likely be similar to the one used in Example 12.2b. Try a 7-1/4×3/4 plate. The gross area is

$$A_g = 7.25(0.75) = 5.44 \text{ in.}^2$$

and the allowable yield strength is

$$R_n/\Omega = F_y A_g/\Omega = (50)(5.44)/1.67 = 163 > 111 \text{ kips}$$

Step 4: Check the plate for tension rupture.

The net area is

$$A_n = (7.25 - 2(7/8 + 1/8))(0.750) = 3.94 \text{ in.}^2$$

and the allowable rupture strength is

$$R_n/\Omega = F_u A_n/\Omega = (65)(3.94)/2.00 = 128 > 111 \text{ kips}$$

So the plate size is adequate based on tension.

Step 5: Determine the number of bolts required based on the bolt shear rupture strength.

First consider the shear rupture strength of the bolts to determine the minimum number of bolts required.

For a 7/8 in. bolt, from *Manual* Table 7-1, $r_n/\Omega = 16.2$ kips; therefore, the minimum number of bolts is

$$n = \frac{111}{16.2} = 6.85$$

Thus, try an eight-bolt connection with bolt spacing of 3.0 in. and end distances equal to the minimum permitted by Table J3.4 of 1-1/8 in. for the end of the plate and the beam end as illustrated in Figure 12.5b.

Step 6: Determine the bolt bearing and tearout strength on the plate.

For the bolt closest to the end of the plate, the clear distance is

$$l_c = 1.125 - \frac{1}{2}(7/8 + 1/16) = 0.656 \text{ in.} < 2(7/8) = 1.75 \text{ in.}$$

Since the clear distance is less than $2d_b$, tearout controls over bearing. The nominal bolt strength for tearout is

$$R_n = 1.2 l_c t F_u = 1.2(0.656)(0.750)(65) = 38.4 \text{ kips}$$

and the allowable tearout strength is

$$R_n/\Omega = 38.4/2.00 = 19.2 \text{ kips}$$

For the other bolts the spacing is 3.0 in. and the clear distance is

$$l_c = s - d_h = 3.0 - (7/8 + 1/16) = 2.06 > 2(7/8) = 1.75 \text{ in.}$$

Since the clear distance is greater than $2d_b$, bearing controls over tearout. The nominal bolt strength for bearing is

$$R_n = 2.4 d t F_u = 2.4(7/8)(0.750)(65) = 102 \text{ kips}$$

and the allowable strength per bolt for bearing on the flange plate is

$$R_n/\Omega = 102/2.00 = 51.0 > 16.2 \text{ kips}$$

Step 7: Determine the bolt bearing and tearout strength on the beam flange

For the bolt closest to the end of the flange, the clear distance is

$$l_c = 1.125 - \frac{1}{2}(7/8 + 1/16) = 0.656 \text{ in.} < 2(7/8) = 1.75 \text{ in.}$$

Since the clear distance is less than $2d_b$, tearout controls over bearing. The nominal bolt strength for tearout is

$$R_n = 1.2 l_c t F_u = 1.2(0.656)(0.570)(65) = 29.2 \text{ kips}$$

and the allowable tearout strength is

$$R_n/\Omega = 29.2/2 = 14.6 \text{ kips}$$

For the other bolts the spacing is 3.0 in. and the clear distance is

$$l_c = s - d_h = 3.0 - (7/8 + 1/16) = 2.06 > 2(7/8) = 1.75 \text{ in.}$$

Since the clear distance is greater than $2d_b$, bearing controls over tearout. The nominal bolt strength for bearing is

$$R_n = 2.4dtF_u = 2.4(7/8)(0.570)(65) = 77.8 \text{ kips}$$

and the allowable strength per bolt for bearing on the flange plate is

$$R_n/\Omega = 77.8/2.00 = 38.9 \text{ kips}$$

Step 8: The controlling strength for each bolt in the connection must be determined.

2 bolts closest to end of plate	
Bolt shear	16.2 kips
Plate tearout	19.2 kips
Flange bearing	38.9 kips
4 interior bolts	
Bolt shear	16.2 kips
Plate bearing	51.0 kips
Flange bearing	38.9 kips
2 bolts closest to end of flange	
Bolt shear	16.2 kips
Plate bearing	51.0 kips
Flange tearout	14.6 kips

Therefore, the strength of the connection at the tension flange is the sum of bolt shear for 6 bolts and flange tearout for the 2 bolts closest to the end of the beam flange. Thus,

$$R_n/\Omega = 6(16.2) + 2(14.6) = 126 > 111 \text{ kips}$$

Step 9: Check the plate for block shear rupture.

Check the plate for block shear using the geometry shown in Figure 12.5. There are three possible block shear failure patterns shown, one with the center portion failing in tension, Figure 12.5a, one with the two outside portions failing in tension, Figure 12.5b, and one with only one shear failure plane and the tension failure over the middle portion with one outside portion, Figure 12.5c. The worst case must be identified. For the first two possibilities, the critical tension area for block shear will be the one associated with the least tension width since they both have the same tension areas. In this case it will be for the middle 3½ in. section as shown in Figure 12.5a, and the critical net tension area is

$$\begin{aligned} A_{nt} &= \left(b-\left(d_h+1/16\right)\right)t_p \\ &= \left(3.5-\left(7/8+1/8\right)\right)(0.750)=1.88\text{ in.}^2 \end{aligned}$$

and the shear areas for the two shear planes are

$$\begin{aligned} A_{gv} &= 2lt_w \\ &= 2\left(10.125(0.750)\right)=15.2\text{ in.}^2 \\ A_{nv} &= 2\left(l-\left(n-0.5\right)\left(d_h+1/16\right)\right)t_w \\ &= 2\left(10.125-3.5\left(7/8+1/8\right)\right)(0.750)=9.94\text{ in.}^2 \end{aligned}$$

Determine the tension rupture strength

$$F_uA_{nt}=65\left(1.88\right)=122\text{ kips}$$

Consider the shear yield and shear rupture and select the one with least strength; thus,

$$\begin{aligned} 0.6F_yA_{gv} &= 0.6\left(50\right)\left(15.2\right)=456\text{ kips} \\ 0.6F_uA_{nv} &= 0.6\left(65\right)\left(9.94\right)=388\text{ kips} \end{aligned}$$

Selecting the shear rupture term and combining it with the tension rupture term gives a connection allowable block shear strength, with U_{bs} = 1.0, of

$$R_n/\Omega=\left(388+1.0\left(122\right)\right)/2.00=510/2.00=255>111\text{ kips}$$

Since the shear contribution is so much greater than the tension contribution, the pattern shown in Figure 12.5c should be checked. For a tension width equal to 5.38 in., the net tension area is

$$\begin{aligned} A_{nt} &= \left(b-1.5\left(d_h+1/16\right)\right)t_p \\ &= \left(5.38-1.5\left(7/8+1/8\right)\right)(0.750)=2.91\text{ in.}^2 \end{aligned}$$

and the shear areas for a single shear plane are

$$A_{gv} = lt_w$$
$$= 10.125(0.750) = 7.59 \text{ in.}^2$$
$$A_{nv} = \left(l - (n-0.5)(d_h + 1/16)\right)t_w$$
$$= \left(10.125 - 3.5(7/8 + 1/8)\right)(0.750) = 4.97 \text{ in.}^2$$

Determine the tension rupture strength

$$F_u A_{nt} = 65(2.91) = 189 \text{ kips}$$

Consider shear yield and shear rupture and select the one with the least strength; thus

$$0.6F_y A_{gv} = 0.6(50)(7.59) = 228 \text{ kips}$$
$$0.6F_u A_{nv} = 0.6(65)(4.97) = 194 \text{ kips}$$

Selecting the shear rupture term and combining it with the tension rupture term gives a connection allowable block shear strength, with U_{bs} = 1.0, of

$$R_n/\Omega = \left(194 + 1.0(189)\right)/2.00 = 383/2.00 = 192 > 111 \text{ kips}$$

Of the three block shear patterns illustrated, this is clearly the most critical. However, it should be noted from Step 4 that tension rupture of the plate is more critical than any of the block shear failure patterns. When the shear strength of the block shear patterns is larger than all the potential tension pattern strengths, tension rupture of the plate will likely control over any of the potential block shear modes. Thus, for typical flange plate widths, any time that the pattern illustrated in Figure 12.5c would be the controlling block shear pattern, tension rupture of the plate will be more critical.

Step 10: Check the beam flange for block shear.

In this case, the beam web adds enough additional shear area beyond that in the flange that it prevents a block shear failure in a pattern similar to Figure 12.5 a. Thus, check a pattern similar to that shown in Figure 12.5 b where the flange width of 7.5 in. yields two tension width2 of 2.0 in. The net tension area is

$$A_{nt} = 2\left(b - \frac{1}{2}(d_h + 1/16)\right)t_p$$
$$= 2\left(2.00 - \frac{1}{2}(7/8 + 1/8)\right)(0.570) = 1.71 \text{ in.}^2$$

and the shear areas for the two shear planes are

$$A_{gv} = 2lt_w$$
$$= 2(10.125(0.570)) = 11.5 \text{ in.}^2$$
$$A_{nv} = 2(l - (n - 0.5)(d_h + 1/16))t_w$$
$$= 2(10.125 - 3.5(7/8 + 1/8))(0.570) = 7.55 \text{ in.}^2$$

Determine the tension rupture strength

$$F_u A_{nt} = 65(1.71) = 111 \text{ kips}$$

Consider the shear yield and shear rupture and select the one with least strength; thus,

$$0.6F_y A_{gv} = 0.6(50)(11.5) = 345 \text{ kips}$$
$$0.6F_u A_{nv} = 0.6(65)(7.55) = 294 \text{ kips}$$

Selecting the shear rupture term and combining it with the tension rupture term gives a connection allowable block shear strength, with $U_{bs} = 1.0$, of

$$R_n/\Omega = (294 + 1.0(111))/2.00 = 405/2.00 = 203 > 111 \text{ kips}$$

Step 11 Check the compression plate for local buckling.

Try the same plate as was used for the tension plate using the geometry given in Figure 12.5a. Check the plate for local buckling in a similar fashion as for the welded plate. In this case, the stiffened plate width is the distance between the bolt lines, and the unstiffened width is from the bolt line to the free edge. Thus, for the stiffened plate, case 7 in Table B4.1a,

$$\frac{b}{t} = \frac{3.5}{0.750} = 4.67 < 1.40\sqrt{E/F_y} = 1.40\sqrt{29{,}000/50} = 33.7$$

For the unstiffened plate, case 1 in Table B4.1a,

$$\frac{b}{t} = \frac{1.875}{0.750} = 2.50 < 0.56\sqrt{E/F_y} = 0.56\sqrt{29{,}000/50} = 13.5$$

So, the plate strength is not limited by local buckling.

Step 12: Check the compression plate for buckling over its length.

The distance from the column flange to the first bolt is taken as the buckling length of the plate; thus, L = 1.125 + 0.50 = 1.625 in. Assuming the effective length factor of a fixed-fixed column, k = 0.65, is appropriate,

$$r=\sqrt{\frac{h^2}{12}}=\sqrt{\frac{(0.750)^2}{12}}=0.217\text{ in.}$$

and

$$\frac{L_c}{r}=\frac{KL}{r}=\frac{0.65(1.625)}{0.217}=4.87<25$$

Thus, according to Section J4.4, $F_n = F_y$ and the strength is the same as for the tension yield limit state.

Step 13: Determine the welds required to connect the flange plates to the column flange.

The force to be transferred is the same for both plates. A comparison of the plate width with the column flange width shows that they are compatible because the plates are narrower than the column flange width, b_f = 14.5 in. In addition, the force is perpendicular to the weld, so the weld strength can be increased by 1.5 according to Equation J2-5. Thus, for fillet welds on both the top and bottom of the plate,

$$D=\frac{111}{1.5(0.928)(2b_p)}=\frac{39.9}{b_p}$$

For both flange plates

$$D=\frac{39.9}{7.25}=5.50\text{ sixteenths of in.}$$

Therefore, use a pair of 3/8 in. welds, which exceeds the minimum for this plate thickness.

Step 14: Consider the web connection.

The same web shear connection as was used in the welded flange plate connection could be used in this connection.

Step 15: Develop the final design.

The flange plates of this connection are PL3/4×7-1/4×0 ft- 11.75 in. with eight 7/8 in. A325-N bolts and a pair of 3/8 fillet welds, as shown in Figure 12.5d. As was the case for the previous connection, the impact of the connection on the column flange and web must be checked. This will be addressed in Section 12.4.

12.4 COLUMN STIFFENING

The connection designs illustrated in the previous examples treated the beam side of the connection. That is, they looked at only the connecting elements and their influence on the beam to which they were attached. They did not consider the influence of the connection and transfer of forces to the supporting element. Normally, a fully restrained moment connection is made to the flange of a column. This is the most efficient use of the column because the strong axis is resisting the transferred moment.

As with each connecting element, the application of force to a supporting element requires a check of all applicable limit states. The typical moment connection, such as those illustrated in Figure 12.2a through c, results in the transfer of a concentrated force to the column flange. The limit states for flanges and webs with concentrated forces that are applicable to the beam-column connection are defined in *Specification* Section J10 as

1. Flange local bending
2. Web local yielding
3. Web local crippling
4. Web compression buckling
5. Web panel zone shear

Application of these limit states varies depending on whether the applied force is tension or compression and on whether the connection is on one side or both sides of the column. If the limit states are exceeded, either the column section should be changed, or stiffeners and web doubler plates employed.

12.4.1 Flange Local Bending

Flange local bending is addressed in Section J10.1 and is illustrated in Figure 12.6. This limit state is applicable only where a tensile force is applied to the column flange. The primary concern addressed through this limit state is the stress distribution in the weld if the column flange deformation is excessive. Thus, the limit on the applied force is set to prevent excessive deformation. The nominal strength for flange local bending is

$$R_n = 6.25 F_{yf} t_f^2 \qquad \text{(AISC J10-1)}$$

and

$$\phi = 0.9 \text{ (LRFD)} \qquad \Omega = 1.67 \text{ (ASD)}$$

If the force is applied over a small central portion of the column flange, less than 15 percent of the flange width, this limit state does not need to be checked because the force is applied close to the column web and very little flange deformation occurs.

If the force is applied close to the end of the column, the distribution of the force within the flange is limited by the proximity of the end of the column, and the resulting deflection increases. Thus, if the force is applied closer than a distance of $10t_f$ to the end of the member, the nominal strength must be reduced by 50 percent. When this limit state is exceeded, a pair of half-depth transverse stiffeners are needed.

12.4.2 Web Local Yielding

Web local yielding is addressed in Section J10.2. It is the same limit state that was considered for bearing of the web in a seated connection, as discussed in both Sections 6.15 and 11.9. Although previously discussed for compressive forces, this limit state is also applicable to tensile forces. Figure 12.7 illustrates the distribution of stresses with the application of a concentrated force to the web of the column.

This force could be transferred through a direct-welded beam flange or a beam flange plate. The bearing length, l_b, is taken as the thickness of the plate applying the force, or the plate thickness plus the weld width, z_w, when attached with fillet welds. The force is distributed in both directions, provided the connection is at a distance equal to at least the depth of the column away from the column end. This distribution is on a slope of 2.5:1 over the depth given by k_{des} in *Manual* Table 1-1. In this case

$$R_n = F_{yw}t_w\left(5k_{des} + l_b\right) \qquad \text{(AISC J10-2)}$$

and

$$\phi = 1.0 \text{ (LRFD)} \qquad \Omega = 1.50 \text{ (ASD)}$$

If the spacing from the connection to the end of the column is less than the depth of the column, d, the distribution of the force can take place in only one direction and the factor 5 is replaced by 2.5. This is the relationship that was used for the seated connection on the end of a beam and is given by

$$R_n = F_{yw}t_w\left(2.5k_{des} + l_b\right) \qquad \text{(AISC J10-3)}$$

When this limit state is exceeded, a pair of half-depth stiffeners or a web doubler plate is needed.

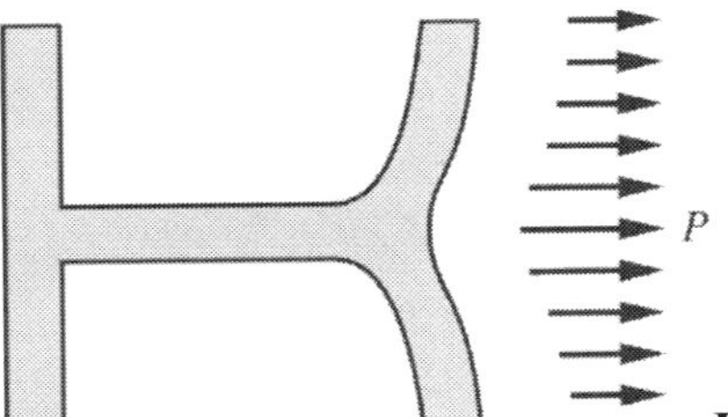

Figure 12.6 Flange Local Bending

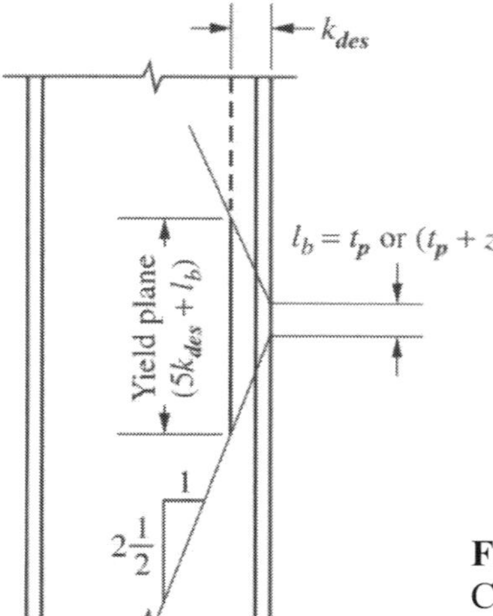

Figure 12.7 Distribution of Concentrated Forces on Column Web

12.4.3 Web local crippling

Web local crippling is addressed in Section J10.3 and applies only to compressive forces. It is the limit state that predicts the crumpling of the web beneath a compressive force. It is similar to local web yielding but occurs in relatively slender webs compared with local web yielding.

Web local crippling strength depends on how close to the column end the force is applied. For illustration here, it is assumed that the force is applied at a point greater than $d/2$ from the column end. When this limit state was considered for the seated connection in Section 11.9, the force was assumed to be applied at a point less than $d/2$ from the end. The nominal strength for that case is also a function of the bearing length to depth ratio. For the case considered here, where the force is assumed to be applied at a distance greater than $d/2$ from the end of the member, the *Specification* gives

$$R_n = 0.80t_w^2\left[1+3\left(\frac{l_b}{d}\right)\left(\frac{t_w}{t_f}\right)^{1.5}\right]\sqrt{\frac{EF_{yw}t_f}{t_w}}Q_f \qquad \text{(AISC J10-4)}$$

and

$$\phi = 0.75 \text{ (LRFD)} \qquad \Omega = 2.00 \text{ (ASD)}$$

where Q_f = 1.0 for wide flange members. If the web local crippling limit state is exceeded, a three quarter-depth stiffener, a pair of three quarter-depth stiffeners, or a three quarter-depth doubler plate is required.

12.4.4 Web Compression Buckling

Web compression buckling is addressed in Section J10.5 and applies only when compressive forces are applied on opposite sides of the column, putting the web into compression. The strength is given by

$$R_n = \frac{24t_w^3\sqrt{EF_{yw}}}{h}Q_f \qquad \text{(AISC J10-8)}$$

and

$$\phi = 0.90 \text{ (LRFD)} \qquad \Omega = 1.67 \text{ (ASD)}$$

where $Q_f = 1.0$ for wide flange members. If the forces are less than $d/2$ from the end of the column, the strength is reduced by 50 percent.

If the web compression buckling limit state is exceeded, a single full-depth stiffener, a pair of full-depth stiffeners, or a full-depth doubler plate are required.

12.4.5 Web Panel Zone Shear

Web panel zone shear, addressed in Section J10.6, within the boundaries of a fully rigid connection may be significant. The strength of the panel zone is based on shear yielding of the web unless a significant axial force also exists. In this case, shear-axial interaction is considered. When the effects of panel zone deformation on frame stability are not considered in the structural analysis, the panel zone is expected to behave elastically. If the behavior of the panel zone is included in the structural analysis, the nonlinear behavior of the panel zone can be included, and its strength increased accordingly. When the effect of inelastic panel zone deformations on stability are not accounted for in the analysis, panel zone strength as given by *Specification* Section J10.6 is

For $\alpha P_r \le 0.4P_y$

$$R_n = 0.6F_y d_c t_w \qquad \text{(AISC J10-9)}$$

For $\alpha P_r > 0.4P_y$

$$R_n = 0.60\,F_y d_c t_w\left(1.4 - \frac{\alpha P_r}{P_y}\right) \qquad \text{(AISC J10-10)}$$

and

$$\phi = 0.90 \text{ (LRFD)} \qquad \Omega = 1.67 \text{ (ASD)}$$

where $\alpha = 1.0$ for LRFD and 1.6 for ASD, P_r is the required column axial strength and P_y is the yield strength. The panel zone strength must be sufficient to resist the total shear in the panel zone, including the story shear carried by the column web. When this limit state is exceeded, a full-depth doubler plate or a pair of diagonal stiffeners is required.

In the discussion of each of these limit states, the concluding statement indicated that if the limit state is exceeded, a stiffener or doubler plate is required. Thus, this is a "go/no-go" decision. It is possible that a stiffener may be required by a very small margin of exceedance of only one of these limit states. Unfortunately, stiffeners and doubler plates are expensive elements to add to a connection, especially if they must be fitted between the column flanges as for a full-depth stiffener. In many cases, it is much more economical to select a column section that may be larger than required for the axial load but avoids the requirement of stiffeners. Thus, stiffener requirements should not be left for the detailing stage, but addressed early in the design process when member sizes can still be revised.

If stiffeners cannot be avoided, they are designed to resist a force calculated as the applied force, either tension or compression, minus the resisting force as defined for each limit state. This net force is resisted by the cross section of the stiffeners, which are sized based on the provisions for tension or compression connecting elements in *Specification* Section J4. The *Specification* provides additional criteria for stiffeners and doubler plates in Sections J10.8 and J10.9.

The dimensional requirements for stiffeners are as follows:

1. The width of each stiffener plus half the column web thickness must be greater than or equal to one-third of the width of the attached flange or plate delivering the force: $b_s \geq (b_p/3 - t_w/2)$.
2. The thickness of the stiffener must be at least half the thickness of the attached plate and at least equal to the flange or plate width delivering the force divided by 16: $t_s \geq t_p/2$ and $t_s \geq b_p/16$.
3. Transverse stiffeners must extend at least one-half the depth of the column.

The strength requirements are also found in these sections. For stiffeners that resist tension forces, the provisions of *Specification* Section J4.1, which refers to Chapter D, must be satisfied. The weld between the loaded flange and stiffener must be sized to transfer the load carried by the stiffener, and the weld to the web must transfer the difference between the forces on the ends of the stiffener.

The strength of a compression stiffener must satisfy the requirements for compression connecting elements found in *Specification* Section J4.4. The stiffener may be designed to bear on the loaded flange or welded to transfer the force that the stiffener is required to resist. The weld to the web is designed to transfer the difference between the forces on the ends of the stiffener.

Doubler plates, when needed, must be designed for the forces they are required to resist according to the provisions for those forces. These include, for compression, the provisions of Chapter E; for tension, the provisions of Chapter D; and for shear, the provisions of Chapter G. Additional limitations are as follows:

1. The plate thickness and size must provide sufficient additional material to equal or exceed the strength requirements.
2. The welds attaching the doubler plate to the column web along the flanges must develop the force transmitted to the doubler plate.

EXAMPLE 12.4a
Column Side Limit States by LRFD

Goal: Check the column side limit states for a moment connection and design any needed stiffeners and doubler plates.

Given: Consider the bolted flange plate connection of Example 12.3a. The flange plates are 7-1/4×3/4 and resist a required force of P_u = 167 kips.

SOLUTION

Step 1: Determine the column flange strength based on flange local bending as addressed in Section J10.1. From *Manual* Table 1-1 $t_f = 0.710$ in.

This limit state is applicable only for a tension force, and the nominal strength is given by Equation J10-1 as

$$R_n = 6.25F_{yf}t_f^2 = 6.25(50)(0.710)^2 = 158 \text{ kips}$$

and the design strength for flange local bending is

$$\phi R_n = 0.9(158) = 142 < 167 \text{ kips}$$

Since the available strength is less than the required strength, a pair of half-depth stiffeners are required.

Step 2: Determine the column web strength based on web local yielding as addressed in Section J10.2. Assume the connection is not near the column end.

This limit state applies to both tension and compression forces applied to the column web. The bearing length, l_b, is the sum of the plate thickness and the 3/8 in. fillet weld on each side of the plate; thus,

$$l_b = 3/4 + 2(3/8) = 1.50 \text{ in.}$$

and from *Manual* Table 1-1, $k_{des} = 1.31$, so the web nominal strength is

$$R_n = F_{yw}t_w(5k_{des} + l_b) = 50(0.440)(5(1.31) + 1.50) = 177 \text{ kips}$$

and the design strength for web local yielding is

$$\phi R_n = 1.00(177) = 177 > 167 \text{ kips}$$

Therefore, this limit state is not exceeded and does not call for stiffeners.

Step 3: Determine the column web strength based on web local crippling as addressed in Section J10.3.

This limit state is applicable only for a compressive force applied to the column.

The column nominal web strength when the force is applied at a distance greater than $d/2$ from the column end is, using Equation J10-4,

$$R_n = 0.80t_w^2\left[1+3\left(\frac{l_b}{d}\right)\left(\frac{t_w}{t_f}\right)^{1.5}\right]\sqrt{\frac{EF_{yw}t_f}{t_w}}Q_f$$

$$R_n = 0.80(0.440)^2\left[1+3\left(\frac{1.50}{14.0}\right)\left(\frac{0.440}{0.710}\right)^{1.5}\right]\sqrt{\frac{29{,}000(50)(0.710)}{0.440}}(1.0)$$

$$= 274 \text{ kips}$$

and the design strength for web local crippling is

$$\phi R_n = 0.75(274) = 206 > 167 \text{ kips}$$

Therefore, no stiffeners are required for this limit state.

Step 4: Determine the column web strength for web compression buckling as addressed in Section J10.5.

This limit state does not need to be checked unless there are opposing compressive forces on opposite sides of the column. The connection described for this example did not mention any connection on the other side of the column. This limit state can be checked to establish any limits on future connections to this column. The value for h is not given explicitly in the *Manual*; however, h/t_w = 25.9 is given. Thus, h = 25.9(0.440) = 11.4 in. The column web nominal strength, based on Equation J10-8, is then

$$R_n = \frac{24t_w^3\sqrt{EF_{yw}}}{h}Q_f$$

$$= \frac{24(0.440)^3\sqrt{29{,}000(50)}}{11.4}(1.0) = 216 \text{ kips}$$

and the design strength for web compression buckling is

$$\phi R_n = 0.9(216) = 194 \text{ kips}$$

Thus, this column web does not experience compression buckling if opposing forces less than 194 kips are applied on opposite sides of the column.

Step 5: Determine the strength of the web for panel zone shear as addressed in Section J10.6.

Based on yielding of the panel zone, without the interaction of any axial force in the column, the nominal panel zone shear strength as given by Equation J10-9 is

$$R_n = 0.6F_y d_c t_w = 0.6(50)(14.0)(0.440) = 185 \text{ kips}$$

and the design strength for web panel zone shear is

$$\phi R_n = 0.9(185) = 167 \text{ kips}$$

This is equal to the force applied by the connection. Note that the story shear from frame action will always reduce the panel zone shear force due to moment connection(s). So, this panel is not likely to have a panel zone shear problem unless the column required axial strength, αP_r, is greater than $0.4P_y$, in which case the panel zone shear strength will be reduced.

Step 6: Determine the force to be transferred by stiffeners.

The only column web limit state that calls for a stiffener in this example is that of flange local bending, which is an issue for the tension flange only. The force to be transferred through the stiffener is the difference between the applied force and that available through the web; thus, from step 1,

$$R_u = 167 - 142 = 25 \text{ kips}$$

This is clearly a small force to be transferred. Careful review of the limit state of flange local bending shows that if the column flange were 0.770 in. thick instead of 0.710 in. thick, no stiffener plates would be required. In this case, a W14×99 would have eliminated the stiffener problem. It is often more economical to provide a heavier column to eliminate stiffeners and web doublers because doing so eliminates the significant labor associated with preparing and welding these items.

Step 7: Determine the required stiffener size. Use A572 Gr. 50 plates.

Based on the dimensional requirements for a stiffener, the minimum width of each stiffener is

$$w = b_s \geq \left(\frac{b_p}{3} - \frac{t_w}{2}\right) = \left(\frac{7.25}{3} - \frac{0.440}{2}\right) = 2.20 \text{ in.}$$

The thickness of the stiffener must be at least

$$t_s \geq \frac{t_p}{2} = \frac{0.750}{2} = 0.375 \text{ in.} \quad \text{and} \quad t_s \geq \frac{b_p}{16} = \frac{7.25}{16} = 0.453 \text{ in.}$$

Transverse stiffeners must also extend at least one-half the depth of the column. Therefore, try a 2.25×1/2 in. stiffener with a 3/4 in. corner cutoff, as shown in Figure 12.8. For the tension stiffener, the gross area of the stiffener is

$$A_g = (2.25 - 0.750)(0.50) = 0.750 \text{ in.}^2$$

and the design tensile strength of one stiffener is

$$\phi R_n = \phi F_y A_g = 0.9(50)(0.750) = 33.8 \text{ kips}$$

Therefore, the pair of stiffeners provide 2(33.8) = 67.6 kips, which is greater than the required strength of 25 kips.

Step 8: Determine the required weld size.

The weld between the loaded flange and stiffener must be sized to transfer the 12.5 kips carried by each stiffener. Fillet welds will be used on the top and bottom of the stiffener. Thus,

$$D = \frac{12.5}{1.5(1.392)\,(2)\,(2.25\text{-}0.750)} = 2.00 \text{ sixteenths of in.}$$

and a minimum 3/16 in. weld is required by *Specification* Table J2.4.

The weld to the web must transfer the difference between the forces on the ends of the stiffener. Because this is a half-depth stiffener, the total force in the stiffener must be transferred to the column web; thus,

$$D = \frac{12.5}{(1.392)(2)(6.50 - 0.750)} = 0.781 \text{ sixteenths of in.}$$

and a minimum 3/16 in. weld is required by *Specification* Table J2.4.

Step 9: Conclusion.

For the conditions as checked in this example, excluding those in steps 4 and 5, the PL1/2×2-1/4×0 ft – 6-1/2 in. stiffeners shown in Figure 12.8 with 3/16 in. fillet welds will be adequate.

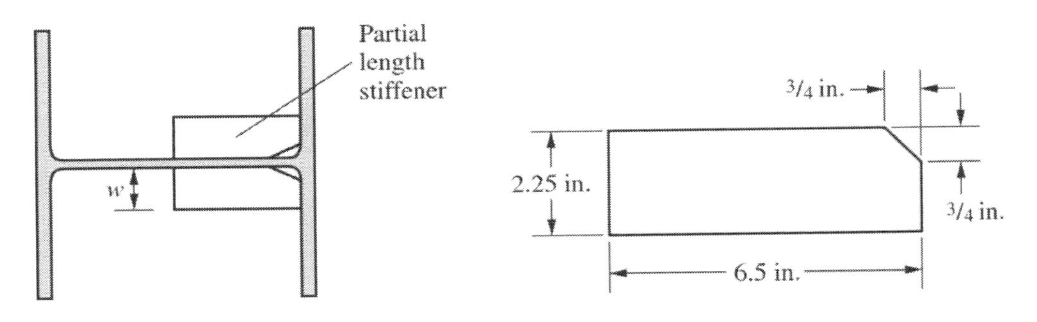

Figure 12.8 Column Stiffener for Example 12.4

EXAMPLE 12.4b
Column Side Limit States by ASD

Goal: Check the column side limit states for a moment connection and design any needed stiffeners and doubler plates.

Given: Consider the bolted flange plate connection of Example 12.3b. The flange plates are 7-1/4×3/4 and resist a required force of P_a = 111 kips.

SOLUTION

Step 1: Determine the column flange strength based on flange local bending as addressed in Section J10.1. From *Manual* Table 1-1 t_f = 0.710 in.

This limit state is applicable only for a tension force, and the nominal strength is given by Equation J10-1 as

$$R_n = 6.25F_{yf}t_f^2 = 6.25(50)(0.710)^2 = 158 \text{ kips}$$

and the allowable strength for flange local bending is

$$R_n/\Omega = 158/1.67 = 94.6 < 111 \text{ kips}$$

Since the available strength is less than the required strength, a pair of half-depth stiffeners are required.

Step 2: Determine the column web strength based on web local yielding as addressed in Section J10.2. Assume the connection is not near the column end.

This limit state applies to both tension and compression forces applied to the column web. The bearing length, l_b, is the sum of the plate thickness and the 3/8 in. fillet weld on each side of the plate; thus,

$$l_b = 3/4 + 2(3/8) = 1.50 \text{ in.}$$

and from *Manual* Table 1-1, k_{des} = 1.31, so the web nominal strength is

$$R_n = F_{yw}t_w(5k_{des} + l_b) = 50(0.440)(5(1.31) + 1.50) = 177 \text{ kips}$$

and the allowable strength for web local yielding is

$$R_n/\Omega = 177/1.50 = 118 > 111 \text{ kips}$$

Therefore, this limit state is not exceeded and does not call for stiffeners.

Step 3: Determine the column web strength based on web local crippling as addressed in Section J10.3.

This limit state is applicable only for a compressive force applied to the column.

The column nominal web strength when the force is applied at a distance greater than $d/2$ from the column end is, using Equation J10-4,

$$R_n = 0.80t_w^2\left[1 + 3\left(\frac{l_b}{d}\right)\left(\frac{t_w}{t_f}\right)^{1.5}\right]\sqrt{\frac{EF_{yw}t_f}{t_w}}Q_f$$

$$R_n = 0.80(0.440)^2\left[1 + 3\left(\frac{1.50}{14.0}\right)\left(\frac{0.440}{0.710}\right)^{1.5}\right]\sqrt{\frac{29{,}000(50)(0.710)}{0.440}}(1.0)$$

$$= 274 \text{ kips}$$

and the allowable strength for web local crippling is

$$R_n/\Omega = 274/2.00 = 137 > 111 \text{ kips}$$

Therefore, no stiffeners are required for this limit state.

Step 4: Determine the column web strength for web compression buckling as addressed in Section J10.5.

This limit state does not need to be checked unless there are opposing compressive forces on opposite sides of the column. The connection described for this example did not mention any connection on the other side of the column. This limit state can be checked to establish any limits on future connections to this column. The value for h is not given explicitly in the *Manual*; however, h/t_w = 25.9, is given. Thus, h = 25.9(0.440) = 11.4 in. The column web nominal strength, based on Equation J10-8, is then

$$R_n = \frac{24t_w^3\sqrt{EF_{yw}}}{h}Q_f$$

$$= \frac{24(0.440)^3\sqrt{29,000(50)}}{11.4}(1.0) = 216 \text{ kips}$$

and the allowable strength for web compression buckling is

$$R_n/\Omega = 216/1.67 = 129 \text{ kips}$$

Thus, this column web does not experience compression buckling if opposing forces less than 129 kips are applied on opposite sides of the column.

Step 5: Determine the strength of the web for panel zone shear as addressed in Section J10.6.

Based on yielding of the panel zone, without the interaction of any axial force in the column, the nominal panel zone shear strength as given by Equation J10-9 is

$$R_n = 0.6F_y d_c t_w = 0.6(50)(14.0)(0.440) = 185 \text{ kips}$$

and the allowable strength for web panel zone shear is

$$R_n/\Omega = 185/1.67 = 111 \text{ kips}$$

Because this is equal to the force applied by the connection, the panel zone cannot accommodate any additive story shear. For a typical exterior column connection, the story shear and the shear from the connection forces are not additive, so this panel is not likely to have a panel zone shear problem unless the column required axial strength, αP_r, is greater than $0.4P_y$, in which case the panel zone shear strength will be reduced.

Step 6: Determine the force to be transferred by stiffeners.

The only column web limit state that calls for a stiffener in this example is that of flange local bending, which is an issue for the tension flange only. The force to be transferred through the stiffener is the difference between the applied force and that available through the web; thus, from step 1,

$$R_a = 111 - 94.6 = 16.4 \text{ kips}$$

This is clearly a small force to be transferred. Careful review of the limit state of flange local bending shows that if the column flange were 0.770 in. thick instead of 0.710 in. thick, no stiffener plates would be required. In this case, a W14×99 would have eliminated the stiffener problem. It is often more economical to provide a heavier column to eliminate stiffeners and web doublers because doing so eliminates the significant labor associated with preparing and welding these items.

Step 7: Determine the required stiffener size. Use A572 Gr. 50 plates.

Based on the dimensional requirements for a stiffener, the minimum width of each stiffener is

$$w = b_s \geq \left(\frac{b_p}{3} - \frac{t_w}{2}\right) = \left(\frac{7.25}{3} - \frac{0.440}{2}\right) = 2.20 \text{ in.}$$

The thickness of the stiffener must be at least

$$t_s \geq \frac{t_p}{2} = \frac{0.750}{2} = 0.375 \text{ in. and } t_s \geq \frac{b_p}{16} = \frac{7.25}{16} = 0.453 \text{ in.}$$

Transverse stiffeners must also extend at least one-half the depth of the column. Therefore, try a 2.25×1/2 in. stiffener with a 3/4 in. corner cutoff, as shown in Figure 12.8. For the tension stiffener, the gross area of the stiffener is

$$A_g = (2.25 - 0.750)(0.50) = 0.750 \text{ in.}^2$$

and the allowable tensile strength of one stiffener is

$$R_n/\Omega = F_y A_g/\Omega = (50)(0.750)/1.67 = 22.5 \text{ kips}$$

Therefore, the pair of stiffeners provide 2(22.5) = 45.0 kips, which is greater than the required strength of 16.4 kips.

Step 8: Determine the required weld size.

The weld between the loaded flange and stiffener must be sized to transfer the 8.20 kips carried by each stiffener. Fillet welds will be used on the top and bottom of the stiffener. Thus,

$$D = \frac{8.20}{1.5(0.928)\ (2)\ (2.25\text{-}0.750)} = 1.96 \text{ sixteenths of in.}$$

and a minimum 3/16 in. weld is required by *Specification* Table J2.4.

The weld to the web must transfer the difference between the forces on the ends of the stiffener. Because this is a half-depth stiffener, the total force in the stiffener must be transferred to the column web; thus,

$$D = \frac{8.20}{(0.928)(2)(6.50 - 0.750)} = 0.768 \text{ sixteenths of in.}$$

and a minimum 3/16 in. weld is required by *Specification* Table J2.4.

Step 9: Conclusion.

For the conditions as checked in this example, excluding those in steps 4 and 5, the PL1/2×2-1/4×0 ft – 6-1/2 in. stiffeners shown in Figure 12.8 with 3/16 in. fillet welds will be adequate.

12.5 PROBLEMS

For all moment connections, design a bolted single-plate shear connection.

1. Design a bolted flange plate connection to connect a W21×44 beam to the flange of a W14×99 column. The connection must transfer a dead load moment of 28 ft-kips and a live load moment of 84 ft-kips, and a dead load shear of 7.2 kips and a live load shear of 22 kips. The plates are A572 Gr. 50 steel and welded to the column with E70 electrodes. Use 3/4 in. A325-N bolts. The beam and column are A992 steel. Design by (a) LRFD and (b) ASD.

2. For the design from Problem 1, determine the column stiffening requirements. If stiffeners or doubler plates are required, design the stiffeners and doublers using A572 Gr. 50 plates by (a) LRFD and (b) ASD.

3. Design a welded flange plate connection to connect a W30×99 beam to the flange of a W14×176 column. The connection must transfer a dead load moment of 167 ft-kips and a live load moment of 500 ft-kips, and a dead load shear of 12 kips and a live load shear of 35 kips. The beam and column are A992 steel, and the plates are A572 Gr. 50. Use E70 electrodes and 3/4 in. A325-N bolts. Design by (a) LRFD and (b) ASD.

4. For the design from Problem 3, determine the column stiffening requirements. If stiffeners or doubler plates are required, design the stiffeners and doublers using A572 Gr. 50 plates by (a) LRFD and (b) ASD.

5. Design a bolted flange plate connection to connect a W18×60 beam to the flange of a W14×109 column. The connection must transfer a dead load moment of 40 ft-kips and a live load moment of 120 ft-kips, and a dead load shear of 8 kips and a live load shear of 24 kips. The plates are A572 Gr. 50 steel and welded to the column with E70 electrodes. Use 7/8 in. A325-N bolts. The beam and column are A992 steel. Design by (a) LRFD and (b) ASD.

6. For the design from Problem 5, determine the column stiffening requirements. If stiffeners or doubler plates are required, design the stiffeners and doublers using A572 Gr. 50 plates by (a) LRFD and (b) ASD.

7. Design a welded flange plate connection to connect a W24×68 beam to the flange of a W14×132 column. The connection must transfer a dead load moment of 35 ft-kips and a live load moment of 105 ft-kips, and a dead load shear of 10 kips and a live load shear of 30 kips. The beam and column are A992 steel, and the plates are A572 Gr. 50. Use E70 electrodes and 3/4 in. A325-N bolts. Design by (a) LRFD and (b) ASD.

8. For the design from Problem 7, determine the column stiffening requirements. If stiffeners or doubler plates are required, design the stiffeners and doublers using A572 Gr. 50 plates by (a) LRFD and (b) ASD.

9. Design a bolted flange plate connection to connect a W21×83 beam to the flange of a W14×120 column. The connection must transfer a dead load moment of 55 ft-kips and a live load moment of 165 ft-kips, and a dead load shear of 12 kips and a live load shear of 36 kips. The plates are A572 Gr. 50 steel and welded to the column with E70 electrodes. Use 3/4 in. A325-N bolts. The beam and column are A992 steel. Design by (a) LRFD and (b) ASD.

10. For the design from Problem 9, determine the column stiffening requirements. If stiffeners or doubler plates are required, design the stiffeners and doublers using A572 Gr. 50 plates by (a) LRFD and (b) ASD.

11. Design a welded flange plate connection to connect a W21×132 beam to the flange of a W14×176 column. The connection must transfer a dead load moment of 150 ft-kips and a live load moment of 450 ft-kips, and a dead load shear of 10 kips and a live load shear of 30 kips. The beam and column are A992 steel, and the plates are A572 Gr. 50. Use E70 electrodes and 3/4 in. A325-N bolts. Design by (a) LRFD and (b) ASD.

12. For the design from Problem 11, determine the column stiffening requirements. If stiffeners or doubler plates are required, design the stiffeners and doublers using A572 Gr. 50 plates by (a) LRFD and (b) ASD.

13. Design a direct-welded flange moment connection to connect a W21×83 beam to the flange of a W14×90 column. The connection must transfer a dead load moment of 80 ft-kips and a live load moment of 240 ft-kips, and a dead load shear of 15 kips and a live load shear of 45 kips. The beam and column are A992 steel, and the web plate is A572 Gr. 50. Use E70 electrodes and 3/4 in. A325-X bolts. Design by (a) LRFD and (b) ASD.

14. For the design from Problem 13, determine the column stiffening requirements. If stiffeners or doubler plates are required, design the stiffeners and doublers using A572 Gr. 50 plates by (a) LRFD and (b) ASD.

15. Design a direct-welded flange moment connection for the conditions of Problem 1. Design by (a) LRFD and (b) ASD.

16. For the design from Problem 15, determine the column stiffening requirements. If stiffeners or doubler plates are required, design the stiffeners and doublers using A572 Gr. 50 plates by (a) LRFD and (b) ASD.

17. Design a direct-welded flange moment connection for the conditions of Problem 3. Design by (a) LRFD and (b) ASD.

18. For the design from Problem 17, determine the column stiffening requirements. If stiffeners or doubler plates are required, design the stiffeners and doublers using A572 Gr. 50 plates by (a) LRFD and (b) ASD.

19. Integrated Design Project. Design bolted flange plate beam-to-column moment connections for the frame designed in Problem 37 or Problem 38 of Chapter 8.

Reconsider the type of moment connections and design using welded flange plate connections.

Chapter 13

Steel Systems for Seismic Resistance

Rainer Square, Seattle, Washington
Photo courtesy of Magnusson Klemencic Associates
© Magnusson Klemencic Associates

13.1 INTRODUCTION

For wind and gravity loads, structural analysis and design are normally performed by assuming that the structural response remains elastic. In seismic design this assumption is too restrictive, particularly for applications that involve significant ground motion. That is, the structural response in a strong earthquake is naturally inelastic, and an elastic analysis may unnecessarily overestimate the resulting forces and incorrectly underestimate the deformations.

This chapter provides an introduction to the design of steel building structures for seismic resistance. The requirements of the *Specification* are supplemented by the AISC *Seismic Provisions for Structural Steel Buildings*, ANSI/AISC 341-22, to provide appropriate guidance in designing for seismic loads. In this chapter, this standard is referred to as the *Seismic Provisions*.

To account for the inelasticity that is expected in the response of a structure to a seismic event, the approach used in the *Seismic Provisions*, the National Earthquake Hazard Reduction Program (NEHRP) Provisions, ASCE 7, and the International Building Code incorporate a seismic response modification coefficient, R, a deflection amplification factor, C_d, and a system overstrength factor, Ω_o. Although more advanced analyses can be used, this approach permits the use of an elastic analysis to predict an inelastic response. These factors are incorporated as follows:

- R is used as a divisor in determining the seismic force for which the structure will be designed. Higher R values represent higher ductility levels in the structural system, which results in seismic forces being reduced in proportion to this ductility.

- C_d is used as a multiplier in determining the story drift. Lower C_d values represent higher levels of structural stiffness and therefore lower story drift.
- Ω_o is used as a multiplier in seismic load combinations. It increases the design loads to account for the level of overstrength present in the seismic force resisting system so that the analysis results in a more accurate prediction of the inelastic behavior.

To determine the values of R, C_d, and Ω_o that are appropriate for a design, buildings are categorized based on occupancy and use. In the NEHRP Provisions, buildings are assigned to one of four risk categories and then to a seismic design category based on the expected acceleration, soil characteristics, and the period of the building. The NEHRP Provisions are a prestandard that is used to develop the provisions in ASCE 7 and the International Building Code, which vary slightly but are similar.

Seismic design categories A, B, and C generally correspond to a classification of low to moderate seismicity. For category A, there are no special seismic design requirements, and design may proceed without reference to the *Seismic Provisions*. In seismic design categories B and C, the engineer can choose to use a basic steel structure with no special detailing, for which $R = 3$, $C_d = 3$, and $\Omega_o = 3$, if permitted by the building code; such building designs can be based upon the *Specification* and do not require consideration of requirements in the *Seismic Provisions*. Alternatively, the engineer can choose to use a system defined in the *Seismic Provisions* and take advantage of a higher R factor.

Seismic design categories D, E, and F generally correspond to a classification of high seismicity. In such cases, the engineer must use a structural system defined in the *Seismic Provisions*. The remainder of this chapter discusses the structural systems provided in the *Seismic Provisions* for resisting seismic forces—those in which R is taken greater than 3.

13.2 EXPECTED BEHAVIOR

For gravity loads, wind loads, and seismic loads associated with smaller earthquakes, it is expected that the structural response will be elastic. However, for larger earthquakes, it is recognized that it may be impractical or impossible to prevent some inelastic behavior. For this reason, and because there is no guarantee that an actual earthquake will be less severe than that defined for design purposes in the building code, the *Seismic Provisions* are based on a capacity design methodology. Accordingly, the provisions included for each system are intended to result in a structure in which controlled inelastic deformations can occur during a strong earthquake to dissipate the energy imparted to the building by the ground motion. These inelastic deformations are forced to occur in a predictable manner and to be restricted to specific elements and/or locations in the structural system. The remainder of the structure remains elastic as these deformations occur, protected in much the same way that a fuse protects the wiring in an electric circuit from overload.

Given this basic premise of the capacity design methodology, the "fuse" elements often establish the design requirements for the members and connections that surround them. This has varying implications for different types of systems.

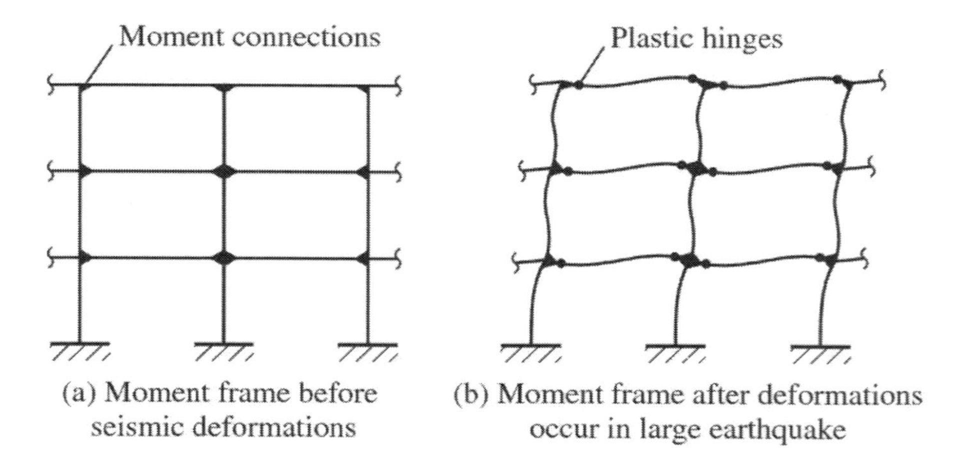

Figure 13.1 Moment-Frame Systems

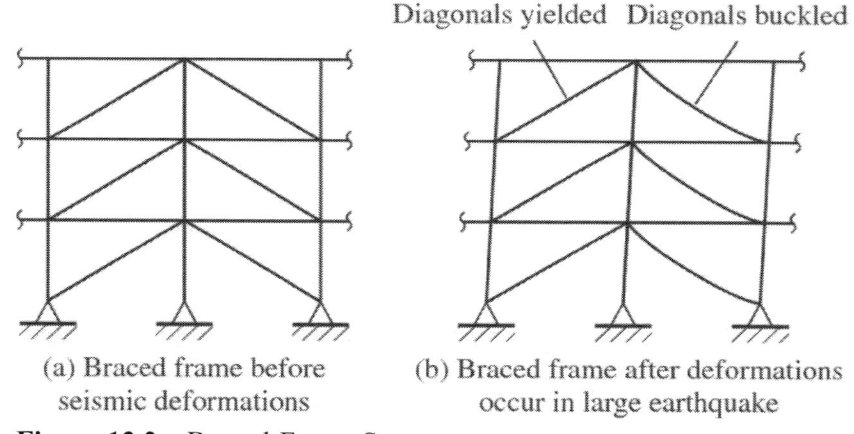

Figure 13.2 Braced-Frame Systems

- As illustrated in Figure 13.1, in a moment frame, the fuse element is typically a plastic hinge that forms in the girders just outside the girder-to-column connection. Accordingly, the girder-to-column connections, column panel zones, and columns must all be designed to develop the flexural strength of the girders connected to them.
- As illustrated in Figure 13.2, in a concentrically braced frame, the fuse element is usually a compression buckling–tension yielding mechanism formed in the diagonal bracing member itself. Accordingly, the brace-to-gusset connections, gussets, gusset connections, beams, and columns must all be designed to develop the tension yield strength and compression buckling strength of the braces that connect to them.

- Other systems exist, including wall and core systems using steel plate shear walls or composite steel and concrete walls. Wall systems use shear in the wall panels as the fuse element. Boundary elements must be designed to remain elastic as the panels deform. Wall systems tend to be stiffer but also can provide significant ductility.

Regardless of the system chosen, fuse elements must deform in a predictable and controlled manner and must provide a ductility that exceeds the level of deformation anticipated. Thus, the systems are configured so that limit states with higher ductility, such as yielding, have control over limit states with lesser ductility, such as rupture.

The actual material properties, such as steel yield strength and strain hardening effects, can influence the behavior of the system. As discussed throughout this book, steel is specified by ASTM designation, which identifies the specified minimum yield strength, among other characteristics. The actual yield strength, however, is most likely higher than the specified value. Also, once strain hardening begins to take place, the effects of load reversals will tend to further elevate the apparent yield strength. The difference between the actual yield strength and specified minimum yield strength and strain hardening effects are important in the capacity design methodology because they increase the strength required in the remainder of the structure to promote yielding in the fuse elements.

These effects are treated directly with multipliers in the *Seismic Provisions*. A multiplier, R_y, is given for each grade of steel. When applied to the specified minimum yield strength, F_y, the resulting quantity is the expected yield strength, R_yF_y. Second, an allowance is made for the effects of strain hardening, generally with a factor of 1.1. Thus, the *Seismic Provisions* use an elevated yield strength, generally equal to $1.1R_yF_y$, in calculating the expected strength of fuse elements in order to determine the design forces for connections and members surrounding the fuse elements. Table 13.1 lists selected examples of ASTM steel with the corresponding R_y.

Table 13.1 Sample Expected Yield Strength Multipliers

Hot-rolled shapes	R_y
A36	1.5
A1043 Gr. 36	1.3
A992	1.1
A913	1.1
HSS	
A500 Gr. C	1.3
A53	1.6
A1085	1.25

13.3 MOMENT-FRAME SYSTEMS

The moment-frame systems, addressed in Chapter E of the *Seismic Provisions*, generally use flexural fuse elements, usually plastic hinges that form in the girders just outside the fully restrained (FR) girder-to-column moment connections. Three types of moment-frame systems are addressed in the *Seismic Provisions*: special moment frames (SMF), in Section E3; intermediate moment frames (IMF), in Section E2; and ordinary moment frames (OMF), in Section E1. Three other systems are addressed in Chapter E: special truss moment frames (STMF), ordinary cantilever column systems (OCCS), and special cantilever column systems (SCCS). STMF are discussed further in Section 13.5. SMF and IMF use connections that have demonstrated inelastic rotation capacities of at least 0.03 radians and 0.01 radians, respectively, in testing.

OMF use a prescriptive connection that provides for small inelastic demands. Assuming that the elastic drift of a moment frame is 0.01 radians, and the inelastic drift is equal to the inelastic rotation at the connections, SMF, IMF, and OMF provide for interstory drifts of 0.04, 0.02, and 0.01 radians, respectively. Some typical seismic moment connections are shown in Figures 13.3 and 13.4.

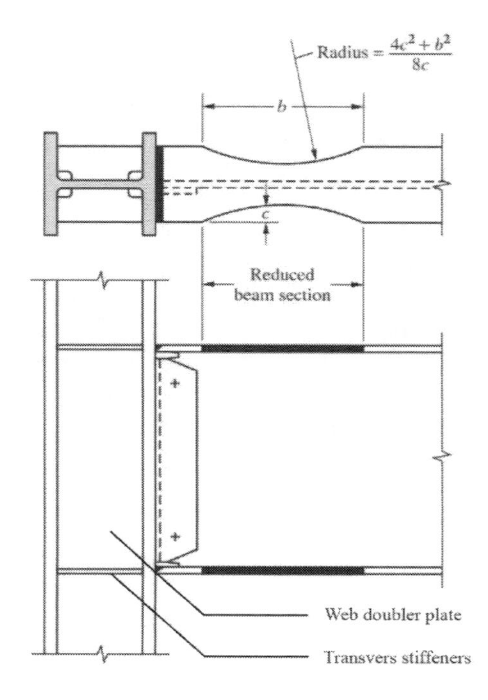

Figure 13.3 Typical Seismic Reduced Beam Section (RBS) Moment Connection

The values of R, C_d, and Ω_o provided in the ASCE 7 for the three systems are given in Table 13.2.

Table 13.2 Seismic Coefficients for Moment Frames

System	R	Ω_o	C_d
SMF	8	3	5-1/2
IMF	4-1/2	3	4
OMF	3-1/2	3	3

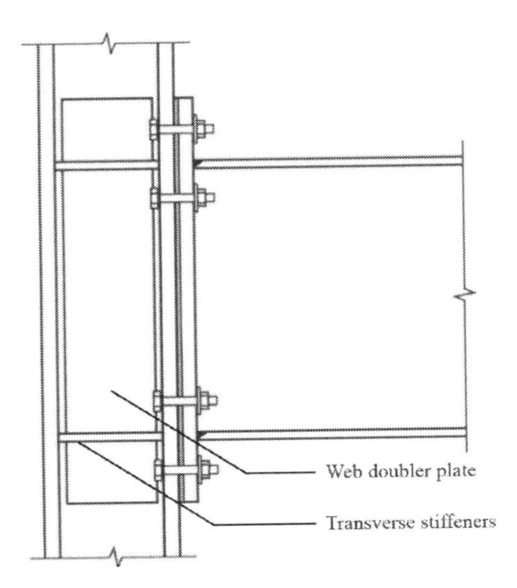

Figure 13.4 Typical Seismic End-Plate Moment Connection

The use of SMF is not limited in any seismic design categories, whereas IMF or OMF usage is restricted according to seismic design category, building height, and structural configuration.

13.3.1 Special Moment Frames (SMF)

SMF are configured to form fuses through plastic hinging in the beams, usually adjacent to the beam-to-column connection, to accommodate significant inelastic deformation during large seismic events. There may also be some inelastic deformation in the column panel zone. Several requirements are included in the *Seismic Provisions* to promote this behavior, as described in the ensuing sections.

Fuse Strength

With the plastic hinges forming in the beams, the fuse flexural strength is $1.1R_yM_p$. The girder-to-column connections, column panel zones, and columns must all be designed to allow the fuse to develop, or achieve, this flexural strength. Alternative approaches recognized in the *Seismic Provisions* include moment-frame systems with partially restrained (PR) connections and weak panel-zone systems, wherein the fuses would form through connection deformations and panel-zone shear deformations, respectively.

Beam-to-Column Connections

The moment connections used in SMF must have supporting tests demonstrating conformance with the ductility requirements, such as through the use of a connection listed in the AISC *Prequalified Connections for Special and Intermediate Steel Moment Frames for Seismic Applications* (AISC 358-22). Alternatively, the use of connections qualified by prior testing or project-specific testing is acceptable. The FR connections shown in Figures 13.3 and 13.4 are prequalified connections according to AISC 358-22. Figure 13.5 shows a reduced beam section (RBS) connection in a special moment frame.

Figure 13.5 Reduced Beam Section (RBS) Moment Connection

Panel Zone Requirements
Some inelastic deformation in the column panel zone is permitted and is in many cases beneficial to the system performance. A panel zone consistent with tested assemblies is required, and the requirements in the *Seismic Provisions* generally result in stiff panel zones with limited yielding. It may also be necessary to reinforce the column with a web doubler plate for shear, Figure 13.4, and/or transverse stiffeners, also called continuity plates, in the column at the beam flanges, Figure 13.3, to transfer the beam flange forces into the column.

Beam and Column Compactness
The compactness criteria in the *Specification* are based on a required ductility level of 3, whereas the expected member ductility demands for beams and columns in SMF can be on the order of 6 or 7. Accordingly, beams and columns in SMF must meet the more stringent width-to-thickness limits of highly ductile members set forth in the *Seismic Provisions*.

Prevention of Story Mechanisms
In SMF, a strong-column, weak-beam relationship must be satisfied in proportioning the columns. This requirement is formulated as a check of the moment ratio between the beam(s) and column(s) at each moment-connected joint in the structure. However, this check is not intended to eliminate all column yielding. Rather, it is a simplified approach that results in a framing system with columns strong enough to force flexural yielding in beams at multiple levels of the frame. This reduces the likelihood of developing a story mechanism, as shown in Figure 13.6, and achieves a higher level of energy dissipation. Some exceptions are permitted, as in the case for a one-story building, where it would not increase energy dissipation if the beams yielded instead of the columns. Because gravity loads accumulate, story mechanisms are a greater concern lower in the building.

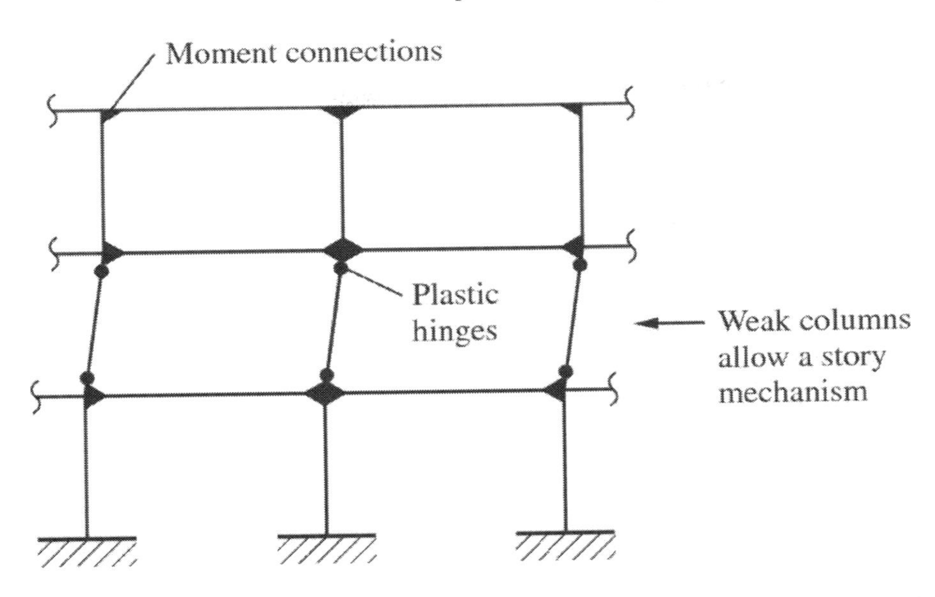

Figure 13.6 Moment Frames without Strong-Column, Weak-Beam Relationship

Stability Bracing Requirements

Special stability bracing requirements apply in SMF because the bracing must be suitable to maintain the position of the braced elements well into the inelastic range. For beams, the permitted unbraced length is generally reduced, and bracing is required near the location where the plastic hinge is expected to form. The *Seismic Provisions* provide several options for bracing the beams at the beam-to-column connection and the column. Often, the configuration of the gravity framing, and interconnection of the floor slab and the beam can be used to satisfy these requirements.

Protected Zones

The fuse regions in SMF—the plastic hinge regions in the beams—are expected to undergo significant inelastic deformations. Accordingly, these regions are designated as protected zones, and attachments and other potential notch effect–inducing conditions are prohibited in these areas.

13.3.2 Intermediate Moment Frames (IMF) and Ordinary Moment Frames (OMF)

IMF and OMF systems are similar in configuration to SMF with lower inelastic deformation capacity for seismic events The *Seismic Provisions* emphasize that IMF and OMF are subject to less stringent special requirements compared with SMF. In fact, OMF are subject to relatively few requirements beyond those in the *Specification.*

IMF are based on the use of a tested connection design with a qualifying interstory drift angle of 0.02 radians. That is, IMF are subject to the same connection testing requirements as SMF, but with a lesser required interstory drift angle. OMF are based on a prescriptive design procedure and an expected interstory drift angle of 0.01 radians, which corresponds to a nominally elastic response.

13.4 BRACED-FRAME SYSTEMS

The braced-frame systems covered in Chapter F of the *Seismic Provisions* fall into two categories: concentric and eccentric. Concentrically braced frames generally use axial fuse elements—usually the braces themselves, which yield in tension and/or buckle in compression. Eccentrically braced frames generally use shear and/or flexural fuse elements—usually a segment, called a link, in the beams themselves adjacent to the braces.

Three types of braced-frame systems are addressed in Chapter F of the *Seismic Provisions*: special concentrically braced frames (SCBF), in Section F2; ordinary concentrically braced frames (OCBF), in Section F1; and eccentrically braced frames (EBF), in Section F3. Two other systems are addressed in Chapter F: buckling restrained braced frames (BRBF) and special plate shear walls (SPSW). These are discussed further in Section 13.5. The values of R, C_d, and Ω_o provided in ASCE 7 for the first three systems are given in Table 13.3.

Table 13.3 Seismic Coefficients for Braced Frames

System	R	Ω_o	C_d
SCBF	6	2	5
OCBF	3-1/4	2	3-1/4
EBF	8	2	4

All braced-frame systems have building height restrictions that vary based on the seismic design category.

13.4.1 Special Concentrically Braced Frames (SCBF)

SCBF are configured to form fuses through tension yielding and compression buckling of the braces between the end connections, to accommodate significant inelastic deformation during major seismic events. Several requirements are included in the *Seismic Provisions* to promote this behavior, as described in the ensuing sections.

Fuse Strength

The fuse axial strength in tension is equal to $R_yF_yA_g$, a value that is usually higher than the fuse axial strength in compression. In compression the fuse axial strength is the lesser of $R_yF_yA_g$ and $(1.0/0.877)F_{ne}A_g$ where F_{ne} is the expected critical buckling strength. The brace-to-gusset connections, gussets, gusset-to-beam and gusset-to-column connections, beams, and columns must all be designed to permit the brace to develop this full axial strength in tension.

Gusset Requirements

Most braces and gussets are detailed such that out-of-plane brace buckling occurs before in-plane buckling. When this is the case, weak-axis bending is induced in the gusset by the end rotations and the gusset must be detailed to accommodate these rotations. One method is to use a free length equal to twice the plate thickness provided between the end of the brace and the bend line in the gusset plate, as illustrated in Figure 13.7. The bend line in the gusset is a line perpendicular to the brace axis that passes through the point on the gusset edge connection that is nearest to the brace end. Alternatively, the bracing connection can be detailed to force the deformation into the bracing member, with either in-plane or out-of-plane buckling occurring.

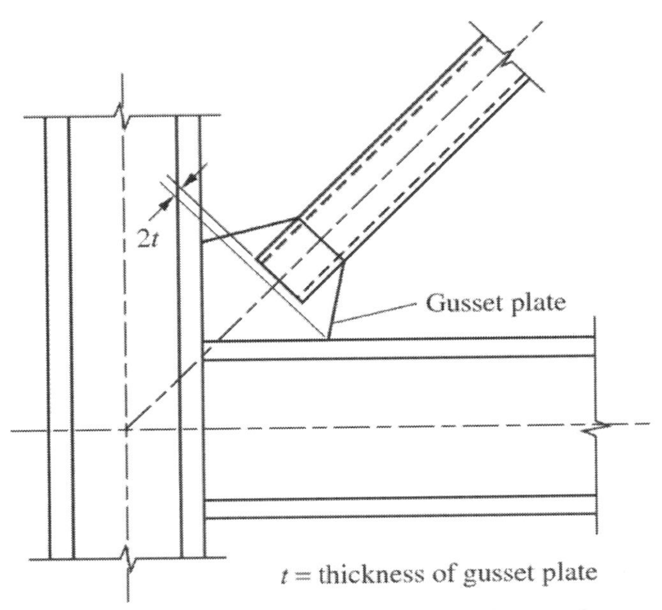

Figure 13.7 Typical Seismic Bracing Connection

Brace Slenderness

The slenderness ratio, L_c/r, of the brace affects post-buckling cyclic performance of the system. Accordingly, a maximum brace slenderness ratio of 200 is permitted.

Brace Net Section Limitations

The effective net area of the brace must not be less than the brace gross area to ensure tension yielding will control over tension rupture. Most end connections include a net section that requires reinforcement to satisfy this requirement.

Distribution of Bracing

Braces must be used in a manner such that the lateral forces in all stories are resisted by a combination of tension yielding and compression buckling of the brace members. Although a 50-50 distribution is considered ideal, the provisions allow up to 70 percent of the lateral force to be resisted by tension or compression braces, unless it can be shown that the system response is essentially elastic. The mixing of tension and compression braces improves the buckling and post-buckling strength of the system and helps prevent the accumulation of inelastic drifts in one direction.

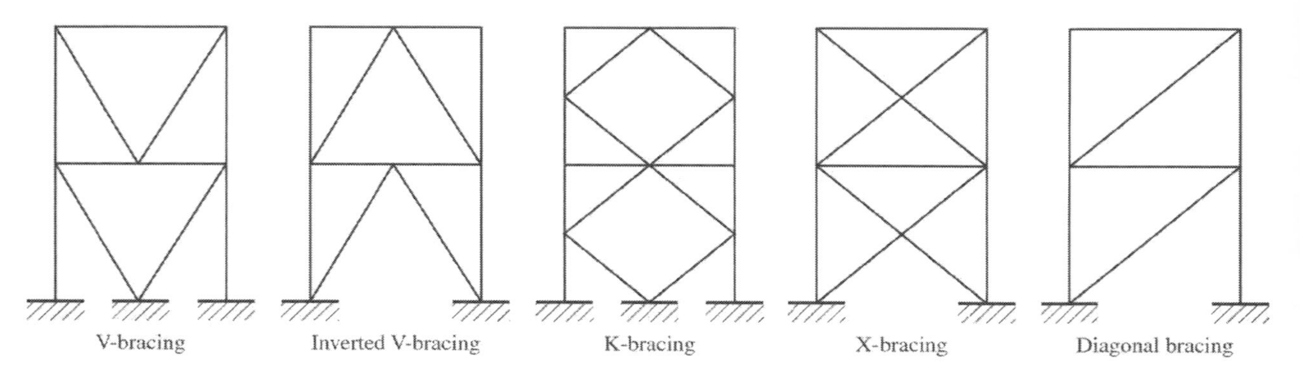

Figure 13.8 Bracing Configurations of Concentrically Braced Frames (K-bracing is not permitted for SFRS)

Beam, Column, and Brace Compactness
The compactness criteria in the *Specification* are based on a required ductility level of 3, whereas the expected member ductility demands for beams and columns in SMF can be on the order of 6 or 7. Accordingly, beams and columns in SCBF must meet the more stringent width-to-thickness ratio limits of highly ductile members in the *Seismic Provisions*. These more stringent limitations also improve the fracture resistance and post-buckling cyclic performance of the braces.

Figure 13.9 Two-Stories of a Special Concentrically Braced Frame (SCBF) Photo courtesy Degenkolb Engineers

Bracing Configurations
A variety of bracing configurations are available. Some configurations require special considerations; others—such as K-bracing, where a column is loaded laterally along its length by bracing—are not permitted. Figure 13.8 illustrates several bracing configurations. Figure 13.9 shows two stories of a special concentrically braced frame.

In V-braced and inverted-V-braced frames, the expected yielding and buckling behavior of the braces creates an unbalanced vertical force because the tension brace remains effective as it yields but the compression brace is ineffective after buckling. This unbalanced force must be resisted by the intersecting beam, as well as by its connections and supporting members. That is, the beam must be designed for the corresponding load redistribution in addition to the gravity loads. Alternatively, the bracing configuration can be altered to provide another path for unbalanced loading. For example, the V and inverted-V configurations can be alternated to form a two-story X configuration. Another approach involves the addition of a zipper column. These bracing configurations are illustrated in Figure 13.10.

The use of K-bracing (and knee bracing) is prohibited in SCBF because the configuration results in an unbalanced force in the columns.

X-bracing is allowed. However, the common tension-only design approach used for wind forces is not permitted. Both diagonals of the X-bracing must be designed to resist the tension and compression forces that result from cyclic load reversals.

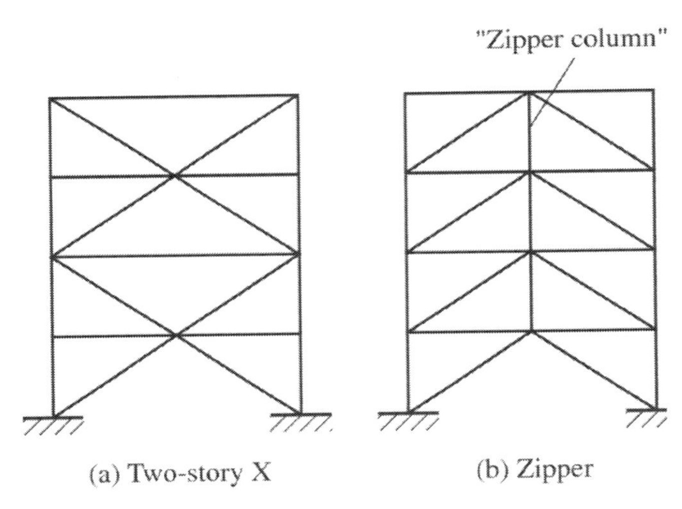

Figure 13.10 Two-Story X and Zipper Configurations

Single-diagonal braced frames, where all braces are oriented in the same direction, are not permitted because all braces in this configuration are called upon to resist the same type of force, either tension or compression, at the same time. In this configuration, for one direction of loading, 30 to 70 percent of the braces would not be in tension. Thus, the diagonal braced frame should be implemented as previously shown in Figure 13.2.

Stability Bracing Requirements

Special stability bracing requirements apply in SCBF because the bracing must be suitable to maintain the position of the braced elements well into the inelastic range. Often, the configuration of the gravity framing and interconnection of the floor slab to the beam can be used to satisfy these requirements.

Protected Zones

The fuse regions in SCBF—the braces and gussets—are expected to undergo significant inelastic deformation. Accordingly, attachments and other potential notch effect–inducing conditions are prohibited in these areas.

13.4.2 Ordinary Concentrically Braced Frames (OCBF)

OCBF are similar in configuration to SCBF but are limited in application because they have lower inelastic deformation capacity for seismic events. The *Seismic Provisions* emphasize that OCBF are subject to less stringent requirements than SCBF.

13.4.3 Eccentrically Braced Frames (EBF)

EBF are configured to form fuses through shear yielding, flexural yielding, or a combination of the two in the EBF link, in order to accommodate significant inelastic deformation during major seismic events. Unlike the behavior of SCBF and OCBF, the braces in EBF are intended to remain nominally elastic and are not expected to buckle. The requirements in the *Seismic Provisions* to promote this behavior are described in the ensuing sections.

Fuse Strength

The fuse strength in shear for I-shaped links is $1.25R_yV_n$, where V_n is the nominal shear strength of the link, which is the lesser of the nominal plastic shear strength and the shear associated with the plastic moment capacity of the link. The strain hardening multiplier for EBF links of 1.25 is higher than the corresponding multiplier for other systems because EBF links are associated with higher deformation demands. The beam segments outside the link, gussets, gusset-to-beam and gusset-to-column connections, and columns must all be designed to develop the shear and/or flexural yielding mechanism in the links.

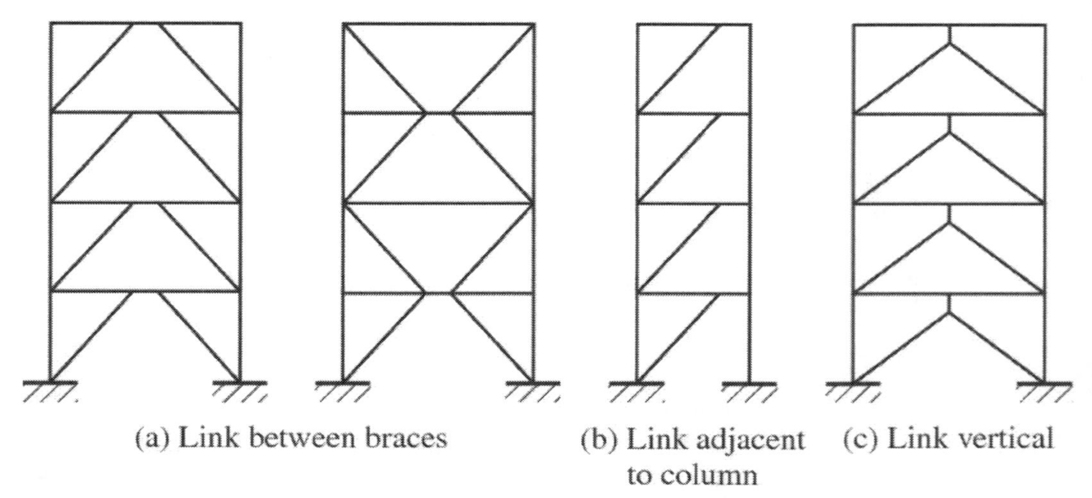

Figure 13.11 Example Configurations of Eccentrically Braced Frames

Figure 13.12 Eccentrically Braced Frame (EBF)
Photo courtesy Degenkolb Engineers

Link Location
EBF links are usually located as segments within the length of the beams, either between braces or between a brace and a beam-to-column connection. Alternatively, links can be provided as vertical elements between beams and V or inverted-V bracing. EBF configurations are shown in Figure 13.11. Figure 13.12 shows an eccentrically braced frame. In the foreground the EBF link is in the upper beam and in the background the link is in the lower beam.

When links are located as segments within the length of the beams, it is preferable to locate the links between the ends of the braces. When the links are located between a brace and a beam-to-column connection, the beam-to-column connection requires special consideration because the rotational demands are substantially higher than those at a beam-to-column connection in SMF. In applications involving a significant flexural demand, a prequalified connection or a connection qualified by testing must be used. When the links used are short enough that shear yielding dominates, the need for qualification testing is eliminated if the beam-to-column connection is reinforced with haunches or other suitable reinforcement designed to preclude inelastic action in the reinforced zone adjacent to the column.

Link Rotations
Plastic rotation of link beams in EBF systems is limited to 0.02 radians for flexural-yielding links and 0.08 radians for shear-yielding links. For links that deform in combined shear and flexure yielding, the rotation limit is determined by linear interpolation between these limits.

Link Stiffening and Bracing
To ensure that the required rotations can be achieved and that yielding occurs without local buckling well into the inelastic range, links are stiffened as prescribed in *Seismic Provisions* Section F3.5(b). The use of web doubler plates to stiffen links is not permitted because this type of reinforcement does not deform consistently with the web

deformations and would increase the strength of the fuse. Additionally, beam web penetrations within the link are not permitted and the link must be braced against out-of-plane displacement and twist at the ends of the link.

Braces and Beam Segments Outside of Links
Because the inelastic action in EBF is intended to occur primarily within the links, the braces and beam segments outside of the links must be designed to remain nominally elastic as the links deform. Limited yielding outside of the links is allowed, as long as the beam segments outside the links and braces have sufficient strength to develop the fully yielded and strain-hardened strength of the links. The braces and beam segments outside the links are normally designed as members subject to the combined effects of axial force and flexure.

13.5 OTHER FRAMING SYSTEMS

Several other systems are provided for in the *Seismic Provisions*, including special truss moment frames (STMF), in Section E4; buckling-restrained braced frames (BRBF), in Section F4; and special plate shear walls (SPSW), in Section F5. The values of R, C_d, and Ω_o provided in ASCE 7 for each of these three systems are given in Table 13.4.

Table 13.4 Seismic Coefficients for Other Framing Systems

System	R	Ω_o	C_d
STMF	7	3	5-1/2
BRBF	8	2-1/2	5
SPSW	7	2	6

Each of these systems has building height restrictions that vary based on the seismic design category.

Composite steel and reinforced concrete systems are also provided for in Chapters G and H of the *Seismic Provisions*.

13.5.1 Special Truss Moment Frames (STMF)

STMF are configured to form fuses through yielding in a special segment of the truss, to accommodate significant inelastic deformation during large seismic events. The special segment can be either a truss panel with diagonals or a Vierendeel truss panel. The remainder of the truss and framing in the system is designed to remain elastic as the special segment deforms. A schematic STMF is illustrated in Figure 13.13.

When diagonals are used, the yielding of the special segment occurs by axial tension yielding and compression buckling of the diagonals. Diagonal web members used in the special segments of STMF systems are limited to flat bars only and must meet a limiting width-to-thickness ratio of 2.5. When a Vierendeel panel—a panel without diagonals—is used, yielding of the special segment occurs by flexural yielding of the chord members.

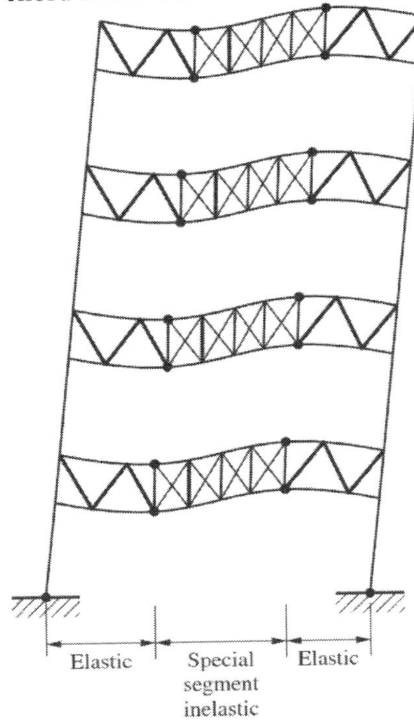

Figure 13.13 Special Truss Moment Frame Configuration

The size of the truss and the size and location of the special segment are limited to correspond with the research on which the system is based. It is required that the STMF special segment be located within the middle half of the truss span because shear due to gravity loads is generally lower in that region. Other than the normal gravity loads carried by the frame, no major structural loading is permitted in the special segment.

13.5.2 Buckling-Restrained Braced Frames (BRBF)

BRBF represent a type of concentrically braced frame system that has special bracing elements, as shown in Figure 13.14. These bracing elements provide essentially the same response in compression as they do in tension. The bracing elements are composed of a load-bearing core and a surrounding sleeve element that restrains the global buckling of the core, forcing yielding in compression rather than buckling. BRBF are configured to form fuses through tension yielding and compression yielding of these special bracing elements, to accommodate significant inelastic deformation during large seismic events. A three-story buckling restrained braced frame is shown in Figure 13.15.

The bracing elements must be qualified by testing to ensure that the braces used provide the necessary strength and deformation capacity. The required deformation capacity during testing is amplified beyond what is required by an analysis in recognition that actual deformations can be larger than those predicted by analysis.

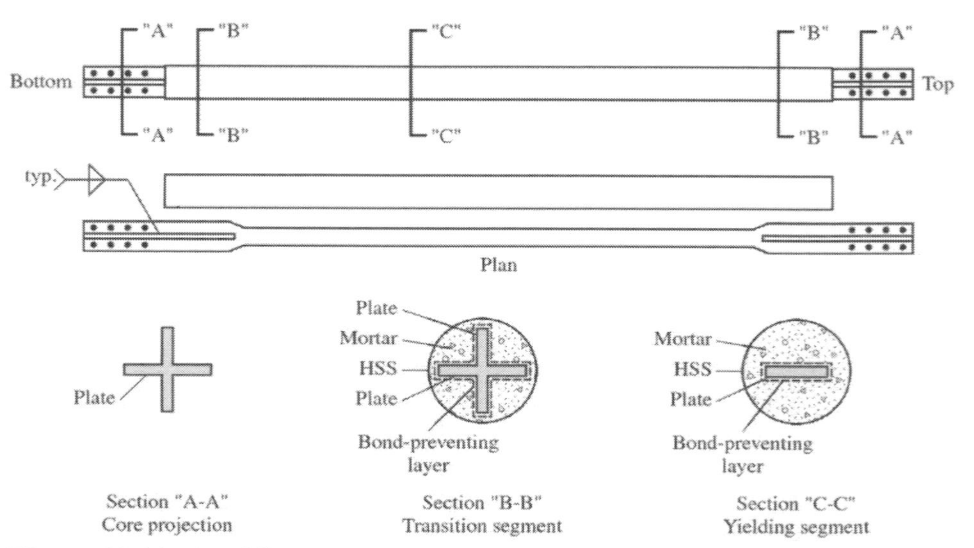

Figure 13.14 Buckling Restrained Braced Frame Bracing Element

Figure 13.15 Application of a Buckling Restrained Brace in a Concentrically Braced Frame (BRBF)
Photo courtesy Star Seismic

The steel core has a yielding segment that is designed with a cross-sectional area and length based on strength, stiffness, and strain demands. Because each bracing element is a manufactured item, the designer can specify an array of braces that promote distributed yielding throughout the frame.

The steel core projections beyond the yielding segment are designed to provide the transition from the core and connection to the remainder of the framing system. The projections are designed so that they remain nominally elastic like the rest of the frame as the yielding segment deforms.

13.5.3 Special Plate Shear Walls (SPSW)

SPSW have slender, unstiffened plate elements surrounded by and connected to horizontal and vertical boundary elements that are rigidly interconnected. A schematic SPSW system is illustrated in Figure 13.16 and an example building application is shown in Figure 13.17.

SPSW are configured to form fuses through web plate yielding and buckling (tension field action in the web element), along with plastic-hinge formation in the ends of horizontal boundary elements to accommodate significant inelastic deformation during major seismic events.

The tension-field action in SPSW is analogous to that in a plate girder, but the behavior and strength of SPSW differs from that of plate girders. Accordingly, the design requirements in the *Seismic Provisions* for SPSW differ from those in the *Specification* for plate girders. A more detailed discussion of SPSW systems is found in AISC Design Guide 20, *Steel Plate Shear Walls*.

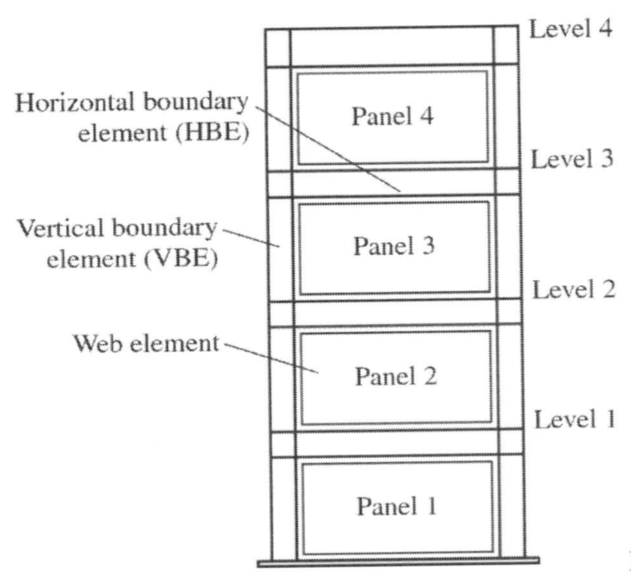

Figure 13.16 Special Plate Shear Wall System

13.5.4 Composite Systems

A variety of composite structural systems are provided for in Chapters G and H of the *Seismic Provisions*. These systems include, in Chapter G, composite moment frames: composite partially restrained moment frames (C-PRMF), composite special moment frames (C-SMF), composite intermediate moment frames (C-IMF), and composite ordinary moment frames (C-OMF). Chapter H includes composite ordinary braced frames (C-OBF), composite special concentrically braced frames (C-SCBF), composite eccentrically braced frames (C-EBF), composite ordinary shear walls (C-OSW), composite special shear walls (C-SSW), composite plate shear walls-concrete encased (C-PSW/CE), composite plate shear walls-concrete filled (C-PSW/CF), and coupled composite plate shear walls-concrete filled (CC-PSW/CF).

The last two systems mentioned above are more commonly known as uncoupled and coupled SpeedCore systems, respectively. These systems offer a rapidly constructable approach to core and shear wall systems that offer high lateral stiffness, strength, and ductility. These systems are treated extensively in AISC Design Guide 38: SpeedCore Systems for Steel Structures.

13.6 OTHER GENERAL REQUIREMENTS

13.6.1 Bolted and Welded Connections

Connections in the seismic force resisting system must be configured such that a ductile limit state in the fuse controls—that is, the deformations occur in the fuse elements before failure occurs in the connections. This generally means that connections in the seismic force resisting system are much larger than they would be if designed for gravity, wind, and low-seismicity applications. There are additional special requirements for the use of bolts and welds in the *Seismic Provisions*.

Bolted joints in shear are designed as pre-tensioned bearing joints with faying surfaces prepared as for Class A or better slip-critical connections. These are not slip-critical connections; they are bearing joints with some slip resistance. Because slip cannot and need not be prevented during large ground motions, the intent is to control slip during lesser ground motions. Pre-tensioned bolts are used because large ground motions can cause full reversal of design load and thus have a negative effect on snug-tight bolt performance.

Hole type is restricted to standard holes and short-slotted holes perpendicular to the loading direction, unless another hole type is shown to be acceptable by testing. One exception provided is that oversized holes are permitted in brace diagonals within certain limits.

For design purposes, bolt bearing checks are required to be made at the *deformation considered* level to prevent excessive deformations of bolted joints due to bearing on the connected material, primarily in order to minimize damage in lesser ground motions.

Figure 13.17 Special Plate Shear Wall for a High-Rise Building
Photo courtesy Michael Dickter/Magnusson Klemencic Associates

In welded connections, filler metal with a minimum specified Charpy V-notch toughness of 20 ft-lbf at 0° F is required in all welds involved in the seismic load path, except for *demand-critical* welded joints, which have more stringent notch toughness requirements.

Sharing of a common force between bolts and welds is prohibited because seismic deformation demands generally exceed the deformation compatibility required for such load sharing. If this load sharing were used, the weld may fail before engaging the full strength of the bolt.

13.6.2 Protected Zones

The fuse elements in the various systems covered in the *Seismic Provisions* may undergo significant inelastic deformations when subjected to large ground motions. Accordingly, construction operations that might cause discontinuities must be barred from these areas. Thus, the *Seismic Provisions* designate protected zones in each system that must be kept free of sharp transitions, penetrations, notches, and so forth. An example of a protected zone is the expected hinge region in beams of SMF, where web openings or studs on the top flange are not allowed. Discontinuities that are inadvertently created in these zones must generally be repaired or elements replaced.

13.6.3 Local Buckling

The yielding of fuse elements requires a member ductility of 6 or 7, which is more than the normal ductility of 3 used in the development of the compactness criteria in the *Specification*. Thus, in the *Seismic Provisions*, more stringent seismic compactness criteria for highly ductile and moderately ductile members are provided in Table D1.1. The highest-performing systems usually require members that meet the criteria called highly ductile. Limitations imposed on use of other systems mean they can meet lesser criteria called moderately ductile (or normal ductility criteria, in some cases).

13.6.4 Column Requirements

Special requirements for columns and column splices in the seismic force resisting system are stipulated in Section D1.5 of the *Seismic Provisions*. Minimum design forces are specified to preclude column and column splice failure in compression or tension. This approach does not necessarily preclude yielding of the column, and some guidance is provided in the Commentary for cases in which yielding of the column might be of concern.

Column splices must be located away from the beam-to-column connections—generally within the middle third of the story height in which the splice occurs—to reduce the effects of flexure. Additionally, if partial-joint-penetration groove welds are used to make column splices, a 100 percent increase in required strength is specified and the use of notch-tough filler metal is required.

There are also requirements for columns that are not a part of the seismic load resisting system, because these columns must still distribute a fraction of the seismic shear between the floors.

13.6.5 Column Bases

To increase frame stiffness, column bases are normally treated similarly to beam-to-column moment connections, accounting for the inherent differences, such as the increased flexibility due to deformations in the base plate, longer anchor rods, compressibility of the grout and concrete, and foundation rocking effects.

13.7 CONCLUSIONS

This introduction to the design of steel structures for seismic force resistance is intended to provide a starting point for further study. The detailed provisions are found in the *Seismic Provisions for Steel Buildings*, ANSI/AISC 341-22. Additional guidance is found in the AISC *Seismic Design Manual* and AISC Facts for Steel Buildings 3, Earthquake and Seismic Design. The interested student is encouraged to study these documents for a more in-depth treatment of seismic design of steel structures.

13.8 PROBLEMS

1. What is the major difference between the analysis and design of a structure for wind and gravity loads and the analysis and design of a structure for seismic loads?

2. Explain the use of the R, C_d, and Ω_o factors. What do these factors account for?

3. How are the R, C_d, and Ω_o factors determined for a particular analysis?

4. What is the purpose of fuse elements in seismic design? Provide some examples of structural fuse elements.

5. How is the R factor used to influence the seismic forces a structure is designed to resist?

6. What are the seismic design categories specified by the building codes and how do they relate to the level of expected seismicity?

7. Explain the capacity design method in relation to material yield strength.

8. What types of fuse elements are typically used in moment frame systems?

9. What are the three primary types of moment frames considered in the *Seismic Provisions*? What are the respective values for R, C_d, and Ω_o for each of these systems?

10. For special moment frames, what type of relationship should exist between the column and beams to reduce the likelihood of a story mechanism?

11. Which type of moment frame is required to meet an interstory drift angle of 0.02 radians?

12. Name the two categories of braced frames provided for in the *Seismic Provisions*. What type of fuse element is used by each of these?

13. List the three primary types of braced frame systems addressed in the *Seismic Provisions* and their corresponding values for R, C_d, and Ω_o.

14. Where are the fuse elements located for SCBF and OCBF?

15. List some examples of CBF configurations.

16. How do EBF differ from CBF in their performance during large seismic events?

17. Where are the fuse elements located for eccentrically braced frames?

18. How do BRBF resist seismic forces and what is the fuse element in this system?

19. Describe how SPSW resist seismic forces.

20. Compare the expected behavior of STMF with other moment frame systems.

21. How does the size of connections in seismic force resisting systems differ from that of connections designed for gravity and wind systems? Why?

22. Is it permissible to share a force between bolts and welds in seismic design? Why or why not?

23. For seismic design, where should column splices be located and why?

24. **Integrated Design Project.** The same lateral load resisting system is used for resisting wind forces as is used for seismic forces. The lateral load resistance in the north-south direction is provided by a concentrically braced frame with a chevron configuration as depicted in Figure 1.24. Before the forces in these bracing members can be determined, the specified seismic load must be determined. For this system, R = 3 may be selected and special detailing according to the AISC *Seismic Provisions* is not required. At this stage in the design, a simplified approach to seismic load calculation might yield the following loads at each level in the north-south direction.

Roof	56.0 kips
Fourth floor	162.0 kips
Third floor	108.0 kips
Second floor	54.0 kips
Total seismic load	380.0 kips

Lateral load resistance in the east-west direction is provided by two perimeter moment frames as shown in Figure 1.24. For these frames, $R = 3$ will also be used. Thus, no special detailing will be required. A simplified approach to seismic load calculation might yield the following loads.

Roof	54.0 kips
Fourth floor	152.0 kips
Third floor	98.0 kips
Second floor	46.0 kips
Total seismic load	350.0 kips

Using the member sizes determined in earlier work, carry out an analysis of both lateral load resisting systems and check the members to be sure that they are adequate to resist the seismic forces.

Reconsider the analysis for the use of special concentrically braced frames and special moment frames. Determine the appropriate loading according to ASCE 7-22 using an approximate building weight as given below and compare the required member strengths to the previous analysis.

Roof	550 kips
Fourth floor	2150 kips
Third floor	2150 kips
Second floor	2150 kips
Total building weight	7000 kips

Index

Made in the USA
Middletown, DE
29 August 2024